SECOND EDITION

DISCRETE MATHEMATICS

Kenneth A. Ross • Charles R.B. Wright

Department of Mathematics, University of Oregon

PRENTICE HALL   *Englewood Cliffs, New Jersey 07632*

Library of Congress Cataloging-in-Publication Data

Ross, Kenneth A.
    Discrete mathematics/Kenneth A. Ross, Charles R. B. Wright.—
2nd ed.
        p.   cm.
    Includes index.
    ISBN 0-13-215427-7
    1. Electronic data processing—Mathematics.   I. Wright, Charles
R. B. (data)   II. Title.
QA76.9.M35R67   1988
511—dc19   87-23586
                                                                    CIP

Editiorial/production supervision: *Eleanor Henshaw Hiatt*
Interior and cover design: *Lorraine Mullaney*
Manufacturing buyers: *Carol Bystrom and Paula Benevento*
Cover photo: *Courtesy of BDM International, Inc.*

Printed in the United States of America

10 9 8 7 6 5 4 3 2

ISBN 0-13-215427-7 01

Prentice-Hall International (UK) Limited, *London*
Prentice-Hall of Australia Pty. Limited, *Sydney*
Prentice-Hall Canada Inc., *Toronto*
Prentice-Hall Hispanoamericana, S.A., *Mexico*
Prentice-Hall of India Private Limited, *New Delhi*
Prentice-Hall of Japan, Inc., *Tokyo*
Prentice-Hall of Southeast Asia Pte. Ltd., *Singapore*
Editora Prentice-Hall do Brasil, Ltda., *Rio de Janeiro*

TO OUR DAUGHTERS

Emily and Laurel Ross
Lisa Madsen
Allyson Wright

CONTENTS

8

GRAPHS   322

9

TREES   407

10

BOOLEAN ALGEBRA   464

11

ALGEBRAIC SYSTEMS   512

PREFACE TO THE
SECOND EDITION

This edition differs from the previous one in two significant respects: the treatment of algorithms has been completely revised, and the account of algebraic topics is now less theoretical.

We have added several algorithms from the standard computer science repertoire and have deleted some that were less central or less efficient. More fundamentally, however, all algorithms after Chapter 0 now come equipped with time complexity analyses. We hope in this way to develop in the reader the habit of automatically considering the running time of any algorithm. In addition, our analyses illustrate some of the basic tools for estimating the efficiency of algorithms.

The shift in our approach to algebraic material is primarily one of concentrating more on concrete examples, with less discussion of the most general settings. Semigroups are in some sense less complicated than groups but the theory of homomorphisms works much better for groups. Rather than presenting the abstract semigroup version first and then specializing to groups, we have gone directly to the material we want to emphasize.

Instructors who are familiar with the first edition may be interested in a brief account of where the major changes occur.

In Chapter 2 we have split the original § 2.1 into two sections, the first of which is an informal introduction to the themes of logic. The intent is to make the reason for studying the predicate calculus clearer and to reduce the emphasis on the mechanical aspects of formal symbolic logic.

We have cut out the section on inverses of matrices in Chapter 4 and we have added a brief introduction to groups of permutations. Later, we have combined §§ 11.3 and 11.4 and have revised the treatment of permutation groups to give beginning students an appreciation for the basic ideas by the study of concrete examples. We use the terminology that a group of permutations of a set *acts on* the set. Readers who know this terminology from a more advanced standpoint will see that we have used it correctly even in the more general context, though we have deliberately not mentioned permutation representations in our discussion.

The new treatment of algorithms begins in § 3.3, where we introduce the "big oh" notation for sequences, a notation which we use from then on to describe the time complexity of our algorithms. Section 3.4 contains a new discussion of iterative and recursive calculation for sequences that are defined recursively. Chapter 3 also has a new section on the solution of elementary recursion relations. The former section on the Division Algorithm is gone; the algorithm itself has been postponed to § 4.3, on modular arithmetic, which we have largely rewritten.

The chapters on graphs and trees have extensive changes in material and emphasis. The account of min-weight algorithms in §§ 8.3, 8.8 and 8.9 is completely redone and now highlights Dijkstra's algorithm as well as Warshall's. We have dropped Hu's and Bavel's algorithms. The treatment of algorithms for finding minimal spanning trees in Chapter 9 is now more streamlined, and includes Prim's algorithm. More significantly, we have completely rewritten the account of tree traversal algorithms to bring out the idea of depth-first search, which we then use to give a fast recursive algorithm for topologically sorting and labeling the vertices of an acyclic digraph. The new treatment of Huffman's algorithm is also more obviously recursive.

These are the highlights. Of course there are also a number of minor changes throughout the text, typically revisions and additions to the exercise sets or clarifications of difficult points. Many of these are in response to the reactions of our students and colleagues who continue to help us to make the book better.

We hope that instructors are aware of the existence of the *Teacher's Manual*, which contains solutions to all exercises that are unanswered in the text. It also gives tips such as "Don't bog down in Chapter 0," and "§ 9.3 is a long one; take two days on it," as well as suggestions for emphasis in classroom presentation.

Our editor for this second edition has been David Ostrow. It is a pleasure also to acknowledge Kevin Johnson's enthusiastic support and Eleanor Hiatt's outstanding editorial work on all phases of production. Finally, we want to give our special thanks to Composition House for their excellent typography. Yes, we know they do it with computers, but we also know that it takes people with talent and experience to produce a math book that looks as good as this.

K.A. Ross / C.R.B. Wright

PREFACE TO THE FIRST EDITION

The term "discrete mathematics" is used broadly to describe the kind of mathematics in which properties such as nearness and smoothness—the key ideas of calculus—are not at issue. In this book the term means the basic noncalculus mathematics a computer science student needs. Some of the material is algebra, some is logic, some is combinatorics or graph theory. Some of the sets we deal with are finite and others are infinite.

Although we do draw on computer science for motivation, we assume no previous experience in the subject. When it is necessary to discuss an application such as logical circuit design in some detail, we provide the needed background, but otherwise we have avoided including topics which are more properly taught in courses in computer science. Of course the choice of topics is a matter of judgment. We have left out a discussion of automata, because the power of mathematics is not very clear in an elementary treatment of the subject. Coding theory is exciting mathematics, but to say anything useful requires more algebra than we develop here. Our treatments of algorithm verification and of time and space complexity are woven into the accounts of other subjects—for instance the division algorithm—rather than set off separately. Algorithm verification is one of the main applications of logic, induction and recursion. We have provided the tools for it, and we have also provided a number of algorithms which come up naturally in the study of relations and graphs. Some of them are written in English and some in a self-explanatory pseudo-code, with the choice of format made for maximum clarity.

We have included more abstract algebra than is absolutely necessary for the undergraduate computer science curriculum today. All of the evidence indicates that in the years to come, computer science will require a higher level of mathematical sophistication, involving more and more advanced mathematics. To give students some of the tools they will need in the future, we have developed enough of the theory of groups to treat nontrivial applications involving symmetry. The discussion of groups also serves as a model for other algebraic systems and for the general algebraic approach which we have taken in much of the book.

One of our main goals is the development of mathematical maturity. We and our colleagues have used this material successfully for several years with average students at the level of beginning calculus, and we find that by the end of two terms they are ready for upperclass work. The presentation begins with an intuitive approach which becomes more and more rigorous as the students' appreciation for proofs and skill at building them increase. Our account is careful but informal. As we go along we illustrate the way mathematicians attack problems, and show the power of an abstract approach. We have aimed to make the account simple enough so the students can learn it and complete enough so they won't have to learn it again.

Chapter 0 gives an introduction to graphs and algorithms and suggests some of the kinds of problems we will consider later. It can be covered quickly, but should not be skipped. Chapters 1, 2, 3, 4 and 7 contain the core material on sets, logic, functions, relations and algebra. The remaining chapters are independent of each other, except that some of the material on digraphs in Sections 8.1 and 8.2 is required in Sections 10.5, 11.2, 11.5 and 11.6 [now 11.4 and 11.5].

We have broken the discussion of logic into two parts: Chapter 2 contains the propositional calculus and an introduction to induction, and Chapter 6 presents the predicate calculus, recursion and generalized induction. We have found that the material goes down better in two bites, and that revisiting logic helps reinforce Chapter 2.

Various choices of topics are possible for a one-semester course. One program which ties together relations, graphs and algorithms consists of Chapters 0, 1, 2, 3, 7, Sections 4.1, 4.2, 4.3 on matrices, algebra and relations, and Sections 8.1, 8.4 and 8.7 on graphs, with Sections 8.2, 8.8 and 8.9 as time permits. A two-semester course can cover the whole book. We strongly recommend coordinating the choice and sequence of topics covered with the computer science courses which the students are taking concurrently. The interactive potential is enormous.

It is a pleasure to acknowledge the helpful comments and suggestions given by our students and colleagues. In particular, we thank our colleagues Frank Anderson, Micheal Dyer, James Harper, David Harrison, William Kantor, Richard Koch, Ivan Niven, Margaret Owens, Andrzej Proskurowski, Stephen Prothero, Mark Reeder and Jerry Wolfe. From among our students, the most helpful comments came from Joyce Eaton, Kenneth Peale and Cathy Phillips.

It is a pleasure also to thank our editor, Bob Sickles, and the members of the staff at Prentice-Hall. In particular, the professional wisdom and experience of Nicholas C. Romanelli are largely responsible for the successful transformation from manuscript to book.

TO THE STUDENT ESPECIALLY

We know that words like "obviously" and "clearly" can be very annoying; they have sometimes bothered us too. When you see them occasionally in this book they are intended as hints. If you don't find the passage obvious or clear, you are probably making the situation too complicated or reading something unintended into the text. Take a break; then back up and read the material again. Similarly, the examples are meant to be helpful. If you are pretty sure you know the ideas involved, but an example seems much too hard, skip over it on first reading and then come back later. If you aren't very sure of the ideas, though, take a more careful look at the example.

Exercises are an important part of the book. They give you a chance to check your understanding and to practice thinking and writing clearly and mathematically. As the book goes on, more and more exercises ask you for proofs. We use the words "show" and "prove" interchangeably, though "show" is more common when a calculation is enough of an answer and "prove" suggests some reasoning is called for. "Prove" means "give a convincing argument or discussion to show why the assertion is true." What you write should be convincing to an instructor, to a fellow student, and to yourself the next day. Proofs should include words and sentences, not just computations, so that the reader can follow your thought processes. Use the proofs in the book as models, especially at first. The discussion of logical proofs in Chapter 2 will also help. Perfecting the ability to write a "good" proof is like perfecting the ability to write a "good" essay or give a "good" oral presentation. They all take practice. Don't be discouraged when one of your proofs fails to convince an expert (say a teacher or a grader). Instead, try to see what failed to be convincing.

Each chapter ends with a list of the main points it covers and with some suggestions for how to use the list to review. One of the best ways to learn material which you plan to use again is to tie each new idea to as many familiar concepts and situations as you can, and to visualize settings in which the new fact would be helpful to you. We have included lots of examples in the text to make this process easier. The review lists can be used to go over the material in the same way by yourself or with fellow students.

Answers or hints to most odd-numbered exercises are given in the back of the book. Wise students will look at the answers only after trying seriously to do the problems. When a proof is called for we usually give a hint or an outline of a proof, which you should first understand and then expand upon. A symbols index appears on the inside of the front cover and the Greek alphabet on the inside of the back cover. At the back of the book there is an index of topics. After Chapter 11 there is a brief dictionary of terms that we use in the text without explanation, but which some readers may have forgotten or never encountered. Look at these items right now to see where they are and what they contain, and then join us for Chapter 0.

K.A. Ross / C.R.B. Wright

0

INTRODUCTION TO GRAPHS AND TREES

The purpose of this text is to present an introduction to several topics in discrete mathematics that are used in or are related to topics in computer science. The presentation will stress precise and mathematical thought, an ingredient essential to both disciplines and in fact to all sciences. However, in practice, such precision is preceded by intuition, which is often obtained by an analysis of examples.

In this chapter we give an informal introduction to graphs and trees. These are topics that are easily grasped, and for which the concepts are easily illustrated with pictures. Our discussion is often intuitive, but it is not careless. The arguments will show up again in later chapters, where we fill in the mathematical formality that is left out for now. This early treatment is intended simply to provide some hands-on practice and a feeling for the sorts of problems dealt with in discrete mathematics. While reading and studying our small easy examples, you should try to imagine large complex ones. Ask yourself whether the thought processes remain valid. We understand complex situations by first understanding simple ones.

§ 0.1   Graphs

You are already undoubtedly familiar with the idea of a graph as a picture of a function. The word "graph" is also used to describe a different kind of structure which arises in a variety of natural settings and which is the subject of this section. In a loose sense these new graphs are diagrams which,

1

properly interpreted, contain information. The graphs we are concerned with are like road maps, circuit diagrams or flowcharts in the sense that they depict connections or relationships between various parts of the diagram.

The diagrams in Figure 1 are from a variety of settings. Figure 1(a) shows a simple flowchart. Figure 1(b) might represent five warehouses of a trucking firm and truck routes between them, labeled with their distances. Figure 1(c) could be telling us about the probability that a rat located in one of four cages will move to one of the other three or stay in its own cage. Figure 1(d) might depict possible outcomes of a repeated experiment such as coin tossing [Heads or Tails]. What do all these diagrams have in common? Each consists of a collection of objects—boxes, circles or dots —and some lines between the objects. Sometimes the lines are directed; that

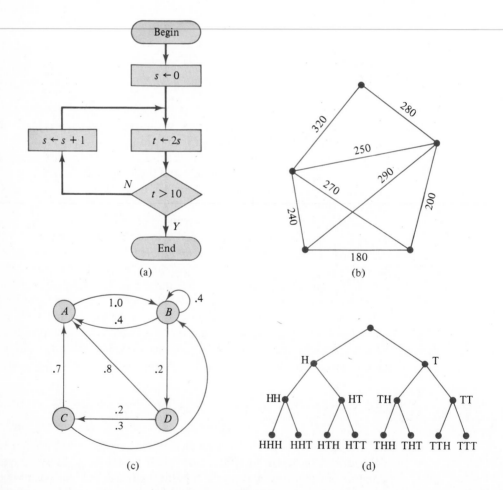

Figure 1

is, they are arrows. Sometimes the objects are labeled, and sometimes the lines are. Later we will worry about arrows and labels, but for now we just concentrate on the objects and lines.

The essential features of a **graph** are its objects and connecting lines; see Figure 2. The objects, drawn here as dots, are called **vertices** [plural of **vertex**] and the connecting lines are called **edges**. The graphs in Figures 2(a) and 2(b) have 5 vertices and 7 edges. The crossing of the two lines in Figure 2(b) is irrelevant and is just a peculiarity of our drawing. The graph in Figure 2(c) has 4 vertices and 6 edges. Two of its edges connect vertex v to vertex w; they are called **multiple edges** or **parallel edges**. One of its edges connects the vertex w to itself; such an edge is called a **loop**.

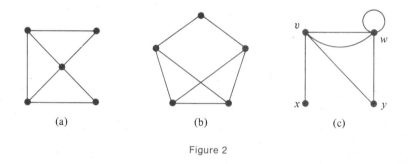

(a)   (b)   (c)

Figure 2

In graph theory we are interested in sequences of edges that link up with each other to form a **path**. To illustrate the idea, we redraw Figure 2(c) in Figure 3(a) and label the edges as well as the vertices. Examples of paths are $d\,b\,f\,e$ [Figure 3(b)] and $c\,f\,e\,b\,a$ [Figure 3(c)]. Note that the drawing alone does not tell us what path we have in mind: Figure 3(c) is also a picture of the paths $b\,a\,f\,e\,c$ and $c\,a\,f\,e\,b$. Paths can repeat edges: $b\,a\,b\,e\,f\,a\,a\,b$ is a path. The **length** of a path is the number of edges in the path. Thus $b\,a\,b\,e\,f\,a\,a\,b$ has length 8.

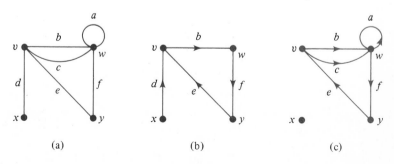

(a)   (b)   (c)

Figure 3

Adjacent edges in a path must have a vertex in common. So a path determines a sequence of vertices. The vertex sequences for the paths discussed above are as follows:

The path	Its vertex sequence
$d\,b\,f\,e$	$x\,v\,w\,y\,v$
$c\,f\,e\,b\,a$	$v\,w\,y\,v\,w\,w$
$b\,a\,f\,e\,c$	$v\,w\,w\,y\,v\,w$
$c\,a\,f\,e\,b$	$v\,w\,w\,y\,v\,w$
$b\,a\,b\,e\,f\,a\,a\,b$	$v\,w\,w\,v\,y\,w\,w\,w\,v$

Note several items. The number of vertices in a vertex sequence is one larger than the number of edges in the path. When a loop appears in a path, its vertex is repeated in the vertex sequence. Vertex sequences treat parallel edges the same and so different paths, such as $b\,a\,f\,e\,c$ and $c\,a\,f\,e\,b$, can have the same vertex sequence. If a graph has no parallel edges or multiple loops, then vertex sequences do uniquely determine paths. In this case the edges can be described by just listing the two vertices which they connect and we may describe a path by its vertex sequence.

A path is a **closed path** if the first and last vertices of its vertex sequence are the same. Some closed paths in Figure 3(a) are $b\,a\,b\,e\,f\,a\,a\,b,\,d\,c\,f\,e\,d$ with vertex sequence $x\,v\,w\,y\,v\,x$, and $f\,e\,b$ with vertex sequence $w\,y\,v\,w$. A **cycle** is a closed path that is efficient in the sense that it repeats no edges and the vertices of its vertex sequence are all distinct except for the first and last ones. Thus $f\,e\,b$ is a cycle. The closed path $d\,c\,f\,e\,d$ repeats the edge d and its vertex sequence $x\,v\,w\,y\,v\,x$ repeats the vertex v. The closed path $c\,a\,f\,e$ is not a cycle; it doesn't repeat any edges but its vertex sequence $v\,w\,w\,y\,v$ repeats the vertex w.

A graph is **acyclic** if it contains no cycles. A path is **acyclic** if the subgraph consisting of the vertices and edges of the path is acyclic. The graph in Figure 4 is not acyclic, since $v\,u\,y\,x\,w\,v$ is a cycle. The path $s\,t\,v\,w\,x\,y\,z$ is acyclic. So is the path $u\,v\,t\,v\,w$; the side trip $v\,t\,v$ is not a

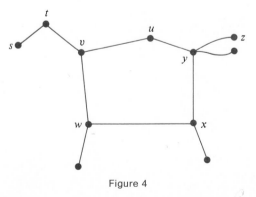

Figure 4

cycle. The path *u v t v w x y z y u* is not acyclic, since the subgraph it determines contains the cycle *u v w x y u*.

One of the oldest problems involving graphs is the Königsberg bridge problem, which asks whether it is possible to take a walk in the town shown in Figure 5(a) crossing each bridge exactly once and returning home. The Swiss mathematician Leonhard Euler [pronounced OIL-er] solved this problem in 1736. He constructed the graph shown in Figure 5(b), replacing the land areas by vertices and the bridges joining them by edges. The question then became: Is there a closed path in this graph which uses each edge exactly once? We call such a path an **Euler circuit**. Euler showed that no such path exists for the graph in Figure 5(b). To see why, we need one more concept. The **degree** of a vertex in a graph is the number of edges connected to that vertex. In Figure 6 we have written the degrees beside the vertices. Note that a loop contributes 2, not 1, to the degree of a vertex. Euler showed that a graph which has an Euler circuit must have all vertices of even

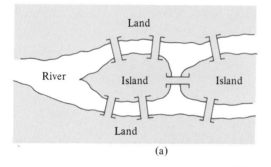

(a)   (b)

Königsberg graph

Figure 5

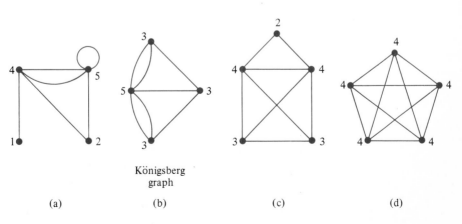

Königsberg
graph

(a)   (b)   (c)   (d)

Figure 6

degree. So the only graph in Figure 6 that could possibly have an Euler circuit is (d). Does it?

Let's see why an Euler circuit forces all vertices to have even degree. Start at some vertex on the circuit and follow the circuit from vertex to vertex, erasing each edge as you go along it. When you go through a vertex you erase one edge going in and one going out, or else you erase a loop. Either way, the erasure reduces the degree of the vertex by 2. Eventually every edge gets erased and all vertices have degree 0. So all vertices must have had even degree to begin with.

Euler also proved that his result almost goes the other way. That is, if every vertex has even degree, this is almost enough to assure us that the graph has an Euler circuit. Almost? Look at Figure 7. This is a picture of a graph since it consists of vertices and edges, even though it's not connected like our earlier examples. [A graph is **connected** if given any two vertices there is a path connecting them.] Every vertex in Figure 7 has even degree, but obviously[1] this graph has no Euler circuit because the graph is not connected. Here is what Euler showed.

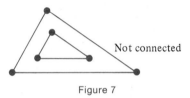
Not connected

Figure 7

Euler's If all the vertices of a connected graph have even degree, then the graph has
Theorem[2] an Euler circuit.

To really understand this theorem we should be able to find a proof, or develop an algorithm or procedure that would always produce an Euler circuit. Indeed, these two approaches are intimately connected. A full understanding of a proof often leads to an algorithm, and behind every algorithm there's a proof. Here's a simple explanation of Euler's theorem, which we illustrate using Figure 8. Start with any vertex, say w, and any edge connected to it, say a. The other vertex, x in this case, has even degree and has been used an odd number of times [once] and so there is an unused edge leaving x. Pick one, say b. Continue in this way. The process won't stop until the starting vertex w is reached since, whenever any other vertex is reached, only an odd number of its edges have been used. In our example, this algorithm might start out with edges $a\ b\ e$ and vertices $w\ x\ w\ y$. At y we can choose any of three edges: d, f, or h. If we select f, the rest of the process

---

[1] See the preface for a discussion concerning this very annoying word.

[2] Some terms, like "theorem," are explained briefly in the dictionary that appears after Chapter 11. Most technical terms can be found by way of the index.

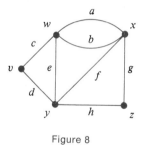

Figure 8

is determined. We end up with the Euler circuit $a\,b\,e\,f\,g\,h\,d\,c$ with vertex sequence $w\,x\,w\,y\,x\,z\,y\,v\,w$.

Simple, wasn't it? Well, it is too simple. What would have happened if, when we first reached vertex y, we had chosen edge d? After choosing edge c, we'd be trapped at vertex w and our path $a\,b\,e\,d\,c$ would have missed edges $f,\,g$ and h. Our explanation and our algorithm must be too simple. In our example it is clear that edge d should have been avoided when we first reached vertex y, but why? What general principle should have warned us to avoid this choice? Think about it. We will continue this discussion in the next section.

EXERCISES 0.1

1. Which of the following vertex sequences describe paths in the graph drawn in Figure 9(a)?
   (a) $s\,t\,u\,v\,w\,x\,y\,z$ (b) $t\,v\,w\,z\,y\,x$
   (c) $s\,t\,u\,s$ (d) $t\,u\,s\,s$
   (e) $v\,w\,v\,w\,v\,w\,v$ (f) $w\,v\,u\,s\,t\,v\,w$

2. Which paths in Exercise 1 are closed paths?

3. Which paths in Exercise 1 are cycles?

4. For the graph in Figure 9(a), give the vertex sequence of a shortest path connecting the following pairs of vertices and give its length.
   (a) s and v (b) s and z
   (c) u and y (d) v and w

5. For each pair of vertices in Exercise 4, give the vertex sequence of a longest path connecting them that repeats no edges. Is there a longest path connecting them?

6. Which graphs in Figure 9 have Euler circuits? For those that don't, give an explanation. For those that do, give an Euler circuit.

7. True or False. "True" means "true in all circumstances under consideration." Consider a graph.
   (a) If there is an edge from a vertex u to a vertex v, then there is an edge from v to u.
   (b) If there is an edge from a vertex u to a vertex v and an edge from v to a vertex w, then there is an edge from u to w.

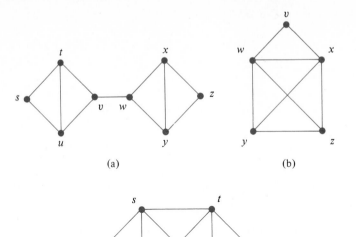

(a)   (b)

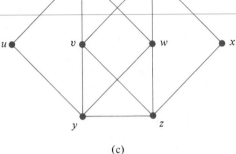

(c)

Figure 9

8. Repeat Exercise 7 with the word "edge" replaced by "path" everywhere.

9. Repeat Exercise 7 with the word "edge" replaced by "path of even length" everywhere.

10. Repeat Exercise 7 with the word "edge" replaced by "path of odd length" everywhere.

11. Give an example of a graph with vertices x, y and z with the following three properties:
    (i)   there is a cycle using vertices x and y;
    (ii) there is a cycle using vertices y and z;
    (iii) no cycle uses vertices x and z.

12. (a) For each graph in Figure 6 calculate
       (i) the sum of the degrees of all of the vertices;
       (ii) the number of edges.
    (b) Compare your answers to part (a) and make a conjecture.
    (c) Use the "erasing edge" idea to justify your conjecture in part (b).
    (d) Can a graph have an odd number of vertices of odd degree?

13. Suppose that a cycle contains a loop. What is its length? Can a cycle contain two loops?

§ 0.2   Special Paths and Trees

We continue our discussion of Euler's theorem and the example in Figure 8 of § 0.1, which we reproduce in Figure 1(a). As we select edges let's remove them from the graph and consider the subgraphs so obtained. Our path started out with edges *a b e*; Figure 1(b) shows the graph with these edges removed. In our successful search for an Euler circuit we next selected *f*, and we noted that if we had selected *d* we were doomed. Figure 1(c) shows the graph if *f* is also removed while Figure 1(d) shows the graph if instead *d* is removed. There is a difference: removal of *d* disconnected the graph while removal of *f* did not. This is the clue for an algorithm that works. At each vertex, FLEURY'S algorithm instructs us to select, if possible, an edge whose removal will not disconnect the graph. If this is not possible, there is exactly one edge available. We select it; then remove it and the vertex from the graph.

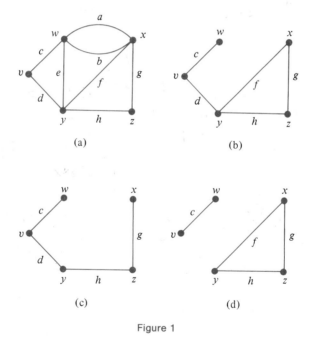

Figure 1

The original choice of edges *a b e f g h d c* satisfied the restrictions of FLEURY'S algorithm. Let's illustrate the procedure again, starting at vertex *z*; see Figure 2(a). Select edges *h* and *f* so that we are at vertex *x*; Figure 2(b) shows these edges removed. Selection of edge *g* would disconnect the graph; vertex *z* would be isolated from the rest of the graph. So we select edge *a* so that we are at vertex *w* [Figure 2(c)]. Now choosing *b* would disconnect the graph, so we select *c* or *e*, say *c*. The rest of the procedure is forced. When we remove *d*, we remove the isolated vertex *v*. Then we remove *e* and the

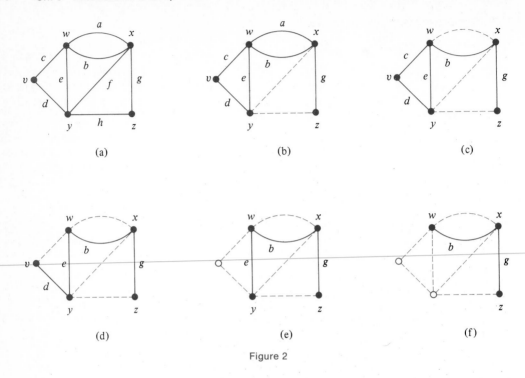

Figure 2

isolated vertex y. Etc. See Figures 2(d)–2(f). The Euler circuit obtained is $h\,f\,a\,c\,d\,e\,b\,g$ with vertex sequence $z\,y\,x\,w\,v\,y\,w\,x\,z$.

Euler's theorem tells us which graphs have closed paths using each edge exactly once, and FLEURY'S algorithm gives a way to construct the paths when they exist. In contrast, much less is known about graphs with paths which go through each vertex exactly once. The Irish mathematician Sir William Hamilton was one of the first to study such graphs, and at one time even marketed a puzzle based on the problem. A closed path is called a **Hamilton circuit** if it uses every vertex of the graph exactly once, except for the last vertex, which duplicates the first vertex. The difference between finding an Euler circuit and a Hamilton circuit is the difference between inspecting streets and inspecting street lights at corners.

Consider the graphs in Figure 3. The graph in Figure 3(a) certainly has a Hamilton circuit which will be an Euler circuit too. The graph in Figure 3(b) has no Hamilton circuit: any path that used every vertex would have to use the central vertex more than once. It does have an Euler circuit. Why? The 8-vertex graph in Figure 3(c) has many Hamilton circuits; the arrows indicate how one of them goes. This graph has no Euler circuit. Why? The 7-vertex graph in Figure 3(d) has no Hamilton circuit. To see this, note that each edge connects an upper vertex to a lower vertex. If there were a Hamilton circuit, its vertex sequence would alternate between upper and lower vertices and would consist of exactly 8 vertices. The first and eighth

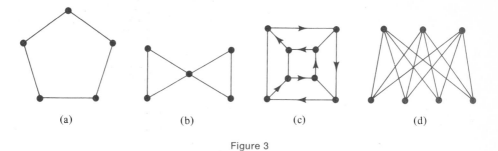

<table>
<tr><td>(a)</td><td>(b)</td><td>(c)</td><td>(d)</td></tr>
</table>

Figure 3

vertices would have to be different [one would be upper and one would be lower] and yet they would have to be the same to complete the circuit. We conclude that no such circuit exists.

With Euler's theorem, the theory of Euler circuits is very nice and complete. What can be proved about Hamilton circuits? Under certain conditions, graphs will have so many edges compared to the number of vertices that they must have Hamilton circuits. But the graph in Figure 3(a) has very few edges and yet it has a Hamilton circuit. And the graph in Figure 3(d) has lots of edges but no Hamilton circuit. It turns out that there is no known simple characterization of those connected graphs possessing Hamilton circuits. The concept of Hamilton circuit seems very close to that of Euler circuit, and yet the theory of Hamilton circuits is vastly more complicated. In particular, no efficient algorithm is known for finding Hamilton circuits. The problem is a special case of the Traveling Salesperson Problem. Here one begins with a graph whose edges are assigned **weights** that may represent mileage, cost, computer time or some other quantity that we wish to minimize. In Figure 4 the weights might represent mileage between cities on a traveling salesperson's route. The goal is to find the shortest round trip that visits each city exactly once. That is, the goal is to find a Hamilton circuit minimizing the sum of the weights of the edges. A nice algorithm solving this problem would also be able to find Hamilton circuits in an unweighted graph, since we could always assign weight 1 to each edge.

We next consider a different problem that resembles the Hamilton circuit problem but which is much easier to solve. Given a connected

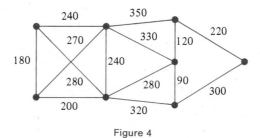

Figure 4

weighted graph, the goal is to find a connected subgraph using all vertices and having minimal total weight. [The **total weight** of a subgraph is the sum of the weights of its edges.] For example, suppose Figure 4 represents distances between cities and that the goal is to connect all cities by pipelines using the minimum possible number of miles of pipeline. It turns out that minimal connected subgraphs are always acyclic, since removal of an edge from a cycle will not disconnect a graph. Acyclic connected graphs are so important that they get a special name: they are called **trees**. We'll devote a whole chapter to them, including applications to such computer science topics as data structures and reverse Polish notation.

Figure 5 contains some examples of trees. Incidentally, we regard the trees in Figures 5(a) and 5(b) as essentially the same. Their pictures are different but the essential structure [vertices and edges] is the same. They share all graph-theoretic properties such as the numbers of vertices and edges, the number of vertices of each degree, etc. To make this clear, we have redrawn them in Figures 6(a) and 6(b) and labeled corresponding vertices. The trees in Figures 6(c) and 6(d) are also essentially the same.

Now back to the problem of minimal connected subgraphs that use all vertices. As we remarked, such a subgraph must be a tree. It is called a **minimal spanning tree**. A connected graph always has at least one minimal spanning tree; let's look at a simple algorithm for finding one. To use the algorithm we first list the edges of the graph in order of their weights. Let's use subscripts and list the edges as $e_1, e_2, e_3, \ldots$ where e_1 is the edge of lowest weight, e_2 is the edge of next lowest weight, etc. If two edges have the same weight, it doesn't matter which comes first. Using this convention we redraw the "pipeline" Figure 4 in Figure 7. Note that the labels on edges e_6 and e_7 could be interchanged. So could the ones on edges e_9 and e_{10}. Our algorithm begins with no edges at all and selects the edges $e_1, e_2, \ldots$ in order, except that at each step we don't use an edge if its selection would create a cycle. Figure 8(a) shows the subgraph obtained after selecting edges e_1, e_2, e_3, e_4, e_5 and e_6. The algorithm rejects edges e_7, e_8 and e_9 because the addition of any of these to the graph in Figure 8(a) would create a cycle. The algorithm next accepts edge e_{10} and rejects the remaining edges. At the end,

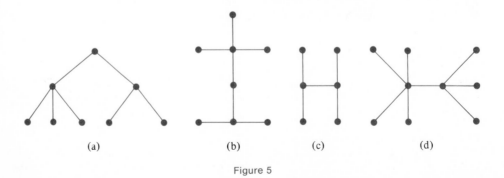

(a) (b) (c) (d)

Figure 5

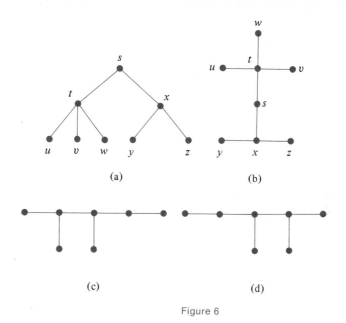

(a)   (b)

(c)   (d)

Figure 6

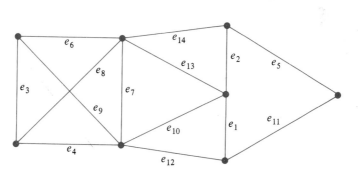

Figure 7

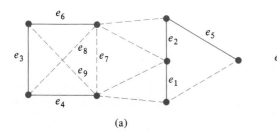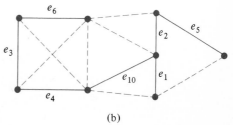

(a)   (b)

Figure 8

the algorithm provides the spanning tree in Figure 8(b). The total weight of this minimal spanning tree is

$$90 + 120 + 180 + 200 + 220 + 240 + 280 = 1330;$$

1330 miles of pipeline would be needed to connect all the cities in Figure 4. This algorithm is often called a **greedy algorithm**. Can you see a reason for the name?

Suppose we want a spanning tree for a connected graph whose edges aren't weighted. No problem. We assign all edges the same weight, say 1, order the edges in any way we like and apply the algorithm described in the last paragraph. Different orderings may provide different spanning trees. If we want *all* spanning trees, then the problem gets more complicated. Of course, we could then apply our algorithm to each ordering, but note, for example, that there are[3]

$$14! = 14 \cdot 13 \cdot 12 \cdot 11 \cdot 10 \cdot 9 \cdot 8 \cdot 7 \cdot 6 \cdot 5 \cdot 4 \cdot 3 \cdot 2 \cdot 1$$

$$\approx 8.72 \cdot 10^{10}$$

possible orderings of the edges for the graph in Figure 7. Even computers don't like to face this many computations. To find all spanning trees one would be advised to look for a better approach, but we will not pursue this matter.

The symbol 14! just used is read "14 factorial." In general, $n!$ is read "n factorial" and is shorthand for

$$n(n-1)(n-2) \cdots 3 \cdot 2 \cdot 1 = 1 \cdot 2 \cdot 3 \cdots (n-2)(n-1)n.$$

Procedures or algorithms that lead to approximately $n!$ steps when one has n vertices or edges are regarded as ineffective because $n!$ grows incredibly rapidly as n gets large. Procedures that lead to something like n^2 steps [or n^3 or $n\sqrt{n}$, say, depending on the circumstances] are regarded as reasonably effective. It is in roughly this sense that we regard FLEURY'S algorithm for finding Euler circuits as acceptable. There are no known effective algorithms for finding Hamilton circuits or even for deciding whether one exists in a particular graph.

EXERCISES 0.2

1. Which of the graphs in Figure 9 have Euler circuits? Hamilton circuits? Give the vertex sequence of a Hamilton or Euler circuit in each case in which one exists.

2. Use FLEURY'S algorithm to find an Euler circuit in the graph of Figure 9(b).

3. Apply FLEURY'S algorithm to the graph of Figure 9(a) until it breaks down. Start at vertex w.

_____

[3] The notation $\approx$ means "approximately equals."

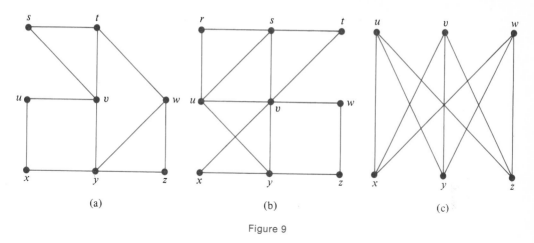

Figure 9

4. Repeat Exercise 3 for the graph in Figure 9(c).

5. Which of the graphs in Figure 2 of § 0.1 have Hamilton circuits?

6. Count the number of spanning trees in the graphs of Figure 10.

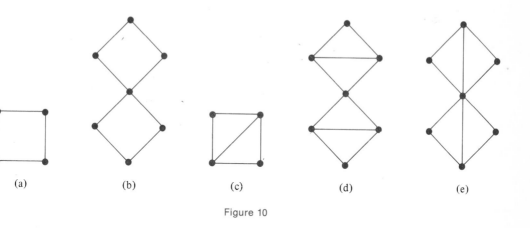

Figure 10

7. In the pipeline problem of Figure 7, the labels on edges e_6 and e_7 could have been interchanged. Interchange them and apply the algorithm for minimal spanning trees to the new sequence of edges. What is the total weight of your new spanning tree?

8. Reverse the order of the edges in Figure 7, i.e., switch the names on e_1 and e_{14}, on e_2 and e_{13}, etc. Redraw the figure and apply the algorithm for minimal spanning trees to this sequence of edges. The tree you get will be a "maximal spanning tree" with respect to the weights in Figure 7. What is its total weight?

9. (a) For each tree in Figure 5, count the number of vertices and the number of edges, and compare the answers.
   (b) On the basis of your work in part (a), make a general conjecture concerning a tree with *n* vertices.
   (c) Why must a connected graph with *n* vertices have at least *n* − 1 edges? *Hint*: It has at least one spanning tree.

10. Vertices of a tree having degree 1 are called **leaves**.
    (a) Count the leaves on the trees in Figure 5.
    (b) Every tree [with more than one vertex] has at least two leaves. Think about this and see if you can give an explanation.
    (c) Draw three trees each of which has exactly two leaves.

11. Consider the tree in Figure 11(a).
    (a) Select several pairs of vertices and, for each pair, give all paths connecting them which repeat no edges.
    (b) Based on your work in part (a), make a general conjecture about trees.

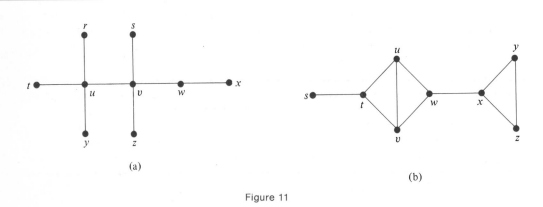

(a)

(b)

Figure 11

12. (a) Consider the graph in Figure 11(b). Here is the vertex sequence for a path from *u* to *z*: *u t s t v w u v w x z y z*. Cut out parts of this path to get an acyclic path from *u* to *z* which repeats no edges. *Hint*: Why visit a vertex twice?
    (b) Describe a general algorithm which, given a path from a vertex *u* to a vertex *v*, produces an acyclic path from *u* to *v*.

13. Consider the graph in Figure 11(b).
    (a) Which edges are parts of cycles? For example, edge *t u* is part of the cycle *t u ẘ v t*.
    (b) Which edges have the property that their removal from the graph would disconnect the graph?
    (c) Can you make a general conjecture based on this example?

14. Consider a tree with *n* vertices. It has exactly *n* − 1 edges [Exercise 9] and so the sum of the degrees of its vertices is 2*n* − 2 [Exercise 13 of § 0.1].
    (a) A certain tree has two vertices of degree 4, one vertex of degree 3 and one vertex of degree 2. If the other vertices have degree 1, how many vertices are

there in the graph? *Hint*: If the tree has n vertices, $n - 4$ of them will have to have degree 1.

(b) Draw a tree as described in part (a).

15. Repeat Exercise 14 for a tree with two vertices of degree 5, three of degree 3, two of degree 2 and the rest of degree 1.

§ 0.3   Matrices for Graphs

Informally, a graph consists of dots and lines, i.e., vertices and edges. How can we, or a computer, describe such a structure in an unambiguous manner? We might list all vertices and all edges and specify for each edge which vertices it connects. This is how we will finally define a graph after we have introduced sets and functions. Another way to completely describe a graph would be to consider all pairs of vertices and to associate with each pair the number of edges connecting them. Since there might be loops in the graph, we allow ourselves to consider pairs where the two vertices are actually the same vertex. By arranging these numbers of edges in a suitable rectangular format we get what's called the matrix of a graph. [The plural of "matrix" is "matrices."]

As an illustration, we find the matrix of the graph in Figures 2(c) and 3(a) of § 0.1, which we have redrawn in Figure 1(a). We will want to list the vertices in order, so we have renamed them v_1, v_2, v_3, v_4, using subscripts to indicate the order. The matrix $\mathbf{M}$ of the graph is given in Figure 1(b). The entries in $\mathbf{M}$ tell how many edges there are between one vertex and another. For example, the entry 1 in the fourth horizontal row and second vertical column of $\mathbf{M}$, designated $\mathbf{M}[4, 2]$, is the number of edges from the fourth vertex v_4 to the second vertex v_2. There is just one such edge, so the entry is 1. [See Figure 2(a).] Similarly, $\mathbf{M}[1, 2] = 2$ because there are two edges connecting v_1 to v_2 [Figure 2(b)], and $\mathbf{M}[3, 4] = 0$ since there are no edges from v_3 to v_4. More generally, the entry $\mathbf{M}[i, j]$ in row i and column j is the number of edges from vertex v_i to vertex v_j. When we describe the location of an entry in a matrix we always tell first which row it's in and then which column.

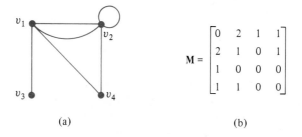

(a)   (b)

Figure 1

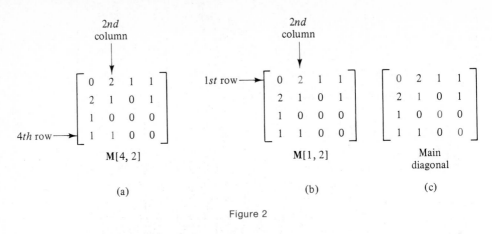

Figure 2

Note that the matrix **M** contains all the information about the connections in the graph, so we can recover the graph from its matrix alone, without the picture. And we can read some information right off the matrix. For example, there are parallel edges if and only if some matrix entry is larger than 1. The loops are all indicated on the main diagonal; see Figure 2(c). The entry $M[2, 2] = 1$ shows that there is a loop at v_2. The other diagonal matrix entries are 0 and so there are no other loops. If a vertex v_i has no loops, its degree is simply the sum of the entries of the ith row [or column]. For example, v_1 has degree 4 since the entries in row 1 add to 4. Finally, observe that the matrix of a graph will be symmetric about the main diagonal, since $M[i, j] = M[j, i]$ for all choices of i and j. Some properties of graphs are not easy to read from their matrices [e.g., the existence of Hamilton circuits], but sometimes we will be able to manipulate the matrices to learn about the graphs.

In § 0.2 we made somewhat vague references to graphs that are "essentially the same." We will not make the notion really precise [the magic word will be "isomorphism"] until after we have a precise set-theoretic definition of graph. But here is an observation that might help give you a sense of what we are looking for. Two graphs are essentially the same [i.e., are "isomorphic"] if their vertices can be ordered so that the matrices of the two graphs are identical. Figure 3 shows all possible trees with 6 vertices. Any tree with 6 vertices must be isomorphic with exactly one of the given trees, and no two trees in the figure are isomorphic with each other.

Isomorphic graphs must share graph-theoretic properties; they must have the same numbers of edges, vertices, loops, cycles of length 4, vertices of even degree, etc. One might hope that there would be a reasonably short list of properties such that any two graphs that shared the properties on the list would have to be isomorphic. No such luck. In general, there are no easy ways to check whether two graphs are isomorphic. Of course, if you can find some graph-theoretic property that they do *not* share, you are done and the graphs *aren't* isomorphic.

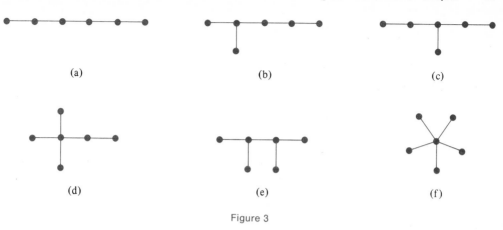

Figure 3

As an illustration, with suitable choices for their orderings the graphs in Figures 3(b) and 3(c) have matrices

$$
\begin{bmatrix}
0 & 1 & 0 & 0 & 0 & 0 \\
1 & 0 & 1 & 0 & 0 & 1 \\
0 & 1 & 0 & 1 & 0 & 0 \\
0 & 0 & 1 & 0 & 1 & 0 \\
0 & 0 & 0 & 1 & 0 & 0 \\
0 & 1 & 0 & 0 & 0 & 0
\end{bmatrix}
\quad \text{and} \quad
\begin{bmatrix}
0 & 1 & 0 & 0 & 0 & 0 \\
1 & 0 & 1 & 0 & 0 & 0 \\
0 & 1 & 0 & 1 & 0 & 1 \\
0 & 0 & 1 & 0 & 1 & 0 \\
0 & 0 & 0 & 1 & 0 & 0 \\
0 & 0 & 1 & 0 & 0 & 0
\end{bmatrix}
$$

We can see that these two graphs are put together differently, so are not isomorphic, since one graph has two leaves [i.e., vertices of degree 1] connected to its vertex of degree 3 while the other graph has only one. This information is contained in the matrices, but we have to be looking for it to find it. At first glance the matrices look different, but not a *lot* different, so we might think we could just rearrange rows and columns to turn one into the other. One way to check this out would be to try all 6! possible orderings of the graph of Figure 3(b) and see if any of the matrices we get are identical with our matrix for Figure 3(c). That factorial sign tells us that in general such a method of checking isomorphism is impractical for large graphs. No method is known which is substantially better than this for arbitrary graphs, though much more effective algorithms are available for graphs of special types, such as trees.

Given a graph, it would be nice to be able to recognize which vertices can be reached from other vertices. Also, it would be useful to know how long the paths between them have to be. Let's tackle a simpler problem. Which vertices can be reached using paths of length 2? While we are at it, we'll keep track of how many such paths there are. We illustrate the ideas using the graph of Figure 1(a), reproduced in Figure 4(a) with the edges labeled. In Figure 4(b) we have created a new graph by connecting pairs of vertices by one edge for each path of length 2 between them in Figure 4(a). There are 6

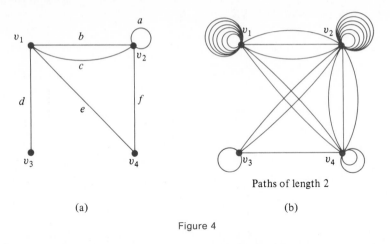

Paths of length 2

(a)   (b)

Figure 4

loops at vertex v_1 because $b\,b, b\,c, c\,b, c\,c, d\,d$ and $e\,e$ are paths of length 2 from v_1 to v_1. The 3 edges from v_2 to v_4 represent the paths $b\,e, c\,e$ and $a\,f$ from v_2 to v_4. No edge in Figure 4(b) connects v_1 and v_3 since there are no paths of length 2 connecting these vertices.

The tedious analysis of the last paragraph for creating Figure 4(b) can be avoided by using the matrix **M** for the original graph. Let's count again the paths of length 2 from v_1 to v_2. Such a path passes through v_1, v_2, v_3 or v_4 on the way, so we can count the number of paths from v_1 to v_2 through v_1, the number through v_2, through v_3 and through v_4 and add these numbers to get the total. Now to get the number of paths with vertex sequence $v_1\,v_2\,v_2$, for example, we count the edges from v_1 to v_2 and the edges from v_2 to v_2 and multiply these numbers to get $2 \cdot 1 = 2$. The corresponding paths are $b\,a$ and $c\,a$. The numbers we multiply, $\mathbf{M}[1, 2]$ and $\mathbf{M}[2, 2]$, are given in the matrix **M**. The general situation for paths from v_1 to v_2 is illustrated in Table 1. The total number of paths of length 2 from v_1 to v_2 is thus

$$\mathbf{M}[1, 1] \cdot \mathbf{M}[1, 2] + \mathbf{M}[1, 2] \cdot \mathbf{M}[2, 2] + \mathbf{M}[1, 3] \cdot \mathbf{M}[3, 2]$$
$$+ \mathbf{M}[1, 4] \cdot \mathbf{M}[4, 2] = 3.$$

This turns out to be the entry in the first row and second column of a matrix called the "product matrix" $\mathbf{MM} = \mathbf{M}^2$. It is the sum of products of entries

TABLE 1.

Vertex v_i	Number of edges from v_1 to v_i	Number of edges from v_i to v_2	Number of paths with vertex sequence $v_1\,v_i\,v_2$
v_1	$\mathbf{M}[1, 1] = 0$	$\mathbf{M}[1, 2] = 2$	$\mathbf{M}[1, 1] \cdot \mathbf{M}[1, 2] = 0 \cdot 2 = 0$
v_2	$\mathbf{M}[1, 2] = 2$	$\mathbf{M}[2, 2] = 1$	$\mathbf{M}[1, 2] \cdot \mathbf{M}[2, 2] = 2 \cdot 1 = 2$
v_3	$\mathbf{M}[1, 3] = 1$	$\mathbf{M}[3, 2] = 0$	$\mathbf{M}[1, 3] \cdot \mathbf{M}[3, 2] = 1 \cdot 0 = 0$
v_4	$\mathbf{M}[1, 4] = 1$	$\mathbf{M}[4, 2] = 1$	$\mathbf{M}[1, 4] \cdot \mathbf{M}[4, 2] = 1 \cdot 1 = 1$

from the first row of **M** and the second column of **M.** We won't pursue this further now, but only want to observe that forming product matrices is a purely mechanical process that computers and many people can perform quickly, and the entries in $\mathbf{M}^2$ tell us the number of paths of length 2 connecting vertices. In other words, $\mathbf{M}^2$ is the matrix for the graph in Figure 4(b). Similarly, the entries in the product $\mathbf{M}^2\mathbf{M} = \mathbf{M}^3$ tell us the number of paths of length 3 connecting vertices. Etc.

There is another matrix of interest related to a graph. It is the **reachability matrix R.** An entry $\mathbf{R}[i, j]$ is 1 if there is some path from vertex v_i to v_j; otherwise it is 0. Thus **R** can be determined once all of the powers $\mathbf{M}^2$, $\mathbf{M}^3$, $\mathbf{M}^4, \ldots$ are. This looks like an infinite task, but fortunately a theorem tells us the following: If a graph has n vertices and there is a path from v_i to v_j, then there is a path from v_i to v_j having length $\leq n$. Hence **R** can be determined once the powers $\mathbf{M}^2$, $\mathbf{M}^3, \ldots, \mathbf{M}^n$ are known. Our example has 4 vertices, so it suffices to examine **M**, $\mathbf{M}^2$, $\mathbf{M}^3$ and $\mathbf{M}^4$. As with any connected graph, the reachability matrix **R** for our example will consist of all 1's. We can tell that our little graph is connected just by looking at its picture, but the matrix procedures would be useful for testing the connectedness of a very large graph or one which was known to us only by its matrix.

We end this section by indicating why the famous Four-Color Theorem is a theorem in graph theory. The theorem asserts that every map in the plane consisting of connected regions without holes can be colored with four colors without coloring two adjacent regions the same color. The theorem was not proved until 1976, after decades of work. The proof finally given used mathematical ingenuity to reduce the problem to a question which could be settled by a computer search. It was one of the first proofs of a significant mathematical result using a computer as a tool.

Given a map, we can construct a graph by viewing regions as vertices and putting an edge between vertices if and only if their corresponding regions are adjacent. We illustrate this in Figure 5. The map coloring problem turns into a vertex coloring problem; vertices connected by an edge must be given different colors. Note that we do not regard Arizona and Colorado (nor Utah and New Mexico) as adjacent.

We said at the beginning that this chapter is meant to be an informal introduction to some of the mathematical topics which are important in computer science and which are the focus of the rest of the book. Our goal has been to provide interesting and concrete examples of problems and methods, so an informal approach was suitable. When we try to generalize from the examples, however, we will need to be careful. It is easy to be fooled into believing that facts about special examples we know and love are true in general. To be sure we can trust our conclusions, we need to start from precisely stated definitions and proceed logically.

In Chapter 1 we lay the foundation and establish some working notation. Chapter 2 begins our study of logic. The material will get harder to read—less like a novel—but the payoff is the knowledge that our conclusions

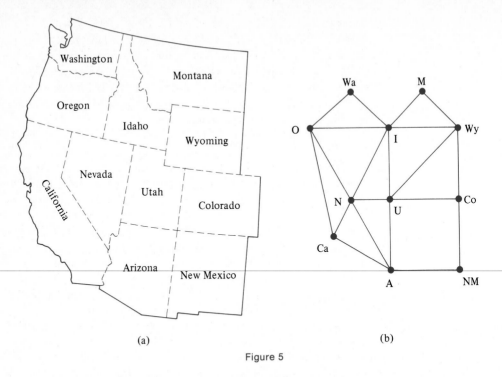

(a)   (b)

Figure 5

are true, and not just plausible assertions. One reason you are reading this book is to learn some facts. Another reason, perhaps even more important, is to get practice working with logical arguments. That practice is about to start.

EXERCISES 0.3

1. Give the matrices for the trees in Figure 3.

2. Write matrices for the graphs in Figure 6.

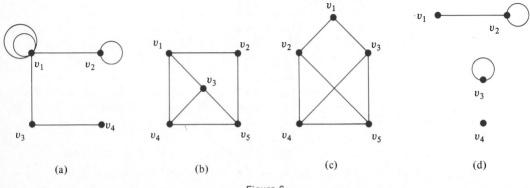

(a)   (b)   (c)   (d)

Figure 6

3. Are the graphs in Figures 6(b) and 6(c) isomorphic? Explain.

4. The trees in Figure 7 all have 6 vertices. Indicate which tree in Figure 3 each is isomorphic with.

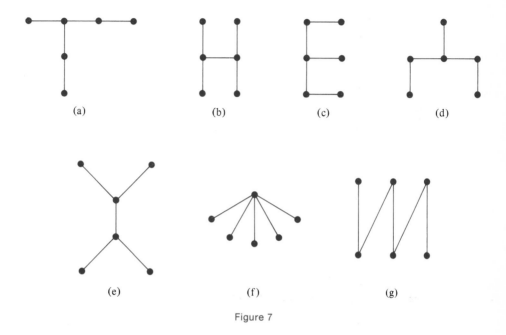

(a)   (b)   (c)   (d)

(e)   (f)   (g)

Figure 7

5. Draw pictures of all connected graphs with 4 edges and 4 vertices. Don't forget loops and parallel edges.

6. For each matrix in Figure 8, draw a graph having the matrix.

$$
\begin{bmatrix} 1 & 1 & 1 & 0 \\ 1 & 1 & 0 & 1 \\ 1 & 0 & 1 & 1 \\ 0 & 1 & 1 & 1 \end{bmatrix}
\begin{bmatrix} 0 & 0 & 1 & 2 \\ 0 & 1 & 0 & 1 \\ 1 & 0 & 1 & 0 \\ 2 & 1 & 0 & 0 \end{bmatrix}
\begin{bmatrix} 0 & 2 & 1 & 0 \\ 2 & 0 & 0 & 1 \\ 1 & 0 & 1 & 0 \\ 0 & 1 & 0 & 1 \end{bmatrix}
\begin{bmatrix} 2 & 0 & 0 & 0 & 0 \\ 0 & 2 & 0 & 0 & 0 \\ 0 & 0 & 1 & 0 & 0 \\ 0 & 0 & 0 & 0 & 1 \\ 0 & 0 & 0 & 1 & 0 \end{bmatrix}
$$

(a)   (b)   (c)   (d)

Figure 8

7. Show that two of the graphs with matrices in Figure 8 are isomorphic.

8. Make a table similar to Table 1 to count the paths of length 2 from v_1 to itself in Figure 4(a).

9. Consider the graph in Figure 6(d).
   (a) Draw the graph that shows all paths of length 2, as we did in Figure 4(b).
   (b) Is the graph in part (a) connected?
   (c) Give the matrix for the graph in part (a).
   (d) Give the reachability matrix **R**.

10. Repeat Exercise 9 for the graph in Figure 3(a); see Exercise 1.

11. Repeat Exercise 9 for the graph in Figure 3(c); see Exercise 1.

12. Use Figure 5 to show that some maps require 4 different colors. *Hint*: Color Nevada and all its neighbors.

13. Let **M** be the matrix for a graph. Give a formula for the degree of a vertex v_i in terms of **M** that works even if there are loops at v_i. *Suggestion*: Look at the matrices for the graphs in Figures 1, 6(a) or 6(d).

14. Consider the matrix of some graph.
   (a) Suppose all the numbers in some row are 0. What does this tell us about the corresponding vertex?
   (b) Suppose all the numbers in some row are nonzero. What does this tell us about the vertex?
   (c) What can you conclude about connectedness of the graph in the situations described in parts (a) and (b)?

CHAPTER HIGHLIGHTS

To check your understanding of the material in this chapter, we recommend that you consider each item listed below and:
   (a) Satisfy yourself that you can define each concept and describe each method.
   (b) Give at least one reason why the item was included in the chapter.
   (c) Think of at least one example of each concept and at least one situation in which the fact or method would be useful.

This chapter is intended to be informal. More precise definitions and justifications of some of the material appear in later chapters.

Concepts

graph
   vertex, edge, degree
   path
      length, vertex sequence
      cycle
      Euler circuit
      Hamilton circuit
   connected
   tree

   weighted graph
     minimal spanning tree
  matrix description of graph
    reachability matrix
  graph isomorphism

Facts

A graph has an Euler circuit if and only if it is connected and all its vertices
   have even degree.

The simpleminded approach to constructing Euler circuits can fail.

Hamilton circuits are in general hard to construct.

Graph isomorphism can be hard to detect.

Map coloring can be converted to a question about graphs.

Methods

FLEURY'S algorithm to construct Euler circuits.

Greedy algorithm to construct minimal spanning trees.

Preview of matrix methods for finding number of paths and reachability.
   [More on this later.]

1

SETS

In this chapter we introduce the notation and terminology of set theory that is basic for the remainder of the book.

§ 1.1 Some Special Sets

In the past few decades it has become traditional to use set theory as the underlying basis for mathematics. That is, the concepts of "set" and "membership" are taken as basic undefined terms and the rest of mathematics is defined or described in these terms. A set is a collection of objects; the definition of a set must be unambiguous in the sense that it must be possible to decide whether particular objects belong to the set. We will usually denote sets by capital letters such as A, B, S or X. Objects are usually denoted by lowercase letters such as a, b, s or x. An object a which belongs to a set S is called a **member** of S or **element** of S. If a is an object and A is a set we write $a \in A$ to mean that a is a member of A and $a \notin A$ to mean that a is not a member of A. The symbol $\in$ can be read as a verb phrase "is an element of," "belongs to" or "is in" or as the preposition "in," depending on context.

Specific sets can be written in a variety of ways. A few especially common and important sets will be given their own names, i.e., their own symbols. We will reserve the symbol $\mathbb{N}$ for the set of **natural numbers**:

$$\mathbb{N} = \{0, 1, 2, 3, 4, 5, 6, \ldots\}.$$

Note that we include 0 among the natural numbers.

We write $\mathbb{P}$ for the set of **positive integers**:

$$\mathbb{P} = \{1, 2, 3, 4, 5, 6, 7, \ldots\}.$$

Many mathematics texts write this set as $\mathbb{N}$ instead. The set of all **integers**, positive, zero or negative, will be denoted by $\mathbb{Z}$ [for the German word "Zahl"]. Numbers of the form m/n where $m \in \mathbb{Z}$, $n \in \mathbb{Z}$ and $n \neq 0$ are called **rational numbers** [since they are ratios of integers]. The set of all rational numbers is denoted by $\mathbb{Q}$. The set of all **real numbers**, rational or not, is denoted by $\mathbb{R}$. Thus $\mathbb{R}$ contains all the numbers in $\mathbb{Q}$ and $\mathbb{R}$ also contains $\sqrt{2}, \sqrt{3}, \sqrt[3]{2}, -\pi, e$ and many many other numbers.

Small finite sets can be listed using braces $\{\ \}$ and commas. For example, $\{2, 4, 6, 8, 10\}$ is the set consisting of the five positive even integers less than 12 and $\{2, 3, 5, 7, 11, 13, 17, 19\}$ consists of the eight primes less than 20. Readers who need to be reminded what "even" or "prime" mean may consult the dictionary that appears after Chapter 11. Two sets are **equal** if they contain the same elements. Thus

$$\{2, 4, 6, 8, 10\} = \{10, 8, 6, 4, 2\} = \{2, 8, 2, 6, 2, 10, 4, 2\};$$

the order of the listing is irrelevant and there is no advantage [or harm] in listing elements more than once.

Large finite sets and even infinite sets can be listed with the aid of the mathematician's etcetera, namely three dots $\ldots$, provided the meaning of the three dots is clear. Thus $\{1, 2, 3, \ldots, 1000\}$ represents the set of positive integers less than or equal to 1000 and $\{3, 6, 9, 12, \ldots\}$ presumably represents the infinite set of positive integers that are divisible by 3. On the other hand, the meaning of $\{1, 2, 3, 5, 8, \ldots\}$ may be less than perfectly clear. The somewhat vague use of three dots is not always satisfactory, especially in computer science, and we will develop techniques for unambiguously describing such sets without using dots.

Sets are often described by properties of their elements via the notation

$$\{\quad : \quad\}.$$

Before the colon a variable [n or x, for instance] is indicated and after the colon the properties are given. For example,

$$\{n : n \in \mathbb{N} \text{ and } n \text{ is even}\}$$

represents the set of nonnegative even integers, i.e., the set $\{0, 2, 4, 6, 8, 10, \ldots\}$. The colon is always read "such that" and so the above is read "the set of all n such that n is in $\mathbb{N}$ and n is even." Similarly,

$$\{x : x \in \mathbb{R} \text{ and } 1 \leqq x < 3\}$$

represents the set of all real numbers that are greater than or equal to 1 and less than 3. The number 1 belongs to the set, but 3 does not. Just to streamline notation, the last two sets can be written as

$$\{n \in \mathbb{N} : n \text{ is even}\} \quad \text{and} \quad \{x \in \mathbb{R} : 1 \leq x < 3\}.$$

The first set is then read "the set of all n in $\mathbb{N}$ such that n is even."

Another way to list a set is to specify a rule for obtaining its elements using some other set of elements. For example, $\{n^2 : n \in \mathbb{N}\}$ represents the set of all integers that are the squares of integers in $\mathbb{N}$, i.e.,

$$\{n^2 : n \in \mathbb{N}\} = \{m \in \mathbb{N} : m = n^2 \text{ for some } n \in \mathbb{N}\}$$
$$= \{0, 1, 4, 9, 16, 25, 36, \ldots\}.$$

Note that this set equals $\{n^2 : n \in \mathbb{Z}\}$. Similarly, $\{(-1)^n : n \in \mathbb{N}\}$ represents the set obtained by evaluating $(-1)^n$ for all $n \in \mathbb{N}$, so that

$$\{(-1)^n : n \in \mathbb{N}\} = \{-1, 1\}.$$

This set has only two elements.

Now consider two sets S and T. We say that T is a **subset** of S provided every element of T belongs to S. If T is a subset of S, we write $T \subseteq S$. The symbol $\subseteq$ can be read as "is a subset of." Two sets S and T are **equal** if they contain the same elements. Thus $S = T$ if and only if $T \subseteq S$ and $S \subseteq T$.

EXAMPLE 1   (a) We have $\mathbb{P} \subseteq \mathbb{N}$, $\mathbb{N} \subseteq \mathbb{Z}$, $\mathbb{Z} \subseteq \mathbb{Q}$, $\mathbb{Q} \subseteq \mathbb{R}$. As with the familiar inequality $\leq$, we can run these assertions together:

$$\mathbb{P} \subseteq \mathbb{N} \subseteq \mathbb{Z} \subseteq \mathbb{Q} \subseteq \mathbb{R}.$$

(b) Since 2 is the only even prime, we have

$$\{n \in \mathbb{P} : n \text{ is prime and } n \geq 3\} \subseteq \{n \in \mathbb{P} : n \text{ is odd}\}.$$

(c) Consider again any set S. Obviously $x \in S$ implies $x \in S$ and so $S \subseteq S$. That is, we regard a set as a subset of itself. This is why we use the notation $\subseteq$ rather than $\subset$. This usage is analogous to our usage of $\leq$ for real numbers. The inequality $x \leq 5$ is valid for many numbers, like 3, 1 and -73. It is also valid for $x = 5$, i.e., $5 \leq 5$. This last inequality looks a bit peculiar because we actually know more, namely $5 = 5$. But $5 \leq 5$ says that "5 is less than 5 or else 5 is equal to 5," and this is a true statement. Similarly, $S \subseteq S$ is true even though we know more, namely $S = S$. Statements like "$5 = 5$," "$5 \leq 5$," "$S = S$" or "$S \subseteq S$" do no harm and are often useful to call attention to the fact that a particular case of a more general statement is valid. ∎[1]

Occasionally we will write $T \subset S$ to mean that $T \subseteq S$ and $T \neq S$, i.e., T is a subset of S different from S. This usage of $\subset$ is analogous to our usage of $<$ for real numbers. If $T \subset S$ we say that T is a **proper subset** of S.

---

[1] We will use ∎ to signify the end of an example or proof.

We next introduce notation for some special subsets of $\mathbb{R}$, called **intervals**. For $a, b \in \mathbb{R}$ with $a < b$, we define

$$[a, b] = \{x \in \mathbb{R} : a \leq x \leq b\}; \qquad (a, b) = \{x \in \mathbb{R} : a < x < b\};$$

$$[a, b) = \{x \in \mathbb{R} : a \leq x < b\}; \qquad (a, b] = \{x \in \mathbb{R} : a < x \leq b\}.$$

The general rule is that brackets [,] signify that the endpoints are to be included and parentheses (,) signify that they are to be excluded. Intervals of the form $[a, b]$ are called **closed**; ones of the form (a, b) are **open**. It is also convenient to use the term "interval" for some unbounded sets which we describe using the symbols ∞ and $-\infty$, which do not represent real numbers but are simply part of the notation for the sets. Thus

$$[a, \infty) = \{x \in \mathbb{R} : a \leq x\}; \qquad (a, \infty) = \{x \in \mathbb{R} : a < x\};$$

$$(-\infty, b] = \{x \in \mathbb{R} : x \leq b\}; \qquad (-\infty, b) = \{x \in \mathbb{R} : x < b\}.$$

Set notation must be dealt with carefully. For example, $[0, 1]$, $(0, 1)$ and $\{0, 1\}$ all denote different sets. In fact, the intervals $[0, 1]$ and $(0, 1)$ are infinite sets while $\{0, 1\}$ has only two elements.

Consider the following sets:

$$\{n \in \mathbb{N} : 2 < n < 3\}, \qquad \{x \in \mathbb{R} : x^2 < 0\},$$

$$\{r \in \mathbb{Q} : r^2 = 2\}, \qquad \{x \in \mathbb{R} : x^2 + 1 = 0\}.$$

These sets have one property in common: they contain no elements. From a strictly logical point of view, they contain the same elements and so they are equal in spite of the different descriptions. This unique set having no elements at all is called the **empty set**. We will use two notations for it, the suggestive $\{\ \ \}$ and the standard $\varnothing$. The symbol $\varnothing$ is not a Greek phi ϕ; it is borrowed from the Norwegian alphabet and non-Norwegians should read it as "empty set." We regard $\varnothing$ as a subset of every set S because we regard the statement "$x \in \varnothing$ implies $x \in S$" as logically true in a vacuous sense. You should probably take this explanation on faith until you study § 2.5.

Sets are themselves objects and so can be members of other sets. The set $\{\{1, 2\}, \{1, 3\}, \{2\}, \{3\}\}$ has four members, namely $\{1, 2\}, \{1, 3\}, \{2\}$ and $\{3\}$. If we had a box containing two sacks full of marbles we would consider it to be a box of sacks, rather than a box of marbles, so it would contain two members. Likewise, if A is a set, then $\{A\}$ is a set with one member, namely A, no matter how many members A itself has. A box containing an empty sack contains something, namely a sack, so it is not an empty box. In the same way, $\{\varnothing\}$ is a set with one member, whereas $\varnothing$ is a set with no members, so $\{\varnothing\}$ and $\varnothing$ are different sets. We have $\varnothing \in \{\varnothing\}$ and even $\varnothing \subseteq \{\varnothing\}$, but $\varnothing \notin \varnothing$.

The set of all subsets of a set S is called the **power set** of S and will be denoted $\mathscr{P}(S)$. Clearly the empty set $\varnothing$ and the set S itself are elements of $\mathscr{P}(S)$, i.e., $\varnothing \in \mathscr{P}(S)$ and $S \in \mathscr{P}(S)$.

EXAMPLE 2   (a) We have $\mathscr{P}(\varnothing) = \{\varnothing\}$ since $\varnothing$ is the only subset of $\varnothing$.

(b) Consider a typical one-element set, say $S = \{a\}$. Then $\mathscr{P}(S) = \{\varnothing, \{a\}\}$ has two elements.

(c) If $S = \{a, b\}$ and $a \neq b$, then $\mathscr{P}(S) = \{\varnothing, \{a\}, \{b\}, \{a, b\}\}$ has four elements.

(d) If $S = \{a, b, c\}$ has three elements, then

$$\mathscr{P}(S) = \{\varnothing, \{a\}, \{b\}, \{c\}, \{a, b\}, \{a, c\}, \{b, c\}, \{a, b, c\}\}$$

has eight elements.

(e) Let S be a finite set. Note that if S has n elements and if $n \leq 3$, then $\mathscr{P}(S)$ has 2^n elements, as shown in parts (a)–(d) above. This is not an accident, as we show in Example 6 of § 2.6.

(f) If S is infinite, then $\mathscr{P}(S)$ is also infinite, of course. ∎

We introduce one more special kind of set, denoted by Σ^*. Sets like this, which may be unfamiliar to many readers, will recur throughout this book. The idea is to allow a rather general, but precise, mathematical treatment of languages. First we need an alphabet. An **alphabet** is a finite nonempty set Σ [Greek capital sigma] whose members are symbols, often called **letters** of Σ, and which is subject to some minor restrictions which we will discuss at the end of this section. Given an alphabet Σ, a **word** is any finite string of letters from Σ. We denote the set of all words using letters from Σ by Σ^* [sigma-star]. Any subset of Σ^* is called a **language** over Σ.

EXAMPLE 3   (a) Let $\Sigma = \{a, b, c, d, \ldots, z\}$ consist of the twenty six letters of the "English" alphabet. *Any* string of letters from Σ belongs to Σ^*. Thus Σ^* contains *math, is, fun, aint, lieblich, amour, zzyzzoomph, etcetera*, etc. Since Σ^* contains *a, aa, aaa, aaaa, aaaaa*, etc., Σ^* is clearly an infinite set. To be definite, we could define the **American language** L to be the subset of Σ^* consisting of words in the latest edition of *Webster's New World Dictionary of the American Language.* Thus

$$L = \{a, aachen, aardvark, aardwolf, \ldots, zymurgy\},$$

a large but finite set.

(b) To get simple examples and yet illustrate the ideas, we will frequently take Σ to be a 2-element set $\{a, b\}$. In this case Σ^* contains *a, b, ab, ba, bab, babbabb*, etc.; again Σ^* is infinite.

(c) If $\Sigma = \{0, 1\}$, then the set B of words in Σ^* that begin with 1 is exactly the set of binary notations for positive integers. That is,

$$B = \{1, 10, 11, 100, 101, 110, 111, 1000, 1001, \ldots\}.$$ ∎

There is a special word in Σ^* somewhat analogous to the empty set, called the **empty word** or **null word**; it is the string with no letters at all and is denoted by ϵ [Greek lowercase epsilon].

EXAMPLE 4   (a) If $\Sigma = \{a, b\}$, then

$$\Sigma^* = \{\epsilon, a, b, aa, ab, ba, bb, aaa, aab, aba, abb, baa, bab, bba, \ldots\}.$$

(b) If $\Sigma = \{0, 1, 2\}$, then

$$\Sigma^* = \{\epsilon, 0, 1, 2, 00, 01, 02, 10, 11, 12, 20, 21, 22, 000, 001, 002, \ldots\}.$$

(c) If $\Sigma = \{a\}$, then

$$\Sigma^* = \{\epsilon, a, aa, aaa, aaaa, aaaaa, aaaaaa, \ldots\}.$$

This example doesn't contain very useful languages, but it will serve to illustrate various concepts.

(d) Various computer languages fit our definition of language. For example, the alphabet Σ for one version of ALGOL has 113 elements. Σ includes letters, the digits $0, 1, 2, \ldots, 9$ and a variety of operators, including sequential operators like "go to" and "if." As usual, Σ^* contains all possible finite strings of letters from Σ, without regard to meaning. The subset of Σ^* consisting of those strings accepted for execution by an ALGOL compiler on a given computer is a well-defined and useful subset of Σ^*; we could call it the ALGOL language determined by the compiler. ∎

As promised, we now discuss the restrictions needed for Σ. Suppose that Σ contains not only the symbols a and b but also the symbol ab. Is the string aab a string of three letters a, a and b in Σ or a string of two letters a and ab? There is no way to tell. So if a, b and ab have some particular significance in an application, the ambiguity here may make it impossible to attach a meaning to the word aab. As another example, we would not want Σ to contain ab, aba and bab; if it did, the letters in the word

$$ababab = (ab)(ab)(ab) = (aba)(bab)$$

would be ambiguous. To avoid these problems, we do not allow Σ to contain letters which are themselves strings beginning with other letters of Σ. Thus $\Sigma = \{a, b, c\}$, $\Sigma = \{a, b, ca\}$ and $\Sigma = \{a, b, Ab\}$ are allowed, but $\Sigma = \{a, b, c, ca\}$ is not. With this agreement, we can unambiguously define **length**(w) for a word w in Σ^* to be the number of letters from Σ in w, counting each appearance of a letter. For example, if $\Sigma = \{a, b\}$, then length$(aab) =$ length$(bab) = 3$. We also define length$(\epsilon) = 0$. A more precise definition is given in § 3.6.

One final word: We will use w, w_1, etc. as variable names for words. This should cause no confusion even though w also happens to be a letter of the English alphabet.

EXAMPLE 5   If $\Sigma = \{a, b\}$ and $A = \{w \in \Sigma^* : \text{length}(w) = 2\}$, then $A = \{aa, ab, ba, bb\}$. If

$$B = \{w \in \Sigma^* : \text{length}(w) \text{ is even}\},$$

then B is the infinite set $\{\epsilon, aa, ab, ba, bb, aaaa, aaab, aaba, aabb, \ldots\}$. Note that A is a subset of B. ∎

EXERCISES 1.1

Terms such as "divisible," "prime" and "even" are defined in the dictionary that appears after Chapter 11.

1. List five elements in each of the following sets.
   (a) $\{n \in \mathbb{N} : n$ is divisible by $5\}$ (b) $\{2n + 1 : n \in \mathbb{P}\}$
   (c) $\mathcal{P}(\{1, 2, 3, 4, 5\})$ (d) $\{2^n : n \in \mathbb{N}\}$
   (e) $\{1/n : n \in \mathbb{P}\}$ (f) $\{r \in \mathbb{Q} : 0 < r < 1\}$
   (g) $\{n \in \mathbb{N} : n + 1$ is prime$\}$

2. List the elements in the following sets.
   (a) $\{1/n : n = 1, 2, 3, 4\}$
   (b) $\{n^2 - n : n = 0, 1, 2, 3, 4\}$
   (c) $\{1/n^2 : n \in \mathbb{P}, n$ is even and $n < 11\}$
   (d) $\{2 + (-1)^n : n \in \mathbb{N}\}$

3. List five elements in each of the following sets.
   (a) Σ^* where $\Sigma = \{a, b, c\}$
   (b) $\{w \in \Sigma^* : \text{length}(w) \leq 2\}$ where $\Sigma = \{a, b\}$
   (c) $\{w \in \Sigma^* : \text{length}(w) = 4\}$ where $\Sigma = \{a, b\}$
   Which sets above contain the empty word ϵ?

4. Determine the following sets, i.e., list their elements if they are nonempty and write $\varnothing$ if they are empty.
   (a) $\{n \in \mathbb{N} : n^2 = 9\}$ (b) $\{n \in \mathbb{Z} : n^2 = 9\}$
   (c) $\{x \in \mathbb{R} : x^2 = 9\}$ (d) $\{n \in \mathbb{N} : 3 < n < 7\}$
   (e) $\{n \in \mathbb{Z} : 3 < |n| < 7\}$ (f) $\{x \in \mathbb{R} : x^2 < 0\}$
   (g) $\{n \in \mathbb{N} : n^2 = 3\}$ (h) $\{x \in \mathbb{Q} : x^2 = 3\}$
   (i) $\{x \in \mathbb{R} : x < 1$ and $x \geq 2\}$ (j) $\{3n + 1 : n \in \mathbb{N}$ and $n \leq 6\}$
   (k) $\{n \in \mathbb{P} : n$ is prime and $n \leq 15\}$ [Recall that 1 isn't prime.]

5. How many elements are there in the following sets? Write ∞ if the set is infinite.
   (a) $\{n \in \mathbb{N} : n^2 = 2\}$ (b) $\{n \in \mathbb{Z} : 0 \leq n \leq 73\}$
   (c) $\{n \in \mathbb{Z} : 5 \leq |n| \leq 73\}$ (d) $\{n \in \mathbb{Z} : 5 < n < 73\}$
   (e) $\{n \in \mathbb{Z} : n$ is even and $|n| \leq 73\}$ (f) $\{x \in \mathbb{Q} : 0 \leq x \leq 73\}$
   (g) $\{x \in \mathbb{Q} : x^2 = 2\}$ (h) $\{x \in \mathbb{R} : x^2 = 2\}$
   (i) $\{x \in \mathbb{R} : .99 < x < 1.00\}$ (j) $\mathcal{P}(\{0, 1, 2, 3\})$
   (k) $\mathcal{P}(\mathbb{N})$ (l) $\{n \in \mathbb{N} : n$ is even$\}$
   (m) $\{n \in \mathbb{N} : n$ is prime$\}$ (n) $\{n \in \mathbb{N} : n$ is even and prime$\}$
   (o) $\{n \in \mathbb{N} : n$ is even or prime$\}$

6. How many elements are there in the following sets? Write ∞ if the set is infinite.
   (a) $\{-1, 1\}$ (b) $[-1, 1]$
   (c) $(-1, 1)$ (d) $\{n \in \mathbb{Z} : -1 \leq n \leq 1\}$
   (e) Σ^* where $\Sigma = \{a, b, c\}$
   (f) $\{w \in \Sigma^* : \text{length}(w) \leq 4\}$ where $\Sigma = \{a, b, c\}$

7. Consider the sets

   $$A = \{n \in \mathbb{P} : n \text{ is odd}\}, \qquad B = \{n \in \mathbb{P} : n \text{ is prime}\},$$

   $$C = \{4n + 3 : n \in \mathbb{P}\}, \qquad D = \{x \in \mathbb{R} : x^2 - 8x + 15 = 0\}.$$

   Which of these sets are subsets of which? Consider all sixteen possibilities.

8. Consider the sets $\{0, 1\}$, $(0, 1)$ and $[0, 1]$. True or False.
   (a) $\{0, 1\} \subseteq (0, 1)$ (b) $\{0, 1\} \subseteq [0, 1]$
   (c) $(0, 1) \subseteq [0, 1]$ (d) $\{0, 1\} \subseteq \mathbb{Z}$
   (e) $[0, 1] \subseteq \mathbb{Z}$ (f) $[0, 1] \subseteq \mathbb{Q}$
   (g) $1/2$ and $\pi/4$ are in $\{0, 1\}$ (h) $1/2$ and $\pi/4$ are in $(0, 1)$
   (i) $1/2$ and $\pi/4$ are in $[0, 1]$

9. Consider the following three alphabets: $\Sigma_1 = \{a, b, c\}$, $\Sigma_2 = \{a, b, ca\}$ and $\Sigma_3 = \{a, b, Ab\}$. Determine to which of Σ_1^*, Σ_2^* and Σ_3^* each word below belongs, and give its length as a member of each set to which it belongs.
   (a) *aba*   (b) *bAb*   (c) *cba*
   (d) *cab*   (e) *caab*   (f) *baAb*

10. Here is a question to think about. Let $\Sigma = \{a, b\}$ and imagine, if you can, a dictionary for all the nonempty words of Σ^* with the words arranged in the usual alphabetical order. All the words a, aa, aaa, $aaaa$, etc. must appear before the word ba. How far into the dictionary will you have to dig to find the word ba? How would the answer change if the dictionary contained only those words in Σ^* of length 5 or less?

11. Suppose that w is a nonempty word in Σ^*.
    (a) If the first [i.e., leftmost] letter of w is deleted, is the resulting string in Σ^*?
    (b) How about deleting letters from both ends of w? Are the resulting strings still in Σ^*?
    (c) If you had a device which could recognize letters in Σ and could delete letters from strings, how could you use it to determine if an arbitrary string of symbols is in Σ^*?

§ 1.2   Set Operations

In this section we introduce operations that allow us to create new sets from old sets. We define the **union** $A \cup B$ and **intersection** $A \cap B$ of sets A and B as follows:

$$A \cup B = \{x : x \in A \text{ or } x \in B \text{ or both}\};$$
$$A \cap B = \{x : x \in A \text{ and } x \in B\}.$$

We added "or both" to the definition of $A \cup B$ for emphasis and clarity. In ordinary English, the word "or" has two interpretations. Sometimes it is the **inclusive or** and means one or the other or both. This is the interpretation when a college catalog asserts: A student's program must include 2 years of science or 2 years of mathematics. At other times, "or" is the **exclusive or** and means one or the other but not both. This is the interpretation when a menu offers soup or salad. In mathematics we always interpret **or** as the "inclusive or" unless explicitly specified to the contrary. Sets A and B are said to be **disjoint** if they have no elements in common, i.e., if $A \cap B = \varnothing$.

For sets A and B, the **relative complement** $A \setminus B$ is the set of objects that are in A and not in B:

$$A \setminus B = \{x : x \in A \text{ and } x \notin B\} = \{x \in A : x \notin B\}.$$

It is the set obtained by removing from A all the elements of B that happen to be in A.

The **symmetric difference** $A \oplus B$ of the sets A and B is the set

$$A \oplus B = \{x : x \in A \text{ or } x \in B \text{ but not both}\}.$$

Note the use of the "exclusive or" here. It follows from the definition that

$$A \oplus B = (A \cup B) \backslash (A \cap B) = (A \backslash B) \cup (B \backslash A).$$

It is sometimes convenient to illustrate relations between sets with pictures called **Venn diagrams**, in which sets correspond to subsets of the plane. See Figure 1, where the indicated sets have been shaded in.

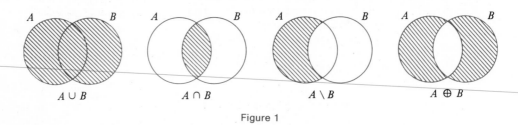

Figure 1

EXAMPLE 1   (a) Let $A = \{n \in \mathbb{N} : n \leq 11\}$, $B = \{n \in \mathbb{N} : n \text{ is even and } n \leq 20\}$ and $E = \{n \in \mathbb{N} : n \text{ is even}\}$. Then we have

$$A \cup B = \{0, 1, 2, 3, 4, 5, 6, 7, 8, 9, 10, 11, 12, 14, 16, 18, 20\},$$

$$A \cap B = \{0, 2, 4, 6, 8, 10\},$$

$$A \backslash B = \{1, 3, 5, 7, 9, 11\},$$

$$B \backslash A = \{12, 14, 16, 18, 20\},$$

$$A \oplus B = \{1, 3, 5, 7, 9, 11, 12, 14, 16, 18, 20\}.$$

We also have $E \cap B = B$, $B \backslash E = \{ \ \ \}$,

$$E \backslash B = \{n \in \mathbb{N} : n \text{ is even} \quad \text{and} \quad n \geq 22\} = \{22, 24, 26, 28, \ldots\},$$

$$\mathbb{N} \backslash E = \{n \in \mathbb{N} : n \text{ is odd}\} = \{1, 3, 5, 7, 9, 11, \ldots\},$$

$$A \oplus E = \{1, 3, 5, 7, 9, 11\} \cup \{n \in \mathbb{N} : n \text{ is even} \quad \text{and} \quad n \geq 12\}$$

$$= \{1, 3, 5, 7, 9, 11, 12, 14, 16, 18, 20, 22, \ldots\}.$$

(b) Consider the intervals $[0, 2]$ and $(0, 1]$. Then $(0, 1] \subseteq [0, 2]$ and so

$$(0, 1] \cup [0, 2] = [0, 2] \quad \text{and} \quad (0, 1] \cap [0, 2] = (0, 1].$$

Moreover, we have

$$(0, 1] \backslash [0, 2] = \{ \ \ \},$$

$$[0, 2] \backslash (0, 1] = \{0\} \cup (1, 2] \quad \text{and} \quad [0, 2] \backslash (0, 2) = \{0, 2\}.$$

(c) Let $\Sigma = \{a, b\}$, $A = \{\epsilon, a, aa, aaa\}$, $B = \{\epsilon, b, bb, bbb\}$ and $C = \{w \in \Sigma^* : \text{length}(w) \leq 2\}$. Then we have

$$A \cup B = \{\epsilon, a, b, aa, bb, aaa, bbb\}, \qquad A \cap B = \{\epsilon\},$$

$$A \setminus B = \{a, aa, aaa\}, \qquad\qquad B \setminus A = \{b, bb, bbb\},$$

$$A \cap C = \{\epsilon, a, aa\}, \qquad\qquad B \setminus C = \{bbb\},$$

$$C \setminus A = \{b, ab, ba, bb\}, \qquad\qquad A \setminus \Sigma = \{\epsilon, aa, aaa\}. \quad \blacksquare$$

It is often convenient to work within some fixed set such as $\mathbb{N}$, $\mathbb{R}$ or Σ^*. That is, it is convenient to fix some set U, which we call the **universe** or **universal set**, and to consider only elements in U and subsets of U. For $A \subseteq U$ the relative complement $U \setminus A$ is called the **absolute complement** or simply the **complement** of A and is denoted by A^c. Note that the relative complement $A \setminus B$ can be written in terms of the absolute complement: $A \setminus B = A \cap B^c$. In the Venn diagrams in Figure 2 we have drawn the universe U as a rectangle and shaded in the indicated sets.

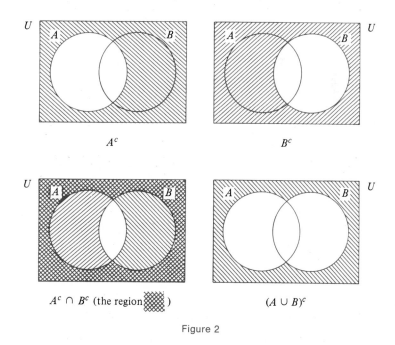

Figure 2

EXAMPLE 2   (a) If the universe is $\mathbb{N}$, and A and E are as in Example 1(a), then

$$A^c = \{n \in \mathbb{N} : n \geq 12\} \quad \text{and} \quad E^c = \{n \in \mathbb{N} : n \text{ is odd}\}.$$

(b) If the universe is $\mathbb{R}$, then $[0, 1]^c = (-\infty, 0) \cup (1, \infty)$, $(0, 1)^c = (-\infty, 0] \cup [1, \infty)$ and $\{0, 1\}^c = (-\infty, 0) \cup (0, 1) \cup (1, \infty)$. For any $a \in \mathbb{R}$, $[a, \infty)^c = (-\infty, a)$ and $(a, \infty)^c = (-\infty, a]$. $\quad \blacksquare$

Note that the last two Venn diagrams in Figure 2 show that $A^c \cap B^c = (A \cup B)^c$. This set identity and many others are true in general. Table 1 lists some basic identities for sets and set operations. Don't be overwhelmed by them; look at them one at a time. As some of the names of the laws suggest, many of them are analogues of laws from algebra. The idempotent laws are new [certainly $a + a = a$ fails for most numbers], and there is only one distributive law for numbers. Of course, the laws involving complementation are new. All sets in Table 1 are presumed to be subsets of some universal set U.

TABLE 1. Laws of Algebra of Sets

1a. $A \cup B = B \cup A$ b. $A \cap B = B \cap A$	commutative laws
2a. $(A \cup B) \cup C = A \cup (B \cup C)$ b. $(A \cap B) \cap C = A \cap (B \cap C)$	associative laws
3a. $A \cup (B \cap C) = (A \cup B) \cap (A \cup C)$ b. $A \cap (B \cup C) = (A \cap B) \cup (A \cap C)$	distributive laws
4a. $A \cup A = A$ b. $A \cap A = A$	idempotent laws
5a. $A \cup \emptyset = A$ b. $A \cup U = U$ c. $A \cap \emptyset = \emptyset$ d. $A \cap U = A$	identity laws
6. $(A^c)^c = A$	double complementation
7a. $A \cup A^c = U$ b. $A \cap A^c = \emptyset$	
8a. $U^c = \emptyset$ b. $\emptyset^c = U$	
9a. $(A \cup B)^c = A^c \cap B^c$ b. $(A \cap B)^c = A^c \cup B^c$	DeMorgan laws

Because of the associative laws, we can write the sets $A \cup B \cup C$ and $A \cap B \cap C$ without any parentheses and cause no confusion.

The identities in Table 1 can be verified in one of two ways. One can shade in the corresponding sets of a Venn diagram and observe that they are equal. Alternatively, one can show that sets S and T are equal by showing that $S \subseteq T$ and $T \subseteq S$; these inclusions can be verified by showing that $x \in S$ implies $x \in T$ and by showing that $x \in T$ implies $x \in S$. We give examples of both sorts of arguments, leaving most of the verifications to the interested reader.

EXAMPLE 3   The DeMorgan law 9a is illustrated by Venn diagrams in Figure 2. Here is a proof in which we first show $(A \cup B)^c \subseteq A^c \cap B^c$ and then we show $A^c \cap B^c \subseteq (A \cup B)^c$.

To show $(A \cup B)^c \subseteq A^c \cap B^c$, we consider an element x in $(A \cup B)^c$. Then $x \notin A \cup B$. In particular, $x \notin A$ and so we must have $x \in A^c$. Similarly, $x \notin B$ and so $x \in B^c$. Therefore $x \in A^c \cap B^c$. We have shown that $x \in (A \cup B)^c$ implies $x \in A^c \cap B^c$; hence $(A \cup B)^c \subseteq A^c \cap B^c$.

To show the reverse inclusion $A^c \cap B^c \subseteq (A \cup B)^c$, we consider x in $A^c \cap B^c$. Then $x \in A^c$ and so $x \notin A$. Also $x \in B^c$ and so $x \notin B$. Since $x \notin A$ and $x \notin B$, we conclude that $x \notin A \cup B$, i.e., $x \in (A \cup B)^c$. Hence $A^c \cap B^c \subseteq (A \cup B)^c$. ∎

EXAMPLE 4 The distributive law 3b is demonstrated in Figure 3. The picture of the set $A \cap (B \cup C)$ is double-hatched ▨.

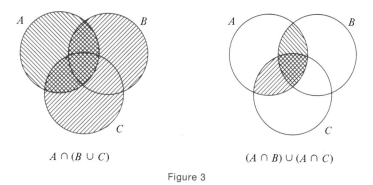

$$A \cap (B \cup C) \qquad\qquad (A \cap B) \cup (A \cap C)$$

Figure 3

Here is a proof where we show that the sets are subsets of each other. First consider $x \in A \cap (B \cup C)$. Then x is in A for sure. Also x is in $B \cup C$. So either $x \in B$, in which case $x \in A \cap B$, or else $x \in C$, in which case $x \in A \cap C$. In either case, we have $x \in (A \cap B) \cup (A \cap C)$. This shows that $A \cap (B \cup C) \subseteq (A \cap B) \cup (A \cap C)$.

Now consider $y \in (A \cap B) \cup (A \cap C)$. Either $y \in A \cap B$ or $y \in A \cap C$; we consider the two cases separately. If $y \in A \cap B$, then $y \in A$ and $y \in B$, so $y \in B \cup C$ and hence $y \in A \cap (B \cup C)$. Similarly, if $y \in A \cap C$ then $y \in A$ and $y \in C$, so $y \in B \cup C$ and thus $y \in A \cap (B \cup C)$. Since $y \in A \cap (B \cup C)$ in both cases, we've shown that $(A \cap B) \cup (A \cap C) \subseteq A \cap (B \cup C)$. We already proved the opposite inclusion and so the two sets are equal. ∎

The proofs using Venn diagrams seem much easier than the proofs where we analyze inclusions elementwise. Proofs by picture make many people nervous; on the other hand, the Venn diagram for A, B, C has eight regions [see Figure 4] and these comprise all the logical possibilities, so that proofs using Venn diagrams are in fact valid. A much more serious objection to proofs via Venn diagrams is that they hide the thought process, i.e., the logic used to shade the diagrams is not specified. If we had written out the reasoning behind the diagrams in Figure 3, the proof would have been as long

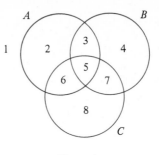

Figure 4

as the elementwise proof in Example 4. The latter proof relies only on logic and would be easier to communicate to a computer or to intelligent life in some far-off galaxy. Another reason for avoiding Venn diagrams is that they are hard to draw whenever there are more than three sets. Still, nearly everyone who works with mathematics uses pictures, including Venn diagrams, to help understand mathematical situations.

Table 1 gives a few of the basic relationships in set theory. Many other relationships exist. They can be verified using one of three methods: (1) Venn diagrams; (2) elementwise arguments as in Examples 3 and 4; (3) applying the laws in Table 1. Sometimes proofs will combine methods (2) and (3).

EXAMPLE 5   We give three proofs for the relationship

$$(A \cup B) \cap A^c \subseteq B.$$

Proof 1. See Figure 5. The picture for $(A \cup B) \cap A^c$ is double-hatched and is clearly a subset of B.

Proof 2. We show that $x \in (A \cup B) \cap A^c$ implies $x \in B$. Consider x in $(A \cup B) \cap A^c$. Then $x \in A^c$ and so $x \notin A$. Since x is also in $A \cup B$, it is in A or B, so it follows that x must be in B.

Proof 3. Using the laws of algebra in Table 1 we obtain

$$
\begin{aligned}
(A \cup B) \cap A^c &= A^c \cap (A \cup B) && \text{commutativity 1b} \\
&= (A^c \cap A) \cup (A^c \cap B) && \text{distributivity 3b} \\
&= (A \cap A^c) \cup (A^c \cap B) && \text{commutativity 1b} \\
&= \varnothing \cup (A^c \cap B) && \text{7b} \\
&= (A^c \cap B) \cup \varnothing && \text{commutativity 1a} \\
&= A^c \cap B. && \text{identity law 5a}
\end{aligned}
$$

This identity agrees, of course, with the picture on the left in Figure 5. Now it is clear that $A^c \cap B \subseteq B$, since if $x \in A^c \cap B$ then x must be in B. ∎

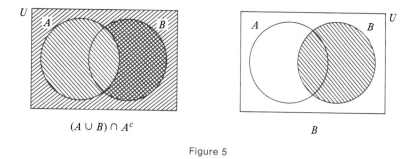

$$(A \cup B) \cap A^c \qquad\qquad\qquad B$$

Figure 5

The symmetric difference $\oplus$ is also an associative operation:

$$(A \oplus B) \oplus C = A \oplus (B \oplus C).$$

We can see this by looking at the Venn diagrams in Figure 6. On the left we have hatched $A \oplus B$ one way and C the other. Then $(A \oplus B) \oplus C$ is the set hatched one way or the other but not both. Doing the same sort of thing with A and $B \oplus C$ gives us the same set, so the sets $(A \oplus B) \oplus C$ and $A \oplus (B \oplus C)$ are equal.

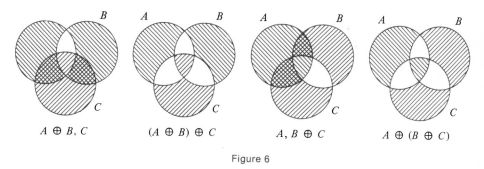

$$A \oplus B, C \qquad (A \oplus B) \oplus C \qquad A, B \oplus C \qquad A \oplus (B \oplus C)$$

Figure 6

Of course it is also possible to prove this fact without appealing to the pictures. You may want to construct such a proof yourself. Be warned, though, that a detailed argument will be fairly complicated.   ∎

Since $\oplus$ is associative, the expression $A \oplus B \oplus C$ is unambiguous. Note that an element belongs to this set provided it belongs to exactly one or to all three of the sets A, B and C.

EXERCISES 1.2

1. Let $U = \{1, 2, 3, 4, 5, \ldots, 12\}$, $A = \{1, 3, 5, 7, 9, 11\}$, $B = \{2, 3, 5, 7, 11\}$, $C = \{2, 3, 6, 12\}$ and $D = \{2, 4, 8\}$. Determine the sets

(a) $A \cup B$

(b) $A \cap C$

(c) $(A \cup B) \cap C^c$

(d) $A \setminus B$

(e) $C \setminus D$

(f) $B \oplus D$

(g) How many subsets are there of C?

2. Let $A = \{1, 2, 3\}$, $B = \{n \in \mathbb{P} : n \text{ is even}\}$ and $C = \{n \in \mathbb{P} : n \text{ is odd}\}$.
   (a) Determine $A \cap B$, $B \cap C$, $B \cup C$ and $B \oplus C$.
   (b) List all subsets of A.
   (c) Which of the following sets are infinite? $A \oplus B$, $A \oplus C$, $A \backslash C$, $C \backslash A$.

3. In this exercise the universe is $\mathbb{R}$. Determine the following sets.
   (a) $[0, 3] \cap [2, 6]$
   (b) $[0, 3] \cup [2, 6]$
   (c) $[0, 3] \backslash [2, 6]$
   (d) $[0, 3] \oplus [2, 6]$
   (e) $[0, 3]^c$
   (f) $[0, 3] \cap \varnothing$

4. Let $\Sigma = \{a, b\}$, $A = \{a, b, aa, bb, aaa, bbb\}$, $B = \{w \in \Sigma^* : \text{length}(w) \geqq 2\}$ and $C = \{w \in \Sigma^* : \text{length}(w) \leqq 2\}$.
   (a) Determine $A \cap C$, $A \backslash C$, $C \backslash A$ and $A \oplus C$.
   (b) Determine $A \cap B$, $B \cap C$, $B \cup C$ and $B \backslash A$.
   (c) Determine $\Sigma^* \backslash B$, $\Sigma \backslash B$ and $\Sigma \backslash C$.
   (d) List all subsets of Σ.
   (e) How many sets are there in $\mathscr{P}(\Sigma)$?

5. In this exercise the universe is Σ^* where $\Sigma = \{a, b\}$. Let A, B and C be as in Exercise 4. Determine the following sets.
   (a) $B^c \cap C^c$
   (b) $(B \cap C)^c$
   (c) $(B \cup C)^c$
   (d) $B^c \cup C^c$
   (e) $A^c \cap C$
   (f) $A^c \cap B^c$
   (g) Which of these sets are equal? Why?

6. The following statements involve subsets of some nonempty universal set U. Tell whether each is true or false. For each false one, give an example for which the statement is false.
   (a) $A \cap (B \cup C) = (A \cap B) \cup C$ for all A, B C.
   (b) $A \cup B \subseteq A \cap B$ implies $A = B$.
   (c) $(A \cap \varnothing) \cup B = B$ for all A, B.
   (d) $A \cap (\varnothing \cup B) = A$ whenever $A \subseteq B$.
   (e) $A \cap B = A^c \cup B^c$ for all A, B.

7. For any set A, what is $A \oplus A$? $A \oplus \varnothing$?

8. Use Venn diagrams to prove the following.
   (a) $A \cap (B \oplus C) = (A \cap B) \oplus (A \cap C)$
   (b) $A \oplus B \subseteq (A \oplus C) \cup (B \oplus C)$

9. Prove the generalized DeMorgan law $(A \cap B \cap C)^c = A^c \cup B^c \cup C^c$. *Hint:* First apply the DeMorgan law 9b to the sets A and $B \cap C$. The elementwise method can be avoided.

10. Prove the following without using Venn diagrams.
    (a) $A \cap B \subseteq A$ and $A \subseteq A \cup B$ for all sets A and B.
    (b) If $A \subseteq B$ and $A \subseteq C$, then $A \subseteq B \cap C$.
    (c) If $A \subseteq C$ and $B \subseteq C$, then $A \cup B \subseteq C$.
    (d) $A \subseteq B$ if and only if $B^c \subseteq A^c$.

Note. In the remaining exercises, you may use any method of proof.

11. Prove or disprove. [A proof needs to be a general argument, but a single counterexample is sufficient for a disproof.]
   (a) $A \cap B = A \cap C$ implies $B = C$.
   (b) $A \cup B = A \cup C$ implies $B = C$.
   (c) $A \cap B = A \cap C$ and $A \cup B = A \cup C$ imply $B = C$.
   (d) $A \cup B \subseteq A \cap B$ implies $A = B$.
   (e) $A \oplus B = A \oplus C$ implies $B = C$.

12. (a) Prove that $A \subseteq B$ if and only if $A \cup B = B$. This is really two assertions:

   "$A \subseteq B$ implies $A \cup B = B$" and "$A \cup B = B$ implies $A \subseteq B$."

   (b) Prove that $A \subseteq B$ if and only if $A \cap B = A$.

13. Draw a Venn diagram for four sets A, B, C and D. Be sure to have a region for all sixteen possible sets such as $A \cap B^c \cap C^c \cap D$.

14. (a) Show that relative complementation is not commutative; that is, $A \setminus B = B \setminus A$ can fail.
   (b) Show that relative complementation is not associative: $(A \setminus B) \setminus C = A \setminus (B \setminus C)$ can fail.
   (c) Show, however, that $(A \setminus B) \setminus C \subseteq A \setminus (B \setminus C)$ for all A, B and C.

§ 1.3   Subscripts and Indexing

Subscript notation comes in handy when we are dealing with a large collection of objects, especially if they are similar; here large often means "more than 3 or 4." For example, letters x, y and z are adequate when dealing with equations involving three or fewer unknowns. But if there are ten unknowns or if we wish to discuss the general situation of n unknowns [n some unspecified integer in $\mathbb{P}$], then $x_1, x_2, \ldots, x_n$ would be a good choice for the names of the unknowns. Here the unknowns are distinguished by the little numbers $1, 2, \ldots, n$ which are called **subscripts**. As another example, a general nonzero polynomial has the form

$$a_n x^n + a_{n-1} x^{n-1} + \cdots + a_2 x^2 + a_1 x + a_0,$$

where $a_n \neq 0$. Here n is the degree of the polynomial and the $n + 1$ possible coefficients are labeled $a_0, a_1, \ldots, a_n$ using subscripts. For example, the polynomial $x^3 + 4x^2 - 73$ fits this general scheme; to see this, let $n = 3$, $a_3 = 1, a_2 = 4, a_1 = 0$ and $a_0 = -73$.

A similar situation arises in set theory. A simple way to handle n sets is to name them $A_1, A_2, \ldots, A_n$. An infinite sequence of sets is also possible: A_1, $A_2, \ldots$ or $\{A_k : k \in \mathbb{P}\}$. Of course the subscripts could begin with 0 or some other integer. We sometimes think of the subscripts as indexing [or labeling] the collection of sets. In this case, we may say the collection of sets is **indexed** by the set of subscripts. For example, $\{A_k : k \in \mathbb{P}\}$ is a family of sets indexed by $\mathbb{P}$ while $\{B_0, B_1, B_2, \ldots\}$ represents a family of sets indexed by $\mathbb{N}$.

EXAMPLE 1   (a) For each n in $\mathbb{P}$ let
$$D_n = \{k \in \mathbb{Z} : k \text{ is divisible by } n\}.$$
Recall that 0 is divisible by all integers. Some of the sets D_n defined above are

$D_1 = \mathbb{Z}$,

$D_2 = \{k \in \mathbb{Z} : k \text{ is divisible by } 2\} = \{k \in \mathbb{Z} : k \text{ is even}\}$

$\quad = \{\ldots, -10, -8, -6, -4, -2, 0, 2, 4, 6, 8, 10, \ldots\}$,

$D_5 = \{\ldots, -30, -25, -20, -15, -10, -5, 0, 5, 10, 15, 20, 25, 30, \ldots\}$.

By a slight abuse of notation, we might also write
$$D_n = \{0, \pm n, \pm 2n, \pm 3n, \ldots\}.$$
The family of sets $\{D_n : n \in \mathbb{P}\}$ is indexed by $\mathbb{P}$.

(b) For $n \in \mathbb{P}$ we define the following intervals in $\mathbb{R}$:
$$A_n = [-n, n] \quad \text{and} \quad B_n = [n, 2n].$$
Thus $A_3 = [-3, 3]$ and $B_3 = [3, 6]$, for instance. Both collections $\{A_n : n \in \mathbb{P}\}$ and $\{B_n : n \in \mathbb{P}\}$ are indexed by $\mathbb{P}$.

(c) Let Σ be an alphabet. For each $k \in \mathbb{N}$, Σ^k is defined to be the set of all words in Σ^* having length k. In symbols,
$$\Sigma^k = \{w \in \Sigma^* : \text{length } (w) = k\}.$$
This "power notation" will seem quite appropriate in the context of § 4.4. Note that the sets Σ^k are disjoint, that $\Sigma^0 = \{\epsilon\}$, and that $\Sigma^1 = \Sigma$. The family of sets $\{\Sigma^k : k \in \mathbb{N}\}$ is indexed by $\mathbb{N}$.

In the case that $\Sigma = \{a, b\}$, we have $\Sigma^0 = \{\epsilon\}$, $\Sigma^1 = \Sigma = \{a, b\}$, $\Sigma^2 = \{aa, ab, ba, bb\}$, etc.

(d) We allow the possibility that two sets with different indices are the same. For instance, consider the set M of memory locations in a given computer and for $n \in \mathbb{N}$ let
$$M_n = \{x \in M : x \text{ contains } n\}.$$
Then $\{M_0, M_1, M_2, \ldots\}$ is a family of sets indexed by $\mathbb{N}$ and $M_n = \varnothing$ for all large enough values of n. It could also happen that $M_n = \varnothing$ for certain small values of n, too.

If we let $L_n = \{x \in M : x \text{ contains a number less than } n\}$, then $L_0, L_1, L_2, \ldots$ is indexed by $\mathbb{N}$, $L_0 \subseteq L_1 \subseteq L_2 \subseteq \cdots$ and $L_n = L_{n+1}$ is possible. ∎

We have used the symbol Σ as a name for an alphabet. In mathematics the big Greek sigma $\sum$ has a standard use as a general summation sign. The terms following it are to be summed according to how $\sum$ is decorated. For example, consider the expression

$$\sum_{k=1}^{10} k^2.$$

The decorations "$k = 1$" and "10" tell us to sum the numbers k^2 obtained by successively setting $k = 1$, then $k = 2$, then $k = 3$, etc. on up to $k = 10$. That is,

$$\sum_{k=1}^{10} k^2 = 1 + 4 + 9 + 16 + 25 + 36 + 49 + 64 + 81 + 100 = 385.$$

The letter k is a variable [it varies from 1 to 10] and it could be replaced by any other variable. Thus

$$\sum_{k=1}^{10} k^2 = \sum_{j=1}^{10} j^2 = \sum_{r=1}^{10} r^2.$$

We can also consider more general sums like

$$\sum_{k=1}^{n} k^2$$

in which the stopping point n can take on different values. Each value of n gives a particular value of the sum; for each choice of n the variable k takes on the values from 1 to n. Here are some of the sums represented by $\sum_{k=1}^{n} k^2$.

Value of n	The sum
$n = 1$	$1^2 = 1$
$n = 2$	$1^2 + 2^2 = 1 + 4 = 5$
$n = 3$	$1^2 + 2^2 + 3^2 = 14$
$n = 4$	$1^2 + 2^2 + 3^2 + 4^2 = 30$
$n = 10$	$1^2 + 2^2 + 3^2 + 4^2 + 5^2 + 6^2 + 7^2 + 8^2 + 9^2 + 10^2$ $= 385$
$n = 73$	$1^2 + 2^2 + 3^2 + 4^2 + \cdots + 73^2 = 132{,}349$

We can also discuss even more general sums like

$$\sum_{k=1}^{n} x_k \quad \text{and} \quad \sum_{j=m}^{n} a_j.$$

Here it is understood that $\{x_k : 1 \leq k \leq n\}$ represents some collection of numbers indexed by $\{k \in \mathbb{N} : 1 \leq k \leq n\}$ and that $\{a_j : m \leq j \leq n\}$ represents a collection of numbers indexed by $\{j \in \mathbb{N} : m \leq j \leq n\}$. Presumably $m \leq n$ since otherwise there would be nothing to sum.

In analogy with $\sum$, the big Greek pi $\prod$ is a general product sign. As explained in § 0.2, for $n \in \mathbb{P}$ the product of the first n integers is called n **factorial** and written $n!$ Thus

$$n! = 1 \cdot 2 \cdot 3 \cdots n = \prod_{k=1}^{n} k.$$

The expression $1 \cdot 2 \cdot 3 \cdots n$ is somewhat confusing for small values of n like 1 and 2; it really means "multiply consecutive integers until you reach n." The expression $\prod_{k=1}^{n} k$ is less ambiguous. Here are the first few values of $n!$

$$1! = 1 \qquad\qquad 5! = 1 \cdot 2 \cdot 3 \cdot 4 \cdot 5 = 120$$

$$2! = 1 \cdot 2 = 2 \qquad\qquad 6! = 720$$

$$3! = 1 \cdot 2 \cdot 3 = 6 \qquad\qquad 7! = 5040$$

$$4! = 1 \cdot 2 \cdot 3 \cdot 4 = 24 \qquad 8! = 40{,}320$$

For technical reasons $n!$ is also defined for $n = 0$; $0!$ is defined to be 1. The definition of $n!$ will be reexamined in § 3.4.

Next consider a family of sets A_k indexed by some nonempty set I of integers; I may be finite or infinite. The **union** $\bigcup_{k \in I} A_k$ is defined to be the set of elements that belong to at least one of the sets A_k. That is,

$$\bigcup_{k \in I} A_k = \{x : x \in A_k \text{ for at least one value of } k \in I\}.$$

Likewise, the **intersection** $\bigcap_{k \in I} A_k$ is defined by

$$\bigcap_{k \in I} A_k = \{x : x \in A_k \text{ for all } k \in I\}.$$

If I has the form $\{k \in \mathbb{Z} : m \leq k \leq n\}$, we write these sets as $\bigcup_{k=m}^{n} A_k$ and $\bigcap_{k=m}^{n} A_k$. For example,

$$\bigcup_{k=0}^{20} A_k = \{x : x \in A_k \text{ for some } k \in \mathbb{N} \text{ such that } 0 \leq k \leq 20\}$$

$$= A_0 \cup A_1 \cup A_2 \cup A_3 \cup \cdots \cup A_{19} \cup A_{20}.$$

If I is infinite and has the form $\{k \in \mathbb{Z} : k \geq m\}$ we write

$$\bigcup_{k=m}^{\infty} A_k \quad \text{for} \quad \bigcup_{k \in I} A_k \quad \text{and} \quad \bigcap_{k=m}^{\infty} A_k \quad \text{for} \quad \bigcap_{k \in I} A_k.$$

Note that the symbol ∞ here is convenient, but does not represent one of the values of the subscripts.

EXAMPLE 2   (a) For $n \in \mathbb{P}$, let

$$D_n = \{k \in \mathbb{Z} : k \text{ is divisible by } n\}.$$

Then we have

$$D_3 \cap D_5 = \{k \in \mathbb{Z} : k \text{ is divisible by both 3 and 5}\}$$
$$= \{k \in \mathbb{Z} : k \text{ is divisible by 15}\} = D_{15}$$
$$= \{0, \pm 15, \pm 30, \pm 45, \ldots\},$$

and

$$D_3 \cup D_5 = \{k \in \mathbb{Z} : k \text{ is divisible by 3 or 5}\}$$
$$= \{0, \pm 3, \pm 5, \pm 6, \pm 9, \pm 10, \pm 12, \pm 15, \pm 18, \pm 20, \ldots\}.$$

Note that

$$\bigcap_{n \in \mathbb{P}} D_n = \bigcap_{n=1}^{\infty} D_n = \{0\}.$$

In fact, if I is any infinite subset of $\mathbb{P}$, then

$$\bigcap_{n \in I} D_n = \{k \in \mathbb{Z} : k \text{ is divisible by all } n \text{ in } I\} = \{0\}.$$

(b) Let $A_n = [-n, n]$ and $B_n = [n, 2n]$ for $n \in \mathbb{P}$. For example, $A_5 = [-5, 5]$ and $B_7 = [7, 14]$. Observe that

$$\bigcap_{n=4}^{73} A_n = [-4, 4], \quad \bigcup_{n=4}^{73} A_n = [-73, 73] \quad \text{and} \quad \bigcup_{n=1}^{\infty} A_n = \mathbb{R}.$$

Note that the sets B_n "slide along $\mathbb{R}$" as n increases:

So clearly $B_2 \cap B_6 = \{ \ \}$ and $\bigcap_{n=1}^{\infty} B_n = \{ \ \}$. But not all pairs of sets B_n, B_m are disjoint. For example,

$$B_5 \cap B_8 = [8, 10] \quad \text{and} \quad \bigcap_{n=4}^{8} B_n = \{8\}.$$

(c) If Σ is an alphabet, then

$$\bigcup_{k=0}^{\infty} \Sigma^k = \bigcup_{k \in \mathbb{N}} \Sigma^k = \Sigma^*. \quad \blacksquare$$

EXAMPLE 3 The DeMorgan laws in § 1.2 extend as follows. Let $\{A_k : k \in I\}$ be a collection of subsets of some universe U that is indexed by I. Then

$$\left(\bigcup_{k \in I} A_k \right)^c = \bigcap_{k \in I} A_k^c \quad \text{and} \quad \left(\bigcap_{k \in I} A_k \right)^c = \bigcup_{k \in I} A_k^c.$$

We verify the first equality. Suppose that $x \in \left(\bigcup_{k \in I} A_k \right)^c$. Then $x \notin \bigcup_{k \in I} A_k$ and

so $x \notin A_k$ for all $k \in I$. Hence we have $x \in A_k^c$ for all $k \in I$ and so

$x \in \bigcap_{k \in I} A_k^c$. Thus we have $\left(\bigcup_{k \in I} A_k \right)^c \subseteq \bigcap_{k \in I} A_k^c$. For the reverse inclusion, con-

sider $x \in \bigcap_{k \in I} A_k^c$. Then $x \in A_k^c$ for all $k \in I$. So $x \notin A_k$ for all $k \in I$. Conse-

quently $x \notin \bigcup_{k \in I} A_k$, i.e., $x \in \left(\bigcup_{k \in I} A_k \right)^c$. This shows that $\bigcap_{k \in I} A_k^c \subseteq \left(\bigcup_{k \in I} A_k \right)^c$ and so

the sets must be equal. ∎

So far we have only considered indexed families of sets, but we can use indexing to provide labels for the objects in any collection.

EXAMPLE 4   In § 0.2 we considered a graph with 14 edges, and named the edges e_1, $e_2, \ldots, e_{14}$. We could say that we indexed the set E of edges using the set $\{1, 2, \ldots, 14\}$. That is,

$$E = \{e_i : i \in \{1, 2, \ldots, 14\}\}. \quad \blacksquare$$

In § 2.6 we will be concerned with families of propositions that are indexed by the set $\mathbb{P}$. By a **proposition** we mean an unambiguous sentence that is either true or false, but not both.

EXAMPLE 5   (a) Here are some propositions: "New York is the capital of the United States." "Every graph has an even number of edges." "73 is a prime number." "$3 + 3 = 5$." "More examples are presented in § 2.1."

(b) Here are some nonpropositions: "Do not walk on the grass!" "Give to the mathematics department of your choice." "Mathematics courses are easier than computer science courses." "$4x - 1 = 2x$." The third example is ambiguous since its truth value [True or False] varies from person to person. The fourth example is ambiguous since x isn't specified, but it becomes a proposition if we write "$4x - 1 = 2x$ for all $x \in \mathbb{R}$" or "$4x - 1 = 2x$ for some $x \in \mathbb{R}$." More examples of nonpropositions appear in Examples 3 and 4 of § 2.1. ∎

EXAMPLE 6   For each choice of n in $\mathbb{P}$ the equation "$n^2 = 2^n$" is a proposition. For $n = 2$ and 4 it is true; for other values of n it is false. If we let $A(n)$ be the proposition "$n^2 = 2^n$" then the set $\{A(1), A(2), \ldots\}$ is a set of propositions indexed by $\mathbb{P}$. Similarly, if we let $B(n)$ be "$n^3 + 1$ is divisible by 3," $C(n)$ be "$1 + 2 + \cdots + n = 1 \cdot 2 \cdots n$," and $D(n)$ be "$n^3 - 4n + 6$ is divisible by 3," then the sets of propositions $\{B(1), B(2), \ldots\}$, $\{C(1), C(2), \ldots\}$ and

$\{D(1), D(2), \ldots\}$ are indexed by $\mathbb{P}$. The propositions $B(n)$ are true for some values of n [for example, $n = 2$, 5 and 11] but not for others. Note that the propositions $C(n)$ could have been written " $\sum_{k=1}^{n} k = n!$ " The first few $C(n)$'s are

Value of n	Proposition C(n)
$n = 1$	$1 = 1$
$n = 2$	$1 + 2 = 1 \cdot 2$
$n = 3$	$1 + 2 + 3 = 1 \cdot 2 \cdot 3$
$n = 4$	$1 + 2 + 3 + 4 = 1 \cdot 2 \cdot 3 \cdot 4$
$n = 5$	$1 + 2 + 3 + 4 + 5 = 1 \cdot 2 \cdot 3 \cdot 4 \cdot 5$

For $n = 1$ or 3, $C(n)$ is true, but otherwise $C(n)$ is false. The first few $D(n)$'s are

Value of n	Proposition D(n)
$n = 1$	$1^3 - 4 \cdot 1 + 6 = 3$ is divisible by 3
$n = 2$	$2^3 - 4 \cdot 2 + 6 = 6$ is divisible by 3
$n = 3$	$3^3 - 4 \cdot 3 + 6 = 21$ is divisible by 3
$n = 4$	$4^3 - 4 \cdot 4 + 6 = 54$ is divisible by 3
$n = 5$	$5^3 - 4 \cdot 5 + 6 = 111$ is divisible by 3

Note that the first five values of n yield true propositions. Is this accidental or a pattern? It turns out that the proposition $D(n)$ is true for every value of n. With enough determination and/or hardware, one could check this for thousands of values of n. But this would not be a proof; conceivably the result fails for some gigantic n. In § 2.6 we will see how to settle this sort of question once and for all; this particular example is dealt with in Exercise 8 of that section. ∎

Frequently propositions indexed by $\mathbb{P}$ will be written as $p(n)$. Thus $p(1)$ is the first proposition, $p(2)$ is the second proposition, $\ldots$, $p(n)$ is the nth proposition.

EXAMPLE 7 Consider the propositions

$$p(n) = \text{``}11^n - 4^n \text{ is divisible by 7.''}$$

Then $p(1)$ is the proposition "$11 - 4$ is divisible by 7," $p(4)$ is the proposition "$11^4 - 4^4$ is divisible by 7," etc. ∎

EXERCISES 1.3

1. Calculate

   (a) $\dfrac{7!}{5!}$

   (b) $\dfrac{10!}{6!4!}$

   (c) $\dfrac{9!}{0!9!}$

   (d) $\dfrac{8!}{4!}$

   (e) $\displaystyle\sum_{k=0}^{5} k!$

   (f) $\displaystyle\prod_{j=3}^{6} j$

2. Simplify

   (a) $\dfrac{n!}{(n-1)!}$

   (b) $\dfrac{(n!)^2}{(n+1)!(n-1)!}$

3. Calculate

   (a) $\displaystyle\sum_{k=1}^{n} 3^k$ for $n = 1, 2, 3$ and 4,

   (b) $\displaystyle\sum_{k=3}^{n} k^3$ for $n = 3, 4$ and 5,

   (c) $\displaystyle\sum_{j=n}^{2n} j$ for $n = 1, 2$ and 5.

4. Calculate

   (a) $\displaystyle\sum_{i=1}^{10} (-1)^i$

   (b) $\displaystyle\sum_{k=0}^{3} (k^2 + 1)$

   (c) $\left(\displaystyle\sum_{k=0}^{3} k^2\right) + 1$

   (d) $\displaystyle\prod_{n=1}^{5} (2n + 1)$

   (e) $\displaystyle\prod_{j=4}^{8} (j - 1)$

5. (a) Calculate $\displaystyle\prod_{r=1}^{n} (r - 3)$ for $n = 1, 2, 3, 4$ and 73.

   (b) Calculate $\displaystyle\prod_{k=1}^{m} \dfrac{k + 1}{k}$ for $m = 1, 2$ and 3. Give a formula for this product for all

   $m \in \mathbb{P}$.

6. (a) Calculate $\displaystyle\sum_{k=0}^{n} 2^k$ for $n = 1, 2, 3, 4$ and 5.

   (b) Use your answers to part (a) to guess a general formula for this sum.

7. For $n \in \mathbb{P}$, let $D_n = \{k \in \mathbb{P} : k \text{ is a multiple of } n\}$. Determine

   (a) $D_2 \cap D_5$

   (b) $D_2 \cap D_3 \cap D_5$

   (c) $D_2 \cap D_4$

   (d) $D_4 \cap D_6$

   (e) $D_4 \oplus D_6$

   (f) $D_2 \oplus D_6$

8. Repeat Exercise 7 for the following sets.
   (a) D_1^c
   (b) D_2^c
   (c) $D_2^c \cap D_3^c$
   (d) $D_2^c \cap D_4^c$
   (e) $D_4^c \cup D_6^c$
   (f) $D_4^c \oplus D_6^c$

9. In this exercise the universe is $\mathbb{N}$. For each $n \in \mathbb{P}$, let $A_n = \{n, n+1, n+2, \ldots\} = \{k \in \mathbb{N} : k \geq n\}$ and $B_n = \{0, 1, 2, \ldots, 2n\} = \{k \in \mathbb{N} : k \leq 2n\}$.
   (a) Write down A_n and B_n for $n = 2$ and $n = 4$.
   (b) Write down A_1^c, A_2^c and A_4^c.
   (c) Determine $A_n \cap B_n$ and A_n^c for $n = 1, 2, 3$ and 7.
   (d) Determine $\bigcup_{n=3}^{6} A_n$, $\bigcup_{n=3}^{6} B_n$, $\bigcap_{n=3}^{6} B_n$, $\bigcap_{n=3}^{6} A_n$.
   (e) Determine $\bigcup_{n=3}^{\infty} A_n$, $\bigcup_{n=3}^{\infty} B_n$, $\bigcap_{n=3}^{\infty} B_n$, $\bigcap_{n=3}^{\infty} A_n$.
   (f) Determine $\bigcup_{n=1}^{5} A_n^c$ and $\bigcup_{n=1}^{\infty} (A_n \cup B_n)^c$.

10. Let $A_0 = \{n \in \mathbb{Z} : n \text{ is divisible by } 5\}$ and for $k \in \mathbb{P}$ let $A_k = \{n + k : n \in A_0\}$.
    (a) List several elements in each of the sets A_1, A_2, A_3, A_4, A_5 and A_6.
    (b) What is the relationship between A_5 and A_0? A_6 and A_1? A_{30} and A_0?
    (c) Generalize your answers to part (b).
    (d) What is $\bigcup_{k=0}^{4} A_k$? $\bigcup_{k=1}^{5} A_k$?

11. Let Σ be an alphabet and, for $k \in \mathbb{N}$, let $\Sigma^k = \{w \in \Sigma^* : \text{length}(w) = k\}$.
    (a) What is $\bigcup_{k=0}^{n} \Sigma^k$ for each $n \in \mathbb{N}$?
    (b) Describe the set $\bigcup_{k \in \mathbb{N}} \Sigma^{2k}$.

12. Let $\Sigma = \{a, b\}$ and for $n \in \mathbb{N}$ let
    $$A_n = \{w \in \Sigma^* : \text{the letter } a \text{ occurs in } w \text{ exactly } n \text{ times}\}.$$
    (a) Describe the members of A_0.
    (b) List five elements in A_1 and five elements in A_4.
    (c) What is $\bigcup_{n \in \mathbb{N}} A_n$?
    (d) Explain briefly why the inclusion $A_n \subseteq \bigcup_{k=n}^{\infty} \Sigma^k$ is valid for all n.
    (e) For $m \in \mathbb{N}$, let B_m be the set of all words w in Σ^* in which the letter b occurs exactly m times. Explain why $A_n \cap B_m \subseteq \Sigma^{n+m}$ for all m and n.

13. Prove the second generalized DeMorgan law stated in Example 3.

14. For $n \in \mathbb{P}$, let $A(n)$ be the proposition "$\sum_{k=1}^{n} k = \frac{1}{2}n(n+1)$." Verify that $A(n)$ is true for $n \leq 5$.

15. For $n \in \mathbb{P}$, let $p(n)$ be the proposition "$11^n - 4^n$ is divisible by 7." Verify that $p(n)$ is true for $n = 1, 2, 3$.

16. For $n \in \mathbb{P}$, let $p(n)$ be the proposition "$n^3 - n$ is divisible by 6." Check whether this is true for each $n \le 5$. Do you have a conjecture for $n \ge 6$?

17. For $n \in \mathbb{P}$, let $p(n)$ be the proposition "$n! > n^2$." Which of the propositions $p(1)$, $p(2)$, $p(3)$, $p(4)$, $p(5)$, $p(6)$ are true? Do you have a conjecture for $n \ge 7$?

18. For $n \in \mathbb{P}$, let $p(n)$ be the proposition "$2n^2 - 2n + 7$ is prime." Determine whether this is true for each $n \le 5$. Do you have a conjecture for $n \ge 6$?

§ 1.4 Ordered Pairs, Matrix Notation

Consider two sets S and T. For each element s in S and each element t in T, we form an **ordered pair** $\langle s, t \rangle$. Here s is the first element of the ordered pair, t is the second element and the order is important. Thus $\langle s_1, t_1 \rangle = \langle s_2, t_2 \rangle$ if and only if $s_1 = s_2$ and $t_1 = t_2$. The set of all ordered pairs $\langle s, t \rangle$ is called the **product of S and** T and written $S \times T$:

$$S \times T = \{\langle s, t \rangle : s \in S \quad \text{and} \quad t \in T\}.$$

If $S = T$ we sometimes write S^2 for $S \times S$.

EXAMPLE 1   (a) Let $S = \{1, 2, 3, 4\}$ and $T = \{a, b, c\}$. Then $S \times T$ consists of the twelve ordered pairs listed on the left in Figure 1. We could also depict these pairs as corresponding points in labeled rows and columns, in the manner shown on the right in the figure. The reader should list or draw $T \times S$ and note that $T \times S \ne S \times T$.

$\langle 1, c \rangle$	$\langle 2, c \rangle$	$\langle 3, c \rangle$	$\langle 4, c \rangle$
$\langle 1, b \rangle$	$\langle 2, b \rangle$	$\langle 3, b \rangle$	$\langle 4, b \rangle$
$\langle 1, a \rangle$	$\langle 2, a \rangle$	$\langle 3, a \rangle$	$\langle 4, a \rangle$

c O O O O
b O O O O
a O O O O
    1   2   3   4

List of $\{1, 2, 3, 4\} \times \{a, b, c\}$ Picture of $\{1, 2, 3, 4\} \times \{a, b, c\}$

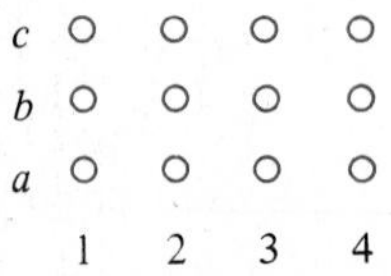

Figure 1

(b) If $S = \{1, 2, 3, 4\}$, then $S^2 = S \times S$ has sixteen ordered pairs; see Figure 2. Note that $\langle 2, 4 \rangle \ne \langle 4, 2 \rangle$; these ordered pairs involve the same two numbers, but in different orders. In contrast, the *sets* $\{2, 4\}$ and $\{4, 2\}$ are

$\langle 1, 4 \rangle$	$\langle 2, 4 \rangle$	$\langle 3, 4 \rangle$	$\langle 4, 4 \rangle$
$\langle 1, 3 \rangle$	$\langle 2, 3 \rangle$	$\langle 3, 3 \rangle$	$\langle 4, 3 \rangle$
$\langle 1, 2 \rangle$	$\langle 2, 2 \rangle$	$\langle 3, 2 \rangle$	$\langle 4, 2 \rangle$
$\langle 1, 1 \rangle$	$\langle 2, 1 \rangle$	$\langle 3, 1 \rangle$	$\langle 4, 1 \rangle$

4 O O O O
3 O O O O
2 O O O O
1 O O O O
   1   2   3   4

List of $\{1, 2, 3, 4\}^2$ Picture of $\{1, 2, 3, 4\}^2$

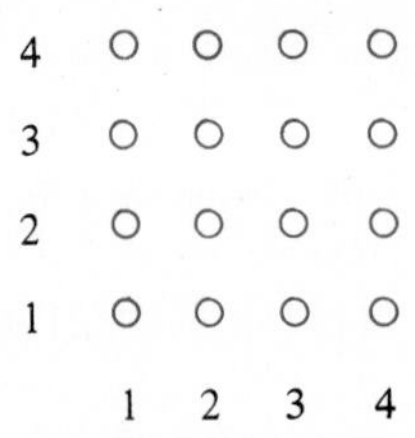

Figure 2

the same. Also note that $\langle 2, 2\rangle$ is a perfectly good ordered pair in which the first element happens to equal the second element. On the other hand, the set $\{2, 2\}$ is just the set $\{2\}$ in which 2 happens to be written twice. If you are used to seeing ordered pairs written (a, b), you may wonder why we do it differently. We do so because we don't want to confuse ordered pairs with open intervals. For us $(2, 4)$ is the set $\{x \in \mathbb{R} : 2 < x < 4\}$.

(c) The set $\mathbb{N}^2 = \mathbb{N} \times \mathbb{N}$ is infinite but we can draw part of it; see Figure 3. The solid points represent $\langle 3, 0\rangle$, $\langle 1, 5\rangle$ and $\langle 4, 2\rangle$. Note that $\mathbb{N}^2 = \{\langle m, n\rangle : m, n \in \mathbb{N}\}$.

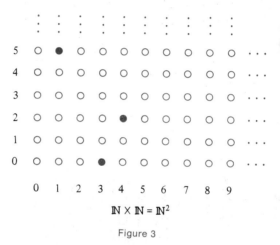

IN X IN = IN²

Figure 3

(d) Part of the set $\mathbb{R}^2 = \mathbb{R} \times \mathbb{R} = \{\langle x, y\rangle : x, y \in \mathbb{R}\}$ is sketched in Figure 4. The solid dots represent the points $\langle 3, 0\rangle$, $\langle \frac{1}{2}, 1\rangle$ and $\langle -2, 2\rangle$. The set $\mathbb{R}^2$ gives a coordinate system for the plane, since every point in the plane

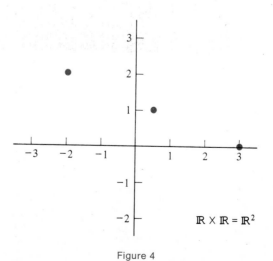

IR X IR = IR²

Figure 4

corresponds to exactly one ordered pair of real numbers, i.e., to an element of $\mathbb{R}^2$. ∎

We are often interested in subsets of product sets.

EXAMPLE 2   (a) Let $S = \{1, 2, 3\}$ and $T = \{0, 1, 2, 3\}$. The pictures of the following subsets of $S \times T$ are indicated in Figure 5:

$$A = \{\langle m, n \rangle \in S \times T : m + n \leq 3\},$$

$$B = \{\langle m, n \rangle \in S \times T : m - n = 2\},$$

$$C = \{\langle m, n \rangle \in S \times T : \max \{m, n\} = 3\},$$

$$D = \{\langle m, n \rangle \in S \times T : m = 1\}.$$

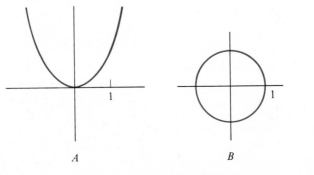

Figure 5

(b) The pictures of the following subsets of $\mathbb{R}^2$ are sketched in Figure 6:

$$A = \{\langle x, y \rangle \in \mathbb{R}^2 : y = x^2\}, \qquad B = \{\langle x, y \rangle \in \mathbb{R}^2 : x^2 + y^2 = 1\},$$

$$C = \{\langle x, y \rangle \in \mathbb{R}^2 : x^2 + y^2 \leq 1\}. \quad ∎$$

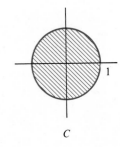

Figure 6

We next define the product of any finite number of sets. Thus, consider n sets $S_1, S_2, \ldots, S_n$. The **product set** $S_1 \times S_2 \times \cdots \times S_n$ consists of all **ordered n-tuples** $\langle s_1, s_2, \ldots, s_n \rangle$ where $s_1 \in S_1$, $s_2 \in S_2$, etc. That is,

$$S_1 \times S_2 \times \cdots \times S_n = \{\langle s_1, s_2, \ldots, s_n \rangle : s_k \in S_k \text{ for } k = 1, 2, \ldots, n\}.$$

Just as with ordered pairs, two ordered n-tuples $\langle s_1, s_2, \ldots, s_n \rangle$ and $\langle t_1, t_2, \ldots, t_n \rangle$ are regarded as equal if all the corresponding entries are equal: $s_k = t_k$ for $k = 1, 2, \ldots, n$. If the sets $S_1, S_2, \ldots, S_n$ are all equal, to S say, we may write S^n for the product $S_1 \times S_2 \times \cdots \times S_n$.

EXAMPLE 3   Just as $\mathbb{R}^2$ gives a coordinate system for the plane, $\mathbb{R}^3$ gives a coordinate system for the usual 3-dimensional space in which we think we live. The set

$$\{\langle x, y, z \rangle \in \mathbb{R}^3 : x^2 + y^2 + z^2 \leq 1\}$$

corresponds to the solid ball with center $\langle 0, 0, 0 \rangle$ and radius 1.   ∎

The set $\mathbb{R}^n$ is called n-**dimensional real space**, and many geometrical notions can be extended to $\mathbb{R}^n$ for $n \geq 4$ even though it is difficult to visualize and draw suitable pictures. Our interest in $\mathbb{R}^n$ will be more algebraic, though we will rely on our intuition from $\mathbb{R}^3$ to give us some rough feeling for geometry in $\mathbb{R}^n$. For example, $\mathbb{R}^n$ is the suitable universal set when considering equations with n unknowns. For instance, the solution set S for the system

$$x^2 + 2xy - y^2 - z - w = 0$$
$$x + 2xz - z^2 - 2yw = 0$$
$$x^2 + 2xw - y^2 - 3z^2 + w^3 = 0$$

is some [horrible] subset of $\mathbb{R}^4$. It is clear that S contains $\langle 0, 0, 0, 0 \rangle$ and $\langle 1, 1, 1, 1 \rangle$, but not much else is clear about S.

Systems of equations with several unknowns are much easier to handle if all the equations are **linear**; this means that each equation has the form

$$a_1 x_1 + a_2 x_2 + \cdots + a_n x_n = b$$

where the coefficients $a_1, a_2, \ldots, a_n, b$ are constants. Hence $x_1, x_2, \ldots, x_n$ represent the n unknown real values; equivalently, the unknown is an element $\langle x_1, x_2, \ldots, x_n \rangle$ of $\mathbb{R}^n$. The equation is called "linear" from the fact that if $n = 2$ its graph is a straight line. Systems of linear equations can be solved and understood best with the aid of matrices. We will not solve any systems of linear equations here since this skill is covered in texts on matrix theory and linear algebra. Our goal here is simply to observe the link between systems of linear equations and matrices.

EXAMPLE 4   The system

$$2x - 3y + \ \ z = 4$$

$$x + \ \ y - 5z = 7$$

is completely described by the coefficients 2, $-3, 1, 4, 1, 1, -5$ and 7, provided we know which coefficient goes where. We can take care of that by presenting them in their proper location,

$$\begin{bmatrix} 2 & -3 & 1 \\ 1 & 1 & -5 \end{bmatrix}$$

for the left-hand side and $\begin{bmatrix} 4 \\ 7 \end{bmatrix}$ for the right-hand side.   ▮

More generally, if we have m linear equations in n unknowns, it is convenient to index the coefficients by double subscripts as follows:

$$a_{11}x_1 + a_{12}x_2 + \cdots + a_{1n}x_n = b_1$$

$$a_{21}x_1 + a_{22}x_2 + \cdots + a_{2n}x_n = b_2$$

$$a_{31}x_1 + a_{32}x_2 + \cdots + a_{3n}x_n = b_3$$

$$\vdots$$

$$a_{m1}x_1 + a_{m2}x_2 + \cdots + a_{mn}x_n = b_m.$$

The rectangular array of coefficients

$$\mathbf{A} = \begin{bmatrix} a_{11} & a_{12} & \cdots & a_{1n} \\ a_{21} & a_{22} & \cdots & a_{2n} \\ a_{31} & a_{32} & \cdots & a_{3n} \\ \vdots & \vdots & & \vdots \\ a_{m1} & a_{m2} & \cdots & a_{mn} \end{bmatrix}$$

is called the **coefficient matrix** for the system. In general, a **matrix** is a rectangular array. Many computer languages allow the definitions of n-dimensional arrays for $n = 2, 3, 4, \ldots$, and with such languages one can describe matrices as 2-dimensional arrays. In many instances special algorithms are also available for performing operations with matrices which would not be appropriate for higher-dimensional arrays. The matrix $\mathbf{A}$ above has m horizontal rows and n vertical columns and is called an $m \times n$ **matrix**. It is traditional to use capital letters, such as $\mathbf{A}$, for matrices. The entry in the ith row and jth column is denoted by a_{ij} and we sometimes write $\mathbf{A}$ as $[a_{ij}]$. Whenever double indexes are used in matrix theory, rows precede columns! Sometimes, as in § 0.3, we denote the entry in the ith row and jth column by $\mathbf{A}[i, j]$; this notation is preferable in computer science since it avoids subscripts. In this text a **matrix** has real entries unless otherwise specified. Note that an $m \times n$ matrix can be viewed as a collection of real numbers indexed by the product set $\{1, 2, \ldots, m\} \times \{1, 2, \ldots, n\}$.

EXAMPLE 5   (a) The matrix

$$A = \begin{bmatrix} 2 & -1 & 0 & 3 & 2 \\ 1 & -2 & 1 & -1 & 3 \\ 3 & 0 & 1 & 2 & -3 \end{bmatrix}$$

is a 3×5 matrix. If we write $A = [a_{ij}]$, then $a_{11} = 2$, $a_{31} = 3$, $a_{13} = 0$, $a_{35} = -3$, etc. If we use the notation $A[i, j]$, then $A[1, 2] = -1$, $A[2, 1] = 1$, $A[2, 2] = -2$, etc.

(b) If B is the 3×4 matrix defined by $B[i, j] = i - j$, then $B[1, 1] = 1 - 1 = 0$, $B[1, 2] = 1 - 2 = -1$, etc. and so

$$B = \begin{bmatrix} 0 & -1 & -2 & -3 \\ 1 & 0 & -1 & -2 \\ 2 & 1 & 0 & -1 \end{bmatrix}. \quad \blacksquare$$

There are at least five reasons why matrices are important in the mathematical sciences.

1. They arise in solving systems of linear equations, as we hinted at in the previous paragraph.
2. Matrices are a convenient device for storing information that is naturally indexed by two variables. This is especially true in business, economics and computer science.
3. Many physical phenomena are linear or very nearly linear in nature and so matrices arise in the mathematical descriptions of these phenomena.
4. As hinted at in § 0.3, matrices are a valuable tool in graph theory, which we will study in Chapter 8.
5. The set of $n \times n$ matrices has a very rich algebraic structure, which is of interest in itself and is also a source of inspiration in the study of more abstract algebraic structures. We will introduce addition of matrices later in this section and we will introduce other algebraic operations on matrices in Chapter 4. More general algebraic systems will be studied in Chapter 11.

For positive integers m and n, we write $\mathfrak{M}_{m,n}$ for the set of all $m \times n$ matrices. Two matrices A and B in $\mathfrak{M}_{m,n}$ are **equal** provided all their corresponding entries are equal, i.e., $A = B$ provided $a_{ij} = b_{ij}$ for all i and j where $1 \leq i \leq m$ and $1 \leq j \leq n$. Matrices that have the same number of rows as columns are called **square matrices**. Thus A is a square matrix if A belongs to $\mathfrak{M}_{n,n}$ for some $n \in \mathbb{P}$. The **transpose** A^T of a matrix $A = [a_{ij}]$ in $\mathfrak{M}_{m,n}$ is the matrix in $\mathfrak{M}_{n,m}$ whose entry in the ith row and jth column is a_{ji}. That is, $\mathbf{A}^T[i, j] = \mathbf{A}[j, i]$. For example, if

$$\mathbf{A} = \begin{bmatrix} 2 & -1 & 0 & 4 \\ 3 & 2 & -1 & 2 \\ 4 & 0 & 1 & 3 \end{bmatrix}, \quad \text{then} \quad \mathbf{A}^T = \begin{bmatrix} 2 & 3 & 4 \\ -1 & 2 & 0 \\ 0 & -1 & 1 \\ 4 & 2 & 3 \end{bmatrix}.$$

The first row in $\mathbf{A}$ becomes the first column in $\mathbf{A}^T$, etc.

Matrices that have only one row, i.e., $1 \times n$ matrices, are often called **row vectors**, while matrices that have only one column, i.e., $m \times 1$ matrices, are called **column vectors**. The transpose of a row vector is a column vector and the transpose of a column vector is a row vector. Thus $[2 \quad 4 \quad -3 \quad -1]$ is a row vector and its transpose

$$\begin{bmatrix} 2 \\ 4 \\ -3 \\ -1 \end{bmatrix}$$

is a column vector. We sometimes view an $m \times n$ matrix as composed of m row vectors or of n column vectors.

Two matrices $\mathbf{A}$ and $\mathbf{B}$ can be added if they are the same size, that is, if they belong to the same set $\mathfrak{M}_{m,n}$. In this case, the **sum** is obtained by adding corresponding entries. More explicitly, if $\mathbf{A} = [a_{ij}]$ and $\mathbf{B} = [b_{ij}]$ are in $\mathfrak{M}_{m,n}$, then $\mathbf{A} + \mathbf{B}$ is the matrix $\mathbf{C} = [c_{ij}]$ in $\mathfrak{M}_{m,n}$ defined by

$$c_{ij} = a_{ij} + b_{ij} \quad \text{for} \quad 1 \leq i \leq m \quad \text{and} \quad 1 \leq j \leq n.$$

Equivalently, we define

$$(\mathbf{A} + \mathbf{B})[i, j] = \mathbf{A}[i, j] + \mathbf{B}[i, j] \quad \text{for} \quad 1 \leq i \leq m \quad \text{and} \quad 1 \leq j \leq n.$$

Since m or n can be 1, this definition applies in particular to row vectors and to column vectors.

EXAMPLE 6   (a) Consider

$$\mathbf{A} = \begin{bmatrix} 2 & 4 & 0 \\ -1 & 3 & 2 \\ -3 & 1 & 2 \end{bmatrix}, \quad \mathbf{B} = \begin{bmatrix} 1 & 0 & 5 & 3 \\ 2 & 3 & -2 & 1 \\ 4 & -2 & 0 & 2 \end{bmatrix}, \quad \mathbf{C} = \begin{bmatrix} 3 & 1 & -2 \\ -5 & 0 & 2 \\ -2 & 4 & 1 \end{bmatrix}.$$

Then we have

$$\mathbf{A} + \mathbf{C} = \begin{bmatrix} 5 & 5 & -2 \\ -6 & 3 & 4 \\ -5 & 5 & 3 \end{bmatrix},$$

but $\mathbf{A} + \mathbf{B}$ and $\mathbf{B} + \mathbf{C}$ are not defined. Of course the sums $\mathbf{A} + \mathbf{A}$, $\mathbf{B} + \mathbf{B}$ and $\mathbf{C} + \mathbf{C}$ are also defined; for example,

$$\mathbf{B} + \mathbf{B} = \begin{bmatrix} 2 & 0 & 10 & 6 \\ 4 & 6 & -4 & 2 \\ 8 & -4 & 0 & 4 \end{bmatrix}.$$

(b) Consider the row vectors

$$\mathbf{v}_1 = [-2 \quad 1 \quad 2 \quad 3], \quad \mathbf{v}_2 = [4 \quad 0 \quad 3 \quad -2], \quad \mathbf{v}_3 = [1 \quad 3 \quad 5]$$

and the column vectors

$$\mathbf{v}_4 = \begin{bmatrix} 1 \\ 2 \\ -3 \\ 2 \end{bmatrix}, \quad \mathbf{v}_5 = \begin{bmatrix} 0 \\ 3 \\ -2 \end{bmatrix} \quad \text{and} \quad \mathbf{v}_6 = \begin{bmatrix} 4 \\ 1 \\ 5 \end{bmatrix}.$$

The only sums of distinct vectors here that are defined are

$$\mathbf{v}_1 + \mathbf{v}_2 = [2 \quad 1 \quad 5 \quad 1] \quad \text{and} \quad \mathbf{v}_5 + \mathbf{v}_6 = \begin{bmatrix} 4 \\ 4 \\ 3 \end{bmatrix}. \quad ∎$$

Elements in $\mathbb{R}^n$ are also often called **vectors**. We add them just as if they were row vectors:

$$\langle x_1, x_2, \ldots, x_n \rangle + \langle y_1, y_2, \ldots, y_n \rangle = \langle x_1 + y_1, x_2 + y_2, \ldots, x_n + y_n \rangle.$$

EXERCISES 1.4

1. Let $A = \{a, b, c\}$ and $B = \{a, b, d\}$.
   (a) List or draw the ordered pairs in $A \times A$.
   (b) List or draw the ordered pairs in $A \times B$.
   (c) List or draw the set $\{\langle x, y \rangle \in A \times B : x = y\}$.

2. Let $S = \{0, 1, 2, 3, 4\}$ and $T = \{0, 2, 4\}$.
   (a) How many ordered pairs are in $S \times T$? $T \times S$?
   (b) List or draw the elements in $\{\langle m, n \rangle \in S \times T : m < n\}$.
   (c) List or draw the elements in $\{\langle m, n \rangle \in T \times S : m < n\}$.
   (d) List or draw the elements in $\{\langle m, n \rangle \in S \times T : m + n \geq 3\}$.
   (e) List or draw the elements in $\{\langle m, n \rangle \in T \times S : mn \geq 4\}$.
   (f) List or draw the elements in $\{\langle m, n \rangle \in S \times S : m + n = 10\}$.

3. For each of the following sets, list all elements if the set has fewer than seven elements. Otherwise, list exactly seven elements of the set.
   (a) $\{\langle m, n \rangle \in \mathbb{N}^2 : m = n\}$
   (b) $\{\langle m, n \rangle \in \mathbb{N}^2 : m + n \text{ is prime}\}$
   (c) $\{\langle m, n \rangle \in \mathbb{P}^2 : m = 6\}$
   (d) $\{\langle m, n \rangle \in \mathbb{P}^2 : \min\{m, n\} = 3\}$
   (e) $\{\langle m, n \rangle \in \mathbb{P}^2 : \max\{m, n\} = 3\}$
   (f) $\{\langle m, n \rangle \in \mathbb{N}^2 : m^2 = n\}$

4. List five elements in each of the following sets.
(a) $\{\langle m, n, p \rangle \in \mathbb{N}^3 : m + n = p\}$
(b) $\{\langle m, n, p \rangle \in \mathbb{N}^3 : m = p = 1\}$
(c) $\{\langle m, n, p, q \rangle \in \mathbb{N}^4 : mnpq = 0\}$

5. Sketch the following sets.
(a) $\{\langle m, n \rangle \in \mathbb{N}^2 : -1 \leq m - n \leq 1\}$
(b) $\{\langle m, n \rangle \in \mathbb{N}^2 : m - n \leq 2\}$

6. Sketch the following subsets of $\mathbb{R}^2$.
(a) $A = \{\langle x, y \rangle \in \mathbb{R}^2 : x = y^2\}$
(b) $B = \{\langle x, y \rangle \in \mathbb{R}^2 : x \leq y^2\}$
(c) $C = \{\langle x, y \rangle \in \mathbb{R}^2 : x \geq 0, y \geq 0, x + y = 1\}$
(d) $D = \{\langle x, y \rangle \in \mathbb{R}^2 : x \geq 0, y \geq 0, x + y \leq 1\}$

7. Consider $A = \{\langle x, y \rangle \in \mathbb{R}^2 : 2x - y = 4\}$, $B = \{\langle x, y \rangle \in \mathbb{R}^2 : x + 3y = 9\}$ and $C = \{\langle x, y \rangle \in \mathbb{R}^2 : y = 2x\}$. Find
(a) $A \cap B$ (b) $A \cap C$
(c) $B \cap C$ (d) $A^c \cup C^c$

8. Solve $\langle 2x - y, x + 2y \rangle = \langle 3, 4 \rangle$ for x and y.

9. Let $\Sigma = \{a, b, c, d\}$.
(a) List seven elements of the set

$$\{\langle w_1, w_2 \rangle \in \Sigma^* \times \Sigma^* : \text{length}(w_1) = \text{length}(w_2)\}.$$

(b) List all the elements in the set

$$\{\langle x, y, z \rangle \in \Sigma^3 : xyz \text{ is a word in the English language}\}.$$

For example, the ordered triple $\langle b, a, d \rangle$ belongs to this set.

10. Consider the matrix

$$\mathbf{B} = \begin{bmatrix} 1 & 2 & -2 & 1 \\ 3 & 0 & 1 & 2 \\ 2 & -1 & 4 & 1 \\ 0 & -3 & 1 & 3 \end{bmatrix}.$$

Evaluate

(a) b_{12} (b) b_{21} (c) b_{23} (d) $\sum_{i=1}^{4} b_{ii}$

11. Consider the matrices

$$\mathbf{A} = \begin{bmatrix} -1 & 0 & 2 \\ 1 & 3 & -2 \\ 4 & 2 & 3 \end{bmatrix}, \quad \mathbf{B} = \begin{bmatrix} 6 & 8 & 5 \\ 4 & -2 & 7 \\ 3 & 1 & 2 \end{bmatrix}, \quad \mathbf{C} = \begin{bmatrix} 1 & 3 \\ 2 & -4 \\ 5 & -2 \end{bmatrix}.$$

Calculate the following when they exist.
(a) $\mathbf{A}^T$ (b) $\mathbf{C}^T$ (c) $\mathbf{A} + \mathbf{B}$
(d) $\mathbf{A} + \mathbf{C}$ (e) $(\mathbf{A} + \mathbf{B})^T$ (f) $\mathbf{A}^T + \mathbf{B}^T$
(g) $\mathbf{B} + \mathbf{B}^T$ (h) $\mathbf{C} + \mathbf{C}^T$ (i) $(\mathbf{A} + \mathbf{A}) + \mathbf{B}$

12. Consider the following elements in $\mathbb{R}^3$:

$$\mathbf{v}_1 = \langle 1, 0, 0 \rangle, \qquad \mathbf{v}_2 = \langle 0, -1, 1 \rangle, \qquad \langle 1, 0, -1 \rangle.$$

Find

(a) $\mathbf{v}_1 + \mathbf{v}_2$ (b) $\mathbf{v}_1 + \mathbf{v}_3$

(c) $\mathbf{v}_3 + \mathbf{v}_2$ (d) $(\mathbf{v}_1 + \mathbf{v}_2) + \mathbf{v}_1$

13. Let $\mathbf{A} = [a_{ij}]$ and $\mathbf{B} = [b_{ij}]$ be matrices in $\mathfrak{M}_{4,3}$ defined by $a_{ij} = (-1)^{i+j}$ and $b_{ij} = i + j$. Find the following matrices when they exist.

(a) $\mathbf{A}^T$ (b) $\mathbf{A} + \mathbf{B}$ (c) $\mathbf{A}^T + \mathbf{B}$

(d) $\mathbf{A}^T + \mathbf{B}^T$ (e) $(\mathbf{A} + \mathbf{B})^T$ (f) $\mathbf{A} + \mathbf{A}$

14. Let $\mathbf{A}$ and $\mathbf{B}$ be matrices in $\mathfrak{M}_{3,3}$ defined by $\mathbf{A}[i, j] = ij$ and $\mathbf{B}[i, j] = i + j^2$.

(a) Find $\mathbf{A} + \mathbf{B}$.

(b) Calculate $\sum_{i=1}^{3} \mathbf{A}[i, i]$.

(c) Calculate $\sum_{i=1}^{3} \left(\sum_{j=1}^{3} \mathbf{B}[i, j] \right)$ and $\sum_{j=1}^{3} \left(\sum_{i=1}^{3} \mathbf{B}[i, j] \right)$.

(d) Does $\mathbf{A}$ equal its transpose $\mathbf{A}^T$?

(e) Does $\mathbf{B}$ equal its transpose $\mathbf{B}^T$?

15. (a) List the six 3×3 matrices whose rows are row vectors $[1 \quad 0 \quad 0]$, $[0 \quad 1 \quad 0]$ and $[0 \quad 0 \quad 1]$.

(b) Which matrices in part (a) are equal to their transposes?

16. In this exercise, $\mathbf{A}$ and $\mathbf{B}$ represent matrices. True or False.

(a) $(\mathbf{A}^T)^T = \mathbf{A}$ for all $\mathbf{A}$.

(b) If $\mathbf{A}^T = \mathbf{B}^T$, then $\mathbf{A} = \mathbf{B}$.

(c) If $\mathbf{A} = \mathbf{A}^T$, then $\mathbf{A}$ is a square matrix.

(d) If $\mathbf{A}$ and $\mathbf{B}$ are the same size, then $(\mathbf{A} + \mathbf{B})^T = \mathbf{A}^T + \mathbf{B}^T$.

17. For each $n \in \mathbb{N}$, let

$$\mathbf{A}_n = \begin{bmatrix} 1 & n \\ 0 & 1 \end{bmatrix} \quad \text{and} \quad \mathbf{B}_n = \begin{bmatrix} 1 & (-1)^n \\ -1 & 1 \end{bmatrix}.$$

(a) Give $\mathbf{A}_n^T$ for all $n \in \mathbb{N}$.

(b) Find $\{n \in \mathbb{N} : \mathbf{A}_n^T = \mathbf{A}_n\}$.

(c) Find $\{n \in \mathbb{N} : \mathbf{B}_n^T = \mathbf{B}_n\}$.

(d) Find $\{n \in \mathbb{N} : \mathbf{B}_n = \mathbf{B}_0\}$.

18. For sets S, T and V, prove

(a) $(S \cap T) \times V = (S \times V) \cap (T \times V)$

(b) $(S \cup T) \times V = (S \times V) \cup (T \times V)$

19. In § 1.1 we mentioned that all of mathematics can be defined or described in terms of "set" and "membership." For example, the **ordered pair** $\langle s, t \rangle$ with first entry s and second entry t can be defined as the set $\{\{s\}, \{s, t\}\}$, in which s and t obviously play different roles. Show that this definition satisfies the basic property of ordered pairs:

$$\langle s_1, t_1 \rangle = \langle s_2, t_2 \rangle \quad \text{if and only if} \quad s_1 = s_2 \quad \text{and} \quad t_1 = t_2.$$

CHAPTER HIGHLIGHTS

To check your understanding of the material in this chapter, we recommend
that you consider each item listed below and:

(a) Satisfy yourself that you can define each concept and each notation.

(b) Give at least one reason why the item was included in the chapter.

Concepts

set [undefined]
    member = element, subset
    equal, disjoint
    set operations
    universe, complement
    Venn diagram
indexed collection
ordered pair, product of sets
vector, matrix, matrix sum
coefficient matrix of a system of linear equations
alphabet, language, word, length of word

Examples and Notation

$\mathbb{N}, \mathbb{P}, \mathbb{Z}, \mathbb{Q}, \mathbb{R}, \mathfrak{M}_{m,n}$

$\in, \notin, \{\ :\ \}, \subseteq, \subset$

$\varnothing = \{\ \} = $ empty set

$\mathscr{P}(S), \cup, \cap, A \backslash B, A \oplus B$

$\displaystyle\bigcup_{k \in I} A_k, \bigcap_{k \in I} A_k$

$\sum$ [summation notation], $\prod$ [product notation]

$n!$

$\langle s, t \rangle, \langle s_1, \ldots, s_n \rangle, S \times T, S_1 \times \cdots \times S_n$

$S^n, \mathbb{R}^n$

Σ^*

Facts

Laws of set algebra [Table 1 of § 1.2].
Additional laws proved from them [e.g., Examples 5 and 6 of § 1.2].

Methods

Use of Venn diagrams.
Reasoning from definitions and previously established facts.

2

ELEMENTARY LOGIC
AND INDUCTION

In this chapter and in Chapter 6 we give an informal introduction to logic, including symbolic logic. Mathematicians, computer scientists, and in fact all scientists need to be able to recognize valid and invalid arguments, and should be aware of some techniques of logic. Thus our emphasis will be on logic as a working tool, though we will give some hints as to what is involved in a careful formal treatment of the subject.

In § 2.1 we introduce some notation and common terminology. In § 2.2 and § 2.3 we give the basic framework of the propositional calculus. The general concepts are important and should be mastered. The somewhat intimidating tables are provided for easy reference, and need not be memorized. Section 2.4 is devoted to formal proofs, to illustrate the "ideal" logical proof, and in § 2.5 we discuss proofs as encountered in the "real" mathematical world. Ultimately, the purpose of proofs is to communicate by providing convincing arguments; their exact structure is, for us, a secondary matter. In § 2.6 we introduce mathematical induction.

§ 2.1   Informal Introduction

Propositional calculus is the study of the logical relationships between objects called propositions, which are usually interpretable as meaningful assertions in real-life contexts. For us, a **proposition** will be any sentence that is either true or false, but not both. That is, it is a sentence that can be assigned the truth value **true** or the truth value **false**, and not both. We do not need to know what its truth value is in order to consider a proposition.

EXAMPLE 1 The following are propositions:

> (a) Julius Caesar was president of the United States.
> (b) The world is flat.
> (c) The Soviet Union is the world's largest country in area.
> (d) $2 + 2 = 4$.
> (e) $2 + 3 = 7$.
> (f) $2^{89301} + 3^{67258} + 1$ is a prime number.
> (g) The number 4 is positive and the number 3 is negative.
> (h) If a tree has n vertices, then it has exactly $n - 1$ edges.
> (i) $2^n + n$ is a prime number for infinitely many n.
> (j) Every even integer greater than 4 is the sum of two prime numbers.

Note that propositions (d) and (e) are mathematical sentences, where " $=$ " serves as the verb "equals" or "is equal to." Clearly proposition (f) is true or false, even though we have no idea which. Proposition (g) is false, since 3 is not negative. If this is not clear now, it will become clear soon, since (g) is the compound proposition: "4 is positive *and* 3 is negative." We have no idea whether proposition (i) is true or false, though some mathematicians may know the answer. On the other hand, as of the writing of this book *no one knows* whether proposition (j) is true; its truth is known as "Goldbach's conjecture." ∎

EXAMPLE 2 Here are some more propositions:

> (a) Every connected graph has an Euler circuit.
> (b) $x + y = y + x$ for all $x, y \in \mathbb{R}$.
> (c) $\mathbf{A} = \mathbf{A}^T$ for all 2×2 matrices $\mathbf{A}$.
> (d) $2^n = n^2$ for some $n \in \mathbb{N}$.
> (e) It is not true that 3 is an even integer or 7 is a prime.
> (f) If the world is flat, then $2 + 2 = 4$.

Propositions (a), (b) and (c) are really infinite sets of propositions covered by the phrase "Every" or "for all." And proposition (d) is a special sort of proposition because of the phrase "for some." Propositions of these types will be studied systematically in §§ 6.1 and 6.2. Proposition (e) is a somewhat confusing compound proposition whose truth value will be easy to analyze after the study of this chapter. Our propositional calculus will allow us to construct propositions like that in (f), even when they may appear silly or even paradoxical. ∎

EXAMPLE 3 The following sentences are not propositions:

> (a) Your place or mine?
> (b) Why is induction important?
> (c) Go directly to jail.
> (d) Help me, please.
> (e) $x - y = y - x$.

The reason that sentence (e) is not a proposition is that the symbols are not specified. If the intention is

(e′) $x - y = y - x$ for all $x, y \in \mathbb{R}$,

then this is a false proposition. If the intention is

(e″) $x - y = y - x$ for some $x, y \in \mathbb{R}$,

or

(e‴) $x - y = y - x$ for all x, y in $\{0\}$,

then this is a true proposition. The problem of unspecified symbols will be dealt with in § 6.1. ∎

EXAMPLE 4   Of course, in the real world there are ambiguous propositions:

(a) Teachers are overpaid.
(b) Doctors are rich.
(c) It was cold in Minneapolis in January 1924.
(d) Math is fun.
(e) Trees are more interesting than matrices.
(f) $A^2 = 0$ implies $A = 0$ for all A.

The difficulty with sentence (f) is that the set of allowable A's is not specified. Proposition (f) is true for all $A \in \mathbb{R}$. It turns out that (f) is meaningful, but false, for the set of all 2×2 matrices A. Ambiguous propositions should either be made unambiguous or abandoned. We will not concern ourselves with this process, but assume that our propositions are unambiguous. ∎

In the propositional calculus, we will generally use lower case letters such as $p, q, r, \ldots$ to stand for propositions and we will combine propositions to obtain compound propositions using standard connective symbols:

¬ for "not" or negation;
∧ for "and";
∨ for "or" [inclusive];
→ for "implies" or the conditional implication;
↔ for "if and only if" or the biconditional.

Other connectives, such as ⊕, appear in the exercises of § 2.3.

In § 2.2 we will carefully discuss each of the connective symbols and explain how they affect the truth values of compound propositions. We now illustrate how some of the propositions in Examples 1 and 2 can be viewed as compound propositions.

EXAMPLE 5   (a) Recall proposition (g) of Example 1: "The number 4 is positive and the number 3 is negative." This can be viewed as the compound proposition $p \wedge q$ where $p = $ "4 is positive" and $q = $ "3 is negative."

(b) Proposition (f) of Example 2, "If the world is flat, then $2 + 2 = 4$," can be viewed as the compound proposition $r \rightarrow s$ where $r =$ "the world is flat" and $s =$ "$2 + 2 = 4$."

(c) Proposition (e) of Example 2 says "It is not true that 3 is an even integer or 7 is a prime." This is $\neg(p \vee q)$ where $p =$ "3 is even" and $q =$ "7 is a prime." Actually, proposition (e) is poorly written and can also be interpreted to mean $(\neg p) \vee q$. ∎

The compound proposition $p \rightarrow q$ is read "p implies q," but has several other English language equivalents, such as "if p, then q." In fact, in Example 5(b) the compound proposition $r \rightarrow s$ was a translation of "if r, then s." So that proposition could have been written: "The world is flat implies that $2 + 2 = 4$." Other English language equivalents for $p \rightarrow q$ are: "p only if q," "q if p," "p is a sufficient condition for q," and "q is a necessary condition for p." We will usually avoid these; but see Exercises 15 and 16.

Compound propositions of the form $p \rightarrow q$, $q \rightarrow p$, $\neg p \rightarrow \neg q$, etc., appear to be related and are sometimes confused with each other. It is important to keep them straight. The proposition $q \rightarrow p$ is called the **converse** of the proposition $p \rightarrow q$. As we will see, it has a different meaning from $p \rightarrow q$. It turns out that $p \rightarrow q$ *is* equivalent to $\neg q \rightarrow \neg p$, which is called the **contrapositive** of $p \rightarrow q$.

EXAMPLE 6   Consider the sentence: "If it is raining, then there are clouds in the sky." This is the compound proposition $p \rightarrow q$ where $p =$ "it is raining" and $q =$ "there are clouds in the sky." This is a true proposition. Its converse $q \rightarrow p$ reads: "If there are clouds in the sky, then it is raining." Fortunately, this is a false proposition. The contrapositive $\neg q \rightarrow \neg p$ says: "If there are no clouds in the sky, then it is not raining." Not only is this a true proposition, but most people would agree that this follows "logically" from $p \rightarrow q$, without having to think again about the physical connection between rain and clouds. It does, and this logical connection will be made more precise in § 2.2 [Table 1, item 9]. ∎

In logic we are concerned with determining the truth values of propositions from related propositions. Example 6 illustrates that the truth of $p \rightarrow q$ does not imply that $q \rightarrow p$ is true, but it suggests that the truth of the contrapositive $\neg q \rightarrow \neg p$ follows from that of $p \rightarrow q$. Here is another illustration of why one must be careful in manipulating logical expressions.

EXAMPLE 7   Consider the argument "If Tom doesn't go to work tomorrow, he will not keep his job. He will go to work tomorrow, so he will keep his job." We are not concerned with whether Tom actually keeps his job [that's Tom's problem], but with whether Tom's keeping his job follows *logically* from the previous two assertions: "If Tom doesn't go to work tomorrow, he will not keep his job," and "He will go to work tomorrow." It turns out that this reasoning is not valid. The first sentence only tells us that Tom is in trouble if he doesn't go to work tomorrow; it tells us nothing otherwise. Perhaps Tom will lose his job for incompetence. If the reasoning above were valid, the following would also be: "If Carol doesn't buy a lottery ticket, then she will not win $1,000,000. Carol does buy a lottery ticket, so she will win $1,000,000."

Symbolically, these invalid arguments both take the form: If $\neg p \rightarrow \neg q$ and p are true, then q is true. The propositional calculus we develop in the next three sections will provide a formal framework with which to analyze arguments such as these. ∎

Let's return to compound propositions that include the phrase "for all" or "for every" and involve several propositions, possibly an infinite number.

EXAMPLE 8   Consider again Goldbach's conjecture from Example 1: "Every even integer greater than 4 is the sum of two prime numbers." This proposition turns out to be decomposable as

$$\text{``}p(6) \wedge p(8) \wedge p(10) \wedge \cdots\text{''}$$

or

$$\text{``}p(n) \text{ for every even } n \text{ in } \mathbb{N} \text{ greater than 4''}$$

where $p(n)$ is the simple proposition "n is the sum of two prime numbers." However, our rules for connectives in the propositional calculus will not allow constructions, such as these, which involve more than a finite number of propositions or phrases like "for every" or "for some." In Chapter 6 we develop the predicate calculus to handle these common logical constructions, which the propositional calculus cannot describe. ∎

We will see that a compound proposition connected by "for every" or "for all" is regarded as true if every single one of the propositions is true; otherwise, it is regarded as false. We will often be able to prove such propositions by using an important technique, mathematical induction, which we describe in § 2.6 and in Chapter 6. Such a compound proposition will be false if any one [or more] of its propositions is false. So to **disprove** such a compound proposition it is enough to show that one of its propositions is false. In other words, it is enough to supply an example that is counter to [or contrary to] the general proposition, i.e., a **counterexample**.

Goldbach's conjecture is still unsettled because no one has been able to prove that *every* even integer greater than 4 is the sum of two primes, and no one has found a counterexample. The conjecture has been verified for many even integers.

EXAMPLE 9   (a) The number 2 provides a counterexample to the assertion "All prime numbers are odd numbers."

(b) The number 7 provides a counterexample to the statement "Every positive integer is the sum of three squares of integers." Incidentally, it can be proved that every positive integer is the sum of four squares of integers, e.g., $1 = 1^2 + 0^2 + 0^2 + 0^2$, $7 = 2^2 + 1^2 + 1^2 + 1^2$, $73 = 8^2 + 3^2 + 0^2 + 0^2$.

(c) The number 23 provides a counterexample to the assertion "Every positive integer is the sum of eight cubes." However, it can be proved that every positive integer is the sum of nine cubes, e.g., $73 = 4^3 + 2^3 + 1^3 +$ some 0's.

(d) Gerald Ford is a counterexample to the assertion: "All American presidents have been right-handed."

(e) The value $n = 3$ provides a counterexample to the statement: "$n^2 \leq 2^n$ for all $n \in \mathbb{N}$." There are no other counterexamples, as we show in Example 7 of § 2.6. ∎

Given a general assertion whose truth value is unknown, often the only strategy is to make a guess and go with it. If you guess it's true, then analyze the situation to see why it's always true. This analysis might lead to a proof. If you fail to find a proof and you can see why, then you might discover a counterexample. Then again, if you can't find a counterexample, you might begin to suspect again that the result is true and formulate reasons why it must be. One of the authors spent a good deal of energy searching for a counterexample to a result that he felt was false, only to have a young Englishman later provide a proof.

EXERCISES 2.1

1. Let p, q, r be the following propositions:

$$p = \text{"it is raining,"}$$

$$q = \text{"the sun is shining,"}$$

$$r = \text{"there are clouds in the sky."}$$

Translate the following into logical notation, using p, q, r and logical connectives.
(a) It is raining and the sun is shining.
(b) If it is raining, then there are clouds in the sky.
(c) If it is not raining, then the sun is not shining and there are clouds in the sky.
(d) The sun is shining if and only if it is not raining.
(e) If there are no clouds in the sky, then the sun is shining.

2. Let p, q, r be as in Exercise 1. Translate the following into English sentences.
(a) $(p \land q) \to r$ (b) $(p \to r) \to q$
(c) $\neg p \leftrightarrow (q \lor r)$ (d) $\neg(p \leftrightarrow (q \lor r))$
(e) $\neg(p \lor q) \land r$

3. (a) Give the truth values of the propositions in parts (a)–(e) of Example 1.

(b) Do the same for parts (a)–(d) of Example 2.

4. Which of the following are propositions? Give the truth values of the propositions.

(a) $x^2 = x$ for all $x \in \mathbb{R}$.

(b) $x^2 = x$ for some $x \in \mathbb{R}$.

(c) $x^2 = x$.

(d) $x^2 = x$ for exactly one $x \in \mathbb{R}$.

(e) $xy = xz$ implies $y = z$.

(f) $xy = xz$ implies $y = z$ for all $x, y, z \in \mathbb{R}$.

(g) $w_1 w_2 = w_1 w_3$ implies $w_2 = w_3$ for all words $w_1, w_2, w_3 \in \Sigma^*$.

5. Consider the ambiguous sentence "$x^2 = y^2$ implies $x = y$ for all x, y."

(a) Make the sentence into an unambiguous proposition whose truth value is true.

(b) Make the sentence into an unambiguous proposition whose truth value is false.

6. Give the converses of the following propositions.

(a) $q \to r$.

(b) If I am smart, then I am rich.

(c) If $x^2 = x$, then $x = 0$ or $x = 1$.

(d) If $2 + 2 = 4$, then $2 + 4 = 8$.

7. Give the contrapositives of the propositions in Exercise 6.

8. (a) Verify that Goldbach's conjecture is true for some small values like 6, 8 and 10.

(b) Do the same for 98.

9. (a) Show that $n = 3$ provides a counterexample to the assertion "$n^3 < 3^n$ for all $n \in \mathbb{N}$."

(b) Can you find any other counterexamples?

10. (a) Show that $\langle m, n \rangle = \langle 4, -4 \rangle$ gives a counterexample to the assertion: "If m, n are nonzero integers that divide each other, then $m = n$."

(b) Give another counterexample.

11. (a) Show that $x = -1$ is a counterexample to "$(x + 1)^2 \geq x^2$ for every $x \in \mathbb{R}$."

(b) Find another counterexample.

(c) Can any nonnegative number serve as a counterexample? Explain.

12. Find counterexamples to the following assertions.

(a) $2^n - 1$ is prime for every $n \geq 2$.

(b) $2^n + 3^n$ is prime for all $n \in \mathbb{N}$.

(c) $2^n + n$ is prime for every positive odd integer n.

13. (a) Give a counterexample to: "$x > y$ implies $x^2 > y^2$ for all $x, y \in \mathbb{R}$." Your answer should be an ordered pair $\langle x, y \rangle$.

(b) How might you restrict x and y so that the proposition in part (a) is true?

14. Let S be a nonempty set. Determine which of the following assertions are true. For the true ones, give a reason. For the false ones, provide a counter-example.
    (a) $A \cup B = B \cup A$ for all $A, B \in \mathscr{P}(S)$.
    (b) $(A \setminus B) \cup B = A$ for all $A, B \in \mathscr{P}(S)$.
    (c) $(A \cup B) \setminus A = B$ for all $A, B \in \mathscr{P}(S)$.
    (d) $(A \cap B) \cap C = A \cap (B \cap C)$ for all $A, B, C \in \mathscr{P}(S)$.

15. Even though we will normally use "implies" and "if ..., then" to describe implication, other word orders and phrases often arise in practice, as in the examples below. Let p, q and r be the propositions:

$$p = \text{"the flag is set,"}$$

$$q = \text{"}I = 0\text{,"}$$

$$r = \text{"subroutine } S \text{ is completed."}$$

Translate each of the following propositions into symbols, using the letters p, q, r and the logical connectives.
    (a) If the flag is set, then $I = 0$.
    (b) Subroutine S is completed if the flag is set.
    (c) The flag is set if subroutine S is not completed.
    (d) Whenever $I = 0$ the flag is set.
    (e) Subroutine S is completed only if $I = 0$.
    (f) Subroutine S is completed only if $I = 0$ or the flag is set.
        Note the ambiguity in part (f); there are two different answers, each with its own claim to validity. Would punctuation help?

16. Consider the following propositions:

$$r = \text{"ODD}(N) = T\text{,"}$$

$$m = \text{"the output goes to the monitor,"}$$

$$p = \text{"the output goes to the printer."}$$

Translate the following, as in Exercise 15.
    (a) The output goes to the monitor if $\text{ODD}(N) = T$.
    (b) The output goes to the printer whenever $\text{ODD}(N) = T$ is not true.
    (c) $\text{ODD}(N) = T$ only if the output goes to the monitor.
    (d) The output goes to the monitor if the output goes to the printer.
    (e) $\text{ODD}(N) = T$ or the output goes to the monitor if the output goes to the printer.

§ 2.2   Propositional Calculus

The fundamental assumption in the propositional calculus is that the truth values of a proposition built up from other propositions by using logical connectives are completely determined by the truth values of the original propositions and the way the proposition is built up from them. Thus, given propositions p and q, the truth values of the compound propositions $\neg p$, $p \wedge q$, $p \vee q$, $p \rightarrow q$ and $p \leftrightarrow q$ will be determined by the truth values of p and q. Since there are only four different combinations of truth values for p and q, we can simply give tables to describe the truth values of the compound propositions for all combinations.

The proposition $\neg p$ should be true exactly when p is false. Most mathematicians and many computer scientists symbolize truth values as T or F [for True or False], while others symbolize them as 1 or 0 [for True or False]. The truth tables for $\neg p$ are:

p	$\neg p$		p	$\neg p$
F	T		0	1
T	F		1	0

Henceforth we will use 0's and 1's to signify False and True. The truth table for $p \wedge q$ is

p	q	$p \wedge q$
0	0	0
0	1	0
1	0	0
1	1	1

The table is read horizontally. For example, the third line tells us that if p is true and q is false, then the compound proposition $p \wedge q$ is to have truth value false. Note that $p \wedge q$ has truth value true exactly when both p *and* q are true.

As we explained in § 1.2, the use of "or" in the English language is somewhat ambiguous, but our use of $\vee$ will not be ambiguous. We define $\vee$ as follows:

p	q	$p \vee q$
0	0	0
0	1	1
1	0	1
1	1	1

Most people would agree with the truth value assignments for the first three lines. The fourth line states that we regard $p \vee q$ to be true if both p and q are true. This is the "inclusive or," sometimes written "and/or." Thus $p \vee q$ is true if p is true or q is true *or both*. The "exclusive or," symbolized $\oplus$, means that one or the other is true but not both; see Exercise 5 of § 2.3.

The **conditional implication** $p \rightarrow q$ means that the truth of p implies the truth of q. In other words, if p is true, then q must be true. The only way that this can fail is if p is true while q is false.

p	q	$p \rightarrow q$
0	0	1
0	1	1
1	0	0
1	1	1

The first two lines of the truth table for $p \to q$ may bother some people because it looks as if false propositions imply anything. In fact, we are simply defining the *compound proposition* $p \to q$ to be true if p is false. This usage of implication appears in ordinary English. Suppose that a politician promises "If I am elected, then taxes will be lower next year." If the politician is not elected, we would surely not regard him or her as a liar, no matter how the tax rates changed.

We will discuss the biconditional $p \leftrightarrow q$ after we introduce general truth tables.

A **truth table** for a compound proposition built up from propositions $p, q, r, \ldots$ is a table giving the truth values of the compound proposition in terms of the truth values of $p, q, r, \ldots$. We call $p, q, r, \ldots$ the **variables** of the table and of the compound proposition. One can determine the truth values of the compound proposition by determining the truth values of subpropositions working from the inside out, as we now illustrate.

EXAMPLE 1 Here is a truth table for the compound proposition $(p \wedge q) \vee \neg(p \to q)$. Note that there are still only four rows, because there are still only four distinct combinations of truth values for p and q.

column	1	2	3	4	5	6
	p	q	$p \wedge q$	$p \to q$	$\neg(p \to q)$	$(p \wedge q) \vee \neg(p \to q)$
	0	0	0	1	0	0
	0	1	0	1	0	0
	1	0	0	0	1	1
	1	1	1	1	0	1

The values in columns 3 and 4 are determined by the values in columns 1 and 2. The values in column 5 are determined by the values in column 4. The values in column 6 are determined by the values in columns 3 and 5. The sixth column gives the truth values of the complete compound proposition.

One can use a simpler truth table, with the same thought processes, by writing the truth values under the connectives, as follows:

p	q	$(p \wedge q)$	$\vee$	$\neg(p \to q)$	
0	0	0	0	0	1
0	1	0	0	0	1
1	0	0	1	1	0
1	1	1	1	0	1
step 1	1	2	4	3	2

The values at each step are determined by the values at earlier steps. For example, the values at the third step were determined by the values in the last column. The values at the fourth step were determined by the values in the third and fifth columns. The column created at the last step gives the truth values of the compound proposition. ∎

The simpler truth tables become more advantageous as the compound propositions get more complicated.

EXAMPLE 2 Here is the truth table for

$$(p \to q) \land [(q \land \neg r) \to (p \lor r)].$$

p	q	r	$(p \to q)$	$\land$	$[(q \land \neg r) \to (p \lor r)]$
0	0	0	1	1	0 1 1 0
0	0	1	1	1	0 0 1 1
0	1	0	1	0	1 1 0 0
0	1	1	1	1	0 0 1 1
1	0	0	0	0	0 1 1 1
1	0	1	0	0	0 0 1 1
1	1	0	1	1	1 1 1 1
1	1	1	1	1	0 0 1 1
step 1 1 1			2	5	3 2 4 2

Notice that the rows of a truth table could be given in any order: we've chosen a systematic order for the truth combinations of p, q, r partly to be sure we have listed them all. ∎

The **biconditional** $p \leftrightarrow q$ is defined by the truth table for $(p \to q) \land (q \to p)$:

p	q	$(p \to q)$	$\land$	$(q \to p)$
0	0	1	1	1
0	1	1	0	0
1	0	0	0	1
1	1	1	1	1
step 1 1		2	3	2

That is,

p	q	$p \leftrightarrow q$
0	0	1
0	1	0
1	0	0
1	1	1

Thus $p \leftrightarrow q$ is true if both p and q are true or if both p and q are false. The following are English language equivalents to $p \leftrightarrow q$: "p if and only if q," "p is a necessary and sufficient condition for q" and "p precisely if q."

It is worth emphasizing that the compound proposition $p \rightarrow q$ and its converse $q \rightarrow p$ are quite different; they have different truth tables.

An important class of compound propositions consists of those that are always true no matter what the truth values of the variables p, q, etc., are. Such a compound proposition is called a **tautology**. Why would we ever be interested in a proposition that is always true, and hence is pretty boring? The answer is that we are going to be dealing with some rather complicated-looking propositions which we hope to show are true, and the way that we will show their truth will be by using other propositions that are known to be true always. We begin with a very simple tautology.

EXAMPLE 3 (a) The classical tautology is the compound proposition $p \rightarrow p$:

p	$p \rightarrow p$
0	1
1	1

(b) The compound proposition $[p \wedge (p \rightarrow q)] \rightarrow q$ is a tautology:

p	q	$[p \wedge (p \rightarrow q)]$		$\rightarrow$	q
0	0	0	1	1	
0	1	0	1	1	
1	0	0	0	1	
1	1	1	1	1	
step 1	1	3	2	4	

(c) $\neg(p \vee q) \leftrightarrow (\neg p \wedge \neg q)$ is a tautology:

p	q	$\neg(p \vee q)$		$\leftrightarrow$	$(\neg p \wedge \neg q)$		
0	0	1	0	1	1	1	1
0	1	0	1	1	1	0	0
1	0	0	1	1	0	0	1
1	1	0	1	1	0	0	0
step 1	1	3	2	4	2	3	2	∎

A compound proposition that is always false is called a **contradiction**. Clearly a compound proposition P is a contradiction if and only if $\neg P$ is a tautology.

EXAMPLE 4   The classical contradiction is the compound proposition $p \wedge \neg p$:

p	p	$\wedge$	$\neg p$
0	0	0	1
1	1	0	0

Two compound propositions P and Q are regarded as **logically equivalent** if they have the same truth values for all choices of truth values of the variables p, q, etc. In other words, the final columns of their truth tables must be the same. When this occurs, we write $P \Leftrightarrow Q$. Since $P \leftrightarrow Q$ has truth values true precisely when the truth values of P and Q agree, we see that:

$$P \Leftrightarrow Q \quad \text{if and only if} \quad P \leftrightarrow Q \text{ is a tautology.}$$

The observation that $P \Leftrightarrow Q$ will be especially useful in cases where P and Q look quite different from each other. See, for instance, the formulas in Table 1.

EXAMPLE 5   (a) In view of Example 3(c), the compound propositions $\neg(p \vee q)$ and $\neg p \wedge \neg q$ are logically equivalent. That is, $\neg(p \vee q) \Leftrightarrow (\neg p \wedge \neg q)$.
    (b) The very nature of the connectives $\vee$ and $\wedge$ suggests that $p \vee q \Leftrightarrow q \vee p$ and $p \wedge q \Leftrightarrow q \wedge p$. Of course, one can verify these assertions by showing that $(p \vee q) \leftrightarrow (q \vee p)$ and $(p \wedge q) \leftrightarrow (q \wedge p)$ are tautologies.   ∎

It is worth stressing the difference between $\Leftrightarrow$ and $\leftrightarrow$. The expression "$P \Leftrightarrow Q$" is an assertion, namely that P and Q are logically equivalent, i.e., $P \leftrightarrow Q$ is a tautology. The expression "$P \leftrightarrow Q$" simply represents some compound proposition that might or might not be a tautology.
    In Table 1 we list a number of logical equivalences selected for their usefulness. To obtain a table of tautologies, replace each $\Leftrightarrow$ by $\leftrightarrow$. These tautologies can all be verified by truth tables. However, most of them should be intuitively reasonable. Many of the entries in Table 1 have names, which we have given, but there is no need to memorize them all. In the table, t represents any tautology and c represents any contradiction.
    The logical equivalences 2, 3, 4, 8 and 9 should be recognized by name, especially 9, the **contrapositive** rule.
    Given two compound propositions P and Q, we say that P **logically implies** Q provided Q has truth value true whenever P has truth value true. We write $P \Rightarrow Q$ when this occurs. Note that

$$P \Rightarrow Q \quad \text{if and only if the compound proposition } P \to Q \text{ is a tautology.}$$

Equivalently, $P \Rightarrow Q$ means that P and Q never simultaneously have the truth values 1 and 0, respectively, so when P is true, Q is true and when Q is false, P is false.

TABLE 1. Logical Equivalences

1. $\neg\,\neg p \Leftrightarrow p$	double negation
2a. $(p \vee q) \Leftrightarrow (q \vee p)$	
b. $(p \wedge q) \Leftrightarrow (q \wedge p)$	commutative laws
c. $(p \leftrightarrow q) \Leftrightarrow (q \leftrightarrow p)$	
3a. $[(p \vee q) \vee r] \Leftrightarrow [p \vee (q \vee r)]$	associative laws
b. $[(p \wedge q) \wedge r] \Leftrightarrow [p \wedge (q \wedge r)]$	
4a. $[p \vee (q \wedge r)] \Leftrightarrow [(p \vee q) \wedge (p \vee r)]$	distributive laws
b. $[p \wedge (q \vee r)] \Leftrightarrow [(p \wedge q) \vee (p \wedge r)]$	
5a. $(p \vee p) \Leftrightarrow p$	idempotent laws
b. $(p \wedge p) \Leftrightarrow p$	
6a. $(p \vee c) \Leftrightarrow p$	
b. $(p \vee t) \Leftrightarrow t$	identity laws
c. $(p \wedge c) \Leftrightarrow c$	
d. $(p \wedge t) \Leftrightarrow p$	
7a. $(p \vee \neg p) \Leftrightarrow t$	
b. $(p \wedge \neg p) \Leftrightarrow c$	
8a. $\neg(p \vee q) \Leftrightarrow (\neg p \wedge \neg q)$	
b. $\neg(p \wedge q) \Leftrightarrow (\neg p \vee \neg q)$	DeMorgan laws
c. $(p \vee q) \Leftrightarrow \neg(\neg p \wedge \neg q)$	
d. $(p \wedge q) \Leftrightarrow \neg(\neg p \vee \neg q)$	
9. $(p \rightarrow q) \Leftrightarrow (\neg q \rightarrow \neg p)$	contrapositive
10a. $(p \rightarrow q) \Leftrightarrow (\neg p \vee q)$	implication
b. $(p \rightarrow q) \Leftrightarrow \neg(p \wedge \neg q)$	
11a. $(p \vee q) \Leftrightarrow (\neg p \rightarrow q)$	
b. $(p \wedge q) \Leftrightarrow \neg(p \rightarrow \neg q)$	
12a. $[(p \rightarrow r) \wedge (q \rightarrow r)] \Leftrightarrow [(p \vee q) \rightarrow r]$	
b. $[(p \rightarrow q) \wedge (p \rightarrow r)] \Leftrightarrow [p \rightarrow (q \wedge r)]$	
13. $(p \leftrightarrow q) \Leftrightarrow [(p \rightarrow q) \wedge (q \rightarrow p)]$	equivalence
14. $[(p \wedge q) \rightarrow r] \Leftrightarrow [p \rightarrow (q \rightarrow r)]$	exportation law
15. $(p \rightarrow q) \Leftrightarrow [(p \wedge \neg q) \rightarrow c]$	reductio ad absurdum

EXAMPLE 6   We have $[p \wedge (p \rightarrow q)] \Rightarrow q$ since $[p \wedge (p \rightarrow q)] \rightarrow q$ is a tautology by Example 3(b). ∎

In Table 2 we list some useful logical implications. Each entry becomes a tautology if $\Rightarrow$ is replaced by $\rightarrow$. As with Table 1, many of the implications have names that need not be memorized.

In checking logical implications $P \Rightarrow Q$ it is only necessary to analyze the rows of the truth table where P is true or where Q is false.

TABLE 2. Logical Implications

16.	$p \Rightarrow (p \lor q)$	addition
17.	$(p \land q) \Rightarrow p$	simplification
18.	$(p \to c) \Rightarrow \neg p$	absurdity
19.	$[p \land (p \to q)] \Rightarrow q$	modus ponens
20.	$[(p \to q) \land \neg q] \Rightarrow \neg p$	modus tollens
21.	$[(p \lor q) \land \neg p] \Rightarrow q$	disjunctive syllogism
22.	$p \Rightarrow [q \to (p \land q)]$	
23.	$[(p \leftrightarrow q) \land (q \leftrightarrow r)] \Rightarrow (p \leftrightarrow r)$	transitivity of $\leftrightarrow$
24.	$[(p \to q) \land (q \to r)] \Rightarrow (p \to r)$	transitivity of $\to$ or hypothetical syllogism

25a. $(p \to q) \Rightarrow [(p \lor r) \to (q \lor r)]$
  b. $(p \to q) \Rightarrow [(p \land r) \to (q \land r)]$
  c. $(p \to q) \Rightarrow [(q \to r) \to (p \to r)]$

26a. $[(p \to q) \land (r \to s)] \Rightarrow [(p \lor r) \to (q \lor s)]$
  b. $[(p \to q) \land (r \to s)] \Rightarrow [(p \land r) \to (q \land s)]$ constructive dilemmas

27a. $[(p \to q) \land (r \to s)] \Rightarrow [(\neg q \lor \neg s) \to (\neg p \lor \neg r)]$
  b. $[(p \to q) \land (r \to s)] \Rightarrow [(\neg q \land \neg s) \to (\neg p \land \neg r)]$ destructive dilemmas

EXAMPLE 7 (a) We verify the implication $(p \land q) \Rightarrow p$. We need only consider the case where $p \land q$ is true, i.e., both p and q are true. Thus we consider the truncated table:

p	q	$(p \land q)$	$\to$	p
1	1	1	1	

(b) We verify the implication 26a. The full truth table would require 16 rows. However, we need only consider the cases where the implication $(p \lor r) \to (q \lor s)$ might be false. Thus it is enough to look at the cases for which $q \lor s$ is false, that is, with both q and s false.

p	q	r	s	$[(p \to q) \land (r \to s)]$			$\to$	$[(p \lor r) \to (q \lor s)]$		
0	0	0	0	1	1	1	1	0	1	0
0	0	1	0	1	0	0	1	1	0	0
1	0	0	0	0	0	1	1	1	0	0
1	0	1	0	0	0	0	1	1	0	0
step	1	1	1	1	2	3	2	4	2	3	2

∎

EXERCISES 2.2

1. Give the converse and contrapositive for each of the following propositions.
   (a) $p \rightarrow (q \wedge r)$.
   (b) If $x + y = 1$, then $x^2 + y^2 \geq 1$.
   (c) If $2 + 2 = 4$, then $3 + 3 = 8$.

2. Consider the proposition "if $x > 0$, then $x^2 > 0$ for $x \in \mathbb{R}$."
   (a) Give the converse and contrapositive of the proposition.
   (b) Which of the following are true propositions: the original proposition, its converse, its contrapositive?

3. Consider the following propositions:

$$p \rightarrow q, \qquad \neg p \rightarrow \neg q, \qquad q \rightarrow p, \qquad \neg q \rightarrow \neg p,$$

$$q \wedge \neg p, \qquad \neg p \vee q, \qquad \neg q \vee p, \qquad p \wedge \neg q.$$

   (a) Which proposition is the converse of $p \rightarrow q$?
   (b) Which proposition is the contrapositive of $p \rightarrow q$?
   (c) Which propositions are logically equivalent to $p \rightarrow q$?

4. Determine the truth values of the following compound propositions.
   (a) If $2 + 2 = 4$, then $2 + 4 = 8$.
   (b) If $2 + 2 = 5$, then $2 + 4 = 8$.
   (c) If $2 + 2 = 4$, then $2 + 4 = 6$.
   (d) If $2 + 2 = 5$, then $2 + 4 = 6$.
   (e) If the earth is flat, then Julius Ceasar was the first president of the United States.
   (f) If the earth is flat, then George Washington was the first president of the United States.
   (g) If George Washington was the first president of the United States, then the earth is flat.
   (h) If George Washington was the first president of the United States, then $2 + 2 = 4$.

5. Suppose that $p \rightarrow q$ is known to be false. Give the truth values for
   (a) $p \wedge q$ (b) $p \vee q$ (c) $q \rightarrow p$

6. Construct truth tables for
   (a) $p \wedge \neg p$ (b) $p \vee \neg p$
   (c) $p \leftrightarrow \neg p$ (d) $\neg \neg p$

7. Construct truth tables for
   (a) $\neg (p \wedge q)$ (b) $\neg (p \vee q)$
   (c) $\neg p \wedge \neg q$ (d) $\neg p \vee \neg q$

8. Construct the truth table for $(p \rightarrow q) \rightarrow [(p \vee \neg q) \rightarrow (p \wedge q)]$.

9. Construct the truth table for $[(p \vee q) \wedge r] \rightarrow (p \wedge \neg q)$.

10. Construct the truth table for $[(p \leftrightarrow q) \vee (p \rightarrow r)] \rightarrow (\neg q \wedge p)$.

11. Construct truth tables for

(a) $\neg(p \vee q) \to r$ (b) $\neg((p \vee q) \to r)$

This exercise shows that one must be careful with parentheses. We will discuss this matter further in Example 6 of § 3.6.

12. Verify the following logical equivalences using truth tables.

(a) the distributive law, rule 4a

(b) the identity laws 6a, 6b, 6c, 6d

(c) the contrapositive, rule 9

13. Verify the following logical equivalences using truth tables.

(a) rule 12a

(b) the exportation law, rule 14

(c) rule 15

14. Verify the following logical implications using truth tables.

(a) modus tollens, rule 20

(b) disjunctive syllogism, rule 21

15. Verify the following logical implications using truth tables and shortcuts as in Example 7.

(a) rule 25b

(b) rule 25c

(c) rule 26b

16. Prove or disprove the following. Note that only *one* line of the truth table is needed to show that a proposition is *not* a tautology.

(a) $(q \to p) \Leftrightarrow (p \wedge q)$

(b) $(p \wedge \neg q) \Rightarrow (p \to q)$

(c) $(p \wedge q) \Rightarrow (p \vee q)$

17. A logician told her son "If you don't finish your dinner, you will not get to stay up and watch TV." He finished his dinner and then was sent straight to bed. Discuss.

§ 2.3   More Propositional Calculus

In this section we will see how to obtain logical equivalences and implications without using truth tables. We also explain what we mean by "theorem" and "proof." We begin with two rules that are useful, but must be handled with care.

Substitution Rules

(a) If a compound proposition P is a tautology and if all occurrences of some variable of P, say q, are replaced by the same proposition E, then the resulting compound proposition P^* is also a tautology.

(b) If a compound proposition P contains a proposition Q and if Q is replaced by a logically equivalent proposition Q^*, then the resulting compound proposition P^* is logically equivalent to P.

To see why the substitution rules are true, one can analyze their effects on the corresponding truth tables. For example, in the case of rule (a) the truth table entries for P are all 1 regardless of the truth values for p. If we replaced all occurrences of p by a contradiction, for example, and considered a table for the resulting proposition, we would in effect be looking at just those rows in the table for P for which p is false, but they would still all have the value 1 for P. If we replace p by some other proposition E, then as soon as we determine the truth value of the proposition E we know which rows in the table for P to look at, but they all give the value 1 for P anyway, so who cares whether E is true or false?

For brevity, the equivalences and implications in Tables 1 and 2 of § 2.2 will often be referred to as **rules**. Here are some illustrations of the use of these rules.

EXAMPLE 1   (a) According to the modus ponens rule 19,

$$P = \text{``}[p \wedge (p \rightarrow q)] \rightarrow q\text{''}$$

is a tautology. If we replace each occurrence of p by the proposition $E = \text{``}q \rightarrow r\text{''}$ we obtain the tautology

$$P^* = \text{``}[(q \rightarrow r) \wedge ((q \rightarrow r) \rightarrow q)] \rightarrow q.\text{''}$$

If instead we replace each occurrence of q by E we obtain the tautology

$$[p \wedge (p \rightarrow (q \rightarrow r))] \rightarrow (q \rightarrow r).$$

(b) Consider the proposition

$$P = \text{``}\neg[(p \rightarrow q) \wedge (p \rightarrow r)] \rightarrow [q \rightarrow (p \rightarrow r)]\text{''}$$

which is not a tautology. We obtain logically equivalent propositions if we replace $(p \rightarrow q)$ by the logically equivalent $(\neg p \vee q)$ or if we replace one or both occurrences of $(p \rightarrow r)$ by $(\neg p \vee r)$; see rule 10a. We could also replace $[(p \rightarrow q) \wedge (p \rightarrow r)]$ by $[p \rightarrow (q \wedge r)]$ thanks to rule 12b. Thus P is logically equivalent to the following propositions among others:

$$\neg[(\neg p \vee q) \wedge (p \rightarrow r)] \rightarrow [q \rightarrow (p \rightarrow r)],$$
$$\neg[(p \rightarrow q) \wedge (\neg p \vee r)] \rightarrow [q \rightarrow (p \rightarrow r)],$$
$$\neg[p \rightarrow (q \wedge r)] \rightarrow [q \rightarrow (\neg p \vee r)]. \quad \blacksquare$$

EXAMPLE 2   (a) We illustrate substitution rules (a) and (b) by showing the nearly obvious equivalence

$$[(p \vee q) \vee (p \vee r)] \Leftrightarrow (p \vee q) \vee r$$

by successively substituting equivalent propositions.

Equivalent propositions	Explanations
$(p \lor q) \lor (p \lor r)$	given
$[(p \lor q) \lor p] \lor r$	associative law 3a with substitutions valid by substitution rule (a)
$[p \lor (q \lor p)] \lor r$	same
$[p \lor (p \lor q)] \lor r$	commutative law 2a
$[(p \lor p) \lor q] \lor r$	associative law 3a and substitution rule (a)
$[p \lor q] \lor r$	idempotent law 5a and substitution rule (b)

Here we carefully mentioned each application of substitution rules (a) and (b), but in practice one would explain the substitution involved at a given step only if it appeared that the reader might not see it without help.

(b) Now we derive the useful tautology

$$[(p \to q) \lor (p \to r)] \to [p \to (q \lor r)]$$

starting from a tautology based on the associative law:

$$[(p \lor q) \lor r] \to [p \lor (q \lor r)].$$

By part (a) and substitution rule (b),

$$[(p \lor q) \lor (p \lor r)] \to [p \lor (q \lor r)]$$

is also a tautology. Replacing each occurrence of p by $\neg p$, we get the tautology

$$[(\neg p \lor q) \lor (\neg p \lor r)] \to [\neg p \lor (q \lor r)].$$

Implication rule 10a and substitution rule (a) tell us

$$\neg p \lor q \Leftrightarrow p \to q, \qquad \neg p \lor r \Leftrightarrow p \to r, \qquad \neg p \lor (q \lor r) \Leftrightarrow p \to (q \lor r).$$

Applying substitution rule (b) three times, we conclude that

$$[(p \to q) \lor (p \to r)] \to [p \to (q \lor r)]$$

is a tautology. ∎

As illustrated in Example 2, we can use substitution to transform one compound proposition into a logically equivalent one. Normally one would use this procedure to obtain a more convenient proposition. What is "convenient" depends, of course, on what comes next. These rules can be considered analogous to the rules of ordinary algebra, with which we often rewrite expressions in forms more suited to whatever task we have at hand.

EXAMPLE 3   We find a proposition logically equivalent to $(p \wedge q) \to (\neg p \wedge q)$ that does not use the connective $\wedge$ by using the DeMorgan law 8d and substitution. Since $p \wedge q$ is equivalent to $\neg(\neg p \vee \neg q)$ and $\neg p \wedge q$ is equivalent to $\neg(\neg \neg p \vee \neg q)$, the given proposition is equivalent to

$$\neg(\neg p \vee \neg q) \to \neg(\neg \neg p \vee \neg q)$$

and so to

$$\neg(\neg p \vee \neg q) \to \neg(p \vee \neg q).$$

If desired, we could apply rule 10a to obtain the equivalent

$$\neg(p \to \neg q) \to \neg(q \to p),$$

which uses neither $\wedge$ nor $\vee$. On the other hand, we could avoid the use of the connective $\to$ by applying rule 10a. ∎

With the tautologies of § 2.2 we will be able to specify precisely what we mean by a valid proof in the propositional calculus. A **theorem** consists of some propositions $H_1, H_2, \ldots, H_n$, called its **hypotheses**, and a proposition C, called its **conclusion**. Here $H_1, H_2, \ldots, H_n$ and C represent propositions as discussed in § 2.1. A theorem with hypotheses $H_1, H_2, \ldots, H_n$ and conclusion C is **true** provided

$$H_1 \wedge H_2 \wedge \cdots \wedge H_n \Rightarrow C.$$

Thus the theorem is true if and only if $H_1 \wedge H_2 \wedge \cdots \wedge H_n \to C$ is a tautology.

A **formal proof** of a theorem consists of a sequence of propositions, ending with the conclusion C, that are regarded as valid for any of several reasons. To be **valid** a proposition may be one of the hypotheses, may be a known tautology, may be derived from propositions earlier in the sequence via the substitution rules or may be inferred from earlier propositions according to certain **rules of inference**. A proposition Q can be inferred from propositions $P_1, P_2, \ldots, P_k$ provided $P_1 \wedge P_2 \wedge \cdots \wedge P_k \Rightarrow Q$. We symbolize such a **rule of inference** as

$$
\begin{array}{l}
P_1 \\
P_2 \\
\vdots \\
P_k \\
\hline
\therefore \ Q
\end{array}
\quad [\therefore \text{ is read "hence" or "therefore."}]
$$

EXAMPLE 4   The logical implication modus ponens $[p \wedge (p \to q)] \Rightarrow q$ corresponds to the rule of inference

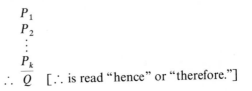

$$
\begin{array}{l}
P \\
P \to Q \\
\hline
\therefore \ Q
\end{array}
$$

TABLE 1. Rules of Inference

28.	P		29.	$P \wedge Q$	
	$\therefore \; \overline{P \vee Q}$	addition		$\therefore \; \overline{P}$	simplification
30.	P		31.	$P \rightarrow Q$	
	$\overline{P \rightarrow Q}$			$\overline{\neg Q}$	
	$\therefore \; Q$	modus ponens		$\therefore \; \neg P$	modus tollens
32.	$P \vee Q$		33.	$P \rightarrow Q$	
	$\overline{\neg P}$	disjunctive		$Q \rightarrow R$	hypothetical
	$\therefore \; Q$	syllogism		$\therefore \; \overline{P \rightarrow R}$	syllogism
34.	P				
	Q				
	$\therefore \; \overline{P \wedge Q}$	conjunction			

where P and Q represent compound propositions. In fact, every logical implication in Table 2 of § 2.2 corresponds to a rule of inference. We list some of them in Table 1. Note that rule 34 is based on the tautology $(p \wedge q) \rightarrow (p \wedge q)$. ∎

Note that if compound propositions P and Q are logically equivalent, then in particular $P \Rightarrow Q$, so we have a corresponding rule of inference.

EXAMPLE 5 (a) From the logical equivalence 4a in Table 1 of § 2.2, we deduce $[(p \vee q) \wedge (p \vee r)] \Rightarrow [p \vee (q \wedge r)]$ and hence we have the rule of inference

$$P \vee Q$$
$$\overline{P \vee R}$$
$$\therefore \; P \vee (Q \wedge R).$$

(b) From the logical equivalence 12b in Table 1 of § 2.2 we obtain the rule of inference

$$P \rightarrow Q$$
$$\overline{P \rightarrow R}$$
$$\therefore \; P \rightarrow (Q \wedge R).$$ ∎

We now have the framework for giving formal proofs, which is what the next section is devoted to. You might wonder, however, why anyone would want to be so formal. We are trying to build a symbolic model to show how logical proofs are constructed and how to tell if a chain of inferences is valid. No one, not even the formal logician, is enchanted about writing out formal proofs. The idea is to learn what kinds of proofs are possible and to see ways the rules of inference apply, since we use these rules implicitly when we write proofs informally.

EXERCISES 2.3

1. Show $[(p \vee r) \wedge (q \to r)] \Leftrightarrow [(p \to q) \to r)]$ by the methods of Example 1.

2. Repeat Exercise 1 for $\neg q \Rightarrow [(p \vee q) \to p]$.

3. Give the rules of inference corresponding to the logical implications 23, 26a and 27b.

4. Verify the following **absorption laws**.
   (a) $[p \vee (p \wedge q)] \Leftrightarrow p$
   (b) $[p \wedge (p \vee q)] \Leftrightarrow p$

5. The "exclusive or" connective $\oplus$ is defined by the truth table

p	q	$p \oplus q$
0	0	0
0	1	1
1	0	1
1	1	0

   (a) Show that $p \oplus q$ has the same truth table as $\neg(p \leftrightarrow q)$.
   (b) Construct a truth table for $p \oplus p$.
   (c) Construct a truth table for $(p \oplus q) \oplus r$.
   (d) Construct a truth table for $(p \oplus p) \oplus p$.

6. (a) Write a compound proposition that is true when exactly one of the three propositions p, q and r is true.
   (b) Write a compound proposition that is true when exactly two of the three propositions p, q and r are true.

7. Show that $(p \oplus q) \Leftrightarrow [(p \vee q) \wedge \neg(p \wedge q)]$, where $\oplus$ is the "exclusive or" introduced in Exercise 5.

8. Prove or disprove.
   (a) $[p \to (q \to r)] \Leftrightarrow [(p \to q) \to (p \to r)]$
   (b) $[p \oplus (q \to r)] \Leftrightarrow [(p \oplus q) \to (p \oplus r)]$

9. Prove or disprove the "associative laws."
   (a) $[(p \to q) \to r] \Leftrightarrow [p \to (q \to r)]$
   (b) $[(p \leftrightarrow q) \leftrightarrow r] \Leftrightarrow [p \leftrightarrow (q \leftrightarrow r)]$
   (c) $(p \oplus q) \oplus r \Leftrightarrow p \oplus (q \oplus r)$

10. Every compound proposition can be written using only the connectives $\neg$ and $\vee$. This fact follows from the equivalences $(p \to q) \Leftrightarrow (\neg p \vee q)$, $(p \wedge q) \Leftrightarrow \neg(\neg p \vee \neg q)$, and $(p \leftrightarrow q) \Leftrightarrow [(p \to q) \wedge (q \to p)]$. Find propositions logically equivalent to the following using only the connectives $\neg$ and $\vee$.
    (a) $p \leftrightarrow q$ (b) $(p \wedge q) \to (\neg q \wedge r)$
    (c) $(p \to q) \wedge (q \vee r)$ (d) $p \oplus q$

11. (a) Show that $p \vee q$ and $p \wedge q$ are logically equivalent to propositions using only the connectives $\neg$ and $\to$.
    (b) Show that $p \vee q$ and $p \to q$ are logically equivalent to propositions using only the connectives $\neg$ and $\wedge$.

(c) Is $p \to q$ logically equivalent to a proposition using only the connectives $\wedge$ and $\vee$?

12. The **Sheffer stroke** is a connective $|$ defined by the truth table:

p	q	$p\vert q$
0	0	1
0	1	1
1	0	1
1	1	0

This connective is interesting because all compound propositions can be written using only this connective. This fact follows from the remarks in Exercise 10 and parts (a) and (b) below.
(a) Show that $\neg p \Leftrightarrow p\vert p$.
(b) Show that $p \vee q \Leftrightarrow (p\vert p)\vert(q\vert q)$.
(c) Find a proposition equivalent to $p \wedge q$ using only the Sheffer stroke.
(d) Do the same for $p \to q$.
(e) Do the same for $p \oplus q$.

13. Consider the tautology $p \to [q \to (p \wedge q)]$. Show that if the first p is replaced by the proposition $p \vee q$, then the new proposition is not a tautology. This shows that substitution rule (a) must be applied with care.

14. Let P be the proposition $[p \wedge (q \vee r)] \vee \neg[p \vee (q \vee r)]$. Replacing all occurrences of $q \vee r$ by $q \wedge r$ yields

$$P^* = \text{``}[p \wedge (q \wedge r)] \vee \neg[p \vee (q \wedge r)].\text{''}$$

Since $q \wedge r \Rightarrow q \vee r$, one might suppose that $P \Rightarrow P^*$ or that $P^* \Rightarrow P$. Show that neither of these is the case.

15. Is the following a proposition? "This sentence is false." If so, what is its truth value?

§ 2.4   Formal Proofs

This section is devoted to formal proofs. The rules governing them are given in § 2.3. The techniques we describe need not be mastered and will not be used later in this book. We turn to practical approaches to proofs in the next section.

A formal proof with a valid sequence of propositions is called a **valid proof** or **valid argument**. Regardless of what the conclusion is, if one or more of the propositions is invalid, then the argument is called a **fallacy**. Here are some examples that begin with propositions given in English.

EXAMPLE 1   We analyze the following argument. "If I study or if I am a genius, then I will pass the course. If I pass the course, then I will be allowed to take the next course. Therefore, if I am not allowed to take the next course, then I am not a genius." We let

$$s = \text{"I study,"}$$

$$g = \text{"I am a genius,"}$$

$$p = \text{"I will pass the course,"}$$

$$a = \text{"I will be allowed to take the next course."}$$

Then the theorem is: if $s \vee g \to p$ and $p \to a$, then $\neg a \to \neg g$. Here is our scratchwork. We want to infer $\neg a \to \neg g$, which is the contrapositive of $g \to a$. From $s \vee g \to p$ we can surely infer $g \to p$ [details in the next sentence] and then combine this with $p \to a$ to get $g \to a$ using rule 33. To get $g \to p$ from $s \vee g \to p$, rule 33 shows that it is enough to observe that $g \to s \vee g$ by the addition rule. Well, almost: the addition rule gives $g \to g \vee s$ and so we need to invoke commutativity too. Here is the formal proof. In the explanations, the numbers prior to the semicolon refer to the earlier propositions from which the proposition is inferred.

Proof	*Explanations*
1. $s \vee g \to p$	hypothesis
2. $p \to a$	hypothesis
3. $g \to g \vee s$	addition (rule 16)
4. $g \to s \vee g$	3; commutative law 2a
5. $g \to p$	4, 1; hypothetical syllogism (rule 33)
6. $g \to a$	5, 2; hypothetical syllogism (rule 33)
7. $\neg a \to \neg g$	6; contrapositive (rule 9)

We stress that this theorem can be proved in several ways. Thus there may be many valid proofs of a single theorem. ∎

EXAMPLE 2   "If I study or if I am a genius, then I will pass the course. I will not be allowed to take the next course. If I pass the course, then I will be allowed to take the next course. Therefore, I did not study." With the notation of Example 1, the theorem is: if $s \vee g \to p$, $\neg a$ and $p \to a$, then $\neg s$. The only rule of inference in Table 1 of §2.3 that might help is rule 31 [modus tollens]. Given the hypothesis $\neg a$, this rule would allow us to infer $\neg s$ if only we had $s \to a$. But this can be inferred from the first and third hypotheses as in Example 1. Here is the formal version.

Proof	*Explanations*
1. $s \vee g \to p$	hypothesis
2. $\neg a$	hypothesis
3. $p \to a$	hypothesis
4. $s \to s \vee g$	addition (rule 16)
5. $s \to p$	4, 1; hypothetical syllogism (rule 33)
6. $s \to a$	5, 3; hypothetical syllogism (rule 33)
7. $\neg s$	6, 2; modus tollens (rule 31) ∎

In general, proving a theorem $H_1 \wedge H_2 \wedge \cdots \wedge H_n \Rightarrow C$ is equivalent to proving

$$H_1 \wedge H_2 \wedge \cdots \wedge H_n \wedge \neg C \Rightarrow \text{a contradiction}$$

in view of the logical equivalence reductio ad absurdum (rule 15). This approach to a proof is called a **proof by contradiction**.

EXAMPLE 3 We prove the assertion in Example 2 by contradiction. We begin by listing the hypotheses and the negation of the conclusion.

Proof	*Explanations*
1. $s \vee g \to p$	hypothesis
2. $\neg a$	hypothesis
3. $p \to a$	hypothesis
4. $\neg(\neg s)$	negation of the conclusion

It will be easy to infer s from line 4. Then we hope to infer a contradiction like $s \wedge (\neg s), a \wedge (\neg a), g \wedge (\neg g)$ or $p \wedge (\neg p)$. Since we already have $\neg a$ and s, the first two contradictions should be easier to reach. We aim for $a \wedge (\neg a)$, because we already have $p \to a$. Since we have s, rule 30 shows that it suffices to get $s \to a$. So all we need is $s \to p$, which we can get from line 1. The proof continues:

5. s	4; double negation (rule 1)
6. $s \to s \vee g$	addition (rule 16)
7. $s \to p$	6, 1; hypothetical syllogism (rule 33)
8. $s \to a$	7, 3; hypothetical syllogism (rule 33)
9. a	5, 8; modus ponens (rule 30)
10. $a \wedge (\neg a)$	9, 2; conjunction (rule 34)
11. contradiction	10; rule 7b

Even though this proof is longer than the proof in Example 2, it may be conceptually more straightforward. In the real world, one resorts to proofs by contradiction when it is easier to use $\neg C$ *in conjunction with* the hypotheses than it is to derive C *from the hypotheses.* We will illustrate this in § 2.5. ∎

Here is another example.

EXAMPLE 4 "If I do not specify the initial conditions, then my program will not begin. If I program an infinite loop, then my program will not terminate. If the program does not begin or if it does not terminate, then the program will fail. Therefore, if the program does not fail, then I specified the initial conditions and I did not program an infinite loop." Let

$$i = \text{"I specified the initial conditions,"}$$

$$b = \text{"the program will begin,"}$$

$$l = \text{"I programmed an infinite loop,"}$$

$$t = \text{"the program will terminate,"}$$

$$f = \text{"the program fails."}$$

The theorem is: if $\neg i \to \neg b$, $l \to \neg t$ and $(\neg b \vee \neg t) \to f$, then $\neg f \to (i \wedge \neg l)$. We will first work for the contrapositive which, by a DeMorgan law, is equivalent to $\neg i \vee l \to f$. [Note that t does not stand for "tautology" here.]

Proof	Explanations
1. $\neg i \to \neg b$	hypothesis
2. $l \to \neg t$	hypothesis
3. $(\neg b \vee \neg t) \to f$	hypothesis
4. $(\neg i \vee l) \to (\neg b \vee \neg t)$	1, 2; the rule of inference based on the constructive dilemma 26a
5. $(\neg i \vee l) \to f$	4, 3; hypothetical syllogism (rule 33)
6. $\neg f \to \neg(\neg i \vee l)$	5; contrapositive (rule 9)
7. $\neg f \to (\neg\neg i \wedge \neg l)$	6; DeMorgan law 8a
8. $\neg f \to (i \wedge \neg l)$	7; double negation (rule 1) $\blacksquare$

EXAMPLE 5   "If the program does not fail, then the program will begin and terminate. The program begins and fails. Therefore the program did not terminate." With the notation of Example 4, the theorem is: if $\neg f \to b \wedge t$ and $b \wedge f$, then $\neg t$.

Proof?	Explanations
1. $\neg f \to b \wedge t$	hypothesis
2. $b \wedge f$	hypothesis
3. $b \wedge t \to t$	simplification (rule 17)
4. $\neg f \to t$	1, 3; hypothetical syllogism (rule 33)
5. f	2; simplification (rule 29)
6. $\neg t$	4, 5; see below

How would you infer proposition 6 from propositions 4 and 5? It appears that the best hope is modus tollens, but a closer look shows that modus tollens does not apply. What is needed is $P \to Q$, $\neg P$, $\therefore$ $\neg Q$ and this is *not* a valid rule of inference. The alleged proof above is invalid and is a fallacy. The proof is invalid, as just indicated, but this failure alone does not show that there is no correct proof; perhaps we simply went about the proof in the wrong way. In fact, no correct proof exists in this case because the theorem itself is not true. That is,

$$\{[\neg f \rightarrow (b \wedge t)] \wedge (b \wedge f)\} \rightarrow \neg t$$

is not a tautology. To see this, consider the logical possibility where b, f and t are all true. In other words, consider the last row of its truth table:

b	f	t	$\{[\neg f \rightarrow (b \wedge t)] \wedge (b \wedge f)\}$						$\rightarrow$	$\neg t$
1	1	1	0	1	1	1	1		0	0

In terms of the original hypotheses, the program might terminate and yet fail for some other reason. ∎

EXERCISES 2.4

1. Complete the following formal proofs by supplying explanations for each step.
   (a) If $p \rightarrow (q \vee r)$, $q \rightarrow s$ and $r \rightarrow t$, then $p \rightarrow (s \vee t)$.

   Proof
   1. $p \rightarrow (q \vee r)$
   2. $q \rightarrow s$
   3. $r \rightarrow t$
   4. $(q \vee r) \rightarrow (s \vee t)$ [See Exercise 3 of § 2.3.]
   5. $p \rightarrow (s \vee t)$

   (b) If $p \rightarrow (q \wedge r)$, $(q \vee s) \rightarrow t$ and $p \vee s$, then t.

   Proof
   1. $p \rightarrow (q \wedge r)$
   2. $(q \vee s) \rightarrow t$
   3. $p \vee s$
   4. $(q \wedge r) \rightarrow q$
   5. $p \rightarrow q$
   6. $(p \vee s) \rightarrow (q \vee s)$
   7. $q \vee s$
   8. t

   (c) If $p \rightarrow (q \rightarrow r)$, $p \vee \neg s$ and q, then $s \rightarrow r$.

   Proof
   1. $p \rightarrow (q \rightarrow r)$
   2. $p \vee \neg s$
   3. q
   4. $\neg s \vee p$
   5. $s \rightarrow p$
   6. $s \rightarrow (q \rightarrow r)$
   7. $(s \wedge q) \rightarrow r$
   8. $q \rightarrow [s \rightarrow (q \wedge s)]$
   9. $s \rightarrow (q \wedge s)$
   10. $s \rightarrow (s \wedge q)$
   11. $s \rightarrow r$

2. (a) Revise the proof in Example 3 to reach the contradiction $s \wedge (\neg s)$.
   Hint: Use lines 2 and 8.
   (b) Revise the proof in Example 3 to reach the contradiction $p \wedge (\neg p)$.
   Hint: Use lines 2 and 3, then lines 5 and 7.

3. Complete the following proof by contradiction, by supplying explanations for each step.

$$\text{If } p \to (q \wedge r), (q \vee s) \to t \text{ and } p \vee s, \text{ then } t.$$

Proof
1. $p \to (q \wedge r)$
2. $(q \vee s) \to t$
3. $p \vee s$
4. $\neg t$
5. $\neg(q \vee s)$
6. $\neg q \wedge \neg s$
7. $\neg q$
8. $\neg s \wedge \neg q$
9. $\neg s$
10. $s \vee p$
11. p
12. $q \wedge r$
13. q
14. $q \wedge (\neg q)$
15. contradiction

4. Complete the following proof by contradiction.

$$\text{If } p \to (q \wedge r), r \to s \text{ and } \neg(q \wedge s), \text{ then } \neg p.$$

Proof
1. $p \to (q \wedge r)$
2. $r \to s$
3. $\neg(q \wedge s)$
4. $\neg(\neg p)$
5. p
6. $q \wedge r$
7. q
8. $r \wedge q$
9. r
10. s
11. $q \wedge s$
12. $(q \wedge s) \wedge \neg(q \wedge s)$
13. contradiction

5. Convert each of the following arguments into logical notation using the suggested variables. Then provide a formal proof.
   (a) "If my computations are correct and I pay the electric bill, then I will run out of money. If I don't pay the electric bill, the power will be turned off. Therefore, if I don't run out of money and the power is still on, then my computations are incorrect." (c, b, r, p)
   (b) "If the weather bureau predicts dry weather, then I will take a hike or go swimming. I will go swimming if and only if the weather bureau predicts

warm weather. Therefore, if I don't go on a hike, then the weather bureau predicts wet or warm weather." (d, h, s, w)

(c) "If I get the job and work hard, then I will get promoted. If I get promoted, then I will be happy. I will not be happy. Therefore, either I will not get the job or I will not work hard." (j, w, p, h)

(d) "If I study law, then I will make a lot of money. If I study archeology, then I will travel a lot. If I make a lot of money or travel a lot, then I will not be disappointed. Therefore, if I am disappointed, then I did not study law and I did not study archeology." (l, m, a, t, d)

6. Construct formal proofs of the following theorems.
   (a) If $p \rightarrow (q \vee r)$ and $q \rightarrow s$, then $p \rightarrow (r \vee s)$.
   (b) If $p \rightarrow q$, $r \rightarrow s$ and $\neg(p \rightarrow s)$, then $q \wedge \neg r$.
   (c) If $p \rightarrow q$, $\neg r \rightarrow s$ and $\neg q \vee \neg s$, then $p \rightarrow r$.

7. Construct formal proofs by contradiction for the following theorems.
   (a) If $p \rightarrow q$, $r \rightarrow (p \wedge s)$, $(q \wedge s) \rightarrow (p \wedge t)$ and $\neg t$, then $p \rightarrow \neg r$.
   (b) If $p \vee (q \rightarrow r)$, $q \vee r$ and $r \rightarrow p$, then p.

8. For each of the following, give a formal proof of the theorem or show that it is false by exhibiting a suitable row of a truth table.
   (a) If $(q \wedge r) \rightarrow p$ and $q \rightarrow \neg r$, then p.
   (b) If $q \vee \neg r$ and $\neg(r \rightarrow q) \rightarrow \neg p$, then p.
   (c) If $p \rightarrow (q \vee r)$, $q \rightarrow s$ and $r \rightarrow \neg p$, then $p \rightarrow s$.

9. For the following sets of hypotheses, state a conclusion that can be inferred and specify the rules of inference used.
   (a) If the TV set is not broken, then I will not study. If I study, then I will pass the course. I will not pass the course.
   (b) If I passed the midterm and the final, then I passed the course, If I passed the course, then I passed the final. I failed the course.
   (c) If I pass the midterm or the final, then I will pass this course. I will take the next course only if I pass this course. I will not take the next course.

10. Consider the following hypotheses. It I take the bus or subway, then I will be late for my appointment. If I take a cab, then I will not be late for my appointment and I will be broke. I will be on time for my appointment.
    Which of the following conclusions *must* follow, i.e., can be inferred from the hypotheses? Justify your answers.
    (a) I will take a cab.
    (b) I will be broke.
    (c) I will not take the subway.
    (d) If I become broke, then I took a cab.
    (e) If I take the bus, then I won't be broke.

§ 2.5 Methods of Proof

The constant emphasis on logic and proofs is what sets mathematics apart from other pursuits. In §§ 2.3 and 2.4 we discussed proofs in the setting and symbolism of the propositional calculus. The proofs used in every-day working mathematics are based on the same logical framework as the propositional calculus but their structure is not usually displayed in the stylized format of § 2.4.

In this section we discuss some common methods of proof and the standard terminology that accompanies them. The most natural sort of proof is the **direct proof** in which the hypotheses $H_1, \ldots, H_n$ are shown to imply the conclusion C:

$$H_1 \wedge H_2 \wedge \cdots \wedge H_n \Rightarrow C.$$

The proofs in § 2.4, except for Example 3, are direct proofs in the propositional calculus.

One type of **indirect proof** is a proof of the **contrapositive** [compare rule 9, Table 1 of § 2.2]:

$$\neg C \Rightarrow \neg(H_1 \wedge H_2 \wedge \cdots \wedge H_n).$$

EXAMPLE 1   Let $m, n \in \mathbb{N}$. We wish to prove that if $m + n \geq 73$ then $m \geq 37$ or $n \geq 37$. To do this, we prove the contrapositive: not "$m \geq 37$ or $n \geq 37$" implies not "$m + n \geq 73$." By DeMorgan's law, the negation of "$m \geq 37$ or $n \geq 37$" is "not $m \geq 37$ and not $n \geq 37$," i.e., "$m \leq 36$ and $n \leq 36$." So the contrapositive proposition is: If $m \leq 36$ and $n \leq 36$, then $m + n \leq 72$. This proposition follows immediately from a general property about inequalities: $a \leq c$ and $b \leq d$ imply $a + b \leq c + d$ for real numbers a, b, c, d. ∎

Another type of **indirect proof** is a **proof by contradiction**:

$$\neg C \wedge H_1 \wedge H_2 \wedge \cdots \wedge H_n \Rightarrow \text{a contradiction}.$$

[Compare rule 15, Table 1 of § 2.2.] Example 3 of § 2.4 contains a proof by contradiction in the propositional calculus.

EXAMPLE 2   We wish to prove that $\sqrt{2}$ is irrational. That is, if x is in $\mathbb{R}$ and $x^2 = 2$, then x is not a rational number. The property of irrationality is a negative sort of property and not easily verified directly. But we can show that x rational and $x^2 = 2$ together lead to a contradiction.

We prove that $\sqrt{2}$ is irrational by contradiction. So assume $x \in \mathbb{R}$, $x^2 = 2$ and x is rational. Then by the definition of a rational number, we have $x = p/q$ where $p, q \in \mathbb{Z}$ and $q \neq 0$. By reducing the fraction if necessary, we may assume that p and q have no common factors. In particular, p and q are not both even. Since $2 = x^2 = p^2/q^2$ we have $p^2 = 2q^2$ and so p^2 is even. This implies that p is even, as we will show in Example 4. Hence $p = 2k$ for some $k \in \mathbb{Z}$. Then $(2k)^2 = 2q^2$ and therefore $q^2 = 2k^2$. Thus q^2 and q are also even. But then p and q are both even, contradicting our earlier statement. Hence $\sqrt{2}$ is irrational. ∎

As already remarked, mathematical proofs could in principle be presented in the formal format of § 2.4 rather than in essay form. However, the proofs would quickly get long and cumbersome. For example, the proof of Example 2 could start out as follows.

Proof	Explanation
1. $x \in \mathbb{R}$	assumption
2. $x^2 = 2$	assumption
3. x is rational	negation of the conclusion
4. $x = p/q, q \neq 0$	definition of rational number
5. p, q have no common factors	fractions can be reduced
6. p and q not both even	5; otherwise 2 would be a common factor
7. $p^2 = 2q^2$	4, 2; algebra
Etc.	

EXAMPLE 3 We prove by contradiction that there are infinitely many primes. Thus assume that there are finitely many primes, say k of them. We write them as $p_1, p_2, \ldots, p_k$ so that $p_1 = 2, p_2 = 3$, etc. Consider $n = 1 + p_1 p_2 \cdots p_k$. Since $n > p_j$ for all $j = 1, 2, \ldots, k$, n itself is not prime. However, n is a product of primes [this believable fact is not trivial to prove and, in fact, we provide a proof later in Example 1 of § 6.3]. Therefore at least one of the p_j's must divide n. Since each p_j divides $n - 1$, at least one p_j divides both $n - 1$ and n, but this is impossible. Indeed, if p_j divides both $n - 1$ and n, then it divides their difference, 1, which is absurd. ∎

One should avoid artificial proofs by contradiction such as in the next example.

EXAMPLE 4 We prove by contradiction that the product of two odd integers is an odd integer. Assume $m, n \in \mathbb{N}$ are odd integers but mn is even. There exist k, l in $\mathbb{N}$ so that $m = 2k + 1$ and $n = 2l + 1$. Then

$$mn = 4kl + 2k + 2l + 1 = 2(2kl + k + l) + 1,$$

an odd number, contradicting the assumption that mn is even.

This proof by contradiction is artificial because we did not use the assumption "mn is even" until after we established that "mn is odd." The following direct proof is far preferable.

Consider odd integers m, n in $\mathbb{N}$. There exist $k, l \in \mathbb{N}$ so that $m = 2k + 1$ and $n = 2l + 1$. Then

$$mn = 4kl + 2k + 2l + 1 = 2(2kl + k + l) + 1,$$

which is odd. ∎

Sometimes a result has the form

$$H_1 \vee H_2 \vee \cdots \vee H_n \Rightarrow C,$$

and sometimes it is convenient to convert a proposition into one of this form. Such a proposition is equivalent to

$$(H_1 \Rightarrow C) \wedge (H_2 \Rightarrow C) \wedge \cdots \wedge (H_n \Rightarrow C).$$

[For $n = 2$, compare rule 12a, Table 1 of § 2.2.] Therefore a theorem like this can be proved by **cases**. The next example illustrates how boring and repetitive a proof by cases can be.

EXAMPLE 5   Recall that the **absolute value** $|x|$ of x in $\mathbb{R}$ is defined by the rule:

$$|x| = \begin{cases} x & \text{if } x \geq 0 \\ -x & \text{if } x < 0 \end{cases}.$$

Assuming the familiar order properties of $\leq$ on $\mathbb{R}$, we prove

$$|x + y| \leq |x| + |y| \quad \text{for} \quad x, y \in \mathbb{R}.$$

We consider four cases: (i) $x \geq 0$ and $y \geq 0$; (ii) $x \geq 0$ and $y < 0$; (iii) $x < 0$ and $y \geq 0$; (iv) $x < 0$ and $y < 0$.

Case (i). If $x \geq 0$ and $y \geq 0$, then $x + y \geq 0$ and so $|x + y| = x + y = |x| + |y|$.

Case (ii). If $x \geq 0$ and $y < 0$, then

$$x + y < x + 0 = |x| \leq |x| + |y|$$

and

$$-(x + y) = -x + (-y) \leq 0 + (-y) = |y| \leq |x| + |y|.$$

Either $|x + y| = x + y$ or $|x + y| = -(x + y)$; either way $|x + y| \leq |x| + |y|$ by the above inequalities.

Case (iii). The case $x < 0$ and $y \geq 0$ is similar to Case (ii).

Case (iv). If $x < 0$ and $y < 0$, then $x + y < 0$ and $|x + y| = -(x + y) = -x + (-y) = |x| + |y|$.
So in all four cases, $|x + y| \leq |x| + |y|$.   ∎

EXAMPLE 6   For every $n \in \mathbb{N}$, $n^3 + n$ is even. This fact can be easily proved by induction [see § 2.6], but here we prove it by cases.

Case (i). Suppose n is even. Then $n = 2k$ for some $k \in \mathbb{N}$ and so

$$n^3 + n = 8k^3 + 2k = 2(4k^3 + k),$$

which is even.

Case (ii). Suppose n is odd; then $n = 2k + 1$ for some $k \in \mathbb{N}$ and so

$$n^3 + n = (8k^3 + 12k^2 + 6k + 1) + (2k + 1) = 2(4k^3 + 6k^2 + 4k + 1),$$

which is even.

Here is a more elegant proof by cases. Given n in $\mathbb{N}$, we have $n^3 + n = n(n^2 + 1)$. If n is even, so is $n(n^2 + 1)$. If n is odd, then n^2 is odd, hence $n^2 + 1$ is even, and so $n(n^2 + 1)$ is even. ∎

An implication $p \to q$ is sometimes said to be **vacuously true** if p is false. This is because we have decreed $p \to q$ to be true whenever p is false and so, in this case, the truth of $p \to q$ tells us nothing about q. A **vacuous proof** is a proof of an implication $p \to q$ in which it is shown that p is false. Such implications rarely have intrinsic interest, but they arise as exceptional cases in proofs of general assertions.

EXAMPLE 7   (a) Consider finite sets A and B, and consider the assertion:

"if A has fewer elements than B, then there is a one-to-one mapping of A onto a proper subset of B."

This is vacuously true if B is the empty set because the hypothesis must be false. A vacuous proof consists of simply observing that in this case the hypothesis is impossible.
    (b) Consider the assertion:

$$n \in \mathbb{N} \text{ and } n \geq 4 \text{ implies } n^2 \leq 2^n.$$

This is vacuously true for $n = 0, 1, 2, 3$. For these values of n, its truth does not depend on whether $n^2 \leq 2^n$ is true. For $n \geq 4$, an induction proof can be given [see § 2.6]. ∎

An implication $p \to q$ is sometimes said to be **trivially true** if q is true. This is because, in this case, the truth value of p is irrelevant. A **trivial proof** of $p \to q$ is one in which q is shown to be true without any reference to p.

EXAMPLE 8   If x and y are real numbers such that $xy = 0$, then $(x + y)^n = x^n + y^n$ for $n \geq 1$. This proposition is trivially true for $n = 1$; $(x + y)^1 = x^1 + y^1$ is obviously true and this fact does not depend on the hypothesis $xy = 0$. For $n \geq 2$, this hypothesis is needed. ∎

One encounters references to **constructive proofs** and **nonconstructive proofs** for the existence of mathematical objects satisfying certain properties. A constructive proof either specifies the object [a number or a matrix, say] or indicates how it [or they] can be determined by some explicit procedure or algorithm. A nonconstructive proof establishes the existence of objects by some indirect means such as a proof by contradiction, without giving directions for how to find them.

EXAMPLE 9   In Example 3 we proved by contradiction that there are infinitely many primes. We did not construct an infinite list of primes. Our proof can be revised to give a constructive procedure for building an arbitrarily long list of distinct primes, provided we have some way of factoring integers. [This is Exercise 16.] ∎

In the next example we use the following axiom which will be discussed further in § 6.3:

Well-Ordering Every nonempty subset of $\mathbb{N}$ has a least element.
Principle

EXAMPLE 10 We prove that there is a least, i.e., smallest, natural number n such that

$$(*) \qquad 1^n + 2^n + 3^n + \cdots + (99)^n = \sum_{k=1}^{99} k^n < (100)^n.$$

Since every nonempty subset of $\mathbb{N}$ has a least element, it suffices to show that the set

$$S = \left\{ n \in \mathbb{N} : \sum_{k=1}^{99} k^n < (100)^n \right\}$$

is nonempty. We simplify the problem by noting that $k^n \leq (99)^n$ for $k \leq 99$ and so

$$\sum_{k=1}^{99} k^n \leq \sum_{k=1}^{99} (99)^n = 99(99)^n.$$

Hence any n that satisfies $99(99)^n < (100)^n$ also satisfies $(*)$ and belongs to S. Thus it suffices to show that some n satisfies the inequality $99(99)^n < (100)^n$ or $99 < (100/99)^n$. A direct calculation shows that this holds if n is sufficiently large. In fact, $n = 458$ works. Thus the set S is nonempty [it contains 458] and has a least element. What is its least element? The proof gives no clue, though we know it is no bigger than 458. The proof is nonconstructive because the well-ordering principle is nonconstructive; it doesn't specify how the least element might be found. ∎

EXAMPLE 11 Every positive integer n has the form $2^k m$ where $k \in \mathbb{N}$ and m is odd. This can be proved in several ways. They all suggest the following constructive procedure. If n is odd, let $k = 0$ and $m = n$. Otherwise, divide n by 2 and apply the procedure to $n/2$. Continue until an odd number is reached. Then k will equal the number of times division by 2 was necessary. Exercise 15 asks you to check out this procedure. ∎

We began this chapter with a painstaking development of a very limited system of logic, namely the propositional calculus. In this section, we relaxed the formality in order to discuss several methods of proof encountered in this book and elsewhere. We hope that you now have a better idea of what a mathematical proof is. In Chapter 6 we will study some more sophisticated aspects of logic.

Outside the realm of logic, and in particular in the remainder of this book, proofs are communications intended to convince the reader of the truths of assertions. Logic will serve as the foundation of the process and will recede into the background except where there is a communications gap. That is, usually it should not be necessary to consciously think of the logic presented in this chapter. But if a particular proof in this book or elsewhere is puzzling, then you may wish to analyze it more closely. What are the exact hypotheses? Is the author using hidden assumptions? Is the author giving an indirect proof?

Finally, there is always the possibility that the author has made an error or has not stated what was intended. Maybe you can show that the assertion is false. Or at least show that the reasoning is fallacious. Even some good mathematicians have made the mistake of trying to prove $P \Leftrightarrow Q$ by showing both $P \Rightarrow Q$ and $\neg Q \Rightarrow \neg P$, probably in some disguise.

EXERCISES 2.5

In all exercises with proofs, indicate the methods of proof used.

1. Prove that the product of two even integers is an even integer.

2. Prove that the product of an even and an odd integer is even.

3. Prove that $|xy| = |x| \cdot |y|$ for $x, y \in \mathbb{R}$.

4. Prove that $n^4 - n^2$ is divisible by 3 for all $n \in \mathbb{N}$.

5. Prove that $n^2 - 2$ is never divisible by 3 for $n \in \mathbb{N}$.

6. (a) Prove that $\sqrt{3}$ is irrational.
   (b) Prove that $\sqrt[3]{2}$ is irrational.

7. Prove or disprove:
   (a) The sum of two even integers is an even integer.
   (b) The sum of two odd integers is an odd integer.
   (c) The sum of two primes is never a prime.
   (d) The sum of three consecutive integers is divisible by 3.
   (e) The sum of four consecutive integers is divisible by 4.
   (f) The sum of five consecutive integers is divisible by 5.

8. (a) It is not known whether there are infinitely many **prime pairs**, i.e., odd primes whose difference is 2. Examples of prime pairs are $\langle 3, 5 \rangle, \langle 5, 7 \rangle, \langle 11, 13 \rangle$ and $\langle 71, 73 \rangle$. Give three more examples of prime pairs.
   (b) Prove that $\langle 3, 5, 7 \rangle$ is the only "prime triple." *Hint:* Given $2k + 1$, $2k + 3$, $2k + 5$ where $k \in \mathbb{N}$, show that one of these must be divisible by 3.

9. Prove the following assertions for a real number x and $n = 1$.
   (a) If $x \geq 0$, then $(1 + x)^n \geq 1 + nx$.
   (b) If $x^n = 0$, then $x = 0$.
   (c) If n is even, then $x^n \geq 0$.

10. Prove the result in Example 8. Use the fact that if $xy = 0$, then $x = 0$ or $y = 0$.

11. Prove that there is a smallest prime that is larger than 10^{21}. Is your proof constructive? If so, produce the prime.

12. Prove that there is a least positive integer n such that

$$\sum_{k=1}^{10} n^k < 2^n.$$

Is the proof constructive? *Hint:* $\sum_{k=1}^{10} n^k \leq \sum_{k=1}^{10} n^{10} = 10 \cdot n^{10}$. Compare $10 \cdot n^{10}$ and 2^n for $n = 70$.

13. (a) Prove that given n in $\mathbb{N}$, there exist n consecutive integers that are not prime, i.e., the set of prime integers has arbitrarily large gaps. *Hint:* Start with $(n + 1)! + 2$.
    (b) Is the proof constructive? If so, use it to give six consecutive nonprimes.
    (c) Give seven consecutive nonprimes.

14. Here is another proof that $\sqrt{2}$ is irrational. Show that if $\sqrt{2}$ is rational, then $\{n \in \mathbb{P} : \sqrt{2}\, n \in \mathbb{Z}\}$ is nonempty. Let m be the least element of the set. Obtain a contradiction by showing that $\sqrt{2}\, m - m$ is also in the set.

15. Use the procedure in Example 11 to write the following positive integers in the form $2^k m$ where $k \in \mathbb{N}$ and m is odd.
    (a) 14   (b) 73   (c) 96   (d) 1168

16. Suppose $p_1, p_2, \ldots, p_k$ is a given list of distinct primes. Explain how one could use an algorithm which factors integers into prime factors to construct a prime which is not in the list. *Suggestion:* Factor $1 + p_1 p_2 \cdots p_k$.

§ 2.6   First Look at Induction

Mathematical induction is a method of proof which we can often use when we want to establish the truth of an infinite list of propositions. The method is a natural one to use in a variety of situations in computer science. Some applications are quite mathematical in flavor, such as verifying that a certain formula holds for all positive integers. Another frequent use of the method is to show that a computer program or algorithm with loops performs as expected.

We will eventually need fairly sophisticated versions of induction, but we feel that the student should first gain experience with the basic version. Accordingly, in this section we give a brief introduction to the subject, with examples and exercises, and a few remarks to indicate why it works. After we have studied some predicate calculus in Chapter 6, we will return to a more thorough and general treatment of induction. At that point it should become clear why we regard mathematical induction as a legitimate method of proof; for now we just see how to use it.

EXAMPLE 1   (a) Consider a video game which begins with a spaceship in the middle of the screen. In 5 seconds an alien appears. Five seconds later the alien splits into two aliens, which appear in two places on the screen. Five seconds later, each of these aliens splits into two, and so on. Every 5 seconds the number of aliens doubles. The player's task is to eliminate the aliens before they fill the screen.

Suppose the player is not very skillful and all the aliens survive. How many will there be 30 seconds after the game begins? By calculating the number of aliens at 5-second intervals we get the table

Time	5	10	15	20	25	30
Aliens	1	2	4	8	16	32

So the answer to our question is that there will be 32 aliens in 30 seconds.

Now suppose we want to find the number of aliens in 5 minutes. We could simply extend the table to the right, doubling the entries in the second row as we go along, but there's clearly an easier way. Let $A(n)$ be the number of aliens on the screen after n 5-second intervals. Then $A(1) = 1$, $A(2) = 2$, $A(3) = 4$, etc. We observe that $A(n) = 2^{n-1}$ for $n = 1, 2, 3, 4, 5, 6$, and it seems reasonable to guess that after 5 minutes, which is $5 \cdot 60$ seconds, the value of $A(60)$ will be 2^{59} [a *lot* of aliens]. How can we be sure this guess is correct without going through the calculations?

The method of mathematical induction applies to just such situations, ones in which

1. we know the answer in the beginning,
2. we know how to determine the answer at one stage from the answer at the previous stage, and
3. we have a guess at the general answer.

Of course, if our guess is wrong that's too bad; we won't be able to prove it's right with this method, or with any other. But if our guess is correct, then mathematical induction often gives us a framework for confirming the guess with a proof. We will return to this example shortly to see how the method applies.

(b) The following simple procedure starts with initial values of quantities called I and S. It calculates and prints successive values of S.

Step 1. Let $I = S = 1$.
Step 2. Print S.
Step 3. Replace S by $S + 2I + 1$.
Step 4. Replace I by $I + 1$ and go back to Step 2.

Figure 1(a) is a flowchart for this procedure.[1] Figure 1(b) lists the first few values of S which are printed. The steps of the program are completely dull; the only interesting feature is the loop caused by the return to Step 2.

---

[1] The reader who is not familiar with flowcharts can simply ignore this one and the few which appear later on. We have included them as possible aids to understanding, and not for their own interest. This particular flowchart gives a simple illustration of a loop in an algorithm. In the flowchart the loop is a chain of arrows that form a closed path.

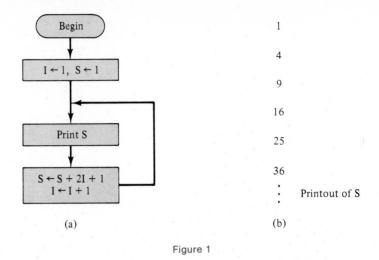

(a)   (b)

Figure 1

A glance at the output shows that the procedure seems to be printing the squares of the positive integers in increasing order. How can we be sure? We check the operation of the loop. When we first come to Step 2, $I = 1$ and $S = 1 = I^2$. Now suppose that at some time when we come to Step 2 we find that I is a positive integer, say n, and $S = I^2$. We will print n^2, replace S by $n^2 + 2n + 1$, replace I by $n + 1$ and go back to Step 2. Since $n^2 + 2n + 1 = (n + 1)^2$, the new S is the square of the new I. We have observed:

$I = 1$ and $S = 1^2$ the first time we are at Step 2, and

If $I = n$ and $S = n^2$ when we are at Step 2, then $I = n + 1$ and $S = (n + 1)^2$ the next time we are at Step 2.

The obvious guess is that for each $n \in \mathbb{P}$ the value of S is n^2 the nth time we are at Step 2. The method of mathematical induction will confirm this guess [Example 2(b)]. ∎

Our first version of mathematical induction is concerned with propositions $p(n)$ that are indexed by $\mathbb{P}$. Section 1.3 contains some preparatory remarks and examples illustrating propositions indexed by $\mathbb{P}$.

Principle of Mathematical Induction   Consider a list $p(1)$, $p(2)$, $p(3),\ldots$ of propositions indexed by $\mathbb{P}$. All the propositions $p(n)$ are true provided

(B) $p(1)$ is true;

(I) $p(n + 1)$ is true whenever $p(n)$ is true.

We will refer to (B), i.e., the fact that $p(1)$ is true, as the **basis for induction** and we will refer to (I) as the **inductive step**. In the notation of the propositional calculus, the inductive step is equivalent to:

the implication $p(n) \to p(n + 1)$ is true for all $n \in \mathbb{P}$.

Notice that the Principle of Mathematical Induction is not itself a proof that $p(n)$ is true for all n, but it tells us that *if* we can somehow show (B) and (I) *then* all $p(n)$'s are true. There is no free lunch. The work goes into showing (B) and (I), which must be verified before the Principle of Mathematical Induction can be applied. In practice, (B) will usually be easy to check.

EXAMPLE 2   (a) In the video game of Example 1(a) we were given that $A(1) = 1$ and that $A(n + 1) = 2 \cdot A(n)$ for $n \geq 1$. We guessed that $A(n) = 2^{n-1}$ for every $n \in \mathbb{P}$, though actually we only cared about $n = 60$.

For each $n \in \mathbb{P}$ let the nth proposition $p(n)$ be "$A(n) = 2^{n-1}$." Then $p(1)$ is "$A(1) = 2^0$," which is true since $2^0 = 1$. Thus the basis (B) holds. We could also verify $p(2)$, i.e., $A(2) = 2^1$, as well as other particular cases, but there is no need to.

For the inductive step (I) we must simply check that for each n if $p(n)$ is true, i.e., if $A(n) = 2^{n-1}$, then $p(n + 1)$ is true, i.e., $A(n + 1) = 2^{(n+1)-1}$. Suppose $A(n) = 2^{n-1}$ for some $n \in \mathbb{P}$. Then

$$A(n + 1) = 2 \cdot A(n) \qquad \text{by the structure of the game}$$

$$= 2 \cdot 2^{n-1} \qquad \text{by the supposition}$$

$$= 2^n = 2^{(n+1)-1},$$

as we wanted to check. Thus (I) holds. By the Principle of Mathematical Induction, all of the propositions $p(n)$ are true. In particular, we have $A(60) = 2^{59}$.

(b) For the looping procedure of Example 1(b) let $p(n)$ be:

"$S = n^2$ the nth time the procedure is at Step 2."

We already checked that $p(1)$ is true and that if $p(n)$ is true then $p(n + 1)$ is true. It follows immediately from the Principle of Mathematical Induction that $p(n)$ is true for every $n \in \mathbb{P}$. Thus the procedure does indeed print the squares of the positive integers in increasing order. ∎

EXAMPLE 3   (a) Here is an illustration of how mathematical induction is useful in mathematics. We calculate

$$2 = 1 \cdot 2, \qquad 2 + 4 = 6 = 2 \cdot 3, \qquad 2 + 4 + 6 = 12 = 3 \cdot 4,$$

$$2 + 4 + 6 + 8 = 20 = 4 \cdot 5,$$

and we begin to suspect that the sum of the first n even positive integers is always $n(n + 1)$. We could verify our guess for many more values of n. We might even call our guess a "conjecture," since that sounds more official. But how do we show our conjecture is true for all n? We use induction. Our nth proposition $p(n)$ is

$$\text{"}2 + 4 + \cdots + (2n) = n(n + 1).\text{"}$$

Thus $p(1)$ asserts that "$2 = 1(1 + 1)$," $p(2)$ asserts that "$2 + 4 = 2(2 + 1)$," $\ldots, p(73)$ asserts that

$$\text{"}2 + 4 + \cdots + 146 = 73(73 + 1) = 5402\text{,"}$$

etc. In particular, $p(1)$ is true by inspection and this establishes the basis for induction.

For the inductive step, suppose that $p(n)$ is true for some n. That is, suppose that

$$2 + 4 + \cdots + (2n) = n(n + 1)$$

is true. We wish to establish $p(n + 1)$:

$$2 + 4 + \cdots + (2n) + (2(n + 1)) = (n + 1)((n + 1) + 1),$$

i.e., we wish to establish

$$2 + 4 + \cdots + (2n) + (2n + 2) = (n + 1)(n + 2).$$

Since $p(n)$ is true by supposition, we have

$$\begin{aligned}
2 + 4 + \cdots + (2n) + (2n + 2) &= [2 + 4 + \cdots + (2n)] + (2n + 2) \\
&= n(n + 1) + (2n + 2) \\
&= n(n + 1) + 2(n + 1) \\
&= (n + 1)(n + 2).
\end{aligned}$$

Thus $p(n + 1)$ holds whenever $p(n)$ holds. By the Principle of Mathematical Induction, we conclude that $p(n)$ is true for all n.

(b) Dividing both sides of the identity proved in part (a) by 2 gives the familiar formula

$$1 + 2 + \cdots + n = \tfrac{1}{2}n(n + 1)$$

for the sum of the first n positive integers. ∎

There are two basic ingredients for a valid induction proof: the basis and the inductive step. In addition, if there is any possible doubt, it should be made clear that one is giving a proof by induction.

It is worth emphasizing that, prior to the last sentence in the preceding proof, we did *not* prove "$p(n + 1)$ is true." We merely proved an implication: "if $p(n)$ is true, then $p(n + 1)$ is true." In a sense we proved an infinite number of assertions, namely: $p(1)$; if $p(1)$ is true then $p(2)$ is true; if $p(2)$ is true then $p(3)$ is true; if $p(3)$ is true than $p(4)$ is true; etc. Then we applied mathematical induction to conclude: $p(1)$ is true; $p(2)$ is true; $p(3)$ is true; $p(4)$ is true; etc.

In Example 3, we could have written $2 + 4 + \cdots + (2n)$ as $\sum\limits_{k=1}^{n} (2k)$ and used summation notation throughout the proof. We illustrate the use of summation notation in the next example.

EXAMPLE 4 We prove

$$\sum_{k=1}^{n} (3k - 2) = \tfrac{1}{2}(3n^2 - n) \quad \text{for all} \quad n \in \mathbb{P}.$$

Proof. Our nth proposition $p(n)$ is

$$\text{`` } \sum_{k=1}^{n} (3k - 2) = \tfrac{1}{2}(3n^2 - n)\text{."}$$

Note that

$$p(1) = \text{`` } 1 = \tfrac{1}{2}(3 \cdot 1^2 - 1)\text{,"}$$
$$p(2) = \text{`` } 1 + 4 = \tfrac{1}{2}(3 \cdot 2^2 - 2)\text{,"}$$
$$p(3) = \text{`` } 1 + 4 + 7 = \tfrac{1}{2}(3 \cdot 3^2 - 3)\text{."}$$

In particular, $p(1)$ is true by inspection and this establishes the basis for induction. Suppose now that $p(n)$ is true for some n:

$$\sum_{k=1}^{n} (3k - 2) = \tfrac{1}{2}(3n^2 - n);$$

we need to show $p(n + 1)$:

$$\sum_{k=1}^{n+1} (3k - 2) = \tfrac{1}{2}[3(n + 1)^2 - (n + 1)].$$

Using $p(n)$, we obtain

$$\sum_{k=1}^{n+1} (3k - 2) = \sum_{k=1}^{n} (3k - 2) + [3(n + 1) - 2] = \tfrac{1}{2}(3n^2 - n) + (3n + 1).$$

To verify $p(n + 1)$, we need to verify that

$$\tfrac{1}{2}(3n^2 - n) + (3n + 1) = \tfrac{1}{2}[3(n + 1)^2 - (n + 1)].$$

This is a purely algebraic matter:

$$\tfrac{1}{2}(3n^2 - n) + (3n + 1) = \tfrac{1}{2}(3n^2 - n + 6n + 2) = \tfrac{1}{2}(3n^2 + 5n + 2)$$
$$= \tfrac{1}{2}(3n + 2)(n + 1) = \tfrac{1}{2}[3(n + 1) - 1](n + 1) = \tfrac{1}{2}[3(n + 1)^2 - (n + 1)].$$

We have shown that $p(n + 1)$ is true whenever $p(n)$ is true. Hence all the propositions $p(n)$ are true by the Principle of Mathematical Induction. We could also have derived this example from the previous one, since

$$\sum_{k=1}^{n} (3k - 2) = 3 \sum_{k=1}^{n} k - \sum_{k=1}^{n} 2 = 3[\tfrac{1}{2}n(n + 1)] - 2n = \tfrac{1}{2}(3n^2 - n). \quad \blacksquare$$

Not all induction proofs in mathematics involve sums.

EXAMPLE 5   All numbers of the form $7^n - 2^n$ are divisible by 5.

Proof. More precisely, we show that $7^n - 2^n$ is divisible by 5 for each $n \in \mathbb{P}$. Our nth proposition is

$$p(n) = \text{``} 7^n - 2^n \quad \text{is divisible by} \quad 5.\text{''}$$

The basis for induction $p(1)$ is clearly true, since $7^1 - 2^1 = 5$. For the inductive step, assume that $p(n)$ is true. Our task is to use this assumption somehow to establish $p(n + 1)$:

$$7^{n+1} - 2^{n+1} \quad \text{is divisible by} \quad 5.$$

Thus we would like to write $7^{n+1} - 2^{n+1}$ somehow in terms of $7^n - 2^n$, in such a way that any remaining terms are easily seen to be divisible by 5. A little trick is to write this as $7(7^n - 2^n)$ plus appropriate terms to preserve the equality:

$$7^{n+1} - 2^{n+1} = 7(7^n - 2^n) + 7 \cdot 2^n - 2^{n+1}$$

$$= 7(7^n - 2^n) + 7 \cdot 2^n - 2 \cdot 2^n = 7(7^n - 2^n) + 5 \cdot 2^n.$$

Now $7^n - 2^n$ is divisible by 5 by assumption and $5 \cdot 2^n$ is obviously divisible by 5, so the same is true for $7^{n+1} - 2^{n+1}$. [In more detail: we can write $7^n - 2^n = 5m$ for some $m \in \mathbb{P}$ so that

$$7^{n+1} - 2^{n+1} = 7 \cdot 5m + 5 \cdot 2^n = 5(7m + 2^n).]$$

We have shown that the inductive step holds and so our proof is complete by the Principle of Mathematical Induction. ∎

The Principle of Mathematical Induction is equally valid if the indexing begins with some integer m other than 1: *All the propositions $p(m)$, $p(m + 1)$, $p(m + 2), \ldots$ are true provided*

(B) $p(m)$ *is true*;

(I) $p(n)$ *implies* $p(n + 1)$ *for all* $n \geq m$.

We will frequently use this with $m = 0$, i.e., when the propositions $p(n)$ are indexed by the set $\mathbb{N}$.

EXAMPLE 6   Let $\mathscr{P}(S)$ be the power set of some finite set S. If S has n elements, then $\mathscr{P}(S)$ has 2^n members. This was shown to be plausible in Example 2 of § 1.1. We prove it now.

Proof. This assertion was verified earlier for $n = 0$, 1, 2 and 3. In particular, the case $n = 0$ establishes the basis for induction. Before proving the inductive step, let's experiment a little and compare $\mathscr{P}(S)$ for $S = \{a, b\}$ and $S = \{a, b, c\}$. Note that

$$\mathscr{P}(\{a, b, c\}) = \{\varnothing, \{a\}, \{b\}, \{a, b\}, \{c\}, \{a, c\}, \{b, c\}, \{a, b, c\}\}.$$

The first four sets comprise $\mathscr{P}(\{a, b\})$; each of the remaining sets is a set in $\mathscr{P}(\{a, b\})$ with c added to it. This is why $\mathscr{P}(\{a, b, c\})$ has twice as many sets as $\mathscr{P}(\{a, b\})$. This argument looks as if it generalizes: every time an element is added to S, the size of $\mathscr{P}(S)$ doubles.

To prove the inductive step, we assume the proposition is valid for n. We consider a set S with $n + 1$ elements; for convenience we use $S = \{1, 2, 3, \ldots, n, n + 1\}$. Let $T = \{1, 2, 3, \ldots, n\}$. The sets in $\mathscr{P}(T)$ are simply the subsets of S that do not contain $n + 1$. By the assumption for n, $\mathscr{P}(T)$ contains exactly 2^n sets. Each remaining subset of S contains $n + 1$, so it is the union of a set in $\mathscr{P}(T)$ with the one-element set $\{n + 1\}$. That is, $\mathscr{P}(S)$ has another 2^n sets that are not subsets of T. It follows that $\mathscr{P}(S)$ has $2^n + 2^n = 2^{n+1}$ members. This completes the inductive step, and hence the result holds for all n by mathematical induction. ∎

Notice that in Example 6 we experimented with a small value of n to get some idea how a general argument might go. Proofs by induction often arise in situations in which we have examined a few cases, think we see the pattern and want to show that our guess is right in general. Sometimes it helps to examine how one case follows from the one before, in order to help us construct an argument for the general inductive step. Such experimentation does not by itself prove anything general, but it can be enormously useful in pointing out what difficulties may arise in a full-dress argument.

EXAMPLE 7 Some experience with inequalities suggests that $n^2 \leq 2^n$ for sufficiently large n. Here is a careful verification of this fact. Some experimentation shows that this inequality holds for $n = 0, 1, 2$ and 4. It appears to hold for all $n \geq 4$. We prove this by induction by observing

(B) $4^2 \leq 2^4$,

and showing

(I) $n^2 \leq 2^n$ implies $(n + 1)^2 \leq 2^{n+1}$ for $n \geq 4$.

Thus we assume that $n^2 \leq 2^n$ for some fixed $n \geq 4$. Here is motivation and "scratch work" for the proof to follow in the next paragraph. We wish to infer $(n + 1)^2 \leq 2^{n+1}$ from $n^2 \leq 2^n$. Since $n^2 \leq 2^n$ implies $2n^2 \leq 2 \cdot 2^n = 2^{n+1}$, it's enough to show $(n + 1)^2 \leq 2n^2$. That is, we want $n^2 + 2n + 1 \leq 2n^2$, i.e., $2n + 1 \leq n^2$, i.e., $2 + (1/n) \leq n$. For $n \geq 3$ [and hence also for $n \geq 4$] this last inequality is true, and we are in business.

Here is the formal proof of (I). Assume that $n^2 \leq 2^n$ and $n \geq 4$. Then $1/n \leq 1$, so $(n + 1)^2 = n^2 + 2n + 1 = n(n + 2 + (1/n)) \leq n(n + 3) \leq n(n + n) = 2n^2 \leq 2 \cdot 2^n = 2^{n+1}$, as desired. We used the assumption $n^2 \leq 2^n$ at the last inequality. Since (B) and (I) hold, the Principle of Mathematical Induction shows that $n^2 \leq 2^n$ for all $n \geq 4$. ∎

EXAMPLE 8   Induction is useful in proving results about graphs and trees. As an example, if you did Exercise 9 of § 0.2 you probably conjectured:

$$p(n) = \text{"Every tree with } n \text{ vertices has exactly } n - 1 \text{ edges,"}$$

for $n \in \mathbb{P}$. For $n = 1$, this says that the trivial one-vertex tree has no edges at all, which is true. Note also that the only tree with 2 vertices has 1 edge. In § 6.4 we'll prove the inductive step, i.e., that $p(n)$ implies $p(n + 1)$ for all $n \in \mathbb{P}$. It will follow that all the propositions $p(n)$ are true. ∎

We close the section with an indication of why induction works. Suppose that

(B) $p(1)$ is true;

(I)   the truth of $p(n)$ implies the truth of $p(n + 1)$, for all $n \in \mathbb{P}$.

Imagine, if possible, that there are some values of n in $\mathbb{P}$ for which $p(n)$ is not true, and start looking for the first such bad n in the list $1, 2, 3, \ldots$. By (B) it's not 1. By (I), if we haven't yet found such a bad n, the next integer we look at won't be bad either. So we'll never find an n for which $p(n)$ is false. This argument will be made more precise in § 6.3.

EXERCISES 2.6

1. Prove

$$\sum_{k=1}^{n} k^2 = 1 + 4 + 9 + \cdots + n^2 = \frac{n(n + 1)(2n + 1)}{6} \quad \text{for} \quad n \in \mathbb{P}.$$

2. Prove

$$4 + 10 + 16 + \cdots + (6n - 2) = n(3n + 1) \quad \text{for all } n \in \mathbb{P}.$$

3. Prove

$$\sum_{k=0}^{n} a^k = \frac{a^{n+1} - 1}{a - 1} \quad \text{for} \quad a \in \mathbb{R}, a \neq 0, a \neq 1, \text{ and } n \in \mathbb{N}.$$

4. Prove

$$\frac{1}{1 \cdot 5} + \frac{1}{5 \cdot 9} + \frac{1}{9 \cdot 13} + \cdots + \frac{1}{(4n - 3)(4n + 1)} = \frac{n}{4n + 1} \quad \text{for} \quad n \in \mathbb{P}.$$

5. Modify the procedure of Example 1(b) by letting $I = 1$ and $S = 2$ in Step 1 but otherwise making no changes.
   (a) List the first four printed values of S.
   (b) Guess the value of S at the nth time the procedure is at Step 2 and prove your guess is correct for all $n \in \mathbb{P}$.

6. Consider the following procedure.

Step 1. Let $S = 1$.
Step 2. Print S.
Step 3. Replace S by $S + 2\sqrt{S} + 1$ and go back to Step 2.

(a) List the first four printed values of S.

(b) Use mathematical induction to show that the value of S is always an integer. [It is easier to prove the stronger statement that the value of S is always the square of an integer and in fact $S = n^2$ the nth time the procedure is at Step 2.]

7. Prove that $11^n - 4^n$ is divisible by 7 for all $n \in \mathbb{P}$.

8. Prove that $n^3 - 4n + 6$ is divisible by 3 for all $n \in \mathbb{N}$.

9. (a) Calculate $1 + 3 + \cdots + (2n - 1)$ for a few values of n, and then guess a general formula for this sum.

(b) Prove the formula obtained in part (a) by induction.

10. (a) Prove that $n^2 > n + 1$ for $n \geq 2$.

(b) Prove that $n! > n^2$ for $n \geq 4$. [$n!$ is defined in § 1.3.]

11. (a) Decide for which positive values of n the inequality $3n < n^2 - 1$ holds.

(b) Prove your claim in part (a).

12. Repeat Exercise 11 for $4n \leq n^2 - 7$.

13. Consider the proposition $p(n) = $ "$n^2 + 5n + 1$ is even."

(a) Prove that the truth of $p(n)$ implies the truth of $p(n + 1)$, for all $n \in \mathbb{P}$.

(b) For which values of n is $p(n)$ actually true? What is the moral of this exercise?

14. Prove $(2n + 1) + (2n + 3) + (2n + 5) + \cdots + (4n - 1) = 3n^2$ for $n \in \mathbb{P}$.

The sum can also be written $\sum\limits_{k=n}^{2n-1} (2k + 1)$.

15. Prove that $5^n - 4n - 1$ is divisible by 16 for $n \in \mathbb{P}$.

16. Prove $\quad 1^3 + 2^3 + \cdots + n^3 = (1 + 2 + \cdots + n)^2, \quad$ i.e., $\quad \sum\limits_{k=1}^{n} k^3 = \left(\sum\limits_{k=1}^{n} k \right)^2 \quad$ for $n \in \mathbb{P}$. *Hint*: Use the identity in Example 3(b).

17. This exercise requires a little knowledge of trigonometric identities. Prove that $|\sin nx| \leq n|\sin x|$ for all $x \in \mathbb{R}$ and all $n \in \mathbb{P}$.

18. Give examples of lists of propositions $p(1)$, $p(2)$, $p(3), \ldots$ for which

(a) (B) holds but (I) fails.

(b) (I) holds but (B) fails.

Can you conclude that $p(n)$ is true for all n in either case?

CHAPTER HIGHLIGHTS

To check your understanding of the material in this chapter, we recommend that you consider each item listed below and:

(a) Satisfy yourself that you can define each concept and describe each method.

(b) Give at least one reason why the item was included in the chapter.

(c) Think of at least one example of each concept and at least one situation in which the fact or method would be useful.

Concepts

propositional calculus
  proposition
  logical connectives $\neg$, $\vee$, $\wedge$, $\rightarrow$, $\leftrightarrow$
  converse, contrapositive, counterexample
  compound proposition
      truth table
      variable
    tautology, contradiction
    logical equivalence, implication, contrapositive
  formal proof
    theorem, hypothesis, conclusion
    rule of inference
    valid, fallacy
  methods of proof
    direct, indirect, by contradiction
    vacuous, trivial
    constructive, nonconstructive
  induction
    basis, inductive step

Facts

Basic logical equivalences [Table 1 of § 2.2].
Basic logical implications [Table 2 of § 2.2].
Basic rules of inference [Table 1 of § 2.3].
Substitution rules [§ 2.3].
Well-ordering principle.
Principle of Mathematical Induction.

Methods

Use of truth tables.
Use of DeMorgan laws to eliminate $\wedge$ or $\vee$.
Use of rules of inference to construct proofs.
Use of mathematical induction to construct proofs.

FUNCTIONS
AND SEQUENCES

This chapter begins with a study of functions, which leads naturally to the study of a special class of functions called sequences. This, in turn, leads to consideration of recursive definitions and algorithms.

§ 3.1   Functions

We begin with a working descriptive definition of "function." A **function** f assigns to each element x in some set S a unique element in a set T. We say such an f is **defined on** S with **values in** T. The set S is called the **domain of** f and is sometimes written $\text{Dom}(f)$. The element assigned to x is usually written $f(x)$. Care should be taken to avoid confusing a function f with its functional values $f(x)$, especially when people write, as we will later, "the function $f(x)$." A function f is completely specified by:

(a) the set on which f is defined, namely $\text{Dom}(f)$;

(b) the assignment, rule or formula giving the value $f(x)$ for each $x \in \text{Dom}(f)$.

For x in $\text{Dom}(f)$, $f(x)$ is called the **image of** x **under** f. The set of all images $f(x)$ is a subset of T called the **image of** f and written $\text{Im}(f)$. Thus we have

$$\text{Im}(f) = \{f(x) : x \in \text{Dom}(f)\}.$$

It is often convenient to specify a set T of allowable images, i.e., a set T containing $\text{Im}(f)$. Such a set is called a **codomain of** f. While a function f has exactly one domain Dom(f) and exactly one image Im(f), any set containing Im(f) can serve as a codomain. Of course, when we specify a codomain we will try to choose one which is useful or informative in context. The notation $f : S \to T$ is shorthand for: "f is a function with domain S and codomain T." We sometimes refer to a function as a **map** or **mapping** and say that f **maps** S into T. When we feel the need of a picture we sometimes draw sketches such as those in Figure 1.

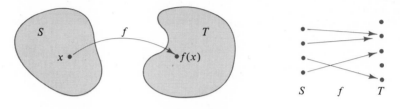

A function f mapping S into T

Figure 1

EXAMPLE 1   (a) Consider a function $f : \mathbb{R} \to \mathbb{R}$. This means that Dom($f$) = $\mathbb{R}$ and, for each $x \in \mathbb{R}$, $f(x)$ represents a unique number in $\mathbb{R}$. Thus $\mathbb{R}$ is a codomain for f but the image Im(f) may be a much smaller set. For example, if $f_1(x) = x^2$ for all $x \in \mathbb{R}$, then Im(f_1) = $[0, \infty)$ and we could write $f_1 : \mathbb{R} \to [0, \infty)$. If f_2 is defined by

$$f_2(x) = \begin{cases} 1 & \text{if } x \geq 0, \\ 0 & \text{if } x < 0, \end{cases}$$

then Im(f_2) = $\{0, 1\}$ and we could write $f_2 : \mathbb{R} \to [0, \infty)$ or $f_2 : \mathbb{R} \to \mathbb{N}$ or $f_2 : \mathbb{R} \to \{0, 1\}$ among other choices.

(b) Consider the function $g : \mathbb{N} \to \mathbb{N}$ defined by $g(n) = n^2 - n$. Here it is useful to specify $\mathbb{N}$ as a codomain, since we might not be interested in the exact set Im(g). ∎

We will avoid the terminology "range of a function f" because many authors use "range" for what we call the image of f and many others use "range" for what we call a codomain.

There is a connection between functions and computer addresses, but they are not the same thing. Suppose that as part of a computer program we want to take whatever number is stored at address X, multiply it by itself and store the result at address SQ. We might write the program line

$$SQ \leftarrow X * X$$

or

$$SQ = X * X.$$

This line describes the squaring function f with rule $f(x) = x^2$ and domain the set of all numbers the computer can store at address X. To help us remember its rule, we might even want to call the function SQ, so that $SQ(x) = x^2$. So far as the computer is concerned, though, SQ is simply the name of a memory location, and we just happen to be storing a single value of the squaring function there.

In Chapter 0 we introduced two important concepts: graphs and trees. Unfortunately, the term "graph" has a somewhat different meaning in connection with functions. To be precise, we consider a function $f : S \to T$. The **graph of** f is the following subset of $S \times T$:

$$\text{Graph}(f) = \{\langle x, y \rangle \in S \times T : f(x) = y\}.$$

This definition is compatible with the use of the term in algebra and calculus. The graphs of the functions in Example 1 are sketched in Figure 2.

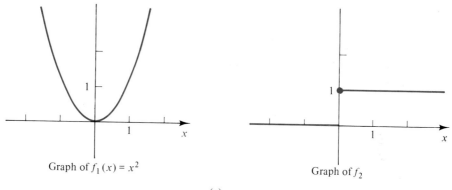

Graph of $f_1(x) = x^2$ Graph of f_2

(a)

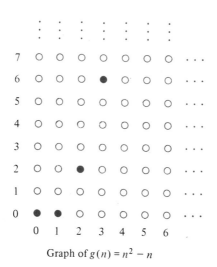

Graph of $g(n) = n^2 - n$

(b)

Figure 2

Unlike the notation for matrices, the labels on the vertical axis of a graph decrease as one goes from top to bottom. This inconsistency is unfortunately completely standard.

Our working definition of "function" is incomplete; in particular, the term "assigns" is undefined. A very precise set-theoretical definition can be given. The key observation is this: Not only does a function determine its graph, but a function can be recovered from its graph. In fact, the graph of a function $f : S \to T$ is a subset G of $S \times T$ with the following property:

for each $x \in S$ there is exactly one $y \in T$ such that $\langle x, y \rangle \in G$.

Given G, we have $\text{Dom}(f) = S$ and, for each $x \in S$, $f(x)$ is the unique element in T such that $\langle x, f(x) \rangle \in G$. The point is that nothing is lost if we regard functions and their graphs as the same, and we gain some precision in the process.

Definition   Let S and T be sets. **A function** with domain S and codomain T is a subset G of $S \times T$ satisfying:

for each $x \in S$ there is exactly one $y \in T$ such that $\langle x, y \rangle \in G$.

If S and T are subsets of $\mathbb{R}$ and if $S \times T$ is graphed so that S is part of the horizontal axis and T is part of the vertical axis, then a subset G of $S \times T$ is a function [or the graph of a function] if every vertical line through a point in S intersects G in exactly one point.

A function $f : S \to T$ is said to be **one-to-one** if distinct elements in S have distinct images in T under f:

$$x_1, x_2 \in S \quad \text{and} \quad x_1 \neq x_2 \quad \text{imply} \quad f(x_1) \neq f(x_2).$$

This is logically equivalent to the contrapositive:

$$x_1, x_2 \in S \quad \text{and} \quad f(x_1) = f(x_2) \quad \text{imply} \quad x_1 = x_2,$$

the form that is most useful in proofs. In terms of the graph G of f, f is one-to-one if:

for each $y \in T$ there is at most one $x \in S$ such that $\langle x, y \rangle \in G$.

If S and T are subsets of $\mathbb{R}$ and f is graphed as above, this condition states that horizontal lines intersect G at most once.

Given $f : S \to T$ we say that f maps **onto** a subset B of T provided $B = \text{Im}(f)$. In particular, we say f maps **onto** T provided $\text{Im}(f) = T$. In terms of the graph G of f, f maps S onto T if and only if:

for each $y \in T$ there is at least one $x \in S$ such that $\langle x, y \rangle \in G$.

A function $f : S \to T$ that is one-to-one and maps onto T is called a **one-to-one correspondence** between S and T. Thus f is a one-to-one correspondence if and only if:

for each $y \in T$ there is exactly one $x \in S$ such that $\langle x, y \rangle \in G$.

These three kinds of special functions are illustrated in Figure 3.

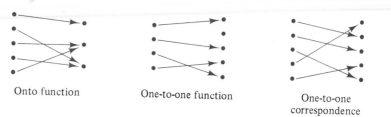

Onto function   One-to-one function   One-to-one
                                                     correspondence

Before we turn to mathematical examples, we illustrate the ideas in a nonmathematical setting.

EXAMPLE 2 Suppose that each student in a class S is assigned a seat number from the set $T = \{1, 2, \ldots, 75\}$. This provides a function $f: S \to T$; thus for each student s, $f(s)$ represents his or her seat number. The function will be one-to-one provided no two students are assigned the same seat number. In this case, the class cannot have more than 75 students. The function will map S onto T provided every number in T has been assigned to at least one student. Note that for this to happen the class must have at least 75 students. The only way f could be a one-to-one correspondence of S onto T is if the class has exactly 75 students.

If we view the function f as a set of ordered pairs, then it will consist of pairs in $S \times T$ like $\langle$ Les Moore, 73 $\rangle$. ∎

EXAMPLE 3 (a) We define $f: \mathbb{N} \to \mathbb{N}$ by the rule $f(n) = 2n$. Then f is one-to-one since

$$f(n_1) = f(n_2) \quad \text{implies} \quad 2n_1 = 2n_2 \quad \text{implies} \quad n_1 = n_2.$$

However, f does not map $\mathbb{N}$ onto $\mathbb{N}$ since $\mathrm{Im}(f)$ consists only of the even natural numbers.

(b) Let Σ be an alphabet. Then length$(w) \in \mathbb{N}$ for each word w in Σ^*; see § 1.1. Thus "length" is a function from Σ^* onto $\mathbb{N}$. [Note that functions can have fancier names than "f."] To see this, recall that Σ is nonempty and so Σ contains some letter, say a. Now $0 = $ length(ϵ), $1 = $ length(a), $2 = $ length(aa), etc. The function length is not one-to-one unless Σ has only one element. ∎

EXAMPLE 4 We prove that $f: \mathbb{R} \to \mathbb{R}$ defined by $f(x) = 3x - 5$ is a one-to-one correspondence of $\mathbb{R}$ onto $\mathbb{R}$. To check that f is one-to-one we need to show

$$f(x) = f(x') \quad \text{implies} \quad x = x',$$

i.e.,

$$3x - 5 = 3x' - 5 \quad \text{implies} \quad x = x'.$$

But $3x - 5 = 3x' - 5$ implies $3x = 3x'$ [add 5 to both sides] and this implies that $x = x'$ [divide both sides by 3].

To show that f maps $\mathbb{R}$ onto $\mathbb{R}$ we consider an element y in $\mathbb{R}$. We need to find an x in $\mathbb{R}$ such that $f(x) = y$, i.e., $3x - 5 = y$. So we solve for x and obtain $x = (y + 5)/3$. Thus, given y in $\mathbb{R}$, $(y + 5)/3$ belongs to $\mathbb{R}$ and $f((y + 5)/3) = 3((y + 5)/3) - 5 = y$. This shows that every $y \in \mathbb{R}$ belongs to Im(f) so that f maps $\mathbb{R}$ onto $\mathbb{R}$. ∎

EXAMPLE 5   Consider the set $\mathfrak{M}_{m,n}$ of all $m \times n$ matrices. For $\mathbf{A} \in \mathfrak{M}_{m,n}$ let TRANS($\mathbf{A}$) = $\mathbf{A}^T$, the transpose of $\mathbf{A}$. Then TRANS is a one-to-one correspondence between $\mathfrak{M}_{m,n}$ and $\mathfrak{M}_{n,m}$. To see this, first observe that

$$(\mathbf{A}^T)^T = \mathbf{A} \quad \text{for all} \quad \mathbf{A} \in \mathfrak{M}_{m,n}.$$

Suppose that $\mathbf{A}$ and $\mathbf{B}$ are in $\mathfrak{M}_{m,n}$ and that TRANS($\mathbf{A}$) = TRANS($\mathbf{B}$). Then $\mathbf{A}^T = \mathbf{B}^T$ and so $\mathbf{A} = (\mathbf{A}^T)^T = (\mathbf{B}^T)^T = \mathbf{B}$. Thus the function TRANS is one-to-one. To see that TRANS maps onto $\mathfrak{M}_{n,m}$, consider an arbitrary $\mathbf{C}$ in $\mathfrak{M}_{n,m}$. Then $\mathbf{C}^T$ belongs to $\mathfrak{M}_{m,n}$ and TRANS($\mathbf{C}^T$) = $(\mathbf{C}^T)^T = \mathbf{C}$. Thus every member of $\mathfrak{M}_{n,m}$ is in Im(TRANS).

Notice how tempting it would have been to write TRANS(TRANS($\mathbf{A}$)) instead of $(\mathbf{A}^T)^T$. Why didn't we? Because the function TRANS under discussion maps $\mathfrak{M}_{m,n}$ to $\mathfrak{M}_{n,m}$, but not the other way back unless $m = n$. There is another function, say SNART, from $\mathfrak{M}_{n,m}$ to $\mathfrak{M}_{m,n}$ defined by

$$\text{SNART}(\mathbf{A}) = \mathbf{A}^T.$$

The rules for defining SNART and TRANS look the same, but because they have different domains the two functions are different. ∎

EXAMPLE 6   (a) We are now in a position to give a precise definition of a graph G. As noted in §0.3, a graph consists of vertices and edges, and each edge connects certain vertices. Thus a graph G consists of a set $V(G)$, whose elements are called **vertices**, and a set $E(G)$, whose elements are called **edges**. In addition, there is a function γ [Greek lowercase gamma] mapping $E(G)$ into the family of two- and one-element subsets of $V(G)$. For each e in $E(G)$, if $\gamma(e) = \{u, v\}$ where $u \neq v$, then we say the edge e **joins** the vertices u and v. If $\gamma(e) = \{u\}$ then we call e a **loop** and say that e **joins** u to itself. The members of $\gamma(e)$ are called the **endpoints** of e. We say that edges e and f are **parallel** if $\gamma(e) = \gamma(f)$, i.e., if e and f have the same endpoints.

This precision is not only comforting, it also makes it clearer how a computer can view a graph as two sets plus a function γ that specifies the endpoints of the edges.

(b) If there are no parallel edges, then γ is one-to-one and the sets $\gamma(e)$ uniquely determine the edges e. That is, there is only one edge for each set $\gamma(e)$. In this case, we often dispense with the set $E(G)$ and the function γ and simply write the edges as sets, like $\{u, v\}$ or $\{u\}$, or as vertex sequences, like $u\,v$, $v\,u$ or $u\,u$. ∎

EXAMPLE 7   In § 1.3 we indicated how various collections of objects can be indexed. For example, $\{A_k : k \in \mathbb{N}\}$ might represent a family of subsets of $\mathbb{R}$ indexed by $\mathbb{N}$, while $\{p(n) : n \in \mathbb{P}\}$ might represent a set of propositions indexed by $\mathbb{P}$. This indexing concept can be made more precise using functions, as we illustrate below.

The indexed family $\{A_k : k \in \mathbb{N}\}$ can be described by the function f with domain $\mathbb{N}$ such that $f(k) = A_k$ for all $k \in \mathbb{N}$. A codomain of f is the set $\mathscr{P}(\mathbb{R})$ of all subsets of $\mathbb{R}$. The function f will be one-to-one if and only if sets A_k and A_j with different subscripts are always different.

For the set $\{p(n) : n \in \mathbb{P}\}$, we can also view the symbol p as a function with domain $\mathbb{P}$, so that p is a proposition-valued function with domain $\mathbb{P}$. As before, this just means that for each $n \in \mathbb{P}$, $p(n)$ represents some proposition.

In general, an indexed family of objects $\{x_i : i \in I\}$ can be described with a function, say g, with domain I such that $g(i) = x_i$ for all $i \in I$. The indexing function g is one-to-one if and only if $x_i \neq x_j$ whenever $i \neq j$. ∎

Some special functions occur so often that they have special names. Let S be a nonempty set. The **identity function** 1_S on S is the function that maps each element of S to itself:

$$1_S(x) = x \quad \text{for all} \quad x \in S.$$

Thus the identity function is a one-to-one correspondence of S onto S.

A function $f : S \to T$ is called a **constant function** if there is some $y_0 \in T$ so that $f(x) = y_0$ for all $x \in S$. The value a constant function takes does not change or vary as x varies over S.

Consider a set S and a subset A of S. The function on S that takes the value 1 at members of A and the value 0 at the other members of S is called the **characteristic function of** A and is denoted χ_A [Greek lowercase chi, sub A]. Thus

$$\chi_A(x) = \begin{cases} 1 & \text{for} \quad x \in A, \\ 0 & \text{for} \quad x \in S \setminus A. \end{cases}$$

Note that $\chi_A : S \to \{0, 1\}$ is rarely one-to-one and is usually an onto map. In fact, χ_A maps S onto $\{0, 1\}$ unless $A = S$ or $A = \varnothing$. If either A or $S \setminus A$ has at least two members then χ_A is not one-to-one.

Now consider functions $f : S \to T$ and $g : T \to U$; see Figure 4. We define the **composition** $g \circ f : S \to U$ by the rule

$$g \circ f(x) = g(f(x)) \quad \text{for all} \quad x \in S.$$

One might read the left side "g circle f of x" or "g of f of x." Complicated operations that are performed in calculus or on a calculator can be viewed as the composition of simpler functions.

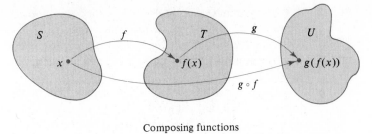

Composing functions

Figure 4

EXAMPLE 8   (a) Consider the function $h: \mathbb{R} \to \mathbb{R}$ given by

$$h(x) = (x^3 + 2x)^7.$$

The value $h(x)$ is obtained by first calculating $x^3 + 2x$ and then taking its seventh power. We write f for the first or inside function: $f(x) = x^3 + 2x$. We write g for the second or outside function: $g(x) = x^7$. The name of the variable x is irrelevant; we could just as well have written $g(y) = y^7$ for $y \in \mathbb{R}$. Either way, we see that

$$g(f(x)) = g(x^3 + 2x) = (x^3 + 2x)^7 = h(x) \quad \text{for} \quad x \in \mathbb{R}.$$

Thus $h = g \circ f$. The ability to view complicated functions as the composition of simpler functions is a critical skill in calculus. Note that the order of f and g is important. In fact,

$$f \circ g(x) = f(x^7) = (x^7)^3 + 2(x^7) = x^{21} + 2x^7 \quad \text{for} \quad x \in \mathbb{R}.$$

(b) Suppose that one wishes to calculate $h(x) = \sqrt{\log x}$ for certain positive values of x on a hand-held calculator. The calculator has the functions $\sqrt{x}$ and $\log x$, which stands for $\log_{10} x$. One works from the inside out. For example, if $x = 73$, one keys in this value, performs $\log x$ to obtain 1.8633, and then performs $\sqrt{x}$ to obtain 1.3650. Note that $h = g \circ f$ where $f(x) = \log x$ and $g(x) = \sqrt{x}$. As in part (a), order is important: $h \neq f \circ g$, i.e., $\sqrt{\log x}$ is not generally equal to $\log \sqrt{x}$. For example, if $x = 73$, then $\sqrt{x}$ is approximately 8.5440 and $\log \sqrt{x}$ is approximately .9317.

(c) Of course, some functions f and g do commute under composition, i.e., satisfy $f \circ g = g \circ f$. For example, if $f(x) = \sqrt{x}$ and $g(x) = 1/x$ for $x \in (0, \infty)$, then $f \circ g = g \circ f$ because

$$\sqrt{\frac{1}{x}} = \frac{1}{\sqrt{x}} \quad \text{for} \quad x \in (0, \infty).$$

For example, for $x = 9$ we have $\sqrt{1/9} = 1/3 = 1/\sqrt{9}$. ∎

We can compose more than two functions if we wish.

EXAMPLE 9 Consider the functions f, g and h that map $\mathbb{R}$ into $\mathbb{R}$ and are defined by

$$f(x) = x^4, \qquad g(y) = \sqrt{y^2 + 1}, \qquad h(z) = z^2 + 72.$$

We've used the different variable names x, y and z to help clarify our computations below. Let's calculate $h \circ (g \circ f)$ and $(h \circ g) \circ f$ and compare the answers. First, for $x \in \mathbb{R}$ we have

$$(h \circ (g \circ f))(x) = h(g \circ f(x)) \qquad \text{by definition of } h \circ (g \circ f)$$

$$= h(g(f(x))) \qquad \text{by definition of } g \circ f$$

$$= h(g(x^4)) \qquad \text{since } f(x) = x^4$$

$$= h(\sqrt{x^8 + 1}) \qquad y = x^4 \text{ in definition of } g$$

$$= (\sqrt{x^8 + 1})^2 + 72 \qquad z = \sqrt{x^8 + 1} \text{ in definition of } h$$

$$= x^8 + 73 \qquad \text{algebra.}$$

On the other hand,

$$((h \circ g) \circ f)(x) = (h \circ g)(f(x)) \qquad \text{by definition of } (h \circ g) \circ f$$

$$= h(g(f(x))) \qquad \text{by definition of } h \circ g$$

$$= x^8 + 73 \qquad \text{exactly as above.}$$

We conclude that

$$(h \circ (g \circ f))(x) = ((h \circ g) \circ f)(x) = x^8 + 73 \quad \text{for all} \quad x \in \mathbb{R},$$

and so the functions $h \circ (g \circ f)$ and $(h \circ g) \circ f$ are exactly the same function. This is no accident, as we observe in the next general theorem. ∎

Associativity of Composition. Consider functions $f: S \to T$, $g: T \to U$ and $h: U \to V$. Then $h \circ (g \circ f) = (h \circ g) \circ f$.

The proof of this basic result amounts to checking that the functions $h \circ (g \circ f)$ and $(h \circ g) \circ f$ both map S into V and that, just as in Example 9, for each $x \in S$ the values $(h \circ (g \circ f))(x)$ and $((h \circ g) \circ f)(x)$ are both equal to $h(g(f(x)))$.

Since composition is associative, we can write $h \circ g \circ f$ unambiguously without any parentheses. We can also compose any finite number of functions without using parentheses.

EXAMPLE 10   (a) If $f(x) = x^4$ for $x \in [0, \infty)$, $g(x) = \sqrt{x + 2}$ for $x \in [0, \infty)$ and $h(x) = x^2 + 1$ for $x \in \mathbb{R}$, then

$$h \circ g \circ f(x) = h(g(x^4)) = h(\sqrt{x^4 + 2}) = (x^4 + 2) + 1$$

$$= x^4 + 3 \quad \text{for} \quad x \in [0, \infty),$$

$$f \circ g \circ h(x) = f(g(x^2 + 1)) = f(\sqrt{x^2 + 1 + 2})$$

$$= (x^2 + 3)^2 \quad \text{for} \quad x \in \mathbb{R},$$

$$f \circ h \circ g(x) = f(h(\sqrt{x + 2})) = f(x + 2 + 1)$$

$$= (x + 3)^4 \quad \text{for} \quad x \in [0, \infty).$$

(b) The function F given by

$$F(x) = (\sqrt{x^2 + 1} + 3)^5 \quad \text{for} \quad x \in \mathbb{R}$$

can be written as $k \circ h \circ g \circ f$ where

$$f(x) = x^2 + 1 \quad \text{for} \quad x \in \mathbb{R},$$

$$g(x) = \sqrt{x} \qquad \text{for} \quad x \in [0, \infty),$$

$$h(x) = x + 3 \quad \text{for} \quad x \in \mathbb{R},$$

$$k(x) = x^5 \qquad \text{for} \quad x \in \mathbb{R}. \quad \blacksquare$$

EXERCISES 3.1

1. We define $f : \mathbb{R} \to \mathbb{R}$ as follows:

$$f(x) = \begin{cases} x^3 & \text{if } x \geq 1, \\ x & \text{if } 0 \leq x < 1, \\ -x^3 & \text{if } x < 0. \end{cases}$$

(a) Calculate $f(3)$, $f(\frac{1}{3})$, $f(-\frac{1}{3})$ and $f(-3)$.
(b) Sketch a graph of f.
(c) Find $\text{Im}(f)$.

2. The functions sketched in Figure 5 have domain and codomain both equal to $[0, 1]$.
(a) Which of these functions are one-to-one?
(b) Which of these functions map $[0, 1]$ onto $[0, 1]$?
(c) Which of these functions are one-to-one correspondences?

3. Let $S = \{1, 2, 3, 4, 5\}$ and $T = \{a, b, c, d\}$. For each question below: if the answer is YES give an example; if the answer is NO explain briefly.
(a) Are there any one-to-one functions from S into T?
(b) Are there any one-to-one functions from T into S?
(c) Are there any functions mapping S onto T?
(d) Are there any functions mapping T onto S?
(e) Are there any one-to-one correspondences between S and T?

4. Let $S = \{1, 2, 3, 4, 5\}$ and consider the following functions from S into S:
$1_S(n) = n$, $f(n) = 6 - n$, $g(n) = \max\{3, n\}$, $h(n) = \max\{1, n - 1\}$.
(a) Write each of these functions as a set of ordered pairs, i.e., list the elements in their graphs.
(b) Sketch a graph of each of these functions.
(c) Which of these functions are one-to-one and onto?

5. Here is a one-to-one function from $\mathbb{N} \times \mathbb{N}$ into $\mathbb{N} : f(\langle m, n \rangle) = 2^m 3^n$.
(a) Calculate $f(\langle m, n \rangle)$ for five different elements $\langle m, n \rangle$ in $\mathbb{N} \times \mathbb{N}$.
(b) Explain why f is one-to-one.
(c) Does f map $\mathbb{N} \times \mathbb{N}$ onto $\mathbb{N}$? Explain.
(d) Show that $g(\langle m, n \rangle) = 2^m 4^n$ defines a function on $\mathbb{N} \times \mathbb{N}$ that is not one-to-one.

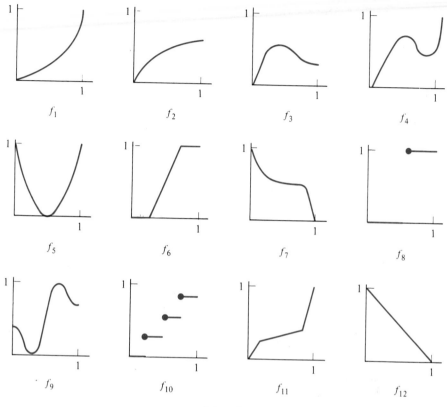

Figure 5

6. Consider the following functions from $\mathbb{N}$ into $\mathbb{N}$: $1_{\mathbb{N}}(n) = n$, $f(n) = 3n$, $g(n) = n + (-1)^n$, $h(n) = \min\{n, 100\}$, $k(n) = \max\{0, n - 5\}$.
   (a) Which of these functions are one-to-one?
   (b) Which of these functions map $\mathbb{N}$ onto $\mathbb{N}$?

7. Let A and B be nonempty sets. The projection map PROJ picks the first element from each pair in $A \times B$, i.e., PROJ: $A \times B \rightarrow A$ where PROJ($\langle a, b \rangle$) $= a$.
   (a) Does this function map $A \times B$ onto A? Justify.
   (b) Is PROJ one-to-one? What if B has only one element?

8. Let $\mathfrak{M}_{n,n}$ be the set of $n \times n$ matrices. Here is an important function in linear algebra: $\text{trace}(\mathbf{A}) = \sum_{i=1}^{n} a_{ii}$ where $\mathbf{A} = [a_{ij}]$. Thus trace maps $\mathfrak{M}_{n,n}$ into $\mathbb{R}$. For this exercise, $n = 2$.
   (a) Calculate trace($\mathbf{A}$) for the following matrices:

   $$\begin{bmatrix} 2 & 3 \\ 4 & 5 \end{bmatrix}, \quad \begin{bmatrix} 1 & 0 \\ 0 & -1 \end{bmatrix}, \quad \begin{bmatrix} 1 & 1 \\ 1 & 1 \end{bmatrix}, \quad \begin{bmatrix} 0 & 17 \\ 8 & 73 \end{bmatrix}.$$

   (b) Show that trace maps $\mathfrak{M}_{2,2}$ onto $\mathbb{R}$.

9. For $n \in \mathbb{Z}$, let $f(n) = \frac{1}{2}[(-1)^n + 1]$. The function f is the characteristic function for some subset of $\mathbb{Z}$. Which subset?

10. In Example 8(b), we compared the functions $\sqrt{\log x}$ and $\log \sqrt{x}$. Show that these functions take the same value for $x = 10{,}000$.

11. We define functions mapping $\mathbb{R}$ into $\mathbb{R}$ as follows: $f(x) = x^3 - 4x$, $g(x) = 1/(x^2 + 1)$, $h(x) = x^4$. Find
   (a) $f \circ g \circ h$ (b) $f \circ h \circ g$ (c) $h \circ g \circ f$
   (d) $f \circ f$ (e) $g \circ g$ (f) $h \circ g$
   (g) $g \circ h$

12. Show that if $f: S \to T$ and $g: T \to U$ are one-to-one, then $g \circ f$ is one-to-one.

13. Prove that the composition of functions is associative.

14. Several important functions can be found on hand-held calculators. Why isn't the identity function, i.e., the function $1_\mathbb{R}$ where $1_\mathbb{R}(x) = x$ for all $x \in \mathbb{R}$, among them?

15. Consider the functions f and g mapping $\mathbb{Z}$ into $\mathbb{Z}$, where $f(n) = n - 1$ for $n \in \mathbb{Z}$ and g is the characteristic function χ_E of $E = \{n \in \mathbb{Z} : n \text{ is even}\}$.
   (a) Calculate $(g \circ f)(5)$, $(g \circ f)(4)$, $(f \circ g)(7)$ and $(f \circ g)(8)$.
   (b) Calculate $(f \circ f)(11)$, $(f \circ f)(12)$, $(g \circ g)(11)$ and $(g \circ g)(12)$.
   (c) Determine the functions $g \circ f$ and $f \circ f$.
   (d) Show that $g \circ g = g \circ f$ and that $f \circ g$ is the negative of $g \circ f$.

§ 3.2 Invertible Functions

Roughly speaking, if a function has an inverse, then the inverse function undoes the action of the function. See Figure 1. Thus if we compose a function and its inverse, we get a function that leaves values unchanged.

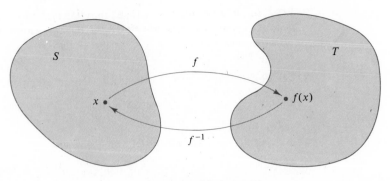

A function and its inverse

Figure 1

EXAMPLE 1
(a) The functions x^2 and $\sqrt{x}$ with domains $[0, \infty)$ are "inverses" to each other. If you apply these operations in either order to some value, the original value is obtained. Try it on a calculator! In other words,

$$\sqrt{x^2} = x \quad \text{and} \quad (\sqrt{x})^2 = x \quad \text{for} \quad x \in [0, \infty).$$

(b) The function $1/x$ is its own "inverse." If you apply the operation twice to some value, the original value is obtained. That is,

$$\frac{1}{1/x} = x \quad \text{for} \quad \text{all nonzero } x \text{ in } \mathbb{R}. \quad \blacksquare$$

To give a precise definition, we consider a function $f : S \to T$. A function $f^{-1} : T \to S$ is said to be the **inverse of** f if $f^{-1} \circ f = 1_S$ and $f \circ f^{-1} = 1_T$. In other words,

$$f^{-1}(f(x)) = x \quad \text{for all} \quad x \in S$$

and

$$f(f^{-1}(y)) = y \quad \text{for all} \quad y \in T.$$

A function f is said to be **invertible** if it has an inverse.

EXAMPLE 2
(a) Consider a positive real number b where $b \neq 1$. Important examples of b are 2, 10 and the number e that appears in calculus and is approximately 2.718. The function f_b given by $f_b(x) = b^x$ for $x \in \mathbb{R}$ has an inverse f_b^{-1} with domain $(0, \infty)$, which is called a **logarithm function**. We write $f_b^{-1}(y) = \log_b y$; by the definition of an inverse we have

$$\log_b b^x = x \quad \text{for} \quad x \in \mathbb{R}$$

and

$$b^{\log_b y} = y \quad \text{for} \quad y \in (0, \infty).$$

In particular, e^x and $\log_e x$ are inverse functions. The function $\log_e x$ is called the **natural logarithm** and is often denoted $\ln x$. The functions 10^x and $\log_{10} x$ are inverses, and so are 2^x and $\log_2 x$. The functions $\log_{10} x = \log$ and $\log_e x = \ln$ appear on many calculators; such calculators also allow one to compute their inverses 10^x and e^x.

(b) The functions $\text{TRANS} : \mathfrak{M}_{m,n} \to \mathfrak{M}_{n,m}$ and $\text{SNART} : \mathfrak{M}_{n,m} \to \mathfrak{M}_{m,n}$ in Example 5 of § 3.1 are inverses to each other, since

$$\text{SNART}(\text{TRANS}(\mathbf{A})) = \mathbf{A} \quad \text{for all} \quad \mathbf{A} \in \mathfrak{M}_{m,n}$$

and

$$\text{TRANS}(\text{SNART}(\mathbf{B})) = \mathbf{B} \quad \text{for all} \quad \mathbf{B} \in \mathfrak{M}_{n,m}. \quad \blacksquare$$

Not all functions have inverses. The next theorem tells us which ones do.

Theorem   Consider $f: S \to T$. The function f is invertible if and only if f is one-to-one and maps S onto T.

Proof. Suppose that f is invertible. Then there is a function $f^{-1}: T \to S$ such that $f^{-1} \circ f = 1_S$ and $f \circ f^{-1} = 1_T$. To see that f is one-to-one, we consider $x_1, x_2 \in S$ such that $f(x_1) = f(x_2)$. Then

$$x_1 = 1_S(x_1) = f^{-1}(f(x_1)) = f^{-1}(f(x_2)) = 1_S(x_2) = x_2.$$

Thus f is one-to-one. To see that f maps S onto T, we consider $y \in T$. Then $f^{-1}(y)$ belongs to S and

$$f(f^{-1}(y)) = f \circ f^{-1}(y) = 1_T(y) = y.$$

Thus $y \in \text{Im}(f)$. Since every y in T is in $\text{Im}(f)$, we conclude that $\text{Im}(f) = T$. That is, f maps S onto T.

Now suppose that f is one-to-one and maps S onto T. Then

for each $y \in T$ there is exactly one $x \in S$ such that $f(x) = y$.

This provides a ready-made formula for f^{-1}, namely: for $y \in T$, $f^{-1}(y) =$ that unique $x \in S$ such that $f(x) = y$. Consider x_0 in S. Then $f(x_0)$ is in T and so

$$f^{-1}(f(x_0)) = \text{that unique } x \in S \text{ such that } f(x) = f(x_0),$$

i.e., $f^{-1}(f(x_0)) = x_0$. This is true for all x_0 in S and so $f^{-1} \circ f = 1_S$. Now consider $y \in T$. Then $f^{-1}(y) = x$ where $f(x) = y$, and so $f(f^{-1}(y)) = f(x) = y$. Since this holds for all y in T, we have $f \circ f^{-1} = 1_T$. Thus f is invertible.   ∎

EXAMPLE 3   Consider the function $f: \mathbb{R} \to \mathbb{R}$ given by $f(x) = x^3 + 1$. Let us check that f is one-to-one and maps $\mathbb{R}$ onto $\mathbb{R}$. To see that it's one-to-one, we note that

$$f(x_1) = f(x_2) \quad \text{implies} \quad x_1^3 + 1 = x_2^3 + 1$$

$$\text{implies} \quad x_1^3 = x_2^3 \quad \text{implies} \quad x_1 = x_2;$$

the last implication holds because each real number has a unique cube root.

To check that f maps onto $\mathbb{R}$, we consider $y \in \mathbb{R}$ and we need to find $x \in \mathbb{R}$ so that $f(x) = y$; i.e., we need to solve $x^3 + 1 = y$ for x. This yields $x = \sqrt[3]{y - 1}$, which belongs to $\mathbb{R}$. Hence f maps $\mathbb{R}$ onto $\mathbb{R}$.

Since f is one-to-one and maps $\mathbb{R}$ onto $\mathbb{R}$, f is invertible, by the theorem. To find its inverse f^{-1}, note that $f^{-1}(y) = x$ if and only if $y = f(x)$. We need to solve for x in terms of y, which we already did in the last paragraph. Thus $f^{-1}(y) = \sqrt[3]{y - 1}$. This formula works for each y in $\mathbb{R}$ and so f^{-1} is completely determined.   ∎

EXAMPLE 4   Consider the function $g : \mathbb{Z} \times \mathbb{Z} \to \mathbb{Z} \times \mathbb{Z}$ given by $g(\langle m, n \rangle) = \langle -n, -m \rangle$. We will check that g is one-to-one and onto, and then we'll find its inverse. To show that g is one-to-one we need to show that

$$g(\langle m, n \rangle) = g(\langle m', n' \rangle) \quad \text{implies} \quad \langle m, n \rangle = \langle m', n' \rangle.$$

First $g(\langle m, n \rangle) = g(\langle m', n' \rangle)$ implies $\langle -n, -m \rangle = \langle -n', -m' \rangle$. Since these ordered pairs are equal we must have $-n = -n'$ and $-m = -m'$. Hence $m = m'$ and $n = n'$ so that $\langle m, n \rangle = \langle m', n' \rangle$ as desired.

   To show that g maps onto $\mathbb{Z} \times \mathbb{Z}$, we consider $\langle p, q \rangle$ in $\mathbb{Z} \times \mathbb{Z}$ and need to find $\langle m, n \rangle$ in $\mathbb{Z} \times \mathbb{Z}$ so that $g(\langle m, n \rangle) = \langle p, q \rangle$. Thus we need $\langle -n, -m \rangle = \langle p, q \rangle$, and this tells us that n should be $-p$ and m should be $-q$. In other words, given $\langle p, q \rangle$ in $\mathbb{Z} \times \mathbb{Z}$ we see that $\langle -q, -p \rangle$ is an element in $\mathbb{Z} \times \mathbb{Z}$ such that $g(\langle -q, -p \rangle) = \langle p, q \rangle$. That is, g maps $\mathbb{Z} \times \mathbb{Z}$ onto $\mathbb{Z} \times \mathbb{Z}$.

   To find the inverse of g we need to consider $\langle p, q \rangle$ in $\mathbb{Z} \times \mathbb{Z}$ and find $g^{-1}(\langle p, q \rangle)$. But this is, by definition, exactly the element in $\mathbb{Z} \times \mathbb{Z}$ that g maps onto $\langle p, q \rangle$. So the work of the last paragraph must be relevant and, in fact, shows that g maps $\langle -q, -p \rangle$ onto $\langle p, q \rangle$. Hence $g^{-1}(\langle p, q \rangle) = \langle -q, -p \rangle$ for all $\langle p, q \rangle$ in $\mathbb{Z} \times \mathbb{Z}$.

   It is interesting to note that $g = g^{-1}$ in this case.   ∎

   Inverses of functions are so useful that we sometimes restrict functions that are not one-to-one to smaller domains on which they are one-to-one. If we then arrange for the codomain to equal the image of the function, we obtain an invertible function.

EXAMPLE 5   (a) Consider $f : \mathbb{R} \to \mathbb{R}$ where $f(x) = x^2$. Then f is not one-to-one, but it is one-to-one if we restrict the domain to $[0, \infty)$. Thus we define a new function F by the same rule $F(x) = x^2$ but having $\text{Dom}(F) = [0, \infty)$. Then F is one-to-one. In fact, $F : [0, \infty) \to [0, \infty)$ is one-to-one and onto. It is this function that has $F^{-1}(x) = \sqrt{x}$ as its inverse; see Example 1(a).

   The function F is called the **restriction** of f to $[0, \infty)$. This sort of restriction is clearly possible and desirable in many settings of interest.

   (b) You should be able to follow this example even if you know no trigonometry. It turns out that none of the trigonometric functions are one-to-one. For example, consider the graph of $\sin x$ in Figure 2. But $\sin x$ is

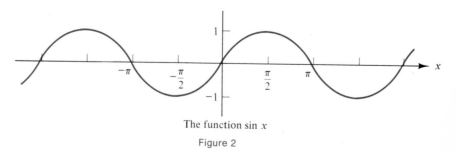

The function $\sin x$

Figure 2

one-to-one if its domain is restricted to, say, $[-\pi/2, \pi/2]$. See Figure 3(a) where we have denoted the restriction by Sin x. With codomain $[-1, 1]$, we obtain an invertible function; the inverse is given in Figure 3(b). This is the inverse sine or Arcsin encountered in trigonometry, calculus and many hand-held calculators. ∎

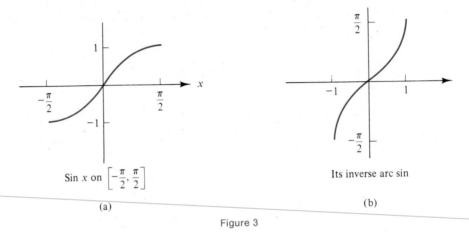

Sin x on $\left[-\dfrac{\pi}{2}, \dfrac{\pi}{2}\right]$ Its inverse arc sin

(a)   (b)

Figure 3

Consider a function $f : S \to T$. For a subset A of S, we define

$$f(A) = \{f(x) : x \in A\}.$$

Thus $f(A)$ is the set of images $f(x)$ as x varies over A. We call $f(A)$ the **image of the set A under** f.

If f is invertible and B is a subset of T, then clearly

$$f^{-1}(B) = \{f^{-1}(y) : y \in B\}.$$

Since $f^{-1}(y) = x$ if and only if $f(x) = y$, we have

$$f^{-1}(B) = \{x \in S : f(x) \in B\}.$$

The last set written makes sense even if f is not invertible. Because of this we use the useful, but misleading, notation $f^{-1}(B)$ for this last set *even when f^{-1} does not represent a function*. This standard notation will appear more natural in the context of relations later on. The reader needs to become comfortable with its usage now. We recapitulate: If $f : S \to T$ and $B \subseteq T$, then we define

$$f^{-1}(B) = \{x \in S : f(x) \in B\}.$$

The set $f^{-1}(B)$ is called the **pre-image of the set B under** f. This notation will be used whether f is invertible or not. If f is invertible, then the pre-image of the subset B of T under f equals the image of B under f^{-1}.

Finally, for $y \in T$ we write $f^{-1}(y)$ for the set $f^{-1}(\{y\})$. That is,

$$f^{-1}(y) = \{x \in S : f(x) = y\}.$$

This set is the **pre-image of the element** y **under** f. Note that solving the equation $f(x) = y$ for x is equivalent to finding the set $f^{-1}(y)$. That is, $f^{-1}(y)$ is the **solution set** for the equation $f(x) = y$. As with equations in algebra, the set $f^{-1}(y)$ might have one element, several elements or no elements at all.

EXAMPLE 6   (a) Consider $f : \mathbb{R} \to \mathbb{R}$ where $f(x) = x^2$. Then we have

$$f^{-1}(4) = \{x \in \mathbb{R} : x^2 = 4\} = \{-2, 2\},$$

the solution set of the equation $x^2 = 4$. The pre-image of the set $[1, 9]$ is

$$f^{-1}([1, 9]) = \{x \in \mathbb{R} : x^2 \in [1, 9]\} = \{x \in \mathbb{R} : 1 \leq x^2 \leq 9\}$$
$$= [-3, -1] \cup [1, 3].$$

Also we have $f^{-1}([-1, 0]) = \{0\}$ and $f^{-1}([-1, 1]) = [-1, 1]$.

(b) Consider the function $g : \mathbb{N} \times \mathbb{N} \to \mathbb{N}$ defined by $g(\langle m, n \rangle) = m^2 + n^2$. Then

$$g^{-1}(0) = \{\langle 0, 0 \rangle\}, \qquad g^{-1}(1) = \{\langle 0, 1 \rangle, \langle 1, 0 \rangle\},$$

$$g^{-1}(2) = \{\langle 1, 1 \rangle\}, \qquad g^{-1}(3) = \varnothing, \qquad g^{-1}(4) = \{\langle 0, 2 \rangle, \langle 2, 0 \rangle\},$$

etc. Note also that $g^{-1}(25) = \{\langle 0, 5 \rangle, \langle 3, 4 \rangle, \langle 4, 3 \rangle, \langle 5, 0 \rangle\}$. ∎

EXAMPLE 7   (a) Let Σ be an alphabet and let L be the length function on $\Sigma^* : L(w) = \text{length}(w)$ for $w \in \Sigma^*$. As we noted in Example 3(b) of § 3.1, L maps Σ^* onto $\mathbb{N}$. For each $k \in \mathbb{N}$,

$$L^{-1}(k) = \{w \in \Sigma^* : L(w) = k\} = \{w \in \Sigma^* : \text{length}(w) = k\};$$

that is, $L^{-1}(k) = \Sigma^k$ in the notation of Example 1(c) of § 1.3. Note that the various sets $L^{-1}(k)$ are disjoint and that their union is Σ^*:

$$\bigcup_{k \in \mathbb{N}} L^{-1}(k) = \bigcup_{k=0}^{\infty} \Sigma^k = \Sigma^*.$$

(b) Consider $h : \mathbb{Z} \to \{-1, 1\}$ where $h(n) = (-1)^n$. Then

$$h^{-1}(1) = \{n \in \mathbb{Z} : n \text{ is even}\}$$

and

$$h^{-1}(-1) = \{n \in \mathbb{Z} : n \text{ is odd}\}.$$

These two sets are disjoint and their union is all of $\mathbb{Z}$:

$$h^{-1}(1) \cup h^{-1}(-1) = \mathbb{Z}. \quad ∎$$

In Example 7 we observed that the pre-images of different elements in the image are disjoint sets whose union is the domain of the function. They cut S up into pieces. There is a technical term for such a family of sets. A **partition** of a nonempty set S is a collection of nonempty subsets which are disjoint and whose union is S.

If the partition is indexed by a set I, say $\{A_i : i \in I\}$, these requirements become

$$A_i \neq \varnothing \text{ for each } i \in I;$$

$$\text{for each } i, j \in I \text{ either } A_i = A_j \text{ or } A_i \cap A_j = \varnothing;$$

$$\bigcup_{i \in I} A_i = S.$$

EXAMPLE 8   (a) For the length function $L: \Sigma^* \to \mathbb{N}$ in Example 7(a), the family $\{L^{-1}(k) : k \in \mathbb{N}\} = \{\Sigma^k : k \in \mathbb{N}\}$ partitions Σ^*.

(b) For the function $h: \mathbb{Z} \to \{-1, 1\}$ where $h(n) = (-1)^n$, the two-subset family $\{h^{-1}(1), h^{-1}(-1)\}$ partitions $\mathbb{Z}$. ∎

Example 8 illustrates a general phenomenon. If f is a function from S onto T, then $\{f^{-1}(y) : y \in T\}$ partitions S. In the first place, each $f^{-1}(y)$ is nonempty, since f maps *onto* T. Every x in S is in exactly one subset of the form $f^{-1}(y)$, namely the set $f^{-1}(f(x))$, which consists of all s in S with $f(s) = f(x)$. If $y \neq z$ then we have $f^{-1}(y) \cap f^{-1}(z) = \varnothing$. Also $\bigcup_{y \in T} f^{-1}(y) = S$, so $\{f^{-1}(y) : y \in T\}$ partitions S.

EXAMPLE 9 For $n \in \mathbb{N}$ let $C_n = \{k \in \mathbb{P} : k - n \text{ is a multiple of } 5\}$. Then

$$C_0 = \{5, 10, 15, 20, \ldots\},$$

$$C_1 = \{1, 6, 11, 16, 21, \ldots\},$$

$$C_2 = \{2, 7, 12, 17, 22, \ldots\},$$

$$\vdots$$

$$C_{73} = \{3, 8, 13, 18, \ldots, 68, 73, 78, \ldots\}$$

$$\vdots$$

Actually, $C_{73} = C_3$. In fact, even though we've used infinitely many indices n, there are only five *distinct* sets, namely C_0, C_1, C_2, C_3 and C_4. Moreover, the family $\{C_n : n \in \mathbb{N}\}$ partitions $\mathbb{P}$ because

(1) $C_n \neq \varnothing$ for each $n \in \mathbb{N}$;

(2)   for each $m, n \in \mathbb{N}$ either $C_m = C_n$ or $C_m \cap C_n = \varnothing$;

(3) $\bigcup_{n \in \mathbb{N}} C_n = \mathbb{P}$.

To verify (2) it's enough to show that

$$C_m \cap C_n \neq \varnothing \quad \text{implies} \quad C_m = C_n.$$

Indeed, if $k_0 \in C_m \cap C_n$, then $k_0 - m$ and $k_0 - n$ are both multiples of 5. Therefore their difference, $n - m$, is also a multiple of 5. If $k \in C_m$, then $k - m$ is a multiple of 5 and so $k - m - (n - m) = k - n$ is too, which means that $k \in C_n$. Thus $C_m \subseteq C_n$, and similarly $C_n \subseteq C_m$. ∎

EXERCISES 3.2

1. Find the inverses of the following functions mapping $\mathbb{R}$ into $\mathbb{R}$.
   (a) $f(x) = 2x + 3$
   (b) $g(x) = x^3 - 2$
   (c) $h(x) = (x - 2)^3$
   (d) $k(x) = \sqrt[3]{x} + 7$

2. Many hand-held calculators have the functions $\log x$, x^2, $\sqrt{x}$ and $1/x$.
   (a) Specify the domains of these functions.
   (b) Which of these functions are inverses to each other?
   (c) Which pairs of these functions commute with respect to composition?
   (d) Some hand-held calculators also have the functions $\sin x$, $\cos x$ and $\tan x$. If you know a little trigonometry, repeat parts (a), (b) and (c) for these functions.

3. Here are some functions from $\mathbb{N} \times \mathbb{N}$ to $\mathbb{N}$: $\text{SUM}(\langle m, n \rangle) = m + n$, $\text{PROD}(\langle m, n \rangle) = m * n$, $\text{MAX}(\langle m, n \rangle) = \max\{m, n\}$, $\text{MIN}(\langle m, n \rangle) = \min\{m, n\}$; $*$ denotes multiplication of integers.
   (a) Which of these functions map $\mathbb{N} \times \mathbb{N}$ onto $\mathbb{N}$?
   (b) Show that none of these functions are one-to-one.
   (c) For each of these functions F, how big is the set $F^{-1}(4)$?

4. Here are some functions mapping $\mathscr{P}(\mathbb{N}) \times \mathscr{P}(\mathbb{N})$ into $\mathscr{P}(\mathbb{N})$: $\text{UNION}(\langle A, B \rangle) = A \cup B$, $\text{INTER}(\langle A, B \rangle) = A \cap B$ and $\text{SYM}(\langle A, B \rangle) = A \oplus B$.
   (a) Show that each of these functions maps $\mathscr{P}(\mathbb{N}) \times \mathscr{P}(\mathbb{N})$ onto $\mathscr{P}(\mathbb{N})$.
   (b) Show that none of these functions are one-to-one.
   (c) For each of these functions F, how big is the set $F^{-1}(\varnothing)$? the set $F^{-1}(\{0\})$?

5. Here are two "shift functions" mapping $\mathbb{N}$ into $\mathbb{N}$: $f(n) = n + 1$ and $g(n) = \max\{0, n - 1\}$ for $n \in \mathbb{N}$.
   (a) Calculate $f(n)$ for $n = 0, 1, 2, 3, 4, 73$.
   (b) Calculate $g(n)$ for $n = 0, 1, 2, 3, 4, 73$.
   (c) Show that f is one-to-one but does not map $\mathbb{N}$ onto $\mathbb{N}$.
   (d) Show that g maps $\mathbb{N}$ onto $\mathbb{N}$ but is not one-to-one.
   (e) Show that $g \circ f = 1_{\mathbb{N}}$ but that $f \circ g \neq 1_{\mathbb{N}}$.

6. We define $f : \mathbb{N} \to \mathbb{N}$ and $g : \mathbb{N} \to \mathbb{N}$ as follows: $f(n) = 2n$ for all $n \in \mathbb{N}$, $g(n) = n/2$ if n is even and $g(n) = (n - 1)/2$ if n is odd.
   (a) Calculate $g(n)$ for $n = 0, 1, 2, 3, 4, 73$.
   (b) Show that $g \circ f = 1_{\mathbb{N}}$ but that $f \circ g \neq 1_{\mathbb{N}}$.

7. If $f : S \to S$ and $f \circ f = 1_S$, then f is its own inverse. Show that the following functions are their own inverses.
   (a) The function $f : (0, \infty) \to (0, \infty)$ where $f(x) = 1/x$.
   (b) Let S be a set and define $\phi : \mathscr{P}(S) \to \mathscr{P}(S)$ by $\phi(A) = A^c$.

(c) Let $\mathfrak{M}_{n,n}$ be the set of $n \times n$ matrices and define $\mathrm{TRANS}(A) = A^T$ for A in $\mathfrak{M}_{n,n}$.

(d) Let C be a set and define $\mathrm{REV} : C \times C \to C \times C$ by $\mathrm{REV}(\langle x, y \rangle) = \langle y, x \rangle$.

8. Let A be a subset of some set S and consider the characteristic function χ_A of A. Find $\chi_A^{-1}(1)$ and $\chi_A^{-1}(0)$.

9. Let $f : S \to T$ and $g : T \to U$ be invertible functions. Show that $g \circ f$ is invertible and that $(g \circ f)^{-1} = f^{-1} \circ g^{-1}$.

10. Let $f : S \to T$ be an invertible function. Show that f^{-1} is invertible and that $(f^{-1})^{-1} = f$.

11. Consider functions $f : S \to T$ and $g : T \to S$ such that $g \circ f = 1_S$. Nontrivial examples of such pairs of functions appear in Exercises 5 and 6.
(a) Prove that f is one-to-one.
(b) Prove that g maps T onto S.

12. Consider the function $f : \mathbb{R} \times \mathbb{R} \to \mathbb{R} \times \mathbb{R}$ defined by

$$f(\langle x, y \rangle) = \langle x + y, x - y \rangle.$$

(a) Prove that f is one-to-one on $\mathbb{R} \times \mathbb{R}$.
(b) Prove that f maps $\mathbb{R} \times \mathbb{R}$ onto $\mathbb{R} \times \mathbb{R}$.
(c) Find the inverse function f^{-1}.
(d) Find the composite functions $f \circ f^{-1}$ and $f \circ f$.

13. Let $f : S \to T$.
(a) Show that $f(f^{-1}(B)) \subseteq B$ for subsets B of T.
(b) Show that $A \subseteq f^{-1}(f(A))$ for subsets A of S.
(c) Show that $f^{-1}(B_1 \cap B_2) = f^{-1}(B_1) \cap f^{-1}(B_2)$ for subsets B_1 and B_2 of T.
(d) Under what conditions on B does equality hold in part (a)?

14. Let $f : S \to T$. Prove or disprove. If false, a single example will suffice.
(a) $f(A_1 \cap A_2) = f(A_1) \cap f(A_2)$ for subsets A_1 and A_2 of S.
(b) $f(A_1 \setminus A_2) = f(A_1) \setminus f(A_2)$ for subsets A_1 and A_2 of S.
(c) $f(A_1) = f(A_2)$ implies $A_1 = A_2$.

15. (a) One can show that if $f : T \to U$ is one-to-one and if $g : S \to T$ and $h : S \to T$ satisfy $f \circ g = f \circ h$, then $g = h$. Give examples of functions f, g and h for which $f \circ g = f \circ h$ but $g \neq h$.
(b) Give examples of functions f, g and h for which $g \circ f = h \circ f$ but $g \neq h$.
(c) Give a condition on f so that $g \circ f = h \circ f$ implies $g = h$.

16. Consider the functions g and h mapping $\mathbb{Z}$ into $\mathbb{N}$ defined as follows : $g(n) = |n|$, $h(n) = 1 + (-1)^n$.
(a) Describe the sets in the partition $\{g^{-1}(k) : k \in \mathbb{N}\}$ of $\mathbb{Z}$. How many sets are there?
(b) Describe the sets in the partition $\{h^{-1}(k) : k \in \mathbb{N}\}$ of $\mathbb{Z}$. How many sets are there?

§ 3.3   Sequences and Big-Oh Notation

An important family of functions consists of those whose domains are the set $\mathbb{N} = \{0, 1, 2, \ldots\}$ of natural numbers [or $\{m, m + 1, m + 2, \ldots\}$ for some integer m]. Such functions are called **sequences.** Thus a sequence on $\mathbb{N}$ is a function that has a specified value for each integer $n \in \mathbb{N}$. It has been traditional in mathematics to denote a sequence by a letter such as s and to denote its value at n as s_n rather than $s(n)$. We frequently call s_n the nth **term** of the sequence. It is often convenient to denote the sequence itself by (s_n) or $(s_n)_{n \in \mathbb{N}}$ or $(s_0, s_1, s_2, \ldots)$.

EXAMPLE 1   (a) Consider the sequence $(s_n)_{n \in \mathbb{N}}$ where $s_n = n^2$. This is the sequence $(0, 1, 4, 9, 16, \ldots)$. Formally, of course, this is the function with domain $\mathbb{N}$ whose value at each n is n^2. The *set* of values is $\{0, 1, 4, 9, 16, \ldots\} = \{n^2 : n \in \mathbb{N}\}$.

(b) Consider the sequence given by $a_n = (-1)^n$ for $n \in \mathbb{N}$, i.e., $(a_n)_{n \in \mathbb{N}}$ where $a_n = (-1)^n$. This is the sequence $(1, -1, 1, -1, 1, -1, 1, \ldots)$. Formally, this is a function whose domain is $\mathbb{N}$ and whose *set* of values is $\{-1, 1\}$. ∎

As the last example suggests, it is important to distinguish between a sequence and its set of values. We will always use parentheses ( ) to signify a sequence and braces { } to signify a set. The sequence given by $a_n = (-1)^n$ has an infinite number of terms even though their values are repeated over and over. On the other hand, the *set* $\{(-1)^n : n \in \mathbb{N}\}$ is exactly the set $\{-1, 1\}$ consisting of two numbers.

EXAMPLE 2   An important sequence is given by $s_n = n!$ for $n \geq 0$, where $n!$ represents "n factorial" introduced in § 1.3. As examples, $s_0 = s_1 = 1$, $s_5 = 120$ and $s_6 = 720$. ∎

As we noted in § 1.4, in computer science it is desirable to communicate information in a linear fashion, without subscripts and other decorations. Accordingly, sequences are frequently written as functions with parentheses around the variable. Moreover, sequences and functions are frequently given suggestive abbreviated names, such as SEQ, FACT, SUM, etc.

EXAMPLE 3   (a) Let $\text{FACT}(n) = n!$ for $n \in \mathbb{N}$. This is exactly the same sequence as in Example 2; only its name [FACT, instead of s] has been changed. Note that $\text{FACT}(n + 1) = (n + 1) * \text{FACT}(n)$ for $n \in \mathbb{N}$, where $*$ denotes multiplication of integers.

(b) For $n \in \mathbb{N}$, let $\text{TWO}(n) = 2^n$. Then TWO is a sequence. Note that $\text{TWO}(n + 1) = 2 * \text{TWO}(n)$ for $n \in \mathbb{N}$. ∎

Our definition of sequence allows the domain to be any set of the form $\{m, m + 1, m + 2, \ldots\}$ where m is an integer.

EXAMPLE 4   Consider the sequence (b_n) given by $b_n = 1/n^2$ for $n \geq 1$. Clearly this sequence needs to have its domain avoid the value $n = 0$. ∎

One way that sequences arise naturally in computer science is as lists of successive computed values; the sequences FACT and TWO in Example 3 are of this sort. Another important application of sequences is to the problem of estimating how long a computation will take for a given input.

For instance, in § 0.2 we looked at the problem of finding a minimal spanning tree for a weighted graph, and we described an algorithm to find such a tree. How long the algorithm takes obviously depends on how many vertices and edges the given graph has, as well as on the sizes of the weights and on other details of the graph. We can define a sequence L by

$L(n) = $ the longest time the algorithm can take for an input graph
with n vertices.

We don't care very much about the actual values of L, which is lucky because to compute $L(n)$ exactly we might have to look at all graphs with n vertices. What we do want to know is how fast $L(n)$ grows as n gets large. Is $L(n)$ roughly proportional to n, or to n^2, or to 10^n, or to what? If the answer is 10^n, then each time we add one more vertex to the graph the time increases by a factor of 10, and a graph with, say, 100 vertices is already much too big to deal with. If $L(n)$ is roughly n^2, though, then going from n to $2n$ vertices only multiplies the time by a factor of 4, and the worst graph with 100 vertices only takes about $5^2 = 25$ times as long as the worst graph with 20 vertices. It turns out, as we will see in Chapter 9, that the algorithm of § 0.2 actually runs in a time no worse than something on the order of n^3.

To state this fact more precisely, we need to develop a new notation, called the **"big-oh" notation**, which is commonly used to describe estimates of this sort, and which we will use consistently from now on to describe the running times of our algorithms.

Suppose that f and g are sequences with nonnegative real values. We say that $f(n)$ is $O(g(n))$ [read "$f(n)$ is big oh of $g(n)$"] in case there is a constant C such that $f(n) \leq C \cdot g(n)$ for all large enough values of n.

EXAMPLE 5   (a) We say that $f(n)$ is $O(n)$ if $f(n)$ is bounded by some constant multiple of n for large n, that is, if $f(n) \leq Cn$ for all large enough n. Here $g(n) = n$ for all $n \in \mathbb{N}$.

(b) We say that $f(n)$ is $O(1)$ if there is a constant C such that $f(n) \leq C$ for all large n, that is, if the values of f are bounded above by some constant. Here we take $g(n) = 1$ for all $n \in \mathbb{N}$.

(c) The number of edges in a graph with n vertices and with no parallel edges is $O(n^2)$. In fact, the largest possible number of edges in such a graph can be shown to be $n(n + 1)/2 = \frac{1}{2}n^2 + \frac{1}{2}n$. In this case, even though we have a precise upper bound, it may be more convenient for many purposes simply to use the rough estimate $O(n^2)$. This bound is correct, since $\frac{1}{2}n^2 + \frac{1}{2}n \leq \frac{1}{2}n^2 + \frac{1}{2}n^2 = n^2$ for all positive n.

(d) The longest-time function L for the minimal spanning tree algorithm in § 0.2 is slightly worse than $O(n^2)$. It's $O(n^2 \cdot \log n)$, though, and since $\log n \leq n$ for $n \geq 2$ the algorithm runs in time $O(n^3)$. In fact, $\log n$ grows *much* more slowly than n for large n, so $O(n^3)$ is a gross overestimate in this case. In § 9.1 we will go through a time analysis of this algorithm and will present a different minimal spanning tree algorithm that does work in time $O(n^2)$.

(e) Let $f(n) = 3 - 400n^2 + 17n^3 + 2n^5$ for $n \in \mathbb{N}$. We claim that $f(n)$ is $O(n^5)$. For small values of n, $f(n)$ is actually negative, but since we only care about large n we follow standard practice and allow f to have a few negative values. To check that $f(n)$ is $O(n^5)$ we observe that $3 \leq n^5$ for $n \geq 2$, $-400n^2 \leq 0$ for all $n \in \mathbb{N}$, $17n^3 \leq n^5$ for $n \geq 5$, and so

$$f(n) \leq n^5 + 0 + n^5 + 2n^5 = 4n^5$$

for all large enough n. ∎

The kind of estimating we just did in Example 5(e) can be done much more generally. One can show [Exercise 8] that if $f(n)$ is $O(n^a)$ and $g(n)$ is $O(n^b)$ then $f(n) \cdot g(n)$ is $O(n^{a+b})$ and $f(n) + g(n)$ is $O(n^{\max\{a, b\}})$. It follows that if $f(n)$ is a polynomial in n in which n^k is the highest power of n that appears, then $f(n)$ is $O(n^k)$. Exercises 11 and 12 give some other basic facts about the big-oh notation.

So far, all of our sequences have had real numbers as values. However, there is no such restriction in the definition and, in fact, we will be interested in sequences with values in other sets.

EXAMPLE 6   (a) For each $n \in \mathbb{P}$, we define the following subsets of $\mathbb{Z}$:

$$A_n = \{m \in \mathbb{Z} : m \text{ divides } n\}.$$

Thus $A_1 = \{1, -1\}$, $A_2 = \{1, -1, 2, -2\}, \ldots, A_6 = \{1, -1, 2, -2, 3, -3, 6, -6\}$, etc. These sets are the terms of a sequence $(A_n)_{n \in \mathbb{P}}$ of subsets of $\mathbb{Z}$.

(b) Another sequence $(D_n)_{n \in \mathbb{N}}$ of subsets of $\mathbb{Z}$ is defined by

$$D_n = \{m \in \mathbb{Z} : m \text{ is a multiple of } n\}$$

$$= \{0, \pm n, \pm 2n, \pm 3n, \ldots\}.$$

(c) Let Σ be an alphabet. For $k \in \mathbb{N}$ we define

$$\Sigma^k = \{w \in \Sigma^* : \text{length}(w) = k\}.$$

The sequence $(\Sigma^k)_{k \in \mathbb{N}}$ is a sequence of subsets of Σ^*.

(d) An example of a matrix-valued sequence $(\mathbf{M}_n)$ is given by

$$\mathbf{M}_n = \begin{bmatrix} n & (-1)^n \\ 1 & -n \end{bmatrix} \quad \text{for} \quad n \in \mathbb{N}.$$

The first few terms are:

$$\mathbf{M}_0 = \begin{bmatrix} 0 & 1 \\ 1 & 0 \end{bmatrix}, \quad \mathbf{M}_1 = \begin{bmatrix} 1 & -1 \\ 1 & -1 \end{bmatrix} \quad \text{and} \quad \mathbf{M}_2 = \begin{bmatrix} 2 & 1 \\ 1 & -2 \end{bmatrix}. \quad \blacksquare$$

Note that a sequence $(s_n)_{n \in \mathbb{N}}$ can always be viewed as a set indexed by $\mathbb{N}$ and that a set indexed by $\mathbb{N}$ can always be viewed as a sequence. See Example 6, parts (b) and (c).

We will also have occasion to use finite sequences. A **finite sequence** is a function whose domain is a finite subset of $\mathbb{Z}$ having the form $\{m, m + 1, \ldots, n\}$. Frequently m will be 0 or 1.

EXAMPLE 7   (a) Consider a finite set S with n elements. The members of S can be listed as a sequence $s_1, s_2, \ldots, s_n$ in many ways, in fact in $n!$ ways. Which listing we choose will often depend on the problem at hand. For instance, if S consists of real numbers we might wish to list its elements in increasing order, so that $s_1 < s_2 < \cdots < s_n$.

(b) The values of a sequence may not all be different. For instance, consider the sequence of values stored at a particular address during the operation of a computer program. One natural way to list such values would be as the sequence $v_1, v_2, \ldots, v_m$ in which v_i is the value at the beginning of step i in the program execution. It might also be natural to list these values as a nondecreasing sequence $s_1, s_2, \ldots, s_m$ with $s_1 \leq s_2 \leq \cdots \leq s_m$ or as a nonincreasing sequence $t_1, t_2, \ldots, t_m$ with $t_1 \geq t_2 \geq \cdots \geq t_m$.

(c) The set of prime integers less than 65,536 can be listed as $p_1, p_2, p_3, p_4, \ldots, p_m$ where $p_1 = 2$, $p_2 = 3$, $p_3 = 5$, etc. The exact value of m could be determined, but is probably not important.

(d) A path in a graph is a finite sequence $e_1, e_2, \ldots, e_n$ of edges that link up with each other. Such a path determines a sequence of vertices $v_1, v_2, \ldots, v_n, v_{n+1}$. The last subscript is $n + 1$ because the number of vertices in a vertex sequence is always one larger than the number of edges. $\quad \blacksquare$

The sequences in Example 3 have the property that the value of the sequence at $n + 1$ can be calculated in terms of the value at n. Indeed, $\text{FACT}(n + 1) = (n + 1) * \text{FACT}(n)$ and $\text{TWO}(n + 1) = 2 * \text{TWO}(n)$. In mathematics, and especially in computer science, it is frequently convenient to define sequences by such means, that is, it is convenient to define their values in terms of previous values. The next section deals with this method of defining sequences.

EXERCISES 3.3

1. Consider the sequence given by $a_n = \dfrac{n - 1}{n + 1}$ for $n \in \mathbb{P}$.

(a) List the first six terms of this sequence.

(b) Calculate $a_{n+1} - a_n$ for $n = 1, 2, 3$.

(c) Show that $a_{n+1} - a_n = \dfrac{2}{(n + 1)(n + 2)}$ for $n \in \mathbb{P}$.

2. Consider the sequence given by $b_n = [1 + (-1)^n]/2$ for $n \in \mathbb{N}$.
   (a) List the first seven terms of this sequence.
   (b) What is its set of values?

3. Consider the matrix-valued sequence $(\mathbf{M}_n)$ given by

$$\mathbf{M}_n = \begin{bmatrix} n & n-1 \\ n+1 & n \end{bmatrix}, \qquad n \in \mathbb{N}.$$

   (a) Give the first four terms of this sequence.
   (b) Caclulate $\mathbf{M}_n + \mathbf{M}_n^T$ for $n = 0, 1$ and 2.

   (c) Show that $\mathbf{M}_n + \mathbf{M}_n^T = \begin{bmatrix} 2n & 2n \\ 2n & 2n \end{bmatrix}$ for all n.

4. Calculate

   (a) $\dfrac{5!}{2!\,3!}$ (b) $\dfrac{6!}{3!\,3!}$ (c) $\dfrac{4!}{0!\,4!}$

5. For $n \in \mathbb{N}$, let $SEQ(n) = n^2 - n$.
   (a) Calculate $SEQ(n)$ for $n \le 6$.
   (b) Show that $SEQ(n+1) = SEQ(n) + 2n$ for all $n \in \mathbb{N}$.
   (c) Show that $SEQ(n+1) = \dfrac{n+1}{n-1} * SEQ(n)$ for $n \ge 2$.

6. For $n = 1, 2, 3, \ldots$, let $SSQ(n) = \sum\limits_{i=1}^{n} i^2$.

   (a) Calculate $SSQ(n)$ for $n = 1, 2, 3$ and 5.
   (b) Observe that $SSQ(n+1) = SSQ(n) + (n+1)^2$ for $n \ge 1$.
   (c) It turns out that $SSQ(73) = 132{,}349$. Use this to calculate $SSQ(74)$ and $SSQ(72)$.

7. For the following sequences, write the first several terms until the behavior of the sequence is clear.
   (a) $a_n = [2n - 1 + (-1)^n]/4$ for $n \in \mathbb{N}$.
   (b) (b_n) where $b_n = a_{n+1}$ for $n \in \mathbb{N}$ and a_n is as in part (a).
   (c) $VEC(n) = \langle a_n, b_n \rangle$ for $n \in \mathbb{N}$.

8. Suppose $f(n)$ is $O(n^a)$ and $g(n)$ is $O(n^b)$. Prove
   (a) $f(n)g(n)$ is $O(n^{a+b})$.
   (b) $f(n) + g(n)$ is $O(n^{\max\{a, b\}})$.
   [In view of these facts we could write $O(n^a) \cdot O(n^b) = O(n^{a+b})$ and $O(n^a) + O(n^b) = O(n^{\max\{a, b\}})$.]

9. In each case find the smallest integer k such that $f(n)$ is $O(n^k)$.
   (a) $f(n) = 13n^2 + 4n - 73$
   (b) $f(n) = 1/(n+1)$
   (c) $f(n) = 1/(n-1)$
   (d) $f(n) = (n-1)^3$
   (e) $f(n) = (n^3 + 2n - 1)/(n+1)$
   (f) $f(n) = \sqrt{n^2 - 1}$
   (g) $f(n) = \sqrt{n^2 + 1}$

10. (a) Show that $n!$ is $O(n^n)$.
    (b) Show that 2^n is $O(n!)$.
    (c) Show that $n!$ is not $O(2^n)$. *Hint:* Show that there is no constant C with

$$\frac{1}{2} \cdot \frac{2}{2} \cdot \frac{3}{2} \cdot \frac{4}{2} \cdots \frac{n}{2} \leqq C$$

   for all $n \in \mathbb{P}$.

11. (a) Show that if $f(n)$ is $O(g(n))$ and $g(n)$ is $O(h(n))$ then $f(n)$ is $O(h(n))$. [Symbolically, $O(O(h(n))) = O(h(n))$.]
    (b) Give examples of $f(n)$, $g(n)$ and $h(n)$ such that $f(n)$ is $O(h(n))$ and $g(n)$ is $O(h(n))$ but $f(n)$ is not $O(g(n))$.

12. Show that if $f(n)$ is $O(g(n))$ and $g(n)$ is $O(f(n))$ then $O(f(n)) = O(g(n))$ in the sense that

$$h(n) \text{ is } O(f(n)) \Leftrightarrow h(n) \text{ is } O(g(n)).$$

13. For each $n \in \mathbb{P}$ let $\mathrm{DIGIT}(n)$ be the number of digits in the decimal expansion of n.
    (a) Show that $10^{\mathrm{DIGIT}(n)-1} \leqq n < 10^{\mathrm{DIGIT}(n)}$.
    (b) Show that $\log_{10} n$ is $O(\mathrm{DIGIT}(n))$.
    (c) Show that $\mathrm{DIGIT}(n)$ is $O(\log_{10} n)$.
        [In the notation of Exercise 12, $O(\mathrm{DIGIT}(n)) = O(\log_{10} n)$.]
    (d) Show that $O(\log_{10} n)$ and $O(\log_2 n)$ mean the same thing. [Recall that $\log_2 x = \log_2 10 \cdot \log_{10} x$.]
    (e) Let $\mathrm{DIGIT2}(n)$ be the number of digits in the binary expansion of n. How are $O(\mathrm{DIGIT}(n))$ and $O(\mathrm{DIGIT2}(n))$ related?

14. (a) Show that

$$\sum_{k=1}^{n} k^2 \text{ is } O(n^3).$$

    (b) More generally, show that for $m \in \mathbb{P}$

$$\sum_{k=1}^{n} k^m \text{ is } O(n^{m+1}).$$

§ 3.4   Recursive Definitions

The values of the terms in a sequence may be given explicitly by formulas such as $s_n = n^3 - 73n$ or by descriptions like "Let t_n be the weight of the nth edge on the path." Terms may also be defined sometimes by descriptions which involve other terms that come before them in the sequence.

We say that a sequence is defined **recursively** provided:

(B) some finite set of values, usually the first one or first few, are specified,

(R) the remaining values of the sequence are defined in terms of previous values of the sequence. A formula which does this is called **a recursion formula** or **relation.**

The requirement (B) will provide the **basis** or starting point for the definition. The remainder of the sequence will be obtained by using the recursion relation (R) repeatedly. [The relation will occur again and again, i.e., it "recurs."]

EXAMPLE 1

(a) We define the sequence FACT via

(B) FACT(0) = 1,

(R) FACT($n + 1$) = ($n + 1$) $*$ FACT(n) for $n \in \mathbb{N}$.

One can use (R) to calculate FACT(1), then FACT(2), then FACT(3), etc. and one quickly sees that FACT(n) = $n!$ for the first several values of n. This identity can be proved for all n by mathematical induction; a proof is requested in Exercise 21. Since we understand the sequence $n!$, the recursive definition above may seem silly, but we will try to convince you in part (b) that recursive definitions of even simple sequences are useful.

(b) Consider the sequence SUM(n) = $\sum\limits_{i=0}^{n} 1/i!$. To write a computer program that calculates the values of SUM for large values of n, one would use the following recursive definition:

(B) SUM(0) = 1,

(R) SUM($n + 1$) = SUM(n) + $\dfrac{1}{(n + 1)!}$.

The added term in (R) is the reciprocal of $(n + 1)!$, so FACT($n + 1$) will be needed as the program progresses. At each n, one could instruct the program to calculate FACT($n + 1$) from scratch or one could store a large number of these values. Clearly, it would be more efficient to alternately calculate FACT($n + 1$) and SUM($n + 1$) using the recursive definition in part (a) for FACT and the recursive definition above for SUM.

(c) Define the sequence SEQ as follows:

(B) SEQ(0) = 1,

(R) SEQ($n + 1$) = ($n + 1$)/SEQ(n) for $n \in \mathbb{N}$.

With $n = 0$, we find SEQ(1) = $1/1 = 1$. Then with $n = 1$, we find SEQ(2) = $2/1 = 2$. Continuing in this fashion, we find that the first few terms are 1, 1, 2, 3/2, 8/3, 15/8, 16/5, 35/16. It is by no means apparent what a general formula for SEQ(n) might be. It is evident that SEQ(73) exists but it would take considerable calculation to find it. ∎

In Example 1, how did we *know* that SEQ(73) exists? Our certainty is based on the belief that recursive definitions do indeed define sequences on all of $\mathbb{N}$, unless some step leads to an illegal computation such as division by 0. We prove that the recursive definition in Example 1 defines a sequence by proving

$$p(n) = \text{``SEQ}(n) \text{ is defined and SEQ}(n) \neq 0\text{''}$$

for all $n \in \mathbb{N}$. We do so by induction. Since $\text{SEQ}(0) = 1$, $p(0)$ is clearly true. Assume that $p(n)$ is true. Then $\text{SEQ}(n) \neq 0$ and so $(n + 1)/\text{SEQ}(n)$ is a well-defined real number. Thus $\text{SEQ}(n + 1)$ is defined by the recursive definition. Moreover, $\text{SEQ}(n + 1) \neq 0$ since $n + 1 \neq 0$. That is, $p(n + 1)$ is true. Since $p(n)$ implies $p(n + 1)$ for all $n \in \mathbb{N}$, the Principle of Mathematical Induction shows that all $p(n)$'s are true.

This proof that $\text{SEQ}(n)$ is defined for each n uses the fact that $\text{SEQ}(n + 1)$ only depends on $\text{SEQ}(n)$. Recursive definitions allow a term to depend on other terms besides the one just before it. In such cases an enhanced version of the Principle of Mathematical Induction which we take up in § 6.3 is the tool that proves the sequences are well defined.

The values of the terms in a recursively defined sequence can be calculated in two ways. An **iterative calculation** finds s_n by computing all of the values $s_1, s_2, \ldots, s_{n-1}$ first, so they are available to use in computing s_n. We had in mind an iterative calculation of the sequences in Example 1. To calculate $\text{FACT}(73)$, for instance, we would first calculate $\text{FACT}(k)$ for $k = 1, 2, \ldots, 72$, even though we might have no interest in these preliminary values themselves.

In the case of FACT there seems really to be no better alternative. Sometimes, though, there is a more clever way to calculate a given value of s_n. A **recursive calculation** finds the value of s_n by looking to see which terms s_n depends on, then which terms those terms depend on, and so on. It may turn out that the value of s_n only depends on the values of a relatively small set of its predecessors, in which case the other previous terms can be ignored.

EXAMPLE 2   (a) The **integer part** of a real number a, denoted by $\lfloor a \rfloor$, is the largest integer m such that $m \leq a$. Define the sequence T by

(B) $T(1) = 1$,
(R) $T(n) = 2 \cdot T(\lfloor n/2 \rfloor)$ for $n \geq 2$.

Then

$$T(73) = 2 \cdot T(\lfloor 73/2 \rfloor) = 2 \cdot T(36) = 2 \cdot 2 \cdot T(18) = 2 \cdot 2 \cdot 2 \cdot T(9)$$
$$= 2 \cdot 2 \cdot 2 \cdot 2 \cdot T(4) = 2 \cdot 2 \cdot 2 \cdot 2 \cdot 2 \cdot T(2) = 2 \cdot 2 \cdot 2 \cdot 2 \cdot 2 \cdot 2 \cdot T(1) = 2^6.$$

For this calculation we only need the values of $T(36)$, $T(18)$, $T(9)$, $T(4)$, $T(2)$ and $T(1)$, and we have no need to compute the other 66 values of $T(n)$ which precede $T(73)$.

(b) The sequence T in part (a) can be described in another way as follows [Exercise 19 of § 6.3]:

$$T(n) \text{ is the largest integer } 2^k \text{ with } 2^k \leq n.$$

Using this description we could compute $T(73)$ by looking at the list of powers of 2 less than 73 and taking the largest one. A slight change in (R) from part (a) gives the sequence Q with

(B) $Q(1) = 1$,
(R) $Q(n) = 2 \cdot Q(\lfloor n/2 \rfloor) + n$ for $n \geq 2$.

Now the general term is not so clear, but we can still find $Q(73)$ recursively from $Q(36)$, $Q(18)$, ..., $Q(2)$, $Q(1)$. ∎

EXAMPLE 3   (a) The **Fibonacci sequence** is defined as follows:

(B) $\text{FIB}(0) = \text{FIB}(1) = 1$,
(R) $\text{FIB}(n) = \text{FIB}(n-1) + \text{FIB}(n-2)$ for $n \ge 2$.

Note that the recursion formula makes no sense for $n = 1$ and so $\text{FIB}(1)$ had to be defined separately in the basis. The first few terms of this sequence are

$$1, 1, 2, 3, 5, 8, 13, 21, 34, 55, 89.$$

(b) Here is an easy way to define the sequence 0, 0, 1, 1, 2, 2, 3, 3, ...

(B) $\text{SEQ}(0) = \text{SEQ}(1) = 0$,
(R) $\text{SEQ}(n) = 1 + \text{SEQ}(n-2)$ for $n \ge 2$.

Compare Exercise 5 of § 3.3. ∎

EXAMPLE 4 Let $\Sigma = \{a, b\}$.
    (a) We are interested in the number s_n of words of length n that do not have consecutive a's, i.e., do not contain the string aa. Let's write A_n for the set of words in Σ^n having no consecutive a's. Then $A_0 = \{\epsilon\}$, $A_1 = \Sigma$, $A_2 = \Sigma^2 \backslash \{aa\}$ and so $s_0 = 1$, $s_1 = 2$ and $s_2 = 4 - 1 = 3$. To get a recursion formula for s_n, we consider $n \ge 2$ and count the number of words in A_n in terms of shorter words. If a word in A_n ends in b it can be preceded by any word in A_{n-1}. So s_{n-1} words in A_n end in b. If a word in A_n ends in a, then the last two letters must be ba and this string can be preceded by any word in A_{n-2}. So s_{n-2} words in A_n end in a. Thus $s_n = s_{n-1} + s_{n-2}$ for $n \ge 2$. This is the recursion relation for the Fibonacci sequence, but note that the basis is different: $s_1 = 2$, while $\text{FIB}(1) = 1$. In fact, $s_n = \text{FIB}(n+1)$ for $n \in \mathbb{N}$.
    (b) Since Σ^n has 2^n words in it, there are $2^n - s_n$ words of length n that contain consecutive a's. ∎

EXAMPLE 5 Let $\Sigma = \{a, b, c\}$, let B_n be the set of words in Σ^n with an even number of a's and let t_n denote the number of words in B_n. Then $B_0 = \{\epsilon\}$, $B_1 = \{b, c\}$, $B_2 = \{aa, bb, bc, cb, cc\}$ and so $t_0 = 1$, $t_1 = 2$ and $t_2 = 5$. We count the number of words in B_n in terms of shorter words by analyzing the last letter. If a word in B_n ends in b, it can be preceded by any word in B_{n-1}. So t_{n-1} words in B_n end in b. Similarly, t_{n-1} words in B_n end in c. If a word in B_n ends in a it must be preceded by a word in Σ^{n-1} with an *odd* number of a's. Since Σ^{n-1} has 3^{n-1} words, $3^{n-1} - t_{n-1}$ of them must have an odd number of a's. Hence $3^{n-1} - t_{n-1}$ words in B_n end in a. Thus

$$t_n = t_{n-1} + t_{n-1} + (3^{n-1} - t_{n-1}) = 3^{n-1} + t_{n-1}$$

for $n \ge 1$. Hence $t_3 = 3^2 + t_2 = 9 + 5 = 14$, $t_4 = 3^3 + t_3 = 27 + 14 = 41$, etc.

In this case, it's relatively easy to find an explicit formula for t_n. First note that

$$t_n = 3^{n-1} + t_{n-1} = 3^{n-1} + 3^{n-2} + t_{n-2} = \cdots$$

$$= 3^{n-1} + 3^{n-2} + \cdots + 3^0 + t_0 = 1 + \sum_{k=0}^{n-1} 3^k.$$

If those three dots make you nervous, you can supply a proof by induction. Now we apply Exercise 3 of § 2.6 to obtain

$$t_n = 1 + \frac{3^n - 1}{3 - 1} = 1 + \frac{3^n - 1}{2} = \frac{3^n + 1}{2}.$$

Our mental process works for $n \geq 1$ but the formula works for $n = 0$ too. The formula for t_n looks right: about half of the words in Σ^n use an even number of a's.   ∎

EXAMPLE 6   (a) Define the sequence S by

(B) $S(0) = 0$, $S(1) = 1$,
(R) $S(n) = S(\lfloor n/2 \rfloor) + S(\lfloor n/5 \rfloor)$ for $n \geq 2$.

[Recall from Example 2 that $\lfloor m \rfloor$ is the integer part of m.] It makes sense to calculate the values of S recursively, rather than iteratively. For instance,

$$S(73) = S(36) + S(14) = [S(18) + S(7)] + [S(7) + S(2)]$$

$$= S(18) + 2S(7) + S(2)$$

$$= S(9) + S(3) + 2[S(3) + S(1)] + S(1) + S(0)$$

$$= \cdots$$

$$= 8S(1) + 6S(0) = 8.$$

The calculation of $S(73)$ involves the values of $S(36)$, $S(18)$, $S(14)$, $S(9)$, $S(7)$, $S(4)$, $S(3)$, $S(2)$, $S(1)$ and $S(0)$, but that's still better than finding all values of $S(k)$ for $k = 1, \ldots, 72$. One can show more generally that the value of $S(n)$ in this example only depends on the values of the terms $S(m)$ for m of the form $\lfloor n/2^a 5^b \rfloor$.

(b) Recursive calculation requires storage addresses for the intermediate values that have been called for but not yet computed. It may be possible, though, to keep the number of storage slots fairly small. For example, the recursive calculation of FACT(6) goes like this:

$$\text{FACT}(6) = 6 \cdot \text{FACT}(5) = 30 \cdot \text{FACT}(4) = 120 \cdot \text{FACT}(3) = \cdots.$$

Only one address is needed for the intermediate [unknown] value FACT(k) for $k < 6$. Similarly,

$$\text{FIB}(6) = \text{FIB}(5) + \text{FIB}(4) = (\text{FIB}(4) + \text{FIB}(3)) + \text{FIB}(4)$$
$$= 2 \cdot \text{FIB}(4) + \text{FIB}(3)$$
$$= 3 \cdot \text{FIB}(3) + 2 \cdot \text{FIB}(2)$$
$$= 5 \cdot \text{FIB}(2) + 3 \cdot \text{FIB}(1)$$
$$= 8 \cdot \text{FIB}(1) + 5 \cdot \text{FIB}(0)$$

only requires two intermediate addresses. ∎

EXAMPLE 7 Consider the graph and its matrix in Figure 1. In § 0.3 we indicated that $\mathbf{M}$ can be used to study this graph. In particular, the entries in the nth power $\mathbf{M}^n$ give the exact number of paths of length n that connect two vertices. Matrix

$$\mathbf{M} = \begin{bmatrix} 1 & 1 \\ 1 & 0 \end{bmatrix}$$

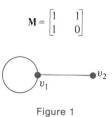

Figure 1

multiplication is discussed in § 4.1, and Exercise 14 of that section asks for a proof that

$$\mathbf{M}^n = \begin{bmatrix} \text{FIB}(n) & \text{FIB}(n-1) \\ \text{FIB}(n-1) & \text{FIB}(n-2) \end{bmatrix} \quad \text{for} \quad n \geq 2.$$

Thus this simple graph leads to matrices whose entries are defined by the recursively defined Fibonacci sequence in Example 3. ∎

Of course we can give recursive definitions even if the sequence is not real-valued.

EXAMPLE 8 Let S be a set and let f be a function from S into S. We define

(B) $f^{(0)} = 1_S$ [the identity function on S],
(R) $f^{(n+1)} = f^{(n)} \circ f$.

Thus

(1) $\qquad f^{(1)} = f, \quad f^{(2)} = f \circ f, \quad f^{(3)} = f \circ f \circ f, \quad \text{etc.}$

In other words,

(2) $\qquad f^{(n)} = f \circ f \circ \cdots \circ f \qquad [n \text{ times}].$

$f^{(n)}$ is simply the composite of the function f, n times. The recursive definition is more precise than the "etc." in (1) or the three dots in (2). ∎

As in Example 8, we will often use recursive definitions to give concise definitions for concepts that we already understand quite well.

EXAMPLE 9   Let a be a nonzero real number.

    (B) $a^0 = 1$,
    (R) $a^{n+1} = a^n \cdot a$ for $n \in \mathbb{N}$.

Equivalently,

    (B) $\text{POW}(0) = 1$,
    (R) $\text{POW}(n + 1) = \text{POW}(n)*a$ for $n \in \mathbb{N}$. ∎

EXAMPLE 10   Let $(a_j)_{j \in \mathbb{P}}$ be a sequence of real numbers. Then

$$(\text{B})\ \prod_{j=1}^{1} a_j = a_1,$$

$$(\text{R})\ \prod_{j=1}^{n+1} a_j = a_{n+1} \cdot \prod_{j=1}^{n} a_j \text{ for } n \geq 1.$$

Equivalently,

    (B) $\text{PROD}(1) = a_1$,
    (R) $\text{PROD}(n + 1) = a_{n+1}*\text{PROD}(n)$ for $n \geq 1$.

These recursive definitions start at $n = 1$. An alternative is to define the "empty product" to be 1, i.e.,

$$(\text{B})\ \prod_{j=1}^{0} a_j = 1 \text{ [which looks peculiar]},$$

or

    (B) $\text{PROD}(0) = 1$.

Then the same recursive relation (R) as before serves to define the remaining terms of the sequence. ∎

EXERCISES 3.4

1. We recursively define $s_0 = 1$ and $s_{n+1} = 2/s_n$ for $n \in \mathbb{N}$.
    (a) List the first few terms of the sequence.
    (b) What is the set of values of s?

2. We recursively define $\text{SEQ}(0) = 0$ and $\text{SEQ}(n + 1) = 1/[1 + \text{SEQ}(n)]$ for $n \in \mathbb{N}$. Calculate $\text{SEQ}(n)$ for $n = 1, 2, 3, 4$ and 6.

3. Consider the sequence $(1, 3, 9, 27, 81, \ldots)$.
   (a) Give a formula for the nth term $SEQ(n)$ where $SEQ(0) = 1$.
   (b) Give a recursive definition for the sequence SEQ.

4. (a) Give a recursive definition for the sequence $(2, 2^2, (2^2)^2, ((2^2)^2)^2, \ldots)$, i.e., $(2, 4, 16, 256, \ldots)$.
   (b) Give a recursive definition for the sequence $(2, 2^2, 2^{(2^2)}, 2^{(2^{(2^2)})}, \ldots)$, i.e., $(2\ 4, 16, 65536, \ldots)$.

5. Is the following a recursive definition for a sequence SEQ? Explain.

   (B) $SEQ(0) = 1$,
   (R) $SEQ(n + 1) = SEQ(n)/[100 - n]$.

6. (a) Calculate $SEQ(9)$ where SEQ is as in Example 1(c).
   (b) Calculate $FIB(11)$ where FIB is as in Example 3(a).

7. Let $\Sigma = \{a, b, c\}$ and let s_n denote the number of words of length n that do not have consecutive a's.
   (a) Calculate s_0, s_1 and s_2.
   (b) Find a recursion formula for s_n.
   (c) Calculate s_3 and s_4.

8. Let $\Sigma = \{a, b\}$ and let s_n denote the number of words of length n that do not contain the string ab.
   (a) Calculate s_0, s_1, s_2 and s_3.
   (b) Find a formula for s_n and prove it is correct.

9. Let $\Sigma = \{a, b\}$ and let t_n denote the number of words of length n with an even number of a's.
   (a) Calculate t_0, t_1, t_2 and t_3.
   (b) Find a formula for t_n and prove it is correct.
   (c) Does your formula for t_n work for $n = 0$?

10. Consider the sequence defined by

   (B) $SEQ(0) = 1$, $SEQ(1) = 0$,
   (R) $SEQ(n) = SEQ(n - 2)$ for $n \geq 2$.

   (a) List the first few terms of this sequence.
   (b) What is the set of values of this sequence?

11. We recursively define $a_0 = a_1 = 1$ and $a_n = a_{n-1} + 2a_{n-2}$ for $n \geq 2$.
   (a) Calculate a_6 recursively.
   (b) Prove that all the terms a_n are odd integers.

12. Recursively define $b_0 = b_1 = 1$ and $b_n = 2b_{n-1} + b_{n-2}$ for $n \geq 2$.
   (a) Calculate b_5 iteratively.
   (b) Explain why all the terms b_n are odd integers.

13. Let $SEQ(0) = 1$ and $SEQ(n) = \sum_{i=0}^{n-1} SEQ(i)$ for $n \geq 1$. This is actually a simple, familiar sequence. What is it?

14. We recursively define $a_0 = 0$, $a_1 = 1$, $a_2 = 2$ and $a_n = a_{n-1} - a_{n-2} + a_{n-3}$ for $n \geq 3$.

   (a) List the first few terms of the sequence until the pattern is clear.

   (b) What is the set of values of the sequence?

15. The process of assigning n children to n classroom seats can be broken down into (1) choosing a child for the first seat and (2) assigning the other $n - 1$ children to the remaining seats. Let $A(n)$ be the number of different assignments of n children to n seats.

   (a) Write a recursive definition of the sequence A.

   (b) Calculate $A(6)$ recursively.

   (c) Is the sequence A familiar?

16. Consider the process of assigning $2n$ children to n cars in a playground train so that two children go in each car. First choose two children for the front car [there are $2n(2n - 1)/2$ ways to do this, as we will see in Chapter 5]. Then distribute the rest of the children to the remaining $n - 1$ cars. Let $B(n)$ be the number of ways to assign $2n$ children to n cars.

   (a) Write a recursive definition of the sequence B.

   (b) Calculate $B(3)$ recursively.

   (c) Calculate $B(5)$ iteratively.

   (d) Give an explicit formula for $B(n)$.

17. Verify the equalities in formula (1) of Example 8.

18. Let $(a_1, a_2, \ldots)$ be a sequence of real numbers.

   (a) Give a recursive definition for $\text{SUM}(n) = \sum_{j=1}^{n} a_j$ for $n \geq 1$.

   (b) Revise your recursive definition for $\text{SUM}(n)$ by starting with $n = 0$. What is the "empty sum"?

19. Let $(A_1, A_2, \ldots)$ be a sequence of subsets of some set S.

   (a) Give a recursive definition for $\bigcup_{j=1}^{n} A_j$.

   (b) How would you define the "empty union"?

   (c) Give a recursive definition for $\bigcap_{j=1}^{n} A_j$.

   (d) How would you define the "empty intersection"?

20. Let $(A_1, A_2, \ldots)$ be a sequence of subsets of some set S. Define

   (B) $\text{SYM}(1) = A_1$,

   (R) $\text{SYM}(n + 1) = A_{n+1} \oplus \text{SYM}(n)$ for $n \geq 1$.

   Recall that $\oplus$ denotes symmetric difference. It turns out that an element x in S belongs to $\text{SYM}(n)$ if and only if the set $\{k : x \in A_k \text{ and } k \leq n\}$ has an odd number of elements. Prove this by mathematical induction.

21. Prove that $\text{FACT}(n) = n!$ for $n \in \mathbb{N}$ where the sequence FACT is defined recursively in Example 1 and $n!$ is defined as in § 1.3.

§ 3.5 Recursion Relations

Sequences that are defined by recursion appear frequently in mathematics and science, and there are several techniques for obtaining explicit formulas for them. We will only give one theorem, which will allow us to solve recursion relations of the form

$$s_n = as_{n-1} + bs_{n-2}.$$

Here a and b are constants, and it is assumed that the two initial values s_0 and s_1 have been specified.

The cases where $a = 0$ or $b = 0$ are especially easy to deal with. If $b = 0$ and $s_n = as_{n-1}$ for $n \geq 1$, then $s_1 = as_0$, $s_2 = as_1 = a^2 s_0$, etc. A simple induction argument shows that $s_n = a^n s_0$ for all $n \in \mathbb{N}$. Now suppose that $a = 0$. Then $s_2 = bs_0$, $s_4 = bs_2 = b^2 s_0$, etc., so that $s_{2n} = b^n s_0$ for all $n \in \mathbb{N}$. Likewise, $s_3 = bs_1$, $s_5 = b^2 s_1$, etc., so that $s_{2n+1} = b^n s_1$ for all $n \in \mathbb{N}$.

EXAMPLE 1   (a) Consider the recursion relation $s_n = 3s_{n-1}$ with $s_0 = 5$. Here $a = 3$ and so $s_n = 5 \cdot 3^n$ for $n \in \mathbb{N}$.

(b) Consider the recursion relation $s_n = 3s_{n-2}$ with $s_0 = 5$ and $s_1 = 2$. Here $b = 3$, so $s_{2n} = 5 \cdot 3^n$ and $s_{2n+1} = 2 \cdot 3^n$ for $n \in \mathbb{N}$. ∎

From now on we will assume that $a \neq 0$ and $b \neq 0$. It is convenient to ignore the specified values of s_0 and s_1 until later. In view of the special cases we've examined, it is reasonable to hope that some solutions have the form $s_n = cr^n$ for some constant c. This hope, if true, would force

$$r^n = ar^{n-1} + br^{n-2}.$$

Dividing by r^{n-2} would then give $r^2 = ar + b$, or $r^2 - ar - b = 0$. In other words, if $s_n = cr^n$ for all n, then r must be a solution of the quadratic equation $x^2 - ax - b = 0$, which is called the **characteristic equation** of the recursion relation. The characteristic equation has one or two solutions, which we'll assume from now on are real numbers.

Theorem   Consider a recursion relation of the form

$$s_n = as_{n-1} + bs_{n-2}$$

with characteristic equation

$$x^2 - ax - b = 0,$$

where a and b are nonzero constants.

(a) If the characteristic equation has distinct solutions r_1 and r_2, then

(*) $$s_n = c_1 r_1^n + c_2 r_2^n$$

for certain constants c_1 and c_2. If s_0 and s_1 are specified, the constants can be determined by setting $n = 0$ and $n = 1$ in (*) and solving the two equations for c_1 and c_2.

(b) If the characteristic equation has only one solution r, then

(**)
$$s_n = c_1 r^n + c_2 \cdot n \cdot r^n$$

for certain constants c_1 and c_2. As in (a), c_1 and c_2 can be determined if s_0 and s_1 are specified.

Warning: This fine theorem only applies to recursion relations of the form $s_n = as_{n-1} + bs_{n-2}$.

EXAMPLE 2 Consider the recursion relation $s_n = s_{n-1} + 2s_{n-2}$ where $s_0 = s_1 = 3$. Here $a = 1$ and $b = 2$. The characteristic equation $x^2 - x - 2 = 0$ has solutions $r_1 = 2$ and $r_2 = -1$ since $x^2 - x - 2 = (x - 2)(x + 1)$, so part (a) of the theorem applies. By the theorem,

$$s_n = c_1 \cdot 2^n + c_2 \cdot (-1)^n$$

for constants c_1 and c_2. By setting $n = 0$ and $n = 1$ we find

$$s_0 = c_1 \cdot 2^0 + c_2 \cdot (-1)^0 \quad \text{and} \quad s_1 = c_1 \cdot 2^1 + c_2 \cdot (-1)^1,$$

i.e.,

$$3 = c_1 + c_2 \quad \text{and} \quad 3 = 2c_1 - c_2.$$

Solving this system of two equations gives $c_1 = 2$ and $c_2 = 1$. We conclude that

$$s_n = 2 \cdot 2^n + 1 \cdot (-1)^n = 2^{n+1} + (-1)^n \quad \text{for} \quad n \in \mathbb{N}. \quad \blacksquare$$

EXAMPLE 3 Consider again the Fibonacci sequence in § 3.4. Writing s_n for FIB(n), we have $s_0 = s_1 = 1$ and $s_n = s_{n-1} + s_{n-2}$ for $n \geq 2$. Here $a = b = 1$, so we solve $x^2 - x - 1 = 0$. By the quadratic formula, the equation has two solutions:

$$r_1 = \frac{1 + \sqrt{5}}{2} \quad \text{and} \quad r_2 = \frac{1 - \sqrt{5}}{2}.$$

Thus part (a) of the theorem applies, and so

$$s_n = c_1 \left(\frac{1 + \sqrt{5}}{2} \right)^n + c_2 \left(\frac{1 - \sqrt{5}}{2} \right)^n \quad \text{for} \quad n \in \mathbb{N}.$$

While solving for c_1 and c_2, it's convenient to retain the notation r_1 and r_2. Setting $n = 0$ and $n = 1$ gives

$$1 = c_1 + c_2 \quad \text{and} \quad 1 = c_1 r_1 + c_2 r_2.$$

If we replace c_2 by $1 - c_1$ in the second equation we get $1 = c_1 r_1 + (1 - c_1)r_2$, so $1 - r_2 = c_1(r_1 - r_2)$ and

$$c_1 = \frac{1 - r_2}{r_1 - r_2}.$$

Since $r_1 + r_2 = 1$ and $r_1 - r_2 = \sqrt{5}$, we conclude that $c_1 = r_1/\sqrt{5}$. Now $c_2 = 1 - c_1 = (\sqrt{5} - r_1)/\sqrt{5} = -r_2/\sqrt{5}$. Finally

$$s_n = c_1 r_1^n + c_2 r_2^n = \frac{r_1}{\sqrt{5}} r_1^n - \frac{r_2}{\sqrt{5}} r_2^n$$

$$= \frac{1}{\sqrt{5}} (r_1^{n+1} - r_2^{n+1})$$

and so

$$\text{FIB}(n) = s_n = \frac{1}{\sqrt{5}} \left[\left(\frac{1 + \sqrt{5}}{2} \right)^{n+1} - \left(\frac{1 - \sqrt{5}}{2} \right)^{n+1} \right]. \quad \blacksquare$$

EXAMPLE 4 Consider the sequence (s_n) defined by $s_0 = 1$, $s_1 = -3$ and $s_n = 6s_{n-1} - 9s_{n-2}$ for $n \geq 2$. Here the characteristic equation is $x^2 - 6x + 9 = 0$, which has exactly one solution, namely $r = 3$. By part (b) of the theorem,

$$s_n = c_1 \cdot 3^n + c_2 \cdot n \cdot 3^n \quad \text{for} \quad n \in \mathbb{N}.$$

Setting $n = 0$ and $n = 1$ we get

$$s_0 = c_1 \cdot 3^0 + 0 \quad \text{and} \quad s_1 = c_1 \cdot 3^1 + c_2 \cdot 3^1$$

or

$$1 = c_1 \quad \text{and} \quad -3 = 3c_1 + 3c_2.$$

So $c_1 = 1$ and $c_2 = -2$. Therefore

$$s_n = 3^n - 2 \cdot n \cdot 3^n \quad \text{for} \quad n \in \mathbb{N}. \quad \blacksquare$$

Proof of the Theorem. (a) No matter what s_0 and s_1 are, the equations

$$s_0 = c_1 + c_2 \quad \text{and} \quad s_1 = c_1 r_1 + c_2 r_2$$

can be solved for c_1 and c_2, since $r_1 \neq r_2$. The original sequence (s_n) is determined by the values s_0 and s_1 and the recurrence condition $s_n = as_{n-1} + bs_{n-2}$, so it suffices to show that the sequence defined by $(*)$ also satisfies this recurrence condition. Since r_1 satisfies $x^2 = ax + b$, we have $r_1^n = ar_1^{n-1} + br_1^{n-2}$ and so the sequence (r_1^n) satisfies the condition $s_n = as_{n-1} + bs_{n-2}$. So does (r_2^n). It is now easy to check that the sequence defined by $(*)$ also satisfies the recurrence condition:

$$as_{n-1} + bs_{n-2} = a(c_1 r_1^{n-1} + c_2 r_2^{n-1}) + b(c_1 r_1^{n-2} + c_2 r_2^{n-2})$$

$$= c_1 [ar_1^{n-1} + br_1^{n-2}] + c_2 [ar_2^{n-1} + br_2^{n-2}]$$

$$= c_1 r_1^n + c_2 r_2^n = s_n.$$

(b) If r is the only solution of the characteristic equation, then the characteristic equation has the form $(x - r)^2 = 0$. Thus $x^2 - 2rx + r^2 = x^2 - ax - b$ and so $a = 2r$ and $b = -r^2$. The recursion relation can now be written

$$s_n = 2rs_{n-1} - r^2 s_{n-2}.$$

Putting $n = 0$ and $n = 1$ in $(**)$ gives the equations

$$s_0 = c_1 \quad \text{and} \quad s_1 = c_1 r + c_2 r.$$

Since $r \neq 0$, these equations have the solutions $c_1 = s_0$ and $c_2 = -s_0 + s_1/r$. As in our proof of part (a), it suffices to show that any sequence defined by $(**)$ satisfies $s_n = 2rs_{n-1} - r^2 s_{n-2}$. But

$$2rs_{n-1} - r^2 s_{n-2}$$
$$= 2r[c_1 r^{n-1} + c_2(n-1)r^{n-1}] - r^2[c_1 r^{n-2} + c_2(n-2)r^{n-2}]$$
$$= 2c_1 r^n + 2c_2(n-1)r^n - c_1 r^n - c_2(n-2)r^n$$
$$= c_1 r^n + c_2 \cdot n \cdot r^n = s_n. \quad \blacksquare$$

The proof of the theorem is still valid if the roots of the characteristic equation are not real, so the theorem is true in that case as well. Finding the values of the terms with formula $(*)$ will then involve complex arithmetic, but the answers will of course all be real if a, b, s_0 and s_1 are real. This situation is analogous to the calculation of the Fibonacci numbers, which are integers, using $\sqrt{5}$, which is not.

EXERCISES 3.5

1. Give an explicit formula for s_n where $s_0 = 3$ and $s_n = -2s_{n-1}$ for $n \geq 1$.

2. (a) Give an explicit formula for $s_n = 4s_{n-2}$ where $s_0 = s_1 = 1$.
   (b) Repeat part (a) for $s_0 = 1$ and $s_1 = 2$.

3. Prove that if $s_n = as_{n-1}$ for $n \geq 1$ and $a \neq 0$, then $s_n = a^n \cdot s_0$ for $n \in \mathbb{N}$.

4. Verify that the sequence given by $s_n = 2^{n+1} + (-1)^n$ in Example 2 satisfies the conditions $s_0 = s_1 = 3$ and $s_n = s_{n-1} + 2s_{n-2}$ for $n \geq 2$.

5. Verify that the sequence $s_n = 3^n - 2 \cdot n \cdot 3^n$ in Example 4 satisfies the condition $s_0 = 1, s_1 = -3$ and $s_n = 6s_{n-1} - 9s_{n-2}$ for $n \geq 2$.

6. Use the formula for FIB(n) in Example 3 and a hand calculator to verify that FIB(5) = 8.

7. Give an explicit formula for s_n where $s_0 = 3$, $s_1 = 6$ and $s_n = s_{n-1} + 2s_{n-2}$ for $n \geq 2$. *Hint:* Imitate Example 2, but note that now $s_1 = 6$.

8. Repeat Exercise 7 with $s_0 = 3$ and $s_1 = -3$.

9. Give an explicit formula for the sequence in Example 4 of § 3.4: $s_0 = 1, s_1 = 2$ and $s_n = s_{n-1} + s_{n-2}$ for $n \geq 2$. *Hint:* Use Example 3.

10. Consider the sequence (s_n) where $s_0 = 2$, $s_1 = 1$ and $s_n = s_{n-1} + s_{n-2}$ for $n \geq 2$.
    (a) Calculate s_n for $n = 2, 3, 4, 5$ and 6.
    (b) Give an explicit formula for s_n.

11. In each of the following cases give an explicit formula for s_n.
    (a) $s_0 = 2$, $s_1 = -1$ and $s_n = -s_{n-1} + 6s_{n-2}$ for $n \geq 2$.
    (b) $s_0 = 2$ and $s_n = 5 \cdot s_{n-1}$ for $n \geq 1$.
    (c) $s_0 = 1$, $s_1 = 8$ and $s_n = 4s_{n-1} - 4s_{n-2}$ for $n \geq 2$.
    (d) $s_0 = c$, $s_1 = d$ and $s_n = 5s_{n-1} - 6s_{n-2}$ for $n \geq 2$. Here c and d are unspecified constants.
    (e) $s_0 = 1$, $s_1 = 4$ and $s_n = s_{n-2}$ for $n \geq 2$.
    (f) $s_0 = 1$, $s_1 = 2$ and $s_n = 3 \cdot s_{n-2}$ for $n \geq 2$.
    (g) $s_0 = 1$, $s_1 = -3$ and $s_n = -2s_{n-1} + 3s_{n-2}$ for $n \geq 2$.
    (h) $s_0 = 1$, $s_1 = 2$ and $s_n = -2s_{n-1} + 3s_{n-2}$ for $n \geq 2$.

12. Recall that if $s_n = bs_{n-2}$ for $n \geq 2$, then $s_{2n} = b^n s_0$ and $s_{2n+1} = b^n s_1$ for $n \in \mathbb{N}$. Show that the theorem holds for $a = 0$ and $b > 0$, and reconcile this assertion with the preceding sentence. That is, specify r_1, r_2, c_1 and c_2 in terms of b, s_0 and s_1.

§ 3.6   General Recursive Definitions

The recursive definitions in § 3.4 allowed us to define sequences, i.e., sets of objects that are conveniently indexed by $\mathbb{N}$. More generally, we say that a set of objects is defined **recursively** provided

(B) some members of the set are specified explicitly,

(R) the remaining members of the set are defined in terms of members already defined.

The specification in (B) is called the **basis** for the definition and the recipe in (R) the **recursive clause**.

We begin with an example that is familiar from algebra.

EXAMPLE 1   We recursively define the set $\text{POLY}(\mathbb{R})$ of **polynomial functions** on $\mathbb{R}$.

(B) The constant functions and the function I, defined by $I(x) = x$ for all $x \in \mathbb{R}$, are polynomial functions.

(R) If f and g are polynomial functions, so are $f + g$ and fg.

We illustrate this definition by showing, step by step, that the function $f(x) = 5 + 2x^2$ is a polynomial function. Let $f_1(x) = 5$ and $f_2(x) = 2$ for all $x \in \mathbb{R}$. Then f_1 and f_2 are polynomial functions by (B). Since I is a polynomial function, so is its square I^2 and hence so is $f_3 = f_2 \cdot I^2$. Note that since $I^2(x) = I(x) \cdot I(x) = x^2$, we have $f_3(x) = 2x^2$ for all x. Finally, since $f = f_1 + f_3$, f is a polynomial function. ∎

A third condition was implicit in the definition in Example 1, namely the so-called **extremal** condition:

(E) No object is a polynomial function unless it can be obtained from (B) and (R) in a finite number of steps.

We will usually omit the extremal condition, but it should be specified if there is any danger of confusion.

EXAMPLE 2 Let Σ be a finite alphabet. We define the set Σ^* as follows:

(B) The empty word ϵ is in Σ^*.
(R) If w is in Σ^* and x is in Σ, then wx is in Σ^*.

By wx we mean the word obtained by attaching x to the right of the word w.
To illustrate this definition, suppose that $\Sigma = \{a, b, c\}$. By (B), ϵ is in Σ^*. Since $c \in \Sigma$, (R) shows that $\epsilon c = c \in \Sigma^*$. Since $a \in \Sigma$, another application of (R) shows that $ca \in \Sigma^*$. Again, since $b \in \Sigma$, we conclude that $cab \in \Sigma^*$. Letters can be repeated, of course, so $cabb \in \Sigma^*$. In fact, any finite string of letters, using only a, b and c, will be in Σ^*. Note that, as in English, we have obtained the word "*cabb*" by building it from left to right. ∎

EXAMPLE 3 (a) In reading about trees in Chapter 0, it may have occurred to you that they are built up from smaller trees. It seems as if one can just keep adding leaves, starting with the trivial tree, to get any tree. In fact, the following recursive definition provides an alternative way to describe the class of [finite] trees.

(B) A single vertex is a [trivial] tree.
(R) If T is a tree and v is some element that is not a vertex of T, then we obtain a new tree T' if we add v to the set of vertices of T and if we add a single edge connecting v to some vertex of T.

(b) To illustrate the recursive definition in part (a), we consider the tree T_1 in Figure 1 and pluck leaves, one by one, in any old way until a single vertex is left. We obtain T_2, T_3, T_4, T_5, T_6 in Figure 1. Now to obtain T_1 using the recursive definition we use Figure 1 *in reverse*. We begin with T_6, which is a tree by the basis (B). By (R), T_5 is a tree because it is obtained from T_6 by adding one vertex, namely w, and one edge, namely $w\,y$, connecting w to a vertex of T_6. By (R) again, T_4 is a tree because it is obtained from T_5 by adding v and the edge $v\,w$ which connects v to a vertex of T_5. Three more

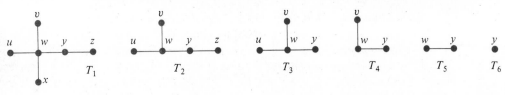

Figure 1

such applications of (R) show that T_1 is a tree according to our recursive definition. ∎

EXAMPLE 4   Sometimes subsets of sets can be defined recursively. For example, consider the following subset of $\mathbb{N} \times \mathbb{N}$:

$$S = \{\langle m, n \rangle : m \leq 2n\}.$$

The set S is drawn in Figure 2. An examination of the picture suggests how membership in S can be based on membership of elements closer to $\langle 0, 0 \rangle$. If an ordered pair is in S, so are the ordered pairs immediately above it illustrated by the arrows in Figure 2. This observation can be converted into a recursive definition:

(B) $\langle 0, 0 \rangle \in S$.
(R) If $\langle m, n \rangle \in S$, then $\langle m, n + 1 \rangle$, $\langle m + 1, n + 1 \rangle$ and $\langle m + 2, n + 1 \rangle$ are in S.

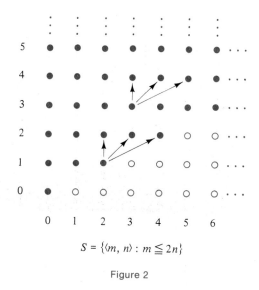

$$S = \{\langle m, n \rangle : m \leq 2n\}$$

Figure 2

We illustrate the above definition by showing that $\langle 5, 3 \rangle$ is in S. By the basis (B), $\langle 0, 0 \rangle \in S$. By (R), $\langle 1, 1 \rangle \in S$. Then by (R), $\langle 3, 2 \rangle \in S$ and again by (R) we have $\langle 5, 3 \rangle \in S$. There are other sequences of arguments that will show that $\langle 5, 3 \rangle \in S$. For example, we could apply (R) to conclude that $\langle 2, 1 \rangle$, then $\langle 4, 2 \rangle$, then $\langle 5, 3 \rangle$ are in S. ∎

To help motivate the next example, consider the following questions that probably did not arise in algebra: What is an acceptable formula? What is it that makes $(x + y)(x - y)$ look good and makes $(x + -(^4/y)$ look worthless? An answer leads to the notion of a **well-formed formula** or **wff**.

EXAMPLE 5   (a) Here is a definition of wff's for algebra.

(B) Numerical constants and variables are wff's.

(R) If f and g are wff's, so are $(f + g)$, $(f - g)$, (fg), (f/g) and (f^g).

Being variables, x and y are wff's. Therefore both $(x + y)$ and $(x - y)$ are wff's. Finally, we conclude that $((x + y)(x - y))$ is a wff. The definition isn't entirely satisfactory, since the outside parentheses here seem extraneous. However, without them the square $(((x + y)(x - y))^2)$ would look like $((x + y)(x - y)^2)$ and these expressions have different meanings. The problem is that in algebra we traditionally allow the omission of parentheses in some circumstances. Taking all the exceptional cases into account, we would be led to a complicated definition. Note also that our definition does not exclude division by 0. Thus $(0/0)$ is a wff even though we would not assign a numerical value to this expression.

(b) In computer science, the symbol $*$ is often used for multiplication and $\uparrow$ used for exponentiation [$a \uparrow b$ means a^b]. With this notation, the definition of wff's can be rewritten as:

(B) Numerical constants and variables are wff's.

(R) If f and g are wff's, so are $(f + g)$, $(f - g)$, $(f * g)$, (f/g) and $(f \uparrow g)$.

For example, $(((((X + Y) \uparrow 2) - (2 * (X * Y))) - (X \uparrow 2)) - (Y \uparrow 2))$ is a wff.

(c) In § 9.4 we will discuss Polish notation, which is a parenthesis-free notation. The preceding examples and related exercises may help you appreciate its value. ∎

EXAMPLE 6   (a) A recursive definition of wff's for the propositional calculus is as follows.

(B) Variables, such as p, q, r, are wff's.

(R) If P and Q are wff's, so are $(P \vee Q)$, $(P \wedge Q)$, $(P \rightarrow Q)$, $(P \leftrightarrow Q)$ and $\neg P$.

Note that we do not require parentheses when we negate a proposition. Consequently, the negation symbol $\neg$ always negates the shortest subexpression following it that is a wff. In practice, we tend to omit the outside parentheses and, for the sake of readability, we may use brackets [ ] or braces { } for parentheses.

(b) In the expression $\neg(p \vee q) \rightarrow r$, the negation sign negates $(p \vee q)$, i.e., the expression means $(\neg(p \vee q)) \rightarrow r$. To emphasize the difference between the expressions $\neg(p \vee q) \rightarrow r$ and $\neg((p \vee q \rightarrow r)$ we describe the recursive procedures that lead to each of them. First, $\neg(p \vee q) \rightarrow r$ is a wff via the following recursive procedure: p, q, r are wff's; so $(p \vee q)$ is a wff; so $\neg(p \vee q)$ is a wff; finally $(\neg(p \vee q) \rightarrow r)$ is a wff. Compare this with the recursive procedure: p, q, r are wff's; $(p \vee q)$ is a wff; so $((p \vee q) \rightarrow r)$ is a wff; and finally $\neg((p \vee q) \rightarrow r)$ is a wff. ∎

Once a set S has been defined recursively, it is often possible and convenient to define a function f recursively on the set. The function definition can sort of ride along with the set definition in the following way:

(B) First define f for the members of the set specified in the basis for defining S.

(R) Whenever a member s of S is defined in terms of previously defined members according to the recursive clause for S, define $f(s)$ appropriately in terms of the values of f assigned to the previously defined members.

EXAMPLE 7   Let Σ^* be as defined in Example 2. Using that definition, we recursively define **length** for words in Σ^*.

(B) $\text{length}(\epsilon) = 0$.

(R) If $\text{length}(w)$ has been defined and $x \in \Sigma$, then $\text{length}(wx) = 1 + \text{length}(w)$.

For example, $\text{length}(c) = \text{length}(\epsilon c) = 1 + \text{length}(\epsilon) = 1 + 0 = 1$, hence $\text{length}(ca) = 1 + \text{length}(c) = 1 + 1 = 2$, and therefore $\text{length}(cab) = 1 + \text{length}(ca) = 1 + 2 = 3$. ∎

One must be extremely careful with this sort of recursive definition. If members of the set can be recursively constructed in more than one way, then we must be certain that the function f is **well-defined**, that is, that the assigned value $f(s)$ does not depend on the way we think of s as constructed.

EXAMPLE 8   Consider the polynomial functions defined in Example 1. Note that if f and g are the polynomial functions defined by $f(x) = 2x^3 - x$ and $g(x) = x^7 + 4x^4 - 8$, then the degree of $f + g$ is 7 and the degree of fg is 10. Because of examples like this, it is tempting to recursively define the **degree** deg of polynomial functions as follows:

(B) $\deg(c) = 0$ for constant functions c and $\deg(I) = 1$.

(R) If $\deg(f)$ and $\deg(g)$ have been defined, then $\deg(f + g) = \max\{\deg(f), \deg(g)\}$ and $\deg(fg) = \deg(f) + \deg(g)$.

This fine-looking definition *does not work*. For example, let $f(x) = x + 3$, $g(x) = x^2 + x - 1$ and $h(x) = 4 - x^2$ for $x \in \mathbb{R}$. Using (B) and (R), one can show that $\deg(f) = 1$ and $\deg(g) = \deg(h) = 2$, as is reasonable. Then by (R), $\deg(g + h)$ would equal 2, but $g + h$ is equal to the function f. The trouble is that f can be recursively constructed in more than one way and our definition of degree is ambiguous.

Incidentally, there seems to be no easy way to repair the unsuccessful definition above. One would have to prove that if f is a polynomial function there exists a *unique n* in $\mathbb{N}$ such that f can be written as

$$f(x) = \sum_{j=0}^{n} a_j x^j, \quad \text{i.e.,} \quad f = \sum_{j=0}^{n} a_j I^j,$$

where $a_0, a_1, \ldots, a_n$ are real numbers and $a_n \neq 0$. [Except for the uniqueness, this is done in Example 4 of § 6.4.] Then one can safely define $\deg(f) = n$. ∎

EXERCISES 3.6

1. Use the definition in Example 1 to show that the following functions on $\mathbb{R}$ are polynomial functions.
   (a) $f(x) = 4 - 3x + 5x^3$
   (b) $g(x) = \pi$
   (c) $h(x) = (x + 1)^3$

2. Use the definition in Example 2 to show that the following objects are in Σ^*, where Σ is the usual English alphabet.
   (a) cat   (b) math
   (c) zzpq   (d) aint

3. Add enough parentheses to the following algebraic expressions so that they are wff's as defined in Example 5(a).
   (a) $x + y + z$ (b) $x + y/z$
   (c) xyz (d) $(x + y)^{x+y}$

4. Add enough parentheses to the following algebraic expressions so that they are wff's as defined in Example 5(b).
   (a) $X + Y + Z$ (b) $X * (Y + Z)$
   (c) $X \uparrow 2 + 2 * X + 1$ (d) $X + Y/Z - Z * X$

5. Use the recursive definition of wff in Example 5 to show that the following are wff's.
   (a) $((x^2) + (y^2))$ (b) $(((X \uparrow 2) + (Y \uparrow 2)) \uparrow 2)$
   (c) $((X + Y) * (X - Y))$

6. Use the definition in Example 6 to show that the following are wff's in the propositional calculus.
   (a) $\neg(p \vee q)$ (b) $(\neg p \wedge \neg q)$
   (c) $((p \leftrightarrow q) \rightarrow ((r \rightarrow p) \vee q))$

7. Modify the definition in Example 6 so that the "exclusive or" connective $\oplus$ is allowable.

8. Show that if f and g are polynomial functions, so is $f - g$. See Example 1.

9. We recursively define **piecewise polynomial functions** in $\mathbb{R}$.
   (B) Polynomial functions are piecewise polynomial functions.
   (R) If f_1 and f_2 are piecewise polynomial functions and $a \in \mathbb{R}$, then so are f_3 and f_4 where

   $$f_3(x) = \begin{cases} f_1(x) & \text{for } x < a \\ f_2(x) & \text{for } x \geq a \end{cases}, \qquad f_4(x) = \begin{cases} f_1(x) & \text{for } x \leq a \\ f_2(x) & \text{for } x > a \end{cases}.$$

   Show that the following functions on $\mathbb{R}$ are piecewise polynomial functions.
   (a) The absolute value function $g_1(x) = |x|$.
   (b) The function $g_2(x) = |x|^3$.

(c) The function f where

$$f(x) = \begin{cases} 0 & \text{for} \quad x < 0 \\ 1 & \text{for} \quad x \geq 0 \end{cases}.$$

(d) $h(x) = |x| + |x - 1|$.

10. Let $\Sigma = \{a, b\}$. We recursively define a subset S of Σ^* as follows:

(B) $a \in S$ and $b \in S$,

(R) if w is in S, then awb is in S.

(a) List four different members of S.

(b) Is $aaabb$ in S? Explain.

(c) Is $aaabbb$ in S? Explain.

11. (a) Give a recursive definition for the following subset of $\mathbb{N} \times \mathbb{N}$:
$S = \{\langle m, n \rangle : n = 3m\}$.

(b) Use the recursive definition to show that $\langle 3, 9 \rangle \in S$.

12. (a) Give a recursive definition for the following subset of $\mathbb{N} \times \mathbb{N}$:
$T = \{\langle m, n \rangle : m \leq n\}$.

(b) Use the recursive definition to show that $\langle 3, 5 \rangle \in T$.

13. Let $\Sigma = \{a, b\}$ and let S be the set of words in Σ^* in which all the a's precede all the b's. For example, aab, $abbb$, a, b and even ϵ belong to S, but bab and ba do not.

(a) Give a recursive definition for the set S.

(b) Use the recursive definition to show that $abbb \in S$.

(c) Use the recursive definition to show that $aab \in S$.

14. Let $\Sigma = \{a, b\}$ and let T be the set of words in Σ^* which have exactly one a.

(a) Give a recursive definition for the set T.

(b) Use the recursive definition to show that $bbab \in T$.

15. We recursively define the **depth** of a wff as follows; see Example 5.

(B) Numerical constants and variables have depth 0.

(R) If $\text{depth}(f)$ and $\text{depth}(g)$ have been defined, then each of $(f + g)$, $(f - g)$, (fg), (f/g) and (f^g) has depth equal to $1 + \max\{\text{depth}(f), \text{depth}(g)\}$.

This turns out to be a well-defined definition; compare Example 8. Calculate the depth of the following algebraic expressions:

(a) $((x^2) + (y^2))$

(b) $(((X \uparrow 2) + (Y \uparrow 2)) \uparrow 2)$

(c) $((X + Y) * (X - Y))$

(d) $(((((X + Y) \uparrow 2) - (2 * (X * Y))) - (X \uparrow 2)) - (Y \uparrow 2))$

(e) $(((x + (x + y)) + z) - y)$

(f) $(((X * Y)/X) - (Y \uparrow 4))$

16. Let Σ^* be as defined in Example 2. We define the **reversal** $\tilde{w}$ of a word w in Σ^* recursively as follows:

(B) $\tilde{\epsilon} = \epsilon$,

(R) if $\tilde{w}$ has been defined and $x \in \Sigma$, and $x \in \Sigma$, then $\widetilde{wx} = x\tilde{w}$.

This is another well-defined definition.

(a) Prove that $\bar{\bar{x}} = x$ for all $x \in \Sigma$.

(b) Use this definition to find the reversal of *cab*.

(c) Use this definition to find the reversal of *abbaa*.

(d) If w_1 and w_2 are in Σ^*, what is $\overline{w_1 w_2}$ in terms of $\overline{w}_1$ and $\overline{w}_2$? What is $\overline{\overline{w}}_1$?

CHAPTER HIGHLIGHTS

Satisfy yourself that you can define each concept and describe each method.
Give at least one reason why the item was included in the chapter.
Think of at least one example of each concept and at least one situation in
  which the fact or method would be useful.

Concepts

function = map = mapping
  domain, codomain
  image of x, image of f = Im(f)
  graph of f
  one-to-one, onto, one-to-one correspondence
  well-defined
  special functions
      identity function, constant function, characteristic function, logarithm
        function, polynomial function
  composition of functions
  restriction, $f(A)$
  pre-image, $f^{-1}(B)$, $f^{-1}(y) = f^{-1}(\{y\})$
partition
sequence, finite sequence
"big-oh" notation
recursive definition [= basis + recursive clause]
  of a sequence, set, function
iterative and recursive calculation of sequences

Facts

Composition of functions is associative.
A function has an inverse if and only if it is one-to-one and onto.
Each function determines a partition of its domain.

Methods

Use of characteristic equation to solve $s_n = as_{n-1} + bs_{n-2}$.

4

MATRICES AND OTHER SEMIGROUPS

The main aim of this chapter is to study situations in which the associative law holds. We begin by discussing addition and multiplication of matrices.in § 4.1. In § 4.2 we fit our earlier results into the more general context of semigroups and monoids. Section 4.3 contains a study of the finite semigroups $\mathbb{Z}(p)$, and § 4.4 is an introduction to the semigroup $\mathscr{P}(\Sigma^*)$.

§ 4.1   Matrices

We first encountered matrices in § 0.3, where we showed that they provide a useful way of studying graphs. In § 1.4 we discussed matrix notation and also considered the operations of transposition and matrix addition. We now want to focus attention on addition, and to do so we require that the matrices we are adding have the same shape and have entries we can add. Accordingly, we consider fixed positive integers m and n and look at members of $\mathfrak{M}_{m,n}$, the set of all $m \times n$ matrices with real entries.

Recall that if $\mathbf{A} = [a_{ij}]$ and $\mathbf{B} = [b_{ij}]$, then $\mathbf{A} + \mathbf{B}$ has (i,j) entry $a_{ij} + b_{ij}$ for each i and j; we simply add corresponding entries of $\mathbf{A}$ and $\mathbf{B}$ to get $\mathbf{A} + \mathbf{B}$. Before listing properties of addition we give a little more notation. Let $\mathbf{0}$ represent the $m \times n$ matrix all entries of which are 0. [Context will always make plain what size this matrix is.] For $\mathbf{A}$ in $\mathfrak{M}_{m,n}$ the matrix $-\mathbf{A}$, called the **negative of A**, is obtained by negating each entry in $\mathbf{A}$. Thus if $\mathbf{A} = [a_{ij}]$, then $-\mathbf{A} = [-a_{ij}]$; equivalently $(-\mathbf{A})[i,j] = -\mathbf{A}[i,j]$.

Theorem   For all $\mathbf{A}$, $\mathbf{B}$ and $\mathbf{C}$ in $\mathfrak{M}_{m,n}$

(a) $\mathbf{A} + (\mathbf{B} + \mathbf{C}) = (\mathbf{A} + \mathbf{B}) + \mathbf{C}$ [associative law]
(b) $\mathbf{A} + \mathbf{B} = \mathbf{B} + \mathbf{A}$ [commutative law]
(c) $\mathbf{A} + \mathbf{0} = \mathbf{0} + \mathbf{A} = \mathbf{A}$ [additive identity]
(d) $\mathbf{A} + (-\mathbf{A}) = (-\mathbf{A}) + \mathbf{A} = \mathbf{0}$ [additive inverses]

Proof. These properties of matrix addition are reflections of corresponding properties of addition of real numbers and are easy to check. We check
(a) and leave the rest to Exercise 18.
    Say $\mathbf{A} = [a_{ij}]$, $\mathbf{B} = [b_{ij}]$ and $\mathbf{C} = [c_{ij}]$. The (i,j) entry of $\mathbf{B} + \mathbf{C}$ is
$b_{ij} + c_{ij}$, so the (i,j) entry of $\mathbf{A} + (\mathbf{B} + \mathbf{C})$ is $a_{ij} + (b_{ij} + c_{ij})$. Similarly, the
(i,j) entry of $(\mathbf{A} + \mathbf{B}) + \mathbf{C}$ is $(a_{ij} + b_{ij}) + c_{ij}$. Since addition of real numbers
is associative, corresponding entries of $\mathbf{A} + (\mathbf{B} + \mathbf{C})$ and $(\mathbf{A} + \mathbf{B}) + \mathbf{C}$ are
equal, and so the matrices are equal. ∎

    Since addition of matrices is associative, we can write $\mathbf{A} + \mathbf{B} + \mathbf{C}$
without causing ambiguity.
    Matrices can be multiplied by real numbers, which in this context are
often called **scalars**. Given $\mathbf{A}$ in $\mathfrak{M}_{m,n}$ and c in $\mathbb{R}$, $c\mathbf{A}$ is the $m \times n$ matrix
whose (i,j) entry is ca_{ij}; thus $(c\mathbf{A})[i,j] = c\mathbf{A}[i,j]$. This multiplication is called
scalar multiplication and $c\mathbf{A}$ is called the **scalar product**.

EXAMPLE 1   (a) If

$$\mathbf{A} = \begin{bmatrix} 2 & 1 & -3 \\ -1 & 0 & 4 \end{bmatrix},$$

then

$$2\mathbf{A} = \begin{bmatrix} 4 & 2 & -6 \\ -2 & 0 & 8 \end{bmatrix} \quad \text{and} \quad -7\mathbf{A} = \begin{bmatrix} -14 & -7 & 21 \\ 7 & 0 & -28 \end{bmatrix}.$$

(b) In general, the scalar product $(-1)\mathbf{A}$ is the negative $-\mathbf{A}$ of $\mathbf{A}$. ∎

    Addition and scalar multiplication of matrices are natural, but the
definition of the product of matrices we are about to make may appear
peculiar and unnatural. The standard linear algebra explanation for choosing this definition shows how matrices correspond to certain functions, called
linear transformations; then multiplication of matrices corresponds to composition of the linear transformations. An explanation can also be given in
terms of systems of linear equations. A treatment along these lines would
take us too far into linear algebra, so we will present the rule for multiplying
matrices without any algebraic justification.
    Recall, however, that in § 0.3 we saw how the powers $\mathbf{M}^n$ of a matrix $\mathbf{M}$
of a graph contain useful information about the graph. Here's another
example that involves two different graphs and matrices.

EXAMPLE 2 Consider the graphs and their matrices in Figure 1; note that they use the same set $\{v_1, v_2, v_3\}$ of vertices. Let $\mathbf{C}$ be the matrix whose (i,j) entry is the number of paths of length 2 from v_i to v_j whose first edge comes from the first

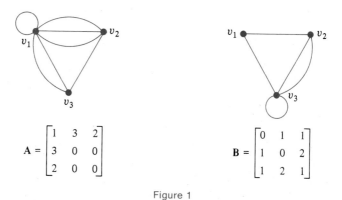

$$\mathbf{A} = \begin{bmatrix} 1 & 3 & 2 \\ 3 & 0 & 0 \\ 2 & 0 & 0 \end{bmatrix} \qquad \mathbf{B} = \begin{bmatrix} 0 & 1 & 1 \\ 1 & 0 & 2 \\ 1 & 2 & 1 \end{bmatrix}$$

Figure 1

graph and whose second edge comes from the second graph. As an example, let's calculate c_{13}, the number of paths like this from v_1 to v_3. The intermediate vertex can be v_1, v_2 or v_3:

Intermediate vertex v_i	Number of paths in first graph from v_1 to v_i	Number of paths in second graph from v_i to v_3	Total number of paths through v_i
v_1	$a_{11} = 1$	$b_{13} = 1$	$a_{11}b_{13} = 1$
v_2	$a_{12} = 3$	$b_{23} = 2$	$a_{12}b_{23} = 6$
v_3	$a_{13} = 2$	$b_{33} = 1$	$a_{13}b_{33} = 2$
			$\overline{9}$

The answer 9 is exactly

$$c_{13} = a_{11}b_{13} + a_{12}b_{23} + a_{13}b_{33},$$

i.e.,

$$\mathbf{C}[1, 3] = \mathbf{A}[1, 1]\mathbf{B}[1, 3] + \mathbf{A}[1, 2]\mathbf{B}[2, 3] + \mathbf{A}[1, 3]\mathbf{B}[3, 3].$$

Similar calculations yield the other entries of $\mathbf{C}$ and we find

$$\mathbf{C} = \begin{bmatrix} 5 & 5 & 9 \\ 0 & 3 & 3 \\ 0 & 2 & 2 \end{bmatrix}.$$

This is the "product matrix" $\mathbf{AB}$ about to be defined. ∎

In general, matrices **A** and **B** can be multiplied to get **AB** provided the number of columns of **A** equals the number of rows of **B**. So consider an $m \times n$ matrix **A** and an $n \times p$ matrix **B**. Thus **A** is in $\mathfrak{M}_{m,n}$ and **B** is in $\mathfrak{M}_{n,p}$. The **product AB** is the $m \times p$ matrix in $\mathfrak{M}_{m,p}$ defined by

$$c_{ik} = \sum_{j=1}^{n} a_{ij} b_{jk} \quad \text{for} \quad 1 \le i \le m \quad \text{and} \quad 1 \le k \le p.$$

In subscript-free notation

$$(\mathbf{AB})[i, k] = \sum_{j=1}^{n} \mathbf{A}[i, j] \mathbf{B}[j, k].$$

Schematically, the (i, k) entry of **AB** is obtained by multiplying terms of the ith row of **A** with corresponding terms of the kth column of **B** and summing. See Figure 2. One can calculate c_{ik} by mentally lifting the ith row of **A**, rotating it clockwise by 90°, placing it on top of the kth column of **B**, and then summing the products of the corresponding terms:

$$c_{ik} = a_{i1} b_{1k} + a_{i2} b_{2k} + \cdots + a_{in} b_{nk}.$$

For this calculation to make sense, the rows of **A** and the columns of **B** must have the same number of entries. If **A** is $m \times n$ and **B** is $r \times p$, the matrix product **AB** is only defined if $n = r$, in which case **AB** is an $m \times p$ matrix.

$$\begin{bmatrix} a_{11} & a_{12} & \cdots & a_{1n} \\ & \vdots & & \\ a_{i1} & a_{i2} & \cdots & a_{in} \\ & \vdots & & \\ a_{m1} & a_{m2} & \cdots & a_{mn} \end{bmatrix} \begin{bmatrix} b_{11} & b_{12} & \cdots & b_{1k} & \cdots & b_{1p} \\ b_{21} & b_{22} & \cdots & b_{2k} & \cdots & b_{2p} \\ \vdots & \vdots & & \vdots & & \vdots \\ b_{n1} & b_{n2} & \cdots & b_{nk} & \cdots & b_{np} \end{bmatrix} = \begin{bmatrix} c_{11} & c_{12} & \cdots & c_{1p} \\ c_{21} & c_{22} & \cdots & c_{2p} \\ \vdots & \vdots & c_{ik} & \vdots \\ c_{m1} & c_{m2} & \cdots & c_{mp} \end{bmatrix}$$

A   B   AB = C

Figure 2

EXAMPLE 3   Consider matrices and vectors

$$\mathbf{A} = \begin{bmatrix} 3 & -1 \\ -2 & 4 \end{bmatrix}, \quad \mathbf{B} = \begin{bmatrix} -1 & 0 & 3 \\ 2 & 1 & -5 \end{bmatrix},$$

$$\mathbf{v}_1 = [2 \quad -3 \quad 4] \quad \text{and} \quad \mathbf{v}_2 = \begin{bmatrix} 1 \\ -3 \end{bmatrix}.$$

(a) To calculate **AB**, we begin by mentally placing the first row of **A** over the first, second and third columns of **B** in turn. These three computations give the first row of **AB**:

$$\mathbf{AB} = \begin{bmatrix} -3 - 2 & 0 - 1 & 9 + 5 \end{bmatrix} = \begin{bmatrix} -5 & -1 & 14 \end{bmatrix}.$$

Using the second row of **A** in the same way, we obtain the second row of **AB**:

$$\mathbf{AB} = \begin{bmatrix} -5 & -1 & 14 \\ 10 & 4 & -26 \end{bmatrix}.$$

(b) The product **BA** is not defined, since **B** is a 2×3 matrix, **A** is a 2×2 matrix and $3 \neq 2$. Furthermore, our schematic procedure breaks down, since the rows of **B** have three terms and the columns of **A** have two terms; so it is not clear how we would mentally place the rows of **B** on top of the columns of **A**.

(c) We have

$$\mathbf{A}^2 = \mathbf{AA} = \begin{bmatrix} 3 & -1 \\ -2 & 4 \end{bmatrix} \begin{bmatrix} 3 & -1 \\ -2 & 4 \end{bmatrix} = \begin{bmatrix} 11 & -7 \\ -14 & 18 \end{bmatrix}.$$

(d) We have

$$\mathbf{Av}_2 = \begin{bmatrix} 3 & -1 \\ -2 & 4 \end{bmatrix} \begin{bmatrix} 1 \\ -3 \end{bmatrix} = \begin{bmatrix} 6 \\ -14 \end{bmatrix}.$$

(e) Neither $\mathbf{Bv}_1$ nor $\mathbf{v}_1\mathbf{B}$ is defined. But $\mathbf{v}_1\mathbf{B}^T$ and $\mathbf{Bv}_1^T$ are:

$$\mathbf{v}_1\mathbf{B}^T = \begin{bmatrix} 2 & -3 & 4 \end{bmatrix} \begin{bmatrix} -1 & 2 \\ 0 & 1 \\ 3 & -5 \end{bmatrix} = \begin{bmatrix} 10 & -19 \end{bmatrix}$$

and

$$\mathbf{Bv}_1^T = \begin{bmatrix} -1 & 0 & 3 \\ 2 & 1 & -5 \end{bmatrix} \begin{bmatrix} 2 \\ -3 \\ 4 \end{bmatrix} = \begin{bmatrix} 10 \\ -19 \end{bmatrix}.$$

The similarity of these two products is not an accident, as noted in Exercise 19. ∎

EXAMPLE 4 Consider

$$\mathbf{A} = \begin{bmatrix} 3 & -1 \\ -2 & 4 \end{bmatrix}, \qquad \mathbf{B} = \begin{bmatrix} 1 & 2 \\ -3 & 1 \end{bmatrix}, \qquad \mathbf{C} = \begin{bmatrix} 1 & 0 \\ 2 & 3 \end{bmatrix}.$$

(a) We have

$$\mathbf{AB} = \begin{bmatrix} 3 & -1 \\ -2 & 4 \end{bmatrix} \begin{bmatrix} 1 & 2 \\ -3 & 1 \end{bmatrix} = \begin{bmatrix} 6 & 5 \\ -14 & 0 \end{bmatrix}$$

and

$$\mathbf{BA} = \begin{bmatrix} 1 & 2 \\ -3 & 1 \end{bmatrix} \begin{bmatrix} 3 & -1 \\ -2 & 4 \end{bmatrix} = \begin{bmatrix} -1 & 7 \\ -11 & 7 \end{bmatrix}.$$

This example shows that multiplication of matrices is not commutative! Even when both **AB** and **BA** are defined they may or may not be equal.

(b) Since

$$AB = \begin{bmatrix} 6 & 5 \\ -14 & 0 \end{bmatrix},$$

we have

$$(AB)C = \begin{bmatrix} 6 & 5 \\ -14 & 0 \end{bmatrix}\begin{bmatrix} 1 & 0 \\ 2 & 3 \end{bmatrix} = \begin{bmatrix} 16 & 15 \\ -14 & 0 \end{bmatrix}.$$

On the other hand,

$$BC = \begin{bmatrix} 1 & 2 \\ -3 & 1 \end{bmatrix}\begin{bmatrix} 1 & 0 \\ 2 & 3 \end{bmatrix} = \begin{bmatrix} 5 & 6 \\ -1 & 3 \end{bmatrix}$$

and so

$$A(BC) = \begin{bmatrix} 3 & -1 \\ -2 & 4 \end{bmatrix}\begin{bmatrix} 5 & 6 \\ -1 & 3 \end{bmatrix} = \begin{bmatrix} 16 & 15 \\ -14 & 0 \end{bmatrix}.$$

These rather different computations show that $(AB)C = A(BC)$ in this case. This is *not* an accident, as we will see later in this section. ∎

EXAMPLE 5   Consider the $n \times n$ matrix

$$I = \begin{bmatrix} 1 & 0 & 0 & \cdots & 0 \\ 0 & 1 & 0 & \cdots & 0 \\ 0 & 0 & 1 & \cdots & 0 \\ \vdots & \vdots & \vdots & & \vdots \\ 0 & 0 & 0 & \cdots & 1 \end{bmatrix};$$

thus $I[i, i] = 1$ for $i = 1, 2, \ldots, n$ and $I[i, j] = 0$ for $i \neq j$. This special matrix is called the $n \times n$ **identity matrix**. Whenever we wish to specify its size explicitly we will denote it by I_n. Thus for example

$$I_2 = \begin{bmatrix} 1 & 0 \\ 0 & 1 \end{bmatrix} \quad \text{and} \quad I_4 = \begin{bmatrix} 1 & 0 & 0 & 0 \\ 0 & 1 & 0 & 0 \\ 0 & 0 & 1 & 0 \\ 0 & 0 & 0 & 1 \end{bmatrix}.$$

Now consider any $m \times n$ matrix A. Then the product AI_n is defined and

$$(AI_n)[i, k] = \sum_{j=1}^{n} A[i, j]I_n[j, k]$$

for $1 \leq i \leq m$ and $1 \leq k \leq n$. For $j \neq k$, we have $I_n[j, k] = 0$ and so this sum collapses to $A[i, k]I_n[k, k] = A[i, k]$. This is true for all i and k, and hence

(1) $AI_n = A$ for all $A \in \mathfrak{M}_{m, n}.$

Next consider an $n \times p$ matrix $\mathbf{B}$. Then $\mathbf{I}_n\mathbf{B}$ is defined and

$$(\mathbf{I}_n\mathbf{B})[i, k] = \sum_{j=1}^{n} \mathbf{I}_n[i, j]\mathbf{B}[j, k] = \mathbf{B}[i, k].$$

Thus

(2) $$\mathbf{I}_n\mathbf{B} = \mathbf{B} \quad \text{for all} \quad \mathbf{B} \in \mathfrak{M}_{n, p}.$$

Assertions (1) and (2) both apply to square $n \times n$ matrices and so

(3) $$\mathbf{AI}_n = \mathbf{I}_n\mathbf{A} = \mathbf{A} \quad \text{for all} \quad \mathbf{A} \in \mathfrak{M}_{n, n}. \quad \blacksquare$$

Consider a fixed n in $\mathbb{P}$. We claim that the set $\mathbb{R}^n$, the set $\mathfrak{M}_{n, 1}$ of column vectors and the set $\mathfrak{M}_{1, n}$ of row vectors are in one-to-one correspondence with each other. In fact,

$$f(\langle x_1, x_2, \ldots, x_n \rangle) = [x_1 \quad x_2 \quad \cdots \quad x_n]$$

defines a one-to-one correspondence f between $\mathbb{R}^n$ and $\mathfrak{M}_{1, n}$, and $\mathrm{TRANS}(\mathbf{A}) = \mathbf{A}^T$ provides a one-to-one correspondence between $\mathfrak{M}_{1, n}$ and $\mathfrak{M}_{n, 1}$. Composition of these correspondences provides a one-to-one correspondence between $\mathbb{R}^n$ and $\mathfrak{M}_{n, 1}$.

Multiplication of matrices is associative [that is, $(\mathbf{AB})\mathbf{C} = \mathbf{A}(\mathbf{BC})$ whenever either side makes sense] as we illustrated in Example 4. At the computational level, this fact is rather mysterious; see Exercise 22. The mystery vanishes when we relate matrices to linear transformations, since the composition of functions is associative. However, we have not explained the connection between matrices and linear transformations, and so we state the general associative law without proof.

Associative Law If $\mathbf{A}$ is an $m \times n$ matrix, $\mathbf{B}$ is an $n \times p$ matrix and $\mathbf{C}$ is a $p \times q$ matrix, then
for Matrices $(\mathbf{AB})\mathbf{C} = \mathbf{A}(\mathbf{BC})$.

Since multiplication of matrices is associative, we can write $\mathbf{ABC}$ without ambiguity. Also, powers such as $\mathbf{A}^3 = \mathbf{AAA}$ are unambiguous.

To conclude this section we stress again that although *multiplication of matrices is associative* it is *not commutative*: $\mathbf{AB}$ need not equal $\mathbf{BA}$ even if both products are defined. Example 4(a) illustrates noncommutativity. Examples of other laws of arithmetic not satisfied by matrices appear in Exercises 17 and 20(b).

EXERCISES 4.1

1. Let

$$\mathbf{A} = \begin{bmatrix} 1 & 2 & 4 \\ 3 & 0 & 2 \end{bmatrix} \quad \text{and} \quad \mathbf{B} = \begin{bmatrix} 2 & 1 \\ -1 & 0 \\ -2 & 3 \end{bmatrix}.$$

Find the following when they exist.
(a) **AB**   (b) **BA**   (c) **ABA**
(d) **A** + **B**^{T} (e) 3**A**^{T} − 2**B**   (f) (**AB**)^2

2. Let

$$C = \begin{bmatrix} 1 \\ 0 \\ 1 \end{bmatrix}$$

and let **A** and **B** be as in Exercise 1. Find the following when they exist.
(a) **AC**   (b) **BC**   (c) **C**^{2}
(d) **C**^{T}**C**   (e) **CC**^{T} (f) 73**C**

3. Let

$$A = \begin{bmatrix} 3 & -4 & 3 & 1 \\ 2 & 0 & 1 & -2 \\ -1 & 1 & 2 & 0 \end{bmatrix} \quad \text{and} \quad B = \begin{bmatrix} -1 & 1 & 0 \\ 1 & 2 & 1 \\ 0 & 1 & -1 \end{bmatrix}.$$

Find the following when they exist.
(a) **A**^{2} (b) **B**^{2} (c) **AB**
(d) **BA**   (e) **BA**^{T} (f) **A**^{T}**B**

4. Let **A** and **B** be as in Exercise 3, and let **v** = [−2   1   −1]. Find the following
   when they exist.
(a) **vA**   (b) **vB**   (c) **Bv**^{T}
(d) (**vB**)^T (e) 5(**vB**)^T − 3**Bv**^{T}

5. (a) Calculate both (**AB**)**C** and **A**(**BC**) for

$$A = \begin{bmatrix} -1 & 4 \\ 2 & 5 \end{bmatrix}, \quad B = \begin{bmatrix} 1 & 1 \\ 0 & 1 \end{bmatrix} \quad \text{and} \quad C = \begin{bmatrix} 2 & -1 \\ 1 & 3 \end{bmatrix}.$$

   (b) Calculate both **B**(**AC**) and (**BA**)**C**.

6. Let **A**, **B** and **C** be as in Exercise 5. Calculate
   (a) both **AB** and **BA**
   (b) both **AC** and **CA**
   (c) **A**^{2}

7. Let

$$A = \begin{bmatrix} 3 & -1 \\ 2 & 1 \\ -2 & 4 \end{bmatrix}, \quad B = \begin{bmatrix} 1 & 2 \\ 0 & 1 \end{bmatrix} \quad \text{and} \quad C = \begin{bmatrix} -1 & 3 \\ 2 & 1 \end{bmatrix}.$$

   (a) Calculate **A**(**BC**) and (**AB**)**C**.
   (b) Calculate **A**(**B**^{2}) and (**AB**)**B**.

8. For **A** and **B** in $\mathfrak{M}_{m,n}$ let **A** − **B** = **A** + (−**B**). Show that
   (a) (**A** − **B**) + **B** = **A**
   (b) −(**A** − **B**) = **B** − **A**
   (c) (**A** − **B**) − **C** ≠ **A** − (**B** − **C**) in general

9. Consider **A**, **B** in $\mathfrak{M}_{m,n}$ and a, b, c in $\mathbb{R}$. Show
   (a) $c(a\mathbf{A} + b\mathbf{B}) = (ca)\mathbf{A} + (cb)\mathbf{B}$
   (b) $-a\mathbf{A} = (-a)\mathbf{A} = a(-\mathbf{A})$
   (c) $(a\mathbf{A})^T = a\mathbf{A}^T$
   (d) $(a\mathbf{A})\mathbf{B} = a(\mathbf{A}\mathbf{B}) = \mathbf{A}(a\mathbf{B})$

10. Let **A** be a square matrix. Give a recursive definition for $\mathbf{A}^n$.

11. Let

$$\mathbf{A} = \begin{bmatrix} 1 & 0 \\ 0 & 0 \end{bmatrix}, \quad \mathbf{B} = \begin{bmatrix} 0 & 1 \\ 0 & 0 \end{bmatrix} \quad \text{and} \quad \mathbf{C} = \begin{bmatrix} 0 & 0 \\ 1 & 0 \end{bmatrix}.$$

   Calculate
   (a) **AB**   (b) **BA**   (c) **AC**
   (d) **CA**   (e) **BC**   (f) **CB**

12. Let **A**, **B** and **C** be as in Exercise 11. Calculate
   (a) $\mathbf{A}^n$ for all $n \in \mathbb{P}$,
   (b) $\mathbf{B}^n$ for all $n \in \mathbb{P}$,
   (c) $\mathbf{C}^n$ for all $n \in \mathbb{P}$.

13. Let

$$\mathbf{A} = \begin{bmatrix} 1 & 0 \\ 1 & 1 \end{bmatrix}.$$

   (a) Calculate $\mathbf{A}^n$ for $n = 1, 2, 3, 4$.
   (b) Guess a general formula for $\mathbf{A}^n$ and prove your guess is correct by induction.

14. Let

$$\mathbf{M} = \begin{bmatrix} 1 & 1 \\ 1 & 0 \end{bmatrix}$$

   and let FIB be the Fibonacci sequence introduced in Example 3 of § 3.4. Prove that

$$\mathbf{M}^n = \begin{bmatrix} \text{FIB}(n) & \text{FIB}(n-1) \\ \text{FIB}(n-1) & \text{FIB}(n-2) \end{bmatrix} \quad \text{for} \quad n \geq 2.$$

15. (a) Let

$$\mathbf{A} = \begin{bmatrix} a & 0 \\ 0 & a \end{bmatrix}$$

   for some fixed a in $\mathbb{R}$. Show that $\mathbf{AB} = \mathbf{BA}$ for all **B** in $\mathfrak{M}_{2,2}$.
   (b) Consider a fixed matrix **A** in $\mathfrak{M}_{2,2}$ that satisfies $\mathbf{AB} = \mathbf{BA}$ for all $\mathbf{B} \in \mathfrak{M}_{2,2}$. Show that

$$\mathbf{A} = \begin{bmatrix} a & 0 \\ 0 & a \end{bmatrix} \quad \text{for some} \quad a \in \mathbb{R}.$$

   Hint: Write

$$\mathbf{A} = \begin{bmatrix} a & b \\ c & d \end{bmatrix} \quad \text{and try} \quad \mathbf{B} = \begin{bmatrix} 1 & 0 \\ 0 & 0 \end{bmatrix} \quad \text{and} \quad \begin{bmatrix} 0 & 1 \\ 0 & 0 \end{bmatrix}.$$

16. Let

$$M = \begin{bmatrix} 2 & 1 & 1 \\ 1 & 0 & 2 \\ 1 & 2 & 0 \end{bmatrix}.$$

    (a) Draw a graph having **M** as its matrix, as in § 0.3.

    (b) Calculate the matrix that counts the number of paths of length 2 between pairs of vertices.

17. Find 2×2 matrices that show that $(\mathbf{A} + \mathbf{B})(\mathbf{A} - \mathbf{B}) = \mathbf{A}^2 - \mathbf{B}^2$ does not generally hold.

18. Prove (b), (c) and (d) of the theorem.

19. Show that if **A** is an $m \times n$ matrix and **B** is an $n \times p$ matrix, then $(\mathbf{AB})^T = \mathbf{B}^T\mathbf{A}^T$. Note that both sides of the equality represent $p \times m$ matrices.

20. (a) Prove the cancellation law for $\mathfrak{M}_{m,n}$ under addition, i.e., prove that if **A**, **B**, **C** are in $\mathfrak{M}_{m,n}$ and $\mathbf{A} + \mathbf{C} = \mathbf{B} + \mathbf{C}$, then $\mathbf{A} = \mathbf{B}$.

    (b) Show that the cancellation law for $\mathfrak{M}_{n,n}$ under multiplication fails, i.e., show that $\mathbf{AC} = \mathbf{BC}$ need not imply $\mathbf{A} = \mathbf{B}$.

21. (a) Let **A** and **B** be $m \times n$ matrices and let **C** be an $n \times p$ matrix. Show that the distributive law holds: $(\mathbf{A} + \mathbf{B})\mathbf{C} = \mathbf{AC} + \mathbf{BC}$.

    (b) Verify the distributive law $\mathbf{A}(\mathbf{B} + \mathbf{C}) = \mathbf{AB} + \mathbf{AC}$. First specify the sizes of the matrices for which this makes sense.

22. (a) Show directly that $\mathbf{A}(\mathbf{BC}) = (\mathbf{AB})\mathbf{C}$ for matrices **A**, **B** and **C** in $\mathfrak{M}_{2,2}$.

    (b) Did you enjoy part (a)? If yes, give a direct proof of the general associative law for matrices.

23. Let **A** and **B** be $n \times n$ matrices such that $\mathbf{AB} = \mathbf{BA}$.

    (a) Prove that $\mathbf{BA}^k = \mathbf{A}^k\mathbf{B}$ for all $k \in \mathbb{P}$.

    (b) Prove that $(\mathbf{AB})^k = \mathbf{A}^k\mathbf{B}^k$ for all $k \in \mathbb{P}$.

24. Let **A** and **B** be $n \times n$ matrices such that $\mathbf{AB} = \mathbf{BA} = \mathbf{0}$. Prove that $(\mathbf{A} + \mathbf{B})^k = \mathbf{A}^k + \mathbf{B}^k$ for all $k \in \mathbb{P}$.

§ 4.2   Semigroups

We have already seen several ways of combining mathematical objects, such as numbers and matrices, two at a time. In this section we look at associative operations for combining two elements of a set, and we consider what sorts of general statements are true in such a context. One merit of such a general abstract approach is that it allows one to focus on fundamental properties, without being distracted by the details of different settings. It also means that one can prove theorems once and for all, rather than again and again in various situations. Later on, in Chapter 11, we will give a much more complete treatment of associative operations.

Consider a nonempty set S. A function from $S \times S$ into S is sometimes called a **binary operation**. In this case, we usually have in mind some sort of addition or multiplication and we use notation such as $+$ or $*$ instead of functional notation. Moreover, we normally place the operation symbol between the variables. Thus we would write $s_1 + s_2$, not $+(\langle s_1, s_2 \rangle)$, for example.

EXAMPLE 1   (a) Addition on $\mathbb{R}$ is a binary operation. Formally, the operation is the function $\mathrm{SUM} : \mathbb{R} \times \mathbb{R} \to \mathbb{R}$ defined by $\mathrm{SUM}(\langle x, y \rangle) = x + y$. We will usually use the symbol $+$, not the formal functional notation SUM.

(b) The function SUM also gives a binary operation on $\mathbb{Z}$, $\mathbb{N}$ and $\mathbb{Q}$. All it takes for SUM to give a binary operation on a subset S of $\mathbb{R}$ is for it to map $S \times S$ into S, i.e.,

$$x + y \in S \quad \text{whenever} \quad x, y \in S.$$

It is common practice to say that S is **closed under** $+$, or under SUM, in this case. Note that $\mathbb{Z}$, $\mathbb{N}$ and $\mathbb{Q}$ are closed under $+$. The subset $\{1, 2, 3, 4, 5\}$ of $\mathbb{R}$ is not closed since, for example, $3 + 3 \notin \{1, 2, 3, 4, 5\}$; $+$ is not a binary operation on this subset.

(c) Multiplication $*$ on $\mathbb{R}$ is a binary operation. Formally, the operation is the function $\mathrm{PROD} : \mathbb{R} \times \mathbb{R} \to \mathbb{R}$ defined by $\mathrm{PROD}(\langle x, y \rangle) = x * y$. Multiplication is also a binary operation on $\mathbb{Z}$, $\mathbb{N}$ and $\mathbb{Q}$.

(d) Let $\mathfrak{M}_{m,n}$ be the set of all $m \times n$ matrices. Matrix addition $+$ is a binary operation on $\mathfrak{M}_{m,n}$ since

$$A, B \in \mathfrak{M}_{m,n} \quad \text{imply} \quad A + B \in \mathfrak{M}_{m,n}.$$

(e) Matrix multiplication is a binary operation on the set $\mathfrak{M}_{n,n}$ of $n \times n$ matrices.

(f) Let $\mathscr{P}(S)$ be the set of all subsets of some set S. Then union $\cup$ and intersection $\cap$ are binary operations on $\mathscr{P}(S)$. Formally, these are the functions UNION and INTER mapping $\mathscr{P}(S) \times \mathscr{P}(S)$ into $\mathscr{P}(S)$ and defined by

$$\mathrm{UNION}(\langle A, B \rangle) = A \cup B \quad \text{and} \quad \mathrm{INTER}(\langle A, B \rangle) = A \cap B. \quad \blacksquare$$

We will sometimes use a little square $\square$ for a generic binary operation. This is a neutral symbol that does not suggest addition or multiplication or any other familiar operation like union or intersection. So consider a binary operation $\square$ on a nonempty set S. The operation $\square$ is said to be **associative** if

$$s_1 \square (s_2 \square s_3) = (s_1 \square s_2) \square s_3 \quad \text{for all} \quad s_1, s_2, s_3 \in S.$$

A set S with an associative binary operation $\square$ is said to be a **semigroup under** $\square$. To be brief, we will say that $(S, \square)$ is a **semigroup**. If the operation is understood, we may simply call S a **semigroup**. All the binary operations in Example 1 are associative and so all the sets in Example 1 are semigroups under the indicated binary operations. Of course not all useful binary operations are associative—subtraction and division are not, for example —but most of them are. By limiting our study to associative operations we will impose enough structure to be able to state some general facts without excluding the main examples. We will look at some operations in detail after a few basic definitions.

A semigroup $(S, \square)$ is **commutative** if

$$s_1 \square s_2 = s_2 \square s_1 \quad \text{for all} \quad s_1, s_2 \in S.$$

An element e in S will be called an **identity for** S provided

$$s \square e = e \square s = s \quad \text{for all} \quad s \in S.$$

A semigroup with an identity is called a **monoid**. A monoid can't have more than one identity, because if e and f are identities then $e = e \square f = f$. If S has an identity e, an element t such that $s \square t = e = t \square s$ is called an **inverse** of s, often written as s^{-1}. Inverses are unique, when they exist, because if r and t are inverses of s then $s \square t = e$ and $r \square s = e$ and so $r = r \square e = r \square (s \square t) = (r \square s) \square t = e \square t = t$. A monoid in which every element has an inverse is called a **group**. Thus a group is a semigroup with an identity and inverses.

In view of associativity, the expression $s_1 \square s_2 \square s_3$ is unambiguous in a semigroup.

EXAMPLE 2   (a) $(\mathbb{R}, +)$ is a commutative group. The identity is 0 since

$$x + 0 = 0 + x = x \quad \text{for all} \quad x \in \mathbb{R}.$$

For emphasis and clarity, 0 is often referred to as the **additive identity in** $\mathbb{R}$. Inverses with respect to the operation $+$ are called **additive inverses** and written $-x$ not x^{-1}, since the inverse law in this case says

for each $x \in \mathbb{R}$ there is an element $-x \in \mathbb{R}$ such that
$$x + (-x) = (-x) + x = 0.$$

(b) $(\mathbb{R}, *)$ is a commutative monoid. The identity [or **multiplicative identity**] is 1 since

$$x * 1 = 1 * x = x \quad \text{for all} \quad x \in \mathbb{R}.$$

Each nonzero x in $\mathbb{R}$ has an inverse [or **multiplicative inverse**], i.e., a number x^{-1} such that

$$x * x^{-1} = x^{-1} * x = 1.$$

Sometimes x^{-1} is written $1/x$ and is called the **reciprocal** of x. $(\mathbb{R}, *)$ is not quite a group because 0 has no multiplicative inverse: $0 * x$ never equals 1.

(c) Since the product of nonzero real numbers is nonzero, $*$ is also a binary operation on $\mathbb{R} \setminus \{0\}$. Moreover, $(\mathbb{R} \setminus \{0\}, *)$ is a bona fide commutative group. ∎

EXAMPLE 3   (a) $(\mathbb{Q}, +)$ and $(\mathbb{Q}\backslash\{0\}, *)$ are commutative groups. Note that $-x$ is in $\mathbb{Q}$ whenever x is in $\mathbb{Q}$ and that x^{-1} is in $\mathbb{Q}$ whenever x is a nonzero number in $\mathbb{Q}$.

(b) $(\mathbb{Z}, +)$ is a commutative group but $(\mathbb{Z}\backslash\{0\}, *)$ is not a group, since $n \in \mathbb{Z}\backslash\{0\}$ does not imply that $n^{-1} \in \mathbb{Z}\backslash\{0\}$. In fact, if n and n^{-1} are both in $\mathbb{Z}\backslash\{0\}$, then n must be 1 or -1. We can assert that $(\mathbb{Z}\backslash\{0\}, *)$ is a commutative monoid; its identity is 1.

(c) $(\mathbb{N}, +)$ is not a group, since $n \in \mathbb{N}$ does not imply that $-n \in \mathbb{N}$. $(\mathbb{N}, +)$ is a commutative monoid with additive identity 0. ∎

EXAMPLE 4   (a) The set $\mathfrak{M}_{m,n}$ of all $m \times n$ matrices is a commutative group under addition. This fact is spelled out in the theorem in § 4.1.

(b) The set $\mathfrak{M}_{n,n}$ is a monoid under multiplication, with identity $\mathbf{I}_n$. The associative law is discussed at the end of § 4.1. Except for the trivial case $n = 1$, this monoid is not commutative: $\mathbf{AB}$ does not necessarily equal $\mathbf{BA}$. Some, but not all, matrices in $\mathfrak{M}_{n,n}$ have inverses. See Exercises 12 and 13. ∎

EXAMPLE 5   (a) Let $\mathscr{P}(S)$ be the set of all subsets of some set S. With the operation $\cup$, $\mathscr{P}(S)$ is a commutative semigroup with identity $\varnothing$; see laws 1a, 2a and 5a in Table 1 of § 1.2. Only the empty set itself has an inverse, since $A \cup B \neq \varnothing$ whenever $A \neq \varnothing$.

(b) $\mathscr{P}(S)$ is also a semigroup under the operations $\cap$ and $\oplus$; see Exercises 4 and 5. ∎

EXAMPLE 6   Consider a set $T = \{a, b, \ldots\}$ with at least two members. The set $\mathrm{FUN}(T, T)$ of all functions mapping T into T is a semigroup under composition; the associative law is discussed in § 3.1. The identity function 1_T on T is the identity for this semigroup, since

$$(1_T \circ f)(t) = 1_T(f(t)) = f(t) \quad \text{for all} \quad t \in T,$$

so that $1_T \circ f = f$ and similarly $f \circ 1_T = f$. This monoid is not commutative. For example, let f and g be the constant functions defined by $f(t) = a$ and $g(t) = b$ for all t in T. Then $(f \circ g)(t) = f(g(t)) = a$ for all t and $(g \circ f)(t) = b$ for all t. That is, $f \circ g = f \neq g = g \circ f$. For a function f in $\mathrm{FUN}(T, T)$ we write f^n for the composition of f with itself n times. Thus $f^2 = f \circ f$, $f^3 = f \circ f \circ f$, etc. Also, we agree that $f^0 = 1_T$. ∎

EXAMPLE 7   (a) For a nonempty set T, let $\mathrm{PERM}(T)$ be the set of all permutations of T, i.e., all one-to-one functions from T onto T. Then $\mathrm{PERM}(T)$ is a subset of $\mathrm{FUN}(T, T)$, and it is a semigroup under composition because

$$f, g \in \mathrm{PERM}(T) \quad \text{imply} \quad g \circ f \in \mathrm{PERM}(T)$$

[Exercise 9 of § 3.2]. This semigroup has an identity, namely 1_T, and every element has an inverse, by the theorem in § 3.2. Thus $\mathrm{PERM}(T)$ is a group.

(b) If T is a finite set with n elements, PERM(T) is often denoted by S_n and called "the **symmetric group** on n letters."

(c) The symmetric group S_5 of all permutations of $T = \{1, 2, 3, 4, 5\}$ contains the permutation f defined in Figure 1(a). A schematic picture of f is given in Figure 1(b). The permutations 1_T, f, f^2, f^3, f^4 and f^5 are all different, and $f^6 = 1_T$. The set consisting of these six permutations is a subset of S_5 that is itself a group. ∎

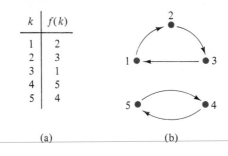

k	$f(k)$
1	2
2	3
3	1
4	5
5	4

(a)   (b)

Figure 1

EXAMPLE 8   (a) Let T be a nonempty set. The set FUN(T, $\mathbb{N}$) of functions mapping T into $\mathbb{N}$ is a semigroup under the operation $+$ defined by

$$(f + g)(t) = f(t) + g(t) \quad \text{for all} \quad t \in T.$$

[Observe, for instance, that $(f + g) + h$ is defined by $((f + g) + h)(t) = (f + g)(t) + h(t) = (f(t) + g(t)) + h(t)$ for all $t \in T$.] The identity element of FUN(T, $\mathbb{N}$) is the constant function z defined by $z(t) = 0$ for all $t \in T$. An inverse for f would be a function g with $f(t) + g(t) = 0$ for all t, i.e., with $g(t) = -f(t)$ for all t. Since f and g have values in $\mathbb{N}$, no such g exists if $f(t) > 0$ for some t.

(b) Replace $\mathbb{N}$ by $\mathbb{Z}$ in (a). Now inverses exist; the inverse of f is the function $-f$ defined by $(-f)(t) = -f(t)$ for all $t \in T$. Thus FUN(T, $\mathbb{Z}$) is a group under $+$.

(c) The set FUN($\mathbb{R}$, $\mathbb{R}$) of real-valued functions on $\mathbb{R}$ is a group under the operation $+$ defined by $(f + g)(x) = f(x) + g(x)$ for all $x \in \mathbb{R}$. It is a monoid under the operation $\cdot$ defined by $(f \cdot g)(x) = f(x) * g(x)$ for all $x \in \mathbb{R}$. The identity now is the constant function c_1 where $c_1(x) = 1$ for all $x \in \mathbb{R}$. Both the additive group and the multiplicative monoid are commutative, since $\mathbb{R}$ itself has commutative addition and multiplication. ∎

EXAMPLE 9   Let Σ be an alphabet. The set Σ^* of all words is informally defined in § 1.1 and is recursively defined in Example 2 of § 3.6. Two words w_1 and w_2 in Σ^* are multiplied by **concatenation**; that is, $w_1 w_2$ is the word obtained by placing the string w_2 right after the string w_1. In other words, if $w_1 = a_1 a_2 \cdots a_m$

and $w_2 = b_1 b_2 \cdots b_n$ where the a_j's and b_k's are in Σ, then $w_1 w_2 = a_1 a_2 \cdots a_m b_1 b_2 \cdots b_n$. For example, if $w_1 = cat$ and $w_2 = nip$, then $w_1 w_2 = catnip$ and $w_2 w_1 = nipcat$. Multiplication by the empty word ϵ leaves the word unchanged:

$$w\epsilon = \epsilon w = w \quad \text{for all} \quad w \in \Sigma^*.$$

It is evident from the definition that concatenation is an associative binary operation. Since the empty word ϵ serves as an identity for Σ^*, Σ^* is a monoid. ∎

Consider once again an arbitrary semigroup $(S, \square)$. For $s \in S$ and $n \in \mathbb{P}$, we continue a familiar convention and write s^n for the $\square$-product of s with itself n times. A precise recursive definition for s^n is:

(B) $s^1 = s$,
(R) $s^{n+1} = s^n \square s$ for $n \in \mathbb{P}$.

If $(S, \square)$ is a monoid, i.e., if S has an identity e, then we also define $s^0 = e$. Alternatively, we could modify the recursive definition to read:

(B) $s^0 = e$,
(R) $s^{n+1} = s^n \square s$ for $n \in \mathbb{N}$.

With these definitions we can prove, once and for all, some familiar algebraic properties of exponents.

Theorem Let $(S, \square)$ be a semigroup. For $s \in S$ and $m, n \in \mathbb{P}$, we have

(a) $s^m \square s^n = s^{m+n}$
(b) $(s^m)^n = s^{mn}$.

For a monoid, these formulas hold for $m, n \in \mathbb{N}$.

Proof. (a) We fix $m \in \mathbb{P}$ and prove

(1) $$s^m \square s^n = s^{m+n} \quad \text{for} \quad n \in \mathbb{P},$$

by induction on n. For $n = 1$, this follows from the recursive definition. Assume the equality is valid for some n. Then

$$
\begin{aligned}
s^m \square s^{n+1} &= s^m \square (s^n \square s) &&\text{definition of } s^{n+1} \\
&= (s^m \square s^n) \square s &&\text{associativity} \\
&= s^{m+n} \square s &&\text{inductive assumption} \\
&= s^{(m+n)+1} &&\text{definition of } s^{(m+n)+1} \\
&= s^{m+(n+1)}.
\end{aligned}
$$

The Principle of Mathematical Induction now shows that (1) holds for all n in $\mathbb{P}$.

(b) Again we fix m and prove

(2) $$(s^m)^n = s^{mn} \quad \text{for} \quad n \in \mathbb{P},$$

by induction on n. For $n = 1$, this is clear since $(s^m)^1 = s^m$ by the basis part of the recursive definition. If the equality holds for n, then

$$
\begin{aligned}
(s^m)^{n+1} &= ((s^m)^n) \square s^m & \text{definition} \\
&= s^{mn} \square s^m & \text{inductive assumption} \\
&= s^{mn+m} & \text{part (a)} \\
&= s^{m(n+1)}.
\end{aligned}
$$

Hence (2) holds for all n by induction.

If S is a monoid and m or n is 0, then (a) and (b) are simple to verify. ∎

We have treated the generic operation $\square$ the same way we would treat multiplication. Let's see how the foregoing would look when $\square$ is replaced by $+$. The general product s^n is replaced by the general sum ns and the first recursive definition becomes:

(B) $1s = s$,

(R) $(n+1)s = ns + s$ for $n \in \mathbb{P}$.

The theorem now says

(a) $ms + ns = (m+n)s$

(b) $n(ms) = (mn)s$

for $s \in S$ and $m, n \in \mathbb{P}$. It might be a useful exercise to rewrite the proof of the last theorem using this notation.

EXERCISES 4.2

1. $(\mathbb{N}, *)$ is a semigroup.
   (a) Is this semigroup commutative?
   (b) Is there an identity for this semigroup?
   (c) If yes, do inverses exist? If yes, specify them.
   (d) Is this a monoid? A group?

2. $(\mathbb{P}, +)$ is a semigroup. Repeat Exercise 1.

3. $(\mathbb{P}, *)$ is a semigroup. Repeat Exercise 1.

4. Let $\mathscr{P}(S)$ be the set of all subsets of a nonempty set S. $\mathscr{P}(S)$ is a semigroup with respect to $\cap$. Repeat Exercise 1 for this semigroup.

5. $\mathscr{P}(S)$ is a semigroup with respect to symmetric difference $\oplus$. Repeat Exercise 1 for this semigroup.

6. Repeat Exercise 1 for the semigroup $\text{PERM}(T)$ of permutations of T.

7. Repeat Exercise 1 for the semigroup $\text{FUN}(T, \mathbb{R})$ of real-valued functions on T under addition.

8. Repeat Exercise 1 for the semigroup $\text{FUN}(T, \mathbb{R})$ under multiplication.

9. The set $\mathfrak{M}_{2,2}$ of 2×2 matrices is a semigroup with respect to matrix multiplication.
   (a) Is this semigroup commutative?
   (b) Is there an identity for this semigroup?
   (c) If yes, do inverses exist?
   (d) Is this semigroup a monoid? A group?

10. Let $\Sigma = \{a, b, c, d\}$ and consider the words: $w_1 = bad$, $w_2 = cab$ and $w_3 = abcd$.
    (a) Determine $w_1 w_2$, $w_2 w_1$, $w_2 w_3 w_2 w_1$ and $w_3 w_2 w_3$.
    (b) Determine w_1^2, w_2^3 and ϵ^4.

11. Let Σ be the usual English alphabet and consider the words: $w_1 = break$, $w_2 = fast$, $w_3 = lunch$ and $w_4 = food$.
    (a) Determine ϵw_1, $w_2 \epsilon$, $w_2 w_4$, $w_3 w_1$ and $w_4 \epsilon w_4$.
    (b) Compare $w_1 w_2$ and $w_2 w_1$.
    (c) Determine w_2^2, w_4^2, $w_2^2 w_4 w_1^2$ and ϵ^{73}.

12. Show that a 2×2 matrix $\mathbf{A} = \begin{bmatrix} a & b \\ c & d \end{bmatrix}$ has an inverse if and only if $ad - bc \neq 0$, in which case the inverse is

$$\mathbf{A}^{-1} = \frac{1}{ad - bc} \begin{bmatrix} d & -b \\ -c & a \end{bmatrix}.$$

   Hint: Try to solve $\begin{bmatrix} a & b \\ c & d \end{bmatrix} \begin{bmatrix} x & y \\ z & w \end{bmatrix} = \begin{bmatrix} 1 & 0 \\ 0 & 1 \end{bmatrix}$ for x, y, z, w.

13. Use Exercise 12 to determine which of the following matrices have inverses. Find the inverses when they exist and check your answers.

   (a) $\mathbf{I} = \begin{bmatrix} 1 & 0 \\ 0 & 1 \end{bmatrix}$
   (b) $\mathbf{A} = \begin{bmatrix} 1 & 1 \\ 0 & 1 \end{bmatrix}$
   (c) $\mathbf{B} = \begin{bmatrix} 1 & 1 \\ 1 & 1 \end{bmatrix}$

   (d) $\mathbf{C} = \begin{bmatrix} 2 & -3 \\ 5 & 8 \end{bmatrix}$
   (e) $\mathbf{D} = \begin{bmatrix} 0 & 1 \\ 1 & 0 \end{bmatrix}$

14. Verify the claims in Example 7. That is,
    (a) Determine f^2, f^3, f^4 and f^5, either by giving a table as in Figure 1(a) or by giving a schematic picture as in Figure 1(b).
    (b) Show that $f^6 = 1_T$.
    (c) Give inverses of f, f^2, f^3, f^4 and f^5.

15. Give examples of nonidentity permutations g and h in S_5 [see Example 7] with $g^3 = 1_T$ and $h^5 = 1_T$.

16. Show that $\mathbb{R}^+ = \{x \in \mathbb{R} : x > 0\}$ is not a semigroup with the binary operation $\langle x, y \rangle \to x/y$.

17. (a) Convince yourself that $\mathbb{N}$ is a semigroup under the binary operation $\langle m, n \rangle \to \min\{m, n\}$ and also under $\langle m, n \rangle \to \max\{m, n\}$.
    (b) Are the semigroups in part (a) monoids?

18. (a) Show that $\mathbb{P}$ is a semigroup with respect to $\langle m, n \rangle \to \gcd(m, n)$ where $\gcd(m, n)$ represents the greatest common divisor of m and n.

(b) Show that $\mathbb{P}$ is a semigroup with respect to $\langle m, n \rangle \rightarrow \mathrm{lcm}(m, n)$ where $\mathrm{lcm}(m, n)$ represents the least common multiple of m and n.

(c) Are the semigroups in parts (a) and (b) monoids?

19. Consider a binary operation on a nonempty set S and use functional notation f for the operation.

(a) State the associative law for f.

(b) State the commutative law for f.

20. Let $(S, \square)$ be a group. How would you define s^n for s in S and negative integers n?

21. Prove that an element of a monoid cannot have two different inverses.

22. Fill in the details of the argument that $\mathrm{FUN}(\mathbb{R}, \mathbb{R})$ is a commutative group under addition; see Example 8(c).

23. Let A and B be subsets of $\mathbb{R}$.

(a) Show that $\chi_{A \cap B} = \chi_A \cdot \chi_B$ where the right side of the equation is the ordinary product of functions as in Example 8(c).

(b) Show that $\chi_{A \cup B} = \chi_A + \chi_B - \chi_{A \cap B}$.

§ 4.3   The Division Algorithm and $\mathbb{Z}(p)$

The examples of semigroups we have looked at so far have almost all been infinite. They are familiar examples, but they are not the ones a computer works with. When we label a quantity as INTEGER or REAL in a computer program, we mean that it should be an integer or a real number within the capacity of the computer. Such notions as "infinite precision arithmetic" notwithstanding, too big is just too big. There is no way to deal with an enormously large integer or with π carried out to more digits than the number of bits of memory available.

One strategy that an arithmetic unit of a computer can use to add or multiply reasonable-sized numbers amounts to carrying out the operations in a very large finite semigroup, rather than in $\mathbb{R}$ or $\mathbb{Z}$. In this section we discuss some important finite semigroups which arise in machine calculations and in a variety of applications, including message encryption schemes and error-correcting codes.

The key idea is the notion of **congruence**. Let p be a fixed integer greater than 1. Consider integers m and n. We say that m **is congruent to** n **modulo** p and we write $m \equiv n \pmod{p}$ provided $m - n$ is a multiple of p.

There is another way to describe congruence using remainders after division by p. The following fact about integers is probably familiar.

The Division Let $p \in \mathbb{P}$. For each integer n there are unique integers q and r satisfying
Algorithm

$$n = p \cdot q + r \quad \text{and} \quad 0 \le r < p.$$

The numbers q and r are called the **quotient** and **remainder**, respectively, **when** n **is divided by** p. For example, if $p = 7$ and $n = 31$ then $31 = 7 \cdot 4 + 3$, so $q = 4$ and $r = 3$.

It may seem odd to call this theorem an algorithm, since the statement doesn't explain a procedure for finding either q or r. The name for the theorem is traditional, however, and in most applications the actual method of computation is unimportant.

We will not prove the Division Algorithm here. It can be proved fairly quickly using a nonconstructive argument, or one can simply list the steps of an algorithm that produces q and r. Given a computer, or even a calculator, it is easy to find the quotient and remainder, as follows. Rewrite the equation $n = p \cdot q + r$ as

$$\frac{n}{p} = q + \frac{r}{p}.$$

Since $0 \leq r < p$, $0 \leq r/p < 1$ and so q must be the largest integer less than or equal to n/p. For n positive this means that q is the **integer part** of n/p, as defined in Example 2 of § 3.4 and denoted by $\lfloor n/p \rfloor$, and r/p is the **fractional part**, i.e., the number to the right of the decimal point in n/p. Once we have q, we know $r = n - p \cdot q$.

EXAMPLE 1   (a) Consider $p = 7$ and $n = 31$ as above. A pocket calculator gives $31 \div 7 \approx 4.429$, so $q = 4$ and $r = 31 - 4 \cdot 7 = 3$, as we observed earlier.

(b) Now consider $p = 7$ and $n = -31$. Then $-31/7 \approx -4.429$. Does this mean that $q = -4$ and $r = -3$? No, because r must be nonnegative. Recall that q is always the largest integer less than or equal to n/m. In our present case, $q = -5$ since $-5 < -4.429 < -4$ and so $r = -31 - (-5) \cdot 7 = 4$. ∎

Now back to congruences. Consider two integers m and n. By the Division Algorithm $m = k_1 p + r_1$ and $n = k_2 p + r_2$ for some $k_1, k_2 \in \mathbb{Z}$ and $r_1, r_2 \in \{0, 1, 2, \ldots, p - 1\}$. Then $m - n = (k_1 - k_2)p + r_1 - r_2$, so $m - n$ is a multiple of p if and only if $r_1 - r_2$ is a multiple of p. But $-p < r_1 - r_2 < p$ [why?], so $r_1 - r_2$ can only be a multiple of p if $r_1 = r_2$. Thus $m \equiv n \,(\mathrm{mod}\; p)$ if and only if m and n have the same remainders when divided by p. We have thus proved the following basic fact.

Theorem 1   Fix $p \geq 2$. For each $m \in \mathbb{Z}$ there is exactly one r in $\{0, 1, 2, \ldots, p - 1\}$ such that $m \equiv r \,(\mathrm{mod}\; p)$.

EXAMPLE 2   (a) Two integers are congruent modulo 2 if they are both even or if they are both odd.

(b) The integers which are multiples of 5, namely

$$\ldots, -25, -20, -15, -10, -5, 0, 5, 10, 15, 20, 25, \ldots,$$

are all congruent to each other modulo 5 since the difference between any two numbers on this list is a multiple of 5. These numbers all have remainders 0 when divided by 5.

If we add 1 to each member of this list we get a new list

$$\ldots, -24, -19, -14, -9, -4, 1, 6, 11, 16, 21, 26, \ldots.$$

The *differences* between numbers haven't changed, so the differences are still multiples of 5. For instance

$$16 - (-14) = (15 + 1) - (-15 + 1) = 15 + 1 - (-15) - 1$$
$$= 15 - (-15) = 30.$$

Thus the numbers on the new list are also congruent to each other modulo 5. They all have remainder 1 when divided by 5.

The integers

$$\ldots, -23, -18, -13, -8, -3, 2, 7, 12, 17, 22, 27, \ldots$$

are also congruent to each other modulo 5. So are the integers

$$\ldots, -22, -17, -12, -7, -2, 3, 8, 13, 18, 23, 28, \ldots,$$

and the integers

$$\ldots, -21, -16, -11, -6, -1, 4, 9, 14, 19, 24, 29, \ldots.$$

Each integer belongs to exactly one of these five lists, i.e., the lists form a partition of $\mathbb{Z}$, and each of the lists contains exactly one of the numbers 0, 1, 2, 3, 4, as stated in Theorem 1. ∎

Congruences appear in real life.

EXAMPLE 3 An **odometer** is a device that measures the distance traveled by a vehicle. In most American automobiles, the odometer only goes up to 99,999 miles. In an old clunker an odometer reading of, say, 28,802 only tells us the true mileage modulo 100,000. That is, the clunker may have traveled 28,802 brutal miles, or 128,802 miles, or 228,802 miles, etc. ∎

The next theorem lists some basic properties of congruence. The labels R, S and T stand for reflexivity, symmetry and transitivity, notions that will be studied in Chapter 7.

Theorem 2 Fix $p \geq 2$.

(a) For integers a, b and c we have

(R) $a \equiv a \pmod{p}$;
(S) $a \equiv b \pmod{p}$ implies $b \equiv a \pmod{p}$;
(T) $a \equiv b \pmod{p}$ and $b \equiv c \pmod{p}$ imply $a \equiv c \pmod{p}$.

(b) If $a \equiv b \pmod{p}$ and $c \equiv d \pmod{p}$, then

$$a + c \equiv b + d \pmod{p}$$

and

$$ac \equiv bd \ (\mathrm{mod}\ p).$$

Proof. (a) Properties (R), (S) and (T) are immediate if we remember that "$a \equiv b \ (\mathrm{mod}\ p)$" means "$a$ and b have the same remainders on division by p."

(b) By assumption, $a - b = kp$ and $c - d = lp$ for some $k, l \in \mathbb{Z}$. Then

$$(a + c) - (b + d) = (a - b) + (c - d) = kp + lp = (k + l)p$$

and so $a + c \equiv b + d \ (\mathrm{mod}\ p)$. To deal with the product, we resort to some algebraic chicanery:

$$ac - bd = ac - ad + ad - bd = a(c - d) + (a - b)d$$
$$= alp + kpd = (al + kd)p.$$

Since $al + kd$ is in $\mathbb{Z}$, $ac - bd$ is divisible by p and so $ac \equiv bd \ (\mathrm{mod}\ p)$. ∎

We are now prepared to define the promised finite semigroups. For fixed $p \geq 2$, we write $\mathbb{Z}(p)$ for the set $\{0, 1, 2, \ldots, p - 1\}$. We are going to define an addition and multiplication for $\mathbb{Z}(p)$. It is traditional to use the symbols $+$ and $\cdot$ or $*$, but to avoid confusion with the ordinary operations on $\mathbb{Z}$ we will employ $+_p$ and $*_p$ in this section. For $m, n \in \mathbb{Z}(p)$, $m +_p n$ is defined to be the unique number in $\mathbb{Z}(p)$ satisfying

$$m +_p n \equiv m + n \ (\mathrm{mod}\ p).$$

Theorem 1 states that such a unique number exists; in fact, it is the remainder on dividing $m + n$ by p. Similarly, $m *_p n$ is the unique number in $\mathbb{Z}(p)$ satisfying

$$m *_p n \equiv mn \ (\mathrm{mod}\ p).$$

To get $m *_p n$, multiply m by n, divide by p and take the remainder.

EXAMPLE 4   (a) For $\mathbb{Z}(5) = \{0, 1, 2, 3, 4\}$, we have $4 +_5 4 = 3$ since $4 + 4 \equiv 3$ (mod 5). Similarly, $4 *_5 4 = 1$ since $4 \cdot 4 \equiv 1$ (mod 5). The complete addition and multiplication tables for $\mathbb{Z}(5)$ are given in Figure 1.

$+_5$	0	1	2	3	4
0	0	1	2	3	4
1	1	2	3	4	0
2	2	3	4	0	1
3	3	4	0	1	2
4	4	0	1	2	3

$*_5$	0	1	2	3	4
0	0	0	0	0	0
1	0	1	2	3	4
2	0	2	4	1	3
3	0	3	1	4	2
4	0	4	3	2	1

$\mathbb{Z}(5)$

Figure 1

(b) Very simple but very important semigroups are obtained using $\mathbb{Z}(2)$ as a starting point. The addition and multiplication tables for $\mathbb{Z}(2)$ are given in Figure 2. ∎

$+_2$	0	1
0	0	1
1	1	0

$*_2$	0	1
0	0	0
1	0	1

$\mathbb{Z}(2)$

Figure 2

Theorem 3   The operations $+_p$ and $*_p$ are associative, so $\mathbb{Z}(p)$ is a semigroup under each of these binary operations.

Proof. Consider m, n and r in $\mathbb{Z}(p)$. By definition of $+_p$, we have

$$(m +_p n) +_p r \equiv (m +_p n) + r \,(\text{mod } p)$$

and

$$m +_p n \equiv m + n \,(\text{mod } p).$$

By Theorem 2(b) with $a = m +_p n$, $b = m + n$ and $c = d = r$, we see that

$$(m +_p n) +_p r \equiv (m + n) + r \,(\text{mod } p).$$

Similarly

$$m +_p (n +_p r) \equiv m + (n + r) \,(\text{mod } p).$$

Since addition in $\mathbb{Z}$ is associative, property (T) of Theorem 2 shows that

$$(m +_p n) +_p r \equiv m +_p (n +_p r) \,(\text{mod } p).$$

Since both sides of this congruence are in $\mathbb{Z}(p)$, we conclude that

$$(m +_p n) +_p r = m +_p (n +_p r).$$

The associativity of $*_p$ is proved in exactly the same way. ∎

In fact [Exercise 12] $(\mathbb{Z}(p), +_p)$ is a group; $(\mathbb{Z}(p), *_p)$ is not, though it does have an identity element.

Applications of congruence arithmetic to computing are outside the scope of this book, but a few comments may serve to give perspective. Computers commonly do arithmetic in $\mathbb{Z}(2)$ at the most primitive hardware level and in $\mathbb{Z}(L)$ where L is very large at the working register level. Really fast computers and really fast arithmetic programs have been designed to do arithmetic in several different $\mathbb{Z}(p)$'s, either all at once or in succession, and then fit the results together at the end. Many high-level programming languages have provisions, such as the MOD function in Pascal, for carrying out congruence arithmetic operations.

EXERCISES 4.3

1. Use any method to find q and r for the following values of n and m.
   (a) $n = 20,\ m = 3$ (b) $n = 20,\ m = 4$
   (c) $n = -20,\ m = 3$ (d) $n = -20,\ m = 4$
   (e) $n = 371{,}246,\ m = 65$ (f) $n = -371{,}246,\ m = 65$

2. Find an integer s and a rational number t such that $0 \le t < 1$ and $s + t =:$
   (a) $20/3$ (b) $20/4$ (c) $-20/3$
   (d) $-20/4$ (e) $371{,}246/65$ (f) $-371{,}246/65$

3. List three integers that are congruent modulo 4 to each of the following.
   (a) 0   (b) 1   (c) 2   (d) 3   (e) 4

4. Repeat Exercise 3 for
   (a) -3 (b) 73   (c) -73

5. For each of the following integers m find the unique integer r in $\{0, 1, 2, 3\}$ such that $m \equiv r \pmod 4$. See Theorem 1.
   (a) 17   (b) 7   (c) -7 (d) 2   (e) -88

6. Calculate
   (a) $4 +_7 4$ (b) $5 +_7 6$ (c) $4 *_7 4$
   (d) $0 +_7 k$ for any $k \in \mathbb{Z}(7)$ (e) $1 *_7 k$ for any $k \in \mathbb{Z}(7)$

7. (a) Calculate $6 +_{10} 7$ and $6 *_{10} 7$.
   (b) Describe in words $m +_{10} k$ for any $m,\ k \in \mathbb{Z}(10)$.
   (c) Do the same for $m *_{10} k$.

8. (a) List the elements in the sets A_0, A_1 and A_2 where

$$A_k = \{m \in \mathbb{Z} : -10 \le m \le 10 \text{ and } m \equiv k \,(\mathrm{mod}\ 3)\}.$$

   (b) What is A_3? A_4? A_{73}?

9. Give the complete addition and multiplication tables for $\mathbb{Z}(4)$. Compare the tables with those for $\mathbb{Z}(5)$ in Figure 1.

10. Use Figure 1 to solve the following equations in $\mathbb{Z}(5)$.
   (a) $1 +_5 x = 0$ (b) $2 +_5 x = 0$
   (c) $3 +_5 x = 0$ (d) $4 +_5 x = 0$

11. Use Figure 1 to solve the following equations in $\mathbb{Z}(5)$.
   (a) $1 *_5 x = 1$ (b) $2 *_5 x = 1$
   (c) $3 *_5 x = 1$ (d) $4 *_5 x = 1$

12. Prove that the semigroup $(\mathbb{Z}(p),\ +_p)$ is a commutative group.

13. Show that the semigroup $(\mathbb{Z}(p),\ *_p)$ is a commutative monoid.

14. (a) Use Exercise 11 to show that $\mathbb{Z}(5)\backslash\{0\}$ is a group under multiplication $*_5$.
   (b) Show that $\mathbb{Z}(4)\backslash\{0\}$ is not a group under $*_4$.

15. Prove that $a \equiv b \pmod p$ implies $a^2 \equiv b^2 \pmod p$ for $a, b \in \mathbb{Z}$.

16. Let $\mathscr{S}$ be a set of propositions such that $p, q \in \mathscr{S}$ implies $p \wedge q$ and $p \oplus q$ are in $\mathscr{S}$; $p \oplus q$ is defined in Exercise 5 of § 2.3. Let $\phi(p) = 1$ if p is true and $\phi(p) = 0$ if p is false.

(a) Show that $\phi(p \wedge q) = \phi(p)\phi(q)$ for all $p, q \in \mathscr{S}$.

(b) Show that $\phi(p \oplus q) \equiv \phi(p) + \phi(q) \pmod 2$ for all $p, q \in \mathscr{S}$.

§ 4.4   The Semigroup $\mathscr{P}(\Sigma^*)$

In this section we look at an example that is important in the study of formal languages and finite automata. Consider the set $\mathscr{P}(\Sigma^*)$ of all subsets of Σ^*, i.e., all languages using letters from Σ. Of course, $\mathscr{P}(\Sigma^*)$ is a semigroup under binary operations such as $\cup, \cap$ and $\oplus$. But none of these operations reflects the special structure of Σ^*. When we speak of **the semigroup** $\mathscr{P}(\Sigma^*)$ we refer to the binary operation in the next definition.

Definition   For $A, B \in \mathscr{P}(\Sigma^*)$, the **set-product** AB is defined to be the set

$$AB = \{w_1 w_2 : w_1 \in A \text{ and } w_2 \in B\}.$$

Thus AB is the set of all words obtained by concatenating a word in A with a word in B.

Using the associativity of concatenation in Σ^*, it is easy to see that set-product is associative:

$$(AB)C = A(BC) \quad \text{for all} \quad A, B, C \in \mathscr{P}(\Sigma^*).$$

Thus $\mathscr{P}(\Sigma^*)$ is a semigroup.

EXAMPLE 1   Here we compare and contrast the empty set $\varnothing$ and the set $\{\epsilon\}$ consisting of the empty word. These sets are different and they both belong to $\mathscr{P}(\Sigma^*)$. Since

$$A\{\epsilon\} = \{\epsilon\}A = A \quad \text{for all} \quad A \in \mathscr{P}(\Sigma^*),$$

$\{\epsilon\}$ is an identity for $\mathscr{P}(\Sigma^*)$. Thus $\mathscr{P}(\Sigma^*)$ is a monoid. For any set A,

$$A\varnothing = \{w_1 w_2 : w_1 \in A \text{ and } w_2 \in \varnothing\};$$

since $w_2 \in \varnothing$ never occurs, this set is empty and so

$$A\varnothing = \varnothing A = \varnothing \quad \text{for all} \quad A \in \mathscr{P}(\Sigma^*). \quad \blacksquare$$

Powers A^n for $A \in \mathscr{P}(\Sigma^*)$ and $n \in \mathbb{N}$ are defined just as for general monoids. In particular, $A^0 = \{\epsilon\}$ for all $A \in \mathscr{P}(\Sigma^*)$. Note that for $k \in \mathbb{N}$, Σ^k consists of all words obtained by concatenating k letters from Σ, i.e.,

$$\Sigma^k = \{w \in \Sigma^* : \text{length}(w) = k\},$$

so that the new power notation is consistent with the definition in Example 1(c) of § 1.3.

EXAMPLE 2   Let $\Sigma = \{a, b, c\}$, $A = \{a, ab\}$, $B = \{\epsilon, b, bb\}$ and $C = \{w \in \Sigma^* : \text{length}(w)$ is even$\}$. Note that C contains the empty word ϵ. Then

$$AB = \{a, ab, abb, abbb\},$$
$$BA = \{a, ab, ba, bab, bba, bbab\},$$
$$A^2 = \{aa, aab, aba, abab\},$$
$$B^2 = \{\epsilon, b, bb, bbb, bbbb\},$$
$$A\Sigma = \{aa, ab, ac, aba, abb, abc\},$$
$$C\Sigma = \Sigma C = \{w \in \Sigma^* : \text{length}(w) \text{ is odd}\}.$$

Note that

$$C = \bigcup_{n=0}^{\infty} \Sigma^{2n}. \quad \blacksquare$$

For any language $A \subseteq \Sigma^*$, the set A^* defined by

$$A^* = \bigcup_{n=0}^{\infty} A^n$$

is called the **Kleene closure** of A. Note that A^* consists of all the possible words obtained by concatenating words from A, including the empty word ϵ since $A^0 = \{\epsilon\}$. For $A = \Sigma$ this notation is consistent with our earlier usage of Σ^* since $\Sigma^* = \bigcup_{n=0}^{\infty} \Sigma^n$. We also define the **positive closure** A^+ by

$$A^+ = \bigcup_{n=1}^{\infty} A^n.$$

In particular,

$$\Sigma^+ = \bigcup_{n=1}^{\infty} \Sigma^n = \{w \in \Sigma^* : \text{length}(w) > 0\} = \Sigma^* \setminus \{\epsilon\}.$$

Note that we always have $A^* = A^+ \cup \{\epsilon\}$ and that $A^* = A^+$ if and only if ϵ belongs to A [Exercise 9].

EXAMPLE 3   Let A, B and C be as in Example 2: $A = \{a, ab\}$, $B = \{\epsilon, b, bb\}$, $C = \{w \in \Sigma^* : \text{length}(w)$ is even$\}$. It is easy to verify that

$$C = \bigcup_{n=0}^{\infty} C^n = C^* = C^+$$

since $C^n \subseteq C$ for all $n \in \mathbb{N}$. It is also easy to show that

$$B^* = B^+ = \{w \in \Sigma^* : w \text{ uses only the letter } b\}$$
$$= \{\epsilon, b, bb, bbb, bbbb, \ldots\} = \{b^n : n \in \mathbb{N}\}.$$

The exact nature of the set A^* is less clear, though it is a subset of

$$D = \{w \in \Sigma^* : \text{the number of occurrences of } b \text{ is } \leq \text{ the} \\ \text{number of occurrences of } a\}.$$

This can be seen by proving that $A^n \subseteq D$ for all $n \in \mathbb{N}$ by induction. The inclusion $A^* \subseteq D$ is proper; for example, $aabb$ is not in A^*. Observe that $\epsilon \in A^*$ but $\epsilon \notin A^+$ so that $A^+ \neq A^*$. ∎

Here are some other basic properties about the Kleene and positive closures.

Theorem 1   For a subset A of Σ^* we have

(a) $A^*A^* = A^*$,
(b) $(A^*)^* = A^*$,
(c) $A^+ = AA^* = A^*A = A^+A^* = A^*A^+$,
(d) $(A^+)^+ = A^+$,
(e) $(A^*)^+ = (A^+)^* = A^*$.

Proof. We freely use the following obvious facts:

(1)   if $A \subseteq B$ and $C \subseteq D$, then $AC \subseteq BD$;

(2)   if $A \subseteq B$, then $A^n \subseteq B^n$ for all n, $A^+ \subseteq B^+$ and $A^* \subseteq B^*$;

(3) $A \subseteq A^+ \subseteq A^*$.

(a) Clearly $A^* = \{\epsilon\}A^* \subseteq A^*A^*$. For the other inclusion, consider $w \in A^*A^*$. Then $w = w_1w_2$ where $w_1 \in A^*$ and $w_2 \in A^*$. Hence for some $m, n \in \mathbb{N}$ we have $w_1 \in A^m$ and $w_2 \in A^n$. Consequently $w = w_1w_2 \in A^mA^n = A^{m+n} \subseteq A^*$. This shows that $A^*A^* \subseteq A^*$ and so (a) is established.

(b) Since $A \subseteq A^*$, we have $A^* \subseteq (A^*)^*$. To show

$$(A^*)^* = \bigcup_{n=0}^{\infty} (A^*)^n \subseteq A^*,$$

it suffices to prove

(4) $(A^*)^n \subseteq A^*$ for all $n \in \mathbb{N}$.

This is obvious for $n = 0$ and $n = 1$. With induction in mind, we assume the inclusion for some $n \geq 1$ and then obtain

$$(A^*)^{n+1} = (A^*)^nA^* \subseteq A^*A^* \quad \text{inductive assumption} \\ = A^* \qquad \text{by part (a).}$$

By the Principle of Mathematical Induction, (4) holds.
(c) We will prove

(5) $A^+ \subseteq AA^* \subseteq A^+A^* \subseteq A^+.$

This will show that $A^+ = AA^* = A^+A^*$ and the other equalities in (c) have similar proofs. For each $n \geq 1$, we have $A^n = AA^{n-1} \subseteq AA^*$ and so $A^+ = \bigcup_{n=1}^{\infty} A^n \subseteq AA^*$. This establishes the first inclusion in (5) and the second inclusion is clear. If $w \in A^+A^*$, then $w = w_1w_2$ with $w_1 \in A^+$ and $w_2 \in A^*$ so that $w_1 \in A^m$ and $w_2 \in A^n$ for some $m \in \mathbb{P}$ and $n \in \mathbb{N}$. Since $m + n$ is in $\mathbb{P}$, we see that $w = w_1w_2 \in A^{m+n} \subseteq A^+$. This establishes the last inclusion in (5).

The proofs of parts (d) and (e) are left to Exercise 5. ∎

Parts (a) and (c) can be given quick and intuitive proofs provided one is comfortable with manipulating infinite unions as in Exercise 8.

Loosely speaking, a closure operator is something which, if repeated, gives you nothing new. For example, A^* will usually be larger than A. But applying the star operator again, this time to A^*, gives nothing new since $(A^*)^* = A^*$. This is why $*$ is called a closure operator. The $^+$ operator is another example of a closure operator.

The next theorem gives various representations of the Kleene closure of the union of two languages. The proof is made easy by applying Theorem 1; elementwise arguments can be avoided.

Theorem 2   For subsets A and B of Σ^* we have

$$(A \cup B)^* = (A^* \cup B^*)^* = (A^*B^*)^* = (B^*A^*)^*.$$

Proof. Since $A \cup B \subseteq A^* \cup B^* \subseteq A^*B^*$, we have

(1) $$(A \cup B)^* \subseteq (A^* \cup B^*)^* \subseteq (A^*B^*)^*.$$

Since $A \subseteq A \cup B$ and $B \subseteq A \cup B$, we have

$$A^*B^* \subseteq (A \cup B)^*(A \cup B)^* = (A \cup B)^*,$$

the equality following from Theorem 1(a). Hence

$$(A^*B^*)^* \subseteq ((A \cup B)^*)^* = (A \cup B)^*$$

where the equality follows from Theorem 1(b). This inclusion and (1) show that the first three sets of the theorem are equal. Now reversing the roles of A and B gives

$$(B^*A^*)^* = (B \cup A)^* = (A \cup B)^*.$$ ∎

EXAMPLE 4   To remove some of the mystery from the equalities in Theorem 2, we illustrate the equality $(A \cup B)^* = (A^*B^*)^*$ for the sets $A = \{a, ad\}$ and $B = \{b, bdb\}$.

Each word in $(A \cup B)^*$ has the form $c_1c_2 \cdots c_n$, where each c_i belongs to A or to B. For example,

$$w = aabdbadaababbdbada$$
$$= aa(bdb)(ad)aabab(bdb)(ad)a$$

is such a word. Neighboring elements in A can be grouped together to give a product in A^*. Similarly, adjacent elements of B give products in B^*. Thus one way to break up w is

$$w = \quad aa \quad (bdb) \quad (ad)aa \quad b \quad a \quad b(bdb) \quad (ad)a.$$
$$\uparrow \qquad \uparrow \qquad \uparrow \qquad \uparrow \quad \uparrow \qquad \uparrow \qquad \uparrow$$
$$\text{in } A^* \ \text{ in } B^* \ \text{ in } A^* \ \text{ in } B^* \ \text{ in } A^* \ \text{ in } B^* \ \text{ in } A^*$$

The empty word ϵ is in B^*. If we attach it to the end of w, we see that w belongs to $(A^*B^*)^4$ and hence to $(A^*B^*)^*$.

Now consider a typical word in $(A^*B^*)^*$, i.e., one of the form $w_1w_2 \cdots w_m$ where each w_i is in A^*B^*. Each w_i is itself of the form u_iv_i with $u_i \in A^*$ and $v_i \in B^*$. One example of such a word is $u_1v_1u_2v_2u_3v_3$ with $u_1 = (ad)a(ad)$, $u_2 = aa$, $u_3 = (ad)(ad)a$ and $v_1 = bbbb$, $v_2 = \epsilon$, $v_3 = b(bdb)$. This is the word

$$adaadbbbbaaadadabbdb,$$

which is clearly in $(A \cup B)^*$. ∎

For any subset A of Σ^* the sets A^+ and A^* are closed under the concatenation operation on Σ^* by Theorem 1(a) and the obvious inclusion $A^+A^+ \subseteq A^+$. Hence each of these sets is a semigroup containing A; indeed A^* is a monoid, with identity ϵ.

In general, a **subsemigroup** of a semigroup $(S, \square)$ is a subset T of S which is closed under the operation $\square$, i.e., satisfies $t \square t' \in T$ for all $t, t' \in T$. A **submonoid** of a monoid $(M, \square)$ is a subsemigroup of M which contains the identity of M, and a **subgroup** of a group $(G, \square)$ is a submonoid of G which contains the inverses of all of its members. We will return to these ideas in Chapter 11.

In our present situation A^+ is a subsemigroup and A^* a submonoid of the monoid Σ^*. In fact, A^+ is the smallest subsemigroup of Σ^* containing A, and A^* is the smallest submonoid containing A. For surely any subsemigroup of Σ^* which contains A must contain A^n for $n = 1, 2, \ldots$ [an induction is hiding here] and so must contain A^+. Throw in ϵ to get the smallest submonoid.

EXERCISES 4.4

1. Let $\Sigma = \{a, b\}$, $A = \{ab, ba\}$ and $B = \{\epsilon, b^2\}$. Calculate
(a) A^2
(b) AB
(c) BA
(d) B^2
(e) $A\Sigma$
(f) $\Sigma^2 \cap B$

2. Let $\Sigma = \{a, b, c\}$, $A = \{a, ab, aa\}$, $B = \{cab, bac, cc\}$ and $E = \{\epsilon\}$. Determine
(a) AB
(b) BA
(c) AE
(d) EA
(e) $A\Sigma$
(f) ΣA
(g) A^2
(h) B^2
(i) E^{73}

3. Let Σ, A and B be as in Exercise 1.
   (a) Describe B^*.
   (b) Show that $A^* \subseteq \bigcup_{n=0}^{\infty} \Sigma^{2n}$.
   (c) Give an example of a word in $\Sigma^* \backslash A^*$ having even length.
   (d) Show that $A^*B \neq AB^*$.

4. (a) Explain why Σ^* is not a group. That is, exhibit a word having no inverse.
   (b) When is the monoid Σ^* commutative?

5. (a) Prove part (d) of Theorem 1.
   (b) Prove part (e) of Theorem 1.

6. Let A and D be as in Example 3. Prove that $A^n \subseteq D$ for all $n \in \mathbb{N}$.

7. Let A, B and C be languages over an alphabet Σ. Specify which statements below are true and which are false. For each false statement, provide an example that shows it is false.
   (a) $A \cap B \subseteq AB$.
   (b) $A^* = B^*$ implies $A = B$.
   (c) Σ^* is infinite.
   (d) ϵ is in A^*.
   (e) $AB = BA$.
   (f) $AB = A$ implies $B = \{\epsilon\}$.
   (g) $AB = AC$ implies $B = C$.
   (h) $A \subseteq A^2$.
   (i) $A \subseteq B$ implies $A^2 \subseteq B^2$.
   (j) $\epsilon \in AB$ implies $\epsilon \in A \cap B$.
   (k) $AA^* = A^*$.
   (l) $A^2 = B^2$ implies $A = B$.
   (m) $A^* = \Sigma^*$ implies $\Sigma \subseteq A$.
   (n) Every word in Σ has length 1.

8. (a) Let $\{C_i : i \in I\}$ be a family of subsets of Σ^* indexed by some set I. Prove that if B is also a subset of Σ^* then $B(\bigcup_{i \in I} C_i) = \bigcup_{i \in I} BC_i$.

   (b) Use part (a) to prove parts (a) and (c) of Theorem 1.

9. For $A \subseteq \Sigma^*$, prove that $A^* = A^+$ if and only if ϵ is in A.

10. Let A be a nonempty language over an alphabet Σ such that $A^2 = A$. Prove that A contains the empty word ϵ. *Hint:* Consider a word in A having least length.

11. Consider languages A and B over an alphabet Σ.
    (a) Prove that $(BA)^{n+1} = B(AB)^n A$ for all $n \in \mathbb{N}$ by induction.
    (b) Prove that $(BA)^+ = B(AB)^* A$.
    (c) Prove that if $AB = BA$, then $(BA)^n = B^n A^n$ for all $n \in \mathbb{N}$. *Hint:* Use induction and part (a).

12. Generalize parts (a) and (c) of Exercise 11 to an arbitrary monoid $(S, \square)$.

13. Let A and B be languages over an alphabet Σ.
    (a) Show that $(AB)^n \subseteq (A \cup B)^{2n}$ for $n = 0$, 1 and 2.
    (b) Prove, by induction, that $(AB)^n \subseteq (A \cup B)^{2n}$ for all $n \in \mathbb{N}$.
    (c) Show that $(AB)^* \subseteq (A \cup B)^*$.

14. Let A, B and C be languages over an alphabet Σ.
    (a) Prove $A(B \cup C) = AB \cup AC$.
    (b) Prove $A(B \cap C) \subseteq AB \cap AC$.
    (c) Show that the inclusion in part (b) can be proper.

15. Let $\Sigma = \{a, b\}$.
 (a) Find the smallest submonoid of Σ^* containing the set $\{a, ab\}$. That is, describe its members. Compare with the set A^* in Example 3.
 (b) Find the smallest submonoid of Σ^* containing $\{a, ba\}$.

16. (a) Show that the intersection of two subsemigroups of a semigroup S is a subsemigroup of S or the empty set.
 (b) How about the intersection of two submonoids of a monoid or two subgroups of a group?
 (c) How would the answers to parts (a) and (b) change if we intersected more than two sets?
 (d) Show that if S is a semigroup, if $A \subseteq S$ and if $A \neq \varnothing$, then the intersection of all subsemigroups of S containing A is the smallest subsemigroup of S containing A.

CHAPTER HIGHLIGHTS

These lists of items should produce three kinds of responses. What three? [See the end of Chapter 0 for the answer.]

Concepts

matrix
  transpose, sum, product, scalar multiple
  identity, negative
binary operation
  semigroup, subsemigroup
  monoid = semigroup with identity, submonoid
  group = monoid with inverses, subgroup
  congruence modulo p, $\mathbb{Z}(p)$, $+_p$, $*_p$
Σ^* and $\mathscr{P}(\Sigma^*)$ as monoids

  Kleene closure $= A^* = \bigcup_{n=0}^{\infty} A^n, A^+$

Facts

Matrix multiplication is associative, but not commutative.
In a semigroup $s^m \square s^n = s^{m+n}$ and $(s^m)^n = s^{mn}$.
The Division Algorithm $[n = p \cdot q + r, 0 \leq r < p]$.
Properties of congruence [Theorem 2 of § 4.3].
$(\mathbb{Z}(p), +_p)$ and $(\mathbb{Z}(p), *_p)$ are semigroups.
Properties of A^* and A^+ [Theorems 1 and 2 of § 4.4].

5

COUNTING

The major goal of this chapter is to establish several techniques for counting large finite sets without actually listing their elements. Related techniques, such as the Pigeon-Hole Principle, will also be studied. The last section deals with infinite sets.

§ 5.1 Basic Counting Techniques

For any finite set S, we write $|S|$ for the number of elements in the set. Thus $|S| = |T|$ precisely when the finite sets S and T are of the same size. Observe that

$$|\varnothing| = 0 \quad \text{and} \quad |\{1, 2, \ldots, n\}| = n \quad \text{for} \quad n \in \mathbb{P}.$$

We begin with some counting rules with which you are probably more or less familiar.

Union Rules   Let S and T be finite sets.

(a) If S and T are disjoint, i.e., if $S \cap T = \varnothing$, then $|S \cup T| = |S| + |T|$.
(b) In general, $|S \cup T| = |S| + |T| - |S \cap T|$.

The intuitive reason (b) holds is that, in calculating $|S| + |T|$, elements in $S \cap T$ are counted twice and so $|S \cap T|$ needs to be subtracted from the sum $|S| + |T|$ to obtain $|S \cup T|$. Assertion (b) can be obtained from (a) as follows. Applying (a) two times, we get

$$|S \cup T| = |S| + |T \backslash S| \quad \text{and} \quad |T| = |T \backslash S| + |S \cap T|.$$

Thus we have

$$|S \cup T| + |S \cap T| = |S| + |T \setminus S| + |S \cap T| = |S| + |T|$$

and this implies (b).

A general rule for counting unions of more than two sets, called the Inclusion-Exclusion Principle, appears in § 5.2.

EXAMPLE 1   How many integers in $S = \{1, 2, 3, \ldots, 1000\}$ are divisible by 3 or 5? We let

$$D_3 = \{n \in S : n \text{ is divisible by 3}\}$$

and

$$D_5 = \{n \in S : n \text{ is divisible by 5}\}.$$

We seek the number of elements in $D_3 \cup D_5$, which is not obvious. But $|D_3|$ is easily seen to be 333; just divide 1000 by 3 and round down. Doubters should note that

$$D_3 = \{3m : 1 \leq m \leq 333\}.$$

Likewise, $|D_5| = 200$. Since $D_3 \cap D_5 = \{n \in S : n$ is divisible by 15$\}$ and 1000/15 is 66 2/3, $|D_3 \cap D_5|$ equals 66. By the Union Rule (b),

$$|D_3 \cup D_5| = |D_3| + |D_5| - |D_3 \cap D_5| = 333 + 200 - 66 = 467. \quad \blacksquare$$

For finite sets S and T we have $|S \times T| = |S| \cdot |T|$, since

$$S \times T = \{\langle s, t \rangle : s \in S \text{ and } t \in T\}$$

and for each of the $|S|$ choices of s in S there are $|T|$ choices of t in T to make up the ordered pair $\langle s, t \rangle$. The identity $|S \times T| = |S| \cdot |T|$ is illustrated in Figures 1 and 2 of § 1.4. A similar equality holds for the product of more than two sets.

Product Rules   (a) For finite sets $S_1, S_2, \ldots, S_k$ we have

$$|S_1 \times S_2 \times \cdots \times S_k| = \prod_{j=1}^{k} |S_j|.$$

(b) More generally, suppose that a given set can be viewed as a set of ordered k-tuples $\langle s_1, s_2, \ldots, s_k \rangle$ with the following structure. There are n_1 possible choices of s_1. Given s_1 there are n_2 possible choices of s_2. Given s_1 and s_2 there are n_3 possible choices of s_3. In general, given $s_1, s_2, \ldots, s_{j-1}$ there are n_j choices of s_j. Then the set has $n_1 n_2 \cdots n_k$ elements.

In practice we will often use Product Rule (b), but almost never with the forbidding formalism suggested in its statement.

EXAMPLE 2   (a) We calculate the number of ways of selecting five cards with replacement from a deck of 52 cards. Thus we are counting ordered 5-tuples consisting of cards from the deck. **With replacement** means that each card is returned to the deck before the next card is drawn. The set of ways of selecting five cards with replacement is in one-to-one correspondence with $D \times D \times D \times D \times D = D^5$, where D is the 52-element set of cards. Thus by Product Rule (a), the set has 52^5 elements.

This problem can also be solved by Product Rule (b). There are 52 ways of selecting the first card. After selecting a few cards which have been returned to the deck, there are still 52 ways of selecting the next card. So there are $52 \cdot 52 \cdot 52 \cdot 52 \cdot 52$ ways of selecting five cards with replacement.

(b) Now we calculate the number of ways of selecting five cards without replacement from a deck of 52 cards. **Without replacement** means that, once a card is drawn, it is not returned to the deck. This time Product Rule (a) does *not* apply, since not all ordered 5-tuples in D^5 are allowed. Specifically, ordered 5-tuples with cards repeated are forbidden. But Product Rule (b) does apply. The first card can be selected in 52 ways. Once it is selected the second card can be selected in 51 ways. The third card can be selected in 50 ways, the fourth in 49 ways and the fifth in 48 ways. So five cards can be selected without replacement in $52 \cdot 51 \cdot 50 \cdot 49 \cdot 48$ ways.

So far we have only counted ordered 5-tuples of cards, not 5-card subsets. We will return to the subset question in Example 10. ∎

EXAMPLE 3   (a) Let $\Sigma = \{a, b, c, d, e, f, g\}$. The number of words in Σ^* having length 5 is $7^5 = 16,807$, i.e., $|\Sigma^5| = 16,807$. This is seen by applying either Product Rule just as in Example 2(a). The number of words in Σ^5 that have no letters repeated is $7 \cdot 6 \cdot 5 \cdot 4 \cdot 3 = 2520$. This is because the first letter can be selected in 7 ways, then the second letter can be selected in 6 ways, etc.

(b) Let $\Sigma = \{a, b, c, d\}$. The number of words in Σ^2 without repetitions of letters is $4 \cdot 3 = 12$ by Product Rule (b). We can illustrate this by a picture called a tree; see Figure 1. Each path from the start corresponds to a word in

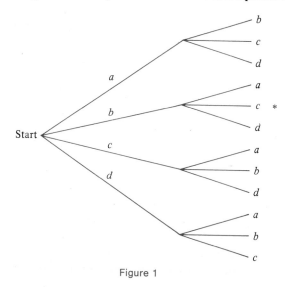

Figure 1

Σ^2 without repetitions. For example, the path ending at ∗ corresponds to the word bc. One can mentally imagine a similar but very large tree for the computation in part (a). ∎

EXAMPLE 4 In Exercise 6 of § 0.2 we asked you to count the number of spanning trees in two graphs, redrawn here in Figures 2(a) and 2(b). A direct count shows that the graph in Figure 2(a) has 8 spanning trees. Any spanning tree in Figure 2(b) can be viewed as a pair of spanning trees, one for the upper half and one

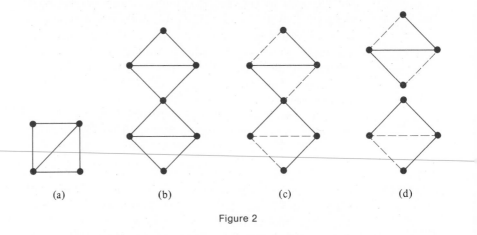

(a)   (b)   (c)   (d)

Figure 2

for the lower half of the graph. Each half of the spanning tree is a spanning tree for the graph in Figure 2(a). For example, the spanning tree in Figure 2(c) is built from the pair of spanning trees in Figure 2(d). By the Product Rule there are $8 \cdot 8 = 64$ such pairs, so the graph in Figure 2(b) has 64 spanning trees. ∎

EXAMPLE 5   (a) Let S and T be finite sets. We will count the number of functions $f : S \to T$. Here it is convenient to write

$$S = \{s_1, s_2, \ldots, s_m\} \quad \text{and} \quad T = \{t_1, t_2, \ldots, t_n\}$$

so that $|S| = m$ and $|T| = n$. A function $f : S \to T$ can be obtained by specifying $f(s_1)$ to be one of the n elements in T, then specifying $f(s_2)$ to be one of the n elements in T, etc. This process leads to $n^m = n \cdot n \cdots n$ [m times] different results, each of which specifies a different function. We conclude that there are n^m functions mapping S into T.

(b) In part (a) we determined $|\text{FUN}(S, T)|$ for the set $\text{FUN}(S, T)$ of all functions from S into T. Some people write T^S in place of $\text{FUN}(S, T)$. This is strange-looking notation, but it does allow one to write

$$|T^S| = |T|^{|S|},$$

giving a "power rule" analogous to our union rules and product rules. ∎

Consider a nonempty finite set S with n elements and consider a positive integer $r \leq n$. An r-**permutation** of S is any ordered listing of r distinct elements of S. More precisely, an r-**permutation** is a one-to-one mapping σ [lowercase Greek sigma] of the set $\{1, 2, \ldots, r\}$ into S. An r-permutation σ is completely described by the ordered r-tuple $\langle \sigma(1), \sigma(2), \ldots, \sigma(r) \rangle$, and we will sometimes use the r-tuple as a notation for the r-permutation itself. An r-permutation can be obtained by assigning 1 to any of the n elements in S. Then 2 can be assigned to any of the $n - 1$ remaining elements, etc. Hence by Product Rule (b), the set S has $n(n - 1)(n - 2) \cdots$ r-permutations where the product consists of exactly r factors. The last factor turns out to be $n - r + 1$. We sometimes abbreviate the product as $P(n, r)$. Thus S has exactly

$$P(n, r) = n(n - 1)(n - 2) \cdots (n - r + 1) = \prod_{j=0}^{r-1} (n - j)$$

r-permutations. We call the n-permutations simply **permutations**. The set S has exactly $P(n, n) = n!$ permutations. Note that $P(n, r) \cdot (n - r)! = n!$ and so

$$P(n, r) = \frac{n!}{(n - r)!} \quad \text{for} \quad 1 \leq r \leq n.$$

It is also convenient to decree $P(n, 0) = 1$, the unique 0-permutation being the "empty permutation."

EXAMPLE 6   (a) The five cards drawn without replacement in Example 2(b) correspond exactly to 5-permutations of the 52-element set of cards. So the number of such drawings is exactly $P(52, 5) = 52 \cdot 51 \cdot 50 \cdot 49 \cdot 48$.

(b) Let Σ be the seven letter alphabet in Example 3(a). The words in Σ^5 that have no letters repeated are 5-permutations of Σ. There are $P(7, 5) = 7 \cdot 6 \cdot 5 \cdot 4 \cdot 3$ such 5-permutations. Note that the empty word ϵ is the empty permutation of Σ.

(c) The words in Σ^2 without repetitions in Example 3(b) are 2-permutations of the 4-element set Σ. There are $P(4, 2) = 4 \cdot 3 = 12$ of them. ∎

EXAMPLE 7   In § 0.2 we were interested in the spanning trees of a graph with 14 edges. We mentioned that there are 14! possible orderings of the edges. This is correct since these orderings are just the various permutations of the 14-element set of edges. ∎

In counting problems where order matters, r-permutations are clearly relevant. Often order is irrelevant, in which case the ability to count sets becomes important. We already know that a set S with n elements has 2^n subsets altogether. For $0 \leq r \leq n$ let $\binom{n}{r}$ be the number of r-element subsets of S. The number $\binom{n}{r}$, called a **binomial coefficient**, is read "n choose r" and is sometimes called the number of **combinations** of n things taken r at a time.

Theorem 1   For $0 \leqq r \leqq n$ we have

$$\binom{n}{r} = \frac{n!}{(n-r)!\, r!}.$$

Proof. Let S be a set with n elements. For each r-element subset T of S, there are $r!$ permutations of S using the elements in T. Hence there are $\binom{n}{r} \cdot r!$ r-permutations of S in all, i.e.,

$$\binom{n}{r} \cdot r! = P(n, r) = \frac{n}{(n-r)!}$$

and so

$$\binom{n}{r} = \frac{n!}{(n-r)!\, r!}. \quad \blacksquare$$

EXAMPLE 8   We count the number of strings of 0's and 1's having length n that contain exactly r 1's. This is equivalent to counting the number of functions from $\{1, 2, \ldots, n\}$ to $\{0, 1\}$ that take the value 1 exactly r times. In other words, we need to count the number of characteristic functions χ_A where $|A| = r$. This is exactly the number of r-element subsets of $\{1, 2, \ldots, n\}$, i.e., $\binom{n}{r}$. $\quad \blacksquare$

EXAMPLE 9   Consider a graph with no loops that is **complete** in the sense that for each pair of distinct vertices there is exactly one edge connecting them. How many edges does the graph have if it has n vertices? Let's assume $n \geq 2$. Each edge determines a 2-element subset of the set V of vertices and, conversely, each 2-element subset of V determines an edge. In other words, the set of edges is in one-to-one correspondence with the set of 2-element subsets of V. Hence there are

$$\binom{n}{2} = \frac{n!}{(n-2)!\, 2!} = \frac{n(n-1)}{2}$$

edges of the graph. $\quad \blacksquare$

An excellent way to illustrate the techniques of this section is to calculate the numbers of various kinds of poker hands. A deck of cards consists of four suits called clubs, diamonds, hearts and spades. Each suit consists of thirteen cards with values A, 2, 3, 4, 5, 6, 7, 8, 9, 10, J, Q, K. Here A stands for ace, J for jack, Q for queen and K for king. There are four cards of each value, one from each suit. A **poker hand** is a set of 5 cards from a 52-card deck of cards. The order in which the cards are chosen is irrelevant. A straight consists of five cards whose values form a consecutive sequence such as 8, 9, 10, J, Q. The ace A can be at the bottom of a sequence A, 2, 3, 4, 5 or at the top of a sequence 10, J, Q, K, A. Poker hands are classified into disjoint sets as follows; they are listed in reverse order of their likelihood.

Royal flush: 10, J, Q, K, A all in the same suit.

Straight flush: A straight all in the same suit that is not a royal flush.

Four of a kind: Four cards in the hand have the same value. For example, four 3's and a 9.

Full house: Three cards of one value and two cards of another value. For example, three jacks and two 8's.

Flush: Five cards all in the same suit, but not a royal or straight flush.

Straight: A straight that is not a royal or straight flush.

Three of a kind: Three cards of one value, a fourth card of a second value and a fifth card of a third value.

Two pairs: Two cards of one value, two more cards of a second value and the remaining card a third value. For example, two queens, two 4's and a 7.

One pair: Two cards of one value, but not classified above. For example, two kings, a jack, a 9 and a 6.

Nothing: None of the above.

EXAMPLE 10
(a) There are $\binom{52}{5}$ poker hands. Note that

$$\binom{52}{5} = \frac{52 \cdot 51 \cdot 50 \cdot 49 \cdot 48}{5 \cdot 4 \cdot 3 \cdot 2 \cdot 1} = 52 \cdot 17 \cdot 10 \cdot 49 \cdot 6 = 2{,}598{,}960.$$

(b) How many poker hands are full houses? Let's call a hand consisting of three jacks and two 8's a full house of type $\langle J, 8 \rangle$, with similar notation for other types of full houses. Order matters, since hands of type $\langle 8, J \rangle$ have three 8's and two jacks. Also, types like $\langle J, J \rangle$ and $\langle 8, 8 \rangle$ are impossible. So types of full houses correspond to 2-permutations of the set of possible values of cards; hence there are $13 \cdot 12$ different types of full houses.

Now we count the number of full houses of each type, say type $\langle J, 8 \rangle$. There are $\binom{4}{3} = 4$ ways to choose three jacks from four jacks, and there are then $\binom{4}{2} = 6$ ways to select two 8's from four 8's. Thus there are $4 \cdot 6 = 24$ hands of type $\langle J, 8 \rangle$. This argument works for all $13 \cdot 12$ types of hands and so there are $13 \cdot 12 \cdot 24 = 3744$ full houses.

(c) How many poker hands are two pairs? Let's say that a hand with two pairs is of type $\{Q, 4\}$ if it consists of two queens and two 4's. This time we have used set notation because order does not matter: hands of type $\{4, Q\}$ are hands of type $\{Q, 4\}$ and we don't want to count them twice. There are $\binom{13}{2}$ types of hands. For each type, say $\{Q, 4\}$, there are $\binom{4}{2}$ ways of choosing two queens, $\binom{4}{2}$ ways of choosing two 4's and $52 - 8 = 44$ ways of choosing the fifth card. Hence there are

$$\binom{13}{2}\binom{4}{2}\binom{4}{2} \cdot 44 = 123{,}552$$

poker hands consisting of two pairs.

(d) How many poker hands are straights? First we count all possible straights even if they are royal or straight flushes. Let's call a straight consisting of the values 8, 9, 10, J, Q a straight of type Q. In general, the type of a straight is the highest value in the straight. Since any of the values 5, 6, 7, 8, 9, 10, J, Q, K, A can be the highest value in a straight, there are 10 types of straights. Given a type of straight, there are 4 choices for each of the 5 values. So there are 4^5 straights of each type and $10 \cdot 4^5 = 10{,}240$ straights altogether. There are 4 royal flushes and 36 straight flushes and so there are 10,200 straights that are not of these exotic varieties.

(e) You are asked to count the remaining kinds of poker hands in Exercise 11, for which all answers are given. ∎

Binomial coefficients get their name from the next theorem.

Binomial Theorem   For real numbers a and b and for $n \in \mathbb{N}$, we have

$$(a + b)^n = \sum_{r=0}^{n} \binom{n}{r} a^r b^{n-r}.$$

This can be proved by induction using the following recursion relation:

$$\binom{n+1}{r} = \binom{n}{r-1} + \binom{n}{r} \qquad \text{for} \quad 1 \leqq r \leqq n;$$

see Exercise 18. This relation, in turn, can be proved by algebraic manipulation, but let us give a set-theoretic explanation.

There are $\binom{n+1}{r}$ r-element subsets of $\{1, 2, \dots, n, n+1\}$. We separate them into two classes. There are $\binom{n}{r}$ subsets which contain only members of $\{1, 2, \dots, n\}$. There are also the remaining subsets, each of which consists of the number $n+1$ and some $r-1$ members of $\{1, 2, \dots, n\}$. Since there are $\binom{n}{r-1}$ ways to choose the elements which aren't $n+1$, this second class of subsets consists of $\binom{n}{r-1}$ subsets. Hence there are exactly $\binom{n}{r} + \binom{n}{r-1}$ r-element subsets of $\{1, 2, \dots, n, n+1\}$, so that $\binom{n}{r} + \binom{n}{r-1} = \binom{n+1}{r}$.

The recursion relation discussed above is the basis for the **Pascal triangle** in Figure 3, which is popular even though its usefulness is limited to small

$$
\begin{array}{ccccccccccccc}
 & & & & & & \binom{0}{0} & & & & & & \\
 & & & & & \binom{1}{0} & & \binom{1}{1} & & & & & \\
 & & & & \binom{2}{0} & & \binom{2}{1} & & \binom{2}{2} & & & & \\
 & & & \binom{3}{0} & & \binom{3}{1} & & \binom{3}{2} & & \binom{3}{3} & & & \\
 & & \binom{4}{0} & & \binom{4}{1} & & \binom{4}{2} & & \binom{4}{3} & & \binom{4}{4} & & \\
 & \binom{5}{0} & & \binom{5}{1} & & \binom{5}{2} & & \binom{5}{3} & & \binom{5}{4} & & \binom{5}{5} & \\
\binom{6}{0} & & \binom{6}{1} & & \binom{6}{2} & & \binom{6}{3} & & \binom{6}{4} & & \binom{6}{5} & & \binom{6}{6} \\
 & & & & & & \vdots & & & & & &
\end{array}
\qquad
\begin{array}{ccccccccccccc}
 & & & & & & 1 & & & & & & \\
 & & & & & 1 & & 1 & & & & & \\
 & & & & 1 & & 2 & & 1 & & & & \\
 & & & 1 & & 3 & & 3 & & 1 & & & \\
 & & 1 & & 4 & & 6 & & 4 & & 1 & & \\
 & 1 & & 5 & & 10 & & 10 & & 5 & & 1 & \\
1 & & 6 & & 15 & & 20 & & 15 & & 6 & & 1 \\
 & & & & & & \vdots & & & & & &
\end{array}
$$

Pascal's triangle

Figure 3

values of n. Each number in the $(n + 1)$-st row of the triangle different from 1 has the form $\binom{n + 1}{r}$ and the nearest numbers just above it are $\binom{n}{r - 1}$ and $\binom{n}{r}$. The recursion relation just says that such an entry is the sum of the two nearest numbers just above it. For example, for $n = 5$ and $r = 2$ the recursion relation says

$$\binom{6}{2} = \binom{5}{1} + \binom{5}{2}, \quad \text{i.e.,} \quad 15 = 5 + 10.$$

EXERCISES 5.1

1. Calculate

(a) $\binom{8}{3}$

(b) $\binom{8}{0}$

(c) $\binom{8}{5}$

(d) $\binom{52}{50}$

(e) $\binom{52}{52}$

(f) $\binom{52}{1}$

2. Let $A = \{1, 2, 3, 4, 5, 6, 7, 8, 9, 10\}$ and $B = \{2, 3, 5, 7, 11, 13, 17, 19\}$.
   (a) Determine the sizes of the sets $A \cup B$, $A \cap B$ and $A \oplus B$.
   (b) How many subsets of A are there?
   (c) How many 4-element subsets of A are there?
   (d) How many 4-element subsets of A consist of 3 even and 1 odd number?

3. Among 150 people, 45 swim, 40 bike and 50 jog. Also, 32 people jog but don't bike, 27 people jog and swim, and 10 people do all three.
   (a) How many people jog but don't swim and don't bike?
   (b) If 21 people bike and swim, how many do none of the three activities?

4. A certain class consists of 12 men and 16 women. How many committees can be chosen from this class consisting of
   (a) seven people?
   (b) three men and four women?
   (c) seven women or seven men?

5. (a) How many committees consisting of 4 people can be chosen from 9 people?
   (b) Redo part (a) if there are two people, Ann and Bob, who will not serve on the same committee.

6. How many committees consisting of 4 men and 4 women can be chosen from a group of 8 men and 6 women?

7. Let $S = \{a, b, c, d\}$ and $T = \{1, 2, 3, 4, 5, 6, 7\}$.
   (a) How many one-to-one functions are there from T into S?
   (b) How many one-to-one functions are there from S into T?
   (c) How many functions are there from S into T?

8. Let $P = \{1, 2, 3, 4, 5, 6, 7, 8, 9\}$ and $Q = \{A, B, C, D, E\}$.
   (a) How many 4-element subsets of P are there?
   (b) How many permutations, i.e., 5-permutations, of Q are there?
   (c) How many license plates are there consisting of three letters from Q followed by two numbers from P? Repetition is allowed; for example, DAD 88 is allowed.

9. Cards are drawn from a deck of 52 cards with replacement.
   (a) In how many ways can ten cards be drawn so that the tenth card is not a repetition?
   (b) In how many ways can ten cards be drawn so that the tenth card is a repetition?

10. Let Σ be the alphabet $\{a, b, c, d, e\}$ and let $\Sigma^k = \{w \in \Sigma^* : \text{length}(w) = k\}$. How many elements are there in each of the following sets?
    (a) Σ^k, for each $k \in \mathbb{N}$
    (b) $\{w \in \Sigma^3 : \text{no letter in } w \text{ is used more than once}\}$
    (c) $\{w \in \Sigma^4 : \text{the letter } c \text{ occurs in } w \text{ exactly once}\}$
    (d) $\{w \in \Sigma^4 : \text{the letter } c \text{ occurs in } w \text{ at least once}\}$

11. Count the number of poker hands of the following kinds:
    (a) four of a kind
    (b) flush [but not straight or royal flush]
    (c) three of a kind
    (d) one pair

12. (a) In how many ways can the letters a, b, c, d, e, f be arranged so that the letters a and b are adjacent?
    (b) In how many ways can the letters a, b, c, d, e, f be arranged so that the letters a and b are not adjacent?

13. Obtain the next two rows of the Pascal triangle in Figure 3.

14. Draw the 8 spanning trees of the graph in Figure 2(a).

15. (a) Give the matrix for a complete graph with n vertices; see Example 9.
    (b) Use the matrix in part (a) to count the number of edges of the graph. *Hint*: How many entries in the matrix are equal to 1?

16. Prove that $\sum_{r=0}^{n} (-1)^r \binom{n}{r} = 0$ for $n \in \mathbb{P}$. *Hint*: Use the binomial theorem.

17. (a) Show that $\binom{n}{r} = \binom{n}{n-r}$ for $0 \le r \le n$.
    (b) Give a set-theoretic interpretation of the identities in part (a).

18. (a) Prove the binomial theorem.
    (b) Prove $\binom{n+1}{r} = \binom{n}{r-1} + \binom{n}{r}$ for $1 \le r \le n$ algebraically.

19. Prove that $2^n = \sum\limits_{r=0}^{n} \binom{n}{r}$

(a) by setting $a = b = 1$ in the binomial theorem.

(b) by counting subsets of an n-element set.

(c) by induction using the recursion relation in Exercise 18(b).

20. Prove that $\sum\limits_{r=0}^{n} \binom{n}{r} 2^r = 3^n$ for $n \in \mathbb{P}$.

21. (a) Verify that $\sum\limits_{k=m}^{n} \binom{k}{m} = \binom{n+1}{m+1}$ for some small values of m and n, like $m = 3$ and $n = 5$.

(b) What does this identity say about Pascal's triangle?

(c) Prove the identity by induction on n.

(d) Prove the identity by counting the $(m+1)$-element subsets of $\{1, 2, \ldots, n+1\}$. *Hint:* How many of these sets are there whose largest element is $k+1$? What can k be?

§ 5.2   More Counting Techniques

In this section we introduce the general Inclusion-Exclusion Principle. We also discuss methods for counting partitions and solving related problems.

It is often easy to count elements in an intersection of sets, where the key connective is "and." On the other hand, it is often difficult to count directly the elements in a union of sets. The Inclusion-Exclusion Principle will tell us the size of a union in terms of the sizes of various intersections. Let $A_1, A_2, \ldots, A_n$ be finite sets. For $n = 2$, the Union Rule (b) asserts that

$$|A_1 \cup A_2| = |A_1| + |A_2| - |A_1 \cap A_2|.$$

For $n = 3$, the Inclusion-Exclusion Principle below asserts that

$$|A_1 \cup A_2 \cup A_3| = |A_1| + |A_2| + |A_3| - |A_1 \cap A_2| - |A_1 \cap A_3|$$
$$- |A_2 \cap A_3| + |A_1 \cap A_2 \cap A_3|$$

and for $n = 4$, it asserts that

$$|A_1 \cup A_2 \cup A_3 \cup A_4| = |A_1| + |A_2| + |A_3| + |A_4| - |A_1 \cap A_2|$$
$$- |A_1 \cap A_3| - |A_1 \cap A_4| - |A_2 \cap A_3|$$
$$- |A_2 \cap A_4| - |A_3 \cap A_4| + |A_1 \cap A_2 \cap A_3|$$
$$+ |A_1 \cap A_2 \cap A_4| + |A_1 \cap A_3 \cap A_4|$$
$$+ |A_2 \cap A_3 \cap A_4| - |A_1 \cap A_2 \cap A_3 \cap A_4|.$$

Inclusion- To calculate the size of $A_1 \cup A_2 \cup \cdots \cup A_n$, calculate the sizes of all possible
Exclusion intersections of sets from $\{A_1, A_2, \ldots, A_n\}$, add the results obtained by
Principle intersecting an odd number of the sets and subtract the results obtained by
intersecting an even number of the sets.

In terms of the phrase "inclusion-exclusion," include or add the sizes of
the sets, then exclude or subtract the sizes of all intersections of two sets, then
include or add the sizes of all intersections of three sets, etc. A concise
statement of the principle is offered in Exercise 16.

EXAMPLE 1 We count the number of integers in $S = \{1, 2, 3, \ldots, 2000\}$ that are divisible
by 9, 11, 13 or 15. For each $k \in \mathbb{P}$, we let $D_k = \{n \in S : n \text{ is divisible by } k\}$ and
we seek $|D_9 \cup D_{11} \cup D_{13} \cup D_{15}|$. Note that $|D_k|$ is the largest integer $\leq$
$2000/k$. Hence

$$|D_9| = 222, \qquad |D_{11}| = 181, \qquad |D_{13}| = 153, \qquad |D_{15}| = 133,$$

$$|D_9 \cap D_{11}| = |D_{99}| = 20, \qquad |D_9 \cap D_{13}| = |D_{117}| = 17,$$

$$|D_9 \cap D_{15}| = |D_{45}| = 44, \qquad |D_{11} \cap D_{13}| = |D_{143}| = 13,$$

$$|D_{11} \cap D_{15}| = |D_{165}| = 12, \qquad |D_{13} \cap D_{15}| = |D_{195}| = 10,$$

$$|D_9 \cap D_{11} \cap D_{13}| = |D_{1287}| = 1, \qquad |D_9 \cap D_{11} \cap D_{15}| = |D_{495}| = 4,$$

$$|D_9 \cap D_{13} \cap D_{15}| = |D_{585}| = 3, \qquad |D_{11} \cap D_{13} \cap D_{15}| = |D_{2145}| = 0,$$

$$|D_9 \cap D_{11} \cap D_{13} \cap D_{15}| = |D_{6435}| = 0.$$

Note that $D_9 \cap D_{15} = D_{45}$ [not D_{135}] since $\operatorname{lcm}(9, 15) = 45$; similar care is
needed in dealing with $D_9 \cap D_{11} \cap D_{15}$, $D_9 \cap D_{13} \cap D_{15}$, etc. Now by the
Inclusion-Exclusion Principle, we have

$$|D_9 \cup D_{11} \cup D_{13} \cup D_{15}| = 222 + 181 + 153 + 133$$

$$-(20 + 17 + 44 + 13 + 12 + 10)$$

$$+(1 + 4 + 3 + 0) - 0 = 581. \quad \blacksquare$$

EXAMPLE 2 We count the number of integers in $T = \{1000, 1001, \ldots, 9999\}$ with at least
one digit that is 0, at least one that is 1 and at least one that is 2. For example,
1072 and 2101 are such numbers. It is easier to count numbers that exclude
certain digits and so we deal with complements. That is, for $k = 0, 1$ and 2
we let

$$A_k = \{n \in T : n \text{ has no digit equal to } k\}.$$

Then each A_k^c consists of those n in T that have at least one digit equal to k
and so $A_0^c \cap A_1^c \cap A_2^c$ consists of those n in T that have at least one 0, one 1
and one 2 among their digits. This is exactly the set whose size we are
after. Since $A_0^c \cap A_1^c \cap A_2^c = (A_0 \cup A_1 \cup A_2)^c$ by DeMorgan's law, we will

first calculate $|A_0 \cup A_1 \cup A_2|$ using the Inclusion-Exclusion Principle. By the Product Rule (a) we have $|A_1| = 8 \cdot 9 \cdot 9 \cdot 9$ since there are 8 choices for the first digit, which cannot be 0 or 1, and 9 choices for the other digits. Similar computations yield

$$|A_0| = 9 \cdot 9 \cdot 9 \cdot 9 = 6561, \qquad |A_1| = |A_2| = 8 \cdot 9 \cdot 9 \cdot 9 = 5832,$$

$$|A_0 \cap A_1| = |A_0 \cap A_2| = 8 \cdot 8 \cdot 8 \cdot 8 = 4096,$$

$$|A_1 \cap A_2| = 7 \cdot 8 \cdot 8 \cdot 8 = 3584,$$

$$|A_0 \cap A_1 \cap A_2| = 7 \cdot 7 \cdot 7 \cdot 7 = 2401.$$

By the Inclusion-Exclusion Principle,

$$|A_0 \cup A_1 \cup A_2| = 6561 + 5832 + 5832 - (4096 + 4096 + 3584)$$
$$+ 2401 = 8850$$

and so

$$|(A_0 \cup A_1 \cup A_2)^c| = |T| - |A_0 \cup A_1 \cup A_2| = 9000 - 8850 = 150.$$

There are 150 integers in T whose digits include at least one 0, 1 and 2.   ∎

An Explanation of the Inclusion-Exclusion Principle. The main barrier to proving the general principle is the notation [cf. Exercise 16]. The principle can be proved by induction on n. We show how the result for $n = 2$ leads to the result for $n = 3$. Using the $n = 2$ case we have

(1) $$|A \cup B \cup C| = |A \cup B| + |C| - |(A \cup B) \cap C|$$

and

(2) $$|A \cup B| = |A| + |B| - |A \cap B|.$$

Applying the distributive law for unions and intersections [rule 3b in Table 1 of § 1.2], we also obtain

(3) $$|(A \cup B) \cap C| = |(A \cap C) \cup (B \cap C)|$$
$$= |A \cap C| + |B \cap C| - |A \cap B \cap C|.$$

Substitution of (2) and (3) into (1) yields

(4) $$|A \cup B \cup C| = |A| + |B| - |A \cap B| + |C| - |A \cap C|$$
$$- |B \cap C| + |A \cap B \cap C|$$

and this is the principle for $n = 3$.   ∎

A much simpler counting principle which we will also find useful is just a formalization of common sense. If a box contains 30 marbles of different colors and if there are 6 marbles of each color, then there must be 5 colors of marbles. This is a special case of the next lemma if we let A be the set of marbles, B the set of colors, ϕ the function which maps each marble to its color, and note that $k = 6$.

Counting If $\phi : A \to B$ maps the finite set A onto B and if the sets
Lemma

$$\phi^{-1}(b) = \{a \in A : \phi(a) = b\}$$

have the same number of elements, say k, for all b in B then

$$|B| = |A|/k.$$

Proof. The set A is the union of the disjoint sets $\phi^{-1}(b)$ and so $|A| = \sum_{b \in B} |\phi^{-1}(b)| = \sum_{b \in B} k = k \cdot |B|$. Therefore $|B| = |A|/k$. ∎

We return to the problem of counting permutations of letters in a word. If all the letters are different, the problem is easy. For example, there are 9! permutations of the letters in ALGORITHM. But how many permutations are there of the letters in CORRESPONDENCE? We solve this by first labeling the repeated letters: $C_1 O_1 R_1 R_2 E_1 SPO_2 N_1 DE_2 N_2 C_2 E_3$ so that all the letters are different. There are 14! permutations of these subscripted letters. For each such permutation σ, we let $\phi(\sigma)$ be the same word without the subscripts. Two permutations yield the same word w if only the two C's are permuted, the two N's are permuted, the two O's are permuted, the two R's are permuted and the three E's are permuted. Thus each $\phi^{-1}(w)$ contains 2! 2! 2! 2! 3! permutations. By the Counting Lemma, we conclude that there are

$$\frac{14!}{2!\,2!\,2!\,2!\,3!}$$

permutations of CORRESPONDENCE. Exactly this sort of argument can be used to prove the following principle.

A Counting Consider a set of n objects of k different types. If $n_1 + n_2 + \cdots + n_k = n$
Principle where n_1 are alike, n_2 are alike, ..., n_k are alike, then there are

$$\frac{n!}{n_1!\,n_2! \cdots n_k!}$$

distinguishable permutations of the n objects, where we distinguish between two permutations in case there is a position in which their entries are of different types.

EXAMPLE 3 Let $\Sigma = \{a, b, c\}$. The number of words in Σ^* having length 10 using 4 a's, 3 b's and 3 c's is

$$\frac{10!}{4!\,3!\,3!} = 4200.$$

The number of words using 5 a's, 3 b's and 2 c's is

$$\frac{10!}{5!\,3!\,2!} = 2520$$

and the number using 5 a's and 5 b's is

$$\frac{10!}{5!\,5!} = 252.$$

For comparison, note that Σ^{10} has $3^{10} = 59{,}049$ words.   ∎

Recall that a partition of a set S is a collection of disjoint subsets whose union is the set S itself. An **ordered partition** is a partition in which the subsets are ordered; in this case, although the sets are ordered, the elements within the sets are not.

EXAMPLE 4   Let $S = \{1, 2, 3, 4, 5, 6, 7\}$.

(a) Here are some ordered partitions of S:

$$\langle\{1, 3, 5\}, \{2, 4, 6, 7\}\rangle, \quad \langle\{2, 4, 6, 7\}, \{1, 3, 5\}\rangle, \quad \langle\{3, 6\}, \{2, 5\}, \{1, 4, 7\}\rangle,$$

$$\langle\{1\}, \{2, 4, 6\}, \{3, 5, 7\}\rangle \quad \text{and} \quad \langle\{1, 6\}, \{2, 5\}, \{3, 4\}, \{7\}\rangle.$$

(b) We count the number of ordered partitions of S of the form $\langle A, B, C\rangle$ where $|A| = 2$, $|B| = 3$ and $|C| = 2$. Given a permutation σ of S, we can form such a partition $\phi(\sigma)$ by letting A consist of the first two elements, B consist of the next three elements, and C the remaining two. There are 7! permutations of S. Given a permutation, say $\langle 3, 7, 4, 1, 6, 5, 2\rangle$, if we permute the first two numbers, the second three and the last two, to obtain say $\langle 7, 3, 4, 6, 1, 5, 2\rangle$, both permutations will be mapped by ϕ to the same ordered partition, namely $\langle\{3, 7\}, \{1, 4, 6\}, \{2, 5\}\rangle$. There are 2! 3! 2! such permutations that ϕ maps to the same ordered partition π. In other words, each $\phi^{-1}(\pi)$ has 2! 3! 2! elements. From the Counting Lemma we see that there are

$$\frac{7!}{2!\,3!\,2!} = 210$$

different ordered partitions of S where $|A| = |C| = 2$ and $|B| = 3$.   ∎

The counting argument in Example 4(b) easily generalizes to establish the following.

Counting
Ordered
Partitions   If a set has n elements and if $n_1 + n_2 + \cdots + n_k = n$, then there are

$$\frac{n!}{n_1!\,n_2!\cdots n_k!}$$

ordered partitions $\langle A_1, A_2, \ldots, A_k\rangle$ of the set with $|A_j| = n_j$ for $j = 1, 2, \ldots, k$.

EXAMPLE 5   In how many ways can three disjoint committees be formed from twenty people if they must have 3, 5 and 7 people, respectively? This is equivalent to counting ordered partitions $\langle A, B, C, D \rangle$ of the set of twenty people where $|A| = 3$, $|B| = 5$, $|C| = 7$ and $|D| = 5$. The set D corresponds to the people with no committee assignment. There are

$$\frac{20!}{3!\,5!\,7!\,5!} \approx 5.587 \cdot 10^9$$

possible ways to form such committees. Note that although $|D| = |B|$, the committee B and the set D play different roles; the ordering of the partition is significant and we are not just interested in ways of breaking the big set up into a 3-element subset, a couple of 5-element subsets and a 7-element subset. ∎

EXAMPLE 6   (a) A **bridge deal** is an ordered partition of 52 cards involving four sets with 13 cards each. Thus there are

$$\frac{52!}{13!\,13!\,13!\,13!} = \frac{52!}{(13!)^4} \approx 5.3645 \cdot 10^{28}$$

bridge deals.

(b) We count the number of bridge deals in which each hand of 13 cards contains one ace. First we deal out the aces; this can be done in $4! = 24$ ways. Just as in part (a), the remaining cards can be partitioned in $48!/(12!)^4$ ways. So $24 \cdot 48!/(12!)^4$ of the bridge deals yield one ace in each hand. The fraction of such deals is

$$24\,\frac{48!}{(12!)^4} \cdot \frac{(13!)^4}{52!} = \frac{(13)^3}{17 \cdot 25 \cdot 49} \approx .1055.$$

(c) A single **bridge hand** consists of 13 cards drawn from a 52-card deck. There are $\binom{52}{13} \approx 6.394 \cdot 10^{11}$ bridge hands. We say that a bridge hand has distribution $n_1 - n_2 - n_3 - n_4$ where $n_1 \geq n_2 \geq n_3 \geq n_4$ and $n_1 + n_2 + n_3 + n_4 = 13$ if there are n_1 cards of some suit, n_2 cards of a second suit, n_3 cards of a third suit and n_4 cards of the remaining suit. To illustrate, we count the number of bridge hands having 4–3–3–3 distribution. There are $\binom{13}{4}\binom{13}{3}\binom{13}{3}\binom{13}{3}$ ways to choose 4 clubs and three of each of the other suits. We get the same result if we replace clubs by one of the other suits. So we conclude that there are

$$4\binom{13}{4}\binom{13}{3}^3 \approx 6.6906 \cdot 10^{10}$$

bridge hands having 4–3–3–3 distribution. Of all bridge hands, about $6.6906/63.94 \approx .1046$ proportion have 4–3–3–3 distribution. ∎

Sometimes problems reduce to counting unordered partitions. When this occurs, count ordered partitions first and then divide by suitable numbers to take into account the lack of order.

EXAMPLE 7   (a) In how many ways can twelve students be divided into three groups, with four students in each group, so that one group studies topic T_1, one studies topic T_2 and one studies topic T_3? Here order matters: if we permuted groups of students, the students would be studying different topics. So we count ordered partitions, of which there are

$$\frac{12!}{4!\,4!\,4!} = 34{,}650.$$

(b) In how many ways can twelve students be divided into three study groups, with four students in each group, so that each group will study the same topic? Now we wish to count unordered partitions, since we regard partitions like $\langle A, B, C \rangle$ and $\langle B, A, C \rangle$ as equivalent. They correspond to the same partition of the twelve students into three equal groups. From part (a), there are 34,650 ordered partitions. If we map each ordered partition $\langle A, B, C \rangle$ to the unordered partition $\phi(\langle A, B, C \rangle) = \{A, B, C\}$, we find that $\phi^{-1}(\{A, B, C\})$ has $3! = 6$ elements, namely $\langle A, B, C \rangle$, $\langle A, C, B \rangle$, $\langle B, A, C \rangle$, $\langle B, C, A \rangle$, $\langle C, A, B \rangle$ and $\langle C, B, A \rangle$. So by the Counting Lemma there are $34{,}650/6 = 5775$ unordered partitions of the desired type. Hence the answer to our question is 5775. ∎

EXAMPLE 8   (a) In how many ways can nineteen students be divided into five groups, two groups of five and three groups of three, so that each group studies a different topic? As in Example 7(a), we count ordered partitions, of which there are

$$\frac{19!}{5!\,5!\,3!\,3!\,3!} \approx 3.911 \cdot 10^{10}.$$

(b) In how many ways can the students in part (a) be divided if all five groups are to study the same topic? In part (a) we counted all ordered partitions $\langle A, B, C, D, E \rangle$ where $|A| = |B| = 5$ and $|C| = |D| = |E| = 3$. If A and B are permuted and C, D, E are permuted, we will get the same study groups, but we cannot permute groups of different sizes like A and D. To count unordered partitions we let $\phi(\langle A, B, C, D, E \rangle) = \langle \{A, B\}, \{C, D, E\} \rangle$. Each inverse image $\phi^{-1}(\langle \{A, B\}, \{C, D, E\} \rangle)$ has $2!\,3!$ elements [such as $\langle B, A, C, E, D \rangle$] and so, by the Counting Lemma, there are

$$\frac{19!}{5!\,5!\,3!\,3!\,3!} \cdot \frac{1}{2!\,3!} \approx 3.26 \cdot 10^9$$

unordered partitions of students into study groups. ∎

EXAMPLE 9   In how many ways can a set of 100 elements be partitioned into 50 sets with 2 elements each? More generally, we can ask in how many ways a set of $2n$ elements can be partitioned into n sets with 2 elements each. To answer the first question, replace each n by 50 in what follows. We are asking for unordered partitions. There are

$$\frac{(2n)!}{2 \cdot 2 \cdots 2} = \frac{(2n)!}{2^n}$$

ordered partitions $\langle A_1, A_2, \ldots, A_n \rangle$ where each set has 2 elements. Any permutation of the n sets gives the same unordered partition and so there are $(2n)!/2^n n!$ unordered partitions of $2n$ elements into n sets with 2 elements each. Note that

$$\frac{(2n)!}{2^n n!} = (2n-1)(2n-3)(2n-5) \cdots 3 \cdot 1$$

since

$$\frac{(2n)!}{(2n-1)(2n-3) \cdots 3 \cdot 1} = (2n)(2n-2)(2n-4) \cdots 6 \cdot 4 \cdot 2$$

$$= 2 \cdot n \cdot 2(n-1) \cdot 2(n-2) \cdots 2 \cdot 3 \cdot 2 \cdot 2 \cdot 2 \cdot 1$$
$$\quad \uparrow \quad \uparrow \qquad\quad \uparrow \qquad\qquad \uparrow \;\; \uparrow \;\; \uparrow$$
$$= 2^n n!.$$

Here is another way to count the unordered partitions above. First list the elements of the set in some order. Pick the first element. There are $2n - 1$ possible elements that can join it to form a 2-element set. Now pick the next unused element. There are $2n - 3$ possible elements that can join it to form a second 2-element set. Continuing in this way, we find that there are $(2n - 1)(2n - 3) \cdots 3 \cdot 1$ ways to construct the desired unordered partitions. ∎

The last counting formula we consider arises in a variety of ways, but we offer it in a form that is easy to remember.

Placing Objects   There are $\binom{n+k-1}{k-1}$ ways to place n indistinguishable objects into k
in Boxes   distinguishable boxes.

Proof. The proof is both elegant and illuminating; we illustrate it for the case $n = 5$ and $k = 4$. We let five 0's represent the objects and then we add three 1's to serve as dividers among the four boxes. Thus we consider the set of all strings consisting of five 0's and three 1's and we claim there is a one-to-one correspondence between this set and the ways to place the five 0's into four boxes. Specifically, a string corresponds to the placement of the 0's before the first 1 into the first box, the 0's between the first and second 1 into the second box, the 0's between the second and third 1 into the third box, and the 0's after the third 1 into the fourth box. For example,

$$0\ 0\ 1\ 1\ 0\ 0\ 0\ 1 \rightarrow 0\ 0\ \|\ 0\ 0\ 0\ |\ \rightarrow \boxed{\begin{array}{c|c|c|c} 0\ 0 & & 0\ 0\ 0 & \end{array}}.$$

box   box   box   box
1   2   3   4

In this example, boxes 2 and 4 are empty because there are no 0's between the first and second dividers and there are no 0's after the last divider. More examples:

$$1\ 0\ 0\ 1\ 0\ 0\ 1\ 0 \rightarrow \boxed{\begin{array}{c|c|c|c} & 0\ 0 & 0\ 0 & 0 \end{array}};$$

$$0\ 0\ 0\ 1\ 1\ 1\ 0\ 0 \rightarrow \boxed{\begin{array}{c|c|c|c} 0\ 0\ 0 & & & 0\ 0 \end{array}}.$$

As noted in Example 8 of § 5.1, there are $\binom{8}{3}$ strings having five 0's and three 1's. This establishes the result for $n = 5$ and $k = 4$.

In the general case, we consider strings of n 0's and $k - 1$ 1's. The 0's correspond to objects and the 1's to dividers. There are $\binom{n + k - 1}{k - 1}$ such strings and, as above, there is a one-to-one correspondence between these strings and the placing of n 0's into k boxes. ∎

EXAMPLE 10   (a) In how many ways can ten red marbles be placed into five distinguishable bags? Here $n = 10$, $k = 5$ and the answer is $\binom{10 + 5 - 1}{5 - 1} = \binom{14}{4} = 1001$.

(b) In how many ways can ten red marbles be placed into five indistinguishable bags? This is much harder. You should be aware that counting problems can get difficult quickly. Here one would like to apply part (a) and the Counting Lemma. The trouble is that if we let ϕ map distinguishable arrangements, like $\langle 1, 1, 3, 2, 3 \rangle$, to indistinguishable arrangements, like $\{1, 1, 2, 3, 3\}$, the inverse images would have *different* sizes. For example, $\phi^{-1}(\{2, 2, 2, 2, 2\})$ would have one element while $\phi^{-1}(\{0, 1, 2, 3, 4\})$ would have 120 elements. We abandon this problem. Any solution we are aware of involves the consideration of several cases. ∎

Sometimes problems need to be manipulated before it is clear how to apply one of our principles.

EXAMPLE 11   How many numbers in $\{1, 2, 3, \ldots, 100000\}$ have the property that the sum of digits is 7? We can ignore the very last number 100000 and we can assume that all the numbers have five digits by placing zeros in front if necessary. So, for example, we replace 1 by 00001 and 73 by 00073. Our question is now: How many strings of five digits have the property that the sum of the digits is 7? We can associate each such string with the placement of seven balls in five boxes; for example,

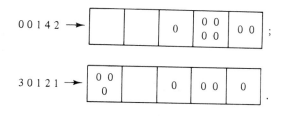

There are $\binom{11}{4} = 330$ such placements and so there are 330 numbers with the desired property. ∎

We now give a little different interpretation of the "Placing Objects in Boxes" principle. Consider the k boxes first and assume that each box contains an unlimited supply of objects labeled according to the boxes they are in. Applying the principle in reverse, we see that there are $\binom{n + k - 1}{k - 1}$ ways to remove n objects from the k boxes. In other words, *the number of ways of selecting n objects from k distinguishable objects, allowing repetitions, is* $\binom{n + k - 1}{k - 1}$.

EXAMPLE 12   In how many ways can ten coins be selected from an unlimited supply of pennies, nickels, dimes and quarters? This is tailor-made for the principle just stated. Let $n = 10$ [for the ten coins] and $k = 4$ [for the four types of coins]. Then the answer is $\binom{10 + 4 - 1}{4 - 1} = \binom{13}{3} = 286$.

The new interpretation can be avoided as follows. The problem is equivalent to counting ordered 4-tuples of nonnegative integers whose sum is 10. For example, $\langle 5, 3, 0, 2 \rangle$ corresponds to the selection of 5 pennies, 3 nickels and 2 quarters. Counting these ordered 4-tuples is equivalent to counting the ways of placing ten indistinguishable objects into 4 boxes and this can be done in $\binom{13}{3}$ ways. ∎

The principles in this section may appear to be a bag of tricks needed to work the problems we set for ourselves. To some extent the subject of counting is like that, though we prefer to think of our techniques as tools rather than tricks. They *are* applicable in a variety of common situations, but won't handle every problem that comes up. The thought processes that we demonstrated in the proofs of the principles are as valuable as the principles themselves, since the same sort of analysis can often be used on problems where the ready-made tools don't apply.

EXERCISES 5.2

1. Among 200 people, 150 either swim or jog or both. If 85 swim and 60 swim and jog, how many jog?

2. Let $S = \{100, 101, 102, \ldots, 999\}$ so that $|S| = 900$.
   (a) How many numbers in S have at least one digit that is a 3 or a 7? Examples: 300, 707, 736, 103, 997.
   (b) How many numbers in S have at least one digit that is a 3 *and* at least one digit that is a 7? Examples: 736 and 377 but not 300, 707, 103, 997.

3. Find the number of integers in $\{1, 2, 3, \ldots, 1000\}$ that are divisible by 4, 5 or 6.

4. How many different mixes of candy are possible if a mix consists of 10 pieces of candy and if there are 4 different kinds of candy available in unlimited quantities?

5. From a total of 15 people, 3 committees consisting of 3, 4 and 5 people respectively are to be chosen.
   (a) How many such sets of committees are possible if no person may serve on more than one committee?
   (b) How many such sets of committees are possible if there is no restriction on the number of committees on which a person may serve?

6. An investor has 7 $1000 bills to distribute by mail among 3 mutual funds.
   (a) In how many ways can she invest her money?
   (b) In how many ways can she invest her money if each fund must get at least $1000?

7. Twelve identical letters are to be placed into four mailboxes.
   (a) In how many ways can this be done?
   (b) How many ways are possible if each mailbox must receive at least two letters?

8. Find the number of permutations that can be formed from all the letters of the following words.
   (a) FLORIDA   (b) CALIFORNIA
   (c) MISSISSIPPI   (d) OHIO

9. How many different signals can be created by lining up nine flags in a vertical column if 3 of them are white, 2 are red and 4 are blue?

10. Among the integers $\{1, 2, 3, \ldots, 1000\}$, how many of them are
    (a) divisible by 7?   (b) divisible by 11?
    (c) not divisible by 7 or 11?   (d) divisible by 7 or 11 but *not* both?

11. (a) How many 4-digit numbers can be formed using only the digits 3, 4, 5, 6 and 7?

(b) How many of the numbers in part (a) have some digit repeated?

(c) How many of the numbers in part (a) are even?

(d) How many of the numbers in part (b) are bigger than 5000?

12. Let $\Sigma = \{a, b, c\}$. If $n_1 \geq n_2 \geq n_3$ and $n_1 + n_2 + n_3 = 6$, we call a word in Σ^6 of type n_1–n_2–n_3 if one of the letters appears in the word n_1 times, another letter appears n_2 times and the other letter appears n_3 times. For example, *accabc* is of type 3–2–1 and *caccca* is of type 4–2–0. The number of words in Σ^6 of each type is:

Type	6-0-0	5-1-0	4-2-0	4-1-1	3-3-0	3-2-1	2-2-2
Number	3	36	90	90	60	360	90

Verify this assertion for three of the types.

13. In how many ways can $2n$ elements be partitioned into two sets with n elements each?

14. The English alphabet consists of 21 consonants and 5 vowels. The vowels are a, e, i, o, u.

(a) Prove that no matter how the letters of the English alphabet are listed in order [e.g. zuvarqlgh $\cdots$] there must be 4 consecutive consonants.

(b) Give a list to show that there need not be 5 consecutive consonants.

(c) Suppose now that the letters of the English alphabet are put in a circular array; for example,

Prove that there must be 5 consecutive consonants in such an array.

15. (a) For how many integers between 1000 and 9999 is the sum of the digits exactly 9? Examples: 1431, 5121, 9000, 4320.

(b) How many of the integers counted in part (a) have all nonzero digits?

16. Consider finite sets $\{A_1, A_2, \ldots, A_n\}$. Let $\mathscr{P}_+(n)$ be the set of nonempty subsets I of $\{1, 2, \ldots, n\}$. Show that the Inclusion-Exclusion Principle says that

$$\left| \bigcup_{i=1}^{n} A_i \right| = \sum_{I \in \mathscr{P}_+(n)} (-1)^{|I|+1} \cdot \left| \bigcap_{i \in I} A_i \right|.$$

17. Let S be the set of all sequences of 0's, 1's and 2's of length ten. For example, S contains 0211012201.
   (a) How many elements are in S?
   (b) How many sequences in S have exactly five 0's and five 1's?
   (c) How many sequences in S have exactly three 0's and seven 1's?
   (d) How many sequences in S have exactly three 0's?
   (e) How many sequences in S have exactly three 0's, four 1's and three 2's?
   (f) How many sequences in S have at least one 0, at least one 1 and at least one 2?

18. Consider finite sets satisfying $\chi_A + \chi_B = \chi_C + \chi_D$. Show that $|A| + |B| = |C| + |D|$.

19. In the proof of the "Placing Objects in Boxes" principle, we set up a one-to-one correspondence between strings of five 0's and three 1's and placements of five 0's in four boxes.
   (a) Give the placements that correspond to the following strings: 1 0 1 0 1 0 0 0, 0 1 0 0 1 0 0 1, 1 0 0 0 0 0 1 1, 1 1 1 0 0 0 0 0.
   (b) Give the strings that correspond to the following placements:

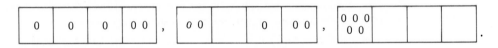

20. Use the Counting Lemma to give another proof of Theorem 1 of § 5.1.

§ 5.3   Pigeon-Hole Principle

The usual Pigeon-Hole Principle asserts that if m objects are placed in n boxes or pigeon-holes and if $m > n$, then some box will receive more than one object. Here is a slight generalization of this fact.

Pigeon-Hole Principle   If a finite set S is partitioned into n sets, then at least one of the sets has $|S|/n$ or more elements.

   Proof. Assume that the set S has a partition $\{A_i : i = 1, 2, \ldots, n\}$ satisfying $|A_i| < |S|/n$ for each i. Then S is the disjoint union of the sets A_i and so

$$|S| = \sum_{i=1}^{n} |A_i| < \sum_{i=1}^{n} \frac{|S|}{n} = |S|,$$

a contradiction.   ∎

   We will often apply this principle when the partition is given by a function. The principle can then be stated as follows.

Pigeon-Hole   Consider a function $f : S \to T$ where S and T are finite sets satisfying
Principle $|S| > k \cdot |T|$. Then at least one of the sets $f^{-1}(t)$ has more than k elements.

Proof. The family $\{f^{-1}(t) : t \in T\}$ partitions S into n sets with $n \leq |T|$. By the principle just proved, some set $f^{-1}(t)$ has at least $|S|/n$ members. Since $|S|/n \geq |S|/|T| > k$ by hypothesis, such a set $f^{-1}(t)$ has more than k elements.   ∎

When $k = 1$, this principle tells us that if $f : S \to T$ and $|S| > |T|$, then at least one of the sets $f^{-1}(t)$ has more than one element. It is remarkable how often this simple observation is helpful in problem-solving.

EXAMPLE 1   Given three integers, there must be two of them whose sum is even. This is because either two of the integers are even or else two are odd, and in either case their sum must be even. Here is a tighter argument. Let S be the set of three integers and for $m \in S$ let $f(m) = 0$ if m is even and $f(m) = 1$ if m is odd. Then $f : S \to \{0, 1\}$ and by the Pigeon-Hole Principle one of the sets $f^{-1}(0)$ or $f^{-1}(1)$ has more than one element. That is, either S contains two [or more] even integers or else S contains two [or more] odd integers.   ∎

EXAMPLE 2   We show that given p integers $a_1, a_2, \ldots, a_p$, not necessarily distinct, some of them add up to a number that is a multiple of p. By Theorem 1 in § 4.3, for each integer m there exists an integer $f(m)$ in $\mathbb{Z}(p)$ such that $m \equiv f(m)$ (mod p). We now consider this function restricted to the set

$$S = \{0, a_1, a_1 + a_2, a_1 + a_2 + a_3, \ldots, a_1 + a_2 + a_3 + \cdots + a_p\}.$$

Since $|S| = p + 1 > p = |\mathbb{Z}(p)|$, the Pigeon-Hole Principle shows that two distinct numbers m and n in S have the same image in $\mathbb{Z}(p)$. Since $m \equiv f(m)$ (mod p), $f(m) = f(n)$ and $f(n) \equiv n$ (mod p), we have $m \equiv n$ (mod p) by (T) of Theorem 2, § 4.3. Therefore the differences $n - m$ and $m - n$ are multiples of p. One of these differences has the form $a_k + a_{k+1} + \cdots + a_l$, i.e., it's a sum of integers from our list that is a multiple of p.

The result just proved is sharp, in the sense that we can give integers $a_1, a_2, \ldots, a_{p-1}$ for which no nonempty subset has sum that is a multiple of p. Simply let $a_j = 1$ for $j = 1, 2, \ldots, p - 1$.   ∎

EXAMPLE 3   Let A be some fixed 10-element subset of $\{1, 2, 3, \ldots, 50\}$. We show that A possesses two different 5-element subsets, the sums of whose elements are equal. Let $\mathcal{S}$ be the family of 5-element subsets B of A. For each B in $\mathcal{S}$, let $f(B)$ be the sum of the numbers in B. Note that we must have $f(B) \geq 1 + 2 + 3 + 4 + 5 = 15$ and $f(B) \leq 50 + 49 + 48 + 47 + 46 = 240$ so that $f : \mathcal{S} \to T$ where $T = \{15, 16, 17, \ldots, 240\}$. Since $|T| = 226$ and $|\mathcal{S}| = \binom{10}{5} = 252$, the Pigeon-Hole Principle shows that $\mathcal{S}$ contains different sets with the same image under f, i.e., different sets the sums of whose elements are equal.   ∎

Some applications of the Pigeon-Hole Principle require considerable ingenuity.

EXAMPLE 4   Here we show that if $a_1, a_2, \ldots, a_{n^2+1}$ is a sequence of $n^2 + 1$ distinct real numbers, then there is a subsequence with $n + 1$ terms that is either increasing or decreasing. This means that there exist subscripts $k_1 < k_2 < \cdots < k_{n+1}$ so that either

$$a_{k_1} < a_{k_2} < \cdots < a_{k_{n+1}}$$

or

$$a_{k_1} > a_{k_2} > \cdots > a_{k_{n+1}}.$$

For each j in $\{1, 2, \ldots, n^2 + 1\}$, let $\text{INC}(j)$ be the length of the longest increasing subsequence stopping at a_j and $\text{DEC}(j)$ be the length of the longest decreasing sequence stopping at a_j. Then define $f(j) = \langle \text{INC}(j), \text{DEC}(j) \rangle$. For example, suppose that $n = 3$ and the original sequence is given by

a_1	a_2	a_3	a_4	a_5	a_6	a_7	a_8	a_9	a_{10}
11	3	15	8	6	12	17	2	7	1 .

Here $a_5 = 6$, $\text{INC}(5) = 2$ since a_2, a_5 is the longest increasing subsequence stopping at a_5 and $\text{DEC}(5) = 3$ since a_1, a_4, a_5 and a_3, a_4, a_5 are longest decreasing subsequences stopping at a_5. Similarly, $\text{INC}(6) = 3$ and $\text{DEC}(6) = 2$ and so $f(5) = \langle 2, 3 \rangle$ and $f(6) = \langle 3, 2 \rangle$. Indeed, in this example

$$f(1) = \langle 1, 1 \rangle \qquad f(2) = \langle 1, 2 \rangle \qquad f(3) = \langle 2, 1 \rangle$$
$$f(4) = \langle 2, 2 \rangle \qquad f(5) = \langle 2, 3 \rangle \qquad f(6) = \langle 3, 2 \rangle$$
$$f(7) = \langle 4, 1 \rangle \qquad f(8) = \langle 1, 4 \rangle \qquad f(9) = \langle 3, 3 \rangle$$
$$f(10) = \langle 1, 5 \rangle.$$

This particular example has increasing subsequences of length 4, such as a_2, a_5, a_6, a_7 since $\text{INC}(7) = 4$, and also decreasing subsequences of length 4, such as a_1, a_4, a_5, a_8 since $\text{DEC}(8) = 4$. Since $\text{DEC}(10) = 5$ it even has a decreasing subsequence of length 5. Note that f is one-to-one in this example so that f cannot map the 10-element set $\{1, 2, 3, \ldots, 10\}$ into the 9-element set $\{1, 2, 3\} \times \{1, 2, 3\}$. In other words, the one-to-oneness of f alone forces at least one $\text{INC}(j)$ or $\text{DEC}(j)$ to exceed 3 and this in turn forces our sequence to have an increasing or decreasing subsequence of length 4.

To prove the general result we first claim that f must be one-to-one, which we prove directly. Consider j, k in $\{1, 2, 3, \ldots, n^2 + 1\}$ with $j < k$. If $a_j < a_k$, then $\text{INC}(j) < \text{INC}(k)$ since a_k could be attached to the longest increasing sequence ending at a_j to get a longer increasing sequence ending at a_k. Similarly, if $a_j > a_k$, then $\text{DEC}(j) < \text{DEC}(k)$. In either case the *ordered pairs* $f(j)$ and $f(k)$ cannot be equal, i.e., $f(j) \neq f(k)$. Since f is one-to-one, f

cannot map $\{1, 2, 3, \ldots, n^2 + 1\}$ into $\{1, 2, \ldots, n\} \times \{1, 2, \ldots, n\}$ by the Pigeon-Hole Principle and so there is a j such that either $\text{INC}(j) \geq n + 1$ or $\text{DEC}(j) \geq n + 1$. Hence the original sequence has an increasing or decreasing subsequence with $n + 1$ terms. ∎

EXAMPLE 5   You probably know that the decimal expansions of rational numbers repeat themselves, but you may never have seen a proof. For example,

$$\frac{29}{54} = .537037037037037 \cdots;$$

check this *by long division* before proceeding further! The general fact can be seen as a consequence of the Pigeon-Hole Principle. We may assume that the given rational number has the form m/n where $0 < m < n$. We analyze the division algorithm. When we divide m by n we obtain $.d_1 d_2 d_3 \cdots$ where

$$10 \cdot m = n \cdot d_1 + r_1 \qquad 0 \leq r_1 < n$$

$$10 \cdot r_1 = n \cdot d_2 + r_2 \qquad 0 \leq r_2 < n$$

$$10 \cdot r_2 = n \cdot d_3 + r_3 \qquad 0 \leq r_3 < n$$

etc. so that $10 \cdot r_j = n \cdot d_{j+1} + r_{j+1}$ where $0 \leq r_{j+1} < n$. We illustrate this in Figure 1. The remainders r_j all take their values in $\{0, 1, 2, \ldots, n - 1\}$. By the Pigeon-Hole Principle, after a while the values must repeat. In fact, two of the numbers $r_1, r_2, \ldots, r_{n+1}$ must be equal. Hence there are k and m in $\{1, 2, \ldots, n + 1\}$ with $k < m$ and $r_k = r_m$. Let $l = m - k$ so that $r_k = r_{k+l}$. [In our carefully selected example, k can be 1 and l can be 3.] We will show that the sequences of r_i's and d_i's repeat every l terms beginning with $i = k + 1$. Since $10r_k = 10r_{k+l}$ we have $nd_{k+1} + r_{k+1} = nd_{k+l+1} + r_{k+l+1}$. Hence $r_{k+1} \equiv r_{k+l+1} \pmod{n}$; since both r_{k+1} and r_{k+l+1} are in $\mathbb{Z}(n)$, they

$$
\begin{array}{llll}
 & .d_1 d_2 d_3 d_4 d_5 \cdots & = .53703 \cdots & \\
54\,\overline{)\,290} & 10 \cdot 29 = 10m & & \\
\underline{270} & 54 \cdot 5 = nd_1 & & \\
200 & 10 \cdot 20 = 10r_1 & & r_1 = 20 \\
\underline{162} & 54 \cdot 3 = nd_2 & & \\
380 & 10 \cdot 38 = 10r_2 & & r_2 = 38 \\
\underline{378} & 54 \cdot 7 = nd_3 & & \\
20 & 10 \cdot 2 = 10r_3 & & r_3 = 2 \\
\underline{0} & 54 \cdot 0 = nd_4 & & \\
200 & 10 \cdot 20 = 10r_4 & & r_4 = 20 \\
\underline{162} & 54 \cdot 3 = nd_5 & & \\
380 & 10 \cdot 38 = 10r_5 & & r_5 = 38 \\
\underline{378} & \text{etc.} & &
\end{array}
$$

Long division

Figure 1

must be equal. Thus $nd_{k+1} = nd_{k+l+1}$ and this implies that $d_{k+1} = d_{k+l+1}$. We have just checked the basis for an induction proof of

(∗) $$d_j = d_{j+l} \quad \text{and} \quad r_j = r_{j+l}$$

for $j \geq k + 1$. For once, the inductive step is no harder to prove. If fact, if (∗) holds for j, then

$$nd_{j+1} + r_{j+1} = 10r_j = 10r_{j+l} = nd_{j+l+1} + r_{j+l+1}$$

so that $r_{j+1} \equiv r_{j+l+1} \pmod{n}$ and hence $r_{j+1} = r_{j+l+1}$. Consequently $nd_{j+1} = nd_{j+l+1}$ and $d_{j+1} = d_{j+l+1}$, which establishes (∗) for $j + 1$. By mathematical induction, (∗) holds for all $j \geq k + 1$.

From (∗) we have

$$d_{k+1} = d_{k+l+1}, \qquad d_{k+2} = d_{k+l+2}, \quad \ldots, \quad d_{k+l} = d_{k+2l}$$

so that

$$d_{k+1} d_{k+2} \cdots d_{k+l} = d_{k+l+1} d_{k+l+2} \cdots d_{k+2l}.$$

In fact, this whole block repeats indefinitely. In other words, the decimal expansion of m/n is a repeating expansion. ∎

The next example doesn't exactly apply the Pigeon-Hole Principle but it is a pigeon-hole problem in spirit.

EXAMPLE 6 Consider nine nonnegative real numbers $a_1, a_2, a_3, \ldots, a_9$ with sum 90.

(a) We show that there must be three of the numbers having sum at least 30. This is easy because

$$90 = (a_1 + a_2 + a_3) + (a_4 + a_5 + a_6) + (a_7 + a_8 + a_9)$$

and so at least one of the sums in parentheses must be at least 30.

(b) We show that there must be four of the numbers having sum at least 40. There are several ways to do this, but none of them is quite as simple as the method in part (a). Our first approach is to note that the sum of all the numbers in Figure 2 is 360 since each row sums to 90. Hence one of the nine columns must have sum at least $360/9 = 40$.

Our second approach is to use part (a) to select three of the numbers having sum $s \geq 30$. One of the remaining six numbers must have value at

a_1	a_2	a_3	a_4	a_5	a_6	a_7	a_8	a_9
a_2	a_3	a_4	a_5	a_6	a_7	a_8	a_9	a_1
a_3	a_4	a_5	a_6	a_7	a_8	a_9	a_1	a_2
a_4	a_5	a_6	a_7	a_8	a_9	a_1	a_2	a_3

Figure 2

least $\frac{1}{6}$ of their sum $90 - s$. Adding this to the selected three gives four numbers with sum at least

$$s + \tfrac{1}{6}(90 - s) = 15 + \tfrac{5}{6}s \geq 15 + \tfrac{5}{6} \cdot 30 = 40.$$

Our third approach is to note that we may as well assume that $a_1 \geq a_2 \geq \cdots \geq a_9$. Then it is clear that $a_1 + a_2 + a_3 + a_4$ is the largest sum using four of the numbers and our task is relatively concrete: to show $a_1 + a_2 + a_3 + a_4 \geq 40$. Moreover, this suggests showing

$$(*) \qquad\qquad a_1 + a_2 + \cdots + a_n \geq 10n$$

for $1 \leq n \leq 9$. We can do this by a finite induction, i.e., by noting that $(*)$ holds for $n = 1$ and showing that if $(*)$ holds for n, $1 \leq n < 9$, then $(*)$ holds for $n + 1$. We will adapt the method of our second approach. Assume $(*)$ holds for n and let $s = a_1 + a_2 + \cdots + a_n$. Since a_{n+1} is the largest of the remaining $9 - n$ numbers, $a_{n+1} \geq (90 - s)/(9 - n)$. Hence

$$a_1 + a_2 + \cdots + a_n + a_{n+1} = s + a_{n+1} \geq s + \frac{90 - s}{9 - n}$$

$$= s + \frac{90}{9 - n} - \frac{s}{9 - n} = s\left(1 - \frac{1}{9 - n}\right) + \frac{90}{9 - n}$$

$$\geq 10n\left(1 - \frac{1}{9 - n}\right) + \frac{90}{9 - n} \qquad \text{[inductive assumption]}$$

$$= 10n + \frac{90 - 10n}{9 - n} = 10n + 10 = 10(n + 1).$$

This finite induction argument shows that $(*)$ holds for $1 \leq n \leq 9$.

For this particular problem, the first and last approaches are far superior because they generalize in an obvious way without further tricks. ∎

Most, but not all, of the following exercises involve the Pigeon-Hole Principle. They also provide more practice using the techniques from § 5.1 and § 5.2. The exercises are not equally difficult and some may require extra ingenuity.

EXERCISES 5.3

1. (a) Given four integers, explain why two of them must be congruent modulo 3.
   (b) Prove that if $a_1, a_2, \ldots, a_{p+1}$ are integers, then two of them must be congruent modulo p.

2. (a) A sack contains 50 marbles of four different colors. Explain why there are at least 13 marbles of the same color.
   (b) If exactly 8 of the marbles are red, explain why there are at least 14 of the same color.

3. Suppose that 73 marbles are placed in eight boxes.
   (a) Show that some box contains at least 10 marbles.
   (b) Show that if two of the boxes are empty, then some box contains at least 13 marbles.

4. (a) Let B be a 12-element subset of $\{1, 2, 3, 4, 5, 6\} \times \{1, 2, 3, 4, 5, 6\}$. Show that B contains two different ordered pairs, the sums of whose entries are equal.
   (b) How many times can a pair of dice be tossed without obtaining the same sum twice?

5. Let A be a 10-element subset of $\{1, 2, 3, \ldots, 50\}$. Show that A possesses two different 4-element subsets, the sums of whose elements are equal.

6. Let S be a 3-element set of integers. Show that S has two different nonempty subsets such that the sums of the numbers in each of the subsets are congruent modulo 6.

7. Let A be a subset of $\{1, 2, 3, \ldots, 149, 150\}$ consisting of 25 numbers. Show that there are two disjoint pairs of numbers from A having the same sum [for example, $\{3, 89\}$ and $\{41, 51\}$ have the same sum, namely 92].

8. For the following sequences, find an increasing or decreasing subsequence of length 5 if you can.
   (a) 4, 3, 2, 1, 8, 7, 6, 5, 12, 11, 10, 9, 16, 15, 14, 13
   (b) 17, 13, 14, 15, 16, 9, 10, 11, 12, 5, 6, 7, 8, 1, 2, 3, 4
   (c) 10, 6, 2, 14, 3, 17, 12, 8, 7, 16, 13, 11, 9, 15, 4, 1, 5

9. Find the decimal expansions for $1/7, 2/7, 3/7, 4/7, 5/7$ and $6/7$ and compare them.

10. (a) Show that if ten nonnegative integers have sum 101, there must be three with sum at least 31.
    (b) Prove a generalization of part (a): If $1 \leq k \leq n$ and if n nonnegative integers have sum m, there must be k with sum at least _____.

11. In this problem the twenty-four numbers 1, 2, 3, 4, $\ldots$, 24 are permuted in some way, say $\langle n_1, n_2, n_3, n_4, \ldots, n_{24} \rangle$.
    (a) Show that there must be four consecutive numbers in the permutation that are less than 20, i.e., ≤ 19.
    (b) Show that $n_1 + n_2 + n_3 + \cdots + n_{24} = 300$.
    (c) Show that there must be three consecutive numbers in the permutation with sum ≥ 38.
    (d) Show that there must be five consecutive numbers in the permutation with sum ≥ 61.

12. A roulette wheel is divided into 36 sectors with numbers 1, 2, 3, $\ldots$, 36. [We are omitting sectors with 0 and 00 that are included in Las Vegas and give the house the edge in gambling.]
    (a) Show that there are four consecutive sectors with sum greater than 74.
    (b) Show that there are five consecutive sectors with sum greater than 94.

13. Let n_1, n_2 and n_3 be distinct positive integers. Show that at least one of n_1, n_2, n_3, $n_1 + n_2$, $n_2 + n_3$ or $n_1 + n_2 + n_3$ is divisible by 3. *Hint*: Map $\{n_1, n_1 + n_2, n_1 + n_2 + n_3\}$ to $\mathbb{Z}(3)$ by f where $f(m) \equiv m \pmod{3}$.

14. A club has six men and nine women members. How many ways can a committee of five be selected if
    (a) there are no restrictions on the committee make-up?
    (b) there must be two men and three women on the committee?
    (c) there must be at least one man and at least one woman on the committee?
    (d) the committee must consist of only men or else it must consist of only women?

15. Six-digit numbers are to be formed using the integers in the set $A = \{1, 2, 3, 4, 5, 6, 7, 8\}$.
    (a) How many such numbers can be formed if repetitions are allowed?
    (b) In part (a), how many of the numbers contain at least one 3 and at least one 5?
    (c) How many six-digit numbers can be formed if each digit in A can be used at most once?
    (d) How many six-digit numbers can be formed that consist of one 2, two 4's and three 5's?

16. How many divisors are there of 6000? *Hint*: $6000 = 2^4 \cdot 3 \cdot 5^3$ and every divisor has the form $2^m 3^n 5^r$ where $m \leq 4$, $n \leq 1$ and $r \leq 3$.

17. Consider n in $\mathbb{P}$ and let S be a subset of $\{1, 2, \ldots, 2n\}$ consisting of $n + 1$ numbers.
    (a) Show that S contains two numbers that are relatively prime.
    (b) Show that S contains two numbers such that one of them divides the other.
    (c) Show that part (a) can fail if S has only n elements.
    (d) Show that part (b) can fail if S has only n elements.

18. (a) Consider a subset A of $\{0, 1, 2, \ldots, p\}$ such that $|A| > (p/2) + 1$. Show that A contains two different numbers whose sum is p.
    (b) For $p = 6$, find A with $|A| = (p/2) + 1$ not satisfying the conclusion in part (a).
    (c) For $p = 7$, find A with $|A| = [(p - 1)/2] + 1$ not satisfying the conclusion in part (a).

§ 5.4   Infinite Sets

Mathematicians have a way of classifying infinite sets according to their "size." First they generalize the concept "two sets are of the same size." The clue to the commonly accepted correct approach is the following elementary observation:

> Two finite sets are of the same size if and only if there exists a one-to-one correspondence between them.

Accordingly, we regard two sets S and T, finite or infinite, to be of the **same size** if there is a one-to-one correspondence between them. In this book we will not study this classification scheme for sets in detail, but we will distinguish between two kinds of infinite sets.

Any set that is the same size as the set $\mathbb{P}$ of positive integers will be called **countably infinite**. Thus a set S is countably infinite if and only if there exists a one-to-one correspondence between $\mathbb{P}$ and S. A set is **countable** if it is finite or countably infinite. One is able to count or list such a nonempty set by matching it with $\{1, 2, \ldots, n\}$ for some $n \in \mathbb{P}$, or with $\mathbb{P}$. In the infinite case, the list will never end. As one would expect, a set is **uncountable** if it is not countable.

EXAMPLE 1   (a) The set $\mathbb{N}$ is countably infinite because $f(n) = n - 1$ defines a one-to-one function f mapping $\mathbb{P}$ onto $\mathbb{N}$. Its inverse f^{-1} is a one-to-one mapping of $\mathbb{N}$ onto $\mathbb{P}$; note $f^{-1}(n) = n + 1$ for $n \in \mathbb{N}$. Even though $\mathbb{P}$ is a proper subset of $\mathbb{N}$, by our definition $\mathbb{P}$ is the same size as $\mathbb{N}$. This may be surprising, since a similar situation does not occur for finite sets. Oh well, $\mathbb{N}$ has only one more element than $\mathbb{P}$.

(b) The set $\mathbb{Z}$ of *all* integers is also countably infinite. A one-to-one function f from $\mathbb{Z}$ onto $\mathbb{P}$ is indicated in Figure 1, where we have found it convenient to bend the picture of $\mathbb{Z}$. This function can be given by a formula, if desired:

$$f(n) = \begin{cases} 2n + 1 & \text{for } n \geq 0, \\ -2n & \text{for } n < 0. \end{cases}$$

Even though $\mathbb{Z}$ looks about twice as big as $\mathbb{P}$, these sets are of the same size. Beware! For infinite sets, your intuition may be unreliable. Or, to take a more positive approach, you may need to refine your intuition when dealing with infinite sets.

(c) Even the set $\mathbb{Q}$ of all rational numbers is countably infinite. This is striking, because the set of rational numbers is distributed evenly throughout $\mathbb{R}$. To give a one-to-one correspondence between $\mathbb{P}$ and $\mathbb{Q}$, a picture is worth

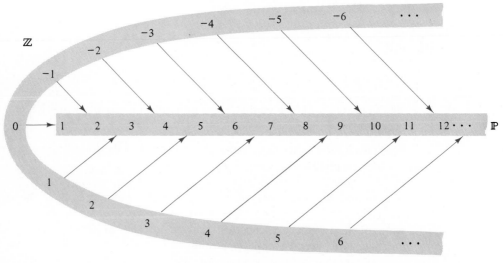

A one-to-one correspondence of $\mathbb{Z}$ onto $\mathbb{P}$

Figure 1

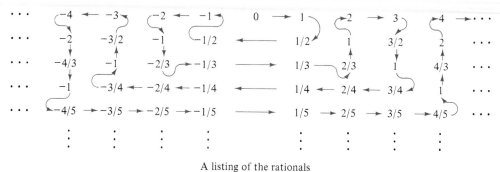

A listing of the rationals

Figure 2

a thousand formulas. See Figure 2. The function f is obtained by following the arrows and skipping over repetitions. Thus $f(1) = 0$, $f(2) = 1$, $f(3) = \frac{1}{2}$, $f(4) = -\frac{1}{2}$, $f(5) = -1$, $f(6) = -2$, $f(7) = -\frac{2}{3}$, etc. ∎

EXAMPLE 2   So far, all of our examples of graphs and trees have had finitely many vertices and edges. However, there is no such restriction in the general definitions. Figure 3 contains partial pictures of some infinite graphs. The set of vertices in Figure 3(a) is $\mathbb{Z}$, and only consecutive integers are connected by an edge. Note that this is an infinite tree with *no* leaves; contrast this with the fact that every finite tree with more than one vertex has leaves. The set of vertices in Figure 3(b) is $\mathbb{Z} \times \{0, 1\}$; this tree has infinitely many leaves. The central vertex in the tree in Figure 3(c) has infinite degree; all its other vertices are leaves. There are only two vertices in the graph of Figure 3(d), but they are connected by infinitely many edges.

In all these examples, the sets of vertices and edges are countable. Graphs don't have to be countable, but it is hard to draw or visualize uncountable ones and we will have no need for them. ∎

In the next example we illustrate a technique that goes back to Georg Cantor, the father of set theory. You may find the result in part (b) more interesting, but the details in part (a) are easier to follow.

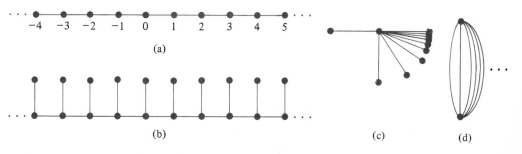

Figure 3

EXAMPLE 3   (a) The set $FUN(\mathbb{P}, \{0,1\})$ of all functions from $\mathbb{P}$ into $\{0,1\}$ is uncountable. Equivalently, the set of all infinite strings of 0's and 1's is uncountable. Obviously $FUN(\mathbb{P}, \{0,1\})$ is infinite, so if it were countable there would exist an infinite listing $\{f_1, f_2, \ldots\}$ of *all* the functions in this set. We define a function f^* on $\mathbb{P}$ as follows:

$$f^*(n) = \begin{cases} 0 & \text{if} \quad f_n(n) = 1 \\ 1 & \text{if} \quad f_n(n) = 0 \end{cases}.$$

For each n in $\mathbb{P}$, $f^*(n) \neq f_n(n)$ and so the functions f^* and f_n must be different. Thus $f^* \neq f_n$ for all $n \in \mathbb{P}$ and so $\{f_1, f_2, \ldots\}$ is not a listing of *all* functions in $FUN(\mathbb{P}, \{0,1\})$. This contradiction shows that $FUN(\mathbb{P}, \{0,1\})$ is uncountable.

(b) The interval $[0, 1)$ is uncountable. If it were countable, there would exist a one-to-one function f mapping $\mathbb{P}$ onto $[0, 1)$. We show that this is impossible. Each number in $[0, 1)$ has a decimal expansion $.d_1 d_2 d_3 \cdots$ where each d_j is a digit in $\{0, 1, 2, 3, 4, 5, 6, 7, 8, 9\}$. In particular, each number $f(k)$ has the form $.d_{1k} d_{2k} d_{3k} \cdots$; here d_{nk} represents the nth digit in $f(k)$. Consider Figure 4 and focus on the indicated "diagonal digits." We define d_n^* for $n \in \mathbb{P}$ as follows: if $d_{nn} \neq 1$, let $d_n^* = 1$ and if $d_{nn} = 1$, let $d_n^* = 2$. The point is that $d_n^* \neq d_{nn}$ for all $n \in \mathbb{P}$. Now $.d_1^* d_2^* d_3^* \cdots$ represents a number x in $[0, 1)$ which is different from $f(n)$ in the nth digit for each $n \in \mathbb{P}$. Thus x cannot be one of the numbers $f(n)$; i.e., $x \notin Im(f)$ and so f does not map $\mathbb{P}$ onto $[0, 1)$.

$$f(1) = .d_{11}\, d_{21}\, d_{31}\, d_{41} \ \cdot\cdot\cdot$$

$$f(2) = .d_{12}\, d_{22}\, d_{32}\, d_{42} \ \cdot\cdot\cdot$$

$$f(3) = .d_{13}\, d_{23}\, d_{33}\, d_{43} \ \cdot\cdot\cdot$$

$$f(4) = .d_{14}\, d_{24}\, d_{34}\, d_{44} \ \cdot\cdot\cdot$$

$$\cdot$$
$$\cdot$$
$$\cdot$$

Cantor's diagonal procedure

Figure 4

Note that we arranged for all of the digits of x to be 1's and 2's. This choice was quite arbitrary, except that we deliberately avoided 0's and 9's since there are some numbers whose expansions involve 0's and 9's that have two decimal expansions. For example, $.250000 \cdots$ and $.249999 \cdots$ represent the same number in $[0, 1)$. ∎

The proof in Example 3(b) can be modified to prove that $\mathbb{R}$ and $(0, 1)$ are uncountable; in fact, all intervals $[a, b]$, $[a, b)$, $(a, b]$ and (a, b) are uncountable for $a < b$. In view of Exercise 9, another way to show that these sets are uncountable is to show that they are in one-to-one correspondence with each other. In fact, they are also in one-to-one correspondence with unbounded intervals. Showing the existence of such one-to-one correspondences can be challenging. We provide a couple of the trickier arguments in the next example, and ask for some easier ones in Exercise 3.

EXAMPLE 4   (a) We show that $\mathbb{R}$ and $(0, 1)$ are in one-to-one correspondence, and hence are of the same size. Though trigonometric functions are not necessary here [Exercise 5], we will use a ready-made function from trigonometry, namely the tangent; see Figure 5(a). The function f given by $f(x) = \tan x$ is one-to-one on $(-\pi/2, \pi/2)$ and maps this interval onto $\mathbb{R}$. It is easy to find a linear function g mapping $(0, 1)$ onto $(-\pi/2, \pi/2)$, namely $g(x) = \pi x - \pi/2$; see Figure 5(b). The composite function $f \circ g$ is a one-to-one correspondence between $(0, 1)$ and $\mathbb{R}$.

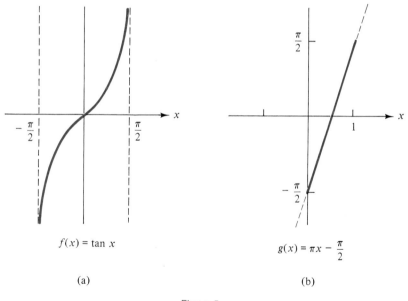

$$f(x) = \tan x$$

$$g(x) = \pi x - \frac{\pi}{2}$$

(a)

(b)

Figure 5

(b) We show that $[0, 1)$ and $(0, 1)$ have the same size. No simple formula provides us with a one-to-one mapping between these sets. The trick is to isolate some infinite sequence in $(0, 1)$, say $\frac{1}{2}, \frac{1}{3}, \frac{1}{4}, \ldots$, and then map this sequence onto $0, \frac{1}{2}, \frac{1}{3}, \frac{1}{4}, \ldots$, while leaving the complement fixed. That is, let

$$C = (0, 1) \setminus \left\{ \frac{1}{n} : n = 2, 3, 4, \ldots \right\}$$

and define

$$f(x) = \begin{cases} 0 & \text{if } x = \frac{1}{2}, \\[2mm] \dfrac{1}{n-1} & \text{if } x = 1/n \text{ for some integer } n \geq 3, \\[2mm] x & \text{if } x \in C. \end{cases}$$

See Figure 6. ▮

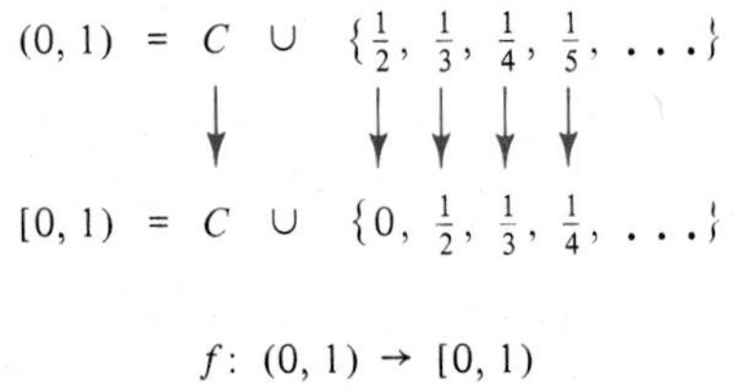

$$(0, 1) \;=\; C \;\cup\; \{\tfrac{1}{2}, \tfrac{1}{3}, \tfrac{1}{4}, \tfrac{1}{5}, \ldots\}$$
$$\downarrow \qquad \downarrow \;\downarrow\; \downarrow \;\downarrow$$
$$[0, 1) \;=\; C \;\cup\; \{0, \tfrac{1}{2}, \tfrac{1}{3}, \tfrac{1}{4}, \ldots\}$$

$$f: (0, 1) \rightarrow [0, 1)$$

Figure 6

We next prove two basic facts about countable sets.

Theorem   (a) Subsets of countable sets are countable.

(b) The countable union of countable sets is countable.

Proof. (a) It is enough to show that subsets of $\mathbb{P}$ are countable. Consider a subset A of $\mathbb{P}$. Clearly A is countable if A is finite. Suppose that A is infinite. We define $f(1)$ to be the least element in A. Then we define $f(2)$ to be the least element in $A \backslash \{f(1)\}$; then $f(3)$ to be the least element in $A \backslash \{f(1), f(2)\}$, etc. This process continues, so that $f(n+1)$ is the least element in $A \backslash \{f(k) : 1 \leq k \leq n\}$ for each $n \in \mathbb{P}$. It is easy to verify that this recursive definition provides a one-to-one function f mapping $\mathbb{P}$ onto A [Exercise 10] and so A is countable.

(b) The statement in part (b) means that if I is a countable index set and if $\{A_i : i \in I\}$ is a family of countable sets, then the union $\bigcup_{i \in I} A_i$ is countable. We may assume that each A_i is nonempty and that $\bigcup_{i \in I} A_i$ is infinite, and we may assume that $I = \mathbb{P}$ or that I has the form $\{1, 2, \ldots, n\}$. If $I = \{1, 2, \ldots, n\}$ and we defined $A_i = A_n$ for $i > n$, we would obtain a family $\{A_i : i \in \mathbb{P}\}$ indexed by $\mathbb{P}$ with the same union. Thus we may assume that $I = \mathbb{P}$. Each set A_i is finite or countably infinite. By repeating elements if A_i is finite, we can list each A_i as follows:

$$A_i = \{a_{1i}, a_{2i}, a_{3i}, a_{4i}, \ldots\}.$$

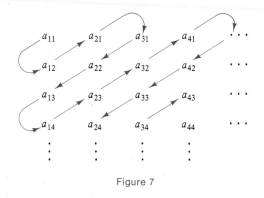

Figure 7

The elements in $\bigcup_{i \in I} A_i$ can be listed in an array as in Figure 7. The arrows in the figure suggest a single listing for $\bigcup_{i \in I} A_i$:

(*) $a_{11}, a_{12}, a_{21}, a_{31}, a_{22}, a_{13}, a_{14}, a_{23}, a_{32}, a_{41}, \ldots .$

Some elements may be repeated, but the list includes infinitely many distinct elements since $\bigcup_{i \in I} A_i$ is infinite. Now a one-to-one mapping f of $\mathbb{P}$ onto $\bigcup_{i \in I} A_i$ is obtained as follows: $f(1) = a_{11}$, $f(2)$ is the next element listed in (*) different from $f(1), f(3)$ is the next element listed in (*) different from $f(1)$ and $f(2)$, etc. ∎

EXAMPLE 5   The argument in Example 1(c) that $\mathbb{Q}$ is countable is similar to the proof of part (b) of the theorem. In fact, we can use the theorem to give another proof that $\mathbb{Q}$ is countable. For each n in $\mathbb{P}$, let

$$A_n = \left\{ \frac{m}{n} : m \in \mathbb{Z} \right\}.$$

Thus A_n consists of all integer multiples of $1/n$. Each A_n is clearly in one-to-one correspondence with $\mathbb{Z}$ [map m to m/n] and so each A_n is countable. By part (b) of the theorem, the union

$$\bigcup_{n \in \mathbb{P}} A_n = \mathbb{Q}$$

is also countable. ∎

EXAMPLE 6   (a) If Σ is a finite alphabet, then the set Σ^* of all words using letters from Σ is countably infinite. Note that Σ is nonempty by definition. We already know that Σ^* is infinite. Recall that

$$\Sigma^* = \bigcup_{k=0}^{\infty} \Sigma^k$$

where each Σ^k is finite. Thus Σ^* is a countable union of countable sets, and hence Σ^* itself is countable by part (b) of the theorem.

(b) Imagine, if you can, a countably infinite alphabet Σ and let Σ^* consist of all words using letters of Σ, i.e., all finite strings of letters from Σ. For each $k \in \mathbb{P}$, the set Σ^k of all words of length k is in one-to-one correspondence with the product set $\Sigma^k = \Sigma \times \Sigma \times \cdots \times \Sigma$. In fact, the correspondence maps each word $a_1 a_2 \cdots a_k$ to the k-tuple $\langle a_1, a_2, \ldots, a_k \rangle$. So each set Σ^k is countable by Exercise 17. The 1-element set $\Sigma^0 = \{\epsilon\}$ is countable too. Hence $\Sigma^* = \bigcup_{k=0}^{\infty} \Sigma^k$ is countable by part (b) of the theorem.   ∎

EXAMPLE 7   Consider any graph and let V and E be the sets of vertices and edges. Even if V or E is infinite, each path has finite length by definition. Let Π [capital Greek pi] be the set of all paths of the graph.

(a) If E is nonempty, Π is infinite. For if e is any edge, then e, $e\,e$, $e\,e\,e$, etc. all describe paths in the graph.

(b) If E is finite, then Π is countable. For purposes of counting, let's view E as an alphabet. Since each path is a sequence of edges, it corresponds to a word in E^*. Of course, not all words in E^* correspond to paths, since endpoints of adjacent edges must match up. But the set Π is in one-to-one correspondence with some *subset* of E^*. The set E^* is countably infinite by Example 6(a) and so Π is countable by part (a) of the theorem.

(c) If E is countably infinite, then Π is still countable. Simply apply Example 6(b) instead of Example 6(a) in the discussion of part (b).   ∎

EXERCISES 5.4

1. Let A and B be finite sets with $|A| < |B|$. True or False.
   (a) There is a one-to-one map of A into B.
   (b) There is a one-to-one map of A onto B.
   (c) There is a one-to-one map of B into A.
   (d) There is a function mapping A onto B.
   (e) There is a function mapping B onto A.

2. True or False.
   (a) The set of positive rational numbers is countably infinite.
   (b) The set of all rational numbers is countably infinite.
   (c) The set of positive real numbers is countably infinite.
   (d) The intersection of two countably infinite sets is countably infinite.
   (e) There is a one-to-one correspondence between the set of all even integers and the set $\mathbb{N}$ of natural numbers.

3. Give one-to-one correspondences between the following pairs of sets.
   (a) $(0, 1)$ and $(-1, 1)$ (b) $[0, 1)$ and $(0, 1]$
   (c) $[0, 1]$ and $[-5, 8]$ (d) $(0, 1)$ and $(1, \infty)$
   (e) $(0, 1)$ and $(0, \infty)$ (f) $\mathbb{R}$ and $(0, \infty)$

4. Let $E = \{n \in \mathbb{N} : n$ is even$\}$. Show that E and $\mathbb{N}\backslash E$ are countable by exhibiting one-to-one functions $f : \mathbb{P} \to E$ and $g : \mathbb{P} \to \mathbb{N}\backslash E$.

5. Here is another one-to-one function f mapping $(0, 1)$ onto $\mathbb{R}$:

$$f(x) = \frac{2x - 1}{x(1 - x)}.$$

(a) Sketch the graph of f.

(b) If you know some calculus, prove that f is one-to-one by showing that its derivative is positive on $(0, 1)$.

6. Which of the following sets are countable? countably infinite?

(a) $\{0, 1, 2, 3, 4\}$ (b) $\{n \in \mathbb{N} : n \le 73\}$

(c) $\{n \in \mathbb{Z} : n \le 73\}$ (d) $\{n \in \mathbb{Z} : |n| \le 73\}$

(e) $\{5, 10, 15, 20, 25, \ldots\}$ (f) $\mathbb{N} \times \mathbb{N}$

(g) $[\frac{1}{4}, \frac{1}{3}]$

7. Let Σ be the alphabet $\{a, b, c\}$. Which of the following sets are countably infinite?

(a) Σ^{73} (b) Σ^*

(c) $\displaystyle\bigcup_{k=0}^{\infty} \Sigma^{2k} = \{w \in \Sigma^* : \text{length}(w) \text{ is even}\}$

(d) $\displaystyle\bigcup_{k=0}^{3} \Sigma^{k}$ (e) $\displaystyle\bigcup_{k=1}^{3} \Sigma^{2k}$

8. A set A has m elements and a set B has n elements. How many functions from A into B are one-to-one? *Hint*: Consider the cases $m \le n$ and $m > n$ separately.

9. Let S be an infinite set.

(a) Show that if there is a one-to-one correspondence of S onto some countable set, then S itself is countable.

(b) Show that if there is a one-to-one correspondence of S onto some uncountable set, then S is uncountable.

10. Complete the proof of part (a) of the theorem by showing that f is one-to-one and that f maps $\mathbb{P}$ onto A.

11. (a) Prove that if S and T are countable, then $S \times T$ is countable.

(b) Prove that if f maps S onto T and S is countable, then T is countable.

(c) Use parts (a) and (b) to give another proof that $\mathbb{Q}$ is countable. *Suggestion*: For $\langle m, n \rangle$ in $\mathbb{Z} \times \mathbb{P}$, define $f(\langle m, n \rangle) = m/n$.

12. Show that if S and T have the same size, so do $\mathscr{P}(S)$ and $\mathscr{P}(T)$.

13. (a) Show that $\text{FUN}(\mathbb{P}, \{0, 1\})$ is in one-to-one correspondence with the set $\mathscr{P}(\mathbb{P})$ of all subsets of $\mathbb{P}$.

(b) Show that $\mathscr{P}(\mathbb{P})$ is uncountable.

14. Show that any disjoint family of nonempty subsets of a countable set is countable.

15. Show that if S is countable, then $S^n = S \times S \times \cdots \times S$ [n times] is countable for each n. *Hint*: Use Exercise 11(a) and induction.

CHAPTER HIGHLIGHTS

See the end of Chapter 0 if you have forgotten how to use the following lists to
review. Think always of examples.

Concepts

number of elements in $S = |S|$
selection with/without replacement
r-permutation, permutation
countable, countably infinite, uncountable

Facts

$|S \cup T| = |S| + |T| - |S \cap T|$ for finite sets.

$$|S_1 \times \cdots \times S_k| = \prod_{j=1}^{k} |S_j|.$$

$$\binom{n}{r} = \frac{n!}{(n-r)!\, r!}.$$

Binomial theorem: $(a + b)^n = \sum\limits_{r=0}^{n} \binom{n}{r} a^r b^{n-r}.$

Counting Lemma.

Formula $\dfrac{n!}{n_1! \cdots n_k!}$ for counting ordered partitions.

Formula $\dbinom{n+k-1}{k-1}$ for ways to place n objects in k boxes.

Inclusion-Exclusion Principle.
Pigeon-Hole Principle.
Every rational number has a repeating decimal expansion.
Subsets and countable unions of countable sets are countable.

Methods

Cantor's diagonalization procedure.
Numerous clever ideas illustrated in examples.

MORE LOGIC
AND INDUCTION

After developing quantifiers in § 6.1, we give in § 6.2 a brief introduction to the predicate calculus. Sections 6.3 and 6.4 concern some sophisticated aspects of mathematical induction, and serve as an elaboration on the treatment given in § 2.6.

§ 6.1   Quantifiers

The propositional calculus is a nice, complete, self-contained theory of logic, but it is totally inadequate for most of mathematics. The problem is that the propositional calculus does not allow the use of an infinite number of propositions. In addition, the notation is awkward for handling a large finite number of propositions. For example, we frequently encounter an infinite sequence of propositions $p(n)$ indexed by $\mathbb{N}$. The informal statement "$p(n)$ is true for all n" means "$p(0)$ is true, $p(1)$ is true, $p(2)$ is true, etc." The only symbolism from the propositional calculus would be $p(0) \wedge p(1) \wedge p(2) \wedge \cdots$, but this is not acceptable in the propositional calculus. Similarly, the informal statement "$p(n)$ is true for some n" would correspond to the unacceptable $p(0) \vee p(1) \vee p(2) \vee \cdots$. To get around this sort of problem we just need some new symbols, a symbol that means "for all" and one that means "for some." Then we need to know the rules for using the new symbols and combining them with the old ones. The augmented system of symbols and rules is called the **predicate calculus**.

The new symbols we introduce are called **quantifiers**. Suppose that $\{p(x) : x \in U\}$ is a family of propositions indexed by a set U that may be

223

infinite; the set U is called the **universe of discourse** or **domain of discourse**. The **universal quantifier** $\forall$, an upside-down A as in "for *A*ll," is used to build compound propositions of the form

$$\forall x \, p(x)$$

which we read as "for all x, $p(x)$." Other translations of $\forall$ are "for each," "for every," "for any." The compound proposition $\forall x \, p(x)$ is assigned truth values as follows:

$\forall x \, p(x)$ is true if $p(x)$ is true for every x in U;
otherwise $\forall x \, p(x)$ is false.

The **existential quantifier** $\exists$, a backward E as in "there *E*xists," is used to form propositions like

$$\exists x \, p(x)$$

which we read as "there exists an x such that $p(x)$," "there is an x such that $p(x)$," or "for some x, $p(x)$." The compound proposition $\exists x \, p(x)$ has these truth values:

$\exists x \, p(x)$ is true if $p(x)$ is true for at least one x in U;
$\exists x \, p(x)$ is false if $p(x)$ is false for every x in U.

EXAMPLE 1
(a) Let $\mathbb{N}$ be the universe of discourse and for each n in $\mathbb{N}$ let $p(n)$ be the proposition "$n^2 = n$." Then $\forall n \, p(n)$ is false because, for example, $p(3)$, i.e., $3^2 = 3$, is false. On the other hand, $\exists n \, p(n)$ is true because at least one proposition $p(n)$ is true; in fact, exactly two of them are true, namely $p(0)$ and $p(1)$.

For the same universe of discourse, let $q(n)$ be the proposition "$(n + 1)^2 = n^2 + 2n + 1$." We can use $p(n)$, $q(n)$ and connectives from the propositional calculus to obtain other quantified propositions. For example, $\exists n[\neg q(n)]$ is false since every proposition $\neg q(n)$ is false. The proposition $\forall n[p(n) \lor q(n)]$ is true because each proposition $p(n) \lor q(n)$ is true. Of course, the weaker proposition $\exists n[p(n) \lor q(n)]$ is also true.

(b) Let $p(x)$ be "$x \le 2x$" and $q(x)$ be "$x^2 \ge 0$," with universe of discourse $\mathbb{R}$. Since $\mathbb{R}$ cannot be listed as a sequence, it would really be impossible to symbolize $\exists x \, p(x)$ or $\forall x \, q(x)$ in the propositional calculus. Clearly $\exists x \, p(x)$ is true; $\forall x \, p(x)$ is false because $p(x)$ is false for negative x. Both $\forall x \, q(x)$ and $\exists x \, q(x)$ are true and $\exists x[\neg q(x)]$ is false.

(c) Quantifiers are useful in mathematics and computer science when the universe of discourse is finite, but large. Suppose, for example, that propositions $p(n)$ are indexed by the universe of discourse $\{n \in \mathbb{N} : 0 \le n \le 65{,}535\}$. The notation $\forall n \, p(n)$ is clearly preferable to

$$p(0) \land p(1) \land p(2) \land p(3) \land \cdots \land p(65{,}535)$$

though we might invent the acceptable $\displaystyle\bigwedge_{n=0}^{65{,}535} p(n)$. ∎

EXAMPLE 2   Occasionally we are confronted with a constant function in algebra, such as $f(x) = 2$ for $x \in \mathbb{R}$, so that the variable x does not appear in the right side of the definition of f. Although the value of $f(x)$ does not depend on the choice of x, we nevertheless regard f as a function of x. Similarly, in logic we occasionally encounter propositions $p(x)$ whose truth values do not depend on the choice of x in U. As artificial examples, consider $p(n) = $ "2 is prime" and $q(n) = $ "16 is prime," with universe of discourse $\mathbb{N}$. Since all propositions $p(n)$ are true, $\exists n\, p(n)$ and $\forall n\, p(n)$ are both true. Since all propositions $q(n)$ are false, $\exists n\, q(n)$ and $\forall n\, q(n)$ are both false. Propositions $p(x)$ whose truth values do not depend on x are essentially the ones we studied earlier in the propositional calculus. In a sense the propositional calculus fills the same place in the predicate calculus that the constant functions fill in the study of all functions.   ∎

Let's analyze the proposition $\forall x\, p(x)$ more closely. The expression $p(x)$ is called a predicate. In ordinary grammatical usage a predicate is the part of a sentence which says something about the subject of the sentence. For example, "_____ went to the moon" and "_____ is bigger than a bread box" are predicates. To make a sentence, we supply the subject. For example, the predicate "_____ is bigger than a bread box" becomes the sentence "This book is bigger than a bread box" if we supply the subject "This book." If we call the predicate p, then the sentence could be denoted $p(\text{This book})$. Each subject yields a sentence.

In our symbolic logic setting a **predicate** is a function which produces a proposition whenever we feed it a member of the universe, that is, a proposition-valued function with domain U. We follow our usual practice and denote such a function by $p(x)$. The variable x in the expression $p(x)$ is called a **free variable** of the predicate. As x varies over U the truth value of $p(x)$ may vary. In contrast, the proposition $\forall x\, p(x)$ has a fixed meaning and truth value and does not vary with x. The variable x in $\forall x\, p(x)$ is called a **bound variable**; it is bound by the quantifier $\forall$. Since $\forall x\, p(x)$ has a fixed meaning and truth value, it would be pointless and unnatural to quantify it again. That is, it would be pointless to introduce $\forall x[\forall x\, p(x)]$ and $\exists x[\forall x\, p(x)]$ since their truth values are the same as that of $\forall x\, p(x)$.

We can also consider predicates which are functions of more than one variable, perhaps from more than one universe of discourse, and in such cases multiple use of quantifiers is natural.

EXAMPLE 3   (a) Let $\mathbb{N}$ be the universe of discourse and for each m and n in $\mathbb{N}$ let $p(m, n)$ be the proposition "$m < n$." We could think of these propositions as being indexed by $\mathbb{N} \times \mathbb{N}$ and think of $\mathbb{N} \times \mathbb{N}$ as the universe of discourse, but for the present we prefer to treat the variables m and n separately. Both variables m and n are free in the sense that the meanings and truth values of $p(m, n)$ vary with both m and n. In the expression $\exists m\, p(m, n)$, the variable m is bound but the variable n is free. The proposition $\exists m\, p(m, n)$ reads "there is an

m in $\mathbb{N}$ with $m < n$," so $\exists m \, p(m, 0)$ is false, $\exists m \, p(m, 1)$ is true, $\exists m \, p(m, 2)$ is true, etc. For each choice of n the proposition $\exists m \, p(m, n)$ is either true or false; its truth value does not depend on m but depends on n alone. It is meaningful to quantify $\exists m \, p(m, n)$ with respect to the free variable n to obtain $\forall n [\exists m \, p(m, n)]$ and $\exists n [\exists m \, p(m, n)]$. The proposition $\forall n [\exists m \, p(m, n)]$ is false because $\exists m \, p(m, 0)$ is false and the proposition $\exists n [\exists m \, p(m, n)]$ is true because, for example, $\exists m \, p(m, 1)$ is true. Henceforth we will usually omit the brackets [ ] and write $\forall n \, \exists m \, p(m, n)$ and $\exists n \, \exists m \, p(m, n)$.

There are eight ways to apply the two quantifiers to the two variables: $\forall m \, \forall n$, $\forall n \, \forall m$, $\exists m \, \exists n$, $\exists n \, \exists m$, $\forall m \, \exists n$, $\exists n \, \forall m$, $\forall n \, \exists m$, $\exists m \, \forall n$. The first two turn out to be logically equivalent, since they have the same meaning as $\forall \langle m, n \rangle \, p(m, n)$ where $\langle m, n \rangle$ varies over the new universe of discourse $\mathbb{N} \times \mathbb{N}$. Similarly, $\exists m \, \exists n \, p(m, n)$ and $\exists n \, \exists m \, p(m, n)$ are logically equivalent. The remaining four must be approached carefully. For our example, we already observed that $\forall n \, \exists m \, p(m, n)$ is false. No matter what m is, $\forall n \, p(m, n)$ is false since $p(m, 0)$ is false. Therefore $\exists m \, \forall n \, p(m, n)$ is false. To analyze $\forall m \, \exists n \, p(m, n)$, note that for each m, $\exists n \, p(m, n)$ is true because $p(m, m + 1)$ is true. Therefore $\forall m \, \exists n \, p(m, n)$ is also true. To analyze $\exists n \, \forall m \, p(m, n)$, note that for each n, $\forall m \, p(m, n)$ is false because, for example, $p(n, n)$ is false. Therefore $\exists n \, \forall m \, p(m, n)$ is false. We repeat:

> for this example,
> $\forall m \, \exists n \, p(m, n)$ is true, while $\exists n \, \forall m \, p(m, n)$ is false.

The left proposition asserts, correctly, that for every m, there is a bigger n. The right proposition asserts, incorrectly, that there is an n bigger than all m.

(b) Here is a less mathematical example illustrating the importance of order when using both quantifiers $\forall$ and $\exists$. Let the universe of discourse consist of all people and consider

$$p(x, y) = \text{``} y \text{ is a mother of } x.\text{''}$$

Then $\forall x \, \exists y \, p(x, y)$ asserts that everyone has a mother, which is true. On the other hand, $\exists y \, \forall x \, p(x, y)$ asserts that someone is the mother of everyone else, which is false.

The proposition $\forall y \, \exists x \, p(x, y)$ asserts that everyone is a mother, and $\exists x \, \forall y \, p(x, y)$ asserts that someone has everyone for his or her mother. These are both clearly false statements. ▮

With these examples in mind we turn now to a more formal account. Let $U_1, U_2, \ldots, U_n$ be nonempty sets. An n-**place predicate** over $U_1 \times U_2 \times \cdots \times U_n$ is a function $p(x_1, x_2, \ldots, x_n)$ with domain $U_1 \times U_2 \times \cdots \times U_n$ which has propositions as its function values. The variables $x_1, x_2, \ldots, x_n$ for $p(x_1, x_2, \ldots, x_n)$ are all **free variables** for the predicate, and each x_j varies over the corresponding universe of discourse U_j. The term "free" is short for "free for substitution," meaning that the variable x_j is available in case we wish to substitute a particular value from U_j for all occurrences of x_j.

If we substitute a value for x_j, say for definiteness we substitute a for x_1 in $p(x_1, x_2, \ldots, x_n)$, we get the predicate $p(a, x_2, \ldots, x_n)$ which is free on the $n - 1$ remaining variables $x_2, \ldots, x_n$ but no longer free on x_1. An application of a quantifier $\forall x_j$ or $\exists x_j$ to a predicate $p(x_1, x_2, \ldots, x_n)$ gives a predicate $\forall x_j\, p(x_1, x_2, \ldots, x_n)$ or $\exists x_j\, p(x_1, x_2, \ldots, x_n)$ whose value depends only on the values of the remaining $n - 1$ variables. We say the quantifier **binds** the variable x_j, making x_j a **bound variable** for the predicate. Application of n quantifiers, one for each variable, makes all variables bound and yields a proposition whose truth value can be determined by applying the rules for $\forall x$ and $\exists x$ specified prior to Example 1 to the universes $U_1, U_2, \ldots, U_n$.

EXAMPLE 4   (a) Consider the proposition

(1) $$\forall m\, \exists n[n > 2^m];$$

here $p(m, n) = $ "$n > 2^m$" is a 2-place predicate over $\mathbb{N} \times \mathbb{N}$. That is, m and n are both allowed to vary over $\mathbb{N}$. Recall our convention that (1) represents

$$\forall m[\exists n[n > 2^m]].$$

Both variables m and n are bound. To decide the truth value of (1), we consider the inside expression $\exists n[n > 2^m]$ in which n is a bound variable and m is a free variable. We mentally fix the free variable m and note that the proposition "$n > 2^m$" is true for some choices of n in $\mathbb{N}$, for example $n = 2^m + 1$. It follows that $\exists n[n > 2^m]$ is true. This thought process is valid for each m in $\mathbb{N}$, so we conclude that $\exists n[n > 2^m]$ is true for all m. That is, (1) is true.

If we reverse the quantifiers in (1), we obtain

(2) $$\exists n\, \forall m[n > 2^m].$$

This is false because $\forall m[n > 2^m]$ is false for each n, since "$n > 2^m$" is false for $m = n$.

(b) Consider the propositions

(3) $$\forall x\, \exists y[x + y = 0],$$

(4) $$\exists y\, \forall x[x + y = 0],$$

(5) $$\forall x\, \exists y\,[xy = 0],$$

(6) $$\exists y\, \forall x[xy = 0],$$

where each universe of discourse is $\mathbb{R}$.

To analyze (3) we consider a fixed x. Then $\exists y[x + y = 0]$ is true, because the choice $y = -x$ makes "$x + y = 0$" true. That is, $\exists y[x + y = 0]$ is true for all x and so (3) is true.

To analyze (4) we consider a fixed y. Then $\forall x[x + y = 0]$ is not true, because the choice $x = 1 - y$ makes "$x + y = 0$" false. That is, for each y, $\forall x[x + y = 0]$ is false and so (4) is false.

Proposition (5) is true, because $\exists y[xy = 0]$ is true for all x. In fact, the choice $y = 0$ makes "$xy = 0$" true.

To deal with (6) we analyze $\forall x[xy = 0]$ in two cases. If $y = 0$, this proposition is clearly true. But if $y \neq 0$ this proposition is false, since the choice $x = 1$ makes "$xy = 0$" false in this case. Thus $\forall x[xy = 0]$ is true if and only if $y = 0$. Since $\forall x[xy = 0]$ is true for some y, namely $y = 0$, the proposition (6) is true.

In the next section we will see that the truth of (6) implies the truth of (5) on purely logical grounds; that is,

$$\exists y \, \forall x \, p(x, y) \rightarrow \forall x \, \exists y \, p(x, y)$$

is always true. ∎

We have already noted that an n-place predicate becomes an $(n - 1)$-place predicate when we bind one of the variables with a quantifier. Its truth value depends on the truth values of the remaining $n - 1$ free variables, and in particular doesn't depend on what name we choose to call the bound variable. Thus if $p(x)$ is a 1-place predicate with universe of discourse U, then $\forall x \, p(x)$, $\forall y \, p(y)$ and $\forall t \, p(t)$ all have the same truth value, namely true if $p(u)$ is true for every u in U and false otherwise. Similarly, if $q(x, y)$ is a 2-place predicate with universes U and V, then $\exists y \, q(x, y)$, $\exists t \, q(x, t)$, and $\exists s \, q(x, s)$ all describe the same 1-place predicate, namely the predicate which has truth value true for a given x in U if and only if $q(x, v)$ is true for some v in the universe V in which the second variable lies. On the other hand, the predicate $\exists x \, q(x, x)$, is *not* the same as these last three. The difference is that the quantifier in this instance binds both of the free variables.

EXAMPLE 5   Let U and V be $\mathbb{N}$ and let $q(x, y) = $ "$x > y$." Then $\exists x \, q(x, y)$ is the 1-place predicate "some member of $\mathbb{N}$ is greater than y," and so is $\exists t \, q(t, y)$. The predicate $\exists y \, q(x, y)$ is the 1-place predicate "there is a member of $\mathbb{N}$ less than x," which is the same predicate as $\exists s \, q(x, s)$ and has the value true for $x > 0$ and false for $x = 0$. But $\exists x \, q(x, x)$ is the proposition "$x > x$ for some x," which has the value false. ∎

In practice, it is common to omit leading universal quantifiers.

EXAMPLE 6   The associative and cancellation laws for $\mathbb{R}$ are often written

(a) $$(x + y) + z = x + (y + z),$$

(b) $$xz = yz \quad \text{and} \quad z \neq 0 \quad \text{imply} \quad x = y.$$

The intended meanings are

$$\forall x \, \forall y \, \forall z[(x + y) + z = x + (y + z)],$$

$$\forall x \, \forall y \, \forall z[(xz = yz \wedge z \neq 0) \rightarrow x = y],$$

where the universe of discourse is $\mathbb{R}$. When writing informally, the universal quantifiers sometimes follow the predicate, so that (a) might be written as

$$(x + y) + z = x + (y + z) \quad \forall x \, \forall y \, \forall z,$$

or

$$(x + y) + z = x + (y + z) \quad \forall x, y, z \in \mathbb{R},$$

or

$$(x + y) + z = x + (y + z) \quad \text{for all} \quad x, y, z \in \mathbb{R}. \quad \blacksquare$$

Another common practice is to build in a description of the universe of discourse just after the variable being quantified. For example, instead of "Let $\mathbb{R}$ be the universe... $\forall x \, p(x)$" one might write $\forall x \in \mathbb{R} \, p(x)$. Similarly, $\exists x \in \mathbb{R} \, \forall n \in \mathbb{P}[x^n > x]$ is read as "there is a real number x such that for every n in $\mathbb{P}$, $x^n > x$" or as "there is a real number x such that $x^n > x$ for every n in $\mathbb{P}$." Along the same lines, one might write $\forall \epsilon > 0 \, \exists \delta > 0$, rather than the cumbersome $\forall \epsilon \in (0, \infty) \, \exists \delta \in (0, \infty)$.

EXERCISES 6.1

1. Determine the truth values of the following, where the universe of discourse is $\mathbb{N}$.
   (a) $\forall m \, \exists n[2n = m]$
   (b) $\exists n \, \forall m[2m = n]$
   (c) $\forall m \, \exists n[2m = n]$
   (d) $\exists n \, \forall m[2n = m]$
   (e) $\forall m \, \forall n[\neg\{2n = m\}]$

2. Determine the truth values of the following, where the universe of discourse is $\mathbb{R}$.
   (a) $\forall x \, \exists y[xy = 1]$
   (b) $\exists y \, \forall x[xy = 1]$
   (c) $\exists x \, \exists y[xy = 1]$
   (d) $\forall x \, \forall y[(x + y)^2 = x^2 + y^2]$
   (e) $\forall x \, \exists y[(x + y)^2 = x^2 + y^2]$
   (f) $\exists y \, \forall x[(x + y)^2 = x^2 + y^2]$
   (g) $\exists x \, \exists y[(x + 2y = 4) \wedge (2x - y = 2)]$
   (h) $\exists x \, \exists y[x^2 + y^2 + 1 = 2xy]$

3. Write the following sentences in logical notation. Be sure to bind all variables. When using quantifiers, specify universes; use $\mathbb{R}$ if no universe is indicated.
   (a) If $x < y$ and $y < z$, then $x < z$.
   (b) For every $x > 0$, there exists an n in $\mathbb{N}$ such that $n > x$ and $x > 1/n$.
   (c) For every $m, n \in \mathbb{N}$ there exists $p \in \mathbb{N}$ such that $m < p$ and $p < n$.
   (d) There exists $u \in \mathbb{N}$ so that $un = n$ for all $n \in \mathbb{N}$.
   (e) For each $n \in \mathbb{N}$, there exists $m \in \mathbb{N}$ such that $m < n$.
   (f) For every $n \in \mathbb{N}$, there exists $m \in \mathbb{N}$ such that $2^m \leq n$ and $n < 2^{m+1}$.

4. Determine the truth values of the propositions in Exercise 3.

5. Write the following sentences in logical notation; the universe of discourse is the set Σ^* of words using letters from a finite alphabet Σ.
   (a) If $w_1 w_2 = w_1 w_3$, then $w_2 = w_3$.
   (b) If length$(w) = 1$, then $w \in \Sigma$.
   (c) $w_1 w_2 = w_2 w_1$ for all $w_1, w_2 \in \Sigma^*$.

6. Determine the truth values of the propositions in Exercise 5.

7. Specify the free and bound variables in the following expressions.
   (a) $\forall x \, \exists z [\sin(x + y) = \cos(z - y)]$
   (b) $\exists x [xy = xz \rightarrow y = z]$
   (c) $\exists x \, \exists z [x^2 + z^2 = y]$

8. Consider the expression $x + y = y + x$.
   (a) Specify the free and bound variables in the expression.
   (b) Apply universal quantifiers over the universe $\mathbb{R}$ to get a proposition. Is the proposition true?
   (c) Apply existential quantifiers over the universe $\mathbb{R}$ to get a proposition. Is the proposition true?

9. Repeat Exercise 8 for the expression $(x - y)^2 = x^2 - y^2$.

10. Consider the proposition $\forall m \, \exists n [m + n = 7]$.
    (a) Is the proposition true for the universes of discourse $\mathbb{N}$?
    (b) Is the proposition true for the universes of discourse $\mathbb{Z}$?

11. Repeat Exercise 10 for $\forall n \, \exists m [m + 1 = n]$.

12. Consider the proposition $\forall x \, \exists y [(x^2 + 1)y = 1]$.
    (a) Is the proposition true for the universes of discourse $\mathbb{N}$?
    (b) Is the proposition true for the universes of discourse $\mathbb{Q}$?
    (c) Is the proposition true for the universes of discourse $\mathbb{R}$?

13. Another useful quantifier is $\exists!$ where $\exists! \, x \, p(x)$ is read "there exists a unique x such that $p(x)$." This compound proposition is assigned truth value true if $p(x)$ is true for exactly one value of x in the universe of discourse; otherwise it is false. Write the following sentences in logical notation.
    (a) There is a unique x in $\mathbb{R}$ such that $x + y = y$ for all $y \in \mathbb{R}$.
    (b) The equation $x^2 = x$ has a unique solution.
    (c) Exactly one set is a subset of all sets in $\mathscr{P}(\mathbb{N})$.
    (d) If $f : A \rightarrow B$, then for each $a \in A$ there is exactly one $b \in B$ such that $f(a) = b$.
    (e) If $f : A \rightarrow B$ is a one-to-one function, then for each $b \in B$ there is exactly one $a \in A$ such that $f(a) = b$.

14. Determine the truth values of the propositions in Exercise 13.

15. In this problem $A = \{0, 2, 4, 6, 8, 10\}$ and the universe of discourse is $\mathbb{N}$. True or False.
    (a) A is the set of even integers in $\mathbb{N}$ less than 12.
    (b) $A = \{0, 2, 4, 6, \ldots\}$
    (c) $A = \{n \in \mathbb{N} : 2n < 24\}$
    (d) $A = \{n \in \mathbb{N} : \forall m [2m = n \rightarrow m < 6]\}$
    (e) $A = \{n \in \mathbb{N} : \forall m [2m = n \wedge m < 6]\}$

(f) $A = \{n \in \mathbb{N} : \exists m[2m = n \to m < 6]\}$

(g) $A = \{n \in \mathbb{N} : \exists m[2m = n \land m < 6]\}$

(h) $A = \{n \in \mathbb{N} : \exists! \, m[2m = n \land m < 6]\}$ [see Exercise 13]

(i) $A = \{n \in \mathbb{N} : n \text{ is even and } n^2 \leq 100\}$

(j) $\forall n[n \in A \to n \leq 10]$

(k) $3 \in A \to 3 < 10$

(l) $12 \in A \to 12 < 10$

(m) $8 \in A \to 8 < 10$

16. With universe of discourse $\mathbb{N}$, let $p(n) = $ "n is prime" and $e(n) = $ "n is even." Write the following in ordinary English.

(a) $\exists m \, \forall n[e(n) \land p(m + n)]$

(b) $\forall n \, \exists m[\neg e(n) \to e(m + n)]$

Translate the following into logical notation using p and e.

(c) There are two prime integers whose sum is even.

(d) If the sum of two primes is even, then neither of them equals 2.

(e) The sum of two prime integers is odd.

17. Determine the truth values of the propositions in Exercise 16.

§ 6.2 Predicate Calculus

The ideas of "proof" and "theorem" which we discussed in § 2.3 for the propositional calculus can be extended to the predicate calculus setting. Not surprisingly, with more possible expressions we also get more complications. A moderately thorough account of the subject would form a substantial part of another book. In this section we limit ourselves to discussing some of the most basic and useful connections between quantifiers and logical operators.

In Chapter 2 we used the term "compound proposition" in an informal way to describe propositions built up out of simpler ones. In § 3.6 we gave a recursive definition of wff's for the propositional calculus; these are the compound propositions of Chapter 2. In the same way, we will use a recursive definition in order to precisely define "compound propositions" for the predicate calculus. At the same time we want to look more closely at what we did in the propositional calculus.

We built up five propositions from symbols p, q, r, etc. which we considered to be names for other propositions. The symbols could be considered to be variables, and a compound proposition such as $p \land (\neg q)$ could be considered a function of the two variables p and q; its value is true or false, depending on the truth values of p and q and the form of the compound proposition. The truth table for the compound proposition is simply a list of the function values as the variables range independently over the set {true, false}. Variables of this sort which can take on just the two values true and false are called **logical variables**. For the propositional calculus they are all we need, but for the predicate calculus we must also consider variables associated with other universes of discourse, such as $\mathbb{R}$, $\mathbb{Z}$ and $\mathfrak{M}_{m,n}$.

Suppose now that we have available a collection of nonempty universes of discourse with which the free variables of all n-place predicates we consider are associated. We define the class of **compound predicates** as follows:

(B_1) Logical variables are compound predicates.

(B_2) n-place predicates are compound predicates for $n \geq 1$.

(R_1) If $p(x)$ is a compound predicate with free variable x, then

$$(\forall x \, p(x)) \quad \text{and} \quad (\exists x \, p(x))$$

are compound predicates for which x is not a free variable.

(R_2) If P and Q are compound predicates, then so are

$$\neg P, \quad P \vee Q, \quad P \wedge Q, \quad P \to Q \quad \text{and} \quad P \leftrightarrow Q.$$

(E) The only compound predicates are those required by (B_1), (B_2), (R_1) and (R_2).

When we write "$p(x)$" here we mean to indicate that the given compound predicate has x as one of its free variables. We admit the possibility that the truth value of $p(x)$ is actually independent of the choice of x. In particular, we can view a proposition as trivially having a free variable, so that $p(x)$ and $q(x)$ in (R_1) might just be propositions p and q. If we delete (B_2) and (R_2) and the reference to free variables in (R_1) we obtain the recursive description of wff's for the propositional calculus in Example 6 of § 3.6. In our present setting there are compound predicates which are propositions besides those made from (B_1) and (R_1). If all of the variables in a compound predicate are bound then the predicate is a proposition. We extend our definition and say that a **compound proposition** is a compound predicate with no free variables. For example,

$$((\exists x(\exists z \, p(x, z))) \to (\forall y(\neg r(y))))$$

is a compound proposition with no free variables. In contrast,

$$(p(x) \vee (\neg \forall y \, q(x, y)))$$

and

$$((\exists z \, p(x, z)) \to (\forall y(\neg r(y))))$$

are compound predicates with free variable x.

The number and the placement of parentheses in a compound predicate are explicitly prescribed by our recursive definition. In practice, for clarity, we may add or suppress some parentheses. For example, we may write $((\forall x \, p(x) \to (\exists x \, p(x)))$ as $\forall x \, p(x) \to \exists x \, p(x)$ and we may write $(\exists x \, \neg p(x))$ as $\exists x(\neg p(x))$. We sometimes also use brackets or braces instead of parentheses.

The truth value of a compound proposition ordinarily depends on the choices of the universes of discourse that the bound variables are quantified over, but there are important instances in which the truth value not only does not depend on the universe choices but is in fact independent of the values of the logical variables as well. A compound proposition which has the value true for all universes of discourse and all values of its logical variables is called a **tautology**. This definition extends the usage in Chapter 2, where there were no universes to worry about.

EXAMPLE 1   (a) An important class of tautologies consists of the generalized DeMorgan laws; compare rules 8a–8d in Table 1 of § 2.2. These are

(1) $$\neg \forall x\, p(x) \leftrightarrow \exists x[\neg p(x)],$$

(2) $$\neg \exists x\, p(x) \leftrightarrow \forall x[\neg p(x)],$$

(3) $$\forall x\, p(x) \leftrightarrow \neg \exists x[\neg p(x)],$$

(4) $$\exists x\, p(x) \leftrightarrow \neg \forall x[\neg p(x)].$$

To see (1), note that $\neg \forall x\, p(x)$ has truth value true exactly when $\forall x\, p(x)$ has truth value false, and this occurs whenever there exists an x in the universe of discourse such that $p(x)$ is false, i.e., $\neg p(x)$ is true. Thus $\neg \forall x\, p(x)$ is true precisely when $\exists x[\neg p(x)]$ is true. This argument does not rely on the choice of universe, so (1) is a tautology. The DeMorgan law (2) can be analyzed in a similar way. Alternatively, we can derive (2) from (1) by applying (1) to the 1-place predicate $\neg p(x)$ to obtain

$$\neg \forall x[\neg p(x)] \leftrightarrow \exists x[\neg\ \neg p(x)].$$

The substitution rules in § 2.3 are still valid and so we may substitute $p(x)$ for $\neg\ \neg p(x)$ and obtain the equivalent expression

$$\neg \forall x[\neg p(x)] \leftrightarrow \exists x[p(x)].$$

This is DeMorgan's law (4) and, if we negate both sides, we obtain (2). An application of (2) to $\neg p(x)$ yields (3).

   (b) The following holds for a 2-place predicate $p(x, y)$:

(5) $$\exists x\, \forall y\, p(x, y) \rightarrow \forall y\, \exists x\, p(x, y).$$

If the left side of (5) has truth value true, then there exists an x_0 in the universe of discourse such that $\forall y\, p(x_0, y)$ is true, and so $p(x_0, y)$ is true for all y. Thus for each y, $\exists x\, p(x, y)$ is true; in fact, the same x_0 works for each y. Since $\exists x\, p(x, y)$ is true for all y, the right side of (5) has truth value true. Therefore (5) is a tautology.

   (c) The converse to (5), namely

$$\forall y\, \exists x\, p(x, y) \rightarrow \exists x\, \forall y\, p(x, y),$$

is not generally true as we noted in Example 3 of § 6.1. To emphasize the difference, we suppose that the x and y vary over a three-element universe U, say $U = \{a, b, c\}$. Then the 2-place predicate $p(x, y)$ takes nine possible values:

$$p(a, a) \qquad p(a, b) \qquad p(a, c)$$
$$p(b, a) \qquad p(b, b) \qquad p(b, c)$$
$$p(c, a) \qquad p(c, b) \qquad p(c, c).$$

As noted in the proof of (5), $\exists x \, \forall y \, p(x, y)$ is true if $\forall y \, p(x_0, y)$ is true for some x_0. Since x_0 must equal a, b or c, we see that $\exists x \, \forall y \, p(x, y)$ is true if and only if all the propositions in some row above are true. In contrast, $\forall y \, \exists x \, p(x, y)$ will be true provided at least one proposition in each column is true. For example, if we consider a predicate $p(x, y)$ with truth values

$$\text{T} \quad \text{F} \quad \text{F}$$
$$\text{F} \quad \text{F} \quad \text{T}$$
$$\text{F} \quad \text{T} \quad \text{T},$$

then $\forall y \, \exists x \, p(x, y)$ will be true, while $\exists x \, \forall y \, p(x, y)$ will be false. For this choice of predicate $p(x, y)$, $\exists x \, p(x, y)$ is true for every y, but the suitable x depends on y; no single x works for all y. ∎

As in the propositional calculus, we say that two compound propositions P and Q are **logically equivalent** if and only if $P \leftrightarrow Q$ is a tautology; and we write $P \Leftrightarrow Q$ in this case. Also, P **logically implies** Q provided $P \to Q$ is a tautology, in which case we write $P \Rightarrow Q$. In Table 1, we give some useful logical equivalences and implications. We begin numbering the rules with 35, since Chapter 2 contains rules 1 through 34.

TABLE 1. Logical Relationships
in the Predicate Calculus

35a.	$\forall x \, \forall y \, p(x, y) \Leftrightarrow \forall y \, \forall x \, p(x, y)$
b.	$\exists x \, \exists y \, p(x, y) \Leftrightarrow \exists y \, \exists x \, p(x, y)$
36.	$\exists x \, \forall y \, p(x, y) \Rightarrow \forall y \, \exists x \, p(x, y)$
37a.	$\neg \forall x \, p(x) \Leftrightarrow \exists x [\neg p(x)]$
b.	$\neg \exists x \, p(x) \Leftrightarrow \forall x [\neg p(x)]$
c.	$\forall x \, p(x) \Leftrightarrow \neg \exists x [\neg p(x)]$
d.	$\exists x \, p(x) \Leftrightarrow \neg \forall x [\neg p(x)]$

DeMorgan Laws

In Example 1 we discussed the tautologies corresponding to rules 36 and 37. The remaining rules are easy to verify.

EXAMPLE 2   To verify rule 35b, that is, to verify that

$$\exists x \, \exists y \, p(x, y) \leftrightarrow \exists y \, \exists x \, p(x, y)$$

is a tautology, we must check that this proposition has the value true for all possible universes of discourse. By the definition of $\leftrightarrow$, we need only check that $\exists x \exists y\, p(x, y)$ has the value true for a given universe if and only if $\exists y \exists x\, p(x, y)$ has the value true for that universe. Suppose $\exists x \exists y\, p(x, y)$ is true. Then $\exists y\, p(x_0, y)$ is true for some x_0 in the universe, so $p(x_0, y_0)$ is true for some y_0 in the universe. Hence $\exists x\, p(x, y_0)$ is true and thus $\exists y \exists x\, p(x, y)$ is true. The implication in the other direction follows similarly. Moreover, both $\exists x \exists y\, p(x, y)$ and $\exists y \exists x\, p(x, y)$ are logically equivalent to the proposition $\exists \langle x, y \rangle\, p(x, y)$ where $\langle x, y \rangle$ varies over $U_1 \times U_2$, with U_1 and U_2 the universes of discourse for the variables x and y. ∎

The DeMorgan laws 37a–37d can be used repeatedly to negate any quantified proposition. For example,

$$\neg\, \exists w\, \forall x\, \exists y\, \exists z\, p(w, x, y, z)$$

is successively logically equivalent to

$$\forall w[\neg\, \forall x\, \exists y\, \exists z\, p(w, x, y, z)]$$

$$\forall w\, \exists x[\neg\, \exists y\, \exists z\, p(w, x, y, z)]$$

$$\forall w\, \exists x\, \forall y[\neg\, \exists z\, p(w, x, y, z)]$$

$$\forall w\, \exists x\, \forall y\, \forall z[\neg\, p(w, x, y, z)].$$

This illustrates the general rule: The negation of a quantified predicate is logically equivalent to the proposition obtained by replacing each $\forall$ by $\exists$, replacing each $\exists$ by $\forall$, and by replacing the predicate itself by its negation.

EXAMPLE 3   (a) The negation of

(1) $$\forall x\, \forall y\, \exists z[x < z < y]$$

is

$$\exists x\, \exists y\, \forall z\{\neg\,[x < z < y]\}.$$

Applying DeMorgan's law, we see that

$$\neg\,[x < z < y] \Leftrightarrow \neg\,[x < z \wedge z < y]$$

$$\Leftrightarrow \neg\,(x < z) \vee \neg\,(z < y) \Leftrightarrow (x \geq z) \vee (z \geq y).$$

Hence the negation of (1) is logically equivalent to

$$\exists x\, \exists y\, \forall z[(z \leq x) \vee (z \geq y)].$$

(b) Consider universes U_1, U_2 and U_3 made up of companies, components and computers, respectively. Let $p(x, y)$ be the predicate "x produces y" and $q(y, z)$ be "y is a component part of z." The predicate

$$p(x, y) \wedge q(y, z)$$

has the meaning "x produces y, which is a component of z." The proposition

$$\forall x\, \forall z\, \exists y[p(x, y) \wedge q(y, z)]$$

means that each company produces some component of each computer. Its negation is

$$\neg \forall x \, \forall z \, \exists y [p(x, y) \wedge q(y, z)]$$

which is logically equivalent, by the DeMorgan laws, to

$$\exists x \, \exists z \, \forall y [\neg p(x, y) \vee \neg q(y, z)].$$

This negation has the interpretation that there exist a company x_0 and a computer z_0 so that for each choice of a component either x_0 does not produce it or it's not a component of z_0. An equivalent form of the negation is

$$\exists x \, \exists z \, \neg \exists y [p(x, y) \wedge q(y, z)],$$

with the interpretation that there exist a company x_0 and a computer z_0 so that no component part is both produced by x_0 and a component of z_0.

Compare this example with part (a). In the case of part (a), the negation is also equivalent to

$$\exists x \, \exists y \, \neg \exists z [x < z < y],$$

i.e., there exist x_0 and y_0 with no z strictly between them.

(c) The negation of

(2) $$\forall x \, \forall y [x < y \rightarrow x^2 < y^2]$$

is

$$\exists x \, \exists y \{ \neg [x < y \rightarrow x^2 < y^2] \}.$$

By rule 10a in Table 1 of § 2.2, and DeMorgan's law $\neg(p \rightarrow q) \Leftrightarrow \neg(\neg p \vee q)$ $\Leftrightarrow p \wedge \neg q$. So $\neg[x < y \rightarrow x^2 < y^2] \Leftrightarrow (x < y) \wedge (x^2 \geq y^2)$. Therefore, the negation of (2) is logically equivalent to

$$\exists x \, \exists y [(x < y) \wedge (x^2 \geq y^2)]. \quad \blacksquare$$

EXAMPLE 4   Let the universe of discourse U consist of a and b. The DeMorgan law 37a then becomes

$$\neg [p(a) \wedge p(b)] \Leftrightarrow [\neg p(a)] \vee [\neg p(b)].$$

Except for the names $p(a)$ and $p(b)$, in place of p and q, this is the DeMorgan law 8b in Table 1 of § 2.2. $\blacksquare$

A general proposition often has the form $\forall x \, p(x)$ where x ranges over some universe of discourse. This is false if and only if $\exists x [\neg p(x)]$ is true, by DeMorgan's law 37a. Thus $\forall x \, p(x)$ is false if some x_0 can be exhibited for which $p(x_0)$ is false. As we pointed out in § 2.1 [after Example 8], such an x_0 is called a **counterexample** to the proposition $\forall x \, p(x)$. Some illustrations are given in Example 9 of that section. Here are a couple more.

EXAMPLE 5   (a) The matrices

$$\begin{bmatrix} 1 & 0 \\ 0 & 0 \end{bmatrix}, \quad \begin{bmatrix} 0 & 0 \\ 1 & 1 \end{bmatrix}$$

provide a counterexample to the assertion "If 2×2 matrices $\mathbf{A}$ and $\mathbf{B}$ satisfy $\mathbf{AB} = \mathbf{0}$, then $\mathbf{A} = \mathbf{0}$ or $\mathbf{B} = \mathbf{0}$." This general assertion could have been written

$$\forall \mathbf{A}\, \forall \mathbf{B}[\mathbf{AB} = \mathbf{0} \rightarrow (\mathbf{A} = \mathbf{0} \vee \mathbf{B} = \mathbf{0})].$$

(b) The proposition "Every connected graph has an Euler circuit" is quite false. In view of Euler's theorem in § 0.1, any connected graph having a vertex of odd degree will serve as a counterexample to this assertion. The simplest counterexample has two vertices and one edge connecting them.   ∎

One compound proposition which is *almost* a tautology is

$$\forall x\, p(x) \rightarrow \exists x\, p(x),$$

with the corresponding logical relationship

$$\forall x\, p(x) \Rightarrow \exists x\, p(x).$$

The proposition has the value true for every nonempty universe, since if $\forall x\, p(x)$ is true for a nonempty U then $p(x_0)$ is true for each member x_0 of U and so $\exists x\, p(x)$ is true. On the other hand, if the universe is empty, then $\forall x\, p(x)$ has the value true vacuously, whereas $\exists x\, p(x)$ is false. It is true that everyone with three heads is rich. [You disagree? Give a counterexample.] But it is not true that there is a rich person with three heads. Here the universe consists of all three-headed people and $p(x)$ denotes "x is rich."

We left this relationship off of the list in Table 1, even though it is an important relationship, because it is not *always* true. We can combine it with other relationships to get implications which hold for all nonempty universes. For instance

$$[\forall x\, p(x) \rightarrow \exists x\, p(x)] \Leftrightarrow (\neg[\exists x\, p(x)] \rightarrow \neg[\forall x\, p(x)])$$

by the contrapositive rule 9. For a nonempty universe the expression on the left is true, so

$$\neg[\exists x\, p(x)] \rightarrow \neg[\forall x\, p(x)]$$

is also true. Using the DeMorgan laws 37b and 37a we find that

$$\forall x[\neg p(x)] \rightarrow \exists x[\neg p(x)]$$

is true, as is

$$\forall x[\neg p(x)] \rightarrow \neg[\forall x\, p(x)].$$

Thus, *for a nonempty universe*

$$\forall x[\neg p(x)] \Rightarrow \exists x[\neg p(x)]$$

$$\Leftrightarrow \neg[\forall x\, p(x)].$$

But note that the fact that every three-headed person is not rich, which is true, does not imply that there is a three-headed person who is not rich.

EXERCISES 6.2

1. Consider a universe U_1 consisting of members of a club, and a universe U_2 of airlines. Let $p(x, y)$ be the predicate "x has been a passenger on y" or equivalently "y has had x as a passenger." Write out the meanings of the following.
   (a) rule 35a   (b) rule 35b   (c) rule 36

2. Consider the universe U of all university professors. Let $p(x)$ be the predicate "x likes punk rock."
   (a) Express the proposition "not all university professors like punk rock" in predicate calculus symbols.
   (b) Do the same for "every university professor does not like punk rock."
   (c) Does either of the propositions in part (a) or (b) imply the other? Explain.
   (d) Write out the meaning of rule 37b for this U and $p(x)$.
   (e) Do the same for rule 37d.

3. Show that the following rules in Table 1 collapse to rules from Table 1 of § 2.2 when the universe of discourse U has two elements, a and b.
   (a) rule 37d   (b) rule 37b

4. (a) Show that the logical implication

$$[\exists x\, p(x)] \wedge [\exists x\, q(x)] \Rightarrow \exists x[p(x) \wedge q(x)]$$

   is false. You may do this either by defining predicates $p(x)$ and $q(x)$ where this implication fails, or by taking a small universe of discourse, say $U = \{a, b\}$, and assigning truth values to the four propositions $p(a)$, $p(b)$, $q(a)$, $q(b)$.
   (b) Do the same for the logical implication

$$\exists x\, \forall y\, p(x, y) \Rightarrow \forall x\, \exists y\, p(x, y).$$

   Compare this with the true implication of rule 36.

5. Write the negation of $\forall n[p(n) \rightarrow p(n + 1)]$ without using the quantifier $\forall$, where the universe of discourse is $\mathbb{N}$.

6. Write the negation of $\exists x\, \forall y\, \exists z[z > y \rightarrow z < x^2]$ without using the connective $\neg$.

7. (a) Write the negation of

$$P = \forall x\, \forall y[x < y \rightarrow \exists z\{x < z < y\}]$$

   without using the connective $\neg$.
   (b) Determine the truth value of P when the universe of discourse is $\mathbb{R}$ or $\mathbb{Q}$.
   (c) Determine the truth value of P when the universe of discourse is $\mathbb{N}$ or $\mathbb{Z}$.

8. Give a counterexample for each of the following assertions.
    (a) Every even integer is the product of two even integers.
    (b) $|S \cup T| = |S| + |T|$ for any two finite sets S and T.
    (c) Every positive integer of the form $6k - 1$ is a prime.
    (d) Every graph has an even number of edges.
    (e) All mathematics courses are fun.

9. Our definition of compound predicate does not permit expressions such as $\exists x \, p(x, x)$ with $p(x, y)$ a 2-place predicate. Describe a predicate $q(x, y)$ such that

$$\exists x \, \exists y [p(x, y) \wedge q(x, y)]$$

is true if and only if $p(x, x)$ is true for some x.

10. In the case that the universe of discourse is empty, $\forall x \, p(x)$ vacuously has the value true regardless of $p(x)$, and $\exists x \, p(x)$ is false. Describe the situation for a universe with exactly one member.

11. The statement "There are arbitrarily large integers n such that $p(n)$ is true" translates into the proposition

$$\forall N \, \exists n [(n \geq N) \wedge p(n)]$$

with universe of discourse $\mathbb{P}$. Write the negation of this proposition using the connective $\rightarrow$ but without using the connective $\neg$. Your answer should translate into a statement which implies that $p(n)$ is true for only a finite set of n's.

§ 6.3 Mathematical Induction

The principle of mathematical induction studied in § 2.6 is often called the First Principle of Mathematical Induction. We restate it here in the form of a rule of inference.

First Principle of Mathematical Induction Let m be an integer and let $p(n)$ be a 1-place predicate over the universe of discourse $\{n \in \mathbb{Z} : n \geq m\}$.

(B) $p(m)$

(I) $\forall_{n \geq m}[p(n) \rightarrow p(n + 1)]$

$\therefore$ $\forall_{n \geq m} p(n)$

In the inductive step (I), each proposition is true provided the proposition immediately preceding it is true. To use this principle as a framework for constructing a proof, we need to check that $p(m)$ is true and that each proposition is true *assuming that the proposition just before it is true*. It is this right to assume the immediately previous case that makes the method of proof by induction so powerful. It turns out that in fact we are permitted to assume *all* previous cases. This apparently stronger assertion is a consequence of the following principle, whose proof we discuss at the end of this section.

Second Principle
of Mathematical
Induction

Let m be an integer and let $p(n)$ be a 1-place predicate over the universe of discourse $\{n \in \mathbb{Z} : n \geq m\}$.

$$\text{(B)} \quad p(m)$$
$$\text{(I)} \quad \forall_{n > m}[p(m) \wedge \cdots \wedge p(n-1) \to p(n)]$$
$$\therefore \qquad \forall_{n \geq m} p(n)$$

The first three implications in (I), corresponding to $n = m + 1$, $m + 2$ and $m + 3$, are

$$p(m) \to p(m+1),$$
$$p(m) \wedge p(m+1) \to p(m+2),$$
$$p(m) \wedge p(m+1) \wedge p(m+2) \to p(m+3).$$

To verify (I) in general one considers an $n > m$, assumes that the propositions $p(k)$ are true for $m \leq k < n$ and shows that $p(n)$ is true. The Second Principle of Mathematical Induction is the appropriate version to use when the truths of the propositions follow from predecessors other than the immediate predecessors.

EXAMPLE 1   Every integer $n \geq 2$ can be written as a product of primes.

Proof. Note that if n is prime the "product of primes" is simply the number n by itself. For $n \geq 2$ let $p(n)$ be the proposition

"n can be written as a product of primes."

Observe that the First Principle of Mathematical Induction is really unsuitable here. The lone fact that 1,311,819, say, happens to be a product of primes is of no help in showing that 1,311,820 is also a product of primes. We apply the Second Principle. Clearly $p(2)$ is true, since 2 is a prime.
Consider $n > 2$ and assume that $p(k)$ is true for all k satisfying $2 \leq k < n$. We need to show that this implies $p(n)$ is true. If n is prime, then $p(n)$ is clearly true. Otherwise, n can be written as a product jk where j and k are integers greater than 1. Thus $2 \leq j < n$ and $2 \leq k < n$. Since both $p(j)$ and $p(k)$ are assumed to be true, we can write j and k as products of primes. Then $n = jk$ is also a product of primes. We have checked the basis and induction step for the Second Principle of Mathematical Induction, and so we infer that all the propositions $p(n)$ are true. ∎

Often the general proof of the inductive step (I) does not work for the first few values of n. In this case, these first few values of n need to be checked separately, so they may serve as part of the basis. We restate the Second Principle of Mathematical Induction in a more general version which applies in such situations.

Second Principle   Let m be an integer, let $p(n)$ be a 1-place predicate over $\{n \in \mathbb{Z} : n \geq m\}$ and let
of Mathematical l be an integer ≥ 0.
Induction

$$\text{(B)} \quad p(m), \ldots, p(m + l)$$
$$\text{(I)} \quad \forall_{n > m + l}[p(m) \wedge \cdots \wedge p(n - 1) \rightarrow p(n)]$$
$$\therefore \qquad \forall_{n \geq m} p(n)$$

If $l = 0$ this is our original version of the Second Principle.

In § 3.4 we saw that many sequences are defined recursively using earlier terms other than the immediate predecessors. The Second Principle is the natural form of induction for proving results about such sequences.

EXAMPLE 2   (a) In Exercise 12 of § 3.4 we recursively defined $b_0 = b_1 = 1$ and $b_n = 2b_{n-1} + b_{n-2}$ for $n \geq 2$. In part (b), we asked for an explanation of why all b_n's are odd integers. We were hoping you would stumble upon the need for the Second Principle of Mathematical Induction. We now give a proof.

The nth proposition is $p(n) = $ "b_n is odd." In the inductive step we will use the relation $b_n = 2b_{n-1} + b_{n-2}$ and so we'll need $n \geq 2$. Hence we'll check the cases $n = 0$ and 1 separately. Thus we will use the Second Principle with $m = 0$ and $l = 1$.

(B) The propositions $p(0)$ and $p(1)$ are obviously true, since $b_0 = b_1 = 1$.

(I) Consider $n \geq 2$ and assume that b_k is odd for all k satisfying $0 \leq k < n$. In particular, b_{n-2} is odd. Clearly $2b_{n-1}$ is even, and so $b_n = 2b_{n-1} + b_{n-2}$ is the sum of an even and an odd integer. Thus b_n is odd. It follows from the Second Principle of Mathematical Induction that all b_n's are odd.

Note that in this proof the oddness of b_n followed from the oddness of b_{n-2}.

(b) For the sequence above, we prove that $b_n < 6b_{n-2}$ for $n \geq 4$. Direct computation shows that $b_2 = 3, b_3 = 7, b_4 = 17$ and $b_5 = 41$. For $n = 4$ the inequality says $b_4 < 6b_2$ or $17 < 6 \cdot 3$ and for $n = 5$ it says $b_5 < 6b_3$ or $41 < 6 \cdot 7$; these are true. We now consider $n \geq 6$ and assume that

$$b_k < 6b_{k-2} \quad \text{for} \quad 4 \leq k < n.$$

Since $n - 1$ and $n - 2$ are both ≥ 4 we have $b_{n-1} < 6b_{n-3}$ and $b_{n-2} < 6b_{n-4}$ by assumption. Hence

$$
\begin{aligned}
b_n &= 2b_{n-1} + b_{n-2} && \text{definition of } b_n \\
&< 2(6b_{n-3}) + 6b_{n-4} && \text{inductive assumption} \\
&= 6[2b_{n-3} + b_{n-4}] && \text{algebra} \\
&= 6b_{n-2} && \text{definition of } b_{n-2}.
\end{aligned}
$$

Hence by the Second Principle of Mathematical Induction the inequality holds for all $n \geq 4$.

Note that we checked the assertion for $n = 4$ and $n = 5$ before going on to the inductive step. Thus we applied the Second Principle with $m = 4$ and $l = 1$. Why did we check the inequality for $n = 5$, as well as for $n = 4$, before proceeding with the inductive step? Before we wrote up the proof we had observed that in the inductive step we were going to need to use $b_k < 6b_{k-2}$ for $k = n - 2$ and so we would need $n - 2 \geq 4$ or $n \geq 6$. In other words, the inductive step wouldn't work for $n = 5$: $b_5 = 2b_4 + b_3$, but b_3 isn't less than $6b_1$. ∎

EXAMPLE 3   We recursively define $a_0 = a_1 = a_2 = 1$ and $a_n = a_{n-2} + a_{n-3}$ for $n \geq 3$. The first few terms of the sequence are 1, 1, 1, 2, 2, 3, 4, 5, 7, 9, 12, 16, 21, 28, 37, 49. We prove that $a_n \leq (\frac{4}{3})^n$ for all $n \in \mathbb{N}$. This inequality is clear for $n = 0, 1$ and 2. So we consider $n \geq 3$ and assume that $a_k < (\frac{4}{3})^k$ for $0 \leq k < n$. In particular, $a_{n-2} \leq (\frac{4}{3})^{n-2}$ and $a_{n-3} \leq (\frac{4}{3})^{n-3}$. Thus we have

$$a_n = a_{n-2} + a_{n-3} \leq \left(\frac{4}{3}\right)^{n-2} + \left(\frac{4}{3}\right)^{n-3} = \left(\frac{4}{3}\right)^{n-3}\left(\frac{4}{3} + 1\right).$$

Since we want to conclude that $a_n \leq (\frac{4}{3})^n$, we observe that if only $(\frac{4}{3} + 1) \leq (\frac{4}{3})^3$ held, then we could conclude

$$a_n \leq \left(\frac{4}{3}\right)^{n-3}\left(\frac{4}{3} + 1\right) \leq \left(\frac{4}{3}\right)^{n-3}\left(\frac{4}{3}\right)^3 = \left(\frac{4}{3}\right)^n,$$

as desired. Direct computation shows that $(\frac{4}{3} + 1) \leq (\frac{4}{3})^3$ [do it!] and so we conclude that $a_n \leq (\frac{4}{3})^n$. This establishes the inductive step. Hence we infer from the Second Principle of Mathematical Induction [with $m = 0$ and $l = 2$] that $a_n \leq (\frac{4}{3})^n$ for all $n \in \mathbb{N}$.

In this proof we were lucky that $(\frac{4}{3} + 1) \leq (\frac{4}{3})^3$ [close, wasn't it?]. If this inequality hadn't held, we would have had to find another proof, prove something else, or abandon the problem. Induction gives us a framework for proofs, but it doesn't provide the details, which are determined by the particular problem at hand. ∎

We have already applied the First Principle of Mathematical Induction to finite sequences, for instance in Example 6 of § 5.3. Both principles can be stated and used for finite sequences. The changes are simple. Suppose that the propositions $p(n)$ are defined for $m \leq n \leq m^*$. The First Principle then reads

(B) $p(m)$
(I) $\forall_{m \leq n < m^*}[p(n) \rightarrow p(n + 1)]$
∴ $\overline{\qquad \forall_{m \leq n \leq m^*} p(n) \qquad}$

and the general Second Principle reads

(B) $p(m), \ldots, p(m + l)$
(I) $\forall_{m+l < n \leq m^*}[p(m) \wedge \cdots \wedge p(n - 1) \rightarrow p(n)]$
∴ $\overline{\qquad \forall_{m \leq n \leq m^*} p(n). \qquad}$

We return to the infinite principles of induction and end this section by discussing the logical relationship between the two principles and explaining why we regard both as valid rules of inference for constructing proofs.

It turns out that each of the two principles implies the other, in the sense that if we accept either as a valid rule of inference then the other is also valid. It is clear that the Second Principle implies the First Principle since, if we are allowed to assume all previous cases, then we are surely allowed to assume the immediately preceding case. A rigorous proof can be given by showing that (B) and (I) of the Second Principle are consequences of (B) and (I) of the First Principle.

It is perhaps more surprising that the First Principle implies the Second. A proof can be given using the propositions

$$q(n) = p(m) \wedge \cdots \wedge p(n) \quad \text{for} \quad n \geq m$$

and showing that if the sequence $p(n)$ satisfies (B) and (I) of the Second Principle, then $q(m)$ and $\forall_{n \geq m}[q(n) \to q(n + 1)]$ are true. Then every $q(n)$ will be true by the First Principle so that $p(n)$ will also be true for every n.

The equivalence of the two principles is of less concern to us than an assurance that they are valid rules. For this we rely on a fundamental property of $\mathbb{N}$:

Well-Ordering Principle   Every nonempty subset of $\mathbb{N}$ has a smallest member.

This property is not a consequence of the rules for arithmetic in $\mathbb{N}$ and is an independent axiom. The Well-Ordering Principle implies that:

Given $m \in \mathbb{Z}$, every nonempty subset of $\{n \in \mathbb{Z} : n \geq m\}$ has a smallest member.

To see this, consider a nonempty subset S of $\{n \in \mathbb{Z} : n \geq m\}$. Then $\{n - m : n \in S\}$ is a nonempty subset of $\mathbb{N}$. If its smallest member is n_0, then $n_0 + m$ is the smallest member of S [Exercise 18].

Proof of the Second Principle. Assume

(B) $p(m), \ldots, p(m + l)$ are all true,
(I) $\forall_{n > m + l}[p(m) \wedge \cdots \wedge p(n - 1) \to p(n)]$ is true,

but that $\forall_{n \geq m} p(n)$ is false. Then the set

$$S = \{n \in \mathbb{Z} : n \geq m \text{ and } p(n) \text{ is false}\}$$

is nonempty. By the Well-Ordering Principle, S has a smallest element n_0. In view of (B) we must have $n_0 > m + l$. Since $p(n)$ is true for $m \leq n < n_0$, the compound proposition $p(m) \wedge \cdots \wedge p(n_0 - 1)$ is true. By (I), so is the implication

$$p(m) \wedge \cdots \wedge p(n_0 - 1) \to p(n_0).$$

Hence $p(n_0)$ is also true [modus ponens, rule 19 in Table 2 of §2.2], contradicting the fact that n_0 belongs to S. It follows that if (B) and (I) hold, then $\forall_{n \geq m} p(n)$ is true. ∎

A similar proof can be given for the First Principle but, since the principles are equivalent, it is not needed.

Some of the exercises for this section require only the First Principle of Mathematical Induction and are included to provide extra practice.

EXERCISES 6.3

1. Prove $3 + 11 + \cdots + (8n - 5) = 4n^2 - n$ for $n \in \mathbb{P}$.

2. For $n \in \mathbb{P}$, prove
   (a) $1 \cdot 2 + 2 \cdot 3 + \cdots + n(n + 1) = \frac{1}{3}n(n + 1)(n + 2)$
   (b) $\dfrac{1}{1 \cdot 2} + \dfrac{1}{2 \cdot 3} + \cdots + \dfrac{1}{n(n + 1)} = \dfrac{n}{n + 1}$

3. Prove that $n^5 - n$ is divisible by 10 for all $n \in \mathbb{P}$.

4. (a) Calculate b_6 for the sequence (b_n) in Example 2.
   (b) Use the recursive definition of (a_n) in Example 3 to calculate a_9.

5. Is the First Principle of Mathematical Induction adequate to prove the fact in Exercise 11(b) of §3.4? Explain.

6. Recursively define $a_0 = a_1 = 1$ and $a_n = 3a_{n-1} - 2a_{n-2}$ for $n \geq 2$.
   (a) Calculate the first few terms of the sequence.
   (b) Using part (a), guess the general formula for a_n.
   (c) Prove the guess in part (b).

7. Recursively define $a_0 = 1$, $a_1 = 2$ and $a_n = 2a_{n-1} - a_{n-2}$ for $n \geq 2$. Repeat Exercise 6 for this sequence.

8. Recursively define $a_0 = 1$, $a_1 = 3$ and $a_n = 2a_{n-1} - a_{n-2}$ for $n \geq 2$. Repeat Exercise 6 for this sequence.

9. Recursively define $a_0 = 1$, $a_1 = 2$ and $a_n = a_{n-1} + 2a_{n-2}$ for $n \geq 2$. Repeat Exercise 6 for this sequence.

10. Recursively define $a_0 = 1$, $a_1 = 2$ and $a_n = 2a_{n-1} + a_{n-2}$ for $n \geq 2$.
    (a) Calculate a_n for $n = 2, 3, 4, 5, 6$.
    (b) Prove that $a_n \leq (\frac{5}{2})^n$ for all $n \in \mathbb{N}$.

11. Recursively define $a_0 = a_1 = a_2 = 1$ and $a_n = a_{n-1} + a_{n-2} + a_{n-3}$ for $n \geq 3$.
    (a) Calculate the first few terms of the sequence.
    (b) Prove that all the a_n's are odd.
    (c) Prove that $a_n \leq 2^{n-1}$ for all $n \geq 1$.

12. Recursively define $a_0 = a_1 = 1$ and $a_n = 2a_{n-1} + 3a_{n-2}$ for $n \geq 2$.
    (a) Calculate a_n for $n = 2, 3, 4, 5, 6$.
    (b) Prove that $a_n > 3^{n-1}$ for $n \geq 2$.
    (c) Prove that $a_n < 2 \cdot 3^{n-1}$ for $n \geq 2$.

13. Recursively define $b_0 = b_1 = b_2 = 1$ and $b_n = b_{n-1} + b_{n-3}$ for $n \geq 3$.
   (a) Calculate b_n for $n = 3, 4, 5, 6$.
   (b) Show that $b_n \geq 2b_{n-2}$ for $n \geq 3$.
   (c) Prove the inequality $b_n \geq (\sqrt{2})^{n-2}$ for $n \geq 2$.

14. As in Exercise 13 of § 3.4, let $\mathrm{SEQ}(0) = 1$ and $\mathrm{SEQ}(n) = \sum\limits_{i=0}^{n-1} \mathrm{SEQ}(i)$ for $n \geq 1$.
   Prove that $\mathrm{SEQ}(n) = 2^{n-1}$ for $n \geq 1$.

15. Recursively define $\mathrm{SEQ}(0) = 0$, $\mathrm{SEQ}(1) = 1$ and

$$\mathrm{SEQ}(n) = \frac{1}{n} * \mathrm{SEQ}(n-1) + \frac{n-1}{n} * \mathrm{SEQ}(n-2)$$

   for $n \geq 2$. Prove that $0 \leq \mathrm{SEQ}(n) \leq 1$ for all $n \in \mathbb{N}$.

16. Recall the Fibonacci sequence in Example 3 of § 3.4:

   (B) $\mathrm{FIB}(0) = \mathrm{FIB}(1) = 1$,
   (R) $\mathrm{FIB}(n) = \mathrm{FIB}(n-1) + \mathrm{FIB}(n-2)$ for $n \geq 2$.

   Prove that

$$\mathrm{FIB}(n) = 1 + \sum_{k=0}^{n-2} \mathrm{FIB}(k) \quad \text{for} \quad n \geq 2.$$

17. The **Lucas sequence** is defined as follows:

   (B) $\mathrm{LUC}(1) = 1$ and $\mathrm{LUC}(2) = 3$,
   (R) $\mathrm{LUC}(n) = \mathrm{LUC}(n-1) + \mathrm{LUC}(n-2)$ for $n \geq 3$.

   (a) List the first eight terms of the Lucas sequence.
   (b) Prove that $\mathrm{LUC}(n) = \mathrm{FIB}(n) + \mathrm{FIB}(n-2)$ for $n \geq 2$, where FIB is the Fibonacci sequence defined in Exercise 16.

18. Let S be a nonempty subset of $\{n \in \mathbb{Z} : n \geq m\}$. Show that n_0 is the smallest member of $S' = \{n - m : n \in S\}$ if and only if $n_0 + m$ is the smallest member of S. *Suggestion:* It might help to first check this out on an example, say with $m = 5$ and $S = \{9, 17, 73\}$.

19. Let the sequence T be defined as in Example 2(a) of § 3.4 by

   (B) $T(1) = 1$,
   (R) $T(n) = 2 \cdot T(\lfloor n/2 \rfloor)$ for $n \geq 2$.

   Show that $T(n)$ is the largest integer of form 2^k with $2^k \leq n$. [I.e., $T(n) = 2^{\lfloor \log n \rfloor}$ where the logarithm is to the base 2.]

20. (a) Show that if T is defined as in Exercise 19 then $T(n)$ is $O(n)$.
   (b) Show that if the sequence Q is defined as in Example 2(b) of § 3.4 by

   (B) $Q(1) = 1$,
   (R) $Q(n) = 2 \cdot Q(\lfloor n/2 \rfloor) + n$ for $n \geq 2$,

   then $Q(n)$ is $O(n^2)$.
   (c) Show that in fact $Q(n)$ is $O(n \log_2 n)$ for Q as in part (b).

21. Show that if S is defined as in Example 6 of § 3.4 by

(B) $S(0) = 0$, $S(1) = 1$,
(R) $S(n) = S(\lfloor n/2 \rfloor) + S(\lfloor n/5 \rfloor)$ for $n \geq 2$,

then $S(n)$ is $O(n)$.

§ 6.4   A Generalization of Induction

Recall from § 3.6 that a set S of objects is defined recursively provided

(B) some members of the set are specified explicitly,
(R) the remaining members of the set are defined in terms of members already defined.

We showed in § 3.6 that once a set itself has been defined recursively, functions cán be recursively defined on it. If the image of such a function consists of propositions, i.e., if the function is a predicate, then it may be possible to prove all the propositions in the image true by a generalized version of induction which we now describe.

Generalized Principle of Induction   Let p be a proposition-valued function on a recursively defined set S. Suppose that

(B) $p(s)$ is true for the members s in S specified in the basis;
(I) whenever a member s of S is defined in terms of a set of previously defined members, say $D_s = \{t_1, \ldots, t_k\}$, then $p(t_1) \wedge \cdots \wedge p(t_k) \rightarrow p(s)$ is true.

Then $p(s)$ is true for every $s \in S$.

To see why this principle is valid, let $T = \{s \in S : p(s) \text{ is true}\}$. Then condition (B) states that all the members of the basis for S are in T, and condition (I) says that every member of S which is defined in terms of members of T is itself in T. Together they say that every member of S which can be built up out of basis members—i.e., every member of S—is in T.

EXAMPLE 1   (a) For a finite alphabet Σ, Σ^* was defined recursively in § 3.6 as follows:

(B) The empty word ϵ is in Σ^*.
(R) If w is in Σ^* and x is in Σ, then wx is in Σ^*.

Each new word wx is defined in terms of a word w in Σ^* already defined and a letter x in Σ. Thus, in terms of the general scheme, we have $D_{wx} = \{w\}$. If p is a proposition-valued function on Σ^*, then $\forall w \, p(w)$ is true provided

(B) $p(\epsilon)$ is true;
(I) $p(w) \rightarrow p(wx)$ is true for all $w \in \Sigma^*$ and $x \in \Sigma$.

This is the principle of induction for Σ^*. We illustrate its use in the next two parts of this example.

(b) Consider any subset S of Σ^*. For $w \in \Sigma^*$ we define $wS = \{ww_1 : w_1 \in S\}$. We claim that if $xS \subseteq S$ for all $x \in \Sigma$, then $wS \subseteq S$ for all $w \in \Sigma^*$.

Proof. We prove the claim by the inductive procedure in part (a). To each w in Σ^* we associate the proposition

$$p(w) = \text{``}wS \subseteq S.\text{''}$$

$p(\epsilon)$ is clearly true since $\epsilon w_1 = w_1$ for all words w_1. This establishes the basis of induction. For the inductive step, consider w in Σ^* and x in Σ; we need to prove $p(w) \to p(wx)$ is true, i.e.,

$$wS \subseteq S \quad \text{implies} \quad wxS \subseteq S.$$

In fact, if $wS \subseteq S$ and $x \in \Sigma$, then

$$wxS \subseteq wS \quad \text{since} \quad xS \subseteq S \quad \text{for } x \in \Sigma,$$
$$\subseteq S \quad \text{since} \quad wS \subseteq S \quad \text{by assumption.}$$

Thus the inductive step (I) has been established, so $p(w)$ is true for all $w \in \Sigma^*$.

(c) For words w in Σ^*, length(w) was recursively defined in § 3.6 as follows:

(B) length$(\epsilon) = 0$;

(R) if length(w) has been defined and $x \in \Sigma$, then

$$\text{length}(wx) = 1 + \text{length}(w).$$

We prove that

$$\text{length}(w_1 w_2) = \text{length}(w_1) + \text{length}(w_2)$$

for $w_1, w_2 \in \Sigma^*$.

Proof. We fix w_1 and we inductively prove all the propositions

$$p(w) = \text{``length}(w_1 w) = \text{length}(w_1) + \text{length}(w).\text{''}$$

Since $w_1 \epsilon = w_1$ and length$(\epsilon) = 0$, the proposition $p(\epsilon)$ is true. For the inductive step, consider w in Σ^* and x in Σ; we prove $p(w) \to p(wx)$. Thus we assume that

(1) $$\text{length}(w_1 w) = \text{length}(w_1) + \text{length}(w)$$

and we want to show

(2) $$\text{length}(w_1 wx) = \text{length}(w_1) + \text{length}(wx).$$

From (1) it is evident that

$$1 + \text{length}(w_1 w) = \text{length}(w_1) + 1 + \text{length}(w).$$

The recursive definition of length shows that

$$\text{length}(w_1 wx) = 1 + \text{length}(w_1 w) \quad \text{and} \quad \text{length}(wx) = 1 + \text{length}(w),$$

and so (2) holds. ∎

EXAMPLE 2   (a) Well-formed formulas [wff's] for algebra were defined recursively in § 3.6 as follows:

(B) Numerical constants and variables are wff's.
(R) If f and g are wff's, so are $(f + g)$, $(f - g)$, (fg), (f/g) and (f^g).

The new wff's are all defined in terms of f and g and so, in terms of the general scheme, $D_{(f+g)} = \{f, g\}$, etc. If p is a proposition-valued function on the set of wff's, then all the propositions $p(f)$ are true provided

(B) $p(f)$ is true for all numerical constants and variables f;
(I) $[p(f) \wedge p(g)] \rightarrow$
$[p((f + g)) \wedge p((f - g)) \wedge p((fg)) \wedge p((f/g)) \wedge p((f^g))]$ is true for all wff's f and g.

(b) We use the principle of induction in part (a) to prove that the number $L(f)$ of left parentheses in a wff f is equal to the number $R(f)$ of right parentheses in f.

Proof. To each wff f we associate the proposition

$$p(f) = \text{``}L(f) = R(f)\text{.''}$$

If f is a numerical constant or variable, then clearly $L(f) = 0$ and $R(f) = 0$, and so $p(f)$ is true. This takes care of the basis of induction. For the inductive step, we consider wff's f and g. We need to prove that if $L(f) = R(f)$ and $L(g) = R(g)$, then similar identities hold for $(f + g)$, $(f - g)$, (fg), (f/g) and (f^g). All five cases are similar, so we deal only with $(f + g)$. It is clear that

$$L((f + g)) = L(f) + L(g) + 1$$

and that

$$R((f + g)) = R(f) + R(g) + 1,$$

from which it follows that $L((f + g)) = R((f + g))$. ∎

EXAMPLE 3   (a) Finite trees were recursively defined in § 3.6:

(B) A single vertex is a [trivial] tree.
(R) If T is a tree and v is not a vertex of T, then T' is a tree if we add v as a vertex and add a single edge connecting v to some vertex of T.

Let p be a proposition-valued function on the class of finite trees. All the propositions are true provided

(B) $p(T_0)$ is true for the trivial tree T_0;

(I) if $p(T)$ is true and T' is obtained from T as in the recursive definition, then $p(T')$ is true.

(b) We use the generalized principle of induction in part (a) to prove that every nontrivial finite tree has at least two leaves, i.e., vertices of degree 1.

Proof. For each tree T the proposition is

$$p(T) = \text{``if } T \text{ is nontrivial, then } T \text{ has at least two leaves.''}$$

For the trivial tree this is vacuously true, and so the basis holds. To prove the inductive step, we consider T and T', and assume $p(T)$ is true. There are two cases: either T is trivial or T has at least two leaves.

If T is trivial, it consists of a single vertex v_0. Then T' has two vertices, v_0 and v, and one edge connecting them. Therefore both v_0 and v are leaves in this case, and so $p(T')$ is true.

If T is nontrivial, then T has at least two leaves by assumption; let v_1 and v_2 be two of them. The new vertex v is a leaf of T' and cannot be connected to both v_1 and v_2. Hence at least one of v_1 and v_2 is also a leaf of T', and so T' has at least two leaves.

This establishes the inductive step and the proof is completed by appealing to the generalized principle of induction in part (a).

(c) In this part and in part (d) we are going to give two proofs of the assertion: If a tree T has n vertices then it has $n - 1$ edges. The proofs will have different structures. One will use induction on n, using part (b) and the old familiar First Principle of Mathematical Induction. The other proof will be more direct and will use induction on T using the principle in part (a).

First we use induction on n. The nth proposition $p(n)$ is

$$\text{``if a tree has } n \text{ vertices, then it has } n - 1 \text{ edges.''}$$

This is clear for $n = 1$. Assume that $p(n)$ is true for some $n \geq 1$ and consider a tree with $n + 1$ vertices. It cannot be trivial, so by part (b) it has leaves. If we remove one leaf and the edge connected to it, we will obtain another tree having only n vertices. By proposition $p(n)$, the new tree has $n - 1$ edges and so the original tree has $(n - 1) + 1 = n$ edges. That is, a tree with $n + 1$ vertices has $(n + 1) - 1$ edges. In other words, $p(n) \to p(n + 1)$ is true for all $n \geq 1$. The result follows from the First Principle of Mathematical Induction.

(d) This time we prove the propositions

$$q(T) = \text{``if the tree } T \text{ has } n \text{ vertices, then it has } n - 1 \text{ edges.''}$$

This is clear for the trivial tree which has 1 vertex and 0 edges. Assume $q(T)$ is true and consider T' recursively defined from T as in part (a). Then T' has one more vertex and one more edge than T has. So $q(T')$ is true if $q(T)$

is. Consequently, all the assertions $q(T)$ hold by the generalized principle of induction in part (a). ∎

EXAMPLE 4   (a) Polynomial functions on $\mathbb{R}$ are defined recursively in Example 1 of § 3.6. Here is the generalized induction procedure for a proposition-valued function p defined on the set of polynomial functions on $\mathbb{R}$. All the propositions $p(f)$ are true provided

(B) $p(f)$ is true for constant functions f,
    and $p(I)$ is true, where $I(x) = x$ for all $x \in \mathbb{R}$;

(I) $p(f) \wedge p(g) \to p(f + g) \wedge p(fg)$ for all f, g.

(b) We prove that if f is a polynomial function on $\mathbb{R}$, then there exist $n \in \mathbb{N}$ and constants $a_0, a_1, \ldots, a_n$ such that

(1) $f(x) = a_n x^n + a_{n-1} x^{n-1} + \cdots + a_1 x + a_0$ for all $x \in \mathbb{R}$.

Proof. If c is a constant and $f(x) = c$ for all $x \in \mathbb{R}$, then (1) holds for f with $n = 0$ and $a_0 = c$. Also, (1) holds for the function I with $n = 1$, $a_1 = 1$ and $a_0 = 0$. These observations establish the basis for induction.

For the inductive step, suppose that f has the form (1) and that g is given by

(2) $g(x) = b_m x^m + b_{m-1} x^{m-1} + \cdots + b_1 x + b_0$ for all $x \in \mathbb{R}$

and some $m \in \mathbb{N}$. Then we have

$$(f + g)(x) = c_p x^p + c_{p-1} x^{p-1} + \cdots + c_1 x + c_0$$

where $p = \max\{m, n\}$ and $c_i = a_i + b_i$ for $0 \leq i \leq p$ where we agree that $c_i = a_i$ if $i > m$ and $c_i = b_i$ if $i > n$. Also

$$(fg)(x) = d_{m+n} x^{m+n} + d_{m+n-1} x^{m+n-1} + \cdots + d_1 x + d_0$$

where each d_i is the sum of all products $a_j b_k$ such that $j + k = i$. [This is seen by multiplying the polynomials f and g and collecting together all the x^i terms.] Thus $f + g$ and fg have the form (1) and so all polynomial functions have this form by the general induction procedure. ∎

As in § 6.3, some of the exercises below require only the First Principle of Mathematical Induction.

EXERCISES 6.4

1. Prove that

$$\frac{1}{n+1} + \frac{1}{n+2} + \cdots + \frac{1}{2n} = 1 - \frac{1}{2} + \frac{1}{3} - \frac{1}{4} + \cdots + \frac{1}{2n-1} - \frac{1}{2n}$$

for $n \in \mathbb{P}$. For $n = 1$ this says $\frac{1}{2} = 1 - \frac{1}{2}$ and for $n = 2$ this says $\frac{1}{3} + \frac{1}{4} = 1 - \frac{1}{2} + \frac{1}{3} - \frac{1}{4}$.

2. For $n \in \mathbb{P}$, prove

(a) $\displaystyle\sum_{k=1}^{n} \frac{1}{\sqrt{k}} \geq \sqrt{n}$

(b) $\displaystyle\sum_{k=1}^{n} \frac{1}{\sqrt{k}} \leq 2\sqrt{n} - 1$

3. Prove that $5^{n+1} + 2 \cdot 3^n + 1$ is divisible by 8 for $n \in \mathbb{N}$.

4. Prove that $8^{n+2} + 9^{2n+1}$ is divisible by 73 for $n \in \mathbb{N}$.

5. Here is a recursive definition for a subset S of $\mathbb{N} \times \mathbb{N}$:

(B) $\langle 0, 0 \rangle \in S$,
(R) if $\langle m, n \rangle \in S$, then $\langle m + 2, n + 3 \rangle \in S$.

(a) List four members of S.
(b) Prove that if $\langle m, n \rangle \in S$, then 5 divides $m + n$.
(c) Is the converse to the assertion in part (b) true?

6. Here is a recursive definition for another subset T of $\mathbb{N} \times \mathbb{N}$:

(B) $\langle 0, 0 \rangle \in T$,
(R) if $\langle m, n \rangle \in T$, then each of $\langle m + 1, n \rangle$, $\langle m + 1, n + 1 \rangle$ and $\langle m + 1, n + 2 \rangle$ is in T.

(a) List six members of T.
(b) Prove that $2m \geq n$ for all $\langle m, n \rangle \in T$.

7. Consider the following recursive definition for a subset A of $\mathbb{N} \times \mathbb{N}$:

(B) $\langle 0, 0 \rangle \in A$,
(R) if $\langle m, n \rangle \in A$, then $\langle m + 1, n \rangle$ and $\langle m, n + 1 \rangle$ are in A.

(a) Show that $A = \mathbb{N} \times \mathbb{N}$.
(b) Let $p(m, n)$ be a proposition-valued function on $\mathbb{N} \times \mathbb{N}$. Use part (a) to devise a general recursive procedure for proving $p(m, n)$ true for all m and n.

8. Let $\Sigma = \{a, b\}$ and let B be the subset of Σ^* defined recursively as follows:

(B) a and b are in B,
(R) if $w \in B$, then abw and baw are in B.

(a) List six members of B.
(b) Prove that if $w \in B$, then length(w) is odd.
(c) Is the converse to the assertion in part (b) true?

9. Let Σ be a finite alphabet. For words w in Σ^*, the reversal $\tilde{w}$ is defined in Exercise 16 of § 3.6.
(a) Prove length$(w) = $ length$(\tilde{w})$ for all $w \in \Sigma^*$.
(b) Prove $\widetilde{w_1 w_2} = \tilde{w}_2 \tilde{w}_1$ for all $w_1, w_2 \in \Sigma^*$.

10. Well-formed formulas P for the propositional calculus are defined recursively in Example 6 of § 3.6.
(a) Give a general recursive procedure for a proposition-valued function on the set of wff's for the propositional calculus.
(b) Prove that the number $L(P)$ of left parentheses in P always equals the number $R(P)$ of right parentheses in P.

11. Throughout this exercise, let p, q be *fixed* propositions. We recursively define the family $\mathfrak{F}$ of compound propositions using only $p, q, \wedge$ and $\vee$ as follows:

(B) $p, q \in \mathfrak{F}$,
(R) if $P, Q \in \mathfrak{F}$, then $(P \wedge Q)$ and $(P \vee Q)$ are in $\mathfrak{F}$.

(a) Use this definition to verify that $(p \wedge (p \vee q))$ is in $\mathfrak{F}$.
(b) Prove that if p and q are false, then all the propositions in $\mathfrak{F}$ are false.
(c) Show that $p \to q$ is not logically equivalent to any proposition in $\mathfrak{F}$.

CHAPTER HIGHLIGHTS

These lists should be used as always to check understanding. See the end of Chapter 0 for a reminder on how to review.

Concepts

predicate calculus
   quantifiers, $\forall, \exists$
   predicate = proposition-valued function
      compound predicate
         free/bound variable
         logical variable
   tautology [extended version for compound predicates]
   logical equivalence, implication
   counterexample

Facts

Basic logical relationships [Table 1 of § 6.2].
$\forall$ and $\exists$ do not commute with each other:
   $\exists x \, \forall y \, p(x, y) \to \forall y \, \exists x \, p(x, y)$ is a tautology,
   but $\forall y \, \exists x \, p(x, y) \to \exists x \, \forall y \, p(x, y)$ is not.

First, Second and Generalized Principles of Mathematical Induction.
   First Principle can be viewed as a special case of Second Principle.

Methods

Use of DeMorgan laws to negate quantified predicates.
Construction of proofs using mathematical induction:
   Second Principle [$n \geqq m$ version] for recursively defined sequences.
   Generalized Principle for predicates on recursively defined sets.

7

RELATIONS

One frequently wants to compare or contrast various members of a set, perhaps to arrange them in some appropriate order or to group together those with similar properties. The mathematical framework to describe this kind of organization of sets is the theory of relations. This chapter begins with order relations, then treats equivalence relations (the ones which group related elements together) and finally examines more general types of relations.

§ 7.1 Partially Ordered Sets

Suppose that there is some natural way to compare the members of a set, so that whenever we are given two members of the set we know how they compare with each other.

EXAMPLE 1   (a) We are used to comparing real numbers. For example, 3 is less than 5, -1 is less than 4 and -1 is greater than -3. We compare two numbers by observing which is larger and which is smaller.

(b) If the set S is indexed by $\mathbb{P}$ or $\mathbb{N}$ so that different elements have different indexes, we can compare two members of S by observing which of them has the smaller index. Different ways of indexing S give different ways of ordering the members. The element which has the lowest subscript of all for one indexing might have many members precede it using another indexing. ∎

A set whose members can be compared in such a way is said to be **ordered**, and the specification of how its members compare with each other is called an **order relation** on the set. To say anything useful about ordered sets we need to make these definitions more precise. First note, however, that in many sets which arise naturally we know how to compare some elements with others but also have pairs which are not comparable.

EXAMPLE 2   (a) If we try to compare makes of automobiles we can perhaps agree that Make X is not as good as Make Y but may not be able to say that either of Make Z or Make W is better than the other.

(b) We can agree to compare two numbers in $\{1, 2, 3, \ldots, 73\}$ in case one is a factor of the other. Then 6 and 72 are comparable and so are 6 and 3, but 6 and 8 are not, since neither 6 nor 8 is a factor of the other.

(c) We can compare two subsets of a set S [i.e., members of $\mathscr{P}(S)$] if one is contained in the other. If S has more than one member it has some incomparable subsets. For example, if $s_1 \neq s_2$ and s_1 and s_2 belong to S, then the sets $\{s_1\}$ and $\{s_2\}$ are incomparable. ■

Sets with comparison relations which allow the possibility of incomparable elements such as those in Example 2 are said to be partially ordered. They form an important class, which we now define precisely.

Suppose we are given a relation on a set S, in the sense that for each pair $\langle x, y \rangle$ in $S \times S$ we know whether or not x is related to y. Write $x \preceq y$ if and only if x is related to y. The relation is called a **partial order** on S if it satisfies the following:

(R) $s \preceq s$ for every s in S;
(AS) $s \preceq t$ and $t \preceq s$ imply $s = t$;
(T) $s \preceq t$ and $t \preceq u$ imply $s \preceq u$.

The conditions (R), (AS) and (T) are the **reflexive**, **antisymmetric** and **transitive** laws, and $\preceq$ is called reflexive, antisymmetric or transitive if it satisfies them. If $\preceq$ is a partial order on S the pair $(S, \preceq)$ is called a **partially ordered set** or **poset** for short. We use the notation "$\preceq$" as a general-purpose name for a partial order. If there is already a notation, such as "$\leq$" or "$\subseteq$," for a particular partial order we will generally use it in preference to "$\preceq$."

In Example 2 the understood relations were "is not as good as," "is a factor of" and "is a subset of." We could just as well have considered the relations "is as good as," "is a multiple of" and "contains," since these relations convey the same comparative information as the chosen ones. Each partial order on a set determines an **inverse** relation in which x and y are related if and only if y and x are related in the original way. The inverse of a partial order $\preceq$ is usually denoted by $\succeq$. The inverse relation is also a partial order [Exercise 12(a)].

Given a partial order $\leq$ on a set S we can define another relation $\prec$ on S by

$$x \prec y \quad \text{if and only if} \quad x \leq y \quad \text{and} \quad x \neq y.$$

For example if $\leq$ is set inclusion $\subseteq$, then $A \prec B$ means A is a proper subset of B, i.e., $A \subset B$. The relation $\prec$ satisfies:

(AR) $s \prec s$ is false for all s in S;
(T) $s \prec t$ and $t \prec u$ imply $s \prec u$.

A relation satisfying (AR) is said to be **antireflexive**. We call a relation which satisfies (AR) and (T) a **quasi-order**. Each partial order on S yields a quasi-order, and conversely if $\prec$ is a quasi-order on S then the relation $\leq$ defined by

$$x \leq y \quad \text{if and only if} \quad x \prec y \quad \text{or} \quad x = y$$

is a partial order on S [Exercise 12(b)]. Whether one considers a partial order or its associated quasi-order to describe comparisons between members of a set depends on the particular problem at hand. We will generally use the partial order but switch back and forth as convenient.

It is possible, at least in principle, to draw a diagram which shows at a glance the order relation on a finite poset. Given a partial order $\leq$ on S, we say the element t **covers** the element s in case $s \prec t$ and there is no u in S with $s \prec u \prec t$. A **Hasse** [pronounced HAH-suh] **diagram** of the poset $(S, \leq)$ is a figure consisting of points [or small circles] labeled by the members of S, with a line segment directed generally upward from s to t whenever t covers s.

EXAMPLE 3   (a) Let $S = \{1, 2, 3, 4, 5, 6\}$. We write $m|n$ in case m divides n, i.e., n is an integer multiple of m. The diagram in Figure 1 is a Hasse diagram of the poset $(S, |)$. There is no segment between 1 and 6 because 6 does not cover 1. We can see from the diagram, though, that $1|6$ because the relation is transitive and there is a chain of segments corresponding to $1|2$ and $2|6$. Similarly, we can see that $1|4$ from the chain $1|2|4$. Note that, in general, transitive relations can be run together without causing confusion: $x \leq y \leq z$ means $x \leq y$, $y \leq z$ and $x \leq z$.

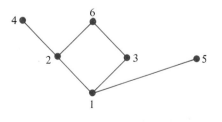

Figure 1

(b) Let S be the power set $\mathscr{P}(\{a, b, c\})$ with $\subseteq$ as partial order. Figure 2 shows a Hasse diagram of $(S, \subseteq)$. Note that the lines from $\{a\}$ to $\{a, c\}$ and from $\{b\}$ to $\{a, b\}$ happen to cross, but this crossing is simply a feature of the drawing and has no significance as far as the partial order is concerned. In particular, the intersection of the two lines does *not* represent an element of the poset.

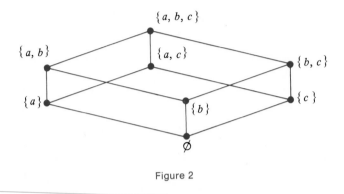

Figure 2

(c) The diagram in Figure 3 is not a Hasse diagram, because u cannot cover x if u also covers y and y covers x. If any of the line segments connecting u, x and y were removed, then the figure would be a Hasse diagram.

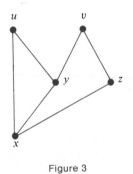

Figure 3

(d) The diagrams in Figure 4 are Hasse diagrams of posets, whose order relations can be read off directly from the diagrams. All elements are related to themselves. In addition:

For $S = \{a, b, c, d, e, f\}$ we have $a \leq b$, $a \leq c$, $a \leq d$, $a \leq e$, $a \leq f$, $b \leq e$, $b \leq f$ and $c \leq f$. We saw this picture before in part (a) of this example.

For $T = \{x, y, z, w\}$ we have $x \leq y$, $x \leq z$, $x \leq w$, $y \leq z$, $y \leq w$ and $z \leq w$. This is the picture we would get for divisors of 8 or of 27 or of 125 with order relation $|$.

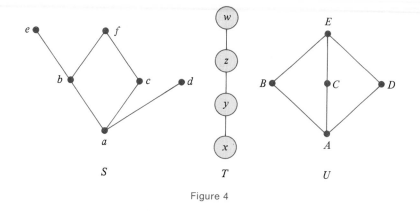

Figure 4

For $U = \{A, B, C, D, E\}$ we have $A \leq B$, $A \leq C$, $A \leq D$, $A \leq E$, $B \leq E$, $C \leq E$ and $D \leq E$. This picture is the Hasse diagram of the poset consisting of the sets $\{1\}$, $\{1, 2\}$, $\{1, 3\}$, $\{1, 4\}$, $\{1, 2, 3, 4\}$ with set inclusion as order relation. ∎

In general, given a Hasse diagram for a poset, we see that $s \leq t$ in case either $s = t$ or there is an upward chain of edges from s to t. The reflexive and transitive laws are understood, and the covering information tells us the rest. The fact that every finite poset has a Hasse diagram is intuitively obvious and is an easy consequence of properties of directed graphs, as we will explain after Theorem 2 of § 8.1. Some infinite posets also have Hasse diagrams. A Hasse diagram of $\mathbb{Z}$ with the usual order $\leq$ is a vertical line with dots spaced along it. On the other hand, no real number covers any other in the usual $\leq$ order, so $(\mathbb{R}, \leq)$ has no Hasse diagram.

EXAMPLE 4 For an alphabet Σ, we make Σ^* into an infinite poset as follows. For words w_1, w_2 in Σ^* define $w_1 \leq w_2$ if w_1 is an **initial segment** of w_2, i.e., if there is a word w in Σ^* with $w_1 w = w_2$. For example, we have $ab \leq abbaa$. This conforms to our definition since, if $w_1 = ab$ and $w_2 = abbaa$, then $w_1 w = w_2$ for $w = baa$. Also, $\epsilon \leq w$ for all words because $w = \epsilon w$. Note that $abbaa$ does not cover ab since $u = abb$ and $u = abba$ both satisfy $ab \prec u \prec abbaa$. However, $abbaa$ covers $abba$, $abba$ covers abb, and abb covers ab. In general, if w_2 covers w_1 then length$(w_2) = 1 + $ length(w_1).

For $\Sigma = \{a, b\}$, part of the Hasse diagram for $(\Sigma^*, \leq)$ is drawn in Figure 5. This Hasse diagram is a tree. In § 9.2 we will view this as a "rooted tree," at which point tradition will force us to draw it upside down. ∎

The elements corresponding to points near the top or bottom of a Hasse diagram often turn out to be important. If $(P, \leq)$ is a poset we call an element x of P **maximal** in case there is no y in P with $x \prec y$, and call x **minimal** if there is no y in P with $y \prec x$. In the posets with Hasse diagrams

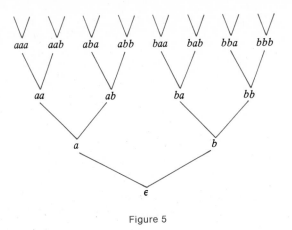

Figure 5

shown in Figure 4 the elements d, e, f, w and E are maximal, while a, x and A are minimal. The infinite poset in Figure 5 has no maximal elements; the empty word ϵ is its only minimal element.

A subset S of a poset P inherits the partial order on P and is itself a poset, since the laws (R), (AS) and (T) apply to all members of P. We call S a **subposet** of P.

EXAMPLE 5   (a) The sets $\{2, 3, 4, 5, 6\}$ and $\{1, 2, 3, 6\}$ are subposets of the poset $\{1, 2, 3, 4, 5, 6\}$ given in Example 3(a), with Hasse diagrams shown in Figure 6. [Notice the placement of the primes in Figure 6(a).]

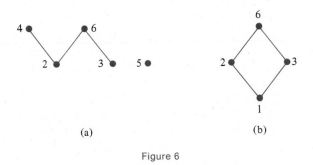

(a)   (b)

Figure 6

(b) The set of nonempty proper subsets of $\{a, b, c\}$ is a subposet of $\mathcal{P}(\{a, b, c\})$ with partial order $\subseteq$. Figure 7 shows a Hasse diagram for it. Compare with Figure 2. ∎

If S is a subposet of a poset $(P, \leq)$, it may happen that S has a member M such that $s \leq M$ for every s in S. In Figure 6(b), $s \leq 6$ for every s, while no such element M exists in Figure 6(a) or 7. An element M with this property is

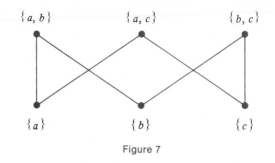

Figure 7

called the **largest member** of S or the **maximum** of S and denoted max(S). [There is at most one such M; why?] This notation is consistent with our usage of max$\{m, n\}$ to denote the larger of the two numbers m and n. Similarly, if S has a member m such that $m \leq s$ for every s in S, then m is called the **smallest member** of S or the **minimum** of S and is denoted min(S).

EXAMPLE 6 Consider again the poset $(\{1, 2, 3, 4, 5, 6\}, |)$ illustrated in Figure 1. This poset has no largest member or maximum even though 4, 6 and 5 are all maximal elements. The element 1 is a minimum of the poset and is the only minimal element. The subset $\{2, 3\}$ has no largest member [3 is larger than 2 in the usual order, but not in the order under discussion]. ∎

    Even if a subposet S of a poset $(P, \leq)$ has no largest member, it may happen that there is an element x in P with $s \leq x$ for every s in S. [For example, both elements in the set $\{2, 3\}$ of Example 6 divide 6.] Such an element is called an **upper bound** for S in P. If x is an upper bound for S in P such that $x \leq y$ for every upper bound y for S in P, x is called a **least upper bound** of S in P, and we write $x = $ lub(S). Similarly, an element z in P such that $z \leq s$ for all s in S is a **lower bound** for S in P. A lower bound z such that $w \leq z$ for every lower bound w is called a **greatest lower bound** of S in P and denoted by glb(S). By the antisymmetric law (AS), a subset of P cannot have two different least upper bounds or two different greatest lower bounds.

EXAMPLE 7   (a) In the poset $(\{1, 2, 3, 4, 5, 6\}, |)$ the subset $\{2, 3\}$ has exactly one upper bound, namely 6, and so lub$\{2, 3\} = 6$. Similarly, glb$\{2, 3\} = 1$. The subset $\{4, 6\}$ has no upper bounds in the poset; 2 and 1 are both lower bounds and so glb$\{4, 6\} = 2$. The subset $\{3, 6\}$ has 6 as an upper bound and 3 and 1 as lower bounds; hence lub$\{3, 6\} = 6$ and glb$\{3, 6\} = 3$. Thus least upper bounds and greatest lower bounds might or might not exist, and if they do they might or might not belong to the subset.
    (b) In the poset P shown in Figure 8 the subset $\{b, c\}$ has d, e, g and h as upper bounds in P, and h is an upper bound for $\{d, f\}$. The set $\{b, c\}$ has no least upper bound in P [why?] but $h = $ lub$\{d, f\}$. The elements a and c are

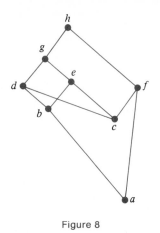

Figure 8

lower bounds for $\{d, e, f\}$, which has no greatest lower bound because a and c are not comparable. Element a is the greatest lower bound of $\{b, d, e, f\}$. ∎

Many of the posets which come up in practice have the property that every 2-element subset has both a least upper bound and a greatest lower bound. Such a poset is called a lattice. Thus a poset $(P, \leq)$ is called a **lattice** if $\text{lub}\{x, y\}$ and $\text{glb}\{x, y\}$ exist for every x and y in P. If $(P, \leq)$ is a lattice, then we introduce the new notation $x \vee y = \text{lub}\{x, y\}$ and $x \wedge y = \text{glb}\{x, y\}$ for $x, y \in P$. We read $x \vee y$ as "x join y" and $x \wedge y$ as "x meet y" or, more informally, as "x bird y" and "x hat y."

Observe that $\vee$ and $\wedge$ are binary operations on P. Also, note that $x \wedge y = x$ if and only if $x \leq y$ if and only if $x \vee y = y$. In particular, we can recover the relation $\leq$ if we know either binary operation $\wedge$ or $\vee$; see Exercise 8. One can show by induction [Exercise 19(b)] that every finite subset of a lattice has both a least upper bound and a greatest lower bound.

EXAMPLE 8   (a) The poset $(\mathcal{P}(\{a, b, c\}), \subseteq)$ shown in Figure 2 is a lattice. For instance

$$\{a\} \vee \{c\} = \{a, c\}, \quad \{a, b\} \vee \{a, c\} = \{a, b, c\}, \quad \{a, b\} \wedge \{c\} = \varnothing$$

and

$$\{a, b\} \wedge \{b, c\} = \{b\}.$$

In general, for any set U, $(\mathcal{P}(U), \subseteq)$ is a lattice with $A \vee B = A \cup B$ and $A \wedge B = A \cap B$ so that

$$\text{lub}\{A, B, \ldots, Z\} = A \cup B \cup \cdots \cup Z$$

and

$$\text{glb}\{A, B, \ldots, Z\} = A \cap B \cap \cdots \cap Z.$$

The poset shown in Figure 7 is not a lattice; for example $\{a, b\}$ and $\{a, c\}$ have no least upper bound in the poset.

(b) Define the partial order $|$ on $\mathbb{P}$ by $m|n$ if and only if m divides n. The subposet $S = \{1, 2, 3, 4, 5, 6\}$ of $\mathbb{P}$, shown in Figure 1, is not a lattice since $\{3, 4\}$ has no upper bound in S. The full poset $(\mathbb{P}, |)$ is a lattice, however. For m and n in $\mathbb{P}$ an upper bound for $\{m, n\}$ is an integer k in $\mathbb{P}$ such that m divides k and n divides k, i.e., a common multiple of m and n. The least upper bound $m \vee n$ is the **least common multiple** of m and n. Similarly the greatest lower bound $m \wedge n$ is the **greatest common divisor** of m and n, the largest positive integer which divides both m and n. For instance, $12 \vee 10 = 60$ and $12 \wedge 10 = 2$. The numbers $m \vee n$ and $m \wedge n$ can be determined from the factorizations of m and n into products of primes. The primes themselves are the minimal members of the subposet obtained by deleting the element 1 from $\mathbb{P}$.

(c) Consider the set $\mathrm{FUN}(\{a, b, c\}, \{0, 1\})$ of all functions from the 3-element set $\{a, b, c\}$ to $\{0, 1\}$. We obtain a partial order $\leq$ on this set by defining

$$f \leq g \quad \text{if and only if} \quad f(x) \leq g(x) \quad \text{for} \quad x = a, b, c.$$

It is convenient to label the eight functions in this poset with subscripts, like 101, that list the values the functions take at a, b and c, respectively. For example, f_{101} represents the function such that $f_{101}(a) = 1$, $f_{101}(b) = 0$ and $f_{101}(c) = 1$. The Hasse diagram for the poset $(\mathrm{FUN}(\{a, b, c\}, \{0, 1\}), \leq)$ is given in Figure 9. ∎

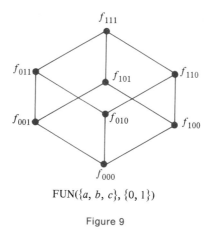

FUN($\{a, b, c\}, \{0, 1\}$)

Figure 9

EXAMPLE 9   Up to this section we have used the symbols $\vee$ and $\wedge$ as logical connectives. Our new usage is consistent with the previous notation. The poset $(\{0, 1\}, \leq)$ with the usual order is a lattice in which

$$0 \vee 0 = 0 \wedge 0 = 0 \wedge 1 = 0 \text{ and } 1 \wedge 1 = 1 \vee 1 = 1 \vee 0 = 1,$$

i.e., $x \vee y = \max\{x, y\}$ and $x \wedge y = \min\{x, y\}$. The table

p	q	$p \vee q$	$p \wedge q$
0	0	0	0
0	1	1	0
1	0	1	0
1	1	1	1

shows that the truth values of $p \vee q$ and $p \wedge q$ are the maximum and minimum values, respectively, of the truth values of p and q. In Chapter 10 we will discuss in detail how to make a set of propositions into a lattice, called a Boolean algebra. ▌

   The definitions of $\vee$ and $\wedge$ imply that a lattice satisfies the following laws:

$$x \vee x = x \qquad\qquad x \wedge x = x$$
$$x \vee y = y \vee x \qquad\qquad x \wedge y = y \wedge x$$
$$(x \vee y) \vee z = x \vee (y \vee z) \qquad (x \wedge y) \wedge z = x \wedge (y \wedge z).$$

A proof of the associativity of the join operation $\vee$ is outlined in Exercise 19. The laws in the left-hand column above match those on the right. The symmetry in the way $\vee$ and $\wedge$ are defined means that we can take the proof of a statement on one side, interchange $\vee$ and $\wedge$ and get a proof of the statement on the other side. This observation can be developed into a formal **principle of duality** for lattices, which says that if we interchange $\vee$ and $\wedge$ in a true theorem about lattices we get another true theorem.

   We began this section by considering sets in which we could compare any two members, but since then we have dealt almost exclusively with posets which have incomparable pairs of elements. In the next section we return to the special case in which each element is related to every other. We also discuss how to use orders on relatively simple sets to produce orders for more complicated ones.

EXERCISES 7.1

1. Draw Hasse diagrams for the following posets.
   (a) $(\{1, 2, 3, 4, 6, 8, 12, 24\}, |)$ where $m|n$ means m is a factor of [i.e., divides] n.
   (b) The set of subsets of $\{3, 7\}$ with $\subseteq$ as partial order.

2. (a) Give examples of two posets which come from everyday life or from other courses.
   (b) Do your examples have maximal or minimal elements? If so, what are they?
   (c) What are the inverses of the partial orders in your examples?

3. Figure 10 shows the Hasse diagrams of three posets.
   (a) What are the maximal members of these posets?
   (b) Which of these posets have minimal elements?

(c) Which of these posets have smallest members?

(d) Which elements cover the element e?

(e) Find each of the following if it exists.

$$\mathrm{lub}\{d, c\}, \qquad \mathrm{lub}\{w, y, v\}, \qquad \mathrm{lub}\{p, m\}, \qquad \mathrm{glb}\{a, g\}.$$

(f) Which of these posets are lattices?

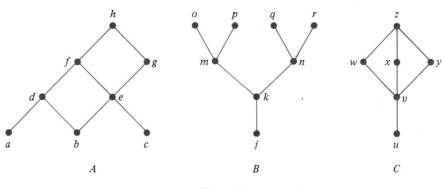

Figure 10

4. Find the maximal proper subsets of the 3-element set $\{a, b, c\}$. That is, find the maximal members of the subposet of $\mathscr{P}(\{a, b, c\})$ consisting of proper subsets of $\{a, b, c\}$.

5. Define the relations $<$, $\leqq$ and $\leq$ on the plane $\mathbb{R} \times \mathbb{R}$ by

$$\begin{aligned}
\langle x, y \rangle < \langle z, w \rangle & \quad \text{if } x^2 + y^2 < z^2 + w^2, \\
\langle x, y \rangle \leqq \langle z, w \rangle & \quad \text{if } \langle x, y \rangle < \langle z, w \rangle \text{ or } \langle x, y \rangle = \langle z, w \rangle, \\
\langle x, y \rangle \leq \langle z, w \rangle & \quad \text{if } x^2 + y^2 \leq z^2 + w^2.
\end{aligned}$$

(a) Which of these relations are partial orders? Explain.

(b) Which are quasi-orders? Explain.

(c) Draw a sketch of $\{\langle x, y \rangle : \langle x, y \rangle \leqq \langle 3, 4 \rangle\}$.

(d) Draw a sketch of $\{\langle x, y \rangle : \langle x, y \rangle \leq \langle 3, 4 \rangle\}$.

6. Let $\mathscr{E}(\mathbb{N})$ be the set of all finite subsets of $\mathbb{N}$ which have an even number of elements, with partial order $\subseteq$.

(a) Let $A = \{1, 2\}$ and $B = \{1, 3\}$. Find four upper bounds for $\{A, B\}$.

(b) Does $\{A, B\}$ have a least upper bound in $\mathscr{E}(\mathbb{N})$? Explain.

(c) Is $\mathscr{E}(\mathbb{N})$ a lattice?

7. Is every subposet of a lattice a lattice? Explain.

8. The table in Figure 11 has been partially filled in. It gives the value of $x \vee y$ for x and y in a certain lattice $(L, \leq)$. For example $b \vee c = d$.

(a) Fill in the rest of the table.

(b) Which are the largest and smallest elements of L?

(c) Show that $f \leq c \leq d \leq e$.

(d) Draw a Hasse diagram for L.

$\vee$	a	b	c	d	e	f
a		e	a	e	e	a
b			d	d	e	b
c				d	e	c
d					e	d
e						e
f						

Figure 11

9. Consider $\mathbb{R}$ with the usual order $\leq$.
   (a) Is $\mathbb{R}$ a lattice? If it is, what are the meanings of $a \vee b$ and $a \wedge b$ in $\mathbb{R}$?
   (b) Give an example of a nonempty subset of $\mathbb{R}$ which has no least upper bound.
   (c) Find $\operatorname{lub}\{x \in \mathbb{R} : x < 73\}$.
   (d) Find $\operatorname{lub}\{x \in \mathbb{R} : x \leq 73\}$.
   (e) Find $\operatorname{lub}\{x \in \mathbb{R} : x^2 < 73\}$.
   (f) Find $\operatorname{glb}\{x \in \mathbb{R} : x^2 < 73\}$.

10. (a) Show that every finite poset has a minimal element. *Hint*: Use induction.
    (b) Give an example of a poset with a maximal element but no minimal element.

11. Consider the poset C whose Hasse diagram is shown in Figure 10. Show that
    $w \vee (x \wedge y) \neq (w \vee x) \wedge (w \vee y)$ and $w \wedge (x \vee y) \neq (w \wedge x) \vee (w \wedge y)$. This
    shows that lattices need not satisfy "distributive laws."

12. (a) Show that if $\leq$ is a partial order on a set S then so is its inverse relation $\geq$.
    (b) Show that if $<$ is a quasi-order on a set S then the relation $\leq$ defined by

$$x \leq y \quad \text{if and only if} \quad x < y \quad \text{or} \quad x = y$$

    is a partial order on S.

13. Let Σ be an alphabet. For w_1, $w_2 \in \Sigma^*$, let $w_1 \leq w_2$ mean $\operatorname{length}(w_1) \leq \operatorname{length}(w_2)$. Is $\leq$ a partial order on Σ^*? Explain.

14. Verify that the partial order $\leq$ on Σ^* in Example 4 is reflexive and transitive.

15. Let Σ be an alphabet.
    (a) For w_1, $w_2 \in \Sigma^*$ define $w_1 \ll w_2$ if there are w and w' in Σ^* with $w_2 = ww_1w'$. Is $\ll$ a partial order on Σ^*? Explain.
    (b) Answer part (a) if w and w' are restricted to belong to Σ.

16. Let S be a set of subroutines of a computer program. For A and B in S write $A \prec B$ if A must be completed before B can be completed. What sort of restriction must be placed on subroutine calls in the program to make $\prec$ a quasi-order on S?

17. Let $\mathscr{F}(\mathbb{N})$ be the collection of all *finite* subsets of $\mathbb{N}$. Then $(\mathscr{F}(\mathbb{N}), \subseteq)$ is a poset.
    (a) Does $\mathscr{F}(\mathbb{N})$ have a maximal element? If yes, give one. If no, explain.
    (b) Does $\mathscr{F}(\mathbb{N})$ have a minimal element? If yes, give one. If no, explain.
    (c) Given A, B in $\mathscr{F}(\mathbb{N})$, does $\{A, B\}$ have a least upper bound in $\mathscr{F}(\mathbb{N})$? If yes, specify it. If no, provide a specific counterexample.

    (d) Given A, B in $\mathscr{F}(\mathbb{N})$, does $\{A, B\}$ have a greatest lower bound in $\mathscr{F}(\mathbb{N})$? If yes, specify it. If no, provide a specific counterexample.

    (e) Is $\mathscr{F}(\mathbb{N})$ a lattice? Explain.

18. Repeat Exercise 17 for the collection $\mathscr{I}(\mathbb{N})$ of all *infinite* subsets of $\mathbb{N}$.

19. (a) Consider elements x, y, z in a poset. Show that if $\text{lub}\{x, y\} = a$ and $\text{lub}\{a, z\} = b$, then $\text{lub}\{x, y, z\} = b$.

    (b) Show that every finite subset of a lattice has a least upper bound.

    (c) Show that if x, y and z are members of a lattice then $(x \vee y) \vee z = x \vee (y \vee z)$.

20. Let $\mathscr{S}$ be a set of compound propositions, and for P and Q in $\mathscr{S}$ define $P \leq Q$ if $P \Rightarrow Q$. Must $(\mathscr{S}, \leq)$ be a poset? Explain.

§ 7.2   Special Orderings

Partially ordered sets arise in a variety of ways, and in many cases the fact that there are pairs of elements which cannot be compared is an essential feature of the context. The important class of data structures called trees, which we will study in detail in Chapter 9, can be thought of as consisting of Hasse diagrams, such as those shown in Figure 1, in which $\text{lub}\{x, y\}$ exists for all x and y but $\text{glb}\{x, y\}$ exists only if $x \leq y$ or $y \leq x$. Trees are useful data structures, even though they have incomparable elements, because it is possible to start at the top and follow the tree to get to any element fairly quickly. Efficient algorithms are also known for examining all elements of a tree.

    An even more common data structure consists of a list or sequence in which, no matter what two elements are chosen, one comes before the other. Such a structure is an example of a **chain** which we define to be a poset

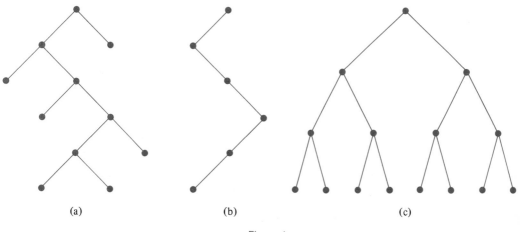

(a)   (b)   (c)

Figure 1

in which every two elements are comparable. A partial order is called a **total order** or **linear order** if for each choice of s and t in S either $s \preceq t$ or $t \preceq s$. Thus a chain is a poset with a total order. The terms "totally ordered set" and "linearly ordered set" are sometimes used as synonyms for "chain."

EXAMPLE 1   (a) The poset of Figure 1(b) is a chain, but the other posets in Figure 1 are not.

(b) The set $\mathbb{R}$ with the usual order $\leq$ is a chain.

(c) The lists of names in a phone book or words in a dictionary are chains if we define $w_1 \preceq w_2$ to mean that $w_1 = w_2$ or w_1 comes before w_2. ∎

Every subposet of a chain is itself a chain. For example the posets $(\mathbb{Z}, \leq)$ and $(\mathbb{Q}, \leq)$ are subposets of $(\mathbb{R}, \leq)$ and are linearly ordered by the orders they inherit from $\mathbb{R}$. The words in the dictionary between "start" and "stop" form a subchain of the chain of all words in Example 1(c).

A poset which is not itself a chain will have subposets which are. For a given poset, it is often useful to know something about the subposets which are chains.

EXAMPLE 2   (a) Let S be a set of people at a family reunion, and write $m \prec n$ in case m is a descendent of n. Then $\prec$ is a quasi-order which defines a partial order $\preceq$ by $m \preceq n$ if and only if $m \prec n$ or $m = n$. A chain in the poset $(S, \preceq)$ is a set of the form $\{m, n, p, \dots, r\}$ in which m is a descendent of n, n a descendent of p, and so on. It would be unusual for such a chain to have more than 5 members, though the set S itself might be quite large.

(b) The Hasse diagram shown in Figure 2 describes a poset with a number of subchains [49 if we count the 1-element chains but not the empty chain]. ∎

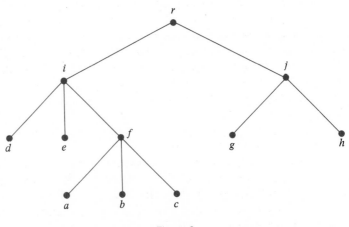

Figure 2

We are frequently interested in the maximal chains of a poset. To define these we first observe that if $(S, \leq)$ is a poset and if $\mathscr{C}(S)$ is the set of chains in S then $(\mathscr{C}(S), \subseteq)$ is also a poset. A **maximal chain** in S is defined to ⊦ a maximal member of $\mathscr{C}(S)$, i.e., a chain which is not properly contai..ed in another chain.

EXAMPLE 3   (a) In the poset of Figure 2 the maximal chains are $\{a, f, i, r\}$, $\{b, f, i, r\}$, $\{c, f, i, r\}$, $\{d, i, r\}$, $\{e, i, r\}$, $\{g, j, r\}$ and $\{h, j, r\}$. Notice that the maximal chains are not all the same size.

(b) In the poset shown in Figure 3 the two maximal chains $\{a, c, d, e\}$ and $\{a, b, e\}$ containing a and e have different numbers of elements. This poset has 4 maximal chains in all. ∎

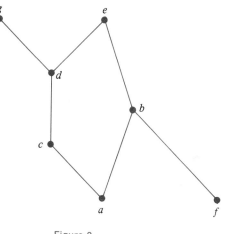

Figure 3

A finite chain must have a smallest member, and so must each of its nonempty subchains. Infinite chains, on the other hand, can exhibit a variety of behaviors. The infinite chains $(\mathbb{R}, \leq)$ and $(\mathbb{Z}, \leq)$ with their usual orders do not have smallest members. The chain $(\{x \in \mathbb{R} : 0 \leq x\}, \leq)$ has a smallest member, namely 0, but has subsets such as $\{x \in \mathbb{R} : 1 < x\}$ without smallest members. The infinite chain $(\mathbb{N}, \leq)$ has a smallest member, and the well-ordering property of $\mathbb{N}$ stated in § 6.3 says that every nonempty subset does too.

We say that a chain C is **well-ordered** in case each nonempty subset of C has a smallest member. If C is well-ordered, and if for each c in C we have a statement $p(c)$, then we can hope to prove all statements $p(c)$ true by supposing the set $\{c \in C : p(c) \text{ is false}\}$ is nonempty, considering the smallest c for which $p(c)$ is false and deriving a contradiction. This was the idea behind our explanation of the validity of the principles of induction in § 6.3. Since every finite chain is well-ordered, the method applies to finite chains, and in particular to any set $\{n \in \mathbb{Z} : m \leq n \leq m^*\}$ with the usual order. Hence it can be used to justify the finite principles of induction stated after Example 3 of §6.3.

In the rest of this section we study how to build new partial orders from known ones. Suppose first that $(S, \leq)$ is a given poset and that T is a nonempty set. We can define a partial order, which we denote $\preccurlyeq$, on the set FUN(T, S) of functions from T to S by defining

$$f \preccurlyeq g \quad \text{if} \quad f(t) \leq g(t) \quad \text{for all} \quad t \text{ in } T.$$

The verification that this new relation is a partial order is straightforward [Exercise 7(a)]. If the order on S is $\leq$, we will write $\leq$ in place of $\preccurlyeq$.

EXAMPLE 4   (a) If $S = T = \mathbb{R}$ with the usual order then $f \leq g$ means that the graph of f lies on or below the graph of g, as in Figure 4.

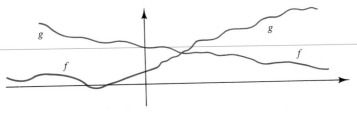

Figure 4

(b) Consider $S = \{0, 1\}$ with $0 < 1$. The functions in FUN($T, \{0, 1\}$) are the characteristic functions of subsets of T. Each subset A of T has a corresponding function χ_A in FUN($T, \{0, 1\}$) with $\chi_A(x) = 1$ if $x \in A$ and 0 if $x \notin A$. Then $\chi_A \leq \chi_B$ if and only if $x \in B$ whenever $x \in A$, hence if and only if $A \subseteq B$. Thus the Hasse diagram for (FUN($T, \{0, 1\}$), $\leq$) is the same as the diagram for ($\mathscr{P}(T), \subseteq$). See Figure 9 of §7.1 for the diagram with $T = \{a, b, c\}$. ∎

Example 4(b) shows that (FUN(T, S), $\preccurlyeq$) need not be a chain even if S is a chain. The poset (FUN(T, S), $\preccurlyeq$) does inherit some properties from S, however. If S has largest or smallest elements so does FUN(T, S), and if S is a lattice so is FUN(T, S). For more in this vein see Exercise 7.

Another way to combine two sets into a new one is to form their product. Suppose that $(S, \leq_1)$ and $(T, \leq_2)$ are posets, where we use the subscripts to keep track of which partial order is which. There is more than one natural way to make $S \times T$ into a poset. Our preference will depend on the problem at hand.

The first partial order we describe for $S \times T$ is called the **product order.** For $s, s' \in S$ and $t, t' \in T$ define

$$\langle s, t \rangle \leq \langle s', t' \rangle \quad \text{if} \quad s \leq_1 s' \quad \text{and} \quad t \leq_2 t'.$$

EXAMPLE 5   Let $S = T = \mathbb{N}$ with the usual order $\leq$ in each case. Then $\langle 2, 5 \rangle \leq \langle 3, 7 \rangle$ since $2 \leq 3$ and $5 \leq 7$. Also $\langle 2, 5 \rangle \leq \langle 3, 5 \rangle$ since $2 \leq 3$ and $5 \leq 5$. But the pairs $\langle 2, 7 \rangle$ and $\langle 3, 5 \rangle$ are not comparable; $\langle 2, 7 \rangle \leq \langle 3, 5 \rangle$ would mean $2 \leq 3$ and $7 \leq 5$, while $\langle 3, 5 \rangle \leq \langle 2, 7 \rangle$ would mean $3 \leq 2$ and $5 \leq 7$. Figure 5 indicates the pairs $\langle m, n \rangle$ in $S \times T = \mathbb{N} \times \mathbb{N}$ with $\langle 2, 1 \rangle \leq \langle m, n \rangle \leq \langle 3, 4 \rangle$.   ∎

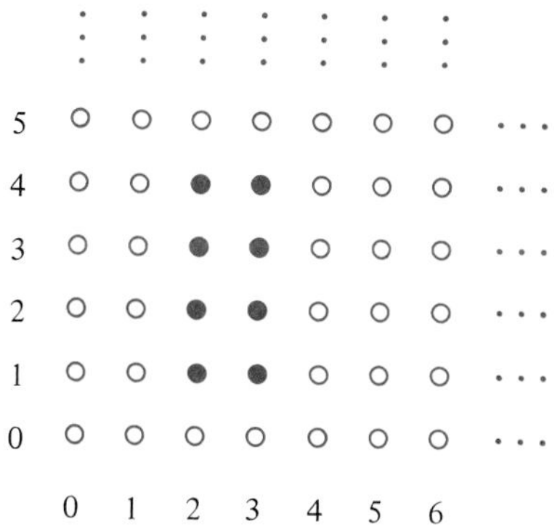

Figure 5

Consider, again, two posets $(S, \leq_1)$ and $(T, \leq_2)$. The fact that the product order $\leq$ is a partial order on $S \times T$ is almost immediate from the definition. For example, if $\langle s, t \rangle \leq \langle s', t' \rangle$ and $\langle s', t' \rangle \leq \langle s, t \rangle$, then $s \leq_1 s'$, $t \leq_2 t'$, $s' \leq_1 s$ and $t' \leq_2 t$. Since $\leq_1$ and $\leq_2$ are antisymmetric $s = s'$ and $t = t'$. So $\langle s, t \rangle = \langle s', t' \rangle$. Thus $\leq$ is antisymmetric.

There is no difficulty in extending this idea to define a partial order on the product $S_1 \times S_2 \times \cdots \times S_n$ of a number of posets. We define

$$\langle s_1, s_2, \ldots, s_n \rangle \leq \langle s_1', s_2', \ldots, s_n' \rangle \quad \text{if} \quad s_i \leq s_i' \quad \text{for all} \quad i.$$

In Example 5, $\mathbb{N}$ is a chain but $\mathbb{N} \times \mathbb{N}$ is not; for instance $\langle 2, 4 \rangle$ and $\langle 3, 1 \rangle$ are not related in the product order. In fact, the product order is almost never a total order [Exercise 12]. On the other hand, if $S_1, S_2, \ldots, S_n$ are chains, there is another natural order on $S_1 \times S_2 \times \cdots \times S_n$ which makes it a chain, which we now illustrate.

EXAMPLE 6   If $S = \{0, 1, 2, \ldots, 9\}$ with the usual order, the set $S \times S$ consists of pairs $\langle m, n \rangle$, which we can identify with the integers $00, 01, 02, \ldots, 98, 99$ from 0 to 99. To make $S \times S$ a chain we can simply define $\langle m, n \rangle \prec \langle m', n' \rangle$ if the corresponding integers are so related, i.e., if $m < m'$ or if $m = m'$ and $n < n'$. For example this order makes $\langle 5, 7 \rangle \prec \langle 6, 3 \rangle$ since $57 < 63$, and $\langle 3, 5 \rangle \prec \langle 3, 7 \rangle$ since $35 < 37$.

In a similar way we can identify the set $S \times S \times S$ with the set of integers from 0 to 999 and make $S \times S \times S$ into a chain by defining $\langle m, n, p \rangle \prec \langle m', n', p' \rangle$ if $m < m'$ or if $m = m'$ and $n < n'$ or if $m = m'$, $n = n'$ and $p < p'$.   ∎

The idea of this example works in general. If $(S_1, \leqq_1), \ldots, (S_n, \leqq_n)$ are posets we can define a relation $\prec$ on $S_1 \times \cdots \times S_n$ by

$$\langle s_1, s_2, \ldots, s_n \rangle \prec \langle t_1, t_2, \ldots, t_n \rangle \text{ if } s_1 \prec_1 t_1 \text{ or if there is an } r \text{ in } \{2, \ldots, n\}$$
such that $s_1 = t_1, \ldots, s_{r-1} = t_{r-1}$ and $s_r \prec_r t_r$.

Then $\prec$ is a quasi-order [Exercise 21] which induces a partial order $\leqq$ on $S_1 \times S_2 \times \cdots \times S_n$ called the **filing order**.

EXAMPLE 7   If $S = \{a, b, c, \ldots, z\}$ and $T = \{0, 1, \ldots, 9\}$ with the usual orders we can identify the members of $S \times T$ with 2-symbol strings such as $a5$ and $x3$. Imagine a device which has subassemblies, labeled by letters, each of which has at most 10 parts. It would be reasonable to label spare parts with letter-number combinations and to file them in bins arranged according to the filing order on $S \times T$. Then bin $a5$ would come before bin $x3$ because a precedes x, but $a5$ would come after $a3$ since $3 < 5$. ∎

The filing order is primarily useful if each S_i is a chain.

Theorem 1   Let $(S_1, \leqq_1), \ldots, (S_n, \leqq_n)$ be chains. Then, with the filing order $\leqq$, the product $S_1 \times \cdots \times S_n$ is a chain.

Proof. We already know that $\leqq$ is a partial order on $S_1 \times \cdots \times S_n$. Now let $\langle s_1, \ldots, s_n \rangle$ and $\langle t_1, \ldots, t_n \rangle$ be distinct elements in $S_1 \times \cdots \times S_n$. Since $s_r \neq t_r$ for some r, there is a first r for which $s_r \neq t_r$. Since $(S_r, \leqq_r)$ is a chain, either $s_r \prec_r t_r$ or $t_r \prec_r s_r$. In the first case $\langle s_1, \ldots, s_n \rangle \prec \langle t_1, \ldots, t_n \rangle$; in the second $\langle t_1, \ldots, t_n \rangle \prec \langle s_1, \ldots, s_n \rangle$. In either case, the two elements of $S_1 \times \cdots \times S_n$ are comparable. ∎

The special case in which all the posets $(S_i, \leqq_i)$ are the same is important enough to warrant a notation of its own. If $(S, \leqq)$ is a poset and $k \in \mathbb{P}$ we write $\leqq^k$ for the filing order on $S^k = S \times \cdots \times S$.

In the remainder of this section we are primarily interested in the case in which S is an alphabet. Accordingly, from now on we go Greek and write Σ instead of S. We also make the natural identification of k-tuples $\langle a_1, \ldots, a_k \rangle$ in the product Σ^k with words $a_1 \cdots a_k$ of length k in Σ^*. We still have in mind some given partial order $\leqq$ on Σ. The set Σ^0 is $\{\epsilon\}$, and we define $\leqq^0$ on Σ^0 in the only possible way: $\epsilon \leqq^0 \epsilon$.

EXAMPLE 8   If $\Sigma = \{a, b, c, \ldots, z\}$ with the usual linear order on the English alphabet then Σ^k consists of all strings of letters of length k and $\leqq^k$ is the usual alphabetical order on Σ^k. For example if $k = 3$, then

$$fed \leqq^3 few \leqq^3 one \leqq^3 six \leqq^3 ten \leqq^3 two \leqq^3 won. \quad ∎$$

Since $\Sigma^* = \Sigma^0 \cup \Sigma^1 \cup \Sigma^2 \cup \cdots$ we can piece the partial orders $\leqq^0, \leqq^1, \leqq^2, \ldots$ together to get an order $\leqq^*$ for Σ^*, called the **standard order**, in which

ϵ comes first, then all words of length 1, then all of length 2, and so on. More precisely

$w_1 \leq^* w_2$ if either $w_1 \in \Sigma^k$ and $w_2 \in \Sigma^r$ and $k < r$
or $w_1, w_2 \in \Sigma^k$ for the same k and $w_1 \leq^k w_2$.

If $(\Sigma, \leq)$ is a chain, as it was in Example 8, then $(\Sigma^*, \leq^*)$ is also a chain.

EXAMPLE 9   (a) Let Σ be the English alphabet in the usual order. The first few terms of Σ^* in the standard order are

$$\epsilon, a, b, \ldots, z, aa, ab, \ldots, az, ba, bb, \ldots, bz, ca, cb, \ldots, cz,$$
$$da, db, \ldots, dz, \ldots, za, zb, \ldots, zz, aaa, aab, aac, \ldots.$$

(b) Let $\Sigma = \{0, 1\}$ with $0 < 1$. The first few terms of Σ^* in the standard order are

$$\epsilon, 0, 1, 00, 01, 10, 11, 000, 001, 010, 011,$$
$$100, 101, 110, 111, 0000, 0001, 0010, \ldots. \quad \blacksquare$$

Note that if a dictionary were constructed using the standard order in Example 9(a) all the short words would be at the beginning of the dictionary, and to find a word it would be essential to know its exact length. [In fact, some dictionaries designed for crossword puzzle solvers are arranged this way.] To find words in a dictionary with the ordinary alphabetical order, one scans words from left to right looking for differences and ignoring the lengths of the words. Thus "aardvark" is listed before "axe" and "break" precedes "breakfast." Alphabetical order is based on the usual alphabet ordered in the usual way. The idea generalizes naturally to an arbitrary Σ with partial order $\leq$.

The **lexicographic** or **dictionary order** $\leq_L$ on Σ^* is defined as follows. For $a_1, \ldots, a_m$ and $b_1, \ldots, b_n$ in Σ let $k = \min\{m, n\}$. We define

$$a_1 \cdots a_m \prec_L b_1 \cdots b_n \quad \text{if} \quad a_1 \cdots a_k \prec^k b_1 \cdots b_k$$
$$\text{or if} \quad k = m < n \quad \text{and} \quad a_1 \cdots a_k = b_1 \cdots b_k.$$

Then $\prec_L$ is a quasi-order which defines the partial order $\leq_L$.

We can describe $\leq_L$ in another way. For words w and z in Σ^*, $w \leq_L z$ if and only if either

(a) w is an initial segment of z, i.e., $z = wu$ for some $u \in \Sigma^*$, or
(b) $w = xu$ and $z = xv$ for words u and v in Σ^* such that the first letter of u precedes the first letter of v in the ordering of Σ.

Note that in (b) x can be any word, possibly the empty word.

Lexicographic order, standard order and filing order all agree if we consider words in some fixed Σ^k, but lexicographic and standard orders differ in the treatment they give to words of different lengths.

EXAMPLE 10   Let $\Sigma = \{a, b\}$ with $a \prec b$. The first few terms of Σ^* in the lexicographic order are

$$\epsilon, a, aa, aaa, aaaa, aaaaa, \ldots.$$

Any word using the letter b is preceded by an *infinite* number of words, including all the words using only the letter a. Moreover, Σ^* contains infinite decreasing sequences of words; for example,

$$\ldots, aaaaab, \ldots, aaaab, \ldots, aaab, \ldots, aab, \ldots, ab, \ldots, b.$$

And there are infinitely many words between these; for example,

$$aaab, aaaba, aaabaa, aaabaaa, aaabaaaa, \ldots$$

all precede *aab*. Thus the lexicographic order on the infinite set Σ^* is very complicated and is difficult to visualize. Nevertheless, it defines a chain, as we show in the next theorem. ∎

Theorem 2   If $(\Sigma, \leq)$ is a chain, then $(\Sigma^*, \leq^*)$ is well-ordered and $(\Sigma^*, \leq_L)$ is a chain.

Proof. We know from Theorem 1 that each $(\Sigma^k, \leq^k)$ is a chain. The standard order $\leq^*$ simply links these chains end to end for $k = 0, 1, 2, \ldots$, so $(\Sigma^*, \leq^*)$ is a chain. Now consider a nonempty subset A of Σ^*. Let k be the shortest length of a word in A. Since $A \cap \Sigma^k$ is nonempty and finite, $A \cap \Sigma^k$ possesses a smallest element w_0 in $(\Sigma^k, \leq^k)$. It follows that $w_0 \leq^* w$ for all w in A so that w_0 is the smallest member of A.

For the lexicographic order $\leq_L$, consider two elements $a_1 \cdots a_m$ and $b_1 \cdots b_n$ in Σ^* where $m \leq n$. If $a_1 \cdots a_m = b_1 \cdots b_m$, then $a_1 \cdots a_m \leq_L b_1 \cdots b_n$ by definition. Otherwise, since $(\Sigma^m, \leq^m)$ is a chain, one of $a_1 \cdots a_m$ and $b_1 \cdots b_m$ precedes the other in $(\Sigma^m, \leq^m)$ and so $a_1 \cdots a_m$ and $b_1 \cdots b_n$ are comparable in $(\Sigma^*, \leq_L)$. Thus every two members of Σ^* are comparable under $\leq_L$, and $\leq_L$ is a total order. ∎

If Σ has more than one element, then $(\Sigma^*, \leq_L)$ is *not* well-ordered. For example, the set $\{b, ab, aab, aaab, aaaab, \ldots\}$ in Example 10 has no smallest member. Of course, every *finite* subset of Σ^* has a smallest member, since it is itself a finite chain.

EXERCISES 7.2

1. Let P be the set of all subsets of $\{1, 2, 3, 4, 5\}$.
   (a) Give two examples of maximal chains in $(P, \subseteq)$.
   (b) How many maximal chains are there in $(P, \subseteq)$?

2. Let $A = \{1, 2, 3, 4\}$ with the usual order and let $S = A \times A$ with the product order.
   (a) Find a chain in S with 7 members.
   (b) Can a chain in S have 8 members? Explain.

3. Let $(S, |)$ be the set $\{2, 3, 4, \ldots, 999, 1000\}$ with partial order "is a factor of."
   (a) There are exactly 500 maximal elements of $(S, |)$. What are they?
   (b) Give two examples of maximal chains in $(S, |)$.
   (c) Does every maximal chain contain a minimal element of S? Explain.

4. (a) Suppose that no chain in the poset $(S, \leq)$ has more than 73 members. Must a chain in S with 73 members be a maximal chain? Explain.
   (b) Give an example of a poset which has two maximal chains with four members and four maximal chains with two members.

5. Is every chain a lattice? Explain.

6. Let $(C_1, \leq_1), (C_2, \leq_2), \ldots, (C_n, \leq_n)$ be a set of disjoint chains. Describe a way to make $C_1 \cup C_2 \cup \cdots \cup C_n$ into a chain.

7. (a) Show that if $(S, \leq)$ is a poset and T is a set, then the relation $\preccurlyeq$ on $\text{FUN}(T, S)$ given by

$$f \preccurlyeq g \quad \text{if} \quad f(t) \leq g(t) \quad \text{for all } t \text{ in } T$$

   is a partial order.
   (b) Show that if m is a maximal element in S then the function f_m defined by $f_m(t) = m$ for all $t \in T$ is a maximal element of $\text{FUN}(T, S)$.
   (c) Show that if S is a lattice and if $f, g \in \text{FUN}(T, S)$, then the function h defined by $h(t) = f(t) \vee g(t)$ for $t \in T$ is the least upper bound of $\{f, g\}$.

8. Let $\mathbb{N} \times \mathbb{N}$ have the product order. Draw a sketch like the one in Figure 5 which shows $\{\langle m, n \rangle : \langle m, n \rangle \leq \langle 5, 2 \rangle\}$.

9. Let $S = \{0, 1, 2\}$ with the usual order and let $T = \{a, b\}$ with $a < b$.
   (a) Draw a Hasse diagram for the poset $(\text{FUN}(T, S), \preccurlyeq)$ with order $\preccurlyeq$ described in Exercise 7. *Hint:* See Example 8(c) of § 7.1.
   (b) Draw a Hasse diagram for the poset $(S \times S, \leq)$ with the product order $\leq$.
   (c) Draw a Hasse diagram for $S \times T$ with the product order.

10. Suppose that $(S, \leq_1)$ and $(T, \leq_2)$ are posets and we define $\leq$ on $S \times T$ by

$$\langle s, t \rangle \leq \langle s', t' \rangle \quad \text{if} \quad s \leq_1 s' \quad \text{or} \quad t \leq_2 t'.$$

Is $\leq$ a partial order? Explain.

11. Let $(S, \leq)$ be a poset and T a nonempty set. Define the partial order $\preccurlyeq$ on $\text{FUN}(T, S)$ as in Exercise 7. Under what conditions is $(\text{FUN}(T, S), \preccurlyeq)$ a chain? Explain.

12. Suppose that $(S, \leq_1)$ and $(T, \leq_2)$ are posets, each with more than one element. Show that $S \times T$ with the product order is not a chain.

13. Let $S = \{0, 1, 2\}$ and $T = \{3, 4\}$ with both sets given the usual order. List the members of the following sets in increasing filing order.
   (a) $S \times S$ (b) $S \times T$ (c) $T \times S$

14. Let $(S, \leq_1)$ and $(T, \leq_2)$ be posets and give $S \times T$ the filing order.
    (a) Show that if m_1 is maximal in S and m_2 is maximal in T, then $\langle m_1, m_2 \rangle$ is maximal in $S \times T$.
    (b) Does $S \times T$ have other maximal elements besides the ones described in part (a)? Explain.
    (c) Suppose $S \times T$ has a largest element. Must S or T have a largest element? Explain.

15. Define the order relation $\geq$ on $\mathbb{N}$ by letting $m \geq n$ if $n \leq m$. [Thus $\geq$ is the inverse of $\leq$.] Let $(S_1, \leq_1) = (\mathbb{N}, \leq)$ and $(S_2, \leq_2) = (\mathbb{N}, \geq)$ and let $S = S_1 \times S_2$ with the filing order. That is, $\langle m, n \rangle \leq \langle m', n' \rangle$ if $m < m'$ or if $m = m'$ and $n \geq n'$.
    (a) Is $(S, \leq)$ a chain?
    (b) Does S have a largest or a smallest element? Explain.
    (c) Is $(S, \leq)$ well-ordered? Explain.

16. Show that in a finite poset $(S, \leq)$ every maximal chain contains a minimal element of S.

17. Let $\mathbb{B} = \{0, 1\}$ with the usual order. List the elements 101, 010, 11, 000, 10, 0010, 1000 of $\mathbb{B}^*$ in increasing order
    (a) for the lexicographic order,
    (b) for the standard order.

18. Let $(\Sigma, \leq)$ be a nonempty chain.
    (a) Does $(\Sigma^*, \leq^*)$ have a maximal member? Explain.
    (b) Does $(\Sigma^*, \leq_L)$ have a maximal member? Explain.

19. Let Σ be the English alphabet with the usual order.
    (a) List the words of this sentence in increasing standard order.
    (b) List the words of this sentence in increasing lexicographic order.

20. Under what conditions on Σ are the lexicographic order and standard order on Σ^* the same?

21. Let $(S_1, \leq_1), \ldots, (S_n, \leq_n)$ be posets and define $\prec$ on $S_1 \times \cdots \times S_n$ by

    $\langle s_1, \ldots, s_n \rangle \prec \langle t_1, \ldots, t_n \rangle$ if $s_1 \prec_1 t_1$ or if there is an r in $\{2, \ldots, n\}$ such that $s_1 = t_1, \ldots, s_{r-1} = t_{r-1}$ and $s_r \prec_r t_r$.

    Show that $\prec$ is a quasi-order.

22. Let $(S_1, \leq_1), \ldots, (S_n, \leq_n)$ be posets and let $S = S_1 \times \cdots \times S_n$ with the product order $\leq$.
    (a) Show that if m_i is a maximal element of S_i for $i = 1, \ldots, n$, then $\langle m_1, \ldots, m_n \rangle$ is maximal in S.
    (b) Show that if x_i and y_i are in S_i for $i = 1, \ldots, n$ and if $\mathrm{lub}\{x_i, y_i\}$ exists for each i, then

    $$\mathrm{lub}\{\langle x_1, \ldots, x_n \rangle, \langle y_1, \ldots, y_n \rangle\} = \langle \mathrm{lub}\{x_1, y_1\}, \ldots, \mathrm{lub}\{x_n, y_n\} \rangle.$$

    (c) Would parts (a) and (b) remain true if "maximal" were replaced by "minimal," "largest" or "smallest" or if "lub" were replaced by "glb?"
    (d) Show that if each S_i is a lattice so is S.

§ 7.3 Equivalence Relations

In this section we study equivalence relations, which are relations that group together elements that have similar characteristics or share some property. These relations occur throughout mathematics and other fields, even though they are not always formally identified as such.

EXAMPLE 1   (a) Let S be a set of marbles. We might regard marbles s and t as equivalent if they have the same color, in which case we might write $s \sim t$. Note that the relation $\sim$ satisfies three properties:

(R) $s \sim s$ for all marbles s,
(S) if $s \sim t$ then $t \sim s$,
(T) if $s \sim t$ and $t \sim u$, then $s \sim u$.

These are nearly obvious; for example (T) asserts that if marbles s and t have the same color and t and u have the same color, then s and u have the same color. Note also that we can partition S into disjoint subsets so that elements belong to the same subset if and only if they are equivalent, i.e., if and only if they have the same color. If we let C be the set of possible colors and define the function $f: S \to C$ by $f(s) =$ "the color of s" for each s in S then the partition is $\{f^{-1}(c): c \in C\}$, a partition of the sort studied in § 3.2.

(b) For the same set S of marbles, we might regard marbles s and t as equivalent if they are of the same size, and write $s \approx t$ in this case. All the comments in part (a) apply to $\approx$, with obvious changes. ∎

Let S be any set and suppose that we have a relation $\sim$ on S, i.e., for each pair $\langle x, y \rangle$ in $S \times S$ we know whether $x \sim y$ or not. The relation $\sim$ is called an **equivalence relation** provided it satisfies:

(R) $s \sim s$ for every $s \in S$;
(S) if $s \sim t$ then $t \sim s$;
(T) if $s \sim t$ and $t \sim u$, then $s \sim u$.

The conditions (R), (S) and (T) are the **reflexive, symmetric** and **transitive** laws. If $s \sim t$ we say that s and t are **equivalent**; depending on the circumstances we might also say that s and t are **similar** or **congruent** or **isomorphic**. Other notations sometimes used for equivalence relations are $s \approx t$, $s \cong t$, $s \equiv t$ and $s \leftrightarrow t$. Unlike the notations for order, these notations all convey the idea that s and t have equal [or equivalent] status. This is reasonable, because of the symmetry law.

EXAMPLE 2   Triangles T_1 and T_2 in the plane are said to be **similar**, and we write $T_1 \approx T_2$, if their angles can be put in one-to-one correspondence so that corresponding angles are equal. If the corresponding sides are also equal, we say that the triangles are **congruent**, and we write $T_1 \cong T_2$. In Figure 1, $T_1 \cong T_2$, $T_1 \approx T_3$ and $T_2 \approx T_3$, but T_3 is not congruent to T_1 or T_2. Both $\approx$ and $\cong$ are

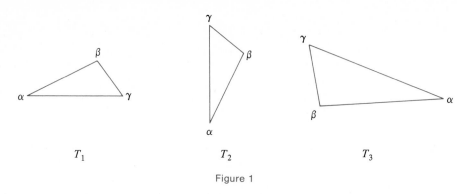

Figure 1

equivalence relations on the set of all triangles in the plane. All the laws (R), (S) and (T) are evident for these relations. ∎

EXAMPLE 3   Consider a graph and let V be the set of its vertices. For vertices u and v we define $u \sim v$ provided $u = v$ or there is a path connecting u to v. Then $\sim$ is an equivalence relation on V. You may have verified symmetry and transitivity back at Exercise 8 of § 0.1. ∎

EXAMPLE 4   (a) Consider a black box, perhaps an automaton, which accepts input strings in Σ^* for some alphabet Σ and generates output strings. We can define an equivalence relation $\sim$ on Σ^* by $w_1 \sim w_2$ if the box generates the same output string for either w_1 or w_2 as input.
        (b) Consider a set S of black boxes which accept inputs in Σ^* and generate outputs. Define relations $\approx_1, \approx_2, \approx_3, \dots$ on S by writing $B \approx_k C$ for boxes B and C in S if B and C produce the same output for every choice of input word of length k. Define $\approx$ on S by letting $B \approx C$ if $B \approx_k C$ for all $k \in \mathbb{P}$. Then all the relations $\approx_k$ and $\approx$ are equivalence relations on S, and two boxes are equivalent under $\approx$ if and only if they produce the same response to all input words with letters from Σ. ∎

EXAMPLE 5   (a) Matrices $\mathbf{A}$ and $\mathbf{B}$ in the set $\mathfrak{M}_{n,n}$ of $n \times n$ matrices are **equivalent** if there are invertible matrices $\mathbf{P}$ and $\mathbf{Q}$ in $\mathfrak{M}_{n,n}$ such that $\mathbf{B} = \mathbf{PAQ}$. If we write $\mathbf{A} \sim \mathbf{B}$, then we obtain an equivalence relation $\sim$ on $\mathfrak{M}_{n,n}$. The (R), (S) and (T) laws are not as obvious as in the previous examples and deserve verification.
        (R) We always have $\mathbf{A} \sim \mathbf{A}$ because $\mathbf{A} = \mathbf{IAI}$, where $\mathbf{I}$ is the $n \times n$ identity matrix.
        (S) Suppose $\mathbf{A} \sim \mathbf{B}$; we need to show $\mathbf{B} \sim \mathbf{A}$. Since $\mathbf{A} \sim \mathbf{B}$ we have $\mathbf{B} = \mathbf{PAQ}$ for invertible matrices $\mathbf{P}$ and $\mathbf{Q}$. Hence $\mathbf{P}^{-1}\mathbf{BQ}^{-1} = \mathbf{P}^{-1}\mathbf{PAQQ}^{-1} = \mathbf{IAI} = \mathbf{A}$. Since $\mathbf{P}^{-1}$ and $\mathbf{Q}^{-1}$ are also invertible, we conclude that $\mathbf{B} \sim \mathbf{A}$.
        (T) Suppose $\mathbf{A} \sim \mathbf{B}$ and $\mathbf{B} \sim \mathbf{C}$; we need to show $\mathbf{A} \sim \mathbf{C}$. Since $\mathbf{A} \sim \mathbf{B}$ and $\mathbf{B} \sim \mathbf{C}$, we have $\mathbf{B} = \mathbf{PAQ}$ and $\mathbf{C} = \mathbf{RBS}$, where $\mathbf{P}$, $\mathbf{Q}$, $\mathbf{R}$ and

S are invertible. Then $\mathbf{C} = \mathbf{RBS} = \mathbf{RPAQS}$. Since $(\mathbf{P}^{-1}\mathbf{R}^{-1})\mathbf{RP} = \mathbf{I} = \mathbf{RP}(\mathbf{P}^{-1}\mathbf{R}^{-1})$, we have $(\mathbf{RP})^{-1} = \mathbf{P}^{-1}\mathbf{R}^{-1}$. Similarly, $(\mathbf{QS})^{-1} = \mathbf{S}^{-1}\mathbf{Q}^{-1}$, so **RP** and **QS** are invertible and so $\mathbf{A} \sim \mathbf{C}$.

(b) It turns out that two matrices are equivalent provided one can be converted into the other via the elementary row and column operations that are encountered in solving systems of equations. In particular, all invertible $n \times n$ matrices are equivalent to each other. ∎

EXAMPLE 6   Consider the set $\mathscr{P}(S)$ of all subsets of some set S. For $A, B \in \mathscr{P}(S)$, we define $A \sim B$ if their symmetric difference $A \oplus B$ is a finite set. Then $\sim$ is an equivalence relation on $\mathscr{P}(S)$, as we now verify.

(R) Since $A \oplus A = \varnothing$ and $\varnothing$ is finite, we have $A \sim A$ for all $A \in \mathscr{P}(S)$.

(S) Symmetry is clear because $A \oplus B = B \oplus A$.

(T) Suppose $A \sim B$ and $B \sim C$ so that $A \oplus B$ and $B \oplus C$ are finite sets. We need to show that $A \sim C$, i.e., that $A \oplus C$ is finite. This is relatively easy *if* one remembers Exercise 8 of § 1.2; part (b) of that exercise asserts that $A \oplus B \subseteq (A \oplus C) \cup (B \oplus C)$. Interchanging B and C and using commutativity, we find

$$A \oplus C \subseteq (A \oplus B) \cup (B \oplus C).$$

Since the right-hand side is a finite set, $A \oplus C$ is also a finite set, and so $A \sim C$. ∎

EXAMPLE 7   (a) Let $\mathscr{S}$ be a set of compound propositions. Recall that $P \Leftrightarrow Q$ signifies that P and Q are logically equivalent, i.e., that P and Q have the same truth table columns. The relation $\Leftrightarrow$ on $\mathscr{S}$ is an equivalence relation:

(R) $P \Leftrightarrow P$ for all $P \in \mathscr{S}$;

(S) if $P \Leftrightarrow Q$, then $Q \Leftrightarrow P$;

(T) if $P \Leftrightarrow Q$ and $Q \Leftrightarrow R$, then $P \Leftrightarrow R$.

For instance, (T) is true because if P and Q have the same truth table columns and if Q and R do, too, then so do P and R.

(b) By the way, when we state a theorem in the form "The following are [logically] equivalent," we really have in mind the equivalence relation in part (a), where the compound propositions are probably propositions in the predicate calculus. The assertion means that for each choice of variables, all the propositions are true or else all the propositions are false. Thanks to the transitivity of $\Rightarrow$, if the propositions are named $P_1, P_2, \ldots, P_n$, it suffices to prove $P_1 \Rightarrow P_2 \Rightarrow \cdots \Rightarrow P_n \Rightarrow P_1$. See the lemma below for a typical illustration of this. ∎

Consider again an equivalence relation $\sim$ on a set S. For each $s \in S$ we define

$$[s] = \{t \in S : s \sim t\};$$

[s] is called the **equivalence class** containing s. For us, "class" and "set" are synonymous, so that [s] could have been called an "equivalence set," but it never is. The set of all equivalence classes of S is denoted by [S], i.e., $[S] = \{[s] : s \in S\}$. We will sometimes attach subscripts to [s] or [S] to clarify exactly which of several possible equivalence relations is being used.

EXAMPLE 8   (a) In the marble setting of Example 1(a) the equivalence class [s] of a given marble s is the set of all marbles that are the same color as s; this includes s itself. The equivalence classes are {blue marbles}, {red marbles}, {green marbles}, etc.

(b) Consider the equivalence relation in Example 3 on the set V of vertices of a graph. Two vertices are equivalent precisely if they belong to the same connected part of the graph. For example, the equivalence classes for the graph in Figure 2 are $\{v_1, v_6, v_8\}$, $\{v_2, v_4, v_{10}\}$ and $\{v_3, v_5, v_7, v_9, v_{11}, v_{12}\}$. If a graph is connected, then the only equivalence class is the set V itself. ∎

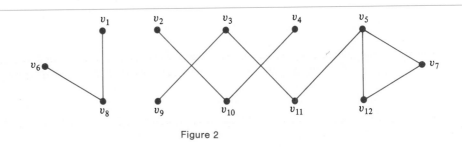

Figure 2

EXAMPLE 9   Let p be an integer ≥ 2. Recall from § 4.3 that for m, $n \in \mathbb{Z}$ we defined $m \equiv n \pmod{p}$ provided $m - n$ is divisible by p. This definition gives an equivalence relation on $\mathbb{Z}$, called **congruence modulo** p, as we noted in Theorem 2(a) of § 4.3.

For the equivalence relation congruence modulo p we denote the equivalence class containing m by $[m]_p$. That is,

$$[m]_p = \{n \in \mathbb{Z} : m \equiv n \pmod{p}\}.$$

For p = 5, we have

$$[0]_5 = \{\ldots, -20, -15, -10, -5, 0, 5, 10, 15, 20, \ldots\},$$
$$[1]_5 = \{\ldots, -19, -14, -9, -4, 1, 6, 11, 16, 21, \ldots\},$$
$$[2]_5 = \{\ldots, -18, -13, -8, -3, 2, 7, 12, 17, 22, \ldots\},$$
$$[3]_5 = \{\ldots, -17, -12, -7, -2, 3, 8, 13, 18, 23, \ldots\},$$
$$[4]_5 = \{\ldots, -16, -11, -6, -1, 4, 9, 14, 19, 24, \ldots\}.$$

Moreover, every equivalence class $[m]_5$ is one of these five sets. For example, $[18]_5 = [3]_5$ and $[-73]_5 = [2]_5$. More generally, we have

$$[\mathbb{Z}]_p = \{[r]_p : r = 0, 1, 2, \ldots, p - 1\}$$

in view of Theorem 1 of §4.3, since $[m]_p = [r]_p$ if and only if $m \equiv r \pmod{p}$. ∎

In Examples 8 and 9 the equivalence classes form a partition of the underlying set. This reflects a general property of equivalence relations, which we give in Theorem 1. First we prove a lemma.

Lemma   Let $\sim$ be an equivalence relation on a set S. For s and t in S the following assertions are logically equivalent:

> (i) $s \sim t$;
> (ii) $[s] = [t]$;
> (iii) $[s] \cap [t] \neq \varnothing$.

Proof. By "logically equivalent" we mean here, as in Example 7(b), that all three assertions are true or else all are false. We prove (i) $\Rightarrow$ (ii), (ii) $\Rightarrow$ (iii) and (iii) $\Rightarrow$ (i).

(i) $\Rightarrow$ (ii). Suppose $s \sim t$ and consider $s' \in [s]$. Then $s \sim s'$. By symmetry $t \sim s$. Since $t \sim s$ and $s \sim s'$, transitivity shows $t \sim s'$. Thus $s' \in [t]$. We've shown that every s' in $[s]$ belongs to $[t]$, and hence $[s] \subseteq [t]$. Similarly $[t] \subseteq [s]$.

(ii) $\Rightarrow$ (iii) is obvious.

(iii) $\Rightarrow$ (i). Select u in $[s] \cap [t]$. Then $s \sim u$ and $t \sim u$. By symmetry $u \sim t$. Since $s \sim u$ and $u \sim t$, we have $s \sim t$ by transitivity. ∎

Theorem 1   (a) If $\sim$ is an equivalence relation on a nonempty set S, then $[S]$ is a partition of S.

(b) Conversely, if $\{A_i : i \in I\}$ is a partition of S, then the sets A_i are the equivalence classes corresponding to some equivalence relation on S.

Proof. (a) To show that $[S]$ partitions S, we need to show

(1) $$\bigcup_{s \in S} [s] = S;$$

(2) $$\text{for } s, t \in S \quad \text{either} \quad [s] = [t] \quad \text{or} \quad [s] \cap [t] = \varnothing.$$

Clearly $[s] \subseteq S$ for each s in S, so that $\bigcup_{s \in S} [s] \subseteq S$. Given s_0 in S, we have $s_0 \sim s_0$ and so $s_0 \in [s_0]$; hence $S \subseteq \bigcup_{s \in S} [s]$. Therefore (1) holds.

Assertion (2) is logically equivalent to

(3) $$[s] \cap [t] \neq \varnothing \quad \text{implies} \quad [s] = [t]$$

and this was proved in the lemma.

(b) Given a partition $\{A_i : i \in I\}$ of S, we define the relation $\sim$ on S by $s \sim t$ if s and t belong to the same set A_i. Properties (R), (S) and (T) are obvious, and so $\sim$ is an equivalence relation on S. Given a nonempty set A_i we have $A_i = [s]$ for all $s \in A_i$, and so the partition consists precisely of all equivalence classes $[s]$. ∎

EXAMPLE 10   Let $\mathscr{S}$ be a family of sets, and for S, $T \in \mathscr{S}$ define $S \sim T$ if there exists a one-to-one function mapping S onto T. The relation $\sim$ is an equivalence relation on $\mathscr{S}$. Indeed:

(R) $S \sim S$ because the identity function $1_S : S \to S$ is a one-to-one mapping of S onto S.

(S) If $S \sim T$, then there is a one-to-one mapping f of S onto T. Its inverse f^{-1} is a one-to-one mapping of T onto S and so $T \sim S$.

(T) If $S \sim T$ and $T \sim U$, then there are one-to-one correspondences $f : S \to T$ and $g : T \to U$. It is easy to check that $g \circ f$ is a one-to-one correspondence of S onto U. This also follows from Exercise 9 of § 3.2 in conjunction with the theorem of that section. In any event, we conclude that $S \sim U$.

Observe that if S is finite, then $[S]$ consists of all sets in $\mathscr{S}$ that are the same size as S. If S is countably infinite, then $[S]$ consists of all countably infinite sets in $\mathscr{S}$. If S is uncountable, then $[S]$ will consist of some of the uncountable sets in $\mathscr{S}$ but probably not all of them, because not all uncountable sets are equivalent to each other. This last assertion is not obvious and is not justified in this book. ∎

Sometimes an equivalence relation is defined in terms of a function on the underlying space. In a sense, which we will make precise in Theorem 2, every equivalence relation arises in this way.

EXAMPLE 11   (a) Consider $\mathbb{N} \times \mathbb{N}$ and define $\langle m, n \rangle \sim \langle j, k \rangle$ provided $m^2 + n^2 = j^2 + k^2$. It is easy to show directly that $\sim$ is an equivalence relation on $\mathbb{N} \times \mathbb{N}$. We take a slightly different approach. We define $f : \mathbb{N} \times \mathbb{N} \to \mathbb{N}$ by the rule

$$f(\langle m, n \rangle) = m^2 + n^2.$$

Then ordered pairs are equivalent exactly when they have equal images under f. The equivalence classes are simply the nonempty sets $f^{-1}(r)$, where $r \in \mathbb{N}$. Some of the sets $f^{-1}(r)$, like $f^{-1}(3)$, are empty, but this does no harm.

(b) Let Σ be an alphabet and let's regard two languages A and B in $\mathscr{P}(\Sigma^*)$ as equivalent if their Kleene closures A^* and B^* are equal. In symbols, $A \sim B$ if $A^* = B^*$. Again it is easy to check directly that $\sim$ is an equivalence relation. One way to see this is to consider the function ϕ where $\phi(A) = A^*$. The domain of ϕ is $\mathscr{P}(\Sigma^*)$ and its image is the set of all submonoids of

Σ^*. Then $A \sim B$ if and only if $\phi(A) = \phi(B)$. The equivalence classes are the sets

$$\phi^{-1}(M) = \{A \in \mathscr{P}(\Sigma^*) : A^* = M\}$$

for submonoids M of Σ^*. ∎

Theorem 2   (a) Let S be a nonempty set. Let f be a function with domain S, and define $s_1 \sim s_2$ if $f(s_1) = f(s_2)$. Then $\sim$ is an equivalence relation on S, and the equivalence classes are the nonempty sets $f^{-1}(t)$, where t is in the codomain T of S.

(b) Every equivalence relation $\sim$ on a set S is determined by a suitable function with domain S as in part (a).

Proof. We check that $\sim$ is an equivalence relation.

(R) $f(s) = f(s)$, so $s \sim s$ for all $s \in S$.

(S) If $f(s_1) = f(s_2)$, then $f(s_2) = f(s_1)$, so $s_1 \sim s_2$ implies $s_2 \sim s_1$.

(T) If $f(s_1) = f(s_2)$ and $f(s_2) = f(s_3)$, then $f(s_1) = f(s_3)$, and so $\sim$ is transitive.

The statement about equivalence classes is just the definition of $f^{-1}(t)$.

To prove (b), we define $\phi(s) = [s]$ for $s \in S$. The function ϕ maps S onto the set $[S]$ of equivalence classes and is called the **natural mapping** of S onto $[S]$. By the lemma before Theorem 1 we have $\phi(s) = \phi(t)$ if and only if $s \sim t$, and so $\sim$ is the equivalence relation determined by ϕ. Note that we have $\phi^{-1}([s]) = [s]$ for all $s \in S$. ∎

Normally we use notation like $\sim$ or $\equiv$ only when we have an equivalence relation. Of course the use of such notation does not automatically guarantee that we have an equivalence relation.

EXAMPLE 12   (a) Let $(P, \leq)$ be a finite poset and consider its Hasse diagram. Define $x \sim y$ if $x = y$ or if x and y are connected by a single line segment. The relation $\sim$ is certainly reflexive and symmetric. But it is probably not transitive: $x \sim y$ and $y \sim z$ need not imply $x \sim z$. The element y might be connected to both x and z by single line segments, and yet x and z might not be connected to each other. See Figure 4 of § 7.1. In the left-hand figure, $a \sim b$ and $b \sim f$ but $a \nsim f$.

(b) For $m, n \in \mathbb{Z}$, define $m \sim n$ in case $m - n$ is odd. The relation is symmetric but is highly nonreflexive and nontransitive. In fact,

$$m \nsim m \quad \text{for all} \quad m \in \mathbb{Z}$$

and

$$m \sim n \quad \text{and} \quad n \sim p \quad \text{always imply} \quad m \nsim p.$$

(c) Consider the set $S = \text{FUN}([0, 1], \mathbb{R})$ of all functions mapping $[0, 1]$ into $\mathbb{R}$. Define $f \sim g$ provided $|f(x) - g(x)| < 1$ for all $x \in [0, 1]$. Then $\sim$ is reflexive and symmetric on S, but it fails to be transitive. For example, if $f(x) = 0$, $g(x) = x/2$ and $h(x) = x$ for $x \in [0, 1]$, then $f \sim g$ and $g \sim h$ but $f \nsim h$. ∎

One of the most common sources of error in dealing with equivalence classes is attempting to define a function or relation on $[S]$ in terms of the members of the equivalence classes.

EXAMPLE 13   (a) Every member of $[\mathbb{Z}]_6$ is an equivalence class of the form $[m]_6$, with m in $\mathbb{Z}$, so let's define a function from $[\mathbb{Z}]_6$ to $\mathbb{Z}$ by the rule

$$f([m]_6) = m^2.$$

For example, $f([2]_6) = 2^2 = 4$. Of course $[2]_6 = [8]_6$ in $[\mathbb{Z}]_6$, so $f([2]_6) = f([8]_6) = 8^2 = 64 \neq 4$. Oops! We have a problem. Our intended function is *not well-defined*; that is, the definition given is ambiguous.

Where did we go wrong? The trouble is that $[2]_6$ contains many integers, including 2 and 8. We tried to base our definition of $f([2]_6)$ on the member of $[2]_6$ we chose, but different choices gave different answers.

(b) Now let's try to define a function g from $[\mathbb{Z}]_6$ to $[\mathbb{Z}]_{12}$ by letting $g([n]_6) = [n^2]_{12}$. This time we can do it. For suppose that $[n]_6 = [m]_6$. Then $n - m$ is a multiple of 6, say $n = m + 6k$; hence

$$n^2 = (m + 6k)^2 = m^2 + 12mk + 36k^2 \equiv m^2 \ (\text{mod } 12)$$

and $[n^2]_{12} = [m^2]_{12}$. In this case we get the same answer regardless of the choice of representative of the equivalence class, so g is well-defined. ∎

In the last illustrations we attempted to define a function value on an equivalence class in terms of the individual members of the class. We ran into trouble if different individuals in a class could give different results. To check that a function, relation or operation is well-defined on a set $[S]$ one must verify that using different members of a class does not change the proposed definition. Exercises 19 and 20 contain further illustrations.

EXAMPLE 14   Let $\mathbb{Q}$ be the set of rational numbers. Each number in $\mathbb{Q}$ has the form m/n with $m, n \in \mathbb{Z}$ and $n \neq 0$. Can we define a function f on $\mathbb{Q}$ by the rule $f(m/n) = m + n$? Since $\frac{2}{3} = \frac{4}{6}$ we would want $f(\frac{2}{3}) = 2 + 3 = 5$ and also $f(\frac{2}{3}) = f(\frac{4}{6}) = 10$. Trouble again, but where is the equivalence relation that causes it? The problem is that we can use different symbols m/n [such as $\frac{2}{3}$ and $\frac{4}{6}$] for the same number in $\mathbb{Q}$.

Let S be the set of all symbols m/n with $m, n \in \mathbb{Z}$ and $n \neq 0$. We define the relation $\approx$ on S by

$$m/n \approx p/q \quad \text{if} \quad mq = np.$$

One can check that $\approx$ is an equivalence relation, and that two symbols m/n and p/q are equivalent if and only if the numerical ratios of m to n and p to q are the same. [Check that $\frac{2}{3} \approx \frac{4}{6}$.] The equivalence classes in S are thus in one-to-one correspondence with the members of $\mathbb{Q}$, and we can [and probably should] think of $\mathbb{Q}$ as the set of equivalence classes. In our fancy notation, we'd write $[\frac{2}{3}] = [\frac{4}{6}]$ rather than $\frac{2}{3} = \frac{4}{6}$.

Now a function on $\mathbb{Q}$ will be well-defined provided the definition does not depend on which form the rationals take, i.e., provided it does not depend on which member of the equivalence class is used. Since $\frac{2}{3} \approx \frac{4}{6}$ and yet $2 + 3 \neq 4 + 6$, our definition $f(m/n) = m + n$ was not well-defined. ∎

EXERCISES 7.3

1. Which of the following describe equivalence relations? For those which are not equivalence relations, specify which of (R), (S) and (T) fail, and illustrate the failures with examples.
    (a) $L_1 \parallel L_2$ for straight lines in the plane if L_1 and L_2 are the same or are parallel.
    (b) $L_1 \perp L_2$ for straight lines in the plane if L_1 and L_2 are perpendicular.
    (c) $p_1 \sim p_2$ for Americans if p_1 and p_2 live in the same state.
    (d) $p_1 \approx p_2$ for Americans if p_1 and p_2 live in the same state or in neighboring states.
    (e) $p_1 \approx p_2$ for people if p_1 and p_2 have a parent in common.
    (f) $p_1 \cong p_2$ for people if p_1 and p_2 have the same mother.

2. (a) List all equivalence classes of $\mathbb{Z}$ for the equivalence relation congruence modulo 4.
    (b) How many different equivalence classes of $\mathbb{Z}$ are there with respect to congruence modulo 73?

3. Let S be a set. Is equality $=$ an equivalence relation?

4. Matrices $\mathbf{A}$ and $\mathbf{B}$ in $\mathfrak{M}_{n,n}$ are **similar** if $\mathbf{B} = \mathbf{PAP}^{-1}$ for an invertible matrix $\mathbf{P}$, in which case we write $\mathbf{A} \approx \mathbf{B}$. Prove that $\approx$ is an equivalence relation on $\mathfrak{M}_{n,n}$.

5. Let S be the set of all sequences (s_n) of real numbers and define $(s_n) \approx (t_n)$ if $\{n \in \mathbb{N} : s_n \neq t_n\}$ is finite. Show that $\approx$ is an equivalence relation on S.

6. Can you think of situations in the real world where you'd use the term "equivalent" and where a natural equivalence relation is involved?

7. Let S be a set and let G be a group of one-to-one functions f mapping S onto S. That is, G satisfies
    (i) the identity function 1_S is in G;
    (ii) if $f, g \in G$ then $f \circ g \in G$;
    (iii) if $f \in G$ then $f^{-1} \in G$.

    For $x, y \in S$ define $x \sim y$ if $f(x) = y$ for some $f \in G$. Show that $\sim$ is an equivalence relation on S. *Note*: Each of the equivalence relations for triangles in Examples 2 is of this sort if G is selected suitably.

8. Verify the claims in Example 12(b).

9. Define the relation $\approx$ on $\mathbb{Z}$ by $m \approx n$ in case $m^2 = n^2$.
   (a) Show that $\approx$ is an equivalence relation on $\mathbb{Z}$.
   (b) Describe the equivalence classes for $\approx$. How many are there?

10. For m, n in $\mathbb{N}$ define $m \sim n$ if $m^2 - n^2$ is a multiple of 3.
   (a) Show that $\sim$ is an equivalence relation on $\mathbb{N}$.
   (b) List four elements in the equivalence class [0].
   (c) List four elements in the equivalence class [1].
   (d) Do you think there are any more equivalence classes?

11. Consider the equivalence relation $\sim$ on $\mathcal{P}(S)$ discussed in Example 6.
   (a) Describe the sets in the equivalence class containing the empty set $\varnothing$.
   (b) Describe the sets in the equivalence class containing S.

12. On the set $\mathbb{N} \times \mathbb{N}$ define $\langle m, n \rangle \sim \langle k, l \rangle$ if $m + l = n + k$.
   (a) Show that $\sim$ is an equivalence relation on $\mathbb{N} \times \mathbb{N}$.
   (b) Draw a sketch of $\mathbb{N} \times \mathbb{N}$ that shows the equivalence classes.

13. The definition of $m \equiv n \pmod{p}$ makes sense even if $p = 1$ or $p = 0$.
   (a) Describe this equivalence relation for $p = 1$ and the corresponding equivalence classes in $\mathbb{Z}$.
   (b) Repeat part (a) for $p = 0$.

14. Let P be a set of computer programs and regard programs p_1 and p_2 as equivalent if they always produce the same outputs for given inputs. Is this an equivalence relation? Explain.

15. Let Σ be an alphabet, and for w_1 and w_2 in Σ^* define $w_1 \sim w_2$ if length(w_1) = length(w_2). Explain why $\sim$ is an equivalence relation, and describe the equivalence classes.

16. Consider $\mathbb{P} \times \mathbb{P}$ and define $\langle m, n \rangle \sim \langle p, q \rangle$ if $mq = np$.
   (a) Show that $\sim$ is an equivalence relation on $\mathbb{P} \times \mathbb{P}$.
   (b) Show that $\sim$ is the equivalence relation corresponding to the function $\mathbb{P} \times \mathbb{P} \to \mathbb{Q}$ given by $f\langle m, n \rangle) = m/n$; see Theorem 2(a).

17. How many equivalence relations are there on $\{0, 1, 2, 3\}$? *Hint*: Count unordered partitions. Why does this solve the problem?

18. In the proof of Theorem 2(b), we obtained the equality $\phi^{-1}([s]) = [s]$. Does this mean that ϕ^{-1} is the identity function on $[S]$? Discuss.

19. As in Exercise 9, define $\approx$ on $\mathbb{Z}$ by $m \approx n$ in case $m^2 = n^2$.
   (a) What is wrong with the following "definition" of $\leq$ on $[\mathbb{Z}]$? Let $[m] \leq [n]$ if and only if $m \leq n$.
   (b) What, if anything, is wrong with the following "definition" of a function $f : [\mathbb{Z}] \to \mathbb{Z}$? Let $f([m]) = m^2 + m + 1$.
   (c) Repeat part (b) with $g([m]) = m^4 + m^2 + 1$.
   (d) What, if anything, is wrong with the following "definition" of the operation $\oplus$ on $[\mathbb{Z}]$? Let $[m] \oplus [n] = [m + n]$.

20. Which of the following are well-defined definitions of functions on $\mathbb{Q}^+ = \{m/n : m, n \in \mathbb{P}\}$?

(a) $f(m/n) = n/m$

(b) $g(m/n) = m^2 + n^2$

(c) $h(m/n) = \dfrac{m^2 + n^2}{mn}$

§ 7.4   General Relations

When we introduced partial orders $\leqq$ and equivalence relations $\sim$ on a set S we indicated that for each pair $\langle x, y \rangle$ in $S \times S$ either the relation $[x \leqq y$ or $x \sim y]$ held or it did not. Also we shrewdly left the term "relation" undefined. Knowing or specifying for which pairs a relationship holds on S is exactly the same as knowing or specifying a subset of $S \times S$. So we now formally define a **binary relation on** S as any subset R of $S \times S$.

By now we hope you have developed a feeling for partial orders and equivalence relations. They are quite different; partial orders sort of "flow" from small to large, while equivalence relations partition the set into unrelated blocks [called equivalence classes]. On the other hand, the notion of binary relation is so general and abstract that there really isn't anything to get a feeling for beyond the obvious fact that binary relations are subsets of $S \times S$. We study such general objects to help us see the common features of the more familiar orderings, equivalence relations and also functions. Viewing binary relations as subsets of $S \times S$ will make them easier to create and manipulate. Finally, machines can store useful information by storing ordered pairs, i.e., members of some relation. Manipulating information, then, corresponds to manipulating relations.

For a binary relation R on S we will often use the notation $x \, R \, y$ [read "x is R-related to y"] to signify $\langle x, y \rangle \in R$. In addition, $x \, R\!\!\!/\, y$ means $\langle x, y \rangle \notin R$. This usage is compatible with our special notation $x \leqq y$ and $x \sim y$. We now define, once and for all, what we mean by a relation R being **reflexive, antireflexive, symmetric, antisymmetric** or **transitive**:

(R) $x \, R \, x$ for all $x \in S$,
(AR) $x \, R\!\!\!/\, x$ for all $x \in S$,
(S) $x \, R \, y$ implies $y \, R \, x$ for all $x, y \in S$,
(AS) $x \, R \, y$ and $y \, R \, x$ imply $x = y$,
(T) $x \, R \, y$ and $y \, R \, z$ imply $x \, R \, z$.

Each of these definitions can be written using ordered-pair notation. For example,

(S) $\langle x, y \rangle \in R$ implies $\langle y, x \rangle \in R$ for all $\langle x, y \rangle \in S \times S$.

From our present point of view, a **partial order on** S is a relation [i.e., subset of $S \times S$] satisfying (R), (AS) and (T). A **quasi-order on** S is a relation satisfying (AR) and (T) [so (AS) is vacuously true]. And an **equivalence relation on** S is a relation satisfying (R), (S) and (T).

It is often convenient to consider the **matrix A** of a relation R, when the set S is finite. The rows and columns of **A** are indexed by S, with entries $A[x, y] = 1$ if $x \, R \, y$ and $A[x, y] = 0$ if $x \, \mathcal{R} \, y$.

EXAMPLE 1   We illustrate several relations on $S = \{1, 2, 3, 4\}$, whose matrices are given in Figure 1.

(a) Let R_1 be the relation defined by $m < n$. That is, $m \, R_1 \, n$ if and only if $m < n$, and $R_1 = \{\langle m, n \rangle \in S \times S : m < n\}$. The relation R_1 is antireflexive and transitive, i.e., a quasi-order as expected. The relation R_1 is also antisymmetric vacuously:

$$m < n \quad \text{and} \quad n < m \quad \text{imply} \quad n = m$$

because the condition "$m < n$ and $n < m$" is always false. Indeed, the same argument shows that all quasi-orders are antisymmetric.

$$
\begin{array}{c}
\begin{array}{cccc}
1 & 2 & 3 & 4
\end{array} \\
\begin{array}{c}
1 \\ 2 \\ 3 \\ 4
\end{array}
\begin{bmatrix}
0 & 1 & 1 & 1 \\
0 & 0 & 1 & 1 \\
0 & 0 & 0 & 1 \\
0 & 0 & 0 & 0
\end{bmatrix}
\end{array}
\qquad
\begin{bmatrix}
1 & 1 & 0 & 0 \\
1 & 1 & 1 & 0 \\
0 & 1 & 1 & 1 \\
0 & 0 & 1 & 1
\end{bmatrix}
\qquad
\begin{bmatrix}
1 & 0 & 0 & 1 \\
0 & 1 & 0 & 0 \\
0 & 0 & 1 & 0 \\
1 & 0 & 0 & 1
\end{bmatrix}
\qquad
\begin{bmatrix}
1 & 0 & 0 & 0 \\
0 & 1 & 0 & 0 \\
0 & 0 & 1 & 0 \\
0 & 0 & 0 & 1
\end{bmatrix}
$$

$$\qquad R_1 \qquad\qquad\qquad R_2 \qquad\qquad\qquad R_3 \qquad\qquad\qquad E$$

Matrices of relations

Figure 1

(b) Let R_2 be the relation defined by $m \, R_2 \, n$ if $|m - n| \leq 1$. This relation is reflexive since $|m - m| \leq 1$ for $m \in S$, and it is symmetric since $|m - n| \leq 1$ implies $|n - m| \leq 1$. It is not transitive since, for instance, $2 \, R_2 \, 3$ and $3 \, R_2 \, 4$ but $2 \, \mathcal{R}_2 \, 4$.

(c) Define R_3 by $m \equiv n \pmod 3$, so that $m \, R_3 \, n$ if and only if $m \equiv n \pmod 3$. This relation is reflexive, symmetric and transitive [Theorem 2 of § 4.3] and so R_3 is an equivalence relation on S.

(d) Let E be the "equality relation" on S:

$$E = \{\langle m, n \rangle \in S \times S : m = n\} = \{\langle m, m \rangle : m \in S\}.$$

In other words, $m \, E \, n$ if and only if $m = n$. It is clear that E is an equivalence relation on S. Note that the matrix for the relation E is the 4×4 identity matrix. ▮

EXAMPLE 2   For $m, n \in \mathbb{Z}$, define $m \, R \, n$ if $m + n$ is a multiple of 3, i.e., $m + n \equiv 0 \pmod 3$. This relation is not reflexive since $1 \, \mathcal{R} \, 1$ for example. It isn't antireflexive either since $3 \, R \, 3$ for example. The relation R is symmetric since

$m + n = n + m$ for $m, n \in \mathbb{Z}$. The relation R is not transitive; for instance $4\ R\ 2$ and $2\ R\ 1$ while $4\ \not\!R\ 1$. ∎

EXAMPLE 3 Consider a graph with no parallel edges or multiple loops. We define a relation R on the set V of vertices by $u\ R\ v$ whenever $\{u, v\}$ is an edge of the graph. The graph determines R, of course, and R also determines the graph since the graph has no parallel edges or multiple loops. More precisely, for every pair $\langle u, v \rangle$ of vertices, exactly one edge connects u to v if $u\ R\ v$, and no edges connect u to v otherwise. The matrix for the graph, as discussed in § 0.3, is exactly the matrix for the relation R.

Note that the relation R is symmetric. It will be reflexive if and only if there is a loop at every vertex. It need not be transitive, generally speaking, since $\{u, v\}$ and $\{v, w\}$ may be edges while $\{u, w\}$ is not. Every symmetric relation is the relation for some graph as described in the last paragraph. In Figure 2 we have drawn the graphs for the symmetric relations whose matrices are in Figure 1. When we study directed graphs in Chapter 8 we will see that every relation, symmetric or not, corresponds to some directed graph. ∎

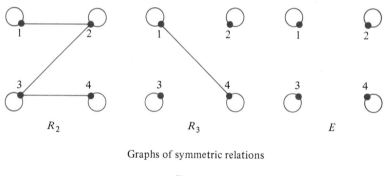

Graphs of symmetric relations

Figure 2

For any set S we will write E for the "equality relation": $E = \{\langle x, x \rangle : x \in S\}$. The next example includes some set-theoretic characterizations of properties (R) and (AR). Set-theoretic characterizations of (S), (AS) and (T) will be given later in this section and in the next section.

EXAMPLE 4 Let R be a binary relation on a set S.

(a) R is reflexive if and only if $E \subseteq R$. This is evident, since R is reflexive precisely in case

$$\langle x, x \rangle \in R \quad \text{for all} \quad x \in S,$$

i.e., in case every ordered pair in E belongs to R.

(b) R is antireflexive if and only if $E \cap R = \varnothing$. In fact, R is antireflexive provided

$$\langle x, x \rangle \notin R \quad \text{for all} \quad x \in S,$$

i.e., provided no ordered pair in E belongs to R.

(c) In § 7.1, prior to Example 3, we indicated how to get quasi-orders from partial orders and vice versa. In short, we observed that if R is a partial order then $R \backslash E$ is a quasi-order, and if R is a quasi-order then $R \cup E$ is a partial order. Exercise 15 contains a slightly more general result. ∎

We have introduced binary relations on a set; it is only the tiniest generalization to define them from one set to another. A **binary relation from a set S to a set T** is a subset of $S \times T$. Matrices for these relations are constructed just as before. The only difference is that the matrices won't be square matrices unless $|S| = |T|$.

EXAMPLE 5   Let $S = \{1, 2, 3, 4, 5\}$ and $T = \{a, b, c\}$, and let

$$R = \{\langle 1, a \rangle, \langle 2, a \rangle, \langle 2, c \rangle, \langle 3, a \rangle, \langle 3, b \rangle, \langle 4, a \rangle, \langle 4, b \rangle, \langle 4, c \rangle, \langle 5, b \rangle\}.$$

The matrix for R is given in Figure 3. ∎

$$
\begin{array}{c c c c}
 & a & b & c \\
1 & 1 & 0 & 0 \\
2 & 1 & 0 & 1 \\
3 & 1 & 1 & 0 \\
4 & 1 & 1 & 1 \\
5 & 0 & 1 & 0 \\
\end{array}
$$

Figure 3

EXAMPLE 6   A mail-order record company has a list L of customers. Each customer indicates interest in certain categories of recordings: classical, easy-listening, Latin, religious, popular, rock, etc. Let C be the set of possible categories. The set of all ordered pairs $\langle$name, selected-category$\rangle$ is a relation R from L to C. This relation might contain such pairs as $\langle$K. A. Ross, classical$\rangle$, $\langle$C. R. B. Wright, classical$\rangle$ and $\langle$C. R. B. Wright, punk rock$\rangle$. ∎

EXAMPLE 7   A university would be interested in the relation given by all ordered pairs whose first entries are students and whose second entries are the courses the students are currently enrolled in. This is a relation R_1 from the set S of university students to the set C of courses offered. Note that if student s in S is fixed, then $\{c \in C : \langle s, c \rangle \in R_1\}$ is the set of courses taken by s. On the other hand, if course c in C is fixed, then $\{s \in S : \langle s, c \rangle \in R_1\}$ is the class list for the course.

Another relation R_2 consists of all ordered pairs whose first entries are courses and whose second entries are the departments for which the course is a major requirement. Thus R_2 is a relation from C to the set D of departments in the university. For fixed $c \in C$, $\{d \in D : \langle c, d \rangle \in R_2\}$ is the set of departments for which c is a major requirement. For fixed $d \in D$, $\{c \in C : \langle c, d \rangle \in R_2\}$ is the list of courses required for that department's majors. A computerized degree-checking program would need to use a data structure which contained enough information to determine relations R_1 and R_2. ∎

EXAMPLE 8   (a) Consider a set P of programs written to be carried out on a computer and a catalog C of canned programs available for use. We get a relation from C to P if we say that a canned program c is related to a program p in P provided p calls c as a subroutine. A frequently used c might be related to a number of p's while a c which is never called is related to no p.

(b) A translator from decimal representations to binary representations can be viewed as the relation consisting of all ordered pairs whose first entries are allowable decimal representations and whose second entries are the corresponding binary representations. Actually this relation is a function. ∎

Recall that in § 3.1 we indicated how functions can be identified with their graphs and hence regarded as sets of ordered pairs. In fact, if $f : S \to T$ we identified f with the set

$$R_f = \{\langle x, y \rangle \in S \times T : y = f(x)\},$$

which is a relation from S to T. Of course, not all relations are functions. From this point of view a **function from S to T** is a relation R from S to T such that

for each $x \in S$ there is exactly one $y \in T$ such that $\langle x, y \rangle \in R$.

This is simply a restatement of the definition in § 3.1.

Consider again an arbitrary relation R from a set S to a set T, i.e., $R \subseteq S \times T$. The **inverse relation** R^{-1} is the relation from T to S defined by:

$$R^{-1} = \{\langle y, x \rangle \in T \times S : \langle x, y \rangle \in R\}.$$

Since every function $f : S \to T$ is a relation, f^{-1} always exists:

$$\textit{as a relation } f^{-1} = \{\langle y, x \rangle \in T \times S : y = f(x)\}.$$

This relation is a function precisely when f is an invertible function as defined in § 3.2.

EXAMPLE 9   (a) Recall that if $f : S \to T$ is a function and $A \subseteq S$, then the image of A under f is

$$f(A) = \{f(x) : x \in A\} = \{y \in T : y = f(x) \text{ for some } x \in A\}.$$

If we view f as the relation R_f, then this set equals

$$\{y \in T : \langle x, y \rangle \in R_f \text{ for some } x \in A\}.$$

Similarly, for any relation R from S to T we can define

$$R(A) = \{y \in T : \langle x, y \rangle \in R \text{ for some } x \in A\}.$$

Since R^{-1} is a relation from T to S, for $B \subseteq T$ we also have

$$\begin{aligned}
R^{-1}(B) &= \{x \in S : \langle y, x \rangle \in R^{-1} \text{ for some } y \in B\} \\
&= \{x \in S : \langle x, y \rangle \in R \text{ for some } y \in B\}.
\end{aligned}$$

If R is actually R_f for a function f from S to T this gives

$$R_f^{-1}(B) = \{x \in S : y = f(x) \text{ for some } y \in B\} = \{x \in S : f(x) \in B\},$$

which is exactly the definition we gave for $f^{-1}(B)$ in § 3.2.

(b) For a concrete example of part (a), let S be a set of suppliers and T a set of products, and define $x \, R \, y$ if supplier x sells product y. For a given set A of suppliers, the set $R(A)$ is the set of products sold by at least one member of A. For a given set B of products, $R^{-1}(B)$ is the set of suppliers who sell at least one product in B. The relation R is R_f for a function f from S to T if and only if each supplier sells exactly one product. ∎

Now consider a binary relation R on a single set S, i.e., $R \subseteq S \times S$. Then

$$R^{-1} = \{\langle y, x \rangle \in S \times S : \langle x, y \rangle \in R\}.$$

This definition of inverse is compatible with the definition of the inverse $\geq$ of a partial order $\leq$ given in § 7.1, since $y \geq x$ if and only if $x \leq y$. We did not introduce the inverse of an equivalence relation in § 7.3 because we would get nothing new. For in general, $R = R^{-1}$ if and only if R is symmetric; in Exercise 19 you are asked to show this and to show that R is antisymmetric if and only if $R \cap R^{-1} \subseteq E$.

EXERCISES 7.4

1. Give matrices for the following relations on $S = \{0, 1, 2, 3\}$ and specify which of the properties (R), (AR), (S), (AS) and (T) the relations satisfy.
   (a) $m \, R_1 \, n$ if $m + n = 3$
   (b) $m \, R_2 \, n$ if $m \equiv n \pmod 2$
   (c) $m \, R_3 \, n$ if $m \leq n$
   (d) $m \, R_4 \, n$ if $m + n \leq 4$
   (e) $m \, R_5 \, n$ if $\max\{m, n\} = 3$

2. Draw graphs for the symmetric relations in Exercise 1.

3. (a) Which of the relations in Exercise 1 are partial orders?
   (b) Which of the relations in Exercise 1 are equivalence relations?

4. Let $A = \{0, 1, 2\}$. Each of the statements below defines a relation R on A by $m \, R \, n$ if the statement is true for m and n. Write each of the relations as a set of ordered pairs.
   (a) $m \leq n$ (b) $m < n$ (c) $m = n$
   (d) $mn = 0$ (e) $mn = m$ (f) $m + n \in A$
   (g) $m^2 + n^2 = 2$ (h) $m^2 + n^2 = 3$ (i) $m = \max\{n, 1\}$

5. Which of the relations in Exercise 4 are reflexive? symmetric?

6. Give a matrix for each of the relations in Exercise 4.

7. The following binary relations are defined on $\mathbb{N}$.
   (a) Write the binary relation R_1 defined by $m + n = 5$ as a set of ordered pairs.
   (b) Do the same for R_2 defined by $\max\{m, n\} = 2$.
   (c) The binary relation R_3 defined by $\min\{m, n\} = 2$ consists of infinitely many ordered pairs. List five of them.

8. For each of the relations in Exercise 7, specify which of the properties (R), (AR), (S), (AS) and (T) it satisfies.

9. Which of the following relations on $\mathbb{Z}$ are equivalence relations? For the equivalence relations, identify the equivalence classes. For the other relations, specify which of (R), (S) and (T) fail.
   (a) $n \equiv m \pmod 4$ (b) $nm = 0$
   (c) $nm > 0$ (d) $n \leq m$

10. Consider the relation R on $\mathbb{Z}$ defined by $m \, R \, n$ if and only if $m^3 - n^3 \equiv 0 \pmod 5$.
    (a) Which of the properties (R), (AR), (S), (AS) and (T) are satisfied by R?
    (b) Is R an equivalence relation? a partial order?

11. Let $\Sigma = \{a, b\}$ and define the binary relation $\leq$ on $\mathcal{P}(\Sigma^*)$ by $A \leq B$ if and only if $A^* \subseteq B^*$.
    (a) Which of the properties (R), (AR), (S), (AS) and (T) does the relation $\leq$ satisfy? Verify your answers.
    (b) Is $\leq$ a partial order? a quasi-order?

12. Let $\Sigma = \{a, b, c, d, e, f, g\}$. Give a matrix for the equivalence relation on Σ determined by the partition $\{\{a, d\}, \{c, e, f\}, \{b, g\}\}$.

13. (a) Consider the **empty relation** $\varnothing$ on a nonempty set S. Which of the properties (R), (AR), (S), (AS) and (T) does $\varnothing$ possess?
    (b) Repeat part (a) for the **universal relation** $U = S \times S$ on S.

14. Give an example of a relation that is:
    (a) antisymmetric and transitive but not reflexive,
    (b) symmetric but not reflexive or transitive.

15. Let R be an antisymmetric and transitive relation on a set S.
    (a) Prove that $R \cup E$ is a partial order on S.
    (b) Prove that $R \setminus E$ is a quasi-order on S.

16. Let R_1 and R_2 be binary relations on a set S.
    (a) Show that $R_1 \cap R_2$ is reflexive if R_1 and R_2 are.
    (b) Show that $R_1 \cap R_2$ is symmetric if R_1 and R_2 are.

(c) Show that $R_1 \cap R_2$ is transitive if R_1 and R_2 are.

(d) Observe that the intersection of equivalence relations is an equivalence relation.

17. Let R_1 and R_2 be binary relations on a set S.
    (a) Must $R_1 \cup R_2$ be reflexive if R_1 and R_2 are?
    (b) Must $R_1 \cup R_2$ be symmetric if R_1 and R_2 are?
    (c) Must $R_1 \cup R_2$ be transitive if R_1 and R_2 are?
    (d) Must $R_1 \cup R_2$ be an equivalence relation if R_1 and R_2 are?

18. If A is the matrix for a relation R from a set S to a set T, what is the matrix for the inverse relation R^{-1}?

19. Let R be a binary relation on a set S.
    (a) Prove that R is symmetric if and only if $R = R^{-1}$.
    (b) Prove that R is antisymmetric if and only if $R \cap R^{-1} \subseteq E$, where $E = \{\langle x, x \rangle : x \in S\}$.

20. Let R_1 and R_2 be binary relations from a set S to a set T.
    (a) Show that $(R_1 \cup R_2)^{-1} = R_1^{-1} \cup R_2^{-1}$.
    (b) Show that $(R_1 \cap R_2)^{-1} = R_1^{-1} \cap R_2^{-1}$.
    (c) Show that if $R_1 \subseteq R_2$ then $R_1^{-1} \subseteq R_2^{-1}$.

§ 7.5 Composition of Relations

In this section we generalize the familiar concept of composition of functions to relations. To help motivate this, we first consider functions $f : S \rightarrow T$ and $g : T \rightarrow U$, and recall that the composite function $g \circ f$ maps S into U and is defined by $g \circ f(x) = g(f(x))$ for all $x \in S$. As in § 7.4 we write R_f for the relation corresponding to f:

$$R_f = \{\langle x, y \rangle \in S \times T : y = f(x)\}.$$

Likewise R_g and $R_{g \circ f}$ represent the relations given by g and $g \circ f$. In particular

$$R_{g \circ f} = \{\langle x, z \rangle \in S \times U : z = g(f(x))\}.$$

Now $z = g(f(x))$ means that $z = g(y)$ where $y = f(x)$, and so $\langle x, z \rangle$ belongs to $R_{g \circ f}$ if and only if there is some $y \in T$ such that $\langle y, z \rangle \in R_g$ and $\langle x, y \rangle \in R_f$. That is,

$$R_{g \circ f} = \{\langle x, z \rangle \in S \times U : \text{for some } y \in T, \langle x, y \rangle \in R_f \text{ and } \langle y, z \rangle \in R_g\}.$$

If we call this set $R_g \circ R_f$, then we can immediately generalize as follows. For a relation R_1 from S to T and a relation R_2 from T to U, we define $R_2 \circ R_1$ from S to U by

$$R_2 \circ R_1 = \{\langle x, z \rangle \in S \times U : \text{for some } y \in T, \langle x, y \rangle \in R_1 \text{ and } \langle y, z \rangle \in R_2\}.$$

The relation $R_2 \circ R_1$ is called the **composite** of R_2 and R_1. Since we think of R_1 first and R_2 second, on esthetic grounds alone this general definition seems backwards. Moreover, it will turn out to be backwards when we observe the connection between composition of relations and products of their matrices. Accordingly, many people write "$R_1 \circ R_2$" in place of $R_2 \circ R_1$; since functions are relations, this usage is inconsistent and a source of confusion. To avoid this confusion, we will write $R_1 R_2$, rather than "$R_1 \circ R_2$," for $R_2 \circ R_1$. We summarize:

Definition   Given relations R_1 from S to T and R_2 from T to U, the **composite relation**

$$\{\langle x, z \rangle \in S \times U : \text{for some } y \in T, \langle x, y \rangle \in R_1 \text{ and } \langle y, z \rangle \in R_2\}$$

will be denoted by either $R_1 R_2$ or $R_2 \circ R_1$. Thus $x \, R_1 R_2 \, z$ precisely if there exists $y \in T$ such that $x \, R_1 \, y$ and $y \, R_2 \, z$.

EXAMPLE 1   Read Example 7 of § 7.4, which concerns university students and courses, again. The relations are

$$R_1 = \{\langle s, c \rangle \in S \times C : s \text{ is enrolled in } c\}$$

and

$$R_2 = \{\langle c, d \rangle \in C \times D : c \text{ is required by } d\}.$$

Observe that

$$R_1 R_2 = \{\langle s, d \rangle \in S \times D : \text{for some } c \in C, \langle s, c \rangle \in R_1 \text{ and } \langle c, d \rangle \in R_2\}.$$

Therefore $\langle s, d \rangle$ belongs to $R_1 R_2$ provided student s is taking some course that is required by department d.

Note that $R_2 R_1$ makes no sense, because the second entries of R_2 lie in the set D while the first entries of R_1 lie in S; it could not happen that $\langle c, t \rangle \in R_2$ and $\langle t, c' \rangle \in R_1$. Of course, $R_2 \circ R_1$ makes sense; this is just another name for $R_1 R_2$. ■

EXAMPLE 2   To illustrate working with general relations, we establish some elementary facts. Consider relations R_1 and R_2 from S to T and relations R_3 and R_4 from T to U.

(a) If $R_1 \subseteq R_2$ and $R_3 \subseteq R_4$, then $R_1 R_3 \subseteq R_2 R_4$. To see this, consider $\langle x, z \rangle \in R_1 R_3$. Then for some $y \in T$ we have $\langle x, y \rangle \in R_1$ and $\langle y, z \rangle \in R_3$. Since $R_1 \subseteq R_2$ and $R_3 \subseteq R_4$ we also have $\langle x, y \rangle \in R_2$ and $\langle y, z \rangle \in R_4$. So $\langle x, z \rangle \in R_2 R_4$. This shows that $R_1 R_3 \subseteq R_2 R_4$.

(b) We show

$$(R_1 \cup R_2)R_3 = R_1 R_3 \cup R_2 R_3.$$

Since $R_1 \subseteq R_1 \cup R_2$ we have $R_1 R_3 \subseteq (R_1 \cup R_2)R_3$ from part (a); likewise $R_2 R_3 \subseteq (R_1 \cup R_2)R_3$ and so

$$R_1 R_3 \cup R_2 R_3 \subseteq (R_1 \cup R_2)R_3.$$

To check the reverse inclusion, consider $\langle x, z \rangle \in (R_1 \cup R_2)R_3$. For some $y \in T$ we have $\langle x, y \rangle \in R_1 \cup R_2$ and $\langle y, z \rangle \in R_3$. Then either $\langle x, y \rangle \in R_1$ so that $\langle x, z \rangle \in R_1 R_3$ or else $\langle x, y \rangle \in R_2$ so that $\langle x, z \rangle \in R_2 R_3$. Either way, $\langle x, z \rangle \in R_1 R_3 \cup R_2 R_3$ and hence

$$(R_1 \cup R_2)R_3 \subseteq R_1 R_3 \cup R_2 R_3. \quad \blacksquare$$

In § 3.1 we observed that composition of functions is associative. So is composition of relations.

Associative   If R_1 is a relation from S to T, R_2 is a relation from T to U and R_3 is a relation
Law for   from U to V, then
Relations

$$(R_1 R_2)R_3 = R_1(R_2 R_3).$$

Proof. We show that an ordered pair $\langle x, v \rangle$ in $S \times V$ belongs to $(R_1 R_2)R_3$ if and only if

(1)   there exist $y \in T$ and $z \in U$ so that
$$\langle x, y \rangle \in R_1, \langle y, z \rangle \in R_2 \text{ and } \langle z, v \rangle \in R_3.$$

A similar argument shows that $\langle x, v \rangle$ belongs to $R_1(R_2 R_3)$ if and only if (1) holds.

Consider $\langle x, v \rangle$ in $(R_1 R_2)R_3$. Since $R_1 R_2$ is a relation from S to U, this means that there exists $z \in U$ such that $\langle x, z \rangle \in R_1 R_2$ and $\langle z, v \rangle \in R_3$. Since $\langle x, z \rangle \in R_1 R_2$ there exists $y \in T$ such that $\langle x, y \rangle \in R_1$ and $\langle y, z \rangle \in R_2$. Thus (1) holds.

Now suppose that (1) holds for an element $\langle x, v \rangle$ in $S \times V$. Then $\langle x, y \rangle \in R_1$ and $\langle y, z \rangle \in R_2$ so that $\langle x, z \rangle \in R_1 R_2$. Since also $\langle z, v \rangle \in R_3$, we conclude that $\langle x, v \rangle \in (R_1 R_2)R_3$. $\quad \blacksquare$

In view of the associative law, we may write $R_1 R_2 R_3$ for either $(R_1 R_2)R_3$ or $R_1(R_2 R_3)$. As shown in the proof, $\langle x, v \rangle$ belongs to $R_1 R_2 R_3$ provided there exist $y \in T$ and $z \in U$ such that $\langle x, y \rangle \in R_1$, $\langle y, z \rangle \in R_2$ and $\langle z, v \rangle \in R_3$.

The sets and relations we have been considering could have been finite or infinite. We now consider relations R_1 from S to T and R_2 from T to U where S, T and U are finite. Let $\mathbf{A}_1$ and $\mathbf{A}_2$ be the matrices for the relations R_1 and R_2. What is the matrix for the composite relation $R_1 R_2$?

EXAMPLE 3   Let $S = \{1, 2, 3, 4, 5\}$, $T = \{a, b, c\}$ and $U = \{e, f, g, h\}$. Consider the relations

$$R_1 = \{\langle 1, a \rangle, \langle 2, a \rangle, \langle 2, c \rangle, \langle 3, a \rangle, \langle 3, b \rangle, \langle 4, a \rangle, \langle 4, b \rangle, \langle 4, c \rangle, \langle 5, b \rangle\},$$

$$R_2 = \{\langle a, e \rangle, \langle a, g \rangle, \langle b, f \rangle, \langle b, g \rangle, \langle b, h \rangle, \langle c, e \rangle, \langle c, g \rangle, \langle c, h \rangle\},$$

so that

$$R_1 R_2 = \{\langle 1, e \rangle, \langle 1, g \rangle, \langle 2, e \rangle, \langle 2, g \rangle, \langle 2, h \rangle, \langle 3, e \rangle, \langle 3, f \rangle, \langle 3, g \rangle, \langle 3, h \rangle,$$
$$\langle 4, e \rangle, \langle 4, f \rangle, \langle 4, g \rangle, \langle 4, h \rangle, \langle 5, f \rangle, \langle 5, g \rangle, \langle 5, h \rangle\}.$$

The matrices $\mathbf{A}_1$, $\mathbf{A}_2$ and $\mathbf{A}$ for these relations are given in Figure 1. Compare the matrix $\mathbf{A}$ with the product $\mathbf{A}_1\mathbf{A}_2$. The 1's in the matrix $\mathbf{A}$ occur where the nonzero entries occur in $\mathbf{A}_1\mathbf{A}_2$. This is not an accident, as we now explain. Consider $x \in \{1, 2, 3, 4, 5\}$ and $z \in \{e, f, g, h\}$. The (x, z)-entry of $\mathbf{A}_1\mathbf{A}_2$ is

$$\sum_{y \in \{a, b, c\}} \mathbf{A}_1[x, y]\mathbf{A}_2[y, z].$$

The sum is positive if any product term is positive. A product $\mathbf{A}_1[x, y]\mathbf{A}_2[y, z]$ is 0 unless both $\mathbf{A}_1[x, y]$ and $\mathbf{A}_2[y, z]$ are 1, in which case we have $\langle x, y \rangle \in R_1$ and $\langle y, z \rangle \in R_2$. Thus the sum is 0 if $\langle x, z \rangle \notin R_1R_2$, and is greater than 0 if $\langle x, z \rangle \in R_1R_2$. More precisely, the sum is exactly the number

$$|\{y \in \{a, b, c\} : \langle x, y \rangle \in R_1 \text{ and } \langle y, z \rangle \in R_2\}|.$$

For example, the $(2, e)$-entry is 2 because

$$\{y \in \{a, b, c\} : \langle 2, y \rangle \in R_1 \text{ and } \langle y, e \rangle \in R_2\} = \{a, c\}.$$

The $(2, f)$-entry is 0 because

$$\{y \in \{a, b, c\} : \langle 2, y \rangle \in R_1 \text{ and } \langle y, f \rangle \in R_2\} = \varnothing,$$

i.e., $\langle 2, f \rangle \notin R_1R_2$. ∎

$$A_1 = \begin{array}{c} \\ 1 \\ 2 \\ 3 \\ 4 \\ 5 \end{array}
\begin{array}{ccc} a & b & c \\ \left[\begin{array}{ccc} 1 & 0 & 0 \\ 1 & 0 & 1 \\ 1 & 1 & 0 \\ 1 & 1 & 1 \\ 0 & 1 & 0 \end{array}\right] \end{array}
\qquad
A_2 = \begin{array}{c} \\ a \\ b \\ c \end{array}
\begin{array}{cccc} e & f & g & h \\ \left[\begin{array}{cccc} 1 & 0 & 1 & 0 \\ 0 & 1 & 1 & 1 \\ 1 & 0 & 1 & 1 \end{array}\right] \end{array}$$

$$A = \begin{array}{c} \\ 1 \\ 2 \\ 3 \\ 4 \\ 5 \end{array}
\begin{array}{cccc} e & f & g & h \\ \left[\begin{array}{cccc} 1 & 0 & 1 & 0 \\ 1 & 0 & 1 & 1 \\ 1 & 1 & 1 & 1 \\ 1 & 1 & 1 & 1 \\ 0 & 1 & 1 & 1 \end{array}\right] \end{array}
\qquad
A_1A_2 = \begin{array}{c} \\ 1 \\ 2 \\ 3 \\ 4 \\ 5 \end{array}
\begin{array}{cccc} e & f & g & h \\ \left[\begin{array}{cccc} 1 & 0 & 1 & 0 \\ 2 & 0 & 2 & 1 \\ 1 & 1 & 2 & 1 \\ 2 & 1 & 3 & 2 \\ 0 & 1 & 1 & 1 \end{array}\right] \end{array}$$

Figure 1

As Example 3 shows, the matrix for the relation R_1R_2 need not be the product $\mathbf{A}_1\mathbf{A}_2$ of the matrices for R_1 and R_2, but there is a connection: both have 0's in the same places. We could define a new product for matrices, letting $\mathbf{A}_1 * \mathbf{A}_2$ be the matrix obtained from $\mathbf{A}_1\mathbf{A}_2$, by replacing each nonzero entry with 1. An equivalent and superior approach involves the use of slightly different operations on the entries which are in $\mathbb{Z}(2) = \{0, 1\}$. We use the **Boolean operations** $\vee$ and $\wedge$ on $\mathbb{Z}(2)$ defined by Table 1. These are the

TABLE 1

familiar logical operations "or" and "and" on the set $\{0, 1\}$ of truth values. Note that for $m, n \in \mathbb{Z}(2)$ we have $m \vee n = \max\{m, n\}$ and $m \wedge n = \min\{m, n\}$. Consider **Boolean matrices** $\mathbf{A}_1$ and $\mathbf{A}_2$, i.e., matrices all of whose entries are 0's and 1's. If $\mathbf{A}_1$ is $m \times n$ and $\mathbf{A}_2$ is $n \times p$, we define the **Boolean product** $\mathbf{A}_1 * \mathbf{A}_2$ to be the usual product, but with the addition and multiplication operations replaced by $\vee$ and $\wedge$. That is, $\mathbf{A}_1 * \mathbf{A}_2$ is the $m \times p$ matrix whose (i, k)-entry is

$$\mathbf{A}_1 * \mathbf{A}_2[i, k] =$$

$$(\mathbf{A}_1[i, 1] \wedge \mathbf{A}_2[1, k]) \vee (\mathbf{A}_1[i, 2] \wedge \mathbf{A}_2[2, k]) \vee \cdots \vee (\mathbf{A}_1[i, n] \wedge \mathbf{A}_2[n, k]);$$

this may be written as

$$\bigvee_{j=1}^{n} (\mathbf{A}_1[i, j] \wedge \mathbf{A}_2[j, k]).$$

EXAMPLE 4   The Boolean product $\mathbf{A}_1 * \mathbf{A}_2$ of the matrices in Figure 1 is the matrix $\mathbf{A}$ in Figure 1. For example, the $(3, g)$-entry of $\mathbf{A}_1 * \mathbf{A}_2$ is $(1 \wedge 1) \vee (1 \wedge 1) \vee (0 \wedge 1) = 1 \vee 1 \vee 0 = 1$. The $(5, e)$-entry is $(0 \wedge 1) \vee (1 \wedge 0) \vee (0 \wedge 1) = 0 \vee 0 \vee 0 = 0$. ∎

The discussion in Example 3 can be modified to prove the following result.

Theorem 1   Consider relations R_1 from S to T and R_2 from T to U where S, T and U are finite. If $\mathbf{A}_1$ and $\mathbf{A}_2$ are the matrices for the relations R_1 and R_2, then the Boolean product $\mathbf{A}_1 * \mathbf{A}_2$ is the matrix for the composite relation $R_1 R_2$.

The Boolean product operation $*$ on Boolean matrices is associative. This can be shown directly or by applying the associativity of the corresponding relations, as advised in Exercise 17.

For the remainder of this section, except for Example 10, we consider relations on a single set S, i.e., subsets of $S \times S$.

Theorem 2   For a set S, let $\mathscr{P}(S \times S)$ denote the set of all relations on S. Under the composition operation, $\mathscr{P}(S \times S)$ is a monoid.

Proof.   First note that if R_1 and R_2 are in $\mathscr{P}(S \times S)$, so is their composite $R_1 R_2$. Associativity of composition has already been verified. Thus $\mathscr{P}(S \times S)$ is a semigroup. The identity for this semigroup is the "equality relation"

$$E = \{\langle x, x\rangle \in S \times S : x \in S\}. \quad \blacksquare$$

The usual notational conventions apply to this monoid. Thus if R is a relation on S, then $R^0 = E$ and, for $n \in \mathbb{P}$, R^n is the composition of R with itself n times. Note that if $n > 1$, then $\langle x, z\rangle$ belongs to R^n provided there exist $y_1, y_2, \ldots, y_{n-1}$ in S such that $\langle x, y_1\rangle, \langle y_1, y_2\rangle, \ldots, \langle y_{n-1}, z\rangle$ are all in R. In other words, $x \, R^n \, z$ if x and z are R-related through a chain of length n.

EXAMPLE 5   (a) Consider a graph with no parallel edges or multiple loops and the relation R on its set V of vertices satisfying $u \, R \, v$ whenever $\{u, v\}$ is an edge. An example is illustrated in Figure 2. We have $u \, R^n \, w$ provided there exist vertices $v_1, \ldots, v_{n-1}$ so that $u \, R \, v_1,\, v_1 \, R \, v_2, \ldots, v_{n-1} \, R \, w$, i.e., provided there is a path of length n connecting u and w. If $\mathbf{A}$ is the Boolean matrix for R, the Boolean power

$$\mathbf{A} * \mathbf{A} * \cdots * \mathbf{A} \qquad [n \text{ times}]$$

is the Boolean matrix for R^n and tells us exactly which pairs of vertices are connected by paths of length n. If we also wanted the number of such paths, we would calculate the ordinary matrix power $\mathbf{A}^n$, as discussed in § 0.3.

(b) If the relation R on $\{1, 2, 3, 4\}$ has the matrix and graph of Figure 2, then the relation R^2 has the matrix and graph of Figure 3. The $(3, 4)$-entry of $\mathbf{A} * \mathbf{A}$ tells us that there is at least one path in Figure 2 of length two from vertex 3 to vertex 4. That is why there is an edge from 3 to 4 in the graph of R^2. The $(3, 4)$-entry of $\mathbf{A}^2$ is 3 and so there are exactly 3 paths of length two in Figure 2 from vertex 3 to vertex 4. What are they? ∎

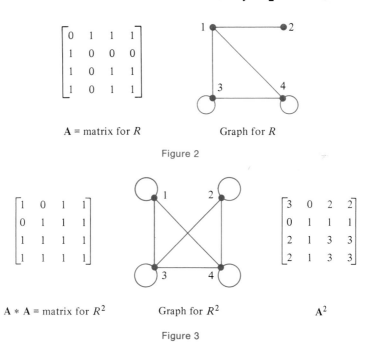

$$\begin{bmatrix} 0 & 1 & 1 & 1 \\ 1 & 0 & 0 & 0 \\ 1 & 0 & 1 & 1 \\ 1 & 0 & 1 & 1 \end{bmatrix}$$

A = matrix for R Graph for R

Figure 2

$$\begin{bmatrix} 1 & 0 & 1 & 1 \\ 0 & 1 & 1 & 1 \\ 1 & 1 & 1 & 1 \\ 1 & 1 & 1 & 1 \end{bmatrix}$$

$$\begin{bmatrix} 3 & 0 & 2 & 2 \\ 0 & 1 & 1 & 1 \\ 2 & 1 & 3 & 3 \\ 2 & 1 & 3 & 3 \end{bmatrix}$$

A * A = matrix for R^2 Graph for R^2 $\mathbf{A}^2$

Figure 3

EXAMPLE 6   Let R be the relation on $\{1, 2, 3\}$ with matrix

$$\mathbf{A} = \begin{bmatrix} 1 & 0 & 0 \\ 1 & 0 & 1 \\ 1 & 1 & 0 \end{bmatrix}.$$

The relation R^2 has matrix

$$\mathbf{A} * \mathbf{A} = \begin{bmatrix} 1 & 0 & 0 \\ 1 & 0 & 1 \\ 1 & 1 & 0 \end{bmatrix} * \begin{bmatrix} 1 & 0 & 0 \\ 1 & 0 & 1 \\ 1 & 1 & 0 \end{bmatrix} = \begin{bmatrix} 1 & 0 & 0 \\ 1 & 1 & 0 \\ 1 & 0 & 1 \end{bmatrix}.$$

The relation R^3 has matrix

$$\mathbf{A} * (\mathbf{A} * \mathbf{A}) = \begin{bmatrix} 1 & 0 & 0 \\ 1 & 0 & 1 \\ 1 & 1 & 0 \end{bmatrix} * \begin{bmatrix} 1 & 0 & 0 \\ 1 & 1 & 0 \\ 1 & 0 & 1 \end{bmatrix} = \begin{bmatrix} 1 & 0 & 0 \\ 1 & 0 & 1 \\ 1 & 1 & 0 \end{bmatrix}.$$

Note that $R = R^3$. In fact, we have $R^n = R^{n+2}$ for $n \geq 1$, since

$$R^{n+2} = R^{(n-1)+3} = R^{n-1}R^3 = R^{n-1}R = R^n. \quad \blacksquare$$

It is sometimes helpful to visualize small relations R by representing the set by points and drawing an arrow from x to y whenever $\langle x, y \rangle \in R$. Figure 4 contains such a picture for the relations R and R^2 in Example 6. Note that R^2 is reflexive; there is an arrow from each point to itself. A relation is symmetric if for every arrow between points x and y there is also an arrow from y to x. Finally, note that R^2 can be drawn using the picture of R: draw

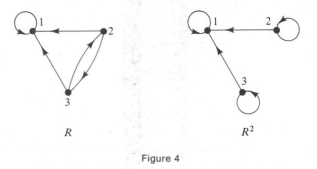

R $\qquad\qquad\qquad\qquad$ R^2

Figure 4

an arrow from x to y if a chain of two arrows of R [one from x to some u and one from u to y] leads from x to y. Thus $1 \to 1$ in R^2 because $1 \to 1 \to 1$ in R, $2 \to 1$ in R^2 because $2 \to 1 \to 1$ in R [also $2 \to 3 \to 1$ in R], $2 \to 2$ in R^2 because $2 \to 3 \to 2$ in R, etc. Note also that $2 \to 3$ is not true in R^2, even though $2 \to 3$ in R, since there is no chain of length 2 in R from 2 to 3.

The monoid $\mathscr{P}(S \times S)$ is almost never commutative.

EXAMPLE 7   Let $S = \{1, 2\}$ and let R_1 and R_2 be the relations with matrices

$$\mathbf{A}_1 = \begin{bmatrix} 1 & 1 \\ 1 & 0 \end{bmatrix} \quad \text{and} \quad \mathbf{A}_2 = \begin{bmatrix} 1 & 1 \\ 0 & 1 \end{bmatrix}.$$

Then $R_1 R_2 \neq R_2 R_1$ since

$$\mathbf{A}_1 * \mathbf{A}_2 = \begin{bmatrix} 1 & 1 \\ 1 & 0 \end{bmatrix} * \begin{bmatrix} 1 & 1 \\ 0 & 1 \end{bmatrix} = \begin{bmatrix} 1 & 1 \\ 1 & 1 \end{bmatrix}$$

$$\neq \begin{bmatrix} 1 & 1 \\ 1 & 0 \end{bmatrix} = \begin{bmatrix} 1 & 1 \\ 0 & 1 \end{bmatrix} * \begin{bmatrix} 1 & 1 \\ 1 & 0 \end{bmatrix} = \mathbf{A}_2 * \mathbf{A}_1.$$

To be explicit, this nonequality shows that $\langle 2, 2 \rangle$ belongs to $R_1 R_2$ but not to $R_2 R_1$. ∎

The next theorem formalizes the fact that we can study finite relations by studying their matrices.

Theorem 3   Let S be a finite set with n elements. There is a one-to-one correspondence between the set $\mathscr{P}(S \times S)$ of relations on S and the set of $n \times n$ Boolean matrices. This correspondence preserves the semigroup operations: if R_1, R_2 and R are relations with Boolean matrices $\mathbf{A}_1$, $\mathbf{A}_2$ and $\mathbf{A}$, then

$$R_1 R_2 = R \quad \text{if and only if} \quad \mathbf{A}_1 * \mathbf{A}_2 = \mathbf{A}.$$

We have proved all of this theorem except the first assertion, which should be obvious upon a moment's reflection.

It should be evident from the definition that composition is somehow related to transitivity. The next theorem shows how simple the connection is.

Theorem 4   If R is a relation on a set S, then R is transitive if and only if $R^2 \subseteq R$.

Proof. Suppose first that R is transitive and consider $\langle x, z \rangle \in R^2$. By definition of R^2 there exists $y \in S$ such that $\langle x, y \rangle \in R$ and $\langle y, z \rangle \in R$. Since R is transitive, $\langle x, z \rangle$ is also in R. We've shown that every $\langle x, z \rangle$ in R^2 is in R, i.e., $R^2 \subseteq R$.

For the converse, suppose that $R^2 \subseteq R$. Consider $\langle x, y \rangle$ and $\langle y, z \rangle$ in R. Then $\langle x, z \rangle$ is in R^2 and hence in R. This proves that R is transitive. ∎

For Boolean $m \times n$ matrices $\mathbf{A}_1$ and $\mathbf{A}_2$, let's write $\mathbf{A}_1 \leq \mathbf{A}_2$ if every entry of $\mathbf{A}_1$ is less than or equal to the corresponding entry of $\mathbf{A}_2$, that is,

$$\mathbf{A}_1[i, j] \leq \mathbf{A}_2[i, j] \quad \text{for} \quad 1 \leq i \leq m \quad \text{and} \quad 1 \leq j \leq n.$$

If R_1 and R_2 are relations from S to T, with matrices $\mathbf{A}_1$ and $\mathbf{A}_2$, then

$$R_1 \subseteq R_2 \quad \text{if and only if} \quad \mathbf{A}_1 \leq \mathbf{A}_2;$$

think about where the 1's and 0's are in $\mathbf{A}_1$ and $\mathbf{A}_2$ [Exercise 16]. In particular, a relation R on a set S satisfies $R^2 \subseteq R$ if and only if its matrix satisfies $\mathbf{A} * \mathbf{A} \leq \mathbf{A}$. So R is transitive if and only if $\mathbf{A} * \mathbf{A} \leq \mathbf{A}$.

EXAMPLE 8   Consider the relation R on $\{1, 2, 3\}$ with matrix

$$\mathbf{A} = \begin{bmatrix} 1 & 0 & 0 \\ 1 & 0 & 0 \\ 1 & 1 & 0 \end{bmatrix}.$$

Since

$$\mathbf{A} * \mathbf{A} = \begin{bmatrix} 1 & 0 & 0 \\ 1 & 0 & 0 \\ 1 & 1 & 0 \end{bmatrix} * \begin{bmatrix} 1 & 0 & 0 \\ 1 & 0 & 0 \\ 1 & 1 & 0 \end{bmatrix} = \begin{bmatrix} 1 & 0 & 0 \\ 1 & 0 & 0 \\ 1 & 0 & 0 \end{bmatrix} \leqq \begin{bmatrix} 1 & 0 & 0 \\ 1 & 0 & 0 \\ 1 & 1 & 0 \end{bmatrix} = \mathbf{A},$$

R is transitive. The transitivity of R can also be seen from its picture in Figure 5. Whenever a chain of two arrows connects x to y, a single arrow also does. For example, $3 \to 2 \to 1$ and also $3 \to 1$. Similarly, $2 \to 1 \to 1$ and also $2 \to 1$. ■

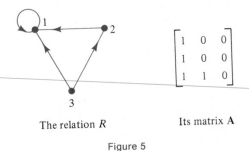

The relation R Its matrix $\mathbf{A}$

Figure 5

We next consider how inverses and compositions of relations interact. We begin with a

Warning. If R is a relation on a set S, then R^{-1} is also a relation on S. However, R^{-1} is not necessarily an inverse in the algebraic sense: RR^{-1} need not equal $R^{-1}R$ or E.

EXAMPLE 9   Let $S = \{1, 2\}$ and let R be the relation with matrix

$$\mathbf{A} = \begin{bmatrix} 1 & 1 \\ 0 & 0 \end{bmatrix}.$$

The matrix for R^{-1} is the transpose

$$\mathbf{A}^T = \begin{bmatrix} 1 & 0 \\ 1 & 0 \end{bmatrix}.$$

So the relations RR^{-1} and $R^{-1}R$ have matrices

$$\mathbf{A} * \mathbf{A}^T = \begin{bmatrix} 1 & 0 \\ 0 & 0 \end{bmatrix} \quad \text{and} \quad \mathbf{A}^T * \mathbf{A} = \begin{bmatrix} 1 & 1 \\ 1 & 1 \end{bmatrix}.$$

Thus $RR^{-1} \neq R^{-1}R$. Since E has matrix

$$\begin{bmatrix} 1 & 0 \\ 0 & 1 \end{bmatrix},$$

neither RR^{-1} nor $R^{-1}R$ equals E. ∎

In spite of Example 9, the inverse relation satisfies some familiar properties.

EXAMPLE 10 Let R_1 be a relation from S to T and R_2 a relation from T to U. Then $(R_1R_2)^{-1}$ and $R_2^{-1}R_1^{-1}$ are relations from U to S. We show that they are equal:

$$(R_1R_2)^{-1} = R_2^{-1}R_1^{-1}.$$

Suppose that $\langle z, x \rangle \in (R_1R_2)^{-1}$. Then $\langle x, z \rangle \in R_1R_2$ and so there exists $y \in T$ such that $\langle x, y \rangle \in R_1$ and $\langle y, z \rangle \in R_2$. It follows that $\langle z, y \rangle \in R_2^{-1}$ and $\langle y, x \rangle \in R_1^{-1}$, and hence $\langle z, x \rangle \in R_2^{-1}R_1^{-1}$. Thus every $\langle z, x \rangle$ in $(R_1R_2)^{-1}$ is in $R_2^{-1}R_1^{-1}$, and so $(R_1R_2)^{-1} \subseteq R_2^{-1}R_1^{-1}$. The reverse inclusion is proved by reversing the steps of the proof above. ∎

Throughout §§ 7.4 and 7.5 we have seen that Boolean matrices help us study finite relations. We will summarize the connections below. First we make two simple and sensible definitions. For Boolean $m \times n$ matrices $\mathbf{A}_1$ and $\mathbf{A}_2$ we define $\mathbf{A}_1 \vee \mathbf{A}_2$ by

$$(\mathbf{A}_1 \vee \mathbf{A}_2)[i, j] = \mathbf{A}_1[i, j] \vee \mathbf{A}_2[i, j] \quad \text{for} \quad 1 \leq i \leq m \quad \text{and} \quad 1 \leq j \leq n.$$

The matrix $\mathbf{A}_1 \wedge \mathbf{A}_2$ has a similar definition. For example, if

$$\mathbf{A}_1 = \begin{bmatrix} 1 & 0 & 1 & 1 \\ 0 & 1 & 1 & 0 \\ 1 & 1 & 0 & 1 \end{bmatrix} \quad \text{and} \quad \mathbf{A}_2 = \begin{bmatrix} 1 & 1 & 1 & 0 \\ 0 & 1 & 1 & 1 \\ 1 & 0 & 1 & 0 \end{bmatrix},$$

then

$$\mathbf{A}_1 \vee \mathbf{A}_2 = \begin{bmatrix} 1 & 1 & 1 & 1 \\ 0 & 1 & 1 & 1 \\ 1 & 1 & 1 & 1 \end{bmatrix} \quad \text{and} \quad \mathbf{A}_1 \wedge \mathbf{A}_2 = \begin{bmatrix} 1 & 0 & 1 & 0 \\ 0 & 1 & 1 & 0 \\ 1 & 0 & 0 & 0 \end{bmatrix}.$$

Summary. Let R be a relation on a finite set S with Boolean matrix $\mathbf{A}$. Then

- (R) R is reflexive if and only if all diagonal entries of $\mathbf{A}$ are 1;
- (AR) R is antireflexive if and only if all diagonal entries of $\mathbf{A}$ are 0;
- (S) R is symmetric if and only if $\mathbf{A} = \mathbf{A}^T$;
- (AS) R is antisymmetric if and only if $\mathbf{A} \wedge \mathbf{A}^T \leq \mathbf{I}$ where $\mathbf{I}$ is the identity matrix;
- (T) R is transitive if and only if $\mathbf{A} * \mathbf{A} \leq \mathbf{A}$.

Let R_1 and R_2 be relations from a finite set S to a finite set T with Boolean matrices A_1 and A_2. Then

(a) $R_1 \subseteq R_2$ if and only if $A_1 \leq A_2$;
(b) $R_1 \cup R_2$ has Boolean matrix $A_1 \vee A_2$;
(c) $R_1 \cap R_2$ has Boolean matrix $A_1 \wedge A_2$.

Finally, composition of relations corresponds to the Boolean product of their matrices as explained in Theorems 1 and 3.

EXERCISES 7.5

1. For each of the following Boolean matrices, consider the corresponding relation R on $\{1, 2, 3\}$. Find the Boolean matrix for R^2 and determine whether R is transitive.

(a) $\begin{bmatrix} 1 & 1 & 0 \\ 0 & 1 & 1 \\ 1 & 0 & 1 \end{bmatrix}$
(b) $\begin{bmatrix} 1 & 0 & 1 \\ 0 & 1 & 0 \\ 1 & 0 & 1 \end{bmatrix}$
(c) $\begin{bmatrix} 0 & 0 & 1 \\ 0 & 1 & 0 \\ 1 & 0 & 0 \end{bmatrix}$

2. Draw pictures of the relations in Exercise 1.

3. Let $S = \{1, 2, 3\}$ and $T = \{a, b, c, d\}$. Let R_1 and R_2 be the relations from S to T with the Boolean matrices

$$A_1 = \begin{bmatrix} 1 & 0 & 1 & 0 \\ 0 & 1 & 0 & 0 \\ 1 & 0 & 0 & 1 \end{bmatrix} \quad \text{and} \quad A_2 = \begin{bmatrix} 0 & 1 & 0 & 0 \\ 1 & 0 & 0 & 1 \\ 0 & 1 & 1 & 0 \end{bmatrix}.$$

(a) Find Boolean matrices for R_1^{-1} and R_2^{-1}.
(b) Find Boolean matrices for $(R_1 \cap R_2)R_1^{-1}$ and $R_1 R_1^{-1} \cap R_2 R_1^{-1}$.
(c) Find Boolean matrices for $R_2(R_1^{-1} \cup R_2^{-1})$ and $R_2 R_1^{-1} \cup R_2 R_2^{-1}$.
(d) Compare your answers to parts (b) and (c) with assertions in Exercise 11.

4. Let $S = \{1, 2, 3\}$ and $R = \{\langle 1, 1\rangle, \langle 1, 2\rangle, \langle 1, 3\rangle, \langle 3, 2\rangle\}$.
(a) Find the matrices for R, RR^{-1} and $R^{-1}R$.
(b) Draw pictures of the relations in part (a).
(c) Show that R is transitive, i.e., $R^2 \subseteq R$, but that $R^2 \neq R$.
(d) Is $R \cup R^{-1}$ transitive? Explain.
(e) Find R^n for all $n = 2, 3, \ldots$.

5. Let $S = \{1, 2, 3\}$ and $R = \{\langle 2, 1\rangle, \langle 2, 3\rangle, \langle 3, 2\rangle\}$.
(a) Find the matrices for R, R^{-1} and R^2.
(b) Draw pictures of the relations in part (a).
(c) Is R transitive?
(d) Is R^2 transitive?
(e) Is $R \cup R^2$ transitive?

6. Let R be the relation on $\{1, 2, 3\}$ with Boolean matrix

$$A = \begin{bmatrix} 0 & 1 & 0 \\ 1 & 1 & 1 \\ 0 & 1 & 0 \end{bmatrix}.$$

(a) Find the Boolean matrix for R^n for $n \in \mathbb{Z}$.

(b) Is R reflexive? symmetric? transitive?

7. Repeat Exercise 6 for

$$A = \begin{bmatrix} 1 & 0 & 0 \\ 0 & 1 & 1 \\ 1 & 0 & 1 \end{bmatrix}.$$

8. Let P be the set of all people and consider the relation R where $p \, R \, q$ if p "likes" q.

(a) Describe in words the relations $R \cap R^{-1}$, $R \cup R^{-1}$ and R^2.

(b) Is R reflexive? symmetric? transitive? Discuss.

9. Consider the functions f and g from $\{1, 2, 3, 4\}$ to itself defined by $f(m) = \max\{2, 4 - m\}$ and $g(m) = 5 - m$.

(a) Find the Boolean matrices A_f and A_g for the relations R_f and R_g corresponding to f and g.

(b) Find the Boolean matrices for $R_f R_g$ and $R_{f \circ g}$ and compare.

(c) Find the Boolean matrices for R_f^{-1} and R_g^{-1}. Do these relations correspond to functions?

10. Let R be a relation on a finite set S and consider its picture. How do you modify the picture to get the picture of R^{-1}?

11. Consider relations R_1 and R_2 from S to T and relations R_3 and R_4 from T to U.

(a) Show that $R_1(R_3 \cup R_4) = R_1 R_3 \cup R_1 R_4$.

(b) Show that $(R_1 \cap R_2)R_3 \subseteq R_1 R_3 \cap R_2 R_3$ and that equality need not hold.

(c) How are the relations $R_1(R_3 \cap R_4)$ and $R_1 R_3 \cap R_1 R_4$ related?

12. Let R_1 and R_2 be relations on a set S. Prove or disprove.

(a) If R_1 and R_2 are reflexive, so is $R_1 R_2$.

(b) If R_1 and R_2 are symmetric, so is $R_1 R_2$.

(c) If R_1 and R_2 are transitive, so is $R_1 R_2$.

13. Let R be the relation from $S = \{1, 2, 3, 4\}$ to $T = \{a, b, c\}$ with Boolean matrix

$$A = \begin{bmatrix} 1 & 0 & 1 \\ 0 & 0 & 1 \\ 1 & 0 & 0 \\ 0 & 1 & 0 \end{bmatrix}.$$

(a) Show that RR^{-1} is a symmetric relation on S.

(b) Show that $R^{-1}R$ is a symmetric relation on T.

(c) Are the relations RR^{-1} and $R^{-1}R$ equivalence relations?

14. Let R be a relation from a set S to a set T.

(a) Prove that RR^{-1} is a symmetric relation on S. Don't use Boolean matrices, since S or T might be infinite.

(b) Use part (a) to quickly infer that $R^{-1}R$ is a symmetric relation on T.

(c) When will RR^{-1} be reflexive?

15. Verify Example 10 for the case that S, T and U are finite, using Boolean matrices.

16. Let R_1 and R_2 be relations from $S = \{1, 2, \ldots, m\}$ to $T = \{1, 2, \ldots, n\}$, with matrices $\mathbf{A}_1$ and $\mathbf{A}_2$. Show that $R_1 \subseteq R_2$ if and only if $\mathbf{A}_1 \leq \mathbf{A}_2$.

17. Use the associative law for relations to prove that the Boolean product is an associative operation.

18. Let S be a set. Is $\mathscr{P}(S \times S)$ a group with inverses R^{-1}? Explain.

§ 7.6   Closures of Relations

Sometimes we may want to form new relations out of ones we already have. For example, we may have two equivalence relations R_1 and R_2 on S and want to find an equivalence relation containing them both. Since R_1 and R_2 are subsets of $S \times S$, the obvious candidate is $R_1 \cup R_2$. Unfortunately, $R_1 \cup R_2$ may not be an equivalence relation; the trouble is that $R_1 \cup R_2$ may not be transitive. Well then, what *is* the smallest transitive relation containing $R_1 \cup R_2$? This turns out to be a loaded question. How do we know there is such a relation? We will see in what follows that if R is a relation on a set S then there is always a smallest transitive relation containing R, which we will denote by $t(R)$, and we will learn how to find it. There are also smallest relations containing R that are reflexive and symmetric; we'll denote them by $r(R)$ and $s(R)$.

EXAMPLE 1   Consider the relation R on $\{1, 2, 3, 4\}$ whose Boolean matrix is

$$
\mathbf{A} = \begin{bmatrix} 0 & 0 & 1 & 1 \\ 0 & 1 & 0 & 0 \\ 0 & 0 & 1 & 0 \\ 1 & 0 & 0 & 0 \end{bmatrix}.
$$

See Figure 1 for the picture of R.

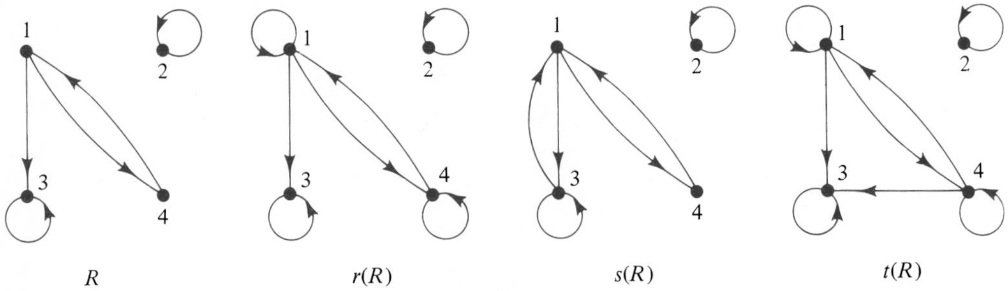

R $r(R)$ $s(R)$ $t(R)$

Figure 1

(a) The relation R is not reflexive, since neither 1 nor 4 is related to itself. To obtain the reflexive relation $r(R)$, all we need to do is add the two ordered pairs $\langle 1, 1 \rangle$ and $\langle 4, 4 \rangle$. The Boolean matrix $\mathbf{r(A)}$ of $r(R)$ is simply the matrix $\mathbf{A}$ with all the diagonal entries set equal to 1:

$$\mathbf{r(A)} = \begin{bmatrix} 1 & 0 & 1 & 1 \\ 0 & 1 & 0 & 0 \\ 0 & 0 & 1 & 0 \\ 1 & 0 & 0 & 1 \end{bmatrix}.$$

To get the picture for $r(R)$ in Figure 1, we simply added all the missing arrows from points to themselves.

(b) The relation R is not symmetric, since $1 \, R \, 3$ but $3 \, \not{R} \, 1$. If we add the ordered pair $\langle 3, 1 \rangle$ to R we get the symmetric relation $s(R)$. Its Boolean matrix is

$$\mathbf{s(A)} = \begin{bmatrix} 0 & 0 & 1 & 1 \\ 0 & 1 & 0 & 0 \\ 1 & 0 & 1 & 0 \\ 1 & 0 & 0 & 0 \end{bmatrix}.$$

To get the picture of $s(R)$ from the picture of R, we simply add the missing reverses of all the arrows.

(c) The relation R isn't transitive either. For example, we have $4 \, R \, 1$ and $1 \, R \, 3$ but $4 \, \not{R} \, 3$. The scheme for finding $t(R)$ [or its Boolean matrix $\mathbf{t(A)}$] is not so simple as those for $r(R)$ and $s(R)$. Since $4 \, R \, 1$ and $1 \, R \, 3$, $t(R)$ will also contain $\langle 4, 1 \rangle$ and $\langle 1, 3 \rangle$. Since $t(R)$ must be transitive, $t(R)$ must contain $\langle 4, 3 \rangle$, so we must put $\langle 4, 3 \rangle$ in $t(R)$. In general, if there is a chain from x to y, i.e., if there are points $x_1, x_2, \ldots, x_{m-1}$ so that

$$x \, R \, x_1, \; x_1 \, R \, x_2, \ldots, x_{m-1} \, R \, y,$$

then $\langle x, y \rangle$ must be in $t(R)$. If there is a chain from x to y and also one from y to z, then there is one from x to z. So the set of all pairs $\langle x, y \rangle$ connected by chains is a transitive relation and is the smallest transitive relation $t(R)$ containing R.

To get the picture of $t(R)$ in Figure 1 from the picture of R, we added an arrow connecting a point x to a point y whenever some sequence of arrows in R connected x to y and there wasn't an arrow from x to y already. For example, we added $4 \to 3$ because $4 \to 1 \to 3$ in R, and we added $1 \to 1$ because $1 \to 4 \to 1$ in R. ∎

The next proposition is nearly obvious. Think about why it's true before you read its proof.

Proposition   Let R be a relation. Then $R = r(R)$ if and only if R is reflexive, $R = s(R)$ if and only if R is symmetric and $R = t(R)$ if and only if R is transitive. Moreover,

$$r(r(R)) = r(R), \quad s(s(R)) = s(R), \quad \text{and} \quad t(t(R)) = t(R).$$

Proof. If R is reflexive then R is clearly the smallest reflexive relation containing R, i.e., $R = r(R)$. Conversely, if $R = r(R)$, then R is reflexive, since $r(R)$ is. Since $r(R)$ is reflexive, $r(R) = r(r(R))$ by what we have just shown. The proofs for $s(R)$ and $t(R)$ are similar. ∎

The next theorem gives explicit descriptions of the relations $r(R)$, $s(R)$ and $t(R)$, which are called the **reflexive**, **symmetric** and **transitive closures** of R.

Theorem 1   If R is a relation on a set S and if $E = \{\langle x, x \rangle : x \in S\}$, as usual, then

(r) $r(R) = R \cup E$;

(s) $s(R) = R \cup R^{-1}$;

(t) $t(R) = \displaystyle\bigcup_{k=1}^{\infty} R^k$.

Proof. (r) As noted in Example 4 of § 7.4, a relation is reflexive if and only if it contains E. Hence $R \cup E$ is reflexive and every reflexive relation that contains R must contain $R \cup E$. So $R \cup E$ is the smallest reflexive relation containing R. This shows that $r(R) = R \cup E$.

(s) Recall that a relation R_1 is symmetric if and only if $R_1^{-1} = R_1$ [Exercise 19 of § 7.4]. If $\langle x, y \rangle \in R \cup R^{-1}$, then $\langle y, x \rangle \in R^{-1} \cup R = R \cup R^{-1}$; thus $R \cup R^{-1}$ is symmetric. Consider any other symmetric relation R_1 that contains R. If $\langle x, y \rangle \in R^{-1}$ then $\langle y, x \rangle \in R \subseteq R_1$ and, since R_1 is symmetric, $\langle x, y \rangle \in R_1$. This shows that $R^{-1} \subseteq R_1$. Since $R \subseteq R_1$ we conclude that $R \cup R^{-1} \subseteq R_1$. This shows that $R \cup R^{-1}$ is the smallest symmetric relation containing R. Hence $s(R) = R \cup R^{-1}$.

(t) First we show that the union $U = \displaystyle\bigcup_{k=1}^{\infty} R^k$ is transitive. Consider x, y, z in S such that $\langle x, y \rangle \in U$ and $\langle y, z \rangle \in U$. Then we must have $\langle x, y \rangle \in R^k$ and $\langle y, z \rangle \in R^j$ for some k and j in $\mathbb{P}$. Then $\langle x, z \rangle$ belongs to $R^k R^j = R^{k+j}$, so that $\langle x, z \rangle \in U$. Thus U is a transitive relation containing R.

Now consider any transitive relation R_1 containing R. To show that $U \subseteq R_1$, we prove $R^k \subseteq R_1$ by induction. This inclusion is obvious for $k = 1$. If the inclusion holds for k then

$$R^{k+1} = R^k R \subseteq R_1 R_1 \subseteq R_1;$$

the last inclusion is valid because R_1 is transitive [Theorem 4 of § 7.5]. The principle of induction shows that $R^k \subseteq R_1$ for all $k \in \mathbb{P}$, and so $U \subseteq R_1$. Thus U is the smallest transitive relation containing R, and

$$t(R) = U = \bigcup_{k=1}^{\infty} R^k. \quad ∎$$

EXAMPLE 2   (a) Suppose that R is a relation on a set S with n elements and that $\mathbf{A}$ is its Boolean matrix. In Theorem 2 we'll learn that

$$t(R) = \bigcup_{k=1}^{n} R^k.$$

The results summarized in § 7.5 show that the Boolean matrices of $t(R)$, $s(R)$ and $r(R)$ are

$$\mathbf{t(A)} = \mathbf{A} \vee \mathbf{A}^2 \vee \cdots \vee \mathbf{A}^n,$$

$$\mathbf{s(A)} = \mathbf{A} \vee \mathbf{A}^T$$

and

$$\mathbf{r(A)} = \mathbf{A} \vee \mathbf{I},$$

where $\mathbf{I}$ is the $n \times n$ identity matrix. Since $\mathbf{t(A)}$, $\mathbf{s(A)}$ and $\mathbf{r(A)}$ determine the relations $t(R)$, $s(R)$ and $r(R)$, these equations provide algorithms for determining transitive, symmetric and reflexive closures. In § 8.9 we will obtain more efficient algorithms for calculating transitive closures.

(b) For the relation R back in Example 1, it is easy to see that $\mathbf{s(A)} = \mathbf{A} \vee \mathbf{A}^T$ and $\mathbf{r(A)} = \mathbf{A} \vee \mathbf{I}$ where $\mathbf{I}$ is the 4×4 identity matrix. One can also verify that

$$\mathbf{t(A)} = \mathbf{A} \vee \mathbf{A}^2 \vee \mathbf{A}^3 \vee \mathbf{A}^4.$$

Of course, for such a simple relation it's easier to find $t(R)$ by using the picture of R. ∎

Theorem 2   If R is a relation on a set S with n elements, then

$$t(R) = \bigcup_{k=1}^{n} R^k.$$

Proof. For motivation we can think of the picture of R. The pair $\langle x, y \rangle$ is in $t(R)$ if and only if there is a path from x to y in the picture. If there is such a path, there's one which doesn't go through the same vertex twice unless $x = y$. It can't involve more than n vertices, so it can't have length more than n.

Now let's write the argument using ordered pairs. Consider $\langle x, y \rangle$ in $t(R)$. If $\langle x, y \rangle \in R$ then clearly $\langle x, y \rangle$ is in $\bigcup_{k=1}^{n} R^k$. Otherwise, there is a chain $x_1, \ldots, x_{m-1}$, with $m \geq 2$, so that $x \, R \, x_1$, $x_1 \, R \, x_2$, $\ldots$, $x_{m-1} \, R \, y$. We can suppose that m is as small as possible for such a chain. Let $x_m = y$. If two of $x_1, \ldots, x_m$ are equal, say $x_i = x_j$ with $1 \leq i < j \leq m$, we can omit $x_i, \ldots, x_{j-1}$ and still get a chain from x to y, contrary to the minimal choice of m. Thus $x_1, \ldots, x_m$ are m different members of the n-element set S, so $m \leq n$. Thus

$$\langle x, y \rangle \in R^m \subseteq \bigcup_{k=1}^{n} R^k. \quad ∎$$

We can think of the mappings which take R to $r(R)$, $s(R)$ or $t(R)$ as closure operators in the sense of § 4.4. Repeating any of these three gives nothing new, but combining two or more of them can lead to other relations.

EXAMPLE 3   For the relation R in Example 1 we obtained

$$\mathbf{r(A)} = \begin{bmatrix} 1 & 0 & 1 & 1 \\ 0 & 1 & 0 & 0 \\ 0 & 0 & 1 & 0 \\ 1 & 0 & 0 & 1 \end{bmatrix} \quad \text{and} \quad \mathbf{s(A)} = \begin{bmatrix} 0 & 0 & 1 & 1 \\ 0 & 1 & 0 & 0 \\ 1 & 0 & 1 & 0 \\ 1 & 0 & 0 & 0 \end{bmatrix}.$$

The Boolean matrix for $sr(R)$ is

$$\mathbf{sr(A)} = \begin{bmatrix} 1 & 0 & 1 & 1 \\ 0 & 1 & 0 & 0 \\ 1 & 0 & 1 & 0 \\ 1 & 0 & 0 & 1 \end{bmatrix}.$$

This is also the matrix $\mathbf{rs(A)}$ for $rs(R)$ and so $rs(R) = sr(R)$. This is not an accident [Exercise 11]. The transitive closure of $sr(R) = rs(R)$ turns out to have matrix

$$\mathbf{tsr(A)} = \mathbf{trs(A)} = \begin{bmatrix} 1 & 0 & 1 & 1 \\ 0 & 1 & 0 & 0 \\ 1 & 0 & 1 & 1 \\ 1 & 0 & 1 & 1 \end{bmatrix}.$$

This is the matrix of the equivalence relation on $\{1, 2, 3, 4\}$ whose equivalence classes are $\{2\}$ and $\{1, 3, 4\}$. Thus $tsr(R)$ is transitive, symmetric and reflexive. This is not obvious from the notation tsr, as you might think. It is conceivable, for example, that applying the transitive closure to a symmetric relation might destroy its symmetry. Actually this does not happen, as we will explain in the lemma to Theorem 3, but applying the symmetric closure to a transitive relation can destroy its transitivity, as we show in the next example. ∎

EXAMPLE 4   Let R be the relation on $\{1, 2, 3\}$ with Boolean matrix

$$\mathbf{A} = \begin{bmatrix} 1 & 1 & 1 \\ 0 & 0 & 0 \\ 0 & 0 & 1 \end{bmatrix}.$$

Since $\mathbf{A} * \mathbf{A} = \mathbf{A}$, R is transitive. The relation $s(R)$ has matrix

$$\mathbf{s(A)} = \begin{bmatrix} 1 & 1 & 1 \\ 1 & 0 & 0 \\ 1 & 0 & 1 \end{bmatrix}$$

and this relation is not transitive. For example, $\langle 2, 1 \rangle$ and $\langle 1, 3 \rangle$ are in $s(R)$ but $\langle 2, 3 \rangle$ is not. ∎

In view of the next lemma, Example 4 illustrates the only way the closure operators r, s and t can destroy reflexivity, symmetry or transitivity; namely, the operator s can destroy transitivity.

Lemma   (a) If R is reflexive, so are $s(R)$ and $t(R)$.
(b) If R is symmetric, so are $r(R)$ and $t(R)$.
(c) If R is transitive, so is $r(R)$.

Proof. (a) This is obvious, because if $E \subseteq R$ then $E \subseteq s(R)$ and $E \subseteq t(R)$. Part (b) is left to Exercise 10.
(c) Suppose that R is transitive and consider $\langle x, y \rangle$ and $\langle y, z \rangle$ in $r(R) = R \cup E$. If $\langle x, y \rangle \in E$ then $x = y$ and so $\langle x, z \rangle = \langle y, z \rangle$ is in $R \cup E$. If $\langle y, z \rangle \in E$ then $y = z$ and so $\langle x, z \rangle = \langle x, y \rangle$ is in $R \cup E$. If neither $\langle x, y \rangle$ nor $\langle y, z \rangle$ is in E, then they are both in R and so $\langle x, z \rangle \in R \subseteq R \cup E$ by the transitivity of R. Hence $\langle x, z \rangle \in R \cup E$ in all cases. ∎

The next theorem answers the basic question with which we began this section.

Theorem 3   For any relation R on a set S, $tsr(R)$ is the smallest equivalence relation containing R.

Proof. Since $r(R)$ is reflexive, two applications of (a) of the lemma show that $tsr(R)$ is reflexive. Since $sr(R)$ is automatically symmetric, one application of (b) of the lemma shows that $tsr(R)$ is symmetric. Finally, $tsr(R)$ is automatically transitive, and so $tsr(R)$ is an equivalence relation.
Consider any equivalence relation R_1 such that $R \subseteq R_1$. Then $r(R) \subseteq r(R_1) = R_1$, hence $sr(R) \subseteq s(R_1) = R_1$ and thus $tsr(R) \subseteq t(R_1) = R_1$. Therefore $tsr(R)$ is the smallest equivalence relation containing R. ∎

EXAMPLE 5   (a) In Example 3, $tsr(R)$ was shown to be the equivalence relation with equivalence classes $\{2\}$ and $\{1, 3, 4\}$.
(b) Let R be the relation on $\{1, 2, 3\}$ in Example 4. Then

$$\mathbf{r}(\mathbf{A}) = \begin{bmatrix} 1 & 1 & 1 \\ 0 & 1 & 0 \\ 0 & 0 & 1 \end{bmatrix}, \mathbf{sr}(\mathbf{A}) = \begin{bmatrix} 1 & 1 & 1 \\ 1 & 1 & 0 \\ 1 & 0 & 1 \end{bmatrix}, \quad \mathbf{tsr}(\mathbf{A}) = \begin{bmatrix} 1 & 1 & 1 \\ 1 & 1 & 1 \\ 1 & 1 & 1 \end{bmatrix}.$$

The smallest equivalence relation containing R is the universal relation $\{1, 2, 3\} \times \{1, 2, 3\}$. These computations can be double-checked by drawing pictures for the corresponding relations. ∎

EXERCISES 7.6

1. Consider the relation R on $\{1,2,3\}$ with Boolean matrix $\mathbf{A} = \begin{bmatrix} 0 & 1 & 0 \\ 0 & 0 & 0 \\ 0 & 0 & 1 \end{bmatrix}$.

   Find the Boolean matrices for
   (a) $r(R)$ (b) $s(R)$ (c) $rs(R)$
   (d) $sr(R)$ (e) $tsr(R)$

2. Repeat Exercise 1 with $\mathbf{A} = \begin{bmatrix} 0 & 1 & 1 \\ 0 & 0 & 1 \\ 0 & 0 & 0 \end{bmatrix}$.

3. For Exercise 1, list the equivalence classes of $tsr(R)$.

4. For Exercise 2, list the equivalence classes of $tsr(R)$.

5. Repeat Exercise 1 for the relation R on $\{1,2,3,4\}$ with Boolean matrix

$$\mathbf{A} = \begin{bmatrix} 0 & 1 & 0 & 1 \\ 1 & 0 & 1 & 0 \\ 0 & 1 & 1 & 0 \\ 1 & 0 & 1 & 0 \end{bmatrix}.$$

6. For Exercise 5, list the equivalence classes of $tsr(R)$.

7. Let R be the usual quasi-order relation on $\mathbb{P}$: $m\,R\,n$ in case $m < n$. Find or describe
   (a) $r(R)$ (b) $sr(R)$ (c) $rs(R)$
   (d) $tsr(R)$ (e) $t(R)$ (f) $st(R)$

8. Repeat Exercise 7 where $m\,R\,n$ means that m divides n.

9. The Fraternal Order of Hostile Hermits is an interesting organization. Hermits know themselves. In addition, everyone knows the High Hermit but neither he nor any of the other members knows any other member. Define the relation R on the F.O.H.H. by $h_1\,R\,h_2$ if h_1 knows h_2. Determine $st(R)$ and $ts(R)$ and compare.

10. (a) Show that if (R_k) is a sequence of symmetric relations on a set S, then the union $\bigcup_{k=1}^{\infty} R_k$ is symmetric.
    (b) Let R be a symmetric relation on S. Show that R^n is symmetric for all $n \in \mathbb{P}$.
    (c) Show that if R is symmetric, so are $r(R)$ and $t(R)$.

11. Consider a relation R on a set S.
    (a) Show that $sr(R) = rs(R)$.
    (b) Show that $tr(R) = rt(R)$.

12. Show that $st(R) \neq ts(R)$ for the relation R in Example 4.

13. Show that there does not exist a smallest antireflexive relation containing the relation R on $\{1,2\}$ whose Boolean matrix is $\begin{bmatrix} 1 & 0 \\ 1 & 0 \end{bmatrix}$.

14. We say that a relation R on a set S is an **onto relation** if for every $y \in S$ there exists $x \in S$ such that $\langle x, y \rangle \in R$. Show that there does not exist a smallest onto relation containing the relation R on $\{1, 2\}$ specified in Exercise 13.

15. Suppose that a property p of relations on a nonempty set S satisfies:
   (i) the universal relation $S \times S$ has property p;
   (ii) p is **closed under intersections**, i.e., if $\{R_i : i \in I\}$ is a nonempty indexed family of relations on S possessing property p, then the intersection $\bigcap_{i \in I} R_i$ also possesses property p.

   (a) Prove that for every relation R there is a smallest relation that contains R and has property p.
   (b) Observe that the properties reflexivity, symmetry and transitivity satisfy both (i) and (ii).
   (c) Which of (i) and (ii) does antireflexivity fail to satisfy?
   (d) Which of (i) and (ii) does property "onto relation" fail to satisfy?

§ 7.7   The Lattice of Partitions

Consider the collection of all equivalence relations on a set S. This collection is partially ordered by inclusion. The minimal element is the equality relation E and the maximal element is the universal relation $U = S \times S$, since $E \subseteq R \subseteq U$ for all equivalence relations R on S. This collection is actually a lattice, as defined in § 7.1, if we let

$$R_1 \wedge R_2 = R_1 \cap R_2 \quad \text{and} \quad R_1 \vee R_2 = tsr(R_1 \cup R_2).$$

The definition of $R_1 \wedge R_2$ is straightforward, since the intersection of equivalence relations is again an equivalence relation [Exercise 16(d) of § 7.4] and so $R_1 \cap R_2$ is the largest equivalence relation contained in both R_1 and R_2. Similarly, $tsr(R_1 \cup R_2)$ is the smallest equivalence relation containing both R_1 and R_2. Note that $tsr(R_1 \cup R_2) = t(R_1 \cup R_2)$ since $R_1 \cup R_2$ is already reflexive and symmetric.

EXAMPLE 1   Consider again a sack S of marbles and the following two equivalence relations on S:

$$\langle s, t \rangle \in R_1 \text{ if } s \text{ and } t \text{ have the same color;}$$

$$\langle s, t \rangle \in R_2 \text{ if } s \text{ and } t \text{ are of the same size.}$$

Then $\langle s, t \rangle \in R_1 \wedge R_2$ if and only if s and t are of the same color *and* size. The pair $\langle s, t \rangle$ will be in $R_1 \vee R_2 = t(R_1 \cup R_2)$ if we can find a sequence of marbles $t_1, \ldots, t_{m-1}$ in S so that

$$\langle s, t_1 \rangle, \langle t_1, t_2 \rangle, \ldots, \langle t_{m-1}, t \rangle \quad \text{are all in} \quad R_1 \cup R_2.$$

For example, if the sack S includes the marbles pictured in Figure 1, then $\langle s, t \rangle$ is in $R_1 \vee R_2$. Actually, the consecutive pairs can be chosen to alternate between R_1 and R_2, so that marble t_2 was not needed. ∎

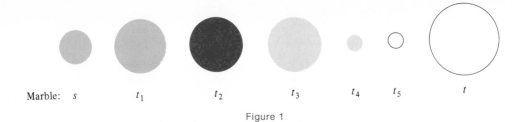

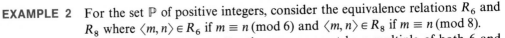

Figure 1

EXAMPLE 2   For the set $\mathbb{P}$ of positive integers, consider the equivalence relations R_6 and R_8 where $\langle m, n \rangle \in R_6$ if $m \equiv n \pmod 6$ and $\langle m, n \rangle \in R_8$ if $m \equiv n \pmod 8$.

(a) If $\langle m, n \rangle \in R_6 \wedge R_8$, then $m - n$ must be a multiple of both 6 and 8. This occurs precisely when $m - n$ is a multiple of their least common multiple 24. So $\langle m, n \rangle \in R_6 \wedge R_8$ if and only if $m \equiv n \pmod{24}$. With obvious notation, $R_6 \wedge R_8 = R_{24}$.

(b) We will show that $R_6 \vee R_8 = R_2$, where $\langle m, n \rangle \in R_2$ if $m \equiv n \pmod 2$. Note that 2 is the greatest common divisor of 6 and 8. Since $R_6 \cup R_8 \subseteq R_2$ and R_2 is an equivalence relation, we have $R_6 \vee R_8 \subseteq R_2$. We show that

$$R_2 \subseteq R_6 \vee R_8 = t(R_6 \cup R_8).$$

First we note that

$(*)$ $\langle k, k + 2 \rangle \in R_6 \vee R_8$ for all $k \in \mathbb{P}$,

since both $\langle k, k + 8 \rangle$ and $\langle k + 8, k + 2 \rangle$ are in $R_6 \cup R_8$. Now consider $\langle m, n \rangle \in R_2$ with, say, $m < n$. Then $n - m$ is even, so that $n = m + 2r$ for some $r \in \mathbb{P}$. Now by $(*)$ all the pairs

$$\langle m, m + 2 \rangle, \langle m + 2, m + 4 \rangle, \dots, \langle m + 2r - 2, m + 2r \rangle$$

are in $R_6 \vee R_8$. By transitivity, $\langle m, m + 2r \rangle = \langle m, n \rangle$ is also in $R_6 \vee R_8$. Every $\langle m, n \rangle$ in R_2 is in $R_6 \vee R_8$, and so $R_2 \subseteq R_6 \vee R_8$. ∎

Theorem 1 of § 7.3 described a one-to-one correspondence between the set of equivalence relations on a set S and the set $\Pi(S)$ of all partitions of S. Statements about equivalence relations on S correspond to statements about partitions of S and conversely we can translate partition statements into relation statements. We have just seen that the set of equivalence relations forms a lattice. There is a corresponding lattice structure on $\Pi(S)$, which we now describe.

Consider equivalence relations R_1 and R_2 with corresponding partitions π_1 and π_2. Then $R_1 \subseteq R_2$ if and only if every two R_1-related elements are also R_2-related, i.e., if and only if two members of the same R_1-class always belong to the same R_2-class. Thus $R_1 \subseteq R_2$ if and only if each set in π_1 is a subset of some set in π_2; we say π_1 **refines** π_2 and write $\pi_1 \precsim \pi_2$ in this case. See Figure 2. The relation $\precsim$ is a partial order on $\Pi(S)$ [Exercise 13] and $\Pi(S)$ becomes a lattice with $\pi_1 \wedge \pi_2$ and $\pi_1 \vee \pi_2$ corresponding to

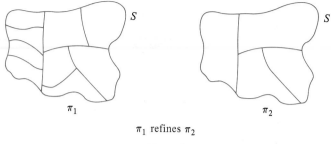

π_1 refines π_2

Figure 2

$R_1 \cap R_2$ and $R_1 \vee R_2$ respectively. The partition $\pi_1 \wedge \pi_2$ is easy to find; it consists of all nonempty sets obtained by intersecting a set in π_1 with a set in π_2. Just as in the case of equivalence relations, it is usually less clear what the partition $\pi_1 \vee \pi_2$ corresponding to $R_1 \vee R_2$ is.

EXAMPLE 3   (a) For the sack of marbles in Example 1, each set in the partition $\pi_1 \wedge \pi_2$ consists of all marbles of a particular color and size. The nature of $\pi_1 \vee \pi_2$ depends on just what marbles are in S and how they are related. It might just consist of the set S by itself. Some possibilities are given in Exercises 1 through 4.

(b) The partition $\pi_6 \wedge \pi_8$ of $\mathbb{P}$ corresponding to $R_6 \wedge R_8 = R_{24}$ in Example 2 consists of the equivalence classes determined by the congruence $m \equiv n \pmod{24}$. There are 24 sets in the partition, one that contains 1, one that contains 2, etc. If this isn't obvious, reread Example 9 of § 7.3.

In this case, the partition $\pi_6 \vee \pi_8$ corresponding to $R_6 \vee R_8$ is also easy to describe, because we have already shown that $R_6 \vee R_8 = R_2$. The corresponding partition of $\mathbb{P}$ consists of two equivalence classes: the set of even numbers in $\mathbb{P}$ and the set of odd numbers in $\mathbb{P}$. ∎

We devote the rest of the section to presenting algorithms for finding $\pi_1 \wedge \pi_2$ and $\pi_1 \vee \pi_2$ when S is finite. For definiteness, consider a partition π of $S = \{1, 2, \ldots, n\}$. For each set A in π we select a fixed element m_A and define $\alpha(k) = m_A$ for $k \in A$; for example, m_A might be taken to be the smallest number in A. Each member of S belongs to some A in π so we obtain a function $\alpha : S \to S$ satisfying:

(1) $\alpha(j) = \alpha(k)$ if and only if j and k belong to the same set in π;

(2) $\alpha(\alpha(k)) = \alpha(k)$ for all k.

For each $k \in S$ the set in π which contains k is the set which contains $\alpha(k)$, so it is $\alpha^{-1}(\alpha(k))$. Thus π is determined by α. If R denotes the equivalence relation for which π is the set of equivalence classes, property (1) asserts:

(1′) $\alpha(j) = \alpha(k)$ if and only if $j \, R \, k$.

Thus α is a function of the sort described in Theorem 2 of § 7.3.

EXAMPLE 4   Let R be an equivalence relation on $S = \{1, 2, 3, \ldots, 10\}$ whose partition π of equivalence classes is $\{\{1, 4, 6\}, \{2\}, \{3, 7, 10\}, \{5, 9\}, \{8\}\}$. The function α that selects the smallest number in each class is given by

k	1	2	3	4	5	6	7	8	9	10
$\alpha(k)$	1	2	3	1	5	1	3	8	5	3

Note that α satisfies (1) and (2). Another function α' that works is given by

k	1	2	3	4	5	6	7	8	9	10
$\alpha'(k)$	4	2	3	4	9	4	3	8	9	3

For partitions π_1 and π_2 of $S = \{1, 2, \ldots, n\}$, let α and β be corresponding functions satisfying (1) and (2). Note that π_1 refines π_2 provided

$$\alpha(i) = \alpha(j) \quad \text{implies} \quad \beta(i) = \beta(j) \quad \text{for all} \quad i, j \in S.$$

We seek the corresponding functions for $\pi_1 \wedge \pi_2$ and $\pi_1 \vee \pi_2$. The first algorithm provides the function γ for $\pi_1 \wedge \pi_2$. It goes through the elements of S one at a time. Whenever it comes to an element which is in a new $\pi_1 \wedge \pi_2$-block it uses that element as the γ-label for the new block.

Algorithm INTERSECT PARTITIONS.

Step 1. Set $\gamma(k) = 0$ for $k = 1, 2, \ldots, n$.
Step 2. Choose $k = 1$.
Step 3. If $\gamma(k) \neq 0$ go to Step 4. Otherwise, for each $j = k, k + 1, \ldots, n$ satisfying $\alpha(j) = \alpha(k)$ and $\beta(j) = \beta(k)$ change $\gamma(j)$ to k.
Step 4. If $k = n$, stop. Otherwise replace k by $k + 1$ and go to Step 3.

After a little experimentation or thought, it is evident that this algorithm works, i.e., produces γ for $\pi_1 \wedge \pi_2$. However, writing out a careful justification is a bit tedious; see Exercise 15. The algorithm clearly takes no worse than $O(n^2)$ time; with a suitable choice of data structures it can be made much faster than that.

EXAMPLE 5   Let π_1 and π_2 be the partitions of $\{1, 2, 3, \ldots, 8\}$ with corresponding functions α and β as follows:

j	1	2	3	4	5	6	7	8
$\alpha(j)$	3	2	3	2	3	7	7	2
$\beta(j)$	5	4	4	4	5	5	5	4

Thus $\pi_1 = \{\{1, 3, 5\}, \{2, 4, 8\}, \{6, 7\}\}$; π_2 consists of two sets. In Table 1 we illustrate Algorithm INTERSECT PARTITIONS step by step. The partition $\pi_1 \wedge \pi_2$ can be read from the last line of Table 1; it consists of four sets.

TABLE 1

k	$\gamma(1)$	$\gamma(2)$	$\gamma(3)$	$\gamma(4)$	$\gamma(5)$	$\gamma(6)$	$\gamma(7)$	$\gamma(8)$
0	0	0	0	0	0	0	0	0
1	1	0	0	0	1	0	0	0
2	1	2	0	2	1	0	0	2
3	1	2	3	2	1	0	0	2
4	1	2	3	2	1	0	0	2
5	1	2	3	2	1	0	0	2
6	1	2	3	2	1	6	6	2
7	1	2	3	2	1	6	6	2
8	1	2	3	2	1	6	6	2

The next algorithm builds a function γ that describes the partition $\pi_1 \vee \pi_2$. The algorithm is quite simple, but its operation may seem mysterious. The idea is to grow γ-classes, i.e., equivalence classes described by the function γ, by taking unions of α-classes until eventually each β-class is contained entirely in a γ-class. The α-classes are contained in γ-classes all along, so when the algorithm stops the γ-classes must contain the classes for $\pi_1 \vee \pi_2$. The algorithm only merges α-classes when it is forced to by β, so the final γ-classes are just barely big enough, and must actually be just the $\pi_1 \vee \pi_2$-classes. The algorithm goes through the list of elements of S. For each element k, it looks at the β-representative $\beta(k)$. When the algorithm comes to an element k which does not live in the same γ-block as its representative $\beta(k)$ it throws the whole γ-block of k into the γ-block containing $\beta(k)$. In particular, then, the α-block of k ends up in the γ-block of $\beta(k)$.

Algorithm MERGE PARTITIONS.

Step 1. For $i = 1, 2, \ldots, n$ set $\gamma(i) = \alpha(i)$.
Step 2. For $k = 1, 2, \ldots, n$ if $\gamma(k) \neq \gamma(\beta(k))$ then find all j with $\gamma(j) = \gamma(k)$ and change $\gamma(j)$ to $\gamma(\beta(k))$ for each such j. ∎

This algorithm, like INTERSECT PARTITIONS, runs in time at worst $O(n^2)$, and much faster with suitable data structures to describe α, β and γ.

EXAMPLE 6   Let π_1 and π_2 be partitions with functions α and β as follows:

j	1	2	3	4	5	6	7	8
$\alpha(j)$	1	2	5	4	5	1	7	4
$\beta(j)$	3	4	3	4	5	6	6	8

Then π_1 and π_2 each have 5 sets, as indicated by the squared-off α-sets and rounded β-sets in Figure 3. We list the values for the operation of the algorithm in Table 2, and show pictorially how it works in Figure 4.

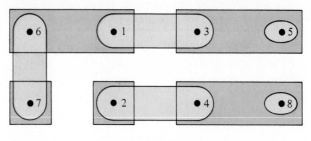

Figure 3

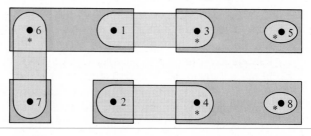

(a)

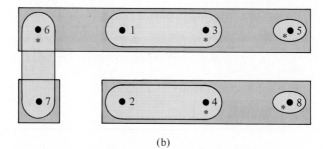

(b)

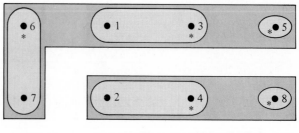

(c)

Figure 4

TABLE 2

k	$\gamma(1)$	$\gamma(2)$	$\gamma(3)$	$\gamma(4)$	$\gamma(5)$	$\gamma(6)$	$\gamma(7)$	$\gamma(8)$	
	1	2	5	4	5	1	7	4	[given α]
1	5	2	5	4	5	5	7	4	
2, 3, 4, 5, 6	5	4	5	4	5	5	7	4	
7, 8	5	4	5	4	5	5	5	4	

Since $\gamma(\beta(1)) = \alpha(3) = 5 \neq \gamma(1)$, each $\gamma(j)$ with $\gamma(j) = \gamma(1) = 1$ is changed to $\gamma(\beta(1)) = 5$ in line $k = 1$. The algorithm has seen that 1 and $\beta(1) = 3$ aren't α-related, so it merges their α-classes as shown in Figure 4(a), where squared-off blocks are γ-classes and stars mark the elements $\beta(k)$. When $k = 2$ the algorithm notes that 2 and $\beta(2) = 4$ aren't yet γ-related, so it merges their γ-classes. Figure 4(b) shows the new blocks. In Table 2 the 2 in the $k = 1$ row changes to a 4. Nothing more happens until $k = 7$, when the algorithm sees that 7 and $\beta(7) = 6$ are not yet γ-related. The final merge is shown in Figure 4(c). Notice that each β-class is now entirely inside a γ-class. ∎

Theorem   Algorithm MERGE PARTITIONS works.

Proof. We restate the algorithm for ease of reference.

Step 1. For $i = 1, 2, \ldots, n$ set $\gamma(i) = \alpha(i)$.
Step 2. For $k = 1, 2, \ldots, n$ if $\gamma(k) \neq \gamma(\beta(k))$ then find all j with $\gamma(j) = \gamma(k)$ and change $\gamma(j)$ to $\gamma(\beta(k))$ for each such j.

For convenience, denote the final function γ by γ^*. Notice first that the only place where $\gamma(j)$ can change during the execution of the algorithm is in Step 2. At that point, if $\gamma(i) = \gamma(j)$ then $\gamma(i)$ and $\gamma(j)$ both change to the same thing, namely $\gamma(\beta(k))$. Thus

(1)   if $\gamma(i)$ and $\gamma(j)$ are ever equal at some stage then they are equal from then on, so $\gamma^*(i) = \gamma^*(j)$.

We next observe that

(2) $\gamma^*(\beta(j)) = \gamma^*(j)$ for every j,

for if $\gamma(\beta(j)) = \gamma(j)$ when we reach Step 2 with $k = j$ then $\gamma^*(\beta(j)) = \gamma^*(j)$ by (1), while if $\gamma(\beta(j)) \neq \gamma(j)$ at that stage then Step 2 leaves $\gamma(\beta(j))$ unchanged and changes $\gamma(j)$ to $\gamma(\beta(j))$ on the spot.

Now if $\alpha(i) = \alpha(j)$ for some i and j then $\gamma(i) = \gamma(j)$ at the start, by Step 1, so $\gamma^*(i) = \gamma^*(j)$ by (1). If $\beta(i) = \beta(j)$ then $\gamma^*(i) = \gamma^*(\beta(i)) = \gamma^*(\beta(j)) = \gamma^*(j)$ by (2). Thus

(3)   if $\alpha(i) = \alpha(j)$ *or*   if $\beta(i) = \beta(j)$ then $\gamma^*(i) = \gamma^*(j)$,

which says that π_1 and π_2 both refine the partition π given by γ^*. To show that $\pi = \pi_1 \vee \pi_2$ we only need to show that π refines $\pi_1 \vee \pi_2$, i.e., that

$$(4) \qquad\qquad \text{if}\quad \gamma^*(i) = \gamma^*(j) \quad\text{then}\quad \langle i, j \rangle \in R$$

for the equivalence relation R corresponding to the partition $\pi_1 \vee \pi_2$.

To show (4) it's enough to show

$$(5) \qquad\qquad \langle j, \gamma^*(j) \rangle \in R \quad\text{for every}\quad j,$$

since if (5) holds and if $\gamma^*(i) = \gamma^*(j)$ then $\langle i, \gamma^*(i) \rangle \in R$ and $\langle \gamma^*(j), j \rangle \in R$ by symmetry of R, so $\langle i, j \rangle \in R$ by transitivity of R.

To show (5), finally, it's enough to show

$$(6) \qquad\qquad \langle j, \gamma(j) \rangle \in R \quad\text{for every}\quad j$$

at all times. Since γ is α at the start, (6) holds then. Assume that (6) holds at some time, and consider a particular j. If $\gamma(j)$ now changes, the change is from $\gamma(k)$ to $\gamma(\beta(k))$ for some k. We want to show that in this case $\langle j, \gamma(\beta(k)) \rangle \in R$. Now $\langle j, \gamma(j) \rangle = \langle j, \gamma(k) \rangle$ before the change, so (6) says that at that time $\langle j, \gamma(k) \rangle \in R$ and also that $\langle k, \gamma(k) \rangle \in R$ and $\langle \beta(k), \gamma(\beta(k)) \rangle \in R$. Since R is symmetric, $\langle \gamma(k), k \rangle \in R$. Thus R contains the chain $\langle j, \gamma(k) \rangle$, $\langle \gamma(k), k \rangle$, $\langle k, \beta(k) \rangle$, $\langle \beta(k), \gamma(\beta(k)) \rangle$. By transitivity, R contains $\langle j, \gamma(\beta(k)) \rangle$ as we wanted to show. [An inductive argument is in the background here, of course.] ▮

EXERCISES 7.7

1. Suppose that the sack in Example 3 has ten marbles: 6 small green ones, 3 large red ones and 1 large green one. Describe $\pi_1 \wedge \pi_2$ and $\pi_1 \vee \pi_2$. How many sets are in each of these partitions?

2. How would your answers to Exercise 1 change if the single large green marble were lost?

3. Repeat Exercise 1 if the sack has ten marbles: 4 small yellow ones, 3 medium blue ones, 2 medium white ones and 1 large yellow one.

4. How would your answer to Exercise 3 change if 1 large blue marble were dropped into the sack?

5. Consider the equivalence relations R_3 and R_5 on $\mathbb{P}$ where $\langle m, n \rangle \in R_3$ if $m \equiv n$ (mod 3) and $\langle m, n \rangle \in R_5$ if $m \equiv n$ (mod 5), with corresponding partitions π_3 and π_5.
   (a) Describe the equivalence relation $R_3 \wedge R_5$.
   (b) Describe the partition $\pi_3 \wedge \pi_5$.
   (c) It turns out [Exercise 6] that $R_3 \vee R_5$ is the universal relation on $\mathbb{P}$, so all numbers in $\mathbb{P}$ are related to each other. Verify that $\langle 1, 2 \rangle$, $\langle 1, 3 \rangle$, $\langle 1, 73 \rangle$, $\langle 47, 73 \rangle$ and $\langle 72, 73 \rangle$ are in $R_3 \vee R_5$.
   (d) Describe the partition $\pi_3 \vee \pi_5$.

6. Prove that the relation $R_3 \vee R_5$ in Exercise 5 is the universal relation.

7. For each partition below of $\{1, 2, 3, 4, 5, 6\}$ give a function α satisfying conditions (1) and (2) stated prior to Example 4.
   (a) $\pi_1 = \{\{1, 3, 5\}, \{2, 6\}, \{4\}\}$
   (b) $\pi_2 = \{\{1, 2, 4\}, \{3, 6\}, \{5\}\}$
   (c) $\pi_3 = \{\{1\}, \{2\}, \{3\}, \{4\}, \{5\}, \{6\}\}$
   (d) $\pi_4 = \{\{1, 2, 3, 4, 5, 6\}\}$
   (e) What equivalence relation corresponds to π_3?
   (f) What equivalence relation corresponds to π_4?

8. Give the partitions $\pi_1, \pi_2, \pi_3, \pi_4$ of $\{1, 2, 3, \ldots, 8\}$ defined by the functions $\alpha_1, \alpha_2,$ α_3 and α_4 below:

k	1	2	3	4	5	6	7	8
$\alpha_1(k)$	1	1	3	1	5	6	3	5
$\alpha_2(k)$	2	2	6	8	5	6	7	8
$\alpha_3(k)$	4	4	3	4	5	3	3	4
$\alpha_4(k)$	3	2	3	8	2	3	7	8

9. Use the algorithms of this section to find functions corresponding to the partitions $\pi_1 \wedge \pi_2$ and $\pi_1 \vee \pi_2$ where π_1 and π_2 are as in Exercise 8.

10. Repeat Exercise 9 for π_3 and π_4.

11. Repeat Exercise 9 for π_2 and π_3.

12. Would Algorithm MERGE PARTITIONS work just as well if the roles of α and β were interchanged?

13. (a) Show that the relation $\leq$ defined on $\Pi(S)$ by $\pi_1 \leq \pi_2$ if and only if π_1 refines π_2 is a partial order on $\Pi(S)$.
   (b) Show that if π_1, π_2 and π_3 are in $\Pi(S)$ and if $\pi_3 \leq \pi_1$ and $\pi_3 \leq \pi_2$ then $\pi_3 \leq \pi_1 \wedge \pi_2$.

14. Analyze the algorithms INTERSECT PARTITIONS and MERGE PARTITIONS in case π_1 refines π_2 by considering the example where $S = \{1, 2, 3, 4, 5, 6, 7\}$ and

k	1	2	3	4	5	6	7
$\alpha(k)$	1	4	3	4	1	6	7
$\beta(k)$	5	4	5	4	5	4	7

15. Verify that Algorithm INTERSECT PARTITIONS works by showing the following.
   (a) The value of $\gamma(j)$ changes at least once for each j during execution of the algorithm.
   (b) If the value of $\gamma(j)$ changes when $k = k_0$ and if $k_0 \leq k'$ with $\alpha(k') = \alpha(j)$ and $\beta(k') = \beta(j)$ then $\gamma(k')$ changes to k_0.

(c) The value of each $\gamma(j)$ changes exactly once during the execution.

(d) If $0 \neq \gamma(i) = \gamma(j)$ then $\alpha(i) = \alpha(j)$ and $\beta(i) = \beta(j)$.

(e) If $\alpha(i) = \alpha(j)$ and $\beta(i) = \beta(j)$, then $\gamma(i) = \gamma(j)$ at the conclusion of the algorithm.

CHAPTER HIGHLIGHTS

See the end of Chapter 0 for a reminder on how to use these lists to review. Think of examples.

Concepts

binary relation on S or from S to T
  reflexive, antireflexive, symmetric, antisymmetric, transitive [for relations on S]
  inverse relation R^{-1}
  composite $R_2 \circ R_1 = R_1 R_2$
  matrix of a relation
partial order, poset, subposet
  quasi-order
  Hasse diagram
  maximal, minimal, largest, smallest elements
  lattice
  chain = totally ordered set = linearly ordered set
  product order on $S_1 \times \cdots \times S_n$
  filing order on $S_1 \times \cdots \times S_n$
  standard order on Σ^*
  lexicographic = dictionary order on Σ^*
equivalence relation
  equivalence class $[s]$
  natural mapping $s \to [s]$
reflexive, symmetric and transitive closures
$R_1 \wedge R_2$, $R_1 \vee R_2$
refine [partitions]

Facts

Composition of relations is associative.

The matrix of a composite $R_1 R_2$ of relations is the Boolean product $\mathbf{A}_1 * \mathbf{A}_2$ of their matrices.

Matrix analogues of relation statements are given in the summary at the end of § 7.5.

Filing order on $S_1 \times \cdots \times S_n$ is linear if each S_i is a chain.

If Σ is a chain then standard order on Σ^* is a well-ordering and lexicographic order is linear but not a well-ordering.

A partition is essentially the same thing as the set of all equivalence classes for an equivalence relation.

Functions define equivalence relations on their domains.

$$r(R) = R \cup E, \ s(R) = R \cup R^{-1} \text{ and } t(R) = \bigcup_{k=1}^{\infty} R^k \ [= \bigcup_{k=1}^{n} R^k \text{ if } |S| = n].$$

The smallest equivalence relation containing R is $tsr(R)$.

Methods

INTERSECT PARTITIONS to give $\pi_1 \wedge \pi_2$.
MERGE PARTITIONS to give $\pi_1 \vee \pi_2$.

8

GRAPHS

Chapter 0 contained an informal introduction to graphs. We discussed and illustrated a number of ideas, but didn't prove theorems or justify algorithms. In this chapter we take another, closer look at the subject and fill in the arguments to support our conclusions. In Chapter 0 we avoided the graphs involving arrows, the so-called directed graphs. This time we begin by studying directed graphs—ones like flowcharts in which the connecting lines have directions associated with them. Then we give the connecting lines "weights," and also consider what happens if we leave off the directions. A road map with mileages as weights can be used to find the shortest highway distance between two points. In § 8.6 we discuss questions such as finding a shortest route for a traveling salesperson. Section 8.7 links graphs with relations and matrices, and §§ 8.8 and 8.9 describe some graph-theoretic algorithms.

This chapter is logically independent of Chapter 0, in the sense that we will develop everything anew. In particular, all the definitions will be repeated here at the appropriate places. We will now stress the theory, however, since motivation has been provided in Chapter 0.

§ 8.1 Directed Graphs

The essential features of a directed graph [digraph for short] are its objects and directed lines. Specifically, a **digraph** G consists of two sets, the non-empty set $V(G)$ of **vertices** of G and the set $E(G)$ of **edges** of G, together with a function γ [Greek lowercase gamma] from $E(G)$ to $V(G) \times V(G)$. If e is an edge of G and $\gamma(e) = \langle p, q \rangle$, then p is called the **initial vertex** of e and q the **terminal vertex** of e and we say e **goes from** p to q. This definition makes sense if $V(G)$ or $E(G)$ is infinite, but because our applications are to finite sets we will assume in this chapter that $V(G)$ and $E(G)$ are finite.

A **picture** of the digraph G is a diagram consisting of points corresponding to the members of $V(G)$ and arrows corresponding to the members of $E(G)$ such that if $\gamma(e) = \langle p, q \rangle$ then the arrow corresponding to e goes from the point labeled p to the point labeled q.

EXAMPLE 1 Consider the digraph G with $V(G) = \{w, x, y, z\}$, $E(G) = \{a, b, c, d, e, f, g, h\}$ and γ given by the table in Figure 1(a). The diagrams in Figures 1(b) and 1(c) are both pictures of G. In Figure 1(b) we labeled the arrows to make the correspondence to $E(G)$ plain. In Figure 1(c) we simply labeled the points and let the arrows take care of themselves. This causes no confusion because, in this case, there are no **parallel edges**, i.e., there is at most one edge with a given initial vertex and terminal vertex. In other words, the function γ is one-to-one. Note also that we omitted the arrow head on edge d since z is clearly both the initial and terminal vertex. ∎

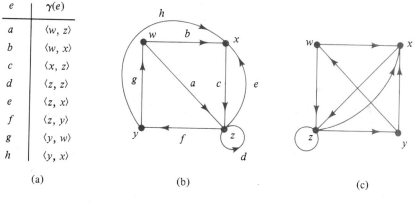

e	$\gamma(e)$
a	$\langle w, z \rangle$
b	$\langle w, x \rangle$
c	$\langle x, z \rangle$
d	$\langle z, z \rangle$
e	$\langle z, x \rangle$
f	$\langle z, y \rangle$
g	$\langle y, w \rangle$
h	$\langle y, x \rangle$

(a)   (b)   (c)

Figure 1

If $\gamma : E(G) \to V(G) \times V(G)$ is one-to-one, then we can identify the edges e with their images $\gamma(e)$ in $V(G) \times V(G)$ and consider $E(G)$ to *be* a subset of $V(G) \times V(G)$. In fact, some people define digraphs to have $E(G) \subseteq V(G) \times V(G)$ and call the more general digraphs we are considering "directed multigraphs."

Given a picture of G we can reconstruct G itself since the arrows tell us all about γ. We will commonly describe digraphs by giving pictures of them rather than tables of γ but the pictorial description is chosen just for human convenience. A computer stores a digraph by storing the function γ in one way or another.

Many of the important questions connected with digraphs can be stated in terms of sequences of edges leading from one vertex to another. A **path** in a digraph G is a sequence of edges such that the terminal vertex of one edge is the initial vertex of the next. Thus if $e_1, \ldots, e_n$ are in $E(G)$, then $e_1 e_2 \cdots e_n$ is a path provided there are vertices $x_1, x_2, \ldots, x_n, x_{n+1}$ so that $\gamma(e_i) = \langle x_i, x_{i+1} \rangle$ for $i = 1, 2, \ldots, n$. We say that $e_1 e_2 \cdots e_n$ is a path of **length** n **from** x_1 **to** x_{n+1}. The path is **closed** if $x_1 = x_{n+1}$.

EXAMPLE 2   In the digraph G in Figure 1 the sequence $f\,g\,a\,e$ is a path of length 4 from z to x. The sequences $c\,e\,c\,e\,c$ and $f\,g\,a\,f\,h\,c$ are also paths, but $f\,a$ is not a path since $\gamma(f) = \langle z, y \rangle$, $\gamma(a) = \langle w, z \rangle$ and $y \neq w$. The paths $f\,g\,a\,f\,h\,c$, $c\,e\,c\,e$ and d are closed; $f\,h\,c\,e$ and $d\,f$ are not. ∎

A path $e_1 \cdots e_n$ with $\gamma(e_i) = \langle x_i, x_{i+1} \rangle$ has an associated sequence of vertices $x_1 x_2 \cdots x_n x_{n+1}$. If each e_i is the only edge from x_i to x_{i+1} then this sequence of vertices uniquely determines the path, and we could describe the path by listing the vertices in succession.

EXAMPLE 3   (a) In Figure 1 the path $f\,g\,a\,e$ has vertex sequence $z\,y\,w\,z\,x$. Observe that this vertex sequence alone determines the path. The path can be recovered from $z\,y\,w\,z\,x$ by looking at Figure 1(b) or 1(c) or using the table of γ in Figure 1(a). Since the digraph has no parallel edges, all its paths are determined by their vertex sequences.

(b) For the digraph pictured in Figure 2 the vertex sequence $y\,z\,z\,z$ corresponds only to the path $f\,g\,g$ but the sequence $y\,v\,w\,z$ belong to both $c\,a\,e$ and $c\,b\,e$. ∎

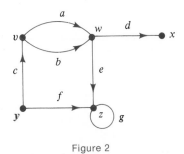

Figure 2

A closed path of length at least 1 with vertex sequence $x_1 x_2 \cdots x_n x_1$ is called a **cycle** if $x_1, \ldots, x_n$ are all different. The language of graph theory has not been standardized: various authors use "circuit" and "loop" for what we call a cycle, and "cycle" is sometimes used as a name for a closed path. A digraph with no cycles is called **acyclic**. Some people call an acyclic digraph a

DAG, short for "directed acyclic graph." A path is **acyclic** if the digraph consisting of the vertices and edges of the path is acyclic.

EXAMPLE 4   In Figure 1 the path $a\,f\,g$ is a cycle since its vertex sequence is $w\,z\,y\,w$. Likewise, the paths $c\,f\,h$ and $c\,f\,g\,b$, with vertex sequences $x\,z\,y\,x$ and $x\,z\,y\,w\,x$, are cycles. The short path $c\,e$ and the loop d are also cycles, since their vertex sequences are $x\,z\,x$ and $z\,z$ respectively. The path $c\,f\,g\,a\,e$ is not a cycle, since its vertex sequence is $x\,z\,y\,w\,z\,x$ and the vertex z is repeated. ∎

EXAMPLE 5   Hasse diagrams may be thought of as digraphs. Consider a poset $(P, \preceq)$ and let H be the digraph with $V(H) = P$ and with an edge from x to y whenever y covers x. Figure 3 shows two pictures of H for the poset $(\{1, 2, 3, 4, 5, 6\}, |)$. The picture on the left is correct but seems less helpful than the one on the right.

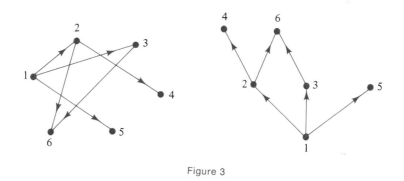

Figure 3

A path in H has a vertex sequence $x_1\, x_2 \cdots x_{n+1}$ in which x_2 covers x_1, x_3 covers x_2, etc., and so $x_1 \prec x_2 \prec \cdots \prec x_{n+1}$. By transitivity and antisymmetry $x_1 \prec x_{n+1}$, so in particular $x_1 \neq x_{n+1}$ and the path is not closed. Hence H is acyclic. ∎

Theorem 1   If u and v are different vertices of a digraph G, and if there is a path in G from u to v, then there is an acyclic path from u to v.

Proof. Among all paths from u to v consider one of smallest length, say having vertex sequence $x_1 \cdots x_n\, x_{n+1}$ with $x_1 = u$ and $x_{n+1} = v$. Suppose that $x_i = x_j$ for some i and j with $1 \leq i < j \leq n+1$. Then the path $x_i\, x_{i+1} \cdots x_j$ from x_i to x_j is closed [see Figure 4 for an illustration] and the path $x_1 \cdots x_i x_{j+1} \cdots x_{n+1}$ obtained by omitting this part still goes from u to v. Since $x_1 \cdots x_n x_{n+1}$ had smallest length, this shorter path is impossible. We conclude that x_i and x_j must be different if $i \neq j$. Thus the path under

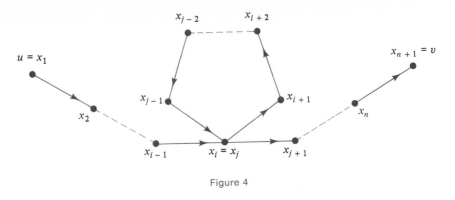

Figure 4

consideration is acyclic. [This is essentially the argument that showed $t(R) \subseteq \bigcup_{k=1}^{n} R^k$ in the proof of Theorem 2 of § 7.6.] ∎

Corollary 1   If there is a closed path from v to v then there is a cycle from v to v.

Proof. If there is an edge e of the graph from v to v, then the one-element sequence e is a cycle from v to v. Otherwise there is a closed path from v to v having the form $v\, x_2 \cdots x_n\, v$ where $x_n \neq v$. Then by Theorem 1 there is an acyclic path from v to x_n. Tacking on the last edge from x_n to v gives the desired cycle. ∎

Corollary 2   A path is acyclic if and only if all its vertices are distinct.

Proof. If a path has no repeated vertex, then it is surely acyclic. If a path has a repeated vertex, then it contains a closed path, so by Corollary 1 it contains a cycle. ∎

If G is an acyclic digraph we can define a natural quasi-order on $V(G)$ by defining $u \prec v$ if there is a path from u to v. This quasi-order gives us a way of comparing vertices of G. In fact, we can do a bit more. We can label the vertices with integers so that smaller vertices have smaller labels. Before we discuss such a labeling we need an observation. Call a vertex of a digraph a **sink** if it is not an initial vertex of any edge. Sinks correspond to points with no arrows leading away from them.

Lemma   Every finite acyclic digraph has at least one sink.

First proof. Since the digraph is acyclic, every path in it is acyclic. Since the digraph is finite, the path lengths are bounded and there must be a path of largest length, say $v_1 v_2 \cdots v_n$. Then v_n must be a sink. [Of course, if the digraph has no edges at all, every vertex is a sink.] ∎

This proof is short and elegant, but it doesn't tell us how to find v_n or any other sink. Our next argument is constructive.

Second proof. Choose any vertex v_1. If v_1 is a sink, we are done. If not, there is an edge from v_1 to some v_2. If v_2 is a sink, we are done. If not, etc. We obtain in this way a sequence $v_1, v_2, v_3, \ldots$ such that $v_1 v_2 \cdots v_k$ is a path for each k. As in the first proof, such paths cannot be arbitrarily long, so at some stage we reach a sink. ∎

Here is an algorithm based on the construction in the second proof, which returns a sink when it is applied to a finite acyclic digraph G. The algorithm uses **immediate successor** sets $\text{SUCC}(v)$ defined by $\text{SUCC}(v) = \{u \in V(G) : \text{there is an edge from } v \text{ to } u\}$. These data sets $\text{SUCC}(v)$ would be supplied in the description of the digraph G when the algorithm is carried out.

Algorithm SINK.

Step 1. Choose v in $V(G)$.
Step 2. If $\text{SUCC}(v) = \varnothing$ go to Step 3. Otherwise choose u in $\text{SUCC}(v)$,
     replace v by u and repeat Step 2.
Step 3. Let $\text{SINK}(G) = v$. Stop. ∎

EXAMPLE 6 Consider the acyclic digraph G shown in Figure 5. The immediate successor sets are $\text{SUCC}(t) = \{u, w, x\}$, $\text{SUCC}(u) = \{v\}$, $\text{SUCC}(v) = \varnothing$, $\text{SUCC}(w) = \{y\}$, $\text{SUCC}(x) = \{y\}$, $\text{SUCC}(y) = \varnothing$ and $\text{SUCC}(z) = \{w\}$. One possible sequence of choices using algorithm SINK on G is t, w, y. Others starting with t are t, x, y and t, u, v. A different first choice could lead to z, w, y. We could even get lucky and choose a sink first time. In any case the value $\text{SINK}(G)$ returns is either v or y. ∎

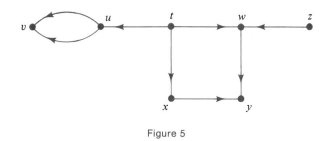

Figure 5

The time algorithm SINK takes is proportional to the number of vertices it chooses before it gets to a sink, so for a digraph with n vertices the algorithm runs in time $O(n)$.

The proof of the next theorem gives one reason for considering sinks.

Theorem 2   An acyclic digraph with n vertices can have its vertices numbered from 1 to n in such a way that $i < j$ whenever there is a path from vertex i to vertex j.

A digraph numbered in the way Theorem 2 describes is said to be **topologically sorted**. For short, we will call the labeling of its vertices a **sorted labeling**. Note that in a sorted labeling $i < j$ does not mean there has to be a path from vertex i to vertex j.

Proof of Theorem 2. We use induction on n, and note that the assertion is obvious for $n = 1$. Assume inductively that acyclic digraphs with fewer than n vertices can be labeled as described, and consider an acyclic digraph G with n vertices. By the last lemma G has a sink, say s. Give s the number n. Form a new graph H with $V(H) = V(G) \backslash \{s\}$ and with edges all those edges of G which do not have s as a vertex. Since G has no cycles, H has no cycles. Since H has only $n - 1$ vertices, by the inductive assumption we can number H with $\{1, 2, \ldots, n - 1\}$ in the way described in the theorem. The vertex s is numbered n, so every vertex in $V(G)$ has a number.

Now suppose there is a path in G from vertex i to vertex j. If the path lies entirely in H then $i < j$ since H is properly numbered. Otherwise, some vertex along the path is s, and since s is a sink it must be the last vertex, vertex j. But then $j = n$, so $i < j$ in this case too. Hence G, with n vertices, has its vertices labeled as in the statement of the theorem. The Principle of Mathematical Induction now shows that the theorem holds for all n. ∎

The idea in the proof of Theorem 2 can be developed into a procedure for constructing sorted labelings.

Algorithm NUMBERING VERTICES.

Step 1.   If $V(G) = \varnothing$, stop. Otherwise, find a sink of G and label it with $|V(G)|$, the number of vertices of G.

Step 2.   Remove the sink and all its attached edges. Replace G by what remains, and go to Step 1. ∎

Each pass through Step 2 removes one vertex from the original set of vertices $V(G)$, so the algorithm must stop, and when it stops each vertex is labeled. Each time the algorithm goes through Step 1 the label goes down by 1, so the labels used are $n, n - 1, \ldots, 2, 1$. In Step 1 we could use algorithm SINK to find the sink of G, in effect calling SINK as a subroutine. You may find it instructive to apply this algorithm to number the graph of Figure 5. Also see Exercise 18 for a procedure which begins numbering with 1.

We can estimate the running time of NUMBERING VERTICES, assuming that it calls SINK to find sinks as needed. Suppose G has n

vertices. When NUMBERING VERTICES first does Step 1 it calls SINK, which takes time $O(n)$, and it attaches a label to a vertex. Then the algorithm goes to Step 2 and removes the labeled vertex and at most $n - 1$ edges. [We assume here that G has no parallel edges.] This step also takes time $O(n)$. The combined time for the two steps is thus $O(n)$. The algorithm repeats this two-step sequence a total of n times, with $n, n - 1, n - 2, \ldots, 1$ vertices, so the total time is at most $nO(n)$, which is $O(n^2)$.

The fact that the number of vertices keeps going down obviously has some significance to the actual running time, but because

$$n + (n - 1) + \cdots + 2 + 1 = n(n + 1)/2 > n^2/2$$

the big oh estimate $O(n^2)$ is as good as we can get. Even if we could somehow cut down the time to remove edges in Step 2, the call to SINK in Step 1 would keep the overall time no better than $O(n^2)$. As a matter of fact, the time it takes to write a label on a vertex depends on how many digits the label has, but this time is at worst $O(n)$, so the time to execute SINK is the most important term in the overall performance time.

Sinks correspond to points with no arrows leading away from them. Points with no arrows leading into them are also special. We call a vertex of a digraph a **source** if it is not a terminal vertex of any edge. Facts and algorithms about sinks have analogues for sources, obtained by reversing all the arrows. In particular, every finite acyclic digraph has at least one source [Exercise 17].

The second proof of the lemma for Theorem 2 consisted of constructing a path from a given vertex v to a sink. We call a vertex u **reachable from** v in G if there is a path of length at least 1 in G from v to u, and we define

$$R(v) = \{u \in V(G) : u \text{ is reachable from } v\}.$$

Then $R(v) = \varnothing$ if and only if v is a sink, and the lemma's second proof showed in effect that in an acyclic digraph each nonempty set $R(v)$ contains at least one sink.

Even if G is not acyclic, the sets $R(v)$ may be important. As we shall see in § 8.7, determining all sets $R(v)$ amounts to finding the transitive closure of a certain relation. In § 8.9 we will study algorithms for finding the $R(v)$'s, as well as for answering other graph-theoretic questions.

One consequence of Theorem 2 is that every finite poset has a Hasse diagram. The discussion in Example 5 showed that if $(P, \leqq)$ is a poset and H is the digraph with $V(H) = P$ and with an edge from x to y whenever y covers x then H is acyclic. We can use Theorem 2 to label H with $\{1, 2, \ldots, n\}$, and then draw a picture of H so that the height of the point corresponding to a vertex increases as its label increases. Since each edge points from a lower-numbered vertex to a higher-numbered one, any edge from an x to a y which covers x points upward. That is, the picture of H is a Hasse diagram for $(P, \leqq)$.

Essentially the same reasoning shows that every acyclic digraph has a picture in which the arrows all go in more or less the same direction, for instance from left to right or from top to bottom. The digraph in Figure 6(b) is acyclic and all its arrows point generally from left to right.

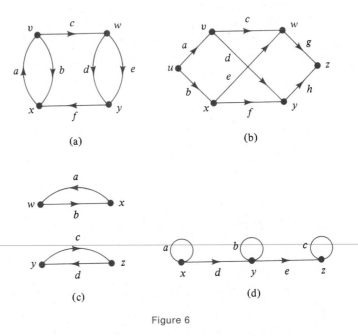

(a)   (b)

(c)   (d)

Figure 6

EXAMPLE 7 According to Theorem 2 it is possible to number college courses so that all prerequisites for a course have lower numbers than the course itself. ∎

EXERCISES 8.1

1. Give a table of the function γ for each of the digraphs pictured in Figure 6.

2. Draw a picture of the digraph G with $V(G) = \{w, x, y, z\}$, $E(G) = \{a, b, c, d, e, f, g\}$ and γ given by the following table.

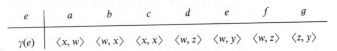

e	a	b	c	d	e	f	g
$\gamma(e)$	$\langle x, w \rangle$	$\langle w, x \rangle$	$\langle x, x \rangle$	$\langle w, z \rangle$	$\langle w, y \rangle$	$\langle w, z \rangle$	$\langle z, y \rangle$

3. Which of the following vertex sequences describe paths in the digraph pictured in Figure 7(a)?
   (a) $z\ y\ v\ w\ t$ (b) $x\ z\ w\ t$ (c) $v\ s\ t\ x$
   (d) $z\ y\ s\ u$ (e) $x\ z\ y\ v\ s$ (f) $s\ u\ x\ t$

4. Find the length of a shortest path from x to w in the digraph shown in Figure 7(a).

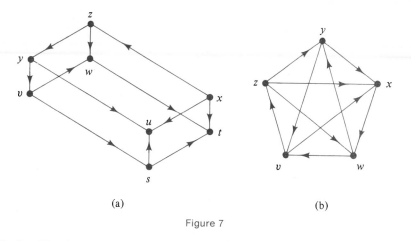

(a)   (b)

Figure 7

5. Consider the digraph pictured in Figure 7(b). Describe an acyclic path
   (a) from x to y (b) from y to z
   (c) from v to w (d) from x to z
   (e) from z to v

6. There are four basic blood types: A, B, AB and O. Type O can donate to any of
   the four types, A and B can donate to AB as well as to their own types but type AB
   can only donate to AB. Draw a digraph which presents this information. Is the
   digraph acyclic?

7. (a) Give the immediate successor sets SUCC(v) for all vertices in the digraph
       shown in Figure 8.
   (b) What value for SINK(G) does an initial choice of vertex w give?
   (c) What sinks of G are in $R(x)$?

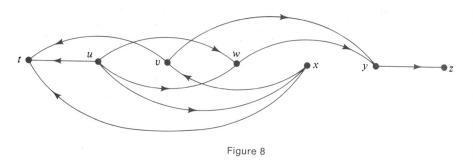

Figure 8

8. Give an example of a digraph with vertices x, y and z in which there is a cycle with
   x and y as vertices and another cycle with y and z, but there is no cycle with x and
   z as vertices.

9. Consider the digraph G pictured in Figure 9.
   (a) Find $R(v)$ for each vertex v in $V(G)$.
   (b) Find all sinks of G.
   (c) Is G acyclic?

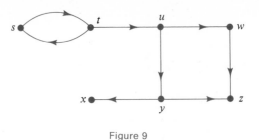

Figure 9

10. Does Algorithm SINK work on digraphs which have cycles? Explain.

11. Consider a digraph G with the following immediate successor sets: $\text{SUCC}(r) =$ $\{s, u\}$, $\text{SUCC}(s) = \varnothing$, $\text{SUCC}(t) = \{r, w\}$, $\text{SUCC}(u) = \varnothing$, $\text{SUCC}(w) = \{r, t, x, y\}$, $\text{SUCC}(x) = \varnothing$, $\text{SUCC}(y) = \{w, z\}$ and $\text{SUCC}(z) = \varnothing$.
    (a) Draw a picture of such a digraph.
    (b) Do these sets $\text{SUCC}(v)$ determine $E(G)$ uniquely? Explain.
    (c) Find all sinks in G.
    (d) Find paths from the vertex w to three different sinks in the digraph.

12. A **tournament** is a digraph in which every two vertices have exactly one edge between them. [Think of $\langle x, y \rangle$ as an edge provided x defeats y.]
    (a) Give an example of a tournament with 4 vertices.
    (b) Show that a tournament cannot have two sinks.
    (c) Can a tournament with a cycle have a sink? Explain.
    (d) Would you like to be the sink of a tournament?

13. Let G be an acyclic digraph. Show that the relation $\prec$, defined by $u \prec v$ if there is a path from u to v, is a quasi-order on $V(G)$.

14. Let G be a digraph and define the relation $\sim$ on $V(G)$ by $x \sim y$ if $x = y$ or if x is reachable from y and y is reachable from x.
    (a) Show that $\sim$ is an equivalence relation.
    (b) Find the equivalence classes for the digraph pictured in Figure 9.
    (c) Describe the relation $\sim$ in the case that G is acyclic.

15. (a) Show that in Theorem 1 and Corollary 1 the path without repeated vertices can be constructed from edges of the given path. Thus every closed path contains at least one cycle.
    (b) Show that if u and v are vertices of a digraph and if there is a path from u to v, then there is a path from u to v in which no edge is repeated. [Consider the case $u = v$, as well as $u \neq v$.]

16. Let G be a digraph.
    (a) Show that if u is reachable from v then $R(u) \subseteq R(v)$.
    (b) Give an alternate proof of the lemma for Theorem 2 by choosing v in $V(G)$ with $|R(v)|$ as small as possible.
    (c) Does your proof in part (b) lead to a useful constructive procedure? Explain.

17. The **reverse** of a digraph G is the digraph $\hat{G}$ obtained by reversing all the arrows of G. That is, $V(\hat{G}) = V(G)$, $E(\hat{G}) = E(G)$ and if $\gamma(e) = \langle x, y \rangle$ then $\hat{\gamma}(e) = \langle y, x \rangle$.

(a) Use $\hat{G}$ and the lemma for Theorem 2 to show that if G is acyclic [and finite] then G has a source.

(b) Find all sources in the digraphs of Figures 5 through 9.

18. (a) Modify Algorithm NUMBERING VERTICES by using sources instead of sinks to produce an algorithm that numbers $V(G)$ in increasing order.

(b) Use your algorithm to number the digraph of Figure 5.

19. Let H be the Hasse diagram of a poset $(P, \leq)$.

(a) Using the terminology of this section, describe the elements of H corresponding to the minimal elements of P.

(b) Answer part (a) with "maximal" instead of "minimal."

20. (a) Suppose a finite acyclic digraph has just one sink. Show that there is a path to the sink from each vertex.

(b) What is the corresponding statement for sources?

21. Show that if a digraph G with n vertices consists of a single chain from a source to a sink then algorithm SINK may make as many as n choices of vertices before it stops.

§ 8.2   Isomorphisms and Invariants

Digraphs are like grains of sand on the beach. A little experimentation shows that the number of different digraphs we can build with just a few vertices and edges is quite large, and in fact this number becomes astronomical in a hurry as we increase the numbers of vertices and edges allowed. If $|V(G)| = n$ and $|E(G)| = m$, then $|V(G) \times V(G)| = n^2$ and there are $(n^2)^m$ functions γ from $E(G)$ to $V(G) \times V(G)$, each of which determines a digraph. For $n = 4$ and $m = 5$ this number is already over a million.

On the other hand, several different functions may determine essentially the same digraph. If we had some way to sift through the million-odd digraphs we get from our $(4^2)^5$ functions and could eliminate essential duplicates, we might hope to get a more manageable list which would still have one sample of each of the kinds of digraphs with 4 vertices and 5 edges. As a practical matter this plan will fail, if not for 4 and 5 then for larger numbers, but it is still a helpful approach when we are looking for digraphs with special properties, as in Exercise 10.

EXAMPLE 1   Figure 1 shows pictures of two digraphs G_1 and G_2 with the same vertex set $\{w, x, y, z\}$ and the same edge set $\{d, f, g, h, i\}$, but described by different functions γ_1 and γ_2, defined as follows.

	d	f	g	h	i
γ_1	$\langle w, x \rangle$	$\langle w, x \rangle$	$\langle w, y \rangle$	$\langle y, y \rangle$	$\langle z, y \rangle$
γ_2	$\langle z, w \rangle$	$\langle z, x \rangle$	$\langle x, x \rangle$	$\langle y, x \rangle$	$\langle z, w \rangle$

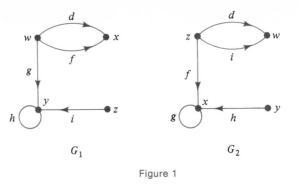

Figure 1

We can see from their pictures that the two digraphs are essentially the same, except that they have been labeled differently. It is clear from Figure 1 how to relabel G_1 to make it into G_2. We can describe the relabeling with the pair of functions $\alpha: \{w, x, y, z\} \rightarrow \{w, x, y, z\}$ and $\beta: \{d, f, g, h, i\} \rightarrow \{d, f, g, h, i\}$ given by the following tables.

v	w	x	y	z		e	d	f	g	h	i
$\alpha(v)$	z	w	x	y		$\beta(e)$	d	i	f	g	h

If we replace each vertex v in the table defining γ_1 by $\alpha(v)$ and each edge e by $\beta(e)$, then starting with

	d	f	g	h	i
γ_1	$\langle w, x \rangle$	$\langle w, x \rangle$	$\langle w, y \rangle$	$\langle y, y \rangle$	$\langle z, y \rangle$

we get the table

d	i	f	g	h
$\langle z, w \rangle$	$\langle z, w \rangle$	$\langle z, x \rangle$	$\langle x, x \rangle$	$\langle y, x \rangle$

which is in fact just the defining table for γ_2, with the entries in another order. ∎

We say that two digraphs G and H are **isomorphic** [pronounced eye-so-MOR-fik] and we write $G \simeq H$ if there are one-to-one correspondences $\alpha: V(G) \rightarrow V(H)$ and $\beta: E(G) \rightarrow E(H)$ such that whenever an edge e of $E(G)$ joins vertices u and v of $V(G)$, the corresponding edge $\beta(e)$ joins the corresponding points $\alpha(u)$ and $\alpha(v)$ in $V(H)$. Figure 2 illustrates the concept. The left-hand arrow is part of the picture of G if and only if the corresponding right-hand arrow is in the picture of H. In symbols, if G and H are described by γ_1 and γ_2, respectively, and if $\gamma_1(e) = \langle u, v \rangle$, then $\gamma_2(\beta(e)) = \langle \alpha(u), \alpha(v) \rangle$. Two isomorphic digraphs are essentially the same except for the

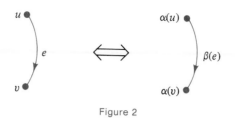

Figure 2

labeling of their vertices and edges. Generally speaking, two sets with some mathematical structure are said to be **isomorphic** if there exist one-to-one correspondences between them which preserve [i.e., are compatible with] the structure. For example, two semigroups $(S, \square)$ and $(T, \triangle)$ are isomorphic if there is a one-to-one correspondence $\phi : S \to T$ such that $\phi(s_1 \square s_2) = \phi(s_1) \triangle \phi(s_2)$ for all s_1, s_2 in S.

It follows from the definition that $G \simeq G$ for every digraph G, and if $G \simeq H$ with correspondences α and β, then α^{-1} and β^{-1} are also one-to-one and so $H \simeq G$. If $G \simeq H$ and $H \simeq K$ then $G \simeq K$ [Exercise 14]. Thus if $\mathscr{S}$ is a set of digraphs, the relation $\simeq$ is an equivalence relation on $\mathscr{S}$. The equivalence classes are called **isomorphism classes**.

If G and H have no parallel edges, we've seen that we can consider $E(G)$ to be a subset of $V(G) \times V(G)$, and $E(H)$ a subset of $V(H) \times V(H)$. In this situation the isomorphism condition is particularly simple: G is isomorphic to H if there is a one-to-one correspondence $\alpha : V(G) \to V(H)$ such that $\langle u, v \rangle$ is in $E(G)$ if and only if $\langle \alpha(u), \alpha(v) \rangle$ is in $E(H)$. Such an α is called an **isomorphism** of G onto H.

EXAMPLE 2 Consider the digraphs G and H pictured in Figure 3, with

$$E(G) = \{\langle w, x \rangle, \langle w, z \rangle, \langle x, y \rangle, \langle x, z \rangle, \langle y, w \rangle, \langle z, x \rangle, \langle z, y \rangle\}$$

and

$$E(H) = \{\langle p, q \rangle, \langle p, r \rangle, \langle q, s \rangle, \langle r, p \rangle, \langle r, q \rangle, \langle s, p \rangle, \langle s, r \rangle\}.$$

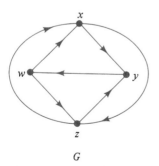

G

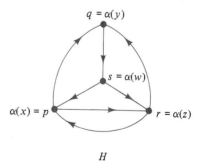

H

Figure 3

It may not be obvious, but these digraphs are isomorphic. To see this, define α by $\alpha(w) = s$, $\alpha(x) = p$, $\alpha(y) = q$, $\alpha(z) = r$. The result of replacing each vertex in the description of $E(G)$ by its image under α is

$$\{\langle s, p \rangle, \langle s, r \rangle, \langle p, q \rangle, \langle p, r \rangle, \langle q, s \rangle, \langle r, p \rangle, \langle r, q \rangle\},$$

which is $E(H)$. Thus $G \simeq H$. The function α we used is not the only possible choice. ∎

Given two digraphs, how do we determine whether they are isomorphic? In general we would probably look first for obvious differences or for the kind of obvious similarity we saw in Figure 1. The problem may not be easy to solve. In fact, one of the main messages of this section could be phrased "The graph isomorphism problem is difficult."

Figure 4 shows pictures of four digraphs. The pictures all look quite different, so we might try to find ways in which the digraphs themselves are essentially different. A human being could attack the problem of the digraphs in Figure 4 in a variety of ways and could shift from one approach to another; a computer program or algorithm must be spelled out in advance and might not be so flexible. One simpleminded algorithm we could write would simply label the vertices of the two candidate digraphs $v_1, v_2, \ldots, v_8$ and $w_1, w_2, \ldots, w_8$, label their edges $e_1, e_2, \ldots, e_{12}$ and $f_1, f_2, \ldots, f_{12}$, and examine each pair of functions

$$\alpha : \{v_1, \ldots, v_8\} \to \{w_1, \ldots, w_8\} \quad \text{and} \quad \beta : \{e_1, \ldots, e_{12}\} \to \{f_1, \ldots, f_{12}\}$$

to see if α and β meet the isomorphism conditions. This would involve $(8!)\,(12!) \approx 1.93 \cdot 10^{13}$ examinations. The computation could be speeded up considerably by taking one α at a time and rejecting whole bunches of β's as impossible for a given α. For example, if e_1 goes from v_1 to v_2 and if $\alpha(v_1) = w_3$ and $\alpha(v_2) = w_7$, we can reject β unless $\beta(e_1)$ goes from w_3 to w_7, without looking at what β does to the other edges. Even this quicker

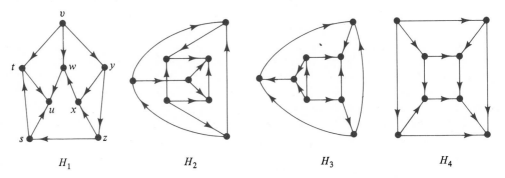

$$H_1 \qquad\qquad H_2 \qquad\qquad H_3 \qquad\qquad H_4$$

Figure 4

algorithm would have to go through all 8! possible α's before concluding that the two digraphs are not isomorphic. If they *were* isomorphic it might also take a while to find an isomorphism by this exhaustive process.

At the present time no isomorphism-detecting algorithm is known which is qualitatively much faster than the simpleminded scheme we just considered. To fully explain this last sentence would require the introduction of technical material which we have no other use for here, but the basic idea is this: The time it takes known algorithms to check for isomorphism of two given digraphs with n vertices can be comparable to $n!$ and in particular is not bounded by some polynomial in n. Polynomial-time algorithms are known to exist for some special classes of digraphs, however. Work in complexity theory has shown that the existence of a polynomial-time algorithm for digraph isomorphism is equivalent to the existence of polynomial-time algorithms to solve a number of other important types of problems, so the search for digraph isomorphism algorithms is an active one.

What the last paragraph means in our present context is that we should not be surprised if checking isomorphism or nonisomorphism turns out to be difficult in specific cases.

One way we could hope to discover quickly that two digraphs are not isomorphic is to show that they differ in some aspect which would be the same for isomorphic graphs. For instance, if $G_1 \simeq G_2$ then obviously $|V(G_1)| = |V(G_2)|$ and $|E(G_1)| = |E(G_2)|$. Thus if our given digraphs have different numbers of vertices or edges, they are surely not isomorphic. A quantity, such as number of vertices or number of edges, is called an **isomorphism invariant** if its value is the same for any two isomorphic digraphs. Invariants are especially useful for showing that digraphs are *not* isomorphic. A set of invariants is **complete** if whenever two digraphs are not isomorphic there is at least one invariant in the set with different values for the two digraphs. A complete set of invariants would let us show that two digraphs *are* isomorphic by showing that each invariant is the same for both digraphs. Unfortunately, no complete set of invariants is known.

All of the digraphs in Figure 4 have 8 vertices and 12 edges, so the invariants $|V(G)|$ and $|E(G)|$ are no help in this case and we must look more closely. First, let's consider some easier examples.

EXAMPLE 3   Figure 5 shows pictures of four digraphs. All have 5 vertices. By counting edges we see that G_2 with 6 edges is not isomorphic to any of the other three, which have 7. Digraphs G_1 and G_3 look somewhat alike though three arrows seem to have changed directions. As we saw in Example 2, it is possible for two digraphs with different-looking pictures to be isomorphic, so G_1 and G_3 could conceivably be isomorphic. They aren't. Digraph G_1 has a vertex v with one arrow coming in and three going out, while G_3 has no such vertex. If G_1 and G_3 were isomorphic, with mappings α and β as above, then β would have to take each of the three edges $\langle v, w \rangle$ with v as initial vertex to an edge $\langle \alpha(v), \alpha(w) \rangle$ with $\alpha(v)$ as initial vertex. No vertex in G_3 could serve as

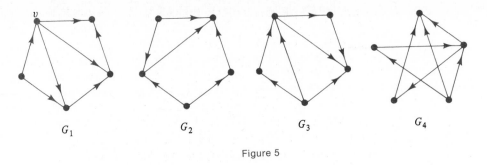

Figure 5

$\alpha(v)$. A similar argument could be based on terminal vertices. We will return to this example shortly and discuss G_4. ∎

The idea of counting arrows is a useful one. Let v be a vertex of a digraph G. The **indegree** of v is the number of edges of G with v as terminal vertex, and the **outdegree** of v is the number with v as initial vertex. The **degree** of v, $\deg(v)$, is the sum of the indegree and outdegree; in symbols

$$\deg(v) = \text{indeg}(v) + \text{outdeg}(v).$$

An edge from v to v gets counted twice here, once going out and once coming in.

As we noticed in Example 3, if α and β describe an isomorphism of G onto H and if $v \in V(G)$ then $\text{indeg}(v) = \text{indeg}(\alpha(v))$ and $\text{outdeg}(v) = \text{outdeg}(\alpha(v))$, so also $\deg(v) = \deg(\alpha(v))$. For each pair $\langle i, j \rangle$ of integers let

$$D_{i,j}(G) = \{v \in V(G) : \text{indeg}(v) = i \text{ and } \text{outdeg}(v) = j\}.$$

Then the one-to-one correspondence α must map $D_{i,j}(G)$ onto $D_{i,j}(H)$; in particular $|D_{i,j}(G)|$ and $|D_{i,j}(H)|$ must be equal. That is, the numbers $|D_{i,j}(G)|$ are isomorphism invariants.

EXAMPLE 4   (a) For the digraphs of Figure 5 these invariants are as follows. [We omit the pairs $\langle i, j \rangle$ for which $D_{i,j}(G)$ is empty for all G's under consideration.]

$\langle i,j \rangle$	$\langle 0,2 \rangle$	$\langle 0,3 \rangle$	$\langle 1,1 \rangle$	$\langle 1,3 \rangle$	$\langle 2,0 \rangle$	$\langle 2,1 \rangle$	$\langle 2,2 \rangle$	$\langle 3,0 \rangle$
G_1	1	0	0	1	1	2	0	0
G_2	2	0	1	0	0	1	0	1
G_3	0	1	2	0	0	0	1	1
G_4	0	1	2	0	0	0	1	1

We can see at once that the only two of these digraphs which can possibly be isomorphic are G_3 and G_4. In fact, just checking one of $|D_{0,2}(G)|$, $|D_{1,1}(G)|$ or $|D_{2,1}(G)|$ would tell us that much. The table of values of $|D_{i,j}(G)|$ does not show that G_3 and G_4 *are* isomorphic. To show isomorphism, one would still need to give a function α. In Exercise 5 you are asked to show that G_3 and G_4 are, in fact, isomorphic.

(b) Now let's go back to the digraphs of Figure 4, drawn again in Figure 6. Their invariants $|D_{i,j}(G)|$ are as follows.

$\langle i,j \rangle$	$\langle 0,3 \rangle$	$\langle 1,2 \rangle$	$\langle 2,1 \rangle$	$\langle 3,0 \rangle$
H_1	1	3	3	1
H_2	1	3	3	1
H_3	1	3	3	1
H_4	1	3	3	1

Not much help here. We must look still more closely. It turns out [Exercise 4] that H_1 and H_2 are, in fact, isomorphic. Exercise 6 suggests one way to show that H_1 and H_3 are not isomorphic by looking hard at these digraphs near their sources. The same idea will show that H_4 is not isomorphic to any of the others so, of the four, only H_1 and H_2 are isomorphic.

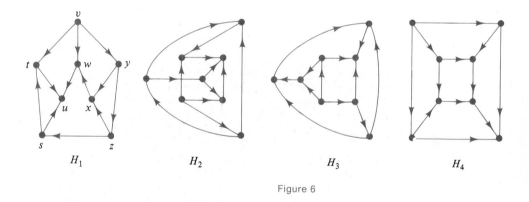

Figure 6

Another way to separate H_4 from the others is to observe that each of the other digraphs has three edges e, f, g so that the path $e\, f$ leads from the initial vertex of g to its terminal vertex. [For instance let $e = y\,z, f = z\,x$ and $g = y\,x$ in H_1.] Digraph H_4 has no such edges.

It turns out that H_1 and H_3 would be isomorphic if we reversed all arrows on one of them. Such digraphs are called **anti-isomorphic**. [See Exercise 6.] ∎

The ideas we used in Example 4 can be extended to get other isomorphism invariants. We could count cycles of a given length, count sources, count triples of edges e, f, g with $e\, f$ a path from initial vertex of g to terminal vertex of g, count vertices with given in- and outdegrees, etc. Some of these invariants are related to others, as the following result illustrates.

Theorem   The sum of all the indegrees of a digraph equals the sum of all the outdegrees, which equals the number of edges of the digraph. In symbols,

$$\sum_{v \in V(G)} \text{indeg}(v) = \sum_{v \in V(G)} \text{outdeg}(v) = |E(G)|.$$

Hence

$$\sum_{v \in V(G)} \deg(v) = 2 \cdot |E(G)|, \text{ an even number.}$$

Proof. Each edge contributes 1 to the indegree of one vertex, so contributes 1 to the indegree sum. Thus the indegree sum is just the total number of edges; similarly for the outdegrees. ∎

We can get the sum of all the indegrees by adding up all the indegrees which are 0, then all which are 1, then 2, and so on. For each $d \in \mathbb{N}$ let $V_d(G) = \{v \in V(G) : \text{indeg}(v) = d\}$. Then $\sum_{v \in V_d(G)} \text{indeg}(v)$ is a sum of $|V_d(G)|$ terms, each of which equals d, so $\sum_{v \in V_d(G)} \text{indeg}(v) = d \cdot |V_d(G)|$. Since $V(G) = \bigcup_{d=0}^{\infty} V_d(G)$, the theorem shows that

$$|E(G)| = \sum_{d \geq 0} \left(\sum_{v \in V_d(G)} \text{indeg}(v) \right) = \sum_{d \geq 0} d \cdot |V_d(G)|$$

$$= 0 \cdot |V_0(G)| + 1 \cdot |V_1(G)| + 2 \cdot |V_2(G)| + \cdots.$$

This gives us a relationship among the invariants $|E(G)|, |V_1(G)|, |V_2(G)|, \ldots.$

EXAMPLE 5   The digraph G shown in Figure 7 has 7 edges and has indegrees and outdegrees as follows.

	w	x	y	z	Sum
indeg	1	1	3	2	7
outdeg	0	3	1	3	7
deg	1	4	4	5	14

The nonempty sets $V_d(G)$ are: $V_1(G) = \{w, x\}$, $V_2(G) = \{z\}$, $V_3(G) = \{y\}$. Sure enough, $7 = 1 \cdot 2 + 2 \cdot 1 + 3 \cdot 1$. We could play the same game with outdegrees,

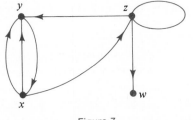

Figure 7

letting $V_d^+ = \{v \in V(G) : \text{outdeg}(v) = d\}$. Then $V_0^+ = \{w\}$, $V_1^+ = \{y\}$, $V_2^+ = \varnothing$, $V_3^+ = \{x, z\}$ and $7 = 0 \cdot 1 + 1 \cdot 1 + 2 \cdot 0 + 3 \cdot 2$.   ∎

The idea of isomorphism is also useful as a means of describing the symmetry of a given digraph. The identity mappings $1_{V(G)}$ and $1_{E(G)}$ are defined by $1_{V(G)}(v) = v$ and $1_{E(G)}(e) = e$ for all v and e. The functions $\alpha = 1_{V(G)}$ and $\beta = 1_{E(G)}$ satisfy the conditions to give $G \simeq G$, but there may be other choices of α and β that work too. Roughly speaking, the more ways we can map G isomorphically back onto itself the more symmetry G has. We call an isomorphism of a digraph onto itself an **automorphism** of the digraph.

EXAMPLE 6   Figure 8 shows examples of digraphs with several different kinds of symmetry. These digraphs have no parallel edges, so we can describe their automorphisms by just giving the mappings α. Digraph D_1 has six different isomorphisms onto itself. Clearly $\alpha(m)$ must be m, but there are three choices for $\alpha(t)$ and then two remaining choices for $\alpha(u)$. The rows in Figure 9(a) list the various possible assignments $\alpha(m)$, $\alpha(t)$, $\alpha(u)$, $\alpha(v)$. Digraph D_2 has four

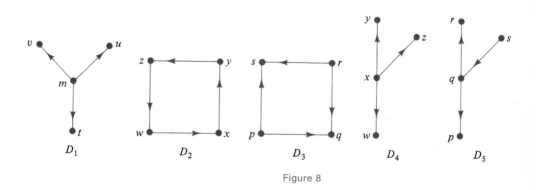

Figure 8

m	t	u	v
m	t	u	v
m	t	v	u
m	u	t	v
m	u	v	t
m	v	t	u
m	v	u	t

w	x	y	z
w	x	y	z
x	y	z	w
y	z	w	x
z	w	x	y

Automorphisms of D_1 Automorphisms of D_2

(a)   (b)

Figure 9

isomorphisms onto itself. There are four choices for $\alpha(w)$, but then $\alpha(x)$, $\alpha(y)$ and $\alpha(z)$ are completely determined; see Figure 9(b). Digraph D_3 also has four automorphisms [Exercise 7], but the digraph has a different kind of symmetry from that of D_2. We can rotate D_2 a quarter turn, while we can flip D_3 over along a diagonal. To describe the difference in terms of isomorphisms we need first to observe that composition of two isomorphisms is again an isomorphism. Then we need to check how the various automorphisms for D_2 and D_3 compose with each other. We return to this question in Chapter 11 where we have the necessary terminology to describe the answer.

The kind of symmetry we are considering may not show up as symmetry of the picture of a digraph. Digraph D_4 doesn't look particularly symmetric as it's drawn, but in fact this digraph is isomorphic to D_1, so D_4 has six isomorphisms onto itself, just as D_1 does. Digraph D_5 also has an asymmetric picture but has two different isomorphisms onto itself [Exercise 7]. We could have drawn more symmetric pictures of these digraphs, of course, but for more complicated digraphs it may be impossible to construct a picture which displays the full symmetry involved, and the techniques of Chapter 11 become more appropriate. ∎

EXERCISES 8.2

1. (a) Assign directions to the edges in the picture in Figure 10(b) to get a digraph isomorphic to the one pictured in Figure 10(a).
   (b) In how many ways could part (a) be anwered correctly? *Hint*: Think of sources and sinks.

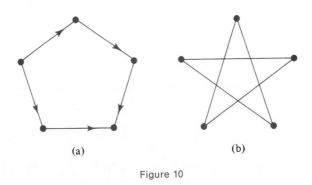

(a)   (b)

Figure 10

2. Draw pictures of all six digraphs with 2 vertices and 2 edges. "All" here means one example from each isomorphism class of such digraphs.

3. Consider the digraphs G_1 and G_2 pictured in Figure 11.
   (a) Find two different isomorphisms of G_1 onto G_2. You need only give the correspondence α.
   (b) Find an isomorphism of G_2 onto G_1.

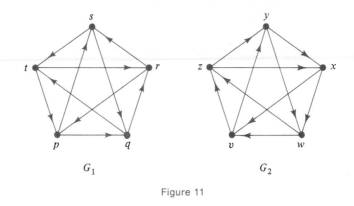

G_1 G_2

Figure 11

4. Label the vertices of digraph H_2 of Figure 6 and find an isomorphism α of H_1 onto H_2. *Hint*: Start with sources and sinks.

5. Label the vertices of digraphs G_3 and G_4 of Figure 5 and find a correspondence α to show that the two digraphs are isomorphic.

6. (a) Show that digraphs H_1 and H_3 of Figure 6 are not isomorphic. *Hint*: Consider edges from the sources of the two digraphs and look at the indegrees and outdegrees of their terminal vertices.
   (b) Label the vertices of H_3 and find a correspondence α to show that H_1 and H_3 are anti-isomorphic. *Hint*: Start with sources and sinks.
   (c) Are digraphs H_2 and H_3 of Figure 6 anti-isomorphic? Explain.

7. (a) Describe the two automorphisms of the digraph D_5 of Figure 8 onto itself.
   (b) List all four automorphisms of the digraph D_3 of Figure 8 onto itself.

8. Consider the digraphs G and H of Example 2. Find an isomorphism of G onto H different from the one chosen in the example.

9. (a) List the nonempty sets $D_{i,j}(H_1)$ for the digraph H_1 in Figure 6.
   (b) Repeat part (a) for the digraph G_1 in Figure 11.

10. A digraph is **regular** if there is some $k \in \mathbb{N}$ such that $\operatorname{indeg}(v) = \operatorname{outdeg}(v) = k$ for all vertices v.
   (a) Draw pictures of all five regular digraphs with 4 vertices and 4 edges. See Exercise 2 for the meaning of "all" in this context.
   (b) How many nonisomorphic regular digraphs are there with 4 vertices and 6 edges? Explain.

11. A student was once given two digraphs G and H and asked if they were isomorphic. In an effort to get the question at least half right, he said that G was isomorphic but H was not. What do you think of this answer?

12. (a) Verify the equations of the theorem in this section for the digraph H_3 of Figure 6. If you haven't done Exercise 6(b), first label the vertices.
   (b) Give the sets $V_d(G)$ for this digraph and verify that $|E(G)| = \sum_{d \geq 0} d \cdot |V_d(G)|$.

13. Suppose that digraphs G and H are isomorphic, with α and β as in the text. If v is a sink of G, must $\alpha(s)$ be a sink of H? Explain.

14. Suppose that digraphs G and H are isomorphic, with correspondences α_1 and β_1 as in the text, and that H and K are isomorphic, with correspondences α_2 and β_2. Verify that G and K are isomorphic, with correspondences $\alpha_2 \circ \alpha_1$ and $\beta_2 \circ \beta_1$.

15. Show that a digraph must have an even number of vertices of odd degree.

16. Show that if G and H are isomorphic digraphs and if H is acyclic then G is acyclic.

17. (a) Show that the number of sources of a digraph is an isomorphism invariant.
(b) Is the number of vertices of indegree 3 an isomorphism invariant? Explain.

18. How many different isomorphisms are there of the digraph G_1 of Figure 11 onto the digraph G_2? Explain.

§ 8.3   Weighted Digraphs

In many applications of digraphs one wants to know if a given vertex v is reachable from another vertex u, that is, if it is possible to get to v from u by following arrows. For instance, suppose each vertex represents a state a machine can be in such as FETCH, DEFER or EXECUTE, and there is an edge from s to t whenever the machine can change from state s to state t in response to some input. If the machine is in state u can it later be in state v? The answer is "yes" if and only if the digraph contains a path from u to v.

Now suppose there is a cost associated with each transition from one state to another, i.e., with each edge in the digraph. Such a cost might be monetary, might be a measure of the time involved to carry out the change, or might have some other meaning. We could now ask for a path from u to v with the smallest total associated cost, obtained by adding all the costs for the edges in the path.

If all edges cost the same amount then the cheapest path is simply the shortest. In general, however, edge costs might differ. A digraph with no parallel edges is called **weighted** if each edge has an associated number, called its **weight**. In a given application it might better be called "cost" or "length" or "capacity" or have some other interpretation. Weights are normally assumed to be nonnegative, but many of the results about weighted digraphs are true without such a limitation. We can describe the weighting of a digraph G with a function W from $E(G)$ to $\mathbb{R}$ where $W(e)$ is the weight of the edge e. The **weight** of a path $e_1 e_2 \cdots e_m$ in G is then the sum $\sum_{i=1}^{m} W(e_i)$. Since a weighted digraph has no parallel edges, we may suppose that $E(G) \subseteq V(G) \times V(G)$, and write $W(u, v)$ for the weight of an edge $\langle u, v \rangle$ from u to v.

EXAMPLE 1   (a) The digraph shown in Figure 1 is stolen from Figure 1 of § 0.1, where it described a rat and some cages. It could just as well describe a machine with states A, B, C and D, and the number next to an arrow [the weight of the

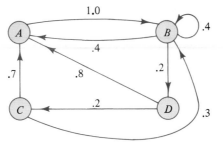

Figure 1

edge] could be the number of microseconds necessary to get from its initial state to its terminal state. With that interpretation it takes .3 microsecond to go from state C to state B, .9 microsecond to go from D to A by way of C, .8 microsecond to go directly from D to A and .4 microsecond to stay in state B in response to an input. What is the shortest time to get from D to B?

(b) Figure 2 shows a more complicated example. In this case the shortest paths $s \, v \, x \, f$ and $s \, w \, x \, f$ from s to f have weights $6 + 7 + 4 = 17$ and $3 + 7 + 4 = 14$, respectively, but the longer path $s \, w \, v \, y \, x \, z \, f$ has weight $3 + 2 + 1 + 3 + 1 + 3 = 13$, which is less than either of these. Thus length is not directly related to weight. This example also shows a path $s \, w \, v$ from s to v which has smaller weight than the edge from s to v.

This digraph has a cycle $w \, v \, y \, w$. Clearly the whole cycle cannot be part of a path of minimum weight, but pieces of it can be. For instance $w \, v \, y$ is the path of smallest weight from w to y and the edge $y \, w$ is the path of smallest weight from y to w. ∎

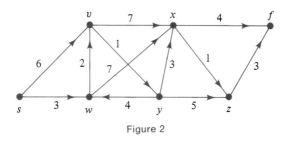

Figure 2

If we wish, we can display the weight function W in a tabular form by indexing rows and columns of an array with the members of $V(G)$ and entering the value of $W(u, v)$ at the intersection of row u and column v.

EXAMPLE 2   The array for the digraph in Figure 2 is given in Figure 3(a). The numbers appear in locations corresponding to edges of the digraph. The table in Figure 3(a) contains enough information to let us reconstruct the weighted

W	s	v	w	x	y	z	f
s		6	3				
v				7	1		
w	2			7			
x						1	4
y		4	3			5	
z							3
f							

W*	s	v	w	x	y	z	f
s		5	3	9	6	10	13
v			5	4	1	5	8
w	2			6	3	7	10
x						1	4
y		6	4	3		4	7
z							3
f							

(a)   (b)

Figure 3

digraph, since from the table we know just where the edges go and what their weights are. Figure 3(b) is a tabulation of the weight function W^* where $W^*(u, v)$ is the smallest weight of a path from u to v if such a path exists. ∎

We will call the smallest weight of a path [of length at least 1] from u to v the **min-weight** from u to v, and we will generally denote it by $W^*(u, v)$, as in Example 2. We also call a path from u to v that has this weight a **min-path**. It is no real restriction to suppose that our weighted digraphs have no loops [why?], so we could just decide not to worry about $W(u, u)$, or we might define $W(u, u) = 0$ for all vertices u. It turns out, though, that we will learn more useful information by another choice, based on the following idea.

Consider vertices u and v with no edge from u to v in G. We can create a fictitious new edge from u to v of enormous weight, so big that the edge would never get chosen in a min-path if a real path from u to v were available. Suppose we create such fictitious edges wherever edges are missing in G. If we ever find that the weight of a path in the enlarged graph is enormous, then we will know that the path includes at least one fictitious edge, so it can't be a path in the original digraph G. A convenient notation is to write $W(u, v) = \infty$ if there is no edge from u to v in G; $W^*(u, v) = \infty$ means that there is no path from u to v. The operating rules for the symbol ∞ are: $\infty + x = x + \infty = \infty$ for every x, and $\infty > a$ for every real number a.

With this notation, we will write $W(u, u) = a$ if there is a loop at u of weight a, and write $W(u, u) = \infty$ otherwise. Then $W^*(u, u) < \infty$ means that there is a path [of length at least 1] from u to itself in G, and $W^*(u, u) = \infty$ means that there is no such path. The digraph G is acyclic if and only if $W^*(v, v) = \infty$ for every vertex v.

EXAMPLE 3   Using this notation, we would fill in all the blanks in the tables of Figures 3(a) and 3(b) with ∞'s. ∎

In Example 2 we simply announced the values of W^*, and the example is small enough that it is easy to check the values given. For more complicated digraphs the determination of W^* and of the min-paths can be nontrivial problems. In §§ 8.8 and 8.9 we will describe algorithms for finding both W^* and the corresponding min-paths, but until then we will stare at the picture until the answer is clear. For small digraphs this method is as good as any.

Attaching one min-path to the end of another might not yield a min-path, but pieces of min-paths are min-paths, as the following proposition shows.

Proposition   If $u \cdots t \cdots v$ is a min-path, then so are the first part $u \cdots t$ and the second part $t \cdots v$. Thus $W^*(u, v) = W^*(u, t) + W^*(t, v)$.

Proof. If $u \cdots t$ weren't already a min-path from u to t, we could replace it with one and get a path from u to t of smaller weight than the one we have. By using the new path, we would reduce the total path weight from u to v, which is impossible since the given path $u \cdots t \cdots v$ is a min-path. Thus $u \cdots t$ and, similarly, $t \cdots v$ are min-paths already.   ▮

In many situations described by weighted graphs the min-weights and min-paths are the important concerns. One class of problems of considerable importance in business and manufacturing, where weighted digraphs are useful, but where min-paths are irrelevant, is the scheduling of processes which involve a number of steps. The following example illustrates the sort of situation which arises.

EXAMPLE 4   Consider a cook preparing a simple meal of curry and rice. The curry recipe calls for the following steps.

  (a) Cut up meat—about 10 minutes.
  (b) Grate onion—about 2 minutes with a food processor.
  (c) Peel and quarter potatoes—about 5 minutes.
  (d) Marinate meat, onions and spices—about 30 minutes.
  (e) Heat oil—4 minutes. Fry potatoes—15 minutes.
      Fry cumin seed—2 minutes.
  (f) Fry marinated meat—4 minutes.
  (g) Bake fried meat and potatoes—60 minutes.

In addition, there is

  (h) Cook rice—20 minutes.

We have grouped three steps together in (e), since they must be done in sequence. Some of the other steps can be done simultaneously if enough help is available. We suppose our cook has all the help needed.

Figure (4a) gives a digraph which shows the sequence of steps and the possibilities for parallel processing. Cutting, grating, peeling and rice cooking can all go on at once. The dotted arrows after cutting and peeling indicate that frying and marinating cannot begin until cutting and peeling are

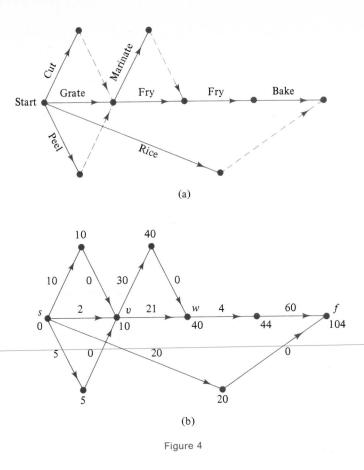

(a)

(b)

Figure 4

completed. The other two dotted arrows have similar meanings. The picture
has been redrawn in Figure 4(b) with weights on the edges to indicate time
involved. [Ignore the numbers on the vertices for the moment.] The vertices
denote stages of partial completion of the overall process, starting at the left
and finishing at the right. In this case the min-path from left to right has
weight 20, but there is much more total time required to prepare the meal
than just the 20 minutes to cook the rice. The min-weight is no help. The
important question here is: What is the smallest total time required to
complete all steps in the process?

To answer the question we first examine vertices from left to right. Suppose we start at s at time 0. What is the earliest time we can have completed
cutting, grating and peeling and arrive at vertex v? Clearly 10 minutes, since
we must wait for the cutting no matter how soon we start grating or
peeling. In fact, 10 is the *largest* weight of a path from s to v. Now what is the
earliest time we can arrive at vertex w? The shortest time to get from v to w is
30 minutes [the largest weight of a path from v to w], so the shortest time to

get from s to w is $10 + 30 = 40$ minutes. Similarly, the earliest we can complete the whole process and arrive at f is $40 + 4 + 60 = 104$ minutes after we start.

In each instance, the smallest time to arrive at a given vertex is the largest weight of a path from s to that vertex. The numbers beside the vertices in Figure 4(b) give these smallest times. ∎

An acyclic digraph with nonnegative weights and with unique source and sink, such as the digraph in Figure 4, is called a **scheduling network**. For the rest of this section we suppose that we are dealing with a scheduling network G with source s [start] and sink f [finish]. For vertices u and v of G a **max-path** from u to v is a path of largest weight, and its weight is the **max-weight** from u to v, which we denote by $M(u, v)$. Max-weights and max-paths can be analyzed in much the same way as min-weights and min-paths. In § 8.8 we describe how to modify an algorithm for W^* to get one for M. For now we determine M by staring at the digraph.

A max-path from s to f is called a **critical path**, and an edge belonging to such a path is a **critical edge**. If $\langle u, v \rangle$ is an edge of G then

$$M(s, u) + W(u, v) + M(v, f) \leqq M(s, f)$$

since a path $s \cdots u\, v \cdots f$ certainly has no greater weight than a critical path from s to f, and equality holds if and only if $\langle u, v \rangle$ is part of a critical path. We define the **float time** $F(u, v)$ of the edge $\langle u, v \rangle$ by

$$F(u, v) = M(s, f) - M(s, u) - W(u, v) - M(v, f).$$

Then $F(u, v) \geqq 0$ for all edges $\langle u, v \rangle$, and $F(u, v) = 0$ if and only if $\langle u, v \rangle$ is a critical edge.

To get a meaningful interpretation of $F(u, v)$ we define two functions A and L on $V(G)$. If we start from s at time 0, the earliest time we can arrive at a vertex v having completed all tasks preceding v is $M(s, v)$. We denote this earliest arrival time by $A(v)$. In particular, $A(f) = M(s, f)$, the time in which the whole process can be completed. Let $L(v) = M(s, f) - M(v, f) = A(f) - M(v, f)$. Since $M(v, f)$ represents the shortest time required to complete all steps from v to f, $L(v)$ is the latest time we can leave v and still complete all remaining steps by time $A(f)$. To calculate $M(v, f)$ we work backwards from f.

EXAMPLE 5   The functions A and L for the scheduling network shown in Figure 5(a) are given in Figure 5(b). Since $A(x) = 5 = L(x)$ we must leave x as soon as we arrive to avoid delaying arrival at f. Similarly, we can't dawdle at u or y. Since $L(v) - A(v) = 9 - 7 = 2$ we may be able to delay departure from v by 2 time units without affecting the overall time from s to f. In this example the path $s\, u\, x\, y\, f$ is the unique critical path. ∎

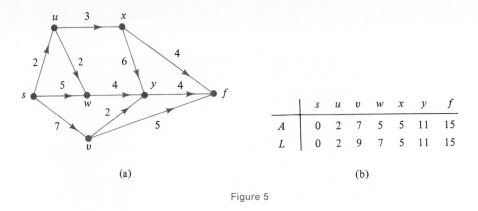

	s	u	v	w	x	y	f
A	0	2	7	5	5	11	15
L	0	2	9	7	5	11	15

(a)   (b)

Figure 5

To describe the float time $F(u, v)$ in terms of A, L and W observe that

$$F(u, v) = M(s, f) - M(v, f) - W(u, v) - M(s, u)$$

$$= L(v) - W(u, v) - A(u).$$

Thus $L(v) - A(u) = W(u, v) + F(u, v)$, i.e., the difference between the earliest we can arrive at u and the latest we can leave v is the time along the edge $\langle u, v \rangle$ plus the float time for that edge. An edge $\langle u, v \rangle$ is critical if and only if $L(v) - A(u) = W(u, v)$.

EXAMPLE 6   (a) For the example in Figure 5 the float time of the edge $\langle u, w \rangle$ is $L(w) - A(u) - W(u, w) = 7 - 2 - 2 = 3$. We could start along this edge as much as 3 time units after $A(u)$ and still arrive at w soon enough not to delay completion of the process. This edge $\langle u, w \rangle$ is not critical. On the other hand, we saw earlier that we must leave u on time in order to get to x on time. The edge $\langle u, x \rangle$ has float time $L(x) - A(u) - W(u, x) = 5 - 2 - 3 = 0$; it is critical.

(b) In the curry dinner problem of Example 4 the steps (a), (d), (f) and (g) are critical. The float time for cooking the rice is $104 - 0 - 20 = 84$ minutes [and in fact one would normally wait about 74 minutes before starting the rice to allow it to "rest" 10 minutes at the end]. The float time for heating the oil and frying potatoes and cumin seed is $40 - 10 - 21 = 9$ minutes. Speeding up these steps will have no effect on the total preparation time and there is no harm done if frying takes as many as 9 more minutes than expected. ∎

As we just saw, shortening the time required for a noncritical edge does not decrease the total time $M(s, f)$ required for the process. Identification of critical edges focuses attention on those steps in a process where improvement may make a difference and where delays will surely be costly. Since its introduction in the 1950s the method of critical path analysis, sometimes called PERT for Program Evaluation and Review Technique, has been a popular way of dealing with industrial management scheduling problems.

EXERCISES 8.3

Use the ∞ notation in all tables of W and W^*.

1. Give tables of W and W^* for the digraph of Figure 1 with the loop at B removed.

2. Give a table of W^* for the digraph of Figure 5(a).

3. Give tables of W and W^* for the digraph of Figure 6.

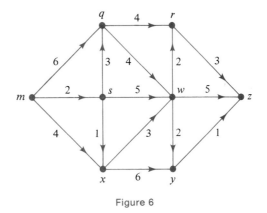

Figure 6

4. The path $s\,w\,v\,y\,x\,z\,f$ is a min-path from s to f in the digraph of Figure 2. Find another min-path from s to f in that digraph.

5. Figure 7 shows a weighted digraph. The directions and weights have been left off the edges, but the number at each vertex v is $W^*(s, v)$.
   (a) Give three different weight functions W which yield these values of $W^*(s, v)$. [An answer could consist of three pictures with appropriate numbers on the edges.]
   (b) Do the different weight assignments yield different min-paths between points? Explain.

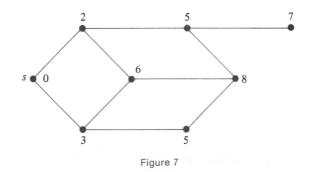

Figure 7

6. Find a critical path from m to z in the digraph of Figure 6.

7. (a) Find a critical path for the digraph of Figure 2 with the edge from y to w removed.
   (b) Why does the critical path method apply only to acyclic digraphs?

8. Suppose that u, v and w are vertices of a weighted digraph with min-weight function W^* and that $W^*(u, v) + W^*(v, w) = W^*(u, w)$. Explain why there is a min-path from u to w through v.

9. (a) Label the vertices in the digraph of Figure 8 and find all critical edges of the digraph.
   (b) How many critical paths does this digraph have?
   (c) What is the largest float time for an edge in this digraph?
   (d) Which edges have the largest float time?

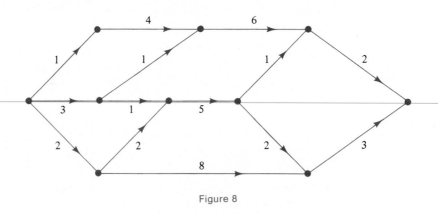

Figure 8

10. If the cook in Example 4 has no helpers, then steps (a), (b) and (c) must be done one after the other, but otherwise the situation is as in the example.
    (a) Draw a scheduling network for the no-helper process.
    (b) Find a critical path for this process.
    (c) Which steps in the process are not critical?

11. In Example 4 we used edges of weight 0 as a device to indicate that some steps could not start until others were finished.
    (a) Explain how to avoid such 0-edges if parallel edges are allowed.
    (b) Draw a digraph for the process of Example 4 to illustrate your answer.

12. (a) Give tables of W and W^* for the digraph of Figure 9.
    (b) Explain how to tell from the table for W^* whether or not this digraph is acyclic.
    (c) How would your answer to part (a) change if the edge of weight -2 had weight -6 instead?
    (d) Explain how to tell which are the sources and sinks of this digraph from the table for W.

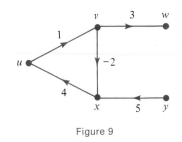

Figure 9

Note. The remaining exercises concern scheduling networks. The functions A, L and F are defined prior to Example 5.

13. (a) Show that $A(u) = \max\{A(w) + W(w, u) : \langle w, u \rangle \in E(G)\}$.
 (b) Show that $L(u) = \min\{L(v) - W(u, v) : \langle u, v \rangle \in E(G)\}$.

14. Define the **slack time** $S(v)$ of a vertex v in a scheduling network by $S(v) = L(v) - A(v)$.
 (a) Show that $S(v) \geq 0$ for all v.
 (b) Show that if $\langle u, v \rangle$ is a critical edge then $S(u) = S(v) = 0$. *Hint:* Use part (a) and Exercise 13.
 (c) If $\langle u, v \rangle$ is an edge with $S(u) = S(v) = 0$, must $\langle u, v \rangle$ be critical? Explain.

15. The float time $F(u, v)$ can be thought of as the amount of time we can delay starting along the edge $\langle u, v \rangle$ without delaying completion of the entire process. Define the **free-float time** $FF(u, v)$ to be the amount of time we can delay starting along $\langle u, v \rangle$ without increasing $A(v)$.
 (a) Find an expression for $FF(u, v)$ in terms of the functions A and W.
 (b) Find $FF(u, v)$ for all edges of the digraph in Figure 5.
 (c) What is the difference between $F(u, v)$ and $FF(u, v)$?

16. (a) Show that increasing the weight of a critical edge in a scheduling network increases the max-weight from the source to the sink.
 (b) Is there any circumstance in which reducing the amount of time for a critical step in a process does not reduce the total time required for the process? Explain.

§ 8.4 Undirected Graphs

We now turn to the study of graphs where the edges have no directions associated with them. These are the sorts of graphs that were introduced in Chapter 0. Readers may wish to reread that chapter to reacquaint themselves with the kinds of problems that arise in this connection. However, the development here will be self-contained. This section is devoted to presenting basic definitions and properties of graphs of this type. Much of the material parallels the theory of digraphs we developed in the last three sections, but the questions we ask and the answers we get have a distinctly different character. The first few definitions were actually given in Example 6 of § 3.1.

Instead of being associated with an ordered pair of vertices, as edges in digraphs are, an undirected edge has an unordered set of vertices. Following the pattern we used for digraphs, we define an [undirected] **graph** G to consist of two sets, the set $V(G)$ of **vertices** of G and the set $E(G)$ of **edges** of G, together with a function γ from $E(G)$ to the set $\{\{u, v\} : u, v \in V(G)\}$ of all subsets of $V(G)$ with one or two members. For an edge e in $E(G)$ the members of $\gamma(e)$ are called the **vertices** of e or the **endpoints** of e; we say that e **joins** its endpoints. A **loop** is an edge with only one endpoint. Distinct edges e and f with $\gamma(e) = \gamma(f)$ are called **parallel** or **multiple** edges.

What we have just described is called a **multigraph** by some authors who reserve the term "graph" for those graphs with no loops or parallel edges. If a graph has no parallel edges we may identify an edge by listing its endpoints, in which case we will commonly write $e = \{u, v\}$ instead of $\gamma(e) = \{u, v\}$; we will also write $e = \{u, u\}$ instead of $e = \{u\}$ if e is a loop with vertex u.

A **picture** of a graph G is a diagram consisting of points corresponding to the vertices of G and arcs corresponding to edges, such that if $\gamma(e) = \{u, v\}$ then the arc for the edge e joins the points labeled u and v.

EXAMPLE 1   If we leave the direction arrows off of the edges in Figure 1 of § 8.1 we get the picture in Figure 1(a). A table of γ for this graph is given in Figure 1(b). This graph has parallel edges c and e joining x and z, and has a loop d with vertex z. The graph pictured in Figure 1(c) has no parallel edges, so for that graph a description such as "the edge $\{x, z\}$" is unambiguous. The same phrase could not be applied to the graph in Figure 1(a). ▮

A **path** of **length** n from the vertex u to the vertex v is a sequence $e_1 \cdots e_n$ of edges together with a sequence $x_1 \cdots x_{n+1}$ of vertices with $\gamma(e_i) = \{x_i, x_{i+1}\}$ for $i = 1, \ldots, n$ and $x_1 = u, x_{n+1} = v$. We will not have occasion to use paths of length 0; in what follows the length will always be a positive integer. If

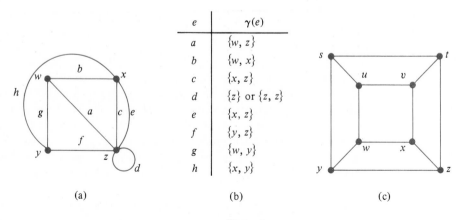

e	$\gamma(e)$
a	$\{w, z\}$
b	$\{w, x\}$
c	$\{x, z\}$
d	$\{z\}$ or $\{z, z\}$
e	$\{x, z\}$
f	$\{y, z\}$
g	$\{w, y\}$
h	$\{x, y\}$

(a)   (b)   (c)

Figure 1

$e_1 e_2 \cdots e_n$ is a path from u to v with vertex sequence $x_1 x_2 \cdots x_{n+1}$, then $e_n \cdots e_2 e_1$ with vertex sequence $x_{n+1} x_n \cdots x_1$ is a path from v to u. We may speak of either of these paths as a **path between** u and v. If $u = v$ the path is **closed**.

The edge sequence of a path usually determines the vertex sequence, and we will sometimes use phrases such as "the path $e_1 e_2 \cdots e_n$" without mentioning vertices. Example 2 illustrates the possible fuzziness in such usage. If a graph has no parallel edges then the vertex sequence completely determines the edge sequence. In that setting, or if the actual choice of edges is unimportant, we will commonly use vertex sequences as descriptions for paths.

A path is called **simple** if all of its edges are different. Thus a simple path cannot use any edge twice, though it may go through the same vertex more than once. A closed simple path with vertex sequence $x_1 \cdots x_n x_1$ is called a **cycle** if the vertices $x_1, \ldots, x_n$ are distinct. A graph is **acyclic** if it contains no cycle. A path is **acyclic** if the "subgraph" consisting of the vertices and edges of the path is acyclic. In general a graph H is a **subgraph** of a graph G if $V(H) \subseteq V(G)$, $E(H) \subseteq E(G)$ and the function γ for G defined on $E(G)$ agrees with the γ for H on $E(H)$. If G has no parallel edges and if we think of $E(G)$ as a set of one- or two-element subsets of $V(G)$, the condition on γ follows from $E(H) \subseteq E(G)$.

EXAMPLE 2   (a) Consider the graph pictured in Figure 2(a). The path $e\,e$ with vertex sequence $x_1 x_2 x_1$ is a closed path, but it is not a cycle because it is not simple. Neither is the path $e\,e$ with vertex sequence $x_2 x_1 x_2$. This graph is acyclic.

(b) The path $e\,f$ with vertex sequence $x_1 x_2 x_1$ in the graph of Figure 2(b) is a cycle. So is the path $e\,f$ with vertex sequence $x_2 x_1 x_2$.

(c) In the graph of Figure 2(c) the path $e\,f\,h\,i\,k\,g$ of length 6 with vertex sequence $u\,v\,w\,x\,y\,w\,u$ is closed and simple but is not a cycle because the first six vertices u, v, w, x, y and w are not all different. The path with vertex sequence $u\,w\,v\,w\,u\,v\,u$ also fails to be a cycle. The graph as a whole is not acyclic, and neither are these two paths, since $u\,v\,w\,u$ is a cycle in both of their subgraphs. ∎

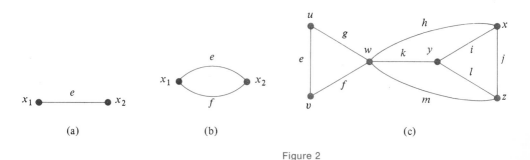

(a)   (b)   (c)

Figure 2

A path $e_1 \cdots e_n$ with all of $x_1, \ldots, x_{n+1}$ distinct must surely be simple since no two edges in it can have the same set of endpoints. Example 2(a) shows, however, that a closed path with $x_1, \ldots, x_n$ distinct need not be simple. That bad example is essentially the only one there is, as the following shows.

Proposition 1   Every closed path $e_1 \cdots e_n$ of length at least 3 with $x_1, \ldots, x_n$ distinct is a cycle.

Proof. We only need to show that $e_1, \ldots, e_n$ are different. Since $x_1, \ldots, x_n$ are distinct, the path $e_1 \cdots e_{n-1}$ is simple. That is, $e_1, \ldots, e_{n-1}$ are all different. But $\gamma(e_n) = \{x_n, x_1\}$ and $\gamma(e_i) = \{x_i, x_{i+1}\}$ for $i < n$. Since $n \geq 3$, $e_n \neq e_i$ for $i < n$ and so the path is simple. ∎

For paths which are not closed, distinctness of vertices can be characterized another way.

Proposition 2   A path has all vertices distinct if and only if it is simple and acyclic.

Proof. Consider first a path with distinct vertices. It must be simple, as we observed earlier. We prove it is acyclic by induction on its length. If the length is 1 the path is clearly acyclic. Suppose the path has length $n > 1$ with vertex sequence $x_1 \cdots x_n x_{n+1}$. By the inductive assumption, the path $x_1 \cdots x_n$ is acyclic, so any cycle in our path must contain the edge joining x_n and x_{n+1}. But x_{n+1} is not a vertex of any other edge in the path, so it's not on any closed path made from edges in the path. Thus our path contains no cycle. The conclusion now follows by induction.

For the converse, suppose a simple path has vertex sequence $x_1 \cdots x_{n+1}$ with some repeated vertices. Consider two such vertices, say x_i and x_j with $i < j$ and with the difference $j - i$ as small as possible. Then $x_i, x_{i+1}, \ldots, x_{j-1}$ are all distinct, so the simple path $x_i x_{i+1} \cdots x_{j-1} x_j$ is a cycle [even if $j = i + 1$] and the original path contains a cycle. ∎

It follows from Proposition 2 that every simple closed path contains a cycle, and in fact the proof shows that such a cycle can be built out of successive edges in the path.

The following result is analogous to Theorem 1 of § 8.1.

Theorem 1   If u and v are distinct vertices of a graph G and if there is a path in G from u to v, then there is a simple acyclic path from u to v.

Proof. Among all paths from u to v in G choose one of smallest length, say with vertex sequence $x_1 \cdots x_{n+1}$, $x_1 = u$ and $x_{n+1} = v$. Just as in the proof of Theorem 1 of § 8.1 the vertices $x_1, \ldots, x_{n+1}$ are distinct. By Proposition 2, the path is simple and acyclic. ∎

Corollary   If there is a simple closed path in G with u as a vertex, then there is a cycle in G with u as a vertex.

> *Proof*. If there is a loop with u as vertex we are done, so suppose there is no such loop. Then the given simple closed path $e_1 e_2 \cdots e_n$ has at least two edges. We may assume the vertex sequence is $x_1 \cdots x_n x_{n+1}$ with $x_1 = x_{n+1} = u$. Then $e_1 \cdots e_{n-1}$ is a simple path from u to x_n. Since $e_1, \ldots, e_n$ are distinct, the path $e_1 \cdots e_{n-1}$ lies in the graph $G \backslash \{e_n\}$ obtained from G by leaving out the edge e_n. Applying Theorem 1 to the graph $G \backslash \{e_n\}$ we get a simple acyclic path $f_1 \cdots f_m$ from u to x_n in $G \backslash \{e_n\}$. Then the path $f_1 \cdots f_m e_n$ is closed. Since $f_1 \cdots f_m$ is simple and acyclic, its vertices are distinct by Proposition 2. Thus only the first and last vertices of $f_1 \cdots f_m e_n$ are equal; i.e., $f_1 \cdots f_m e_n$ is a cycle through u. ∎

We will have other occasions, especially in Chapter 9, to remove an edge e to get a new graph $G \backslash \{e\}$. The graph $G \backslash \{e\}$ is the subgraph of G with $V(G \backslash \{e\}) = V(G)$ and $E(G \backslash \{e\}) = E(G) \backslash \{e\}$.

The portion of § 8.1 after Theorem 1 and its corollary dealt with sinks in acyclic digraphs. For undirected graphs the corrresponding ideas lead to the notion of connectedness, which we examine in the next two sections, and to the study of trees in Chapter 9, where we will need the following fact.

Theorem 2   If the graph G has two distinct vertices u and v and has two different simple paths from u to v then G contains a cycle.

We will most frequently use this theorem in its contrapositive form:

If G is acyclic and if u and v are distinct vertices of G then there is at most one simple path in G from u to v.

> *Proof of Theorem 2*. If G has a loop or has parallel edges it clearly contains a cycle and we are done. Thus suppose G has no loops or parallel edges. Then each path is completely described by its vertex sequence. Suppose $x_0 \cdots x_n$ and $y_0 \cdots y_m$ are vertex sequences for two different simple paths with $x_0 = y_0 = u$ and $x_n = y_m = v$. If either path contains a cycle we are done, so suppose both paths are acyclic. The idea is to build a cycle out of parts of these paths. Since the paths are different, there is an index i such that $x_0 = y_0, \ldots, x_{i-1} = y_{i-1}$, but $x_i \neq y_i$. Let $w = x_{i-1} = y_{i-1}$. Since $x_n = y_m$ there is at least one pair of indices $\langle j, k \rangle$ such that $i \leq j$, $i \leq k$ and $x_j = y_k$. Consider such a pair with $j + k$ as small as possible, and let $z = x_j = y_k$.
>
> Now form the path $w\, x_i \cdots x_{j-1}\, z\, y_{k-1} \cdots y_i\, w$, going out from w to z along the x-path and back from z to w along the y-path. Since $x_i \neq y_i$ this

closed path has length at least 3. We claim it's a cycle. By Proposition 1 we only need to show that $w, x_i, \ldots, x_{j-1}, z, y_{k-1}, \ldots, y_i$ are distinct. Since the x-path and y-path are acyclic, the vertices going out are all different and the ones coming back are all different. By the choice of $\langle j, k \rangle$, none of $x_i, \ldots, x_{j-1}$ can be any of $y_i, \ldots, y_{k-1}$. Thus the path is a cycle. ∎

EXAMPLE 3 For the graph of Figure 2(c), $x\, z\, w\, u$ and $x\, z\, y\, w\, v\, u$ are simple paths from x to u. In the notation of the proof of Theorem 2, $x_0 = x$, $x_1 = z$, $x_2 = w$, $x_3 = u$ and $y_0 = x$, $y_1 = z$, $y_2 = y$, $y_3 = w$, $y_4 = v$, $y_5 = u$. Since $x_1 = y_1$ and $x_2 \neq y_2$, $i = 2$. Since $x_2 = y_3 = w$, $j = 2$ and $k = 3$. The path $x_1\, w\, y_2\, y_1$, i.e., $z\, w\, y\, z$, is the cycle constructed by the proof. The path $x_2\, x_3\, y_4\, y_3$, i.e., $w\, u\, v\, w$, is also a cycle made from parts of the x-path and the y-path. ∎

The ideas of isomorphism and invariant which we encountered in § 8.2 have natural counterparts for graphs which are not directed. If G and H are graphs without parallel edges and we consider edges to be one- or two-element sets of vertices, an **isomorphism** of G onto H is a one-to-one correspondence $\alpha: V(G) \to V(H)$ such that $\{u, v\}$ is in $E(G)$ if and only if $\{\alpha(u), \alpha(v)\}$ is in $E(H)$. Two graphs G and H are **isomorphic**, written $G \simeq H$, if there is an isomorphism α of one onto the other, in which case the inverse correspondence α^{-1} is also an isomorphism. For graphs with parallel edges the situation is slightly more complicated: we require two one-to-one correspondences $\alpha: V(G) \to V(H)$ and $\beta: E(G) \to E(H)$ such that an edge e of $E(G)$ joins vertices u and v in $V(G)$ if and only if the corresponding edge $\beta(e)$ joins $\alpha(u)$ and $\alpha(v)$.

EXAMPLE 4 The correspondence α with $\alpha(t) = t'$, $\alpha(u) = u'$, ..., $\alpha(z) = z'$ is an isomorphism between the graphs pictured in Figures 3(a) and 3(b). The graphs shown in Figures 3(c) and 3(d) are also isomorphic to each other, but not to the graphs in parts (a) and (b). ∎

To tell the graphs of Figures 3(a) and 3(c) apart we can simply count vertices. Just as we observed when dealing with digraphs, isomorphic graphs have the same number of vertices and the same number of edges. These two numbers are examples of **isomorphism invariants** for graphs. Other examples include the number of loops and number of simple paths of a given length.

As in the case of digraphs we can count the number of edges attached to a particular vertex. To get the right count we need to treat loops differently from edges with two distinct vertices. We define deg(v), the **degree** of the vertex v, to be the number of 2-vertex edges with v as a vertex plus twice the number of loops with v as vertex. The number $D_k(G)$ of vertices of degree k in G is an isomorphism invariant, as is the **degree sequence** $(D_0(G), D_1(G), D_2(G), \ldots)$.

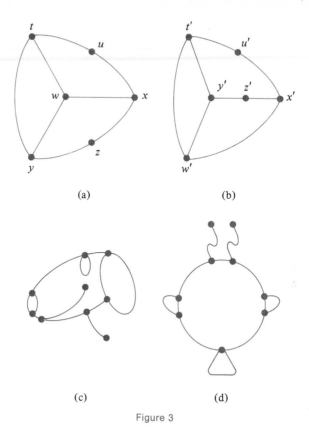

(a)   (b)

(c)   (d)

Figure 3

EXAMPLE 5   (a) The graphs shown in Figures 3(a) and 3(b) each have degree sequence $(0, 0, 2, 4, 0, 0, \ldots)$. Those in Figures 3(c) and 3(d) have degree sequence $(0, 2, 0, 6, 1, 0, 0, \ldots)$.

(b) Erasing the arrows from the digraphs of Figure 6 of § 8.2 gives the graphs pictured in Figure 4. All four graphs have eight vertices of degree 3 and no others. It turns out that $H_1 \simeq H_2 \simeq H_3$, but that none of these three

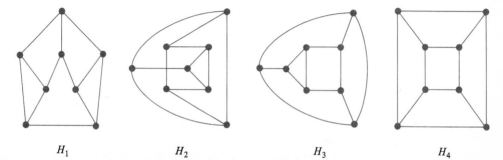

H_1 H_2 H_3 H_4

Figure 4

is isomorphic to H_4. Having the same degree sequence does not guarantee isomorphism.   ∎

Graphs in which all vertices have the same degree, such as those in Figure 4, are called **regular** graphs. As the example shows, regular graphs with the same number of vertices need not be isomorphic. Graphs without loops or multiple edges and in which every vertex is joined to every other by an edge are called **complete** graphs. A complete graph with n vertices has vertices of degree $n - 1$, so such a graph is regular. All complete graphs with n vertices are isomorphic to each other and so we use the symbol K_n for any of them.

EXAMPLE 6   Figure 5(a) shows the first five complete graphs. The graph in Figure 5(b) has four vertices, each of degree 3, but is not complete.   ∎

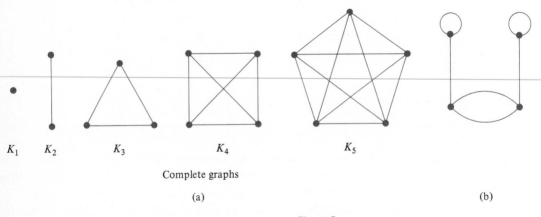

K_1 K_2 K_3 K_4 K_5

Complete graphs

(a)   (b)

Figure 5

A complete graph K_n contains subgraphs isomorphic to the graphs K_m for $m = 1, 2, \ldots, n$. Such a subgraph can be obtained by selecting any m of the n vertices and using all the edges in K_n joining them. Thus K_5 contains $\binom{5}{2} = 10$ subgraphs isomorphic to K_2, $\binom{5}{3} = 10$ subgraphs isomorphic to K_3 [i.e., triangles], and $\binom{5}{4} = 5$ subgraphs isomorphic to K_4. In fact, K_n contains subgraphs isomorphic to all of the graphs with n or fewer vertices and with no loops or parallel edges.

Complete graphs have a high degree of symmetry. Each permutation α of the vertices of a complete graph gives an isomorphism of the graph onto itself, since both $\{u, v\}$ and $\{\alpha(u), \alpha(v)\}$ are edges whenever $u \neq v$.

The next theorem is an analogue to the theorem of § 8.2. You may have anticipated part (a) if you worked Exercise 12 of § 0.1.

Theorem 3   (a) The sum of the degrees of the vertices of a graph is twice the number of edges. That is,

$$\sum_{v \in V(G)} \deg(v) = 2 \cdot |E(G)|.$$

(b) $D_1(G) + 2D_2(G) + 3D_3(G) + 4D_4(G) + \cdots = 2 \cdot |E(G)|.$

Proof. (a) Each edge, whether a loop or not, contributes 2 to the degree sum.

(b) The total degree sum contribution from the $D_k(G)$ vertices of degree k is $k \cdot D_k(G)$. ∎

EXAMPLE 7   (a) The graph of Figure 1(a) has vertices w, x, y and z of degrees 3, 4, 3 and 6, and has eight edges. The degree sequence is $(0, 0, 0, 2, 1, 0, 1, 0, 0, \ldots)$. Sure enough,

$$3 + 4 + 3 + 6 = 2 \cdot 8 = 1 \cdot 0 + 2 \cdot 0 + 3 \cdot 2 + 4 \cdot 1 + 5 \cdot 0 + 6 \cdot 1.$$

(b) The complete graph K_n has n vertices, each of degree $n - 1$, and has $n(n - 1)/2$ edges. ∎

Most of the remaining observations of § 8.2 are still true in the undirected context. In particular, the problem of determining when two graphs are isomorphic is still a difficult one.

EXERCISES 8.4

1. (a) Give a table of the function γ for the graph G pictured in Figure 6.
(b) List the edges of this graph, considered as subsets of $V(G)$. For example, $a = \{w, x\}$.

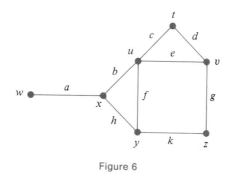

Figure 6

2. Draw a picture of the graph G with $V(G) = \{x, y, z, w\}$, $E(G) = \{a, b, c, d, f, g, h\}$ and γ given by the table:

e	a	b	c	d	f	g	h
$\gamma(e)$	$\{x, y\}$	$\{x, y\}$	$\{w, x\}$	$\{w, y\}$	$\{y, z\}$	$\{y, z\}$	$\{w, z\}$

3. Consider the graph of Figure 6. Carry out the steps in the proof of Theorem 2 to construct cycles starting with the pairs of simple paths with the given vertex sequences.
   (a) $v\ u\ y\ x$ and $v\ z\ y\ u\ x$
   (b) $w\ x\ u\ v\ t\ u\ y$ and $w\ x\ y$
   (c) $w\ x\ u\ t\ v\ z\ y$ and $w\ x\ y$

4. (a) Draw pictures of all 14 graphs with three vertices and three edges. "All" here means one example from each isomorphism class.
   (b) Draw pictures of all graphs with four vertices and four edges which have no loops or parallel edges.
   (c) List the four graphs in parts (a) and (b) which are regular.

5. (a) Draw pictures of all five of the regular graphs with four vertices, each vertex of degree 2.
   (b) Draw pictures of all of the regular graphs with four vertices, each of degree 3, and with no loops or parallel edges.
   (c) Draw pictures of all of the regular graphs with five vertices, each of degree 3.

6. Suppose that a graph H is isomorphic to the graph G of Figure 6.
   (a) How many vertices of degree 1 does H have?
   (b) Give the degree sequence of H.
   (c) How many different isomorphisms are there of G onto G? Explain.
   (d) How many isomorphisms are there of G onto H?

7. Which, if any, of the pairs of graphs shown in Figure 7 are isomorphic? Justify your answer by describing an isomorphism or explaining why one does not exist.

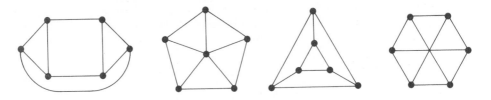

Figure 7

8. Describe an isomorphism between the graphs shown in Figure 8.

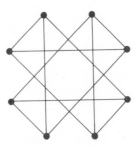

Figure 8

9. Consider the complete graph K_8 with vertices $v_1, v_2, \ldots, v_8$.
   (a) How many subgraphs of K_8 are isomorphic to K_5?
   (b) How many simple paths with 3 or fewer edges are there from v_1 to v_2?
   (c) How many simple paths with 3 or fewer edges are there altogether in K_8?

10. How many different isomorphisms are there of the graph K_n onto itself?

11. (a) A graph with 21 edges has 7 vertices of degree 1, 3 of degree 2, 7 of degree 3 and the rest of degree 4. How many vertices does it have?
    (b) How would your answer to part (a) change if the graph also had 6 vertices of degree 0?

12. Show that the number of vertices of odd degree is even.

13. Which of the following are degree sequences of graphs? In each case either draw a graph with the given degree sequence or explain why no such graph exists.
    (a) $(1, 1, 0, 3, 1, 0, 0, \ldots)$ (b) $(4, 1, 0, 3, 1, 0, 0, \ldots)$
    (c) $(0, 1, 0, 2, 1, 0, 0, \ldots)$ (d) $(0, 0, 2, 2, 1, 0, 0, \ldots)$
    (e) $(0, 0, 1, 2, 1, 0, 0, \ldots)$ (f) $(0, 1, 0, 2, 1, 0, 0, \ldots)$
    (g) $(0, 0, 0, 4, 0, 0, 0, \ldots)$ (h) $(0, 0, 0, 0, 5, 0, 0, \ldots)$

14. Show that a path in a graph G is a cycle if and only if it is possible to assign directions to the edges of G so that the path is a [directed] cycle in the resulting digraph.

15. Show that every finite graph in which each vertex has degree at least 2 contains a cycle.

16. Show that every graph with n vertices and at least n edges contains a cycle. *Hint*: Use induction on n and Exercise 15.

17. Show that
$$2|E(G)| - |V(G)| = -D_0(G) + D_2(G) + 2D_3(G) + \cdots + (k-1)D_k(G) + \cdots.$$

18. (a) Let $\mathscr{S}$ be a set of graphs. Show that isomorphism $\simeq$ is an equivalence relation on $\mathscr{S}$.
    (b) How many equivalence classes are there if $\mathscr{S}$ consists of the four graphs in Figure 4?

§ 8.5   Edge Traversal Problems

Euler's solution to the Königsberg bridge problem is discussed in § 0.1; the Königsberg graph is shown in Figure 5 of that section. FLEURY'S algorithm, which provides a constructive solution to the problem, is discussed in § 0.2. In this section we prove Euler's theorem and some related results and justify FLEURY'S algorithm. A simple path which contains all edges of a graph G is called an **Euler path** of G, and a closed Euler path is called an **Euler circuit**. Euler's solution to the Königsberg bridge problem began with the following elementary observation, whose simple proof is given on page 6.

Theorem 1   A graph which has an Euler circuit must have all vertices of even degree.

Corollary   A graph which has an Euler path has either 2 vertices of odd degree or no vertices of odd degree.

> *Proof.* Suppose G has an Euler path starting at u and ending at v. If $u = v$, the path is closed and Theorem 1 says all vertices have even degree. If $u \neq v$, create a new edge e joining u and v. The new graph $G \cup \{e\}$ has an Euler circuit consisting of the Euler path for G followed by e, so all vertices of $G \cup \{e\}$ have even degree. Remove e. Then u and v are the only vertices of $G = (G \cup \{e\}) \setminus \{e\}$ of odd degree. ∎

EXAMPLE 1   The graph shown in Figure 1(a) has no Euler circuit, since u and v have odd degree, but the path $b\, a\, c\, d\, g\, f\, e$ is an Euler path. The graph in Figure 1(b) has all vertices of even degree and in fact has an Euler circuit. The graph in Figure 1(c) has all vertices of even degree but has no Euler circuit, for the obvious reason that the graph is disconnected into two subgraphs which are not connected to each other. Each of the subgraphs, however, has its own Euler circuit. ∎

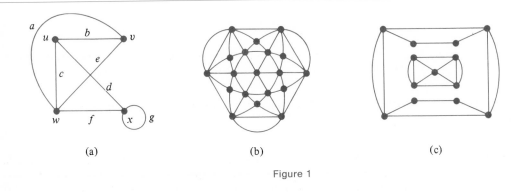

(a)   (b)   (c)

Figure 1

Theorem 1 shows that the even-degree condition is necessary for the existence of an Euler circuit. Euler's major contribution to the problem was his proof that, except for the sort of obvious trouble we ran into in Figure 1(c), the condition is also sufficient to guarantee an Euler path.

We need some terminology to describe the exceptional cases. A graph is called **connected** if each pair of distinct vertices is joined by a path in the graph. The graphs in Figures 1(a) and 1(b) are connected, but the one in Figure 1(c) is not. A connected subgraph of a graph G which is not contained properly in another connected subgraph of G is called a **component** of G. We will show in § 8.7 that the component containing a given vertex v consists of v and all vertices and edges on paths starting at v. For now, we will think of components as maximal connected subgraphs.

EXAMPLE 2   (a) The graphs of Figures 1(a) and 1(b) are connected. In these cases the graph has just one component, namely the graph itself.

(b) The graph of Figure 1(c) has two components, the one drawn on the outside and the one on the inside. Another picture of this graph is shown in Figure 2(a). In this picture there is no "inside" component, but of course there are still two components.

(c) The graph of Figure 2(b) has seven components, two of which are isolated vertices. ∎

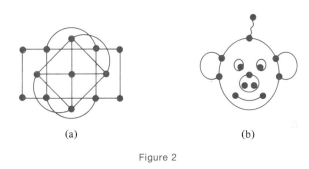

(a)   (b)

Figure 2

We now restate Euler's theorem in § 0.1.

Theorem 2   A finite connected graph in which every vertex has even degree has an Euler circuit.

Before we prove this theorem we illustrate the idea of the proof in a concrete case.

EXAMPLE 3   Consider the graph shown in Figure 3(a). We want a closed simple path which includes all edges, and we observe that $a\,b\,c\,d\,e$ is at least closed and simple, even though it does not contain all edges. We remove the edges a, b, c, d, e. What's left is almost the graph of Figure 3(b). To get Figure 3(b) we also removed the vertex v which would otherwise have been isolated. Now we can look at the two components of Figure 3(b) and see that they have their own Euler circuits $g\,i\,h\,f$ and $j\,k\,m\,n$, as illustrated in Figure 3(c). We can attach those circuits to the cycle $a\,b\,c\,d\,e$ at u and w to get the simple closed path $a\,g\,i\,h\,f\,b\,c\,d\,j\,k\,m\,n\,e$ of Figure 3(d), which is an Euler circuit for the original graph. ∎

Proof of Theorem 2. Suppose G is a finite connected graph with every vertex of even degree. If G has just one vertex the theorem is trivial, so assume $|V(G)| \geq 2$. Since G is connected, each vertex has degree at least 1, and hence at least 2 since all degrees are even.

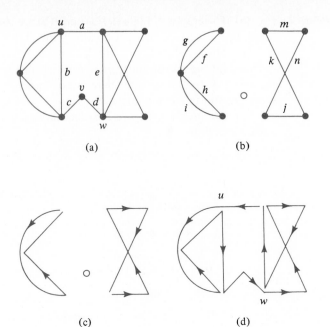

(a)   (b)

(c)   (d)

Figure 3

We first claim that G contains at least one cycle. Let $v_1\ v_2 \cdots v_n$ be the vertex sequence of a path of largest possible length with distinct vertices. Since $\deg(v_n) \geq 2$ there is an edge at v_n different from the last edge in the path. It must join v_n and some v_i with $i \leq n$, or else our chosen path would not be longest possible. Then the path $v_i v_{i+1} \cdots v_n v_i$ with the new edge at the end is a cycle.

Now choose a cycle C in G and remove from G all the edges in C and all the vertices of degree 2 on C, to obtain the subgraph $G \backslash C$. Since C is closed its removal decreases the degree of each vertex by an even number, so every vertex of $G \backslash C$ has even degree. If $G \backslash C$ is empty, then $G = C$ and we are done. Otherwise each of the components of $G \backslash C$ satisfies the hypotheses of the theorem.

It's time for induction. We could have begun with the supposition that the theorem is true for all graphs with fewer edges than G. Then each component of $G \backslash C$ has its own Euler circuit. [To make a constructive algorithm out of this proof, call the algorithm itself recursively on the components.] The components were connected to all the vertices of C before we removed the edges of C, so each has at least one vertex on C. Choose such a vertex in each component. As in Example 3, proceed along C and at each of the chosen vertices attach the corresponding component Euler circuit. The resulting path is closed, simple and contains all the edges of G; it is an Euler circuit for G. ∎

Corollary   A finite connected graph which has exactly two vertices of odd degree has an Euler path.

> *Proof.* Say u and v have odd degree. Create a new edge e joining them. Then $G \cup \{e\}$ has all vertices of even degree and so has an Euler circuit by Theorem 2. Remove e again. What remains of the circuit is an Euler path for G. ∎

The arguments used in the proofs of Theorems 1 and 2 can be easily modified to prove the following analogous result for digraphs.

Digraph   Suppose G is a digraph with more than one edge in which every vertex is
Version   reachable from every other vertex. There is a closed [directed] path in G which contains all edges of G if and only if indeg (v) = outdeg (v) for every vertex v.

We can make the proof of Theorem 2 into an algorithm for actually finding an Euler circuit. The following algorithm gives a different way of constructing Euler paths and circuits, one edge at a time. When the algorithm terminates, the sequence ES is the edge sequence of an Euler path or circuit and VS is its vertex sequence.

FLEURY'S *Algorithm.*

Step 1. Start at any vertex v of odd degree if there is one. Otherwise start at any vertex v. Let VS = v and let ES = ϵ [the empty sequence].

Step 2. If there is no edge remaining at v, stop.

Step 3. If there is exactly one edge remaining at v, say e from v to w, then remove e from $E(G)$ and v from $V(G)$ and go to Step 5.

Step 4. If there is more than one edge remaining at v, choose such an edge, say e from v to w, whose removal will not disconnect the graph; then remove e from $E(G)$.

Step 5. Add w to the end of VS, add e to the end of ES, replace v by w and go to Step 2. ∎

This algorithm was illustrated at the beginning of § 0.2. Before discussing why it works, we give another example to illustrate its use.

EXAMPLE 4   Consider the graph of Figure 4(a). This graph does not have an Euler circuit, but it does have an Euler path joining the vertices z and y of odd degree. We start from z; i.e., let $v = z$ in Step 1. Thus VS = z and ES = ϵ. The only edge at z is i. So we go to Step 3, choose $e = i$ and $w = y$, remove i from $E(G)$, remove z from $V(G)$ and go to Step 5. Then VS = $z\,y$ and ES = i. The new G with both i and z deleted is shown in Figure 4(b). Let $v = y$ and return to Step 2. There are three edges at v, so Step 4 applies.

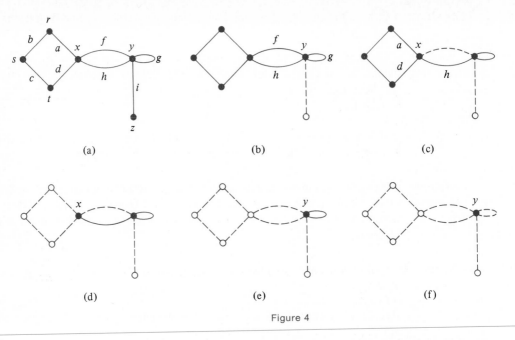

Figure 4

Now e can be f, g or h. Let's take $e = f$ in Step 4. Step 5 gives VS $= z\,y\,x$, ES $= i\,f$, $v = x$ and the new G shown in Figure 4(c). Return to Step 2. Again there are three edges at v and Step 4 applies.

Now e can be either a or d, but not h since removing h would disconnect the graph. Let's choose $e = a$; Step 5 gives VS $= z\,y\,x\,r$, ES $= i\,f\,a$ and $v = r$.

The next three moves are forced. Each leads to Step 3 and the removal of a vertex as well as an edge. We arrive at VS $= z\,y\,x\,r\,s\,t\,x$, ES $= i\,f\,a\,b\,c\,d$, $v = x$ and the graph shown in Figure 4(d).

The remaining two moves are also forced and lead to the graphs in Figures 4(e) and 4(f). The final vertex and edge sequences are VS $= z\,y\,x\,r\,s\,t\,x\,y\,y$ and ES $= i\,f\,a\,b\,c\,d\,h\,g$. ∎

To prove that FLEURY'S algorithm works, we need the next theorem. It will also play a major role in our treatment of trees in Chapter 9.

Theorem 3   Let e be an edge of a connected graph G. The following are equivalent:

(a) $G\backslash\{e\}$ is connected.
(b) e is an edge of some cycle in G.
(c) e is an edge of some simple closed path in G.

Proof. First note that if e is a loop, then $G\backslash\{e\}$ is connected, while e is a cycle all by itself. Since cycles are simple closed paths, the theorem holds in this case, and we may assume that e is not a loop. Thus e connects distinct vertices u and v. If f is another edge connecting u and v, then clearly $G\backslash\{e\}$ is connected and $e\,f$ is a cycle containing e. So the theorem also holds in this case. Hence we may assume that e is the unique edge connecting u and v.

(a) $\Rightarrow$ (b). Suppose that $G\backslash\{e\}$ is connected. By Theorem 1 of § 8.4 there is a simple acyclic path $x_1\,x_2\cdots x_m$ with $u = x_1$ and $x_m = v$. Since there is no edge from u to v in $G\backslash\{e\}$, we have $x_2 \neq v$ and so $m \geq 3$. As noted in Proposition 2 of § 8.4, the vertices $x_1,\,x_2,\ldots,x_m$ are distinct and so $x_1\,x_2\cdots x_m\,u$ is a cycle in G containing the edge e.

(b) $\Rightarrow$ (c). This is obvious since cycles are simple closed paths.

(c) $\Rightarrow$ (a). Now suppose that e is an edge of some simple closed path. Since e is not a loop or parallel edge, e belongs to a simple closed path $v_1\,v_2\cdots v_{n+1}$ where $v_1 = u$, $v_2 = v$, $v_{n+1} = u$ and $n \geq 3$. Consider any vertices x and y of G. Since G is connected, there is some path $w_1\,w_2\cdots w_r$ with $x = w_1$ and $w_r = y$. If u and v appear as consecutive w_i's, replace the edge $u\,v$ in the path from x to y by the path $v_{n+1}\,v_n\cdots v_3\,v$. If v and u appear as consecutive w_i's, replace $v\,u$ by $v\,v_3\cdots v_n\,v_{n+1}$. In this way we obtain a path from x to y that doesn't use e, i.e., a path in $G\backslash\{e\}$. Thus any two vertices of G are connected by a path in $G\backslash\{e\}$ and so $G\backslash\{e\}$ is connected.   ∎

Proof That FLEURY'S Algorithm Works. We consider a finite connected graph all of whose vertices have even degree, and we show that FLEURY'S algorithm produces an Euler circuit. Modifications showing that we get an Euler path in the case of two vertices of odd degree are straightforward.

Each pass through the loop from Step 2 through Step 5 removes an edge from G and adds it to ES in such a way that the edges in ES form a path. Since G has only a finite number of edges to begin with, the algorithm must stop—or break down—sooner or later, and no edge can appear more than once in the path determined by ES. We need to show that the algorithm does not break down, and that when it stops ES contains every edge of G.

The only possible breakdown might occur at Step 4. How do we know that there is an edge we can remove without disconnecting the graph? Say G' is the current value of G and v' the current value of v, and suppose that there is more than one edge at v' in G'. We claim that in fact we can choose *any* of the edges at v', with at most one exception. Since the edges in ES form a path, there are two cases to consider: either G' has two vertices of odd degree, one of which is v' and the other the vertex v_0 chosen in Step 1, or else $v' = v_0$ and G' has no vertices of odd degree.

In the first case, the corollary to Theorem 2 says that G' has an Euler path from v' to v_0. Every time this path returns to v' it completes a simple closed path which includes two of the edges at v' so, except perhaps for the edge that the path uses the last time it leaves v', all edges at v' belong to simple closed paths in G'. In the second case, Theorem 2 says that G' has an Euler circuit, so a similar argument shows that every edge at v' belongs to a simple closed path of G'. By Theorem 3, removing an edge in a simple closed path cannot disconnect G'. Because there is at most one bad edge at v', if the first one we try doesn't belong to a simple closed path then any other choice of an edge at v' will surely work.

Why are there no edges left at the end? At the start the graph is connected, when we execute Step 4 the graph remains connected, and when we execute Step 3 the graph also remains connected, because after we remove the edge e we also remove the isolated vertex v which is left. So at all times the graph is connected. When we're forced to stop because there is no edge left at v' there must not be any other vertices left in G' either. Hence G' has no edges whatever and ES contains every edge of G. ∎

Most of the operations in FLEURY'S algorithm, such as adding or removing edges, take a fixed amount of time independent of how many vertices G has. The operation that takes longer is the test for connectedness of $G \backslash \{e\}$. We will show in § 8.9 that there is a connectedness test based on DIJKSTRA'S algorithm that runs in time $O(|V(G)|^2)$. The proof that FLEURY'S algorithm works shows that we only need to test one edge at Step 4. Since the algorithm makes one pass through the Step 2,..., Step 5 loop for each edge in G and then stops, the total time required by FLEURY'S algorithm is $O(|V(G)|^2 \cdot |E(G)|)$.

EXERCISES 8.5

1. Consider the graph shown in Figure 5(a).
   (a) Describe an Euler path for this graph or explain why there isn't one.
   (b) Describe an Euler circuit for this graph or explain why there isn't one.

2. Repeat Exercise 1 for the graph of Figure 5(b).

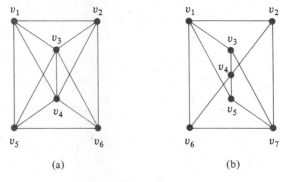

(a)   (b)

Figure 5

3. Is it possible for an insect to crawl along the edges of a cube so as to travel along each edge exactly once? Explain.

4. Draw an Euler circuit in the style of Figure 3(d) for the graph of Figure 1(b).

5. Which complete graphs K_n have Euler circuits?

6. Apply FLEURY'S algorithm as in Example 4 to get an Euler path for the graph of Figure 1(a). Sketch the intermediate graphs obtained in the application of the algorithm, as was done in Figure 4.

7. Construct a graph with vertex set $\{0, 1\}^3$ and with an edge between vertices v and w if v and w differ in two coordinates.
   (a) How many components does the graph have?
   (b) How many vertices does the graph have of each degree?
   (c) Does the graph have an Euler circuit?

8. Answer the same questions as in Exercise 7 for the graph with vertex set $\{0, 1\}^3$ and with an edge between v and w if v and w differ in two or three coordinates.

9. (a) Show that if a connected graph G has exactly $2k$ vertices of odd degree and $k > 0$, then $E(G)$ is the disjoint union of the edge sets of k simple paths. *Hint*: Add more edges, as in the proof of the corollary to Theorem 2.
   (b) Find two disjoint simple paths whose edge set union is $E(G)$ for the Königsberg graph in Figure 5 on page 5.
   (c) Do the same for the graph in Figure 5(b) of Exercise 1.

10. (a) Use the proof of Theorem 2 to create an algorithm for finding Euler circuits.
   (b) How would you modify the algorithm so that it would find Euler paths?

§ 8.6 Vertex Traversal Problems

In § 0.2 we discussed Hamilton circuits and pointed out that characterizing them and finding techniques for discovering them in graphs are very difficult problems. Some facts can be proved, however, and this is our next task. We also discuss a couple of applications.

A path with vertex sequence $x_1 x_2 \cdots x_n$ is called a **Hamilton path** for the graph G if $x_1, x_2, \ldots, x_n$ are distinct and $\{x_1, x_2, \ldots, x_n\} = V(G)$. A closed path $x_1 x_2 \cdots x_n x_1$ is called a **Hamilton circuit** of G if $x_1 x_2 \cdots x_n$ is a Hamilton path, and a graph which has a Hamilton circuit is called a **Hamiltonian graph**. A Hamilton path must clearly be simple, and by Proposition 1 of § 8.4 if G has at least 3 vertices a Hamilton circuit of G must be a cycle.

EXAMPLE 1 (a) The graph shown in Figure 1(a) has Hamilton circuit $v\ w\ x\ y\ z\ v$.
   (b) Adding more edges can't hurt, so the graph K_5 of Figure 1(b) is also Hamiltonian. In fact, every complete graph K_n for $n \geq 3$ is Hamiltonian; we can go from vertex to vertex in any order we please.
   (c) The graph of Figure 1(c) has the Hamilton path $v\ w\ x\ y\ z$ but has no Hamilton circuit since no cycle goes through v.
   (d) The graph of Figure 1(d) has no Hamilton path. ∎

A Hamiltonian graph with n vertices must have at least n edges. This necessary condition may not be sufficient, as Figure 1(d) illustrates. Of

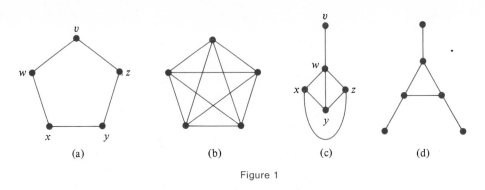

Figure 1

course loops and parallel edges are of no use. The following gives a simple sufficient condition.

Theorem 1   If the graph G has no loops or parallel edges, if $|V(G)| = n \geq 3$ and if $\deg(v) \geq n/2$ for each vertex v of G, then G is Hamiltonian.

EXAMPLE 2   (a) The graph K_5 in Figure 1(b) has $\deg(v) = 4$ for each v and has $|V(G)| = 5$, so it satisfies the condition of Theorem 1.

(b) Each of the graphs in Figure 2 has $|V(G)|/2 = 5/2$ and has a vertex of degree 2. They do not satisfy the hypotheses of Theorem 1, but are nevertheless Hamiltonian.   ∎

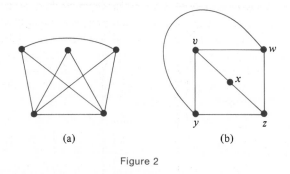

Figure 2

Theorem 1 imposes a uniform condition on all the vertices. Our next theorem requires only that there be enough edges somewhere in the graph. We will establish both of these sufficient conditions as consequences of Theorem 3, which gives a criterion in terms of degrees of pairs of vertices.

Theorem 2   A graph with n vertices and with no loops or parallel edges which has at least $\frac{1}{2}(n - 1)(n - 2) + 2$ edges is Hamiltonian.

EXAMPLE 3   (a) The Hamiltonian graph of Figure 2(a) has $n = 5$, which gives $\frac{1}{2}(n - 1)(n - 2) + 2 = 8$. It has 8 edges and so it satisfies the hypotheses and the conclusion of Theorem 2.

(b) The Hamiltonian graph of Figure 2(b) also has $n = 5$ and so $\frac{1}{2}(n - 1)(n - 2) + 2 = 8$, but it has only 7 edges. It fails to satisfy the hypotheses of Theorem 2, as well as Theorem 1. If there were no vertex in the middle, we would have K_4 with $n = 4$ so $\frac{1}{2}(n - 1)(n - 2) + 2 = 5$, and the 6 edges would be more than enough. As it stands, the graph satisfies the hypotheses of the next theorem. ∎

Theorem 3   Suppose that the graph G has no loops or parallel edges and that $|V(G)| = n \geq 3$. If

$$\deg(v) + \deg(w) \geq n$$

for each two vertices v and w not connected by an edge, then G is Hamiltonian.

EXAMPLE 4   For the graph in Figure 2(b), $n = 5$. There are three pairs of distinct vertices that are not connected by an edge. We verify the hypotheses of Theorem 3 by examining them:

for $\langle v, z \rangle$, $\deg(v) + \deg(z) = 3 + 3 = 6 \geq 5$;

for $\langle w, x \rangle$, $\deg(w) + \deg(x) = 3 + 2 = 5 \geq 5$;

for $\langle x, y \rangle$, $\deg(x) + \deg(y) = 2 + 3 = 5 \geq 5$. ∎

Proof of Theorem 3. Suppose the theorem is false for some n, and let G be a counterexample with $|E(G)|$ as large as possible. Now G is a subgraph of the Hamiltonian graph K_n. Adjoining to G an edge from K_n would give a graph which still satisfies the degree condition but has more than $|E(G)|$ edges. By the choice of G, any such graph would have a Hamilton circuit. This means that G must already have a Hamilton *path*, say with vertex sequence $v_1 v_2 \cdots v_n$. Since G has no Hamilton circuit, v_1 and v_n are not connected by an edge in G, and so $\deg(v_1) + \deg(v_n) \geq n$.

Define subsets S_1 and S_n of $\{2, \ldots, n\}$ by

$$S_1 = \{i : \{v_1, v_i\} \in E(G)\} \quad \text{and} \quad S_n = \{i : \{v_{i-1}, v_n\} \in E(G)\}.$$

Then $|S_1| = \deg(v_1)$ and $|S_n| = \deg(v_n)$. Since $|S_1| + |S_n| \geq n$ and $S_1 \cup S_n$ has at most $n - 1$ elements, $S_1 \cap S_n$ must be nonempty. Thus there is an i for which both $\{v_1, v_i\}$ and $\{v_{i-1}, v_n\}$ are edges of G. Then [see Figure 3] the path

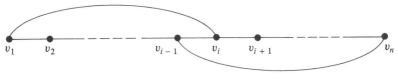

Figure 3

$v_1 \cdots v_{i-1} v_n \cdots v_i v_1$ is a Hamilton circuit in G, contradicting the choice of G as a counterexample. $\blacksquare$

Our first two sufficient conditions follow easily from Theorem 3.

Proofs of Theorems 1 and 2. Suppose G has no loops or parallel edges and $|V(G)| = n \geq 3$.

If $\deg(v) \geq n/2$ for each v then $\deg(v) + \deg(w) \geq n$ for any v and w whether joined by an edge or not, so the hypothesis of Theorem 3 is satisfied and G is Hamiltonian.

Suppose $|E(G)| \geq \frac{1}{2}(n-1)(n-2) + 2 = \binom{n-1}{2} + 2$, and consider vertices u and v with $\{u, v\} \notin E(G)$. Remove from G the vertices u and v and all edges with u or v as a vertex. Since $\{u, v\} \notin E(G)$ we have removed $\deg(u) + \deg(v)$ edges and 2 vertices. The graph G' which is left is a subgraph of K_{n-2}, so

$$\binom{n-2}{2} = |E(K_{n-2})| \geq |E(G')| \geq \binom{n-1}{2} + 2 - \deg(u) - \deg(v).$$

Hence

$$\deg(u) + \deg(v) \geq \binom{n-1}{2} - \binom{n-2}{2} + 2$$

$$= \tfrac{1}{2}(n-1)(n-2) - \tfrac{1}{2}(n-2)(n-3) + 2$$

$$= \tfrac{1}{2}(n-2)[(n-1) - (n-3)] + 2$$

$$= \tfrac{1}{2}(n-2)[2] + 2 = n.$$

Again, G satisfies the hypothesis of Theorem 3. $\blacksquare$

Theorems 1, 2 and 3 are somewhat unsatisfactory in two ways. Not only are their sufficient conditions not necessary, the theorems give no guidance for finding a Hamilton circuit when one is guaranteed to exist. As we explained in §0.2, the problem is related to the Traveling Salesperson Problem. As of this writing no efficient algorithm is known for solving it or for finding Hamilton paths or circuits. On the positive side, a Hamiltonian graph must certainly be connected, so all three theorems give sufficient conditions for a graph to be connected.

EXAMPLE 5   One way to convert the angular position of a rotating pointer into digital form is to divide the circle into 2^n equal segments, label the segments with the binary numbers from 0 to $2^n - 1$ and record the number of the segment the pointer points to. Figure 4 shows two possible ways of assigning labels for such a device with $n = 2$. To read the label electrically, we can provide n segmented concentric rings so that the pointer makes contact with ring i if

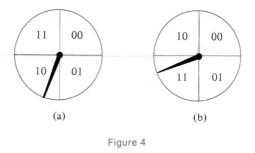

(a)   (b)

Figure 4

and only if the ith digit of the label is 1. Figure 5 shows the rings associated with the labelings of Figure 4. We have numbered the rings from the outside in, so that the outside rings correspond to the first digit and the inside rings correspond to the second digit.

If the pointer is near the boundary between 00 and 11 in Figure 5(a) a slight irregularity in contacts can cause a reading of 01 or 10. The arrangement of Figure 5(b) is better, because a contact error at a boundary can affect only one digit and in case of error the false label is just across the boundary from the true one. Indeed, a misreading of one digit, for whatever reason, yields the label for the segment on one side or the other of the true segment; small mistakes in readings mean small mistakes in interpretation.

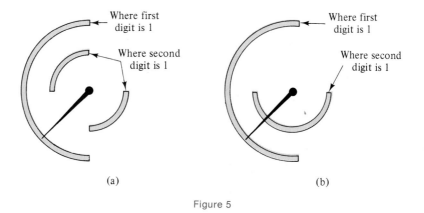

Where first
digit is 1

Where second
digit is 1

Where first
digit is 1

Where second
digit is 1

(a)   (b)

Figure 5

A **Gray code** of length n is a labeling of the 2^n equal segments of a circle with binary strings of length n in such a way that the labels of adjacent segments differ in exactly one digit.

We can view the construction of a Gray code as a graph-theoretic problem. Let $V(G)$ be the set $\{0, 1\}^n$ of binary n-tuples, and join u and v by an edge if u and v differ in exactly one digit. A Gray code of length n is, in effect, a Hamilton circuit of the graph G. Figure 6(a) shows the graph G for $n = 2$. Figure 6(b) shows the same graph redrawn. Compare these pictures

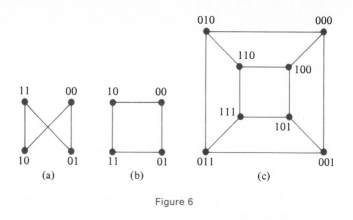

Figure 6

with the labels in Figure 4. This graph has two Hamilton circuits, one in each direction, which shows that there are two [essentially equivalent] Gray codes of length 2. Figure 6(c) shows the graph for $n = 3$. There are 12 Gray codes of length 3. Figure 7(a) indicates the Hamilton path corresponding to one such code, and Figure 7(b) shows the associated segmented rings. Note that a contact error at any boundary would affect only one digit. ∎

The vertices in the graphs we constructed in Example 5 can be partitioned into two sets, those with an even number of 1's and those with an odd number, so that each edge joins a member of one set to a member of the other. We conclude this section with some observations about Hamilton circuits in graphs with this sort of partition.

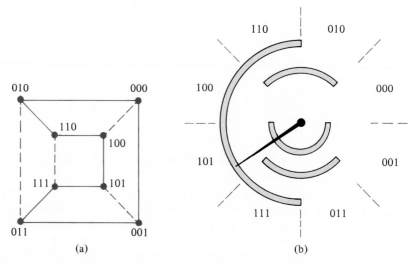

Figure 7

A graph G is called **bipartite** if $V(G)$ is the union of two disjoint nonempty subsets V_1 and V_2 such that every edge of G joins a vertex of V_1 to a vertex of V_2. A graph is called a **complete bipartite** graph if, in addition, every vertex of V_1 is joined to every vertex of V_2 by a unique edge.

EXAMPLE 6 The graphs shown in Figure 8 are all bipartite. All but the one in Figure 8(b) are complete bipartite graphs. ∎

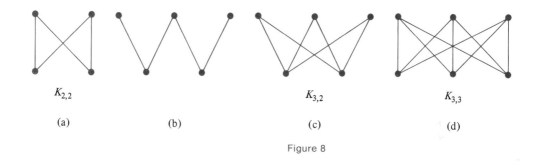

$K_{2,2}$

(a)

(b)

$K_{3,2}$

(c)

$K_{3,3}$

(d)

Figure 8

Given m and n, the complete bipartite graphs with $|V_1| = m$ and $|V_2| = n$ are all isomorphic to each other; we denote them by $K_{m,n}$. Note that $K_{m,n}$ and $K_{n,m}$ are isomorphic.

Theorem 4 Let G be a bipartite graph with partition $V(G) = V_1 \cup V_2$. If G has a Hamilton circuit, then $|V_1| = |V_2|$. If G has a Hamilton path, then the numbers $|V_1|$ and $|V_2|$ differ by at most 1. For complete bipartite graphs with at least 3 vertices the converse statements are also true.

Proof. The vertices on a path in G alternately belong to V_1 and V_2. If $x_1 x_2 \cdots x_n x_1$ is a closed path that goes through each vertex once, then $x_1, x_3, x_5, \ldots$ must belong to one of the sets, say V_1. Since $\{x_n, x_1\}$ is an edge, n must be even and $x_2, x_4, \ldots, x_n$ all belong to V_2. So $|V_1| = |V_2|$. Similar remarks apply to a nonclosed path $x_1 x_2 \cdots x_n$, except that n might be odd, in which case one of V_1 and V_2 will have an extra vertex.

Now suppose $G = K_{m,n}$. If $m = n$ we can simply go back and forth from V_1 to V_2, since edges exist to take us wherever we want. If $m = n + 1$ we should start in V_1 to get a Hamilton path. ∎

A respected computer scientist we know tells the story of how he once spent over two weeks on a computer searching for a Hamilton path in a bipartite graph with 42 vertices before he realized that the graph violated the condition of Theorem 4. The story has two messages: (1) people do have practical applications for bipartite graphs and Hamilton paths, and (2) *thought should precede computation.*

EXERCISES 8.6

1. Consider the graph shown in Figure 5(a) of § 8.5.
   (a) Is this a Hamiltonian graph?
   (b) Is this a complete graph?
   (c) Is this a bipartite graph?
   (d) Is this a complete bipartite graph?

2. Answer the same questions as in Exercise 1 for the graph in Figure 5(b) of § 8.5.

3. (a) How many Hamilton circuits does the graph $K_{n,n}$ have for $n \geq 2$? [Count circuits as different if they have different starting points or vertex sequences.]
   (b) How many Hamilton paths does $K_{n,n-1}$ have for $n \geq 2$?
   (c) Which complete bipartite graphs $K_{m,n}$ have Euler paths?

4. Redraw the graphs in Figure 6 and mark each of the subsets V_1 and V_2 of the bipartite partition of $V(G)$.

5. Arrange eight 0's and 1's in a circle so that each 3-digit binary number occurs as a string of 3 consecutive symbols somewhere in the circle. *Hint*: Find a Hamilton circuit in the graph with vertex set $\{0, 1\}^3$ and with an edge between vertices $\langle v_1, v_2, v_3 \rangle$ and $\langle w_1, w_2, w_3 \rangle$ whenever $\langle v_1 \, v_2 \rangle = \langle w_2, w_3 \rangle$ or $\langle v_2, v_3 \rangle = \langle w_1, w_2 \rangle$.

6. Give two other examples of Gray codes of length 3 besides the one in Example 5. For each code draw the associated segmented rings as in Figure 7(b).

7. Does the graph in Exercise 7 of § 8.5 have a Hamilton circuit? Does it have a Hamilton path?

8. Does the graph in Exercise 8 of § 8.5 have a Hamilton circuit? Does it have a Hamilton path?

9. For $n \geq 4$ build the graph K_n^+ from the complete graph K_{n-1} by adding one more vertex in the middle of an edge of K_{n-1}. [Figure 2(b) shows K_5^+.]
   (a) Show that K_n^+ does not satisfy the condition of Theorem 2.
   (b) Use Theorem 3 to show that K_n^+ is nevertheless Hamiltonian.

10. For $n \geq 4$ build the graph K_n^{++} from the complete graph K_{n-1} by adding one more vertex and an edge from the new vertex to a vertex of K_{n-1}. [Figure 1(c) shows K_5^{++}.] Show that K_n^{++} is not Hamiltonian. Observe that K_n^{++} has n vertices and $\frac{1}{2}(n-1)(n-2) + 1$ edges. This example shows that the number of edges required in Theorem 2 cannot be decreased.

11. The **complement** of a graph G is the graph with vertex set $V(G)$ and with an edge between distinct vertices v and w if G does *not* have an edge joining v and w.
    (a) Draw the complement of the graph of Figure 2(b).
    (b) How many components does the complement in part (a) have?
    (c) Show that if G is not connected then its complement is connected.
    (d) Give an example of a graph which is isomorphic to its complement.
    (e) Is the converse to the statement in part (c) true?

12. Suppose that the graph G is regular of degree $k \geq 1$ [i.e., each vertex has degree k] and has at least $2k + 2$ vertices. Show that the complement of G is Hamiltonian. *Hint*: Use Theorem 1.

13. Show that Gray codes of length n always exist. *Hint:* Use induction on n and consider the graph G_n in which a Hamilton circuit corresponds to a Gray code of length n, as described in Example 5.

14. Explain why none of the theorems in this section can be used to solve Exercise 13.

§ 8.7   Matrices, Relations and Graphs

The pictures of relations in § 7.5 look just like pictures of digraphs in § 8.1. It is not surprising, then, that there is an intimate connection between relations and digraphs. In § 0.3 we briefly examined the connection between matrices and graphs, and in § 7.5 we saw that relations correspond to Boolean matrices. We now tie all these topics together. A relation on a set S determines a digraph G in a natural way: let $V(G) = S$ and put an edge from v to w whenever v is related to w. This was the procedure we used in § 8.1 to get the Hasse diagram of a poset from the covering relation on the poset. In the opposite direction, given a graph or digraph G we can define a relation on $V(G)$ by saying that v is related to w if there is an edge from v to w, or more generally if there is a path of some specified type from v to w.

To begin with, consider an undirected graph G. We call vertices v and w **adjacent** and write $v \, A \, w$ if there is an edge in $E(G)$ from v to w. Thus $v \, A \, w$ if there is a path of length 1 in G from v to w. Then A is a relation on $V(G)$, called the **adjacency relation**. Since G is undirected, the relation A is symmetric. A vertex v is A-related to itself if and only if G has a loop at v, so A is not reflexive in general. Nor is A usually transitive; we can have an edge from u to v and one from v to w without an edge from u to w.

To get a transitive relation from A we must consider chains of edges, from u_1 to u_2, u_2, to u_3, etc. As we saw in § 8.1 the appropriate notion is reachability. Define the **reachable relation** R on $V(G)$ by

$$v \, R \, w \text{ if there is a path of length at least 1 in } G \text{ from } v \text{ to } w.$$

Then R is transitive, and also symmetric since G is undirected. Since we require all paths to have length at least 1, R might not be reflexive. Let $\bar{R} = R \cup E$, the reflexive closure of R, so that $v \, \bar{R} \, w$ if $v = w$ or if $v \, R \, w$. Then $\bar{R}$ is an equivalence relation on $V(G)$. In fact, R is the transitive closure of A, and $\bar{R}$ is the smallest equivalence relation on $V(G)$ which contains A [see the lemma to Theorem 3 of § 7.6].

EXAMPLE 1   (a) Consider the graph G of Figure 1(a). The relation A is $\{\langle v, w \rangle, \langle v, x \rangle, \langle v, y \rangle, \langle w, v \rangle, \langle x, v \rangle, \langle x, x \rangle, \langle x, y \rangle, \langle y, v \rangle, \langle y, x \rangle\}$. It is not reflexive or transitive. The relation R is $V(G) \times V(G)$ since each vertex is reachable from every other and from itself. [For instance, there is a path from w to v to w, so $w \, R \, w$.] Hence $\bar{R} = R$.

(b) The relation A for the graph of Figure 1(b) is $\{\langle u, w \rangle, \langle u, y \rangle, \langle v, x \rangle, \langle v, z \rangle, \langle w, u \rangle, \langle w, y \rangle, \langle x, v \rangle, \langle x, z \rangle, \langle y, u \rangle, \langle y, w \rangle, \langle z, v \rangle, \langle z, x \rangle\}$. The relation R consists of the pairs in A together with the pairs $\langle u, u \rangle, \langle v, v \rangle$,

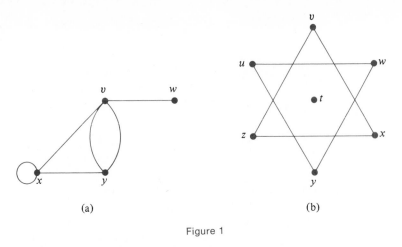

(a)　　　　　　　　　　　(b)

Figure 1

$\langle w, w \rangle$, $\langle x, x \rangle$, $\langle y, y \rangle$ and $\langle z, z \rangle$. The relation $\bar{R}$ is $R \cup \{\langle t, t \rangle\}$. In general, $R = \bar{R}$ if and only if G has no isolated vertices. The $\bar{R}$-equivalence classes of $V(G)$ are $\{u, w, y\}$, $\{v, x, z\}$ and $\{t\}$.　▮

We have seen the $\bar{R}$-equivalence classes of $V(G)$ before; they are the vertex sets of the connected components of G. Recall that a graph is connected if each vertex is reachable from every other vertex, and a component of G is a connected subgraph H which is not contained in any other connected subgraph of G. The next theorem gives us a way of finding components. Simply choose a vertex v, find its $\bar{R}$-class, i.e., all vertices we can get to from v, and then take all edges joining the vertices found. In § 8.9 we will discuss algorithms for finding $\bar{R}$-classes, and hence components. At this point the statement of the theorem is more important to us than its proof.

Theorem　Let G be a graph. Let V be an $\bar{R}$-equivalence class of vertices of G and let H be the subgraph of G with $V(H) = V$ and $E(H) = \{e \in E(G) : e$ joins vertices in $V\}$. Then H is a component of G. Conversely, every component of G is determined by an $\bar{R}$-equivalence class in this way.

Proof. We show first that H is connected, i.e., that there is a path *in H* between any two different vertices in $V(H)$. Let u and v be distinct vertices in V. Since V is an $\bar{R}$-class of G, there is a path in G with vertex sequence $u = u_0 u_1 \cdots u_n = v$. All the vertices $u_1, u_2, \ldots$ on this path are reachable from u in G, so they are all $\bar{R}$-related to u. Thus each is in the $\bar{R}$-class V. Since H contains all edges in G that connect vertices in V, the path is a path in H, as we wished. Thus H is connected.

Now suppose H is contained in a connected subgraph K of G. Choose v in V and let w be any vertex in $V(K)$. Since K is connected and $V \subseteq V(K)$, v and w are joined by a path in K[or $v = w$] and so $w \bar{R} v$ in G. Thus w lies in the $\bar{R}$-class of v, i.e., in V. This shows that $V(K) \subseteq V$, so $V(K) = V$. Since $E(H)$ contains all edges of G between members of V, $E(H) = E(K)$ and thus

$H = K$. We have shown that H is not contained in any other connected subgraph of G, so H is a component.

Now consider a component C of G and a vertex v in $V(C)$. Since C is connected, each vertex of C is in the $\bar{R}$-class of v. We just saw that the graph H with this $\bar{R}$-class as vertex set and with all possible edges is a component of G. Since C is contained in H and C is a component, $C = H$. ∎

As in § 7.5 we can use Boolean matrices to describe the relations A, R and $\bar{R}$ for finite graphs. We index the rows and columns of a matrix by the members of $V(G)$, and we put a 1 in the (u, v)-position if $\langle u, v \rangle$ is in the relation and put a 0 there otherwise.

EXAMPLE 2 (a) The relations A, R and $\bar{R}$ for the graph of Figure 1(a) give the following matrices.

$$\mathbf{M}_A = \begin{array}{c} \\ v \\ w \\ x \\ y \end{array} \begin{array}{cccc} v & w & x & y \\ \left[\begin{matrix} 0 & 1 & 1 & 1 \\ 1 & 0 & 0 & 0 \\ 1 & 0 & 1 & 1 \\ 1 & 0 & 1 & 0 \end{matrix}\right] \end{array}, \qquad \mathbf{M}_R = \mathbf{M}_{\bar{R}} = \begin{bmatrix} 1 & 1 & 1 & 1 \\ 1 & 1 & 1 & 1 \\ 1 & 1 & 1 & 1 \\ 1 & 1 & 1 & 1 \end{bmatrix}.$$

Here we have shown the row and column headings on $\mathbf{M}_A$, but we will often omit them if the indexing is clear.

(b) The graph of Figure 1(b) gives the following.

$$\mathbf{M}_A = \begin{array}{c} \\ u \\ v \\ w \\ x \\ y \\ z \\ t \end{array} \begin{array}{ccccccc} u & v & w & x & y & z & t \\ \left[\begin{matrix} 0 & 0 & 1 & 0 & 1 & 0 & 0 \\ 0 & 0 & 0 & 1 & 0 & 1 & 0 \\ 1 & 0 & 0 & 0 & 1 & 0 & 0 \\ 0 & 1 & 0 & 0 & 0 & 1 & 0 \\ 1 & 0 & 1 & 0 & 0 & 0 & 0 \\ 0 & 1 & 0 & 1 & 0 & 0 & 0 \\ 0 & 0 & 0 & 0 & 0 & 0 & 0 \end{matrix}\right] \end{array}, \qquad \mathbf{M}_{\bar{R}} = \begin{bmatrix} 1 & 0 & 1 & 0 & 1 & 0 & 0 \\ 0 & 1 & 0 & 1 & 0 & 1 & 0 \\ 1 & 0 & 1 & 0 & 1 & 0 & 0 \\ 0 & 1 & 0 & 1 & 0 & 1 & 0 \\ 1 & 0 & 1 & 0 & 1 & 0 & 0 \\ 0 & 1 & 0 & 1 & 0 & 1 & 0 \\ 0 & 0 & 0 & 0 & 0 & 0 & 1 \end{bmatrix}.$$

With a different labeling we get a better idea of the structure.

$$\mathbf{M}_A = \begin{array}{c} \\ u \\ w \\ y \\ v \\ x \\ z \\ t \end{array} \begin{array}{ccccccc} u & w & y & v & x & z & t \\ \left[\begin{matrix} 0 & 1 & 1 & 0 & 0 & 0 & 0 \\ 1 & 0 & 1 & 0 & 0 & 0 & 0 \\ 1 & 1 & 0 & 0 & 0 & 0 & 0 \\ 0 & 0 & 0 & 0 & 1 & 1 & 0 \\ 0 & 0 & 0 & 1 & 0 & 1 & 0 \\ 0 & 0 & 0 & 1 & 1 & 0 & 0 \\ 0 & 0 & 0 & 0 & 0 & 0 & 0 \end{matrix}\right] \end{array}, \qquad \mathbf{M}_{\bar{R}} = \begin{bmatrix} 1 & 1 & 1 & 0 & 0 & 0 & 0 \\ 1 & 1 & 1 & 0 & 0 & 0 & 0 \\ 1 & 1 & 1 & 0 & 0 & 0 & 0 \\ 0 & 0 & 0 & 1 & 1 & 1 & 0 \\ 0 & 0 & 0 & 1 & 1 & 1 & 0 \\ 0 & 0 & 0 & 1 & 1 & 1 & 0 \\ 0 & 0 & 0 & 0 & 0 & 0 & 1 \end{bmatrix}.$$

The blocks of 1's in $\mathbf{M}_{\bar{R}}$ correspond to the components. ∎

Notice that since the relations are symmetric and we have always labeled rows and columns in the same order, the matrices we get are symmetric. Diagonal 1's in M_A correspond to loops. In $M_{\bar{R}}$ all diagonal entries are 1.

The relation R is the transitive closure of A, so by Theorem 2 of § 7.6 we have

$$R = \bigcup_{k=1}^{\infty} A^k.$$

Thus $\bar{R} = E \cup R = \bigcup_{k=0}^{\infty} A^k$. The relations $A^2, A^3, \ldots$ have their own graph-theoretic interpretations. By definition, $u\, A^2\, v$ means there is a vertex z with $u\, A\, z$ and $z\, A\, v$, i.e., there is a path in G of length 2 from u to z to v. Similarly, $u\, A^k\, v$ means there is a path of length k in G joining u and v [Exercise 14]. We saw in Theorem 1 of § 7.5 that if T and U are relations on a set S with matrices M_T and M_U then M_{TU} is the Boolean product $M_T * M_U$. Thus the matrix of A^k is the Boolean product $M_A * M_A * \cdots * M_A$ with k factors.

EXAMPLE 3   (a) Consider the graph of Figure 1(a) and its matrix M_A given in Example 2(a). We compute.

$$M_{A^2} = M_A * M_A = \begin{bmatrix} 0 & 1 & 1 & 1 \\ 1 & 0 & 0 & 0 \\ 1 & 0 & 1 & 1 \\ 1 & 0 & 1 & 0 \end{bmatrix} * \begin{bmatrix} 0 & 1 & 1 & 1 \\ 1 & 0 & 0 & 0 \\ 1 & 0 & 1 & 1 \\ 1 & 0 & 1 & 0 \end{bmatrix} = \begin{bmatrix} 1 & 0 & 1 & 1 \\ 0 & 1 & 1 & 1 \\ 1 & 1 & 1 & 1 \\ 1 & 1 & 1 & 1 \end{bmatrix}.$$

The 0's in M_{A^2} reflect the fact that there is no path of length 2 between v and w. One can easily check that for this example $M_A \vee M_{A^2} = M_{\bar{R}}$.

(b) The graph of Figure 2(a) has the adjacency matrix given in Figure 2(b) with Boolean powers

$$M_{A^2} = \begin{bmatrix} 1 & 0 & 1 & 0 \\ 0 & 1 & 0 & 1 \\ 1 & 0 & 1 & 0 \\ 0 & 1 & 0 & 1 \end{bmatrix}, \quad M_{A^3} = \begin{bmatrix} 0 & 1 & 0 & 1 \\ 1 & 0 & 1 & 0 \\ 0 & 1 & 0 & 1 \\ 1 & 0 & 1 & 0 \end{bmatrix}, \quad M_{A^4} = \begin{bmatrix} 1 & 0 & 1 & 0 \\ 0 & 1 & 0 & 1 \\ 1 & 0 & 1 & 0 \\ 0 & 1 & 0 & 1 \end{bmatrix}.$$

$$M_A = \begin{bmatrix} 0 & 1 & 0 & 0 \\ 1 & 0 & 1 & 0 \\ 0 & 1 & 0 & 1 \\ 0 & 0 & 1 & 0 \end{bmatrix}$$

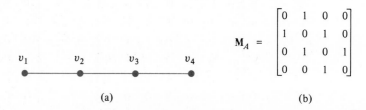

$v_1 \qquad v_2 \qquad v_3 \qquad v_4$

(a)   (b)

Figure 2

Since $\mathbf{M}_{A^2} \vee \mathbf{M}_{A^3}$ has all entries equal to 1, the same is true for $\mathbf{M}_R = \mathbf{M}_{\bar{R}}$ and so R and $\bar{R}$ are the universal relation. Also $\mathbf{M}_{A^2} = \mathbf{M}_{A^4}$ and so $A^2 = A^4$. It follows that $A^3 = A^5$ and in general $A^{2k} = A^2$ and $A^{2k+1} = A^3$ for $k \geq 2$. ∎

Instead of forming the Boolean powers of $\mathbf{M}_A$ we could consider powers using ordinary matrix multiplication. The (u, w)-entry of $\mathbf{M}_A \mathbf{M}_A$ is the sum of all products $\mathbf{M}_A[u, v] \cdot \mathbf{M}_A[v, w]$ for v in $V(G)$. Such a product is 1 if there is a path from u to v to w and is 0 otherwise, since then at least one factor is 0. Thus the (u, w)-entry is the number of products which are 1, i.e., the number of v's in the middle of such paths. If G has no parallel edges this is simply the number of paths of length 2 from u to w.

EXAMPLE 4   (a) For the graph of Example 3(b) we have

$$\mathbf{M}_A \cdot \mathbf{M}_A = \begin{bmatrix} 1 & 0 & 1 & 0 \\ 0 & 2 & 0 & 1 \\ 1 & 0 & 2 & 0 \\ 0 & 1 & 0 & 1 \end{bmatrix} \quad \text{and} \quad \mathbf{M}_A \cdot \mathbf{M}_A \cdot \mathbf{M}_A = \begin{bmatrix} 0 & 2 & 0 & 1 \\ 2 & 0 & 3 & 0 \\ 0 & 3 & 0 & 2 \\ 1 & 0 & 2 & 0 \end{bmatrix}.$$

These matrices have non-0 entries just where $\mathbf{M}_{A^2}$ and $\mathbf{M}_{A^3}$ have them, but some of the entries are 2's and 3's. The diagonal 2's in $\mathbf{M}_A \cdot \mathbf{M}_A$ reflect the fact that there are two paths of length 2 from v_2 and v_3 back to themselves. Similarly, we observe that corresponding to the $(4, 3)$-entry 2 in $\mathbf{M}_A \cdot \mathbf{M}_A \cdot \mathbf{M}_A$ there are 2 paths of length 3 from v_4 to v_3. Their vertex sequences are $v_4 \, v_3 \, v_2 \, v_3$ and $v_4 \, v_3 \, v_4 \, v_3$.

(b) The graph of Figure 1(a) has parallel edges, so $\mathbf{M}_A$ doesn't give the correct number of paths of length 1 between vertices. The matrix

$$\mathbf{M}' = \begin{matrix} v \\ w \\ x \\ y \end{matrix} \begin{bmatrix} 0 & 1 & 1 & 2 \\ 1 & 0 & 0 & 0 \\ 1 & 0 & 1 & 1 \\ 2 & 0 & 1 & 0 \end{bmatrix}$$

does tell the number of edges between vertices of this graph, so its powers yield the number of paths of various lengths. For instance

$$(\mathbf{M}')^2 = \begin{bmatrix} 6 & 0 & 3 & 1 \\ 0 & 1 & 1 & 2 \\ 3 & 1 & 3 & 3 \\ 1 & 2 & 3 & 5 \end{bmatrix}$$

tells us that there are 6 paths of length 2 from v to itself, no paths of length 2 from v to w, 3 paths of length 2 from v to x, etc. ∎

In § 0.3 an example similar to Example 4(b) was analyzed in some detail. There we used the notation $\mathbf{M}$ instead of $\mathbf{M}'$ since we were not also concerned with Boolean matrices. Both examples suggest one way to describe finite graphs with parallel edges using matrices. Choose a linear order for $V(G)$ and let $\mathbf{M}'$ be the matrix with rows and columns indexed by $V(G)$ and with (u, v)-entry the number of edges from u to v. Then $\mathbf{M}'$ completely determines G up to isomorphism and so contains enough information to let us draw a picture of G.

Since

$$\bar{R} = E \cup A \cup A^2 \cup A^3 \cup \cdots,$$

one way to get the relation $\bar{R}$ for a finite graph would be to compute the Boolean power matrices $\mathbf{M}_A, \mathbf{M}_{A^2}, \mathbf{M}_{A^3}, \ldots$ and form their least upper bound to get

$$\mathbf{M}_{\bar{R}} = \mathbf{M}_E \vee \mathbf{M}_A \vee \mathbf{M}_{A^2} \vee \mathbf{M}_{A^3} \vee \cdots.$$

There are only finitely many Boolean matrices of any given size, so only finitely many different matrices $\mathbf{M}_{A^k}$ can arise and this computation can actually be carried out. Indeed, Theorem 2 of § 7.6 points out that we don't need k's greater than $|V(G)|$. It turns out that this matrix procedure is very slow compared with other algorithms. We'll look at this question again in the next section and see some faster methods.

Many of the results we have developed in this section for undirected graphs have analogues for digraphs. Suppose for the rest of this section that G is a digraph. Again we define the **adjacency relation** A by $v \, A \, w$ if there is an edge in $E(G)$ from v to w. In the digraph setting A need not be symmetric. The transitive closure $R = t(A) = \bigcup_{k=1}^{\infty} A^k$ is the **reachable relation**, with $v \, R \, w$ if there is a path in G from v to w, i.e., if w is reachable from v. Even if R is reflexive it might not be an equivalence relation, since it might not be symmetric. The matrices $\mathbf{M}_A$ and $\mathbf{M}_R$ are defined just as in the undirected case. Here again

$$\mathbf{M}_R = \mathbf{M}_A \vee \mathbf{M}_{A^2} \vee \mathbf{M}_{A^3} \vee \cdots.$$

There is no natural analogue for digraphs of the notion of component. The concept of connectedness itself needs to be reexamined. What are we trying to call attention to when we say a digraph is connected? The following two definitions seem to be the most useful. We call a digraph **connected** if the graph we get from it by ignoring directions is connected as an undirected graph, and call a digraph **strongly connected** if $u \, R \, v$ for all vertices u and v. Every strongly connected digraph is connected, and a digraph is strongly connected if and only if the matrix of its reachability relation has all 1's as entries.

EXAMPLE 5   (a) The digraph of Figure 3(a) describes the operation of a **binary up-down counter**. The vertices stand for states of the machine, and the machine shifts from one state to another following arrows in response to inputs. An

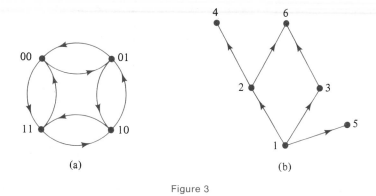

Figure 3

input 1 gives a clockwise shift, and 0 gives a counterclockwise shift. This digraph is strongly connected. Its associated matrices are

$$\mathbf{M}_A = \begin{array}{c} 00 \\ 01 \\ 10 \\ 11 \end{array} \begin{bmatrix} 0 & 1 & 0 & 1 \\ 1 & 0 & 1 & 0 \\ 0 & 1 & 0 & 1 \\ 1 & 0 & 1 & 0 \end{bmatrix} \quad \text{and} \quad \mathbf{M}_R = \begin{bmatrix} 1 & 1 & 1 & 1 \\ 1 & 1 & 1 & 1 \\ 1 & 1 & 1 & 1 \\ 1 & 1 & 1 & 1 \end{bmatrix}.$$

(b) The Hasse diagram in Figure 3(b) has $v\ A\ w$ if w covers v, and has the adjacency and reachability matrices

$$\mathbf{M}_A = \begin{array}{c} 1 \\ 2 \\ 3 \\ 4 \\ 5 \\ 6 \end{array} \begin{bmatrix} 0 & 1 & 1 & 0 & 1 & 0 \\ 0 & 0 & 0 & 1 & 0 & 1 \\ 0 & 0 & 0 & 0 & 0 & 1 \\ 0 & 0 & 0 & 0 & 0 & 0 \\ 0 & 0 & 0 & 0 & 0 & 0 \\ 0 & 0 & 0 & 0 & 0 & 0 \end{bmatrix} \quad \text{and} \quad \mathbf{M}_R = \begin{bmatrix} 0 & 1 & 1 & 1 & 1 & 1 \\ 0 & 0 & 0 & 1 & 0 & 1 \\ 0 & 0 & 0 & 0 & 0 & 1 \\ 0 & 0 & 0 & 0 & 0 & 0 \\ 0 & 0 & 0 & 0 & 0 & 0 \\ 0 & 0 & 0 & 0 & 0 & 0 \end{bmatrix}.$$

This digraph is connected, but not strongly connected. We can't tell that the digraph is connected just by looking at $\mathbf{M}_R$, though the matrix does contain enough information to answer the connectedness question. To check connectedness we could form $\mathbf{M}_A \vee \mathbf{M}_A^T$, the adjacency matrix for the associated undirected graph, and then take its Boolean powers as we did earlier. We could also do the same thing with the matrix $\mathbf{M}_R \vee \mathbf{M}_R^T$. In either case we would arrive at a matrix with all entries 1 and could conclude that the digraph is connected. Of course the easiest method for this example is to stare at the picture. ∎

EXERCISES 8.7

1. (a) Describe the adjacency relation A for an [undirected] graph G in terms of the function γ from $E(G)$ to $\{\{u, v\} : u,\ v \in V(G)\}$.
   (b) Give a description of A in terms of $\gamma : E(G) \to V(G) \times V(G)$ for a digraph G.

2. (a) Draw a picture of a graph with adjacency matrix

$$\begin{bmatrix} 0 & 0 & 1 & 1 & 0 & 1 \\ 0 & 0 & 0 & 0 & 1 & 0 \\ 1 & 0 & 1 & 0 & 0 & 0 \\ 1 & 0 & 0 & 0 & 0 & 1 \\ 0 & 1 & 0 & 0 & 0 & 0 \\ 1 & 0 & 0 & 1 & 0 & 0 \end{bmatrix}.$$

   (b) Find the components of the graph of part (a).

3. Find matrices $\mathbf{M}_A$, $\mathbf{M}_{A^2}$, $\mathbf{M}_{A^3}$, $\mathbf{M}_{A^4}$ and $\mathbf{M}_{\bar{R}}$ for the graph shown in Figure 4(a).

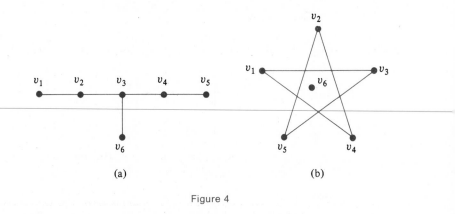

(a)   (b)

Figure 4

4. Find the matrices $\mathbf{M}_A$, $\mathbf{M}_{A^2}$, $\mathbf{M}_{A^4}$ and $\mathbf{M}_{A^{2001}}$ for the graph of Figure 4(b).

5. (a) Give the matrices $\mathbf{M}_A$ and $\mathbf{M}_R$ for the digraph of Figure 5(a).
   (b) Is the digraph connected? Strongly connected?

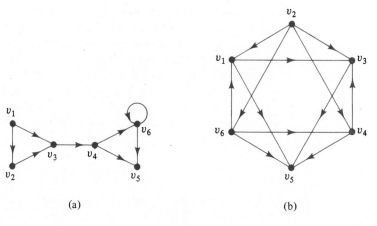

(a)   (b)

Figure 5

6. (a) Find the matrices M_A, M_{A^2}, M_{A^3}, M_{A^4} and M_R for the digraph of Figure 5(b).
   (b) Is this digraph connected? Strongly connected?
   (c) Is this digraph acyclic? [Try to answer this part using your answer to part (a).]

7. (a) How many paths are there of length 3 from v_1 to v_3 in the graph of Figure 4(b)? *Hint*: See Example 4.
   (b) How many paths are there of length 4 from v_1 to itself in this graph?
   (c) How many are there of length 4 from v_1 to v_2?

8. (a) Find $(M_A)^2$ for the digraph of Figure 5(b).
   (b) How many paths are there of length 2 from v_2 to each of the other vertices in this digraph?
   (c) How many paths are there of length 4 from v_2 to other vertices?

9. Define the relation S on $V(G)$ by $v\,S\,w$ if $v = w$ or there is a path of even length from v to w.
   (a) Find the relation S for the graph of Figure 4(a).
   (b) Find S for the graph of Figure 4(b).
   (c) Is S an equivalence relation in general? Explain.
   (d) If G is a bipartite graph, what can you say about S?

10. (a) Describe the matrices M_A and $M_{\bar{R}}$ for a complete graph K_n.
    (b) Describe M_A and $M_{\bar{R}}$ for a complete bipartite graph $K_{m,n}$. [Order the vertices in an intelligent way.]

11. (a) Say as much as you can about a connected graph whose reachability matrix has a 0 entry on the diagonal.
    (b) Do the same for a connected digraph.

12. (a) Show that if all entries of one row of the reachability matrix for a graph are 1, then all entries of M_R are 1.
    (b) Is a statement analogous to that in part (a) true for digraphs? Explain.

13. (a) Can two nonisomorphic graphs have the same adjacency matrix M_A? Explain.
    (b) Give a method for finding the degree of a vertex in a graph without parallel edges by looking at its adjacency matrix.
    (c) Give a method for finding indegrees and outdegrees of a digraph without parallel edges by looking at its adjacency matrix.
    (d) How would your answers to parts (b) and (c) change if parallel edges were allowed?

14. Show that if u and v are vertices of a graph with adjacency relation A and $k \in \mathbb{P}$, then $u\,A^k\,v$ if and only if there is a path of length k from u to v. *Hint*: Use induction on k.

15. (a) Show that it is possible to order the vertices of a finite acyclic digraph so that the non-0 entries of M_A lie on or above the diagonal.
    (b) Will the matrix M_R have the same property?

§ 8.8   Graph Algorithms

> algorithm *n.* Any peculiar method of computing.
> *The American College Dictionary*

Our study of digraphs and graphs has led us to a number of concrete questions. Given a digraph, what is the length of a shortest path from one vertex to another? If the digraph is weighted, what's the minimum or maximum weight of such a path? Is there any path at all? What are the components of a graph? Does removing an edge increase the number of components? Does a given edge belong to a cycle? Do any edges belong to cycles?

This section describes some algorithms for answering these questions and others, algorithms which can be implemented on computers as well as used to organize hand computations. The algorithms we have chosen are reasonably fast, and their workings are comparatively easy to follow. For a more complete discussion we refer the reader to books on the subject, such as *Data Structures and Algorithms* by Aho, Hopcroft and Ullman.

We concentrate first on digraph problems, and deal later with modifications for the undirected case. As we noted in § 8.3, any min-weight algorithm can be used to get a shortest path-length algorithm simply by giving all edges weight 1. We can use such an algorithm to see if there is any path at all from u to v in G by creating fictitious edges of enormous weights between vertices which are not joined by edges in G. If the min-path from u to v in the enlarged graph has an enormous weight, it must be because no path exists made entirely from edges of G.

The min-weight problem is essentially a question about digraphs without loops or parallel edges, so we limit ourselves to that setting. Hence $E(G) \subseteq V(G) \times V(G)$, and we can describe the digraph with a table of the edge-weight function $W(u, v)$, as we did in § 8.3.

The min-weight algorithms we will consider all begin by looking at paths of length 1, i.e., single edges, and then systematically consider longer and longer paths between vertices. As they proceed, the algorithms find smaller and smaller path weights between vertices, and when they stop the weights are the best possible.

Our first algorithm just finds min-weights from a selected vertex to the other vertices in the digraph G. To describe how it works, it will be convenient to suppose that $V(G) = \{1, \dots, n\}$ and that 1 is the selected vertex. Starting with 1, the algorithm looks at additional vertices one at a time, always choosing a new vertex w whose known best path weight from 1 is as small as possible, and updating best path weights from 1 to other vertices by considering paths through w. It keeps track of the set of vertices which it has looked at, by putting them in a set L, and it doesn't look at them again. At any given time, $D(j)$ is the smallest weight of a path from 1 to j whose vertices lie in L. Here is the recipe.

DIJKSTRA'S *Algorithm.*

Set $L = \{1\}$
For $i = 1$ to n
    Set $D(i) = W(1, i)$
End for
While $V(G) \backslash L \neq \emptyset$
    Choose k in $V(G) \backslash L$ with $D(k)$ as small as possible
    Put k in L
    For each j in $V(G) \backslash L$
        If $D(j) > D(k) + W(k, j)$
            Replace $D(j)$ by $D(k) + W(k, j)$
    End for
End while
End ▮

We have written this algorithm in a style similar to a computer program, rather than in our usual "Step 1, Step 2, …" format, to make the looping clearer. Each time the algorithm runs through the While loop, it goes back and checks to see if $V(G) \backslash L$ is empty yet. If it is, the algorithm stops. If not, the algorithm goes through the loop again, with a new k.

We need to check that this algorithm stops, and that the final value of $D(j)$ is $W^*(1, j)$ for each j. We also want to get some estimate of the time it takes the algorithm to run. First, though, let's look at how it works.

EXAMPLE 1   Consider the weighted digraph G shown in Figure 1(a). Its edge-weight table is given in Figure 1(b).

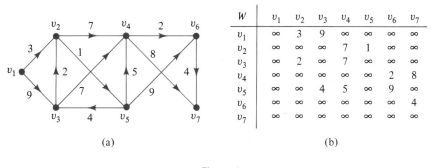

W	v_1	v_2	v_3	v_4	v_5	v_6	v_7
v_1	∞	3	9	∞	∞	∞	∞
v_2	∞	∞	∞	7	1	∞	∞
v_3	∞	2	∞	7	∞	∞	∞
v_4	∞	∞	∞	∞	∞	2	8
v_5	∞	∞	4	5	∞	9	∞
v_6	∞	∞	∞	∞	∞	∞	4
v_7	∞	∞	∞	∞	∞	∞	∞

(a)   (b)

Figure 1

The table in Figure 2 shows how the values of L and of $D(2), \ldots, D(7)$ change as the algorithm progresses, starting with 1.

Notice that the values in the columns decrease with time, and that once j gets into L the value of $D(j)$ doesn't change. ▮

L	D(1)	D(2)	D(3)	D(4)	D(5)	D(6)	D(7)	Comment
$\{1\}$	∞	3	9	∞	∞	∞	∞	Initial data
$\{1, 2\}$	∞	3	9	10	4	∞	∞	Found $v_1 v_2 v_4$ and $v_1 v_2 v_5$
$\{1, 2, 5\}$	∞	3	8	9	4	13	∞	Found $v_1 v_2 v_5 v_3$, $v_1 v_2 v_5 v_4$ and $v_1 v_2 v_5 v_6$
$\{1, 2, 5, 3\}$	∞	3	8	9	4	13	∞	No improvement
$\{1, 2, 5, 3, 4\}$	∞	3	8	9	4	11	17	Found $v_1 v_2 v_5 v_4 v_6$ and $v_1 v_2 v_5 v_4 v_7$
$\{1, 2, 5, 3, 4, 6\}$	∞	3	8	9	4	11	15	Found $v_1 v_2 v_5 v_4 v_6 v_7$

Figure 2

Theorem 1   If the edge weights in G are nonnegative, then DIJKSTRA'S algorithm stops with $D(j) = W^*(1, j)$ for $j = 2, \ldots, n$.

Proof. Since each pass through the While loop in the algorithm adds one more vertex to L, the algorithm makes $n - 1$ trips through the loop and then stops. For convenience, denote the final value of $D(j)$ by $D^*(j)$. It is not at all obvious that $D^*(j) = W^*(1, j)$, because the algorithm makes "greedy" choices of new vertices whenever it gets a chance. Greed does not always pay—consider trying to get 40 cents out of a pile of dimes and quarters by picking a quarter first—but it works this time.

Notice first that $W^*(1, j) \leq D(j)$ at all times and for all j, because $D(j) = W(1, j)$ at the start, and any replacement value is the weight of some path from 1 to j. The following lemma displays the key property of DIJKSTRA'S algorithm.

Lemma   If j is chosen before k, then $D^*(j) \leq D^*(k)$.

Proof. Suppose the lemma is false. Among all pairs $\langle j, k \rangle$ with j chosen before k and with $D^*(j) > D^*(k)$, take a pair with the time interval between the two choices as small as possible.

When j is chosen, k is not yet in L, so that $D^*(j) = D(j) \leq D(k)$. At the end $D^*(k) < D^*(j)$, so there must be an m chosen between j and k for which $D(k)$ is replaced by $D(m) + W(m, k)$ with $D(m) + W(m, k) < D^*(j)$. Then also $D^*(m) \leq D(m) \leq D(m) + W(m, k)$, since $W(m, k) \geq 0$ by the hypothesis of Theorem 1. But this means that $D^*(m) < D^*(j)$ with the time interval between the choices of j and m less than that between j and k, a contradiction. ∎

We return to the proof of Theorem 1. We know that $W^*(1, j) \leq D^*(j)$ for every j, and we want to show equality holds. Suppose not, and among values of j with $W^*(1, j) < D^*(j)$ choose one with $W^*(1, j)$ as small as possible.

Consider a path $1 \cdots kj$ of weight $W^*(1, j)$. For each vertex m in this path, $W^*(1, m)$ is the weight of the initial segment $1 \cdots m$ by the Proposition of §8.3. Hence, in particular, $W^*(1, m) \leq W^*(1, j)$, and we may suppose that j is the first vertex on the path for which $W^*(1, j) < D^*(j)$. If $k = 1$, i.e., if the path has length 1, then $D^*(j) > W^*(1, j) = W(1, j) = D(j)$ at the start, which is absurd. So $k \neq 1$.

Since k precedes j on the path, $W^*(1, k) = D^*(k)$, so

$$D^*(k) \leq D^*(k) + W(k, j) = W^*(1, k) + W(k, j) = W^*(1, j) < D^*(j).$$

By the lemma, this means that k is chosen before j. So at the time k is chosen it must be true that $W^*(1, k) = D^*(k) = D(k)$. At the end of that pass through the loop,

$$W^*(1, j) = W^*(1, k) + W(k, j) = D(k) + W(k, j)$$

$$\geq D(j) \qquad \text{[by replacement, if needed]}$$

$$\geq D^*(j),$$

a final contradiction.   ∎

How long does DIJKSTRA'S algorithm take for a digraph with n vertices? The largest part of the time is spent going through the While loop, removing one of the original n vertices at each pass. Time to find the smallest $D(k)$ is at worst $O(n)$ if we simply examine the vertices in $V(G)$ one by one, and in fact there are sorting algorithms which will do this faster. For each chosen vertex k there are at most n comparisons and replacements, so the total time for one pass through the loop is at most $O(n)$. All told, the algorithm makes n loops, so it takes total time $O(n^2)$.

If the digraph is presented in terms of successor lists, the algorithm can be rewritten so that the replacement/update step only looks at successors of k. During the total operation, each edge is then considered just once in an update step. Such a modification speeds up the overall performance if $|E(G)|$ is much less than n^2.

DIJKSTRA'S algorithm finds the weights of min-paths from a given vertex. To find $W^*(v_i, v_j)$ for all choices of vertices v_i and v_j we could just apply the algorithm n times, starting from each of the n vertices. There is another algorithm, originally due to Warshall and refined by Floyd, which produces all of the values $W^*(v_i, v_j)$ at the end, and which is easy to program. Like DIJKSTRA'S algorithm, it builds an expanding list of examined vertices and looks at paths through vertices on the list.

Suppose that $V(G) = \{v_1, \ldots, v_n\}$. WARSHALL'S algorithm works with an $n \times n$ matrix $\mathbf{W}$, which at the beginning is the edge-weight matrix $\mathbf{W}_0$ with $\mathbf{W}_0[i, j] = W(v_i, v_j)$ for all i and j, and at the end is the min-weight matrix $\mathbf{W}_n = \mathbf{W}^*$ with $\mathbf{W}^*[i, j] = W^*(v_i, v_j)$.

WARSHALL'S *Algorithm.*

For $k = 1$ to n
    For $i = 1$ to n
        For $j = 1$ to n
            If $\mathbf{W}[i, j] > \mathbf{W}[i, k] + \mathbf{W}[k, j]$ then
                Replace $\mathbf{W}[i, j]$ by $\mathbf{W}[i, k] + \mathbf{W}[k, j]$
        End for
    End for
End for
End   ∎

Theorem 2 WARSHALL'S algorithm produces the min-weight matrix $\mathbf{W}^*$.

Proof. We will prove the following assertion for each $k = 1, 2, \ldots, n$. At the conclusion of the execution of the loop for k, each $\mathbf{W}[i, j]$ is the smallest weight of any path from v_i to v_j whose intermediate vertices [if any] are all in $\{v_1, \ldots, v_k\}$. [The smallest weight is ∞ if there is no path of this kind.] It will follow that $\mathbf{W} = \mathbf{W}^*$ at the conclusion of the loop for $k = n$.

First consider the computation of the entry $\mathbf{W}[i, j]$ in the loop for $k = 1$. The computed value of $\mathbf{W}[i, j]$ is either the original value of $\mathbf{W}[i, j]$ or is $\mathbf{W}[i, 1] + \mathbf{W}[1, j]$, so it is either ∞ or the weight of the edge $w_i w_j$ or the weight of the path $w_i w_1 w_j$. It is the smallest of these, so the assertion for $k = 1$ in the last paragraph is true.

Now assume inductively that after execution of the loop for $k = m$ $\mathbf{W}$ is as described, and consider the computation of $\mathbf{W}[i, j]$ in the loop for $k = m + 1$. Suppose first that among all paths from v_i to v_j with intermediate vertices in $\{v_1, \ldots, v_{m+1}\}$ there is one of smallest weight which does not go through v_{m+1}. By the inductive assumption its weight is the current value of $\mathbf{W}[i, j]$. Since $\mathbf{W}[i, m + 1] + \mathbf{W}[m + 1, j]$ is the weight of a path from v_i to v_j through vertices in $\{v_1, \ldots, v_{m+1}\}$, we must have $\mathbf{W}[i, j] \leq \mathbf{W}[i, m + 1] + \mathbf{W}[m + 1, j]$, and so no replacement occurs. The new value of $\mathbf{W}[i, j]$ is the same as the old, and is the smallest weight for a path from v_i to v_j with intermediate vertices in $\{v_1, \ldots, v_{m+1}\}$.

Now suppose every path of smallest weight from v_i to v_j through $\{v_1, \ldots, v_{m+1}\}$ goes through v_{m+1}. Since such a path passes through each vertex at most once, the weight of such a path is $\mathbf{W}[i, m + 1] + \mathbf{W}[m + 1, j]$, where these entries still have the values they attained at the end of the loop for $k = m$. Since this sum is less than $\mathbf{W}[i, j]$ at the end of the loop for $k = m$, the appropriate replacement is made. That is, after executing the loop for $k = m + 1$, $\mathbf{W}$ is as described in the first paragraph.

The claim for all $k = 1, 2, \ldots, n$ follows by finite induction.   ∎

EXAMPLE 2 We apply WARSHALL'S algorithm to the digraph shown in Figure 3. Hand calculations with WARSHALL'S algorithm lead to n new matrices, one for each value of k.

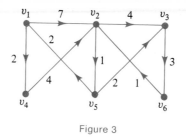

Figure 3

For this example, the matrices are the following.

$$
W = W_0 = \begin{bmatrix}
\infty & 7 & \infty & 2 & \infty & \infty \\
\infty & \infty & 4 & \infty & 1 & \infty \\
\infty & \infty & \infty & \infty & \infty & 3 \\
\infty & 4 & \infty & \infty & \infty & \infty \\
2 & \infty & 2 & \infty & \infty & \infty \\
\infty & 1 & \infty & \infty & \infty & \infty
\end{bmatrix}
\qquad
W_1 = \begin{bmatrix}
\infty & 7 & \infty & 2 & \infty & \infty \\
\infty & \infty & 4 & \infty & 1 & \infty \\
\infty & \infty & \infty & \infty & \infty & 3 \\
\infty & 4 & \infty & \infty & \infty & \infty \\
2 & 9 & 2 & 4 & \infty & \infty \\
\infty & 1 & \infty & \infty & \infty & \infty
\end{bmatrix}
$$

$$
W_2 = \begin{bmatrix}
\infty & 7 & 11 & 2 & 8 & \infty \\
\infty & \infty & 4 & \infty & 1 & \infty \\
\infty & \infty & \infty & \infty & \infty & 3 \\
\infty & 4 & 8 & \infty & 5 & \infty \\
2 & 9 & 2 & 4 & 10 & \infty \\
\infty & 1 & 5 & \infty & 2 & \infty
\end{bmatrix}
\qquad
W_3 = \begin{bmatrix}
\infty & 7 & 11 & 2 & 8 & 14 \\
\infty & \infty & 4 & \infty & 1 & 7 \\
\infty & \infty & \infty & \infty & \infty & 3 \\
\infty & 4 & 8 & \infty & 5 & 11 \\
2 & 9 & 2 & 4 & 10 & 5 \\
\infty & 1 & 5 & \infty & 2 & 8
\end{bmatrix}
$$

$$
W_4 = \begin{bmatrix}
\infty & 6 & 10 & 2 & 7 & 13 \\
\infty & \infty & 4 & \infty & 1 & 7 \\
\infty & \infty & \infty & \infty & \infty & 3 \\
\infty & 4 & 8 & \infty & 5 & 11 \\
2 & 8 & 2 & 4 & 9 & 5 \\
\infty & 1 & 5 & \infty & 2 & 8
\end{bmatrix}
\qquad
W_5 = \begin{bmatrix}
9 & 6 & 9 & 2 & 7 & 12 \\
3 & 9 & 3 & 5 & 1 & 6 \\
\infty & \infty & \infty & \infty & \infty & 3 \\
7 & 4 & 7 & 9 & 5 & 10 \\
2 & 8 & 2 & 4 & 9 & 5 \\
4 & 1 & 4 & 6 & 2 & 7
\end{bmatrix}
$$

$$
W^* = W_6 = \begin{bmatrix}
9 & 6 & 9 & 2 & 7 & 12 \\
3 & 7 & 3 & 5 & 1 & 6 \\
7 & 4 & 7 & 9 & 5 & 3 \\
7 & 4 & 7 & 9 & 5 & 10 \\
2 & 6 & 2 & 4 & 7 & 5 \\
4 & 1 & 4 & 6 & 2 & 7
\end{bmatrix}
$$

Notice that this example is strongly connected, so ∞ does not appear in the final matrix W^*.

We illustrate the computations by calculating $W_4[5, 2]$. Here $k = 4$. The entry $W_3[5, 2]$ is 9, corresponding to the shortest path $v_5 v_1 v_2$ with intermediate vertices from $\{v_1, v_2, v_3\}$. To find $\mathbf{W}_4[5, 2]$ we look at $\mathbf{W}_3[5, 4] + \mathbf{W}_3[4, 2]$, which is $4 + 4 = 8$, corresponding to the pair of paths $v_5 v_1 v_4$ and $v_4 v_2$. Since $9 > 8$, we replace $\mathbf{W}_3[5, 2]$ by $\mathbf{W}_3[5, 4] + \mathbf{W}_3[4, 2] = 8$, corresponding to $v_5 v_1 v_4 v_2$.

A given entry, such as $\mathbf{W}[5, 2]$, may change several times during the calculations, as k runs through all possible values. ∎

It is easy to analyze how long WARSHALL'S algorithm takes. The comparison/replacement step inside the j-loop takes at most some fixed amount of time, say t. The step is done exactly n^3 times, once for each possible choice of the triple $\langle k, i, j \rangle$, so the total time to execute the algorithm is $n^3 t$, which is $O(n^3)$.

The comparison/replacement step for WARSHALL'S algorithm is the same as the one in DIJKSTRA'S algorithm, which includes other sorts of steps as well. Since DIJKSTRA'S algorithm can be done in $O(n^2)$ time, doing it once for each of the n vertices gives an $O(n^3)$ algorithm to find all min-weights. The time constants involved in this multiple DIJKSTRA'S algorithm are different from the ones for WARSHALL'S algorithm, though, so the choice of which algorithm to use may depend on the computer implementation available. If $|E(G)|$ is small compared with n^2, a successor list presentation of the digraph favors choosing DIJKSTRA'S algorithm.

This is probably the time to confess that we are deliberately ignoring one possible complication. Even if G has only a handful of vertices, the weights $W(i, j)$ could still be so large that it would take years to write them down, so both DIJKSTRA'S and WARSHALL'S algorithms might take a long time. In practical situations, though, the numbers that come up in the applications of these two algorithms are of manageable size. In any case, given the same set of weights, our comparison of the relative run times for the algorithms is still valid.

Notice that in the proof of Theorem 2 we only used the fact that $W(v_i, v_j) \geqq 0$ to argue that a min-path cannot visit a vertex twice. If we assume that the digraph is acyclic, then negative weights can be allowed.

WARSHALL'S algorithm can also be adapted to finding max-weights in an acyclic digraph. Replace all the ∞'s by $-\infty$'s, with $-\infty + x = -\infty = x + (-\infty)$ for all x, and $-\infty < a$ for all real a. Then change the inequality in the replacement step to $\mathbf{W}[i, j] < \mathbf{W}[i, k] + \mathbf{W}[k, j]$. The resulting algorithm computes $\mathbf{W}_n[i, j] = M(v_i, v_j)$, where M is the max-weight function of § 8.3.

If we just want max-weights from a single source, as we might for a scheduling network, we can simplify the algorithm somewhat, at the cost of first topologically sorting the digraph, using an algorithm such as NUMBERING VERTICES of § 8.1 or TOPSORT of § 9.3. To find max-weights from a given vertex v_s, for instance v_1, just fix $i = s$ in the max-modified WARSHALL'S algorithm. The sorted labeling also means that $\mathbf{W}[k, j] = -\infty$ if $j \leqq k$, so the j-loop does not need to go all the way from 1 to n. The resulting algorithm looks like this, with $i = 1$.

Algorithm MAX-WEIGHT.

For $k = 2$ to $n - 1$
 For $j = k + 1$ to n
  If $\mathbf{W}[1, j] < \mathbf{W}[1, k] + \mathbf{W}[k, j]$ then
   Replace $\mathbf{W}[1, j]$ by $\mathbf{W}[1, k] + \mathbf{W}[k, j]$
 End for
End for
End ∎

We emphasize that this algorithm is only meant for acyclic digraphs with topologically sorted labelings. The proof that it works in time $O(n^2)$ [Exercise 12] is similar to the argument for WARSHALL'S algorithm.

DIJKSTRA'S algorithm does not work with negative weights [Exercise 10]. Moreover [Exercise 11], there seems to be no natural way to modify it to find max-weights without first sorting the vertices. Given such a labeling, DIJKSTRA'S algorithm for max-weights is essentially the MAX-WEIGHT algorithm we just described. It can be speeded up somewhat if the digraph is given by successor lists and we just consider j in SUCC(k). Then each edge gets examined exactly once to see if it enlarges max-weights. The algorithm runs in a time of $O(\max\{|V(G)|, |E(G)|\})$, which is comparable to the time it takes to sort the vertices using algorithm TOPSORT.

EXAMPLE 3   We apply MAX-WEIGHT to the digraph of Figure 4(a). Figure 4(b) gives the initial matrix $\mathbf{W}_0$. For convenience, we use a row matrix $\mathbf{D}$ with $\mathbf{D}[j] = \mathbf{W}[1, j]$ for $j = 1, \ldots, 6$. The sequence of matrices is as follows.

$$\mathbf{D}_0 = \mathbf{D}_1 = [-\infty \quad 1 \quad 2 \quad -\infty \quad -\infty \quad -\infty]$$
$$\mathbf{D}_2 = [-\infty \quad 1 \quad 2 \quad 3 \quad 4 \quad -\infty]$$
$$\mathbf{D}_3 = [-\infty \quad 1 \quad 2 \quad 7 \quad 4 \quad -\infty]$$
$$\mathbf{D}_4 = \mathbf{D}_5 = [-\infty \quad 1 \quad 2 \quad 7 \quad 4 \quad 9]$$

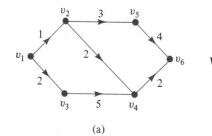

$$\mathbf{W}_0 = \begin{bmatrix} -\infty & 1 & 2 & -\infty & -\infty & -\infty \\ -\infty & -\infty & -\infty & 2 & 3 & -\infty \\ -\infty & -\infty & -\infty & 5 & -\infty & -\infty \\ -\infty & -\infty & -\infty & -\infty & -\infty & 2 \\ -\infty & -\infty & -\infty & -\infty & -\infty & 4 \\ -\infty & -\infty & -\infty & -\infty & -\infty & -\infty \end{bmatrix}$$

(a)   (b)

Figure 4

As an illustration, we compute $\mathbf{D}_4$, assuming that $\mathbf{D}_3$ gives the right values of $\mathbf{W}[1, j]$ for $k = 3$.

The j-loop for $k = 4$ is

For $j = 5$ to 6
　　If $\mathbf{D}[j] < \mathbf{D}[4] + \mathbf{W}[4, j]$ then
　　　　Replace $\mathbf{D}[j]$ by $\mathbf{D}[4] + \mathbf{W}[4, j]$
End for.

Since $\mathbf{D}_3[5] = 4 > -\infty = 7 + (-\infty) = \mathbf{D}_3[4] + \mathbf{W}[4, 5]$, we make no replacement and get $\mathbf{D}_4[5] = \mathbf{D}_3[5] = 4$. Since $\mathbf{D}_3[6] = -\infty < 9 = 7 + 2 = \mathbf{D}_3[4] + \mathbf{W}[4, 6]$, we make a replacement and obtain $\mathbf{D}_4[6] = 9$. The values of $\mathbf{D}_3[1], \ldots, \mathbf{D}_3[4]$ are, of course, unchanged in $\mathbf{D}_4$. ∎

In the next section we consider how to modify the algorithms we have just seen to answer some of our basic questions about graphs, including how to find min-paths or max-paths corresponding to the min-weights or max-weights we have just been computing.

EXERCISES 8.8

1. (a) Give the min-weight matrix $\mathbf{W}^*$ for the digraph shown in Figure 5(a). Any method is allowed, including staring at the picture.
   (b) Repeat part (a) for the digraph in Figure 5(b).

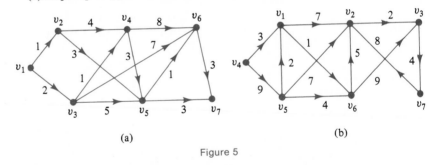

(a)　　　　　　　　　　　　　　　　　(b)

Figure 5

2. Find the max-weight matrix for the digraph of Figure 5(a).

3. (a) Apply DIJKSTRA'S algorithm to the digraph of Figure 5(a). Start at v_1, and use the format of Figure 2.
   (b) Repeat part (a) with the digraph of Figure 5(b).

4. Apply DIJKSTRA'S algorithm to the digraph of Figure 3, starting at v_1. Compare your answer with the answer obtained by WARSHALL'S algorithm in Example 2.

5. (a) Use WARSHALL'S algorithm to find minimum path lengths in the digraph of Figure 6. Give the matrix for $\mathbf{W}$ at the start of each k-loop.
   (b) Use DIJKSTRA'S algorithm on this digraph to find minimum path lengths from v_1. Write your answer in the format of Figure 2.

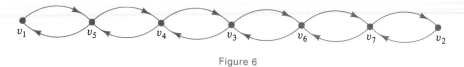

Figure 6

6. (a) Use WARSHALL'S algorithm to find **W*** for the digraph of Figure 4(a).
   (b) Find max-weights for the same digraph using the modified WARSHALL'S algorithm.

7. Give the final matrix **W** if WARSHALL'S algorithm is applied to the matrix

$$\mathbf{W}_0 = \begin{bmatrix} \infty & 11 & 9 & \infty & 2 \\ \infty & \infty & \infty & \infty & \infty \\ \infty & 1 & \infty & \infty & \infty \\ \infty & \infty & 2 & \infty & \infty \\ \infty & \infty & 6 & 3 & \infty \end{bmatrix}.$$

8. Apply DIJKSTRA'S algorithm to find min-weights from v_3 in the digraph of Figure 5(a).

9. (a) Use algorithm MAX-WEIGHT to find max-weights from v_1 to the other vertices in the digraph of Figure 5(a).
   (b) Use MAX-WEIGHT to find max-weights from s to the other vertices in the digraph of Figure 7. [Start by sorting the digraph.]

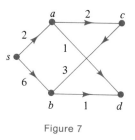

Figure 7

10. (a) Show that DIJKSTRA'S algorithm does not produce correct min-weights for the acyclic digraph shown in Figure 8. [So negative weights can cause trouble.]
    (b) Would DIJKSTRA'S algorithm give the correct min-weights for this digraph if the "For each j in $V(G)\backslash L$" line of the algorithm were replaced by "For each j in $V(G)$"? Explain.

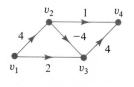

Figure 8

11. This exercise shows some of the difficulties in trying to modify DIJKSTRA'S algorithm to get max-weights. Change the replacement step in DIJKSTRA'S algorithm to the following:

$$\text{If } D(j) < D(k) + W(k,j)$$
$$\qquad \text{Replace } D(j) \text{ by } D(k) + W(k,j)$$

(a) Suppose that this modified algorithm chooses k in $V(G) \backslash L$ with $D(k)$ as large as possible. Show that the new algorithm fails to give the right answer for the digraph in Figure 9(a). [Start at v_1.]

(b) Suppose that the modified algorithm instead chooses k in $V(G) \backslash L$ with $D(k)$ nonnegative but as small as possible. Show that the new algorithm fails for the digraph in Figure 9(b).

(c) Would it help in either part (a) or (b) if "For each j in $V(G) \backslash L$" were replaced by "For each j in $V(G)$"?

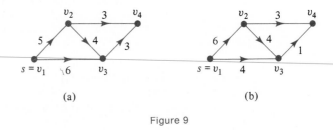

(a)   (b)

Figure 9

12. (a) Show that algorithm MAX-WEIGHT works; it finds max-weights from v_1 to other vertices.

(b) Show that MAX-WEIGHT operates in time $O(n^2)$.

§ 8.9   Modifications and Applications of the Algorithms

The versions of DIJKSTRA'S and WARSHALL'S algorithms that we saw in § 8.8 produce min-weights and max-weights. We can easily modify the algorithms to produce the corresponding min-paths and max-paths as well. The key idea is to think of paths as linked lists.

To describe a path, we associate with each vertex a pointer to the next vertex along the path, until we come to the end. Think, perhaps, of a traveler in a strange city asking for directions to the railway station, going for a block and then asking again, and so on. For a path $x\,y\,z\,w$ we need pointers from x to y, from y to z and from z to w, which we can describe with a function p such that $p(x) = y$, $p(y) = z$ and $p(z) = w$. We can define $p(v)$ in some arbitrary way for other vertices v in $V(G)$ or just define p on $\{x, y, z\}$.

In the case of DIJKSTRA'S algorithm, which finds min-weights from some chosen vertex to each of the other vertices, it turns out to be technically easier to have the pointers point backwards along the corresponding min-paths, rather than forwards. Once the sequence of vertices backwards along the path is known, it is easy, of course, to reverse the order and list the vertices in the forward direction.

To describe the modified DIJKSTRA'S algorithm we take $V(G)$ to be $\{1, 2, \ldots, n\}$ with 1 as the chosen vertex. We create a pointer function P defined on $\{2, \ldots, n\}$ such that at all times either $\langle P(j), j \rangle$ is the final edge on a path of smallest known weight from 1 to j, or $P(j) = 0$ if no path from 1 to j has been discovered. When the algorithm stops, $P(j) = 0$ if there is no path from 1 to j in G. Otherwise, the sequence

$$j, P(j), P(P(j)), P(P(P(j))), \ldots$$

lists the vertices on a min-path from 1 to j in reverse order.

Initially, let $P(j) = 0$ if there is no edge from 1 to j; otherwise let $P(j) = 1$. We add a line to the replacement loop in the original algorithm, so that whenever a path $1 \cdots k j$ is found that is better than the best path from 1 to j previously known, the pointer value $P(j)$ is set equal to k. Thus $P(j)$ is always the next-to-last stop on the best known path to j. Here is the revised algorithm.

DIJKSTRA'S *Algorithm with a Pointer.*

Set $L = \{1\}$
For $i = 1$ to n
    Set $D(i) = W(1, i)$
    If $W(1, i) = \infty$ set $P(i) = 0$
    Otherwise set $P(i) = 1$
End for
While $V(G) \backslash L \neq \varnothing$
    Choose k in $V(G) \backslash L$ with $D(k)$ as small as possible
    Put k in L
    For each j in $V(G) \backslash L$
        If $D(j) > D(k) + W(k, j)$
            Replace $D(j)$ by $D(k) + W(k, j)$
            Replace $P(j)$ by k
    End for
End while
End ∎

EXAMPLE 1 Consider the weighted digraph in Figure 1, which is the same one we looked at in Example 1 of § 8.8. The table in Figure 2 shows the successive values of D and P as each new vertex k is looked at and added to the list L. The reverse listing of a min-path from 1 to 7, derived from the last row of Figure 2, is

$$7, P(7) = 6, P(6) = 4, P(4) = 5, P(5) = 2, P(2) = 1,$$

so the path is 1 2 5 4 6 7. ∎

WARSHALL'S algorithm can also be given a pointer function. In this case, it is just as easy to have the pointer go in the forward direction, so that at

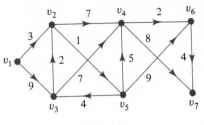

Figure 1

k	D(1)	D(2)	D(3)	D(4)	D(5)	D(6)	D(7)	P(1)	P(2)	P(3)	P(4)	P(5)	P(6)	P(7)
1	∞	3	9	∞	∞	∞	∞	0	1	1	0	0	0	0
2	∞	3	9	10	4	∞	∞	0	1	1	2	2	0	0
5	∞	3	8	9	4	13	∞	0	1	5	5	2	5	0
3	∞	3	8	9	4	13	∞	0	1	5	5	2	5	0
4	∞	3	8	9	4	11	17	0	1	5	5	2	4	4
6	∞	3	8	9	4	11	15	0	1	5	5	2	4	6

Figure 2

all times $\langle v_i, v_{P(i,\,j)} \rangle$ is the first edge in a path of smallest known weight from v_i to v_j, if such a path has been found. When the modified algorithm stops, $P(i, j) = 0$ if there is no path from v_i to v_j. Otherwise the sequence

$$i,\ P(i, j),\ P(P(i, j), j),\ P(P(P(i, j), j), j), \cdots$$

lists the indices of vertices on a min-path from v_i to v_j. Since each subscript k on the list is followed by $P(k, j)$, this sequence is recursively defined in an easy way, using the function P.

We can treat P as an $n \times n$ matrix with entries in $\{0, \ldots, n\}$. At the start, let $\mathbf{P}[i, j] = j$ if there is an edge from v_i to v_j and let $\mathbf{P}[i, j] = 0$ otherwise. Add a line to the replacement loop in WARSHALL'S algorithm so that whenever the algorithm discovers a path $v_i \cdots v_k v_j$ from v_i to v_j which is better than the best path known so far, the pointer value $\mathbf{P}[i, j]$ is set equal to $\mathbf{P}[i, k]$; the right way to head from v_i to v_j is to head for v_k first. The modified WARSHALL'S algorithm looks like this.

WARSHALL'S *Algorithm with a Pointer.*

```
For k = 1 to n
    For i = 1 to n
        For j = 1 to n
            If W[i, j] > W[i, k] + W[k, j] then
                Replace W[i, j] by W[i, k] + W[k, j]
                Replace P[i, j] by P[i, k]
        End for
    End for
End for
End ∎
```

The same modifications will describe max-paths if WARSHALL'S algorithm has been revised to find max-weights.

Either DIJKSTRA'S or WARSHALL'S algorithm can be applied to undirected graphs, in effect by replacing each undirected edge $\{u, v\}$ with $u \neq v$ by two directed edges $\langle u, v \rangle$ and $\langle v, u \rangle$. If the undirected edge has weight w, assign the directed edges the same weight w. If the graph is unweighted, assign weight 1 to all edges. Loops are really irrelevant in the applications of min-paths or max-paths to undirected graphs, so it is convenient when using the algorithms to set $W(i, i) = 0$ for all i.

EXAMPLE 2   Consider the weighted graph shown in Figure 3. We use WARSHALL'S algorithm with a pointer to find min-weights and min-paths between vertices, allowing travel in either direction along the edges. Figure 4 gives the successive values of $\mathbf{W}$ and $\mathbf{P}$. As a sample, we calculate $\mathbf{P}_4[2, 1]$. Since $\mathbf{W}_3[2, 4] + \mathbf{W}_3[4, 1] = 1 + 5 = 6 < 8 = \mathbf{W}_3[2, 1]$, $\mathbf{W}_4[2, 1] = 6$ and also

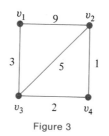

Figure 3

$$\mathbf{W}_0 = \mathbf{W}_1 = \begin{bmatrix} 0 & 9 & 3 & \infty \\ 9 & 0 & 5 & 1 \\ 3 & 5 & 0 & 2 \\ \infty & 1 & 2 & 0 \end{bmatrix} \qquad \mathbf{P}_0 = \mathbf{P}_1 = \begin{bmatrix} 0 & 2 & 3 & 0 \\ 1 & 0 & 3 & 4 \\ 1 & 2 & 0 & 4 \\ 0 & 2 & 3 & 0 \end{bmatrix}$$

$$\mathbf{W}_2 = \begin{bmatrix} 0 & 9 & 3 & 10 \\ 9 & 0 & 5 & 1 \\ 3 & 5 & 0 & 2 \\ 10 & 1 & 2 & 0 \end{bmatrix} \qquad \mathbf{P}_2 = \begin{bmatrix} 0 & 2 & 3 & 2 \\ 1 & 0 & 3 & 4 \\ 1 & 2 & 0 & 4 \\ 2 & 2 & 3 & 0 \end{bmatrix}$$

$$\mathbf{W}_3 = \begin{bmatrix} 0 & 8 & 3 & 5 \\ 8 & 0 & 5 & 1 \\ 3 & 5 & 0 & 2 \\ 5 & 1 & 2 & 0 \end{bmatrix} \qquad \mathbf{P}_3 = \begin{bmatrix} 0 & 3 & 3 & 3 \\ 3 & 0 & 3 & 4 \\ 1 & 2 & 0 & 4 \\ 3 & 2 & 3 & 0 \end{bmatrix}$$

$$\mathbf{W}^* = \mathbf{W}_4 = \begin{bmatrix} 0 & 6 & 3 & 5 \\ 6 & 0 & 3 & 1 \\ 3 & 3 & 0 & 2 \\ 5 & 1 & 2 & 0 \end{bmatrix} \qquad \mathbf{P}^* = \mathbf{P}_4 = \begin{bmatrix} 0 & 3 & 3 & 3 \\ 4 & 0 & 4 & 4 \\ 1 & 4 & 0 & 4 \\ 3 & 2 & 3 & 0 \end{bmatrix}$$

Figure 4

$P_4[2, 1] = P_3[2, 4] = 4$. The min-path from v_2 to v_1 is described by the sequence

$$2, \quad P^*[2, 1] = 4, \quad P^*[4, 1] = 3, \quad P^*[3, 1] = 1.$$

The diagonal entries of P in this example are not especially meaningful. ∎

WARSHALL'S algorithm can be used to find the reachability relation R of a digraph. Simply give all edges, including loops, weight 1. The final value $W^*[i, j]$ is ∞ if there is no path from v_i to v_j and is a positive integer if a path exists. In particular, $W^*[i, i] < \infty$ if and only if v_i is a vertex of a cycle in G. We can test whether a digraph G is acyclic by applying WARSHALL'S algorithm and looking at the diagonal entries of W^*.

In § 8.7 we saw how to view transitive closure in terms of reachability. Given a relation Q on a finite set $\{s_1, \ldots, s_n\}$ we can find $t(Q)$ with WARSHALL'S algorithm by letting $W_0[i, j] = 1$ if $s_i \, Q \, s_j$ and $W_0[i, j] = \infty$ otherwise. Then $W^*[i, j] < \infty$ if and only if $s_i \, t(Q) \, s_j$. See Exercise 10 for a modification with 0 instead of ∞.

FLEURY'S algorithm of § 8.5 involved checking an undirected graph G to see if removing a given edge e increased the number of components, i.e., to see if the endpoints of e were reachable from each other in the graph $G\backslash\{e\}$. WARSHALL'S algorithm can check reachability, but DIJKSTRA'S algorithm is faster in this instance; if e joins v_i and v_j we can apply DIJKSTRA'S algorithm to $G\backslash\{e\}$ with v_i as initial vertex. In fact, by Theorem 3 of § 8.5, in doing this we are simply checking to see if the edge e belongs to a cycle of G, so by testing each edge in turn we can determine whether or not an undirected graph G is acyclic. In § 9.3 we will see an acyclicity check that is even faster.

EXERCISES 8.9

1. (a) Give the initial and final min-path pointer matrices P for WARSHALL'S algorithm applied to the digraph in Figure 5(a). [Compare with Exercise 1 of § 8.8.] You may use any method, including staring at the picture, to get your answer.

   (b) Repeat part (a) for max-paths instead of min-paths.

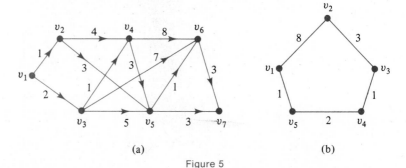

(a)   (b)

Figure 5

2. Use WARSHALL'S algorithm with a pointer to find $\mathbf{W}^*$ and $\mathbf{P}^*$ for the graph of Figure 5(b).

3. (a) Apply DIJKSTRA'S algorithm with a pointer to the digraph shown in Figure 6(a), starting with vertex v_1. Display your answer in the format of Figure 2.

   (b) Repeat part (a) for the digraph of Figure 6(b).

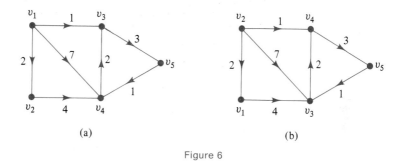

(a)   (b)

Figure 6

4. List the vertices of min-paths from v_1 to v_2, from v_1 to v_4 and from v_1 to v_6 for the digraph of Figure 1. *Hint*: See Figure 2.

5. (a) Use WARSHALL'S algorithm with a pointer to find $\mathbf{P}^*$ for the digraph of Figure 7(a).

   (b) Repeat part (a) for max-paths instead of min-paths.

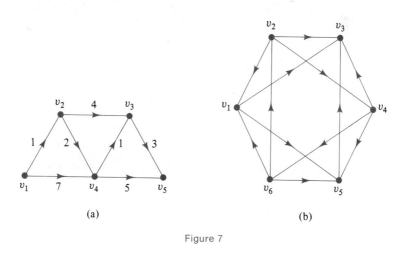

(a)   (b)

Figure 7

6. (a) Give the initial matrix $\mathbf{W}_0$ to find the reachability relation for the digraph of Figure 7(b).

   (b) Find the reachability matrix $\mathbf{M}_R$ for this digraph using WARSHALL'S algorithm.

   (c) Is this digraph strongly connected?

   (d) Is it acyclic?

7. (a) Modify algorithm MAX-WEIGHT of § 8.8 to get an algorithm that finds pointers along max-paths from a single vertex.

(b) Apply your algorithm from part (a) to find max-path pointers from v_1 in the digraph of Figure 4(a) of § 8.8.

8. Give the initial matrix $\mathbf{W}_0$ to find the transitive closure of each of the following relations.

(a) R on $\{1, 2, 3, 4, 5, 6\}$ where iRj if $i = j - 1$,

(b) R on $\{1, 2, 3, 4, 5\}$ where iRj if $|i - j| = 2$.

9. Find the transitive closures of the relations (a) and (b) in Exercise 8.

10. Modify WARSHALL'S algorithm so that $\mathbf{W}[i, j] = 0$ if no path has been discovered from v_i to v_j and $\mathbf{W}[i, j] = 1$ if a path is known. *Hint*: Use $\mathbf{W}[i, k] \wedge \mathbf{W}[k, j]$.

11. (a) Show that DIJKSTRA'S algorithm with a pointer takes $O(n^2)$ time for a digraph with n vertices.

(b) Give the best estimate you can for the running time of WARSHALL'S algorithm with a pointer.

CHAPTER HIGHLIGHTS

For a reminder on how to use this material for review, see the end of Chapter 0.

Some of the material in this chapter was introduced long ago in Chapter 0. A good part of it is first discussed for digraphs and then modified for undirected graphs, with similar definitions and no great surprises. Though there are lots of items on the lists below, there are not really as many new ideas to master as the lists would suggest. Have courage.

Concepts

digraph	**graph** [undirected]
vertex, edge	vertex, edge
initial/terminal vertex	endpoint
path	path
length	length
closed path, cycle	closed path, cycle
acyclic digraph, path	acyclic graph, path
isomorphism	isomorphism
class, invariant	class, invariant
degree, indegree, outdegree	degree
reachable	reachable
connected, strongly connected	connected

digraph (*cont.*)	**graph** [undirected] (*cont.*)
sink, source	component
sorted labeling	Euler path, circuit
weight	Hamilton path, circuit
min-weight, min-path	Gray code
scheduling network	adjacency relation
max-weight, max-path	reachable relation
critical path, edge	

Facts

A path in a digraph is acyclic if and only if its vertices are distinct.

There is at most one simple path between two vertices in an acyclic graph.

Graph and digraph isomorphism may be hard to check, but invariants can sometimes help.

Automorphisms help analyze digraph symmetry.

$\sum_v \text{indeg}(v) = \sum_v \text{outdeg}(v) = |E(G)|$ for digraphs.

$\sum_v \text{deg}(v) = 2|E(G)|$ for graphs and digraphs.

$L(v) - A(u) = W(u, v) + F(u, v)$ for scheduling networks.

Critical edge weights govern max-weight from source to sink in a scheduling network.

A graph has an Euler circuit if and only if it is connected and all vertices have even degree. An analogous statement holds for digraphs.

If e is an edge of a connected graph G then e belongs to some cycle if and only if $G \backslash \{e\}$ is connected. Thus a connectedness algorithm can test for cycles.

Components correspond to $\bar{R}$-classes. $\bar{R} = \bigcup_{k=0}^{\infty} A^k$.

The matrix of A^k is $\mathbf{M}_A * \cdots * \mathbf{M}_A$, with k factors.

If a graph G has no loops or parallel edges, and if $|V(G)| = n \geq 3$, then G is Hamiltonian if any of the following is true:

   (a) $\text{deg}(v) \geq n/2$ for each vertex v [high degrees].
   (b) $|E(G)| \geq \frac{1}{2}(n - 1)(n - 2) + 2$ [lots of edges].
   (c) $\text{deg}(v) + \text{deg}(w) \geq n$ whenever v and w are not connected by an edge.

Theorem 4 of § 8.6 gives information on Hamilton paths in bipartite graphs.

Methods

SINK to find a sink [or source] in a finite acyclic digraph.

NUMBERING VERTICES to assign numbers to vertices in a finite acyclic digraph so all paths lead from smaller to larger numbers.

FLEURY'S algorithm to find an Euler path of a graph.

DIJKSTRA'S algorithm to find min-weights from a selected vertex.

WARSHALL'S algorithm to find min-weights or max-weights between all pairs of vertices, as well as reachability and transitive closure.

MAX-WEIGHT to find max-weights from a selected vertex.

Each of the last three algorithms can be given a pointer to describe the corresponding min-paths or max-paths.

DIJKSTRA'S and WARSHALL'S algorithms can be applied to undirected graphs [but initialize **W** to 0 on the diagonal].

9

TREES

In this chapter we study a class of graphs, called trees, which arise frequently in computer science. Trees are especially suited to represent hierarchical structures and to represent addresses or labels in an organized manner. We give applications involving Polish notation, efficient codes and merging lists. We will also see how to use the structure of trees to process data they represent. This chapter begins with undirected trees and then deals with the most useful directed trees, the so-called "rooted trees," which arise naturally from undirected trees.

§ 9.1 Properties of Trees

A **tree** is a connected acyclic graph. In particular, a tree has no parallel edges and no loops.

EXAMPLE 1   The eleven trees with seven vertices are pictured in Figure 1. More precisely, there are eleven isomorphism classes of trees that have seven vertices, and we have drawn one from each class. ▮

Consider any connected graph G. A subgraph T of G is called a **spanning tree** if T is a tree and if T includes every vertex of G, i.e., $V(T) = V(G)$. In other words, T is a tree obtained by removing some of the edges of G, perhaps, but keeping all of the vertices.

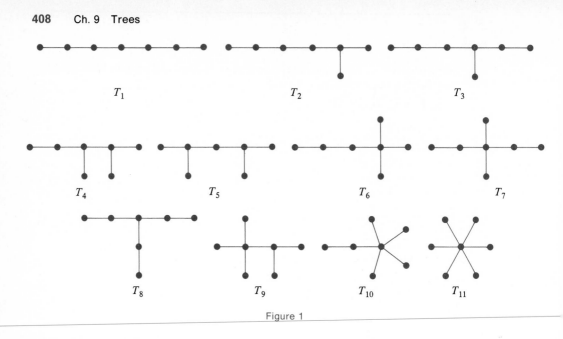

Figure 1

EXAMPLE 2 The graph H in Figure 2 has over 300 spanning trees, of which four have been sketched. They all have 6 edges. ∎

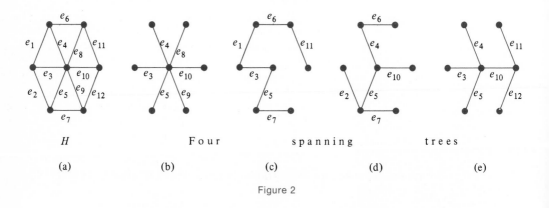

H F o u r   s p a n n i n g   t r e e s

(a)   (b)   (c)   (d)   (e)

Figure 2

The next theorem is a special case of Theorem 3 of § 8.5. Its corollary guarantees that spanning trees exist for all finite connected graphs; later in this section we will see how to construct them.

Theorem 1 Let e be an edge of a connected graph G. Then $G\backslash\{e\}$ is connected if and only if e is in some cycle.

Corollary Every finite connected graph G has a spanning tree.

Proof. Consider a connected subgraph G' of G that uses all the vertices of G and has as few edges as possible. Suppose that G' contains a cycle, say one involving the edge e. By Theorem 1, $G' \setminus \{e\}$ is a connected subgraph of G that has fewer edges than G' has, contradicting the choice of G'. So G' has no cycles. Since it's connected, G' is a tree. ∎

EXAMPLE 3

(a) We illustrate Theorem 1 using the connected graph in Figure 3(a). Note that e_1 does not belong to a cycle and that $G \setminus \{e_1\}$ is disconnected: no path in $G \setminus \{e_1\}$ connects v to the other vertices. Likewise, e_5 belongs to no cycle and $G \setminus \{e_5\}$ is disconnected. The remaining edges belong to cycles. Removal of any *one* of them will not disconnect G.

(b) To illustrate the corollary, note that $G \setminus \{e_{10}\}$ is still connected but has cycles. If we also remove e_8 the resulting graph still has a cycle, namely $e_2 e_3 e_4$. But if we then remove one of the edges in the cycle, say e_4, we obtain an acyclic connected subgraph, i.e., a spanning tree. See Figure 3(b). Clearly, several different spanning trees can be obtained in this way. ∎

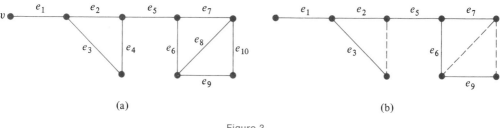

(a)   (b)

Figure 3

In characterizing trees we lose nothing by restricting our attention to graphs with no loops or parallel edges. Our first characterizations hold even if the graph is infinite.

Theorem 2

Let G be a graph with more than one vertex, no loops and no parallel edges. The following are equivalent:

(a) G is a tree.
(b) Each pair of distinct vertices is connected by exactly one simple path.
(c) G is connected, but will not be if any edge is removed.
(d) G is acyclic, but will not be if any edge is added.

Proof. This proof consists of four short proofs.

(a) ⇒ (b). Suppose that G is a tree, so that G is connected and acyclic. By Theorem 1 of § 8.4, each pair of vertices is connected by at least one simple path. By the contrapositive version of Theorem 2 of § 8.4, there is just one simple path.

(b) $\Rightarrow$ (c). If (b) holds, G is clearly connected. Let $e = \{u, v\}$ be an edge of G and assume $G\backslash\{e\}$ is still connected. Note that $u \neq v$ since G has no loops. By Theorem 1 of § 8.4 there is a simple path in $G\backslash\{e\}$ from u to v. Since this path and the one-edge path e are different simple paths in G from u to v, we contradict (b).

(c) $\Rightarrow$ (d). Suppose that (c) holds. If G had a cycle we could remove an edge from G and retain connectedness by Theorem 1. So G is acyclic. Now consider an edge e not in the graph G and let G' denote the graph G with this new edge adjoined. Since $G'\backslash\{e\} = G$ is connected, we apply Theorem 1 to G' to conclude that e belongs to some cycle of G'. In other words, adding e to G destroys acyclicity.

(d) $\Rightarrow$ (a). If (d) holds and G is not a tree, then G is not connected. Then there exist distinct vertices u and v that are not connected by any paths in G. Consider the new edge $e = \{u, v\}$. According to our assumption (d), $G \cup \{e\}$ has a cycle and e must be part of it. The rest of the cycle is a path in G that connects u and v. This contradicts our choice of u and v. Hence G is connected and G is a tree. ∎

In order to appreciate Theorem 2, draw a tree or look at a tree in one of the Figures 1 through 3 and observe that it possesses all the properties (a)-(d) in Theorem 2. Then draw or look at a nontree and observe that it possesses none of the properties (a)-(d).

We need two lemmas for our characterization of finite trees. For a tree, vertices of degree one are called **leaves** [the singular is **leaf**].

EXAMPLE 4   Of the trees in Figure 1: T_1 has two leaves; T_2, T_3 and T_8 have three leaves; T_4, T_5, T_6 and T_7 have four leaves; T_9 and T_{10} have five leaves; and T_{11} has six leaves. ∎

Lemma 1   A finite tree with at least one edge has at least two leaves.

Proof. Consider a longest simple acyclic path, say $v_1 v_2 \cdots v_n$. Then $v_1 \neq v_n$ and both v_1 and v_n are leaves. ∎

Lemma 2   A tree with n vertices has exactly $n - 1$ edges.

Proof. We apply induction. For $n = 2$ the lemma is clear. Assume the result is true for some n, and consider a tree T with $n + 1$ vertices. By Lemma 1, T has a leaf v_0. Let T_0 be the graph obtained by removing v_0 and the edge attached to v_0. Then T_0 is a tree, as is easily checked, and has n vertices. By the inductive assumption T_0 has $n - 1$ edges and so T has n edges. ∎

Lemma 2 holds, of course, for all the trees drawn in this section, and it will hold for any tree we can draw. Try drawing a tree and counting the number of edges as you add new vertices.

Theorem 3   Let G be a finite graph with n vertices, no loops and no parallel edges. The following are equivalent:

    (a) G is a tree.
    (b) G is acyclic and has $n - 1$ edges.
    (c) G is connected and has $n - 1$ edges.

In other words, any two of the properties "connectedness," "acyclicity" and "having $n - 1$ edges" imply the third one.

    Proof. The theorem is obvious for $n = 1$, so we assume $n \geq 2$. Both (a) $\Rightarrow$ (b) and (a) $\Rightarrow$ (c) follow from Lemma 2.
    (b) $\Rightarrow$ (a). Assume that (b) holds but that G is not a tree. Then (d) of Theorem 2 cannot hold. Since G is acyclic we can evidently add some edge and retain acyclicity. Now add as many edges as possible and still retain acyclicity. The graph G' so obtained will satisfy Theorem 2(d) and so G' will be a tree. Since G' has n vertices and at least n edges, this contradicts Lemma 2. Thus G is a tree.
    (c) $\Rightarrow$ (a). Assume (c) holds but that G is not a tree. By the corollary to Theorem 1, G has a spanning tree T, which must have fewer than $n - 1$ edges. This contradicts Lemma 2 and so G is a tree. ∎

    An acyclic graph, whether or not it is connected, is sometimes called a **forest**. Clearly the connected components of a forest are trees. Exercise 15 shows that it is possible to generalize Theorem 3 to characterize forests as well as trees.
    The theorems characterizing trees suggest some methods for finding a spanning tree of a finite connected graph. Using the idea in the proof of the corollary to Theorem 1, we could just remove edges one after another without destroying connectedness, i.e., remove edges that belong to cycles, until we are forced to stop. If G has n vertices and more than $2n$ edges, this procedure will involve examining and throwing out more than half of the edges. It would be faster, if we could do it, to build up a spanning tree by choosing its $n - 1$ edges one at a time so that at each stage the subgraph of chosen edges is part of a spanning tree.
    The question becomes more interesting if the edges are **weighted**, i.e., if each edge e of G is assigned a nonnegative number $W(e)$. The **weight** $W(G')$ of a subgraph G' of G is simply the sum of the weights of the edges of G'. The problem is to find a **minimal spanning tree**, i.e., a spanning tree whose weight is less than or equal to that of any other. If a graph G is not weighted, we can assign each edge the weight 1. Then all spanning trees are minimal since they all have weight $|V(G)| - 1$.

EXAMPLE 5   Figure 4(a) shows a weighted graph with the weights indicated next to the edges. Figure 4(b) shows one possible way to number the edges of the graph so that the weights form a nondecreasing sequence, i.e., with $W(e_i) \leq W(e_j)$ whenever $i < j$. ∎

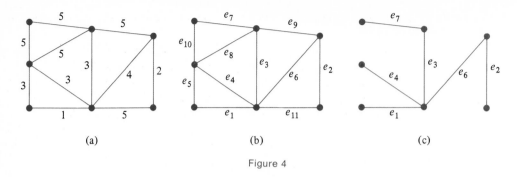

Figure 4

Our first algorithm builds a spanning tree for a weighted graph G whose edges $e_1, \ldots, e_m$ have been sorted so that

$$W(e_1) \leqq W(e_2) \leqq \cdots \leqq W(e_m).$$

This is the procedure that we illustrated in the discussion surrounding Figure 8 of § 0.2 and mentioned in § 3.3. The algorithm proceeds one by one through the list of edges of G, beginning with the smallest weights, choosing edges that do not introduce cycles. When the algorithm stops, the set E is supposed to be the set of edges in a minimal spanning tree for G. The notation $E \cup \{e_j\}$ in the statement of the algorithm stands for the subgraph whose edge set is $E \cup \{e_j\}$ and whose vertex set is $V(G)$.

KRUSKAL'S *Algorithm.*

Set $E = \varnothing$
For $j = 1$ to $|E(G)|$
    If $E \cup \{e_j\}$ is acyclic replace E by $E \cup \{e_j\}$
End for
End ∎

EXAMPLE 6   (a) Applying KRUSKAL'S algorithm to the weighted graph of Figure 4 in Example 5 gives the spanning tree T with edges $e_1, e_2, e_3, e_4, e_6, e_7$ sketched in Figure 4(c). Edge e_5 was rejected because $e_1 \, e_4 \, e_5$ would form a cycle. Edges e_8 through e_{11} were rejected for similar reasons. The spanning tree T has weight 18.

(b) KRUSKAL'S algorithm applied to the unweighted graph of Figure 3(a) gives the spanning tree of Figure 3(b). ∎

Theorem 4   KRUSKAL'S algorithm produces a minimal spanning tree.

Proof. The algorithm will work even if G has loops or parallel edges. It never chooses loops, and it will select the first edge listed in a collection of parallel edges. Thus we may assume that all loops and extra edges have been removed from $E(G)$.

The algorithm will certainly stop. Let E^* be the set of edges E at the time the algorithm stops, and let T be the subgraph with $V(T) = V(G)$ and

$E(T) = E^*$. We show first that T is a tree; since $V(T) = V(G)$ it must then be a spanning tree for G.

Since each new edge added to E gives no cycle, T is acyclic. To show that T is connected, consider any two vertices u and v in $V(G)$. Since G is connected, there is a path from u to v in G. Say e is an edge on such a path. If e is not itself in E^*, then $E^* \cup \{e\}$ must contain a cycle, since otherwise e would have been chosen to be in E^* at some time. But then E^* must contain the edges on a path joining the endpoints of e. Thus the endpoints of every edge on the path from u to v can be joined by a path in E^*, so u is joined to v in T.

To show that T is actually a *minimal* spanning tree, consider a minimal spanning tree S of G that has as many edges as possible in common with T. We will show that $S = T$. Suppose not, and let e_k be the first edge on the list $e_1, \dots, e_m$ that is in T but not in S. Let $S^* = S \cup \{e_k\}$. In view of Theorem 2(d), S^* contains a cycle, say C, which must contain e_k because S itself is acyclic. Since T is also acyclic, there must be some other edge e in C that is not in T. Note that e is an edge of S. Now delete e from S^* to get $U = S^* \setminus \{e\} = (S \cup \{e_k\}) \setminus \{e\}$. By Theorem 1, U is connected, and since it has the same number of edges as S, U is a spanning tree by Theorem 3. Moreover, U has one more edge, namely e_k, in common with T than S has. Because of the choice of S, U is not a *minimal* spanning tree. By comparing the weights of S and U we conclude that $W(e) < W(e_k)$, and so $e = e_i$ for some $i < k$.

Now e is not in T, so it must have been rejected at the $j = i$ stage for the reason that at that time $E \cup \{e\}$ contained a cycle, say C'. All edges in E at that stage were in S, by our choice of e_k as the first edge in S but not in T. Since e is also in S, C' is a cycle in S, which is a contradiction. Thus $S = T$, as we wanted to show. ▮

If G has n vertices, KRUSKAL'S algorithm can't produce more than $n - 1$ edges in E^*. The algorithm could be programmed to stop when $|E| = n - 1$, but it might still need to examine every edge in G before it stopped [see Exercise 11].

Each edge examined requires a test to see if e_j belongs to a cycle. We saw how to make such a check in § 8.9 using DIJKSTRA'S algorithm, which takes $O(n^2)$ time and can be speeded up somewhat by using successor lists. The time KRUSKAL'S algorithm takes is thus at worst $O(n^2 |E(G)|)$, plus the time it takes at the beginning to list the edges of G so that their weights are in increasing order.

The acyclicity check can be handled in a different way. Suppose that G' is the graph when the algorithm is examining e_j, with $V(G') = V(G)$ and $E(G') = E$. If we know which components of G' the endpoints of e_j lie in, then we can add e_j to E if they lie in different components and reject e_j otherwise. This test is quick, provided we keep track of the components. At the start, each component consists of a single vertex, and it's easy to update the component list after accepting e_j; the components of the endpoints of e_j just merge into a single component. The resulting version of KRUSKAL'S algorithm runs in time $O(|E(G)|\log|E(G)|)$, including the time it takes to sort $E(G)$ initially. For complete details see, for example, the account of KRUSKAL'S algorithm in *Data Structures and Algorithms* by Aho, Hopcroft and Ullmann.

It is not actually necessary for G to be connected in order to apply KRUSKAL'S algorithm. In the general case the algorithm produces a **minimal spanning forest** made up of minimal spanning trees for the various components of G.

In both of the cases illustrated in Example 6 it would have been quicker to delete a few bad edges from G than it was to build T up one edge at a time. There is a general algorithm that works by deleting edges: given a connected graph with the edges listed in increasing order of weight, go through the list starting at the big end, throwing out an edge if and only if it belongs to a cycle in the current subgraph of G. The subgraphs that arise during the operation of this algorithm are all connected, and the algorithm only stops when it reaches an acyclic graph, so the final result is a spanning tree for G. It is in fact a minimal spanning tree [Exercise 16]. Indeed, it's the same tree KRUSKAL'S algorithm produces. If $|E(G)| < 2|V(G)| - 1$ this procedure takes less time than KRUSKAL'S algorithm, but of course if G has so few edges then both algorithms work quite quickly.

KRUSKAL'S algorithm makes sure the subgraph being built is always acyclic, while the deletion procedure we have just described keeps all subgraphs connected. Both algorithms are greedy, in the sense that they always choose the smallest edge to add or the largest to delete. Fortunately, greed pays off this time.

The algorithm we next describe is doubly greedy; it makes minimal choices while simultaneously keeping the subgraph both acyclic and connected. Moreover, it does not require the edges of G to be sorted initially. The procedure grows a tree T inside G, with $V(T) = M$ and $E(T) = E$. At each stage the algorithm looks for an edge of smallest weight that joins a vertex in T to some new vertex outside T. Then it adds such an edge and vertex to T and repeats the process.

PRIM'S *Algorithm.*

Set $E = \varnothing$
Choose w in $V(G)$ and set $M = \{w\}$
While $M \neq V(G)$
    Choose an edge $\{u, v\}$ in $E(G)$ of smallest possible weight with $u \in M$
       and $v \in V(G)\backslash M$
    Put $\{u, v\}$ in E
    Put v in M
End while
End ∎

EXAMPLE 7 We apply PRIM'S algorithm to the weighted graph shown in Figure 5(a). Since there are choices possible at several stages of the execution, the resulting tree is not uniquely determined. The solid edges in Figure 5(b) show one possible outcome, with the edges labeled a, b, c, d, e in the order in which they are chosen. The dashed edges b' and d' are alternate choices. Note that KRUSKAL'S algorithm would have chosen edges c and e before edge b. ▌

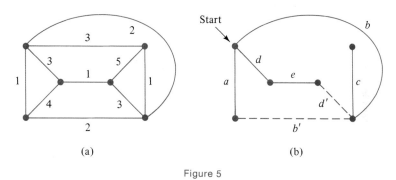

(a)   (b)

Figure 5

Theorem 5 PRIM'S algorithm produces a minimal spanning tree for a connected weighted graph.

    Proof. Each pass through the While loop adds one vertex to M, so the algorithm makes $|V(G)| - 1$ passes through the loop and then stops. At each stage, let T be the subgraph of G with $V(T) = M$ and $E(T) = E$. When the algorithm first reaches the While loop, T is a tree with one vertex. If T is a tree at the start of a pass through the While loop it is also a tree at the end of the pass; the new vertex added is connected to the old tree by the new edge, and the new edge can't be part of a cycle because there's no other path from u to v in T. So T is always a tree. [A finite induction is hiding here.]

    When the algorithm stops, $V(T) = V(G)$, so T is then a spanning tree for G. To show that T is a *minimal* spanning tree at the end, it will be enough to show that at each stage there is some minimal spanning tree that contains T as a subgraph.

    This statement is surely true at the start, when T is just a single vertex. Suppose that at the beginning of some pass through the While loop T is contained in some minimal spanning tree T^* of G. Suppose the algorithm now chooses the edge $\{u, v\}$. If $\{u, v\} \in E(T^*)$, then the new T is still contained in T^*, which is wonderful. Suppose not. Because T^* is a spanning tree, there is a path in T^* from u to v. Now $u \in M$ and $v \notin M$, so there must be some edge in the path that joins a vertex z in M to a vertex w in $V(G) \setminus M$. PRIM'S algorithm chooses $\{u, v\}$ instead of $\{z, w\}$, so $W(u, v) \leq W(z, w)$. Take $\{z, w\}$ out of $E(T^*)$ and replace it with $\{u, v\}$. The new graph T^{**} is still connected, and so it's a tree by Theorem 3. Since $W(T^{**}) \leq W(T^*)$, T^{**} is also a minimal spanning tree, and T^{**} contains the new T. At the end of the loop, T is still contained in some minimal spanning tree. So [induction again] T is always contained in a minimal spanning tree, as we wanted to show. ∎

PRIM'S algorithm makes $n - 1$ passes through the While loop for a graph G with n vertices. Each pass involves choosing a smallest edge subject to a specified condition. A stupid implementation could require looking through all the edges of $E(G)$ to find the right edge. A more clever implementation would keep a record for each vertex x in $V(G) \backslash M$ of the vertex u in M with smallest value $W(u, x)$, and would also store the corresponding value of $W(u, x)$. The algorithm could then simply run through the list of vertices x in $V(G) \backslash M$, find the smallest $W(u, x)$, add $\{u, x\}$ to E and add x to M. Then it could check for each y in $V(G) \backslash M$ whether x is now the vertex in M closest to y, and if so update the record for y. The time to find the closest x to M and then update records is just $O(n)$, so PRIM'S algorithm with the implementation we have just described runs in time $O(n^2)$.

PRIM'S algorithm can easily be modified [Exercise 18] to produce a minimal spanning forest for a graph, whether or not the graph is connected. The algorithm as it stands can also check connectedness; it will break down if the given graph is not connected. KRUSKAL'S algorithm provides another connectedness test, since it can count the components of a graph. We will see in § 9.3, however, that there are even faster methods than these for finding spanning forests if weights don't matter.

EXERCISES 9.1

1. Find all trees with fewer than seven vertices.

2. The trees in Figure 6 have seven vertices. Specify which tree in Figure 1 each is isomorphic to.

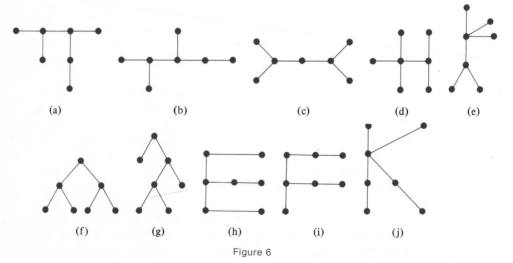

Figure 6

3. (a) Draw a spanning forest for the graph in Figure 7.
   (b) How many spanning forests does this graph have?

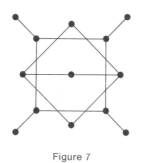

Figure 7

4. Find two nonisomorphic spanning trees of $K_{3,3}$.

5. Sketch a tree with at least one edge and no leaves. *Hint:* See Lemma 1 to Theorem 3.

6. Suppose that the graph in Figure 3(a) is weighted so that $W(e_1) > W(e_2) > \cdots > W(e_{10})$.
   (a) List the edges in a minimal spanning tree for this graph in the order in which KRUSKAL'S algorithm would choose them.
   (b) Repeat part (a) for PRIM'S algorithm, starting at the upper right vertex.

7. (a) Use KRUSKAL'S algorithm to find a minimal spanning tree of the graph in Figure 8(a). Label the edges in alphabetical order as you choose them. Give the weight of the minimal spanning tree.
   (b) Repeat part (a) with PRIM'S algorithm, starting at the lower middle vertex.

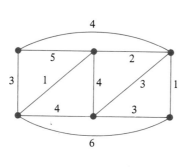

(a)

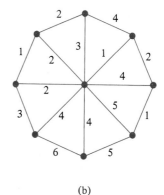

(b)

Figure 8

8. (a) Repeat Exercise 7(a) for the graph in Figure 8(b).
   (b) Repeat Exercise 7(b) for the graph in Figure 8(b), starting at the top vertex.

9. Show that a connected graph with n vertices has at least $n-1$ edges.

10. Does every edge of a finite connected graph with no loops belong to some spanning tree? Justify your answer.

11. Construct a graph with 8 vertices and 11 edges such that KRUSKAL'S algorithm has to examine each edge before producing a minimal spanning tree. Do not use loops or parallel edges.

12. (a) Find all spanning trees of the graph in Figure 9.
    (b) Which edges belong to every spanning tree?
    (c) For a general finite connected graph, characterize the edges which belong to every spanning tree. Prove your assertion.

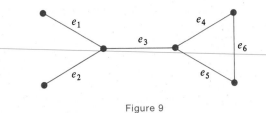

Figure 9

13. (a) Show that the sum of the degrees of the vertices of a tree with n vertices is $2n-2$.
    (b) A tree has three vertices of degree 2, two of degree 3 and one of degree 4. If the remaining vertices have degree 1, how many vertices does the tree have?

14. (a) Show that there is a tree with six vertices of degree 1, one vertex of degree 2, one vertex of degree 3, one vertex of degree 5 and no others.
    (b) For $n \geq 2$, consider n positive integers $d_1, \ldots, d_n$ whose sum is $2n-2$. Show that there is a tree with n vertices whose vertices have degrees $d_1, \ldots, d_n$.
    (c) Show that part (a) illustrates part (b) where $n = 9$.

15. (a) Show that a forest with n vertices and m components has $n-m$ edges.
    (b) Show that a graph with n vertices, m components and $n-m$ edges must be a forest.

16. Show that the edge-deletion algorithm described after KRUSKAL'S algorithm produces a minimal spanning tree. *Hint*: To show that the spanning tree is minimal, follow the proof of Theorem 4 to the point of concluding that $e = e_i$ for some $i < k$. Then show that the deletion criterion implies a contradiction.

17. An oil company wants to connect the cities in the mileage chart below by pipelines going directly between cities. What is the minimal number of miles of pipeline needed?

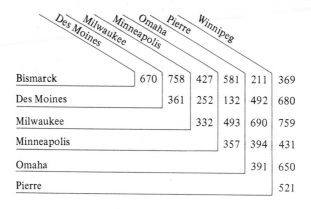

	Des Moines	Milwaukee	Minneapolis	Omaha	Pierre	Winnipeg
Bismarck	670	758	427	581	211	369
Des Moines		361	252	132	492	680
Milwaukee			332	493	690	759
Minneapolis				357	394	431
Omaha					391	650
Pierre						521

18. Modify PRIM'S algorithm to produce a minimal spanning forest.

19. Let G be a finite connected weighted graph in which different edges have different weights. Show that G has exactly one minimal spanning tree. *Hint*: Assume G has more than one minimal spanning tree. Order the edges and consider the smallest subscript k of an edge that belongs to some but not all minimal spanning trees.

20. Let G be a finite connected weighted graph. Show that if e belongs to a cycle and has maximal weight among the edges in the cycle, then $G \backslash \{e\}$ contains a minimal spanning tree of G.

§ 9.2 Rooted Trees

In computer science trees often have distinguished vertices which can be used to give the trees directed structures. We can in general turn any undirected graph into a digraph by associating directions with its edges—putting arrows on them. If the original graph is a tree, the digraph we get is called a **directed tree**. If all of the arrows are directed away from a single vertex, the directed tree is called a rooted tree and that single vertex is its root. Rooted trees are often drawn as in Figure 1(a) so that the top vertex is the root and all the arrows point down and away from it. In this case, we may leave off the arrows, as in Figure 1(b). Thus the same picture will represent an ordinary tree or a rooted tree [or, later, an ordered rooted tree], depending on which we say it represents. We note in passing that our trees are upside-down from the ones in the woods.

EXAMPLE 1   (a) A binary search tree is a structure used to quickly determine the value or location of objects, such as numbers or alphabetized files, that are

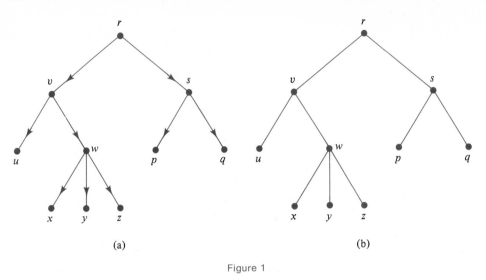

Figure 1

ordered linearly. The binary search tree in Figure 2 would be useful for searching for a number in $\{1, 2, 3, \ldots, 15\}$. The circled numbers represent keys. Given a number in this set, it is first compared with 8: if it is less than 8, proceed to key 4; if it is bigger than 8, proceed to key 12; if it is equal to 8 the search is over. One proceeds down the tree in this manner until the number is found. Note that at most four keys will be needed.

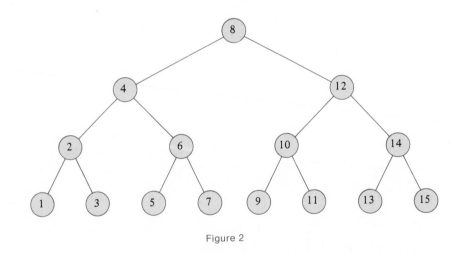

Figure 2

(b) Data structures for diagnosis or identification can frequently be viewed as rooted trees. Figure 3 shows the idea. To use such a data structure we start at the top and proceed from vertex to vertex, taking an appropriate branch in each case to match the symptoms of the patient. The final leaf on

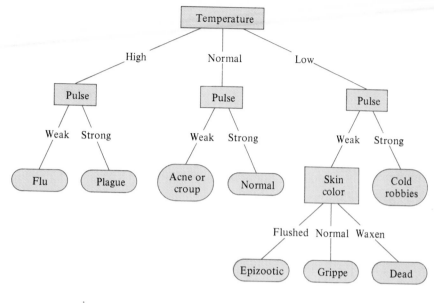

Figure 3

the path gives the name of the most likely condition or conditions for the given symptoms. The same sort of rooted tree structure is the basis for the key system used in field guides for identifying mushrooms, birds, wildflowers and the like.

(c) The chains of command of an organization can often be represented by a rooted tree. Part of the hierarchy of a university is indicated in Figure 4.

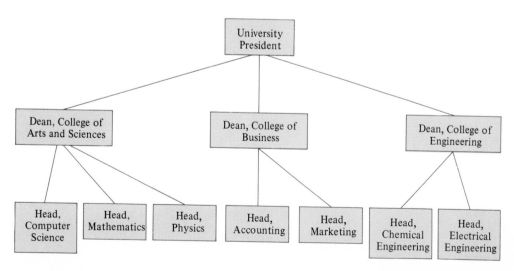

Figure 4

One natural way to get a rooted tree from an undirected tree is to choose a vertex, pick the tree up by that vertex and let gravity direct the edges, with arrows pointing down. We make this idea precise and use it to give the formal definition of a rooted tree. We begin with any [undirected] tree T, as in § 9.1, and select a vertex r which will be its **root**. Theorem 2 of § 9.1 shows that for each vertex $v \neq r$ there is a unique simple path connecting r and v. We direct the edges so that all these paths lead away from the root r. This directed tree is the **rooted tree** with root r made from the tree T; we denote it by T_r. Then an ordered pair $\langle v, w \rangle$ is a [directed] edge of the rooted tree T_r provided $\{v, w\}$ is an edge of the original tree T that is part of the unique simple path from r to the second entry w of the pair $\langle v, w \rangle$. Thus w will be farther from r than v.

EXAMPLE 2   Consider the [undirected] tree in Figure 5(a). If we select v, x and z to be the roots, we obtain the three rooted trees illustrated in Figures 5(b), 5(c) and 5(d). The exact placement of the vertices is unimportant; Figures 5(b) and 5(b′) represent the same rooted tree.

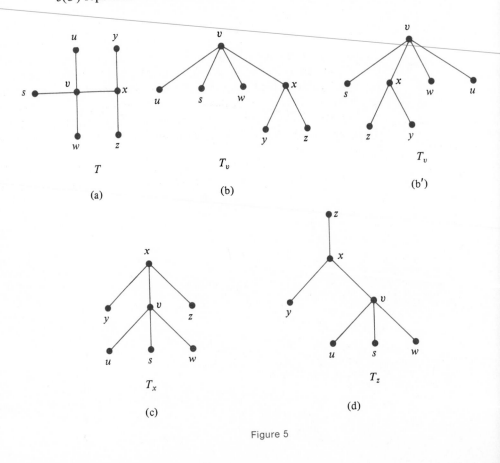

Figure 5

Note that $\langle v, w \rangle$ is an edge in Figure 5(d) because $\{v, w\}$ is an edge of the unique simple path in Figure 5(a) from z to w. On the other hand, $\langle w, v \rangle$ is not an edge of the rooted tree T_z; even though $\{w, v\}$ is an edge of the original tree, it is not an edge of the unique simple path from z to v. Similar remarks apply to all of the other edges. ∎

The terms used to describe various parts of a tree are a curious mixture derived both from the trees in the woods and by family trees. As before, the vertices of degree 1 are called **leaves**; there is one exception: occasionally [as in Figure 5(d)] the root will have degree 1 but we will not call it a leaf. Viewing a tree as a digraph, we see that its root is the sole source while the leaves are all sinks. The remaining vertices are sometimes called "branch nodes" or "interior nodes" and the leaves are sometimes called "terminal nodes." We adopt the convention that if $\langle v, w \rangle$ is an edge of a rooted tree, then v is the **parent** of w and w is a **child** of v. Every vertex except the root has exactly one parent. A parent may have several **children**. More generally, w is a **descendant** of v provided $w \neq v$ and v is a vertex of the unique simple path from r to w. Finally, for any vertex v the **subtree with root** v is precisely the tree consisting of v, all its descendants and all the directed edges connecting them. Whenever v is a leaf, the subtree with root v is a trivial one-vertex tree.

EXAMPLE 3   Consider the rooted tree in Figure 1 redrawn in Figure 6. There are six leaves. The parent v has two children, u and w, and five descendants: u, w, x, y and z. All the vertices except r itself are descendants of r. The whole tree itself is clearly a subtree rooted at r, and there are six trivial subtrees consisting of leaves. The interesting subtrees are given in Figure 6. ∎

Our first theorem gives an intrinsic characterization of rooted trees. As we've seen, rooted trees are viewed both as directed trees and as ordinary

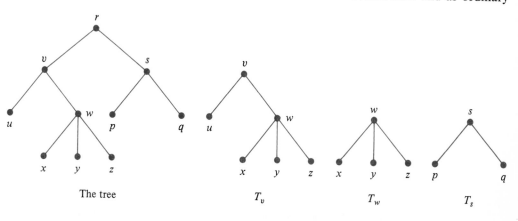

The tree

T_v T_w T_s

Subtrees of the tree

Figure 6

undirected trees. Therefore, for clarity, we will sometimes refer to paths as **directed** or **undirected**.

Theorem 1 A directed tree is a rooted tree if and only if one vertex has indegree 0 and all others have indegree 1.

Proof. Suppose T_r is rooted with root r. If $\langle v, r \rangle$ were a directed edge, then $\{v, r\}$ would be an undirected edge on a simple path from r to r. But T is acyclic, so no such path exists. Thus $\text{indeg}(r) = 0$.

If w is a vertex other than r, there is a unique simple path from r to w, say with vertex sequence $r \cdots u\, w$. Then $\langle u, w \rangle$ is the unique directed edge with w as terminal vertex; thus $\text{indeg}(w) = 1$.

Now suppose T is a directed tree and has one vertex, call it r, of indegree 0 and the rest of indegree 1. Let w be a vertex different from r, and suppose the simple path from r to w has vertex sequence $r = u_0 u_1 u_2 \cdots u_{n-1} u_n = w$. Then $\{r, u_1\}$ is an undirected edge, so either $\langle r, u_1 \rangle$ or $\langle u_1, r \rangle$ is a directed edge. Since $\text{indeg}(r) = 0$, it must be $\langle r, u_1 \rangle$ which is the directed edge. Assume inductively that $\langle u_{k-1}, u_k \rangle$ is a directed edge for some $k < n$. Then $\langle u_{k+1}, u_k \rangle$ is not a directed edge, since $\text{indeg}(u_k) = 1$, so $\langle u_k, u_{k+1} \rangle$ must be a directed edge. By finite induction we conclude

$$\langle r, u_1 \rangle, \langle u_1, u_2 \rangle, \ldots, \langle u_{n-1}, u_n \rangle \text{ are directed edges.}$$

In particular, the unique directed edge of the form $\langle v, w \rangle$ is $\langle u_{n-1}, w \rangle$, and $\{u_{n-1}, w\}$ is the only undirected edge with vertex w on the simple path from r to w. We have shown that T is a rooted tree. ∎

Corollary 1 Every directed path in a rooted tree T is acyclic.

Proof. Consider a directed path $v_0 v_1 \cdots v_m$ in T. We may assume $v_0 = r$ since there is a directed path from r to v_0 which could be adjoined to the original path. Since v_0 has indegree 0, v_0 is not repeated in the sequence. Since no vertex has indegree ≥ 2, none of the other vertices is repeated either. Consequently the path is acyclic. ∎

Corollary 2 A rooted tree is an acyclic digraph.

A rooted tree is an m-**ary tree** if $\text{outdeg}(v) \leq m$ for all vertices v. In other words, each parent has at most m children. The 2-ary trees are called **binary trees**. An m-ary tree [or binary tree] is a **regular** m-**ary tree** if $\text{outdeg}(v) = m$ for all vertices v that are not leaves.

The **level number** of a vertex v is the length of the unique simple path from the root to v. In particular, the root itself has level number 0. The **height** of a rooted tree is the largest level number of a vertex. Only leaves can have their level numbers equal to the height of the tree. A regular m-ary tree is said to be a **full** m-**ary tree** if all the leaves have the same level number, namely the height of the tree.

EXAMPLE 4   (a) The rooted tree in Figures 1 and 6 is a 3-ary tree and, in fact, is an m-ary tree for $m \geq 3$. It is not a regular 3-ary tree since vertices v and s have outdegree 2. Vertices v and s have level number 1, vertices u, w, p and q have level number 2, and the leaves x, y and z have level number 3. The height of the tree is 3.

(b) The labeled tree in Figure 8(a) is a full regular binary tree of height 3. The labeled tree in Figure 8(b) is a regular 3-ary tree of height 3. It is not a full 3-ary tree since one leaf has level number 1 and five leaves have level number 2. ∎

EXAMPLE 5   Consider a full m-ary tree of height h. There are m vertices at level 1. Each parent at level 1 has m children, so there are m^2 vertices at level 2. A simple induction shows that the tree has m^l vertices at level l for each $l \leq h$. Thus it has $1 + m + m^2 + \cdots + m^h$ vertices in all. Since

$$(m - 1)(1 + m + m^2 + \cdots + m^h) = m^{h+1} - 1,$$

as one can check by multiplying and canceling, we have

$$1 + m + m^2 + \cdots + m^h = \frac{m^{h+1} - 1}{m - 1}.$$

Note that the same tree has $p = (m^h - 1)/(m - 1)$ parents and $t = m^h$ leaves. ∎

An **ordered rooted tree** is simply a rooted tree such that the set of children of each parent is linearly ordered. When we draw such a tree, the children of each parent will be ordered from left to right.

EXAMPLE 6   (a) If we view Figure 5(b) as an ordered rooted tree, then the children of v are ordered: $u \prec s \prec w \prec x$. And the children of x are ordered: $y \prec z$. Figure 5(b′) is the picture of a different ordered rooted tree, since $s \prec x \prec w \prec u$ and $z \prec y$.

(b) As soon as we draw a rooted tree it looks like an ordered rooted tree, even if we do not care about the order structure. For example, the important structure in Figure 4 is the rooted tree structure. The ordering of the "children" is not important. The head of the computer science department precedes the head of the mathematics department simply because we chose to list the departments in alphabetical order. ∎

Recall that in general two structures are isomorphic if there is a one-to-one correspondence between them that preserves the structure of concern. Two rooted trees T and T' are **isomorphic rooted trees** provided there exists a one-to-one correspondence α from $V(T)$ onto $V(T')$ so that

(i) if r is the root of T, then $\alpha(r)$ is the root of T';
(ii) if $\{w_1, \ldots, w_k\}$ is the set of children of a vertex v in T, then $\{\alpha(w_1), \ldots, \alpha(w_k)\}$ is the set of children of $\alpha(v)$.

Two ordered rooted trees T and T' are **isomorphic ordered rooted trees** if there is such an α satisfying (i) and

(ii') if $\{w_1, \ldots, w_k\}$ is the set of children of a vertex v in T and if $w_1 \prec w_2 \prec \cdots \prec w_k$, then $\{\alpha(w_1), \ldots, \alpha(w_k)\}$ is the set of children of $\alpha(v)$ and $\alpha(w_1) \prec \alpha(w_2) \prec \cdots \prec \alpha(w_k)$.

Two ordinary trees are **isomorphic trees** if they are isomorphic as [undirected] graphs.

EXAMPLE 7   No two of the trees in Figures 5(b), 5(b'), 5(c) and 5(d) on page 422 are isomorphic as ordered rooted trees. To see this for the trees in Figures 5(b) and 5(b'), assume α is such an isomorphism from (b) to (b'). Then $\alpha(v) = v$ since α must map the root onto the root. Moreover, $\alpha(s) = x$ since α must map the second child of v onto the second child of $\alpha(v)$. But s is childless, while $\alpha(s) = x$ has two children.

   As rooted trees, Figures 5(b) and 5(b') are isomorphic, but they are not isomorphic to the rooted trees in Figures 5(c) and 5(d). As ordinary trees, all five of the trees in Figure 5 are isomorphic.   ∎

EXAMPLE 8   (a) Consider an alphabet Σ. We make Σ^* into a rooted tree as follows. The empty word ϵ will serve as the root. For any word w in Σ^*, its set of children is

$$\{wx : x \in \Sigma\}.$$

If Σ is ordered, then we order each set of children according to the lexicographic order to obtain an ordered rooted tree.

   (b) Let $\Sigma = \{a, b\}$ where $a \prec b$. Each vertex has two children. For instance, the children of $abba$ are $abbaa$ and $abbab$. Part of the infinite ordered rooted tree Σ^* is drawn in Figure 7.   ∎

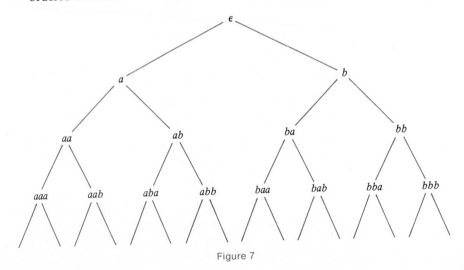

Figure 7

It is often convenient to label the vertices of an ordered rooted tree. This can be done in a variety of ways. One such scheme resembles Example 8: vertices of an *m*-ary tree can be labeled using words from Σ^* where $\Sigma = \mathbb{Z}(m) = \{0, 1, \ldots, m - 1\}$. The ordered children of the root are labeled 0, 1, 2, etc. If a vertex is labeled by the word w, then its ordered children are labeled $w0$, $w1$, $w2$, etc. The label of a vertex tells us the exact location of the vertex in the tree. For example, a vertex labeled 1021 would be the second child of the vertex labeled 102 which, in turn, would be the third child of the vertex labeled 10, etc. The level of the vertex is the length of its label; a vertex labeled 1021 will be at level 4.

EXAMPLE 9   All of the vertices in Figure 8(a) except the root are labeled. In Figure 8(b) we have only labeled the leaves. The labels of the other vertices should be clear.   ∎

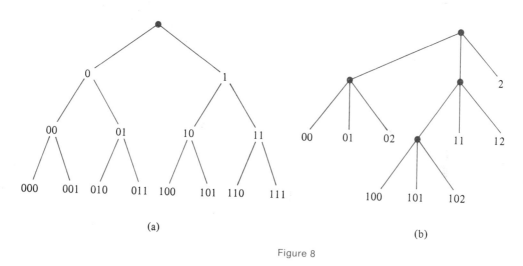

(a)

(b)

Figure 8

Labeled trees are useful for organizing or storing information. In fact, this scheme has been around a long time.

In old mathematics textbooks the paragraphs were frequently numbered using words in Σ^* where $\Sigma = \{1, 2, 3, \ldots\}$. Decimal points were used to set off the letters of the words in Σ^*, i.e., numbers. Thus 3.4.1.2 would refer to the second paragraph of the first subsection of the fourth section of Chapter 3, while 3.4.12 would refer to the twelfth subsection of the fourth section of Chapter 3. This scheme is not very pretty and so modern authors usually avoid it, but it has some real advantages which carry over to present-day uses of trees. One can always insert new paragraphs or sections without disrupting the numbering system. With a little care, paragraphs and sections can be deleted without causing trouble, especially if one doesn't mind gaps. Also the labels, such as 3.4.12, tell you exactly where the subsection or paragraph fits into the book. In contrast, one famous mathematics book has theorems numbered from 1 to 460. All the label "Theorem 303" tells us is that this theorem probably appears about two-thirds of the way through the book. ∎

Another advantage to labeled trees is that vertices can be reached via relatively short paths from the root. For example, a binary tree of height 12 can have up to $2^{12} = 4096$ leaves. Information stored at the leaves can be retrieved in 12 steps. A mindless linear search of the leaves might take thousands of steps.

EXERCISES 9.2

1. (a) For the tree in Figure 1, draw a rooted tree with new root v.
   (b) What is the level number of the vertex r?
   (c) What is the height of the tree?

2. Create a binary search tree for the usual English alphabet $\{a, b, \ldots, z\}$ with its usual order. Arrange it so that at most five keys will ever be needed.

3. (a) For each rooted tree in Figure 5, give the level numbers of the vertices and the height of the tree.
   (b) Which of the trees in Figure 5 are regular m-ary for some m?

4. Discuss why ordinary family trees are not rooted trees.

5. (a) Draw all binary trees of height 2. As usual "all" means up to isomorphism as rooted trees.
   (b) How many trees in part (a) are regular binary trees?
   (c) How many trees in part (a) are full binary trees?

6. Draw all regular binary trees of height 3.

7. (a) Draw full m-ary trees of height h for $m = 2, h = 2$; $m = 2, h = 3$; and $m = 3$, $h = 2$.
   (b) Which trees in part (a) have m^h leaves?

8. Consider a full binary tree T of height h.
   (a) How many leaves does T have?
   (b) How many vertices does T have?

9. Consider a full m-ary tree with p parents and t leaves. Show that $t = (m-1)p + 1$ no matter what the height is.

10. Draw part of the rooted tree Σ^* where $\Sigma = \{a, b, c\}$ and $a < b < c$ as usual.

11. Let $\Sigma = \{a, b\}$ and consider the rooted tree Σ^*; see Example 8. Describe the set of vertices at level k. How big is this set?

12. Give some real-life examples of information storage that can be viewed as labeled trees. *Suggestions*: Federal laws, handbooks of tables, parts catalogs.

13. (a) Show that the rooted trees in Figures 1 and 9 are isomorphic by giving an explicit isomorphism α.

(b) How many such isomorphisms are there?

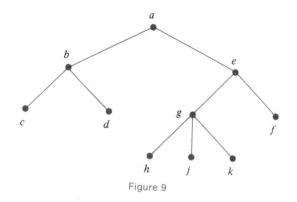

Figure 9

14. Show that the requirement (i) in the definition of isomorphic rooted trees is redundant. That is, show that (i) follows from (ii).

15. (a) Show that the rooted trees in Figures 5(b) and 5(c) are not isomorphic.

(b) Do the same for Figures 5(c) and 5(d).

16. Consider an isomorphism α between rooted trees T and T'. Show that a vertex v in $V(T)$ is a leaf if and only if $\alpha(v)$ is a leaf.

17. Draw an ordered rooted tree that is not isomorphic to the tree in Figure 5(c) viewed as an ordered rooted tree but is isomorphic to that tree viewed simply as a rooted tree.

18. Repeat Exercise 17 for Figure 5(d).

§ 9.3   Depth-First Search Algorithms

We have been thinking of our rooted trees as built from the root down. For finite trees we can also work from the leaves up, building larger and larger subtrees recursively. Such a construction makes rooted trees natural data structures to use for recursive computer programs. It also allows us to prove facts using our generalized principle of induction for recursively defined sets.

Definition 1

(B) A single vertex is a [trivial] rooted tree.

(R) If $T_1, \ldots, T_k$ are rooted trees with roots $r_1, \ldots, r_k$, if $V(T_1), \ldots, V(T_k)$ are disjoint and if r is an element not in $V(T_1) \cup \cdots \cup V(T_k)$, then T is a rooted tree where

$$V(T) = \{r\} \cup V(T_1) \cup \cdots \cup V(T_k),$$

r is a root with children $r_1, \ldots, r_k$ and all other vertices have the same children as before. Thus $T_1, \ldots, T_k$ are subtrees of T with corresponding roots $r_1, \ldots, r_k$.

EXAMPLE 1   Consider T in Figure 1. By (B), the vertices x, y and z each represent trivial trees. By (R), w is the root of a tree T_w with children x, y and z. Since these children were originally childless, they remain childless and are leaves of T_w. By (B) again, the vertex u represents a trivial tree. By (R) again, v is the root of a tree T_v with children u and w. The vertex w still has three children. Similarly, s is the root of a tree T_s with leaves p and q. Finally, by (R) there is a tree with root r whose children v and s are roots of the subtrees T_v and T_s. The process is illustrated in Figure 1.   ∎

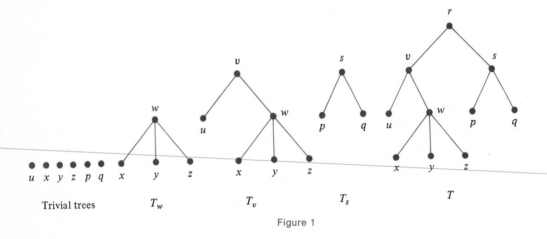

Figure 1

We can also recursively define the height function on the set of rooted trees.

Definition 2   (B) Trivial trees have height 0.
(R) If T is defined as in Definition 1 and the subtrees $T_1, \ldots, T_k$ have heights $h_1, \ldots, h_k$, the height of T is $1 + \max\{h_1, \ldots, h_k\}$.

EXAMPLE 2   We use these recursive definitions to prove that an m-ary tree of height h has at most m^h leaves.
   This is clear for the trivial trees since $m^0 = 1$. Consider an m-ary tree T, defined as in Definition 1, with the property that each subtree T_j has height h_j. Since T is an m-ary tree, each T_j is an m-ary tree and so it has at most m^{h_j} leaves. Let $h^* = \max\{h_1, \ldots, h_k\}$. Then the number of leaves of T is bounded by

$$m^{h_1} + \cdots + m^{h_k} \leq m^{h^*} + \cdots + m^{h^*} = k \cdot m^{h^*}.$$

Since the root has at most m subtrees, $k \leq m$ and so T has at most $m \cdot m^{h^*} = m^{h^*+1}$ leaves. Since $h^* + 1$ equals the height h of T, we are done.   ∎

A tree traversal algorithm is an algorithm for listing [or visiting or searching] all the vertices of a finite ordered rooted tree. The three most common such algorithms provide preorder traversal, inorder traversal [for binary trees *only*] and postorder traversal. All three are most easily described using the recursive definition of ordered rooted trees.

In the **preorder traversal**, the root is listed first and the subtrees are listed in order of their roots. Because of the way we draw ordered rooted trees, we will refer to the order as left to right. Here is the recursive algorithm. Its output is a listing of the tree T_v with root v.

Algorithm PREORDER(v).

Step 1. List the subtrees with the children of v as roots [by using PREORDER(w) to list T_w for each child w of v].

Step 2. List T_v by stringing together v followed by the lists from Step 1 in order, from left to right. ∎

If v has no children, i.e., if v is a leaf, Step 1 is automatically vacuously completed, so the list of T_v is just v in that case.

In the **postorder traversal**, the subtrees are listed in order first, and then the root is listed at the end.

Algorithm POSTORDER(v).

Step 1. List the subtrees with the children of v as roots [using POSTORDER(w) for each child w].

Step 2. List T_v by stringing together the lists from Step 1 in order from left to right *followed by v*. ∎

EXAMPLE 3   (a) Consider the tree T in Figure 1. Algorithm PREORDER gives the following list:

$$T = T_r = r, T_v, T_s$$
$$= r, v, T_u, T_w, s, T_p, T_q$$
$$= r, v, u, w, T_x, T_y, T_z, s, p, q$$
$$= r, v, u, w, x, y, z, s, p, q.$$

This listing can be obtained from the picture of T as illustrated in Figure 2(a). Just follow the dashed line, listing each vertex the first time it is reached.

(b) Applying Algorithm POSTORDER to the trees in Figure 1 gives

$$T = T_r = T_v, T_s, r$$
$$= T_u, T_w, v, T_p, T_q, s, r$$
$$= u, T_x, T_y, T_z, w, v, p, q, s, r$$
$$= u, x, y, z, w, v, p, q, s, r.$$

This listing can be obtained from the picture of T in Figure 2(b), provided the vertices are written in reverse order, i.e., from right to left starting with r, then s, then q, etc. ∎

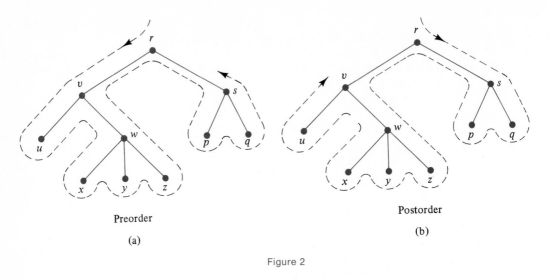

Preorder

(a)

Postorder

(b)

Figure 2

For binary ordered rooted trees, the **inorder traversal** lists the root v in between the left and right subtrees of T_v.

Algorithm INORDER(v).

Step 1. List the left subtree [using INORDER(w) on the left child w of v].
Step 2. List the right subtree [using INORDER(u) on the right child u of v].
Step 3. List T_v by stringing together the result of Step 1, then v, and then the result of Step 2. ▮

If T_v is not regular, then some vertices have just one child. Each such child must be designated as either a left or a right child. A missing child produces an empty list in Step 1 or Step 2.

EXAMPLE 4   (a) Consider the recursively defined binary tree in Figure 3. Algorithm INORDER gives the following list:

$$T = T_r = T_w, r, T_u$$
$$= T_v, w, T_x, r, T_t, u, T_s$$
$$= v, w, T_y, x, T_z, r, t, u, T_p, s, T_q$$
$$= v, w, y, x, z, r, t, u, p, s, q.$$

(b) Consider the labeled tree in Figure 8(a) of § 9.2. The inorder listing of the subtree with root 00 is 000, 00, 001; the other subtrees are also easy to list. The inorder listing of the entire tree is:

000, 00, 001, 0, 010, 01, 011, root, 100, 10, 101, 1, 110, 11, 111. ▮

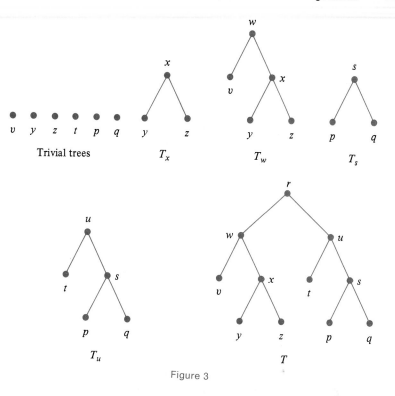

Figure 3

It is of interest to know whether a particular listing determines the tree, that is, whether the tree can be recovered from the listing. It often cannot be, as we'll see in the next example.

EXAMPLE 5   Consider again the tree T in Figure 3. Figure 4 gives two more binary trees having the same inorder listing. The binary tree in Figure 5(a) has the same

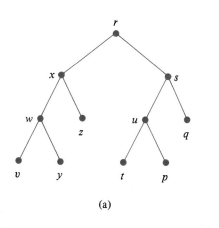

(a)

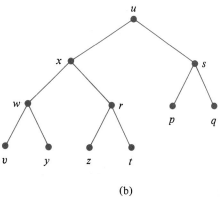

(b)

Figure 4

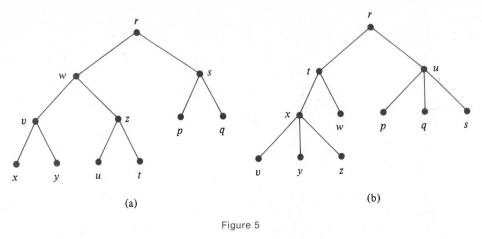

(a)   (b)

Figure 5

preorder listing as T, and the tree in Figure 5(b) has the same postorder listing as T. ∎

In spite of these examples, there are important situations under which trees can be recovered from their listings. Since our applications will be to Polish notation, and since the ideas are easier to grasp in that setting, we will return to this matter at the end of the next section.

We can analyze how long the three traversal algorithms take by using a method of "charges." Let t_1 be the time that it takes to label a vertex, and let t_2 be the time that it takes to attach a list of vertices to another list. [We can assume that the attachment time is constant, by using a linked-list representation for our lists of vertices.] The idea now is to think of the total time for the various operations in the execution of the algorithm as being charged to the individual vertices and edges of the tree. We charge a vertex v with the time t_1 that it takes to label it, and we charge an edge $\langle v, w \rangle$ with the time t_2 that it takes to attach the list of T_w to make it part of the list of T_v. Then the time for every operation gets charged somewhere, every vertex and every edge gets charged, and nobody gets charged twice. By adding up the charges we find that the total time is $t_1|V(T)| + t_2|E(T)|$, which is $O(\max \{|V(T)|, |E(T)|\})$. Since $|E(T)| = |V(T)| - 1$ for the tree T, the time is $O(|V(T)|)$.

The three traversal algorithms above are all examples of what are called **depth-first search** algorithms. Each of them goes as far as it can by following the arrows, then backs up a bit and again goes as far as it can, and so on.

The same idea can be useful in settings in which there is no obvious tree to start with. We illustrate with an algorithm that labels the vertices of an acyclic digraph G, connected or not, with a sorted labeling of the kind we got from NUMBERING VERTICES in § 8.1. For clarity we first describe an algorithm that labels all of the vertices that are accessible from a given vertex v; then we piece together the results for various v's to label all of the vertices of G.

The first algorithm works this way. Starting with an acyclic digraph G, a chosen vertex v and an integer k, the algorithm looks in turn at each immediate successor of v. It calls itself recursively to label the vertices that are accessible from those successors, being careful to avoid labeling the same vertex twice. It marks vertices as it labels them and then doesn't go back to a marked vertex again. Eventually, v and all vertices reachable from v are labeled with integers $k, k - 1, k - 2, \ldots$ in such a way that if u and w are vertices reachable from v and if there is a path from u to w then the label of u is less than the label of w. The algorithm stops with a new value of k which is 1 less than the last label used.

The set L in the algorithm consists of those vertices that have been labeled. For each u in $V(G)$, $\mathrm{SUCC}(u)$ is the set of immediate successors of u, i.e., those vertices w for which there is an edge from u to w. We choose v and k before we start, and set $L = \varnothing$. When the algorithm calls itself recursively it will know what vertex, integer and set L to use.

Algorithm TREESORT(v, k).

For $w \in \mathrm{SUCC}(v) \backslash L$
    Do TREESORT(w, k)
End for
Label v with k, put v in L, decrease k by 1
End ∎

EXAMPLE 6   We apply TREESORT$(z, 8)$ to the digraph shown in Figure 6(a). We need to specify an order in which to examine successor lists. Let's use alphabetical order, as indicated in Figure 6(b). The list of steps in Figure 7 describes what happens. Figure 6(c) shows the final labeling. The algorithm starts at z and calls itself at w, then at t and then at u. Since u has no successors, the algorithm backs up to t and looks for unlabeled successors. There are none, so it backs up to w. The only unlabeled successor at w is x, and from there the algorithm calls itself at y. Since y, x, and w now have no unlabeled successors,

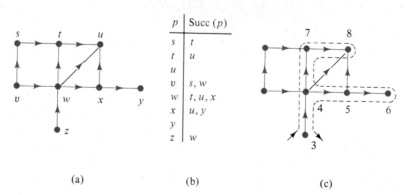

p	Succ (p)
s	t
t	u
u	
v	s, w
w	t, u, x
x	u, y
y	
z	w

(a)   (b)   (c)

Figure 6

TREESORT(z, 8)
  TREESORT(w, 8)
    TREESORT(t, 8)
      TREESORT(u, 8)
        Label u with 8, set $L = \{u\}$, set $k = 7$
        Label t with 7, set $L = \{u, t\}$, set $k = 6$
      TREESORT(x, 6)   [x, not u, since $u \in L$ now]
        TREESORT(y, 6)   [not u now either]
          Label y with 6, set $L = \{u, t, y\}$, set $k = 5$
        Label x with 5, set $L = \{u, t, y, x\}$, set $k = 4$
      Label w with 4, set $L = \{u, t, y, x, w\}$, set $k = 3$
    Label z with 3, set $L = \{u, t, y, x, w, z\}$, set $k = 2$

Figure 7

the algorithm backs up all the way to z, labels it, reduces k by 1 and quits. The dashed curve in Figure 6(c) shows the order in which the vertices are visited. The algorithm has found a tree inside G and has made a depth-first search of it. Note that the two vertices that are not accessible from z are still unlabeled at the end, and the final value of k is 2. ∎

TREESORT only labels the vertices accessible from v. To get all of the vertices of G we just keep applying TREESORT, again being careful not to try to label a vertex twice.

Algorithm TOPSORT.

Set $k = |V(G)|$ and $L = \varnothing$
While $L \neq V(G)$
    Choose $v \in V(G)\backslash L$
    Do TREESORT(v, k)
End while
End ∎

Theorem TOPSORT gives a topologically sorted labeling of the vertices of a finite acyclic digraph.

Proof. We have already seen that TREESORT(v, k) must label v and every vertex reachable from v, so TOPSORT must label every vertex of G. Suppose that there is an edge from some vertex x to another vertex y. Consider what happens when the algorithm is examining SUCC(x). If y is already in L, then y already has a larger label than x will get. If y is not yet in L, then the algorithm will call itself at y and label y before it labels x. Either way, y gets the larger label. Hence the final numbering is a topological sorting of G. ∎

A method of charges argument like the one given earlier lets us estimate the time TOPSORT takes. Charge each vertex with the time it takes to label it, and charge each edge $\langle x, y \rangle$ with the time it takes to check if $y \in L$. [In Figure 7 the checking steps were only hinted at.] As was the case for the traversal algorithms, the total time is $O(\max \{|V(G)|, |E(G)|\})$. NUMBER-ING VERTICES was not this fast. It allowed us to go over the same edges again and again in our search for sinks. In the extreme case in which G is simply a chain of n vertices from a source to a sink and we always start at the source to look for sinks, the time NUMBERING VERTICES takes [see Exercise 21 of § 8.1] will be proportional to

$$\sum_{k=1}^{n} k = \frac{n(n + 1)}{2},$$

whereas TOPSORT works in time $O(n)$.

The ideas in TOPSORT can also be used to give fast algorithms for finding spanning trees or forests, for finding components, and for checking acyclicity in undirected graphs.

If we simply want to grow trees in G, we can leave out the labeling commands in TOPSORT and just keep track of where we've been and which edges got us there. The following recursive algorithm finds as big a tree as possible in a graph, starting at a selected vertex v. Initially, the set V of visited vertices is empty, as is the set E of used edges. We assume that the graph has no loops or multiple edges.

Algorithm TREE(v).

Put v in V
For $w \notin V$ with an edge joining v and w in G
    Put the edge $\{v, w\}$ in E
    Do TREE(w)
End for
End ∎

To get a whole spanning forest, we just keep growing trees.

Algorithm FOREST.

Set $V = \varnothing$ and $E = \varnothing$
While $V \neq V(G)$
    Choose $v \in V(G) \backslash V$
    Do TREE(v)
End while
End ∎

It is easy to keep track of the components as we build them, by using a pointer function C which assigns the value v to each vertex w in the list V produced by TREE(v). Then each pass through the While loop in FOREST produces a different value of C, which is shared by all the vertices in the component for that pass. The modified algorithms look like this.

Algorithm FOREST+.

Set $V = \varnothing$ and $E = \varnothing$
While $V \neq V(G)$
    Choose $v \in V(G) \backslash V$
    Set $C(v) = v$
    Do TREE+(v)
End while
End ▮

Algorithm TREE+(v).

Put v in V
For $w \notin V$ with $\{v, w\} \in E(G)$
    Put $\{v, w\}$ in E and set $C(w) = C(v)$
    Do TREE+(w)
End for
End ▮

EXAMPLE 7   We can illustrate the operation of TREE+ and FOREST+ on the graph shown in Figure 8(a). Figure 9 shows the steps in TREE+(1) and in TREE+(2); then FOREST puts them together. The initial data for TREE+(1) are $V = \varnothing, E = \varnothing$ and $C(1) = 1$. For TREE+(2) the sets V and E start with the output values of TREE+(1) and with $C(2) = 2$. When choices are available we go through the vertices in increasing numerical order. Figure 8(b) shows the two trees that are grown in the spanning forest, with vertices labeled by their output values of C. ▮

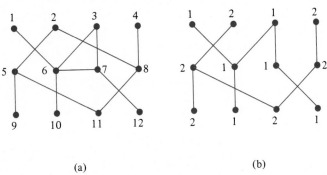

(a)   (b)

Figure 8

TREE + (1):

Put 1 in V
   Put $\{1, 6\}$ in E; set $C(6) = 1$
   Put 6 in V
      Put $\{6, 3\}$ in E; set $C(3) = 1$
      Put 3 in V
         Put $\{3, 7\}$ in E; set $C(7) = 1$
         Put 7 in V
            Put $\{7, 12\}$ in E; set $C(12) = 1$
            Put 12 in V
      Put $\{6, 10\}$ in E; set $C(10) = 1$
      Put 10 in L

TREE + (2):

Put 2 in V
   Put $\{2, 5\}$ in E; set $C(5) = 2$
   Put 5 in V
      Put $\{5, 9\}$ in E; set $C(9) = 2$
      Put 9 in V
      Put $\{5, 11\}$ in E; set $C(11) = 2$
      Put 11 in V
         Put $\{11, 8\}$ in E; set $C(8) = 2$
         Put 8 in V
            Put $\{8, 4\}$ in E; set $C(4) = 2$
            Put 4 in V

Figure 9

Both FOREST and FOREST+, like our other depth-first algorithms, run in time $O(\max\{|V(G)|, |E(G)|\})$. FOREST+ gives a fast test to see if a graph is connected; simply count the components at the end. It also lets us check for cycles, because $E = E(G)$ at the end if and only if G is a forest to begin with.

EXERCISES 9.3

1. Use the schemes illustrated in Figure 2 to give the preorder and postorder listings of the vertices of the tree in Figure 4(a).

2. Repeat Exercise 1 for Figure 5(a).

3. Repeat Exercise 1 for Figure 5(b).

4. Give the inorder listing of the vertices of the labeled tree in Figure 10.

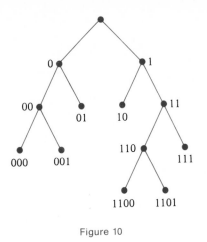

Figure 10

5. Use the PREORDER algorithm to list the vertices of the tree T in Figure 3.

6. Repeat Exercise 5 using the POSTORDER algorithm.

7. (a) Use the recursive definition of rooted tree in Definition 1 to show that Figure 4(b) represents a rooted tree.
   (b) Use the recursive definition of height in Definition 2 to calculate the height of the tree in part (a).

8. Use the recursive definition of height to calculate the height of the tree T in Figure 1.

9. Use the POSTORDER algorithm and the recursive definition in Exercise 7(a) to list the vertices in Figure 4(b).

10. Repeat Exercise 9 using the PREORDER algorithm.

11. Repeat Exercise 9 using the INORDER algorithm.

12. (a) Use the recursive definition of rooted tree to show that Figure 5(b) represents a rooted tree.
    (b) Calculate the height of the tree in part (a) using the recursive definition.

13. Use the POSTORDER algorithm and your answer to Exercise 12(a) to list the vertices of Figure 5(b).

14. Repeat Exercise 13 using the PREORDER algorithm.

15. (a) Apply algorithm TREESORT(a, 6) to the digraph of Figure 11(a). At each choice point in the algorithm choose successors of a vertex in increasing alphabetical order. Draw a dashed curve to show the search pattern and show the labels on the vertices, as in Figure 6(c).
    (b) List the steps for part (a) in the manner of Figure 7.
    (c) Apply algorithm TOPSORT to the digraph of Figure 11(a) and draw a picture which shows the resulting labels on the vertices.

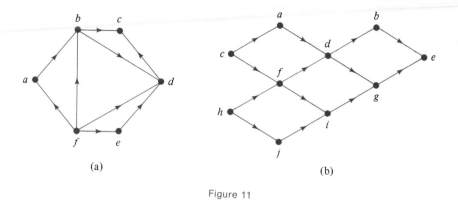

(a)

(b)

Figure 11

16. (a) Repeat Exercise 15(a) for the digraph of Figure 11(b), using TREE-SORT(a, 10).
    (b) Repeat Exercise 15(c) for this digraph.
    (c) How many times does TOPSORT go through the While loop in part (b)?

17. (a) Apply TREE(a) to the graph in Figure 12 with $V = E = \varnothing$ initially. At each choice point in the algorithm choose vertices in increasing alphabetical order. Show the operation in a format like Figure 9. Draw a dashed curve to show the search pattern.
    (b) Repeat part (a) for the algorithm TREE$+(d)$ with $C(d) = d$.

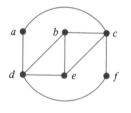

Figure 12

18. Draw a picture with dashed curves which shows the vertex traversal order of the components in Example 7.

19. Show that an acyclic digraph with n vertices cannot have more than $n(n - 1)/2$ edges. [Hence TOPSORT takes time no more than $O(n^2)$.]

20. (a) Give an example of an acyclic digraph without parallel edges that has 4 vertices and $4(4 - 1)/2 = 6$ edges. How many sorted labelings does your example have?
    (b) Show that every acyclic digraph with n vertices and $n(n - 1)/2$ edges has exactly one sorted labeling.

§ 9.4   Polish Notation

Preorder, postorder and inorder traversal give ways to list the vertices of an ordered rooted tree. If we label the vertices, we get a list of labels. If the labels are such things as numbers, addition signs, multiplication signs and the like our list may give a meaningful way to calculate a number; for instance, using ordinary algebraic notation, the list $4 * 3 \div 2$ determines the number 6. The list $4 + 3 * 2$ seems ambiguous; is it the number 14 or 10? Polish notation, which we describe below, is a method for defining algebraic expressions without parentheses, using lists obtained from trees. Of course it is important that the lists completely determine the corresponding labeled trees and their expressions. After we discuss Polish notation we prove that under even more general conditions the lists do determine the trees uniquely.

Polish notation can be used to write expressions which involve objects from some system [of numbers, or matrices, or propositions in the propositional calculus, etc.] and certain operations on the objects. The operations are usually, but not always, binary operations. The corresponding ordered rooted trees have leaves labeled with objects from the system [such as numbers] or by variables representing objects from the system [such as x]. The other vertices are labeled by the operations.

EXAMPLE 1   The algebraic expression

$$((x - 4) \uparrow 2) * ((y + 2)/3)$$

is represented by the tree in Figure 1(a). This expression uses several familiar binary operations on $\mathbb{R}$: $+, -, *, /, \uparrow$. Recall that $*$ represents multiplication and that $\uparrow$ represents exponentiation: $a \uparrow b$ means a^b. Thus our expression is equivalent to

$$(x - 4)^2 \left(\frac{y + 2}{3} \right).$$

Note that this is an *ordered* tree; if x and 4 were interchanged, for example, the tree would represent a different algebraic expression.

It is clear that the ordered rooted tree determines the algebraic expression. Note that the inorder traversal of the vertices yields $x - 4 \uparrow 2 * y + 2 / 3$, and this is exactly the original expression *except for the parentheses*. Moreover, the parentheses are crucial since this expression determines neither the tree nor the original algebraic expression. This listing could just as well come from the algebraic expression

$$((x - (4 \uparrow 2)) * y) + (2/3)$$

whose tree is drawn in Figure 1(b). This algebraic expression is equivalent to $(x - 16)y + \frac{2}{3}$, a far cry from

$$(x - 4)^2 \left(\frac{y + 2}{3} \right).$$

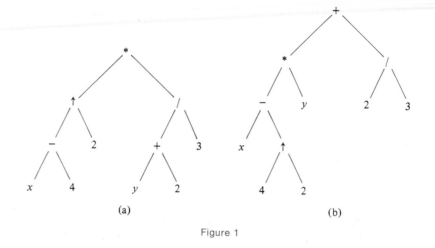

Figure 1

Let us return to our original algebraic expression given by Figure 1(a). Preorder traversal yields

$$* \uparrow - x\ 4\ 2\ /\ +\ y\ 2\ 3$$

and postorder traversal yields

$$x\ 4\ -\ 2 \uparrow y\ 2\ +\ 3\ /\ *.$$

It turns out that each of these listings uniquely determines the tree, and hence the original algebraic expression. Thus these expressions are unambiguous *without parentheses*. This extremely useful observation was made by the Polish logician Łukasiewicz. The preorder listing is known now as **Polish notation** or **prefix notation**. The postorder listing is known as **reverse Polish notation** or **postfix notation**. Our usual algebraic notation, with the necesary parentheses, is known as **infix** notation. ∎

EXAMPLE 2 Consider the compound proposition $(p \to q) \vee (\neg p)$. We treat the binary operations $\to$ and $\vee$ as before. However, $\neg$ is a 1-ary or unary operation. We decree that its child is a right child since the operation precedes the proposition that it operates on. The corresponding ordered tree in Figure 2(a) can be traversed in all three ways. The preorder is $\vee \to p\ q\ \neg\ p$ and the postorder is $p\ q \to p\ \neg\ \vee$. The inorder gives the original expression *without parentheses*. Another tree with the same inorder expression is drawn in Figure 2(b). As in Example 1, the preorder and postorder listings determine the tree and the original compound proposition. ∎

As we illustrated in Example 5 of § 9.3, in general the preorder or postorder list of vertices will not determine a tree, even a regular binary tree. More information is needed. It turns out that if we are provided with the level of each vertex, then the preorder or postorder list determines the tree. We will not pursue this. It also turns out that if we know how many

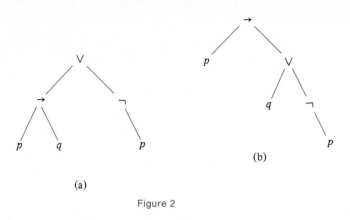

(a)

(b)

Figure 2

children each vertex has, then the tree is determined by its preorder or postorder list.

EXAMPLE 3   We illustrate the last sentence by beginning with the expression

$$x\, 4 - 2 \uparrow y\, 2 + 3 / *$$

in postfix notation. Each vertex $*, \uparrow, -, /$ and $+$ represents a binary operation and so has two children. The other vertices have no children, so we know exactly how many children each vertex has.

   Let us recover the tree, but instead of drawing subtrees we'll determine their corresponding subexpressions. Starting at the left we find the first binary operation, $-$, which must apply to the two preceding subtrees, namely the leaves x and 4. Writing $(x - 4)$ for the corresponding subtree, we obtain the modified sequence $(x - 4)\, 2 \uparrow y\, 2 + 3 / *$. Treating this expression in the same way, we obtain a subtree corresponding to $((x - 4) \uparrow 2)$ which leads us to $((x - 4) \uparrow 2)\, y\, 2 + 3 / *$. Now the operation $+$ acts on the two preceding trees, namely the leaves y and 2. We get $((x - 4) \uparrow 2)(y + 2)\, 3 / *$. Now $/$ acts on the preceding two trees, represented by $(y + 2)$ and 3. We obtain $((x - 4) \uparrow 2)((y + 2) / 3) *$. Now we have one operation, $*$, preceded by two expressions representing subtrees:

$$((x - 4) \uparrow 2) * ((y + 2) / 3).$$

We've recovered the expression for the tree in Figure 1(a).

   We briefly summarize the procedure of the last paragraph:

$$x\, 4 - 2 \uparrow y\, 2 + 3 / *$$
$$(x - 4)\, 2 \uparrow y\, 2 + 3 / *$$
$$((x - 4) \uparrow 2)\, y\, 2 + 3 / *$$
$$((x - 4) \uparrow 2)(y + 2)\, 3 / *$$
$$((x - 4) \uparrow 2)((y + 2) / 3) *$$
$$((x - 4) \uparrow 2) * ((y + 2) / 3).$$

There is another way to view this procedure, which will serve as a model for the proof of the theorem after Example 6. Working left to right, we can pick off the children of each vertex as follows: each binary operation gets the two nearest available children to its left. A child is "available" if it has not already been assigned to a parent. The other vertices get no children.

Vertex	x	4	$-$	2	$\uparrow$	y	2	$+$	3	$/$	$*$
Available children		x	$x, 4$	$-$	$-, 2$	$\uparrow$	$\uparrow, y$	$\uparrow, y, 2$	$\uparrow, +$	$\uparrow, +, 3$	$\uparrow, /$
Assigned children			$x, 4$		$-, 2$			$y, 2$		$+, 3$	$\uparrow, /$

∎

EXAMPLE 4 The same method works for a compound proposition in reverse Polish notation, but when we meet the unary operation $\neg$, it acts on just the immediately preceding subexpression. For example,

$$p\,q \to p\,q \wedge \neg \vee$$
$$(p \to q)\,p\,q \wedge \neg \vee$$
$$(p \to q)\,(p \wedge q)\,\neg \vee \qquad [\neg \text{ just acts on } (p \wedge q)]$$
$$(p \to q)\,(\neg(p \wedge q)) \vee$$
$$(p \to q) \vee (\neg(p \wedge q)).$$

The reader can draw a tree representing this compound proposition. Alternatively, we can pick off the children working from left to right as in Example 3:

Vertex	p	q	$\to$	p	q	$\wedge$	$\neg$	$\vee$
Available children		p	p, q	$\to$	$\to, p$	$\to, p, q$	$\to, \wedge$	$\to, \neg$
Assigned children			p, q			p, q	$\wedge$	$\to, \neg$

∎

Not all strings of operations and symbols lead to meaningful expressions.

EXAMPLE 5 Suppose it is alleged that

$$y + 2\,x * \uparrow 4 \quad \text{and} \quad q \neg p\,q \vee \wedge \to$$

are in reverse Polish notation. The first one is hopeless right away, since $+$ is not preceded by two expressions. The second one breaks down when we attempt to decode it as in Example 4:

$$q \neg p \, q \lor \land \rightarrow$$
$$(\neg q) \, p \, q \lor \land \rightarrow$$
$$(\neg q)(p \lor q) \land \rightarrow$$
$$((\neg q) \land (p \lor q)) \rightarrow .$$

Unfortunately, the operation $\rightarrow$ has only one subexpression preceding it. We conclude that neither of the strings of symbols given above represents a meaningful expression. ▮

Just as we did with ordinary algebraic expressions in §3.6, we can recursively define what we mean by well-formed formulas [wff's] for Polish and reverse Polish notation. We give one example by defining **wff's for reverse Polish notation** of algebraic expressions:

(B) Numerical constants and variables are wff's.
(R) If f and g are wff's, so are $fg +$, $fg -$, $fg *$, $fg /$ and $fg \uparrow$.

EXAMPLE 6   We show that $x \, 2 \uparrow y - x \, y * /$ is a wff. All the variables and constants are wff's by (B). Then $x \, 2 \uparrow$ is a wff by (R). Hence $x \, 2 \uparrow y -$ is a wff where we use (R) with $f = x \, 2 \uparrow$ and $g = y$. Likewise, $x \, y *$ is a wff by (R). Finally, the entire expression is a wff by (R) since it has the form $fg /$ where $f = x \, 2 \uparrow y -$ and $g = x \, y *$. ▮

We end the section by proving the theorem that shows that expressions in Polish and reverse Polish notation uniquely determine the original expression. The proof is based on the second algorithm illustrated in Examples 3 and 4.

Theorem   Let T be a finite ordered rooted tree whose vertices have been listed by a preorder traversal or a postorder traversal. Suppose that the number of children is known for each vertex. Then the tree is determined, that is, the tree can be recovered from the listing.

Proof. We consider only the case of postorder traversal. For each vertex v of T let $S(v)$ be the set of children of v. We'll show that for $m = 1, 2, \ldots, n$ the set $S(v_m)$ and its order are uniquely determined by the postordered list $v_1 v_2 \cdots v_n$ of T and the numbers $c_1, \ldots, c_n$ of children of $v_1, \ldots, v_n$.

Consider some vertex v_m. Then v_m is the root of the subtree T_m consisting of v_m and its descendants. When Algorithm POSTORDER lists v_m the subtrees of T_m have already been listed and their lists immediately precede v_m, in the order determined by the order of $S(v_m)$. Moreover, the algorithm does not insert entries later on into the list of T_m, so the list of T_m appears in the final list $v_1 v_2 \cdots v_n$ of T as an unbroken string with v_m at the right end.

Since v_1 has no predecessors, v_1 is a leaf. Thus $S(v_1) = \varnothing$, and its order is vacuously determined. Assume inductively that for each k with $k < m$ the set $S(v_k)$ and its order are determined, and consider v_m. [We are using the second principle of induction here on the finite set $\{1, 2, \ldots, n\}$.] Then the set $U(m) = \{v_k : k < m\} \setminus \bigcup_{k<m} S(v_k)$ is determined. It consists of the vertices v_k to the left of v_m whose parents are not to the left of v_m. The children of v_m are the members of $U(m)$ which are in the tree T_m. Since the T_m list is immediately to the left of v_m, the children of v_m are the c_m members of $U(m)$ farthest to the right. Since $U(m)$ is determined and c_m is given, the set $S(v_m)$ of children of v_m is determined. Moreover, its order is the order of appearance in the list $v_1 v_2 \cdots v_n$.

By induction, each ordered set $S(v_m)$ of children is determined by the postordered list and the sequence $c_1, c_2, \ldots, c_n$. The root of the tree is, of course, the last vertex v_n. Thus the complete structure of the ordered rooted tree is determined. ∎

EXAMPLE 7 The list $u\ x\ y\ z\ w\ v\ p\ q\ s\ r$ and sequence 0, 0, 0, 0, 3, 2, 0, 0, 2, 2 give the following sets.

v_k	u	x	y	z	w	v	p	q	s	r
$U(k)$	$\varnothing$	$\{u\}$	$\{u, x\}$	$\{u, x, y\}$	$\{u, x, y, z\}$	$\{u, w\}$	$\{v\}$	$\{v, p\}$	$\{v, p, q\}$	$\{v, s\}$
$S(v_k)$	$\varnothing$	$\varnothing$	$\varnothing$	$\varnothing$	$\{x, y, z\}$	$\{u, w\}$	$\varnothing$	$\varnothing$	$\{p, q\}$	$\{v, s\}$

The ordered sets $S(v_k)$ can be assembled recursively into the tree T of Figure 1 of § 9.3. ∎

EXERCISES 9.4

1. Write the algebraic expression given by Figure 1(b) in reverse Polish and in Polish notation.

2. For the ordered rooted tree in Figure 3(a), write the corresponding algebraic expression in reverse Polish notation and also in the usual infix algebraic notation.

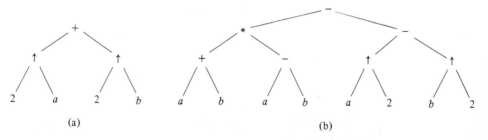

(a) (b)

Figure 3

3. (a) For the ordered rooted tree in Figure 3(b), write the corresponding algebraic expression in Polish notation and also in the usual infix algebraic notation.
   (b) Simplify the algebraic expression obtained in part (a) and then draw the corresponding tree.

4. Calculate the following expressions given in reverse Polish notation.
   (a) $3\,3\,4\,5\,1 - * + +$ (b) $3\,3 + 4 + 5 * 1 -$
   (c) $3\,3\,4 + 5 * 1 - +$

5. Calculate the following expressions given in reverse Polish notation.
   (a) $6\,3 / 3 + 7\,3 - *$ (b) $3\,2 \uparrow 4\,2 \uparrow + 5/2 *$

6. Calculate the following expressions in Polish notation.
   (a) $- * 3 \uparrow 5\,2\,2$ (b) $\uparrow * 3\,5 - 2\,2$
   (c) $- \uparrow * 3\,5\,2\,2$ (d) $/ * 2 + 2\,5 \uparrow + 3\,4\,2$
   (e) $* + / 6\,3\,3 - 7\,3$

7. Write the following algebraic expressions in reverse Polish notation.
   (a) $(3x - 4)^2$ (b) $(a + 2b)/(a - 2b)$
   (c) $x - x^2 + x^3 - x^4$

8. Write the algebraic expressions in Exercise 7 in Polish notation.

9. (a) Write the algebraic expressions $a(bc)$ and $(ab)c$ in reverse Polish notation.
   (b) Do the same for $a(b + c)$ and $ab + ac$.
   (c) What do the associative and distributive laws look like in reverse Polish notation?

10. Write the expression $x\ y + 2 \uparrow x\ y - 2 \uparrow - x\ y * /$ in the usual infix algebraic notation and simplify.

11. Consider the compound proposition represented by Figure 2(b).
    (a) Write the proposition in the usual infix notation [with parentheses].
    (b) Write the proposition in reverse Polish and in Polish notation.

12. The following compound propositions are given in Polish notation. Draw the corresponding rooted trees, and rewrite the expressions in the usual infix notation.
    (a) $\leftrightarrow \neg \wedge \neg p \neg q \vee p\ q$ (b) $\leftrightarrow \wedge p\ q \neg \rightarrow p \neg q$
    [These are laws from Table 1 of § 2.2.]

13. Repeat Exercise 12 for the following.
    (a) $\rightarrow \wedge p \rightarrow p\ q\ q$
    (b) $\rightarrow \wedge \wedge \rightarrow p\ q \rightarrow r\ s \vee p\ r \vee q\ s$

14. Write the following compound propositions in reverse Polish notation.
    (a) $[(p \rightarrow q) \wedge (q \rightarrow r)] \rightarrow (p \rightarrow r)$ (b) $[(p \vee q) \wedge \neg p] \rightarrow q$

15. Illustrate the ambiguity of "parenthesis-free infix notation" by writing the following pairs of expressions in infix notation without parentheses.
    (a) $(a/b) + c$ and $a/(b + c)$ (b) $a + (b^3 + c)$ and $(a + b)^3 + c$

16. Use the recursive definition for wff's for reverse Polish notation to show that the following are wff's.
    (a) $3\ x\ 2 \uparrow *$ (b) $x\ y + 1\ x / 1\ y / + *$
    (c) $4\ x\ 2 \uparrow y\ z + 2 \uparrow / -$

17. (a) Define wff's for Polish notation for algebraic expressions.
    (b) Use the definition in part (a) to show that $\uparrow + x / 4 \, x \, 2$ is a wff.

18. Let $S_1 = x_1 \, 2\uparrow$ and $S_{n+1} = S_n \, x_{n+1} \, 2\uparrow +$ for $n \geq 1$. Here $x_1, x_2, \ldots$ represent variables.
    (a) Show that each S_n is a wff for reverse Polish notation. *Hint*: Use induction.
    (b) What does S_n look like in the usual infix notation?

19. (a) Define wff's for reverse Polish notation for the propositional calculus; see Example 6(a) of § 3.6.
    (b) Use the definition in part (a) to show that $p \, q \, \neg \, \wedge \, \neg \, p \, q \, \neg \, \rightarrow \, \vee$ is a wff.
    (c) Define wff's for Polish notation for the propositional calculus.
    (d) Use the definition in part (c) to show that $\vee \, \neg \, \wedge \, p \, \neg \, q \rightarrow p \, \neg \, q$ is a wff.

20. (a) Draw the tree with postorder vertex sequence $s \, t \, v \, y \, r \, z \, w \, u \, x \, q$ and number of children sequence 0, 0, 0, 2, 2, 0, 0, 0, 2, 3.
    (b) Is there a tree with $s \, t \, v \, y \, r \, z \, w \, u \, x \, q$ as preorder vertex sequence and number of children sequence 0, 0, 0, 2, 2, 0, 0, 0, 2, 3? Explain.

§ 9.5   Weighted Trees

In this section we discuss general weighted trees and we give applications to prefix codes and sorted lists.

Consider a finite rooted tree T. We call T a **weighted tree** if a nonnegative real number is assigned to each leaf. This number is called the **weight** of the leaf. To establish some notation, we assume that T has t leaves whose weights are $w_1, w_2, \ldots, w_t$. We lose no generality if we also assume that $w_1 \leq w_2 \leq \cdots \leq w_t$. It will be convenient to label the leaves by their weights, and so we will often refer to the leaf by referring to its weight. Let $l_1, l_2, \ldots, l_t$ denote the corresponding levels of the leaves, so that l_i is the length of the path from the root to the leaf w_i. The **weight** of the tree T is the number

$$W(T) = \sum_{i=1}^{t} w_i l_i.$$

EXAMPLE 1   (a) Consider the weighted tree in Figure 1(a). The six leaves have weights 2, 4, 6, 7, 7 and 9. Thus $w_1 = 2$, $w_2 = 4$, $w_3 = 6$, $w_4 = 7$, $w_5 = 7$ and $w_6 = 9$. There are two leaves labeled 7, and it does not matter which we regard as w_4 and which we regard as w_5. For definiteness, we let w_4 represent the leaf labeled 7 at level 2. Then the level numbers are $l_1 = 3$, $l_2 = 1$, $l_3 = 3$, $l_4 = 2$, $l_5 = 1$ and $l_6 = 2$. Hence

$$W(T) = \sum_{i=1}^{6} w_i l_i = 2 \cdot 3 + 4 \cdot 1 + 6 \cdot 3 + 7 \cdot 2 + 7 \cdot 1 + 9 \cdot 2 = 67.$$

(b) The same six weights can be placed on a binary tree as in Figure 1(b), for instance. Now the level numbers are $l_1 = 3$, $l_2 = 3$, $l_3 = 2$, $l_4 = 3$, $l_5 = 3$ and $l_6 = 2$ and so

$$W(T) = 2 \cdot 3 + 4 \cdot 3 + 6 \cdot 2 + 7 \cdot 3 + 7 \cdot 3 + 9 \cdot 2 = 90.$$

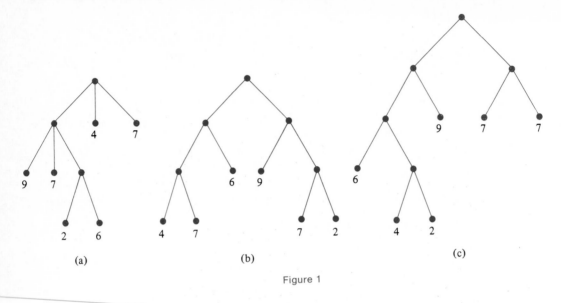

(a)   (b)   (c)

Figure 1

(c) Figure 1(c) shows another binary tree with these weights. Its weight is

$$W(T) = 2 \cdot 4 + 4 \cdot 4 + 6 \cdot 3 + 7 \cdot 2 + 7 \cdot 2 + 9 \cdot 2 = 88.$$

The total weight is less than in part (b), because the heavier leaves are near the root and the lighter ones are farther away. Later in this section we will discuss an algorithm for obtaining a binary tree with minimal weight for any specified sequence of weights $w_1, w_2, \ldots, w_t$. ∎

EXAMPLE 2 As we will explain later in this section, certain sets of binary numbers can serve as codes. An example of such a set is $\{00, 01, 100, 1010, 1011, 11\}$. These numbers are the labels of the leaves in the binary tree of Figure 2(a). This set could serve as a code for the letters in an alphabet Σ which has six letters. Suppose that we also know how frequently each letter in Σ is used. In Figure 2(b) we have placed a weight at each leaf that signifies the percentage of code symbols using that leaf. For example, the letter coded 00 appears 25 percent of the time, the letter coded 1010 appears 20 percent of the time, etc. Since the length of each code symbol as a word in 0's and 1's is exactly its level in the binary tree, the average length of a code message using 100 letters from Σ will just be the weight of the weighted tree, in this case

$$25 \cdot 2 + 10 \cdot 2 + 10 \cdot 3 + 20 \cdot 4 + 15 \cdot 4 + 20 \cdot 2 = 280.$$

This weight measures the efficiency of the code. As we will see in Example 7, there are more efficient codes for this example, i.e., for the set of frequencies 10, 10, 15, 20, 20, 25. ∎

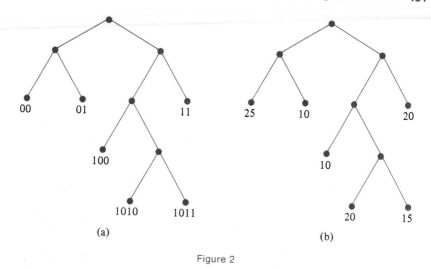

(a)

(b)

Figure 2

EXAMPLE 3   Consider some sorted lists, say $L_1, L_2, \ldots, L_n$, which could be alphabetically ordered files or some other subsets of a given poset, with each listed in increasing order. To illustrate the ideas involved, let's suppose each list is a set of real numbers. Suppose that we can merge lists two at a time to produce new lists. Our problem is to determine how to merge the lists most efficiently to produce a single sorted list.

Two lists are merged by comparing the largest numbers of each set and selecting the larger [either one if they are equal]. The selected number is removed and stored someplace, and the process is repeated for the remaining two lists. The process ends when one of the lists is empty. If the lists contain j and k elements, respectively, this process must end after $j + k - 1$ or fewer comparisons. The goal is to merge $L_1, L_2, \ldots, L_n$ in pairs while minimizing the number of comparisons involved.

Let's consider a more concrete example. Suppose we have five lists L_1, L_2, L_3, L_4, L_5 with 15, 22, 31, 34, 42 items and suppose they are merged as indicated in Figure 3. There are 4 merges indicated by the circled numbers.

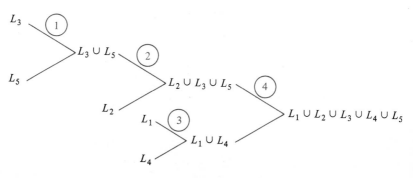

Figure 3

The first merge involves at most $|L_3| + |L_5| - 1 = 72$ comparisons. The second merge involves at most $|L_2| + |L_3| + |L_5| - 1 = 94$ comparisons. The third and fourth merges involve at most $|L_1| + |L_4| - 1 = 48$ and $|L_1| + |L_2| + |L_3| + |L_4| + |L_5| - 1 = 143$ comparisons. The entire process involves at most 357 comparisons. This number isn't very illuminating by itself, but note that

$$357 = 2 \cdot |L_1| + 2 \cdot |L_2| + 3 \cdot |L_3| + 2 \cdot |L_4| + 3 \cdot |L_5| - 4.$$

This is just 4 less than the weight of the tree in Figure 4. Note the intimate connection between Figures 3 and 4. No matter how we merge the five lists in pairs, there will be 4 merges. A computation like the one above shows that the merge will involve at most $W(T) - 4$ comparisons, where T is the tree corresponding to the merge. So finding a merge that minimizes the largest possible number of comparisons is equivalent to finding a binary tree with weights 15, 22, 31, 34, 42 having minimal weight. We return to this problem in Example 8.

The merging of n lists in pairs involves $n - 1$ merges. In general, a merge of n lists will involve at most $W(T) - (n - 1)$ comparisons, where T is the tree corresponding to the merge. ∎

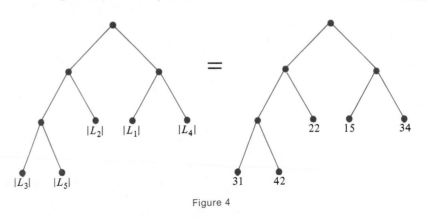

Figure 4

Examples 2 and 3 suggest the following general problem. We are given a list $L = \langle w_1, \ldots, w_t \rangle$ of at least two nonnegative numbers, and we want to construct a binary weighted tree T with the members of L as weights so that $W(T)$ is as small as possible. We call such a tree T an **optimal binary tree** for the weights $w_1, \ldots, w_t$. The following recursive algorithm solves the problem by producing an optimal binary tree $T(L)$.

Algorithm HUFFMAN(L).

If $|L| = 2$ let $T(L)$ be a tree with 2 leaves, of weights w_1 and w_2. Otherwise find the two smallest members of L, say u and v, and let L' be the list obtained from L by removing u and v and inserting $u + v$.

Do HUFFMAN(L').
Form $T(L)$ from $T(L')$ by replacing a leaf of weight $u + v$ in $T(L')$ by
  a subtree with two leaves, of weights u and v.
End ▌

This algorithm ultimately reduces the problem to one of finding optimal
binary trees with two leaves, which is trivial to solve. We will show shortly
that HUFFMAN(L) always produces an optimal binary tree. First we look
at some examples of how it works, and apply the algorithm to the problems
that originally motivated our looking at optimal trees.

EXAMPLE 4   Consider weights 2, 4, 6, 7, 7, 9. First the algorithm repeatedly combines the
smallest two weights to obtain shorter and shorter weight sequences:

$$2, 4, 6, 7, 7, 9 \to 6, 6, 7, 7, 9 \to 7, 7, 9, 12 \to 9, 12, 14 \to 14, 21.$$

When a sequence with two weights is reached, the algorithm proceeds to
construct trees beginning with the shortest weight sequence and working
up. The trees so obtained are given in Figure 5. Note that, for example, the
third tree is obtained from the second tree by replacing the leaf of weight
$14 = 7 + 7$ by a subtree with two leaves of weight 7 each. The final weighted
tree is essentially the tree drawn in Figure 1(c), so that tree is an optimal
binary tree. As noted in Example 1(c), it has weight 88. ▌

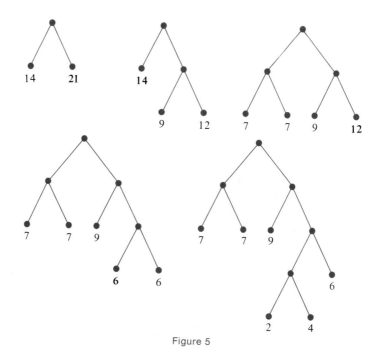

Figure 5

EXAMPLE 5   Let us find an optimal binary tree with weights 2, 3, 5, 7, 10, 13, 19. We repeatedly combine the smallest two weights to obtain the weight sequences

$$\textbf{2, 3}, 5, 7, 10, 13, 19 \rightarrow \textbf{5, 5}, 7, 10, 13, 19$$

$$\rightarrow \textbf{7, 10}, 10, 13, 19 \rightarrow \textbf{10, 13}, 17, 19 \rightarrow \textbf{17, 19}, 23 \rightarrow 23, 36.$$

Then we use HUFFMAN'S algorithm to build the optimal binary trees in Figure 6. After the fourth tree is obtained, either leaf of weight 10 could have been replaced by the subtree with weights 5 and 5. Thus the last two trees

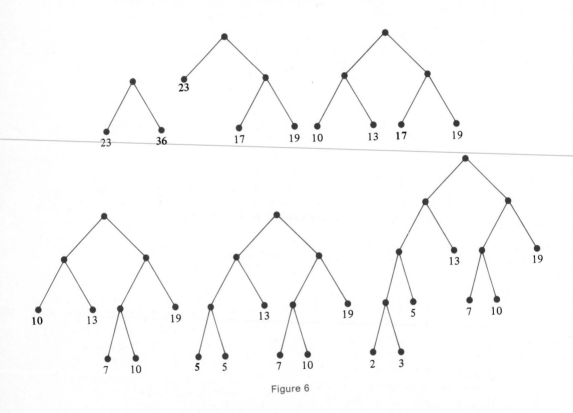

Figure 6

could have been as drawn in Figure 7. Either way, the final tree has weight 150 [Exercise 2]. Note that the optimal tree is by no means unique; the one in Figure 6 has height 4, while the one in Figure 7 has height 5. ∎

We now return to codes using binary numbers. Certain sets of binary numbers must be avoided; for example, a code should not use all three of 10, 01 and 0110, since strings like 100110 would be ambiguous. Actually, it is convenient to require that no code symbol consist of a string of digits which comprises the beginning digits of another code symbol. We can think of strings of 0's and 1's as vertices in a labeled binary tree, as in §9.2. Reading

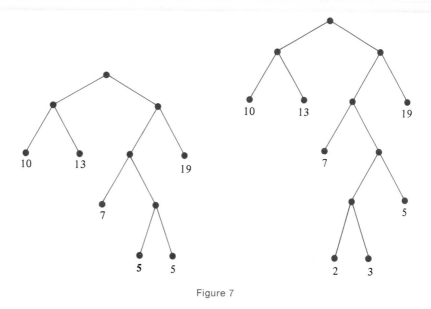

Figure 7

the label of a vertex from left to right tells how to get to the vertex from the root: go left on 0, right on 1. Our restriction simply means that no vertex labeled with a code symbol lies below another in such a labeled binary tree, and means that there is a labeled binary tree whose leaves are exactly the code symbols. We also require that this binary tree be regular, so that every nonleaf has two children. Such a code is called a **prefix code**.

EXAMPLE 6 The set $\{00, 01, 100, 1010, 1011, 11\}$ is a prefix code. It is the set of leaves for the labeled binary tree in Figure 2(a), which we have redrawn in Figure 8(a) with all vertices but the root labeled. Every string of 0's and 1's of length 4 begins with one of these code symbols, since every path of length 4 from the root in the full binary tree in Figure 8(b) runs into one of the code vertices. This means that we can attempt to decode any string of 0's and 1's by proceeding from left to right in the string, finding the first substring that is a code symbol, then the next substring after that, and so on. This procedure either uses up the whole string or it leaves at most three 0's and 1's undecoded at the end.

For example, consider the string

$$1\ 1\ 1\ 0\ 1\ 0\ 1\ 1\ 0\ 1\ 1\ 0\ 0\ 0\ 1\ 0\ 0\ 1\ 1\ 1\ 1\ 1\ 0\ 0\ 1\ 0.$$

We visit vertex 1, then vertex 11. Since vertex 11 is a leaf, we record 11 and return to the root. We next visit vertices 1, 10, 101 and 1010. Since 1010 is a leaf, we record 1010 and return again to the root. Proceeding in this way, we obtain the sequence of code symbols

$$11, 1010, 11, 01, 100, 01, 00, 11, 11, 100$$

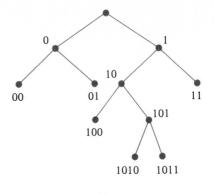

(a)

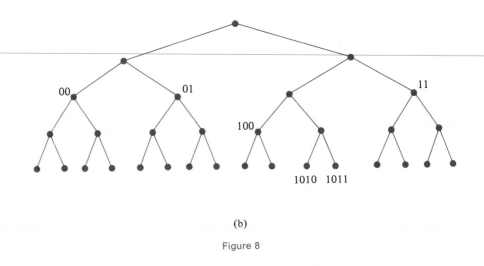

(b)

Figure 8

and have 10 left over. This scheme for decoding arbitrary strings of 0's and 1's will work for any code with the property that every path from the root in a full binary tree runs into a unique code vertex. Prefix codes have this property by their definition. ∎

EXAMPLE 7 We now solve the problem suggested by Example 2; that is, we find a prefix code for the set of frequencies 10, 10, 15, 20, 20, 25 that is as efficient as possible. Hence we want to minimize the average length of a code message using 100 letters from Σ. Thus all we need is an optimal binary tree for these weights. Using the procedure illustrated in Examples 4 and 5, we obtain the weighted tree in Figure 9(a). We label this tree with binary digits in Figure 9(b). Then {00, 01, 10, 110, 1110, 1111} will be a most efficient code for Σ provided we match the letters of Σ to code symbols so that the frequencies of

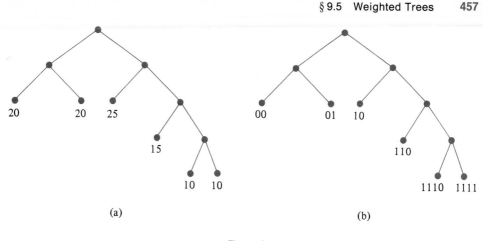

(a)   (b)

Figure 9

the letters are given by Figure 9(a). With this code, the average length of a code message using 100 letters from Σ is

$$20 \cdot 2 + 20 \cdot 2 + 25 \cdot 2 + 15 \cdot 3 + 10 \cdot 4 + 10 \cdot 4 = 255,$$

an improvement over the average length 280 obtained in Example 2. ∎

EXAMPLE 8   We complete the discussion on sorted lists begun in Example 3. There we saw that we needed an optimal binary tree with weights 15, 22, 31, 34, 42. Using the procedure in Examples 4 and 5, we obtain the tree in Figure 10.

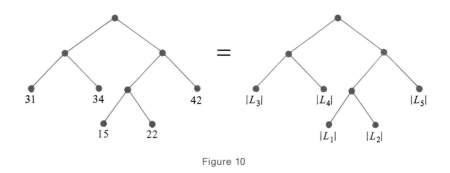

Figure 10

This tree has weight 325. The corresponding merge in pairs given in Figure 11 will require at most $325 - 4 = 321$ comparisons. ∎

To show that HUFFMAN'S algorithm works, we first prove a lemma which tells us that in optimal binary trees the heavy leaves are near the root. The lemma and its corollary are quite straightforward if all the weights are distinct [Exercise 14]. However, we need the more general case. Even if

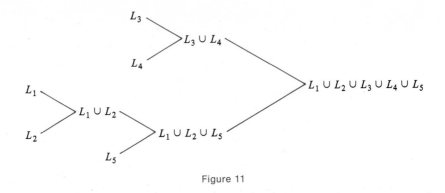

Figure 11

we began with distinct weights, HUFFMAN'S algorithm might lead to the case where the weights are not all distinct, as occurred in Example 5.

Lemma   Let T be an optimal binary tree for weights $w_1, w_2, \ldots, w_t$. For $i = 1, 2, \ldots, t$, let l_i denote the level of w_i. If $w_j < w_k$ then $l_j \geq l_k$.

   Proof. Assume that $w_j < w_k$ and $l_j < l_k$ for some j and k. Let T' be the tree obtained by interchanging the weights w_j and w_k. In calculating $W(T)$ the leaves w_j and w_k contribute $w_j l_j + w_k l_k$, while in calculating $W(T')$ they contribute $w_j l_k + w_k l_j$. Since the other leaves contribute the same to both $W(T)$ and $W(T')$, we have

$$W(T) - W(T') = w_j l_j + w_k l_k - w_j l_k - w_k l_j$$
$$= (w_k - w_j)(l_k - l_j) > 0.$$

Hence $W(T') < W(T)$ and T is not an optimal binary tree, contrary to our hypothesis. ∎

Corollary   There is an optimal binary tree T where the weights w_1 and w_2 are both at the lowest level l.

   Proof. There are at least two leaves at the lowest level, say w_j and w_k. If $w_1 < w_j$ then $l_1 \geq l_j = l$ by the lemma, and so $l_1 = l$ and w_1 is at level l. If $w_1 = w_j$ then conceivably $l_1 < l_j$, but we can interchange w_1 and w_j without changing the total weight of T. Similarly, by interchanging w_2 and w_k if necessary, we may suppose that w_2 is at level l. ∎

   The following result shows that HUFFMAN(L) works.

Theorem   Suppose that $0 \leq w_1 \leq w_2 \leq \cdots \leq w_t$. Let T' be an optimal binary tree with weights $w_1 + w_2, w_3, \ldots, w_t$ and let T be the weighted binary tree obtained from T' by replacing a leaf of weight $w_1 + w_2$ by a subtree with two leaves having weights w_1 and w_2. Then T is an optimal binary tree with weights w_1, $w_2, \ldots, w_t$.

Proof. Since there are only finitely many binary trees with t leaves, there must be an optimal binary tree T_0 with weights $w_1, w_2, \ldots, w_t$. Our task is to show $W(T) = W(T_0)$. By the corollary of the lemma, we may suppose that the weights w_1 and w_2 for T_0 are at the same level. The total weight of T_0 won't change if weights at the same level are interchanged. Thus we may assume that w_1 and w_2 are children of the same parent p. These three vertices form a little subtree T_p.

Now let T_0' be the tree with weights $w_1 + w_2, w_3, \ldots, w_t$ obtained from T_0 by replacing the subtree T_p by a leaf $\bar{p}$ of weight $w_1 + w_2$. Let l be the level of the vertex p and observe that in calculating $W(T_0)$ the subtree T_p contributes $w_1(l+1) + w_2(l+1)$, while in calculating $W(T_0')$, the vertex $\bar{p}$ with weight $w_1 + w_2$ contributes $(w_1 + w_2)l$. Thus

$$W(T_0) = W(T_0') + w_1 + w_2.$$

The same argument shows that

$$W(T) = W(T') + w_1 + w_2.$$

Since T' is optimal for the weights $w_1 + w_2, w_3, \ldots, w_t$, we have $W(T') \leq W(T_0')$ and so

$$W(T) = W(T') + w_1 + w_2 \leq W(T_0') + w_1 + w_2 = W(T_0).$$

Of course $W(T_0) \leq W(T)$, since T_0 is optimal for the weights $w_1, w_2, \ldots, w_t$, so $W(T) = W(T_0)$, as desired. That is, T is an optimal binary tree with weights $w_1, w_2, \ldots, w_t$. $\blacksquare$

HUFFMAN(L) leads recursively to $t - 1$ choices of parents in $T(L)$. Each choice requires a search for the two smallest members of the current list, which can be done in time $O(t)$ by just running through the list. So the total operation of the algorithm takes time at most $O(t^2)$.

There are at least two ways to speed up the algorithm. It is possible to find the two smallest members in time $O(\log t)$, using a binary tree as a data structure for the list L. Alternatively, there are algorithms that can initially sort L into nondecreasing order in time $O(t \log t)$. Then the smallest elements are simply the first two on the list, and after we remove them we can maintain the nondecreasing order on the new list by inserting their sum at the appropriate place, just as we did in Examples 4 and 5. The correct insertion point can be found in time $O(\log t)$, so this scheme, too, works in time at most $O(t \log t)$.

EXERCISES 9.5

1. (a) Calculate the weights of all the trees in Figure 5.
   (b) Calculate the weights of all the trees in Figure 6.

2. Calculate the weights of the two trees in Figures 6 and 7 with weights 2, 3, 5, 7, 10, 13, 19.

3. Construct an optimal binary tree for the following sets of weights and compute the weight of the optimal tree.
   (a) $\{1, 3, 4, 6, 9, 13\}$
   (b) $\{1, 3, 5, 6, 10, 13, 16\}$
   (c) $\{2, 4, 5, 8, 13, 15, 18, 25\}$
   (d) $\{1, 1, 2, 3, 5, 8, 13, 21, 34\}$

4. Find an optimal binary tree for the weights 10, 10, 15, 20, 20, 25 and compare your answer with Figure 9(a).

5. Which of the following sets of sequences are prefix codes? If the set is a prefix code, construct a binary tree whose leaves represent this binary code. Otherwise, explain why the set is not a prefix code.
   (a) $\{000, 001, 01, 10, 11\}$
   (b) $\{00, 01, 110, 101, 0111\}$
   (c) $\{00, 0100, 0101, 011, 100, 101, 11\}$

6. Here is a prefix code: $\{00, 010, 0110, 0111, 10, 11\}$.
   (a) Construct a binary tree whose leaves represent this binary code.
   (b) Decode the string

   $$00100001100100010011111110110$$

   if $00 = A$, $10 = D$, $11 = E$, $010 = H$, $0110 = M$ and 0111 represents the apostrophe '. You will obtain the very short poem titled "Fleas."
   (c) Decode

   $$01011011000101101101101100010.$$

   (d) Decode the following soap opera. $100010010001001100010000110.$
   $01011011000101101100001100010.$ 011000011000101110
   $01010110010.$

7. Suppose we are given a fictitious alphabet Σ of seven letters a, b, c, d, e, f and g with the following frequencies per 100 letters: a-11, b-20, c-4, d-22, e-14, f-8, g-21.
   (a) Design an optimal binary prefix code for this alphabet.
   (b) What is the average length of a code message using 100 letters from Σ?

8. Repeat Exercise 7 for the frequencies: a-25, b-2, c-15, d-10, e-38, f-4, g-6.

9. (a) Show that the code $\{000, 001, 10, 110, 111\}$ satisfies all the requirements of a prefix code, except that the corresponding binary tree is not regular.
   (b) Show that some strings of binary digits are meaningless for this code.
   (c) Show that $\{00, 01, 10, 110, 111\}$ is a prefix code, and compare its binary tree with that of part (a).

10. Repeat Exercise 7 for the frequencies: a-31, d-31, e-12, h-6, m-20.

11. Let L_1, L_2, L_3, L_4 be sorted lists having 23, 31, 61 and 73 elements, respectively. How many comparisons at most are needed if the lists are merged as indicated?

(a)

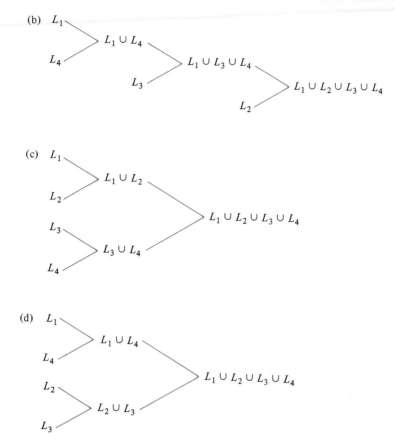

(b) L_1, L_4 → $L_1 \cup L_4$, L_3 → $L_1 \cup L_3 \cup L_4$, L_2 → $L_1 \cup L_2 \cup L_3 \cup L_4$

(c) L_1, L_2 → $L_1 \cup L_2$; L_3, L_4 → $L_3 \cup L_4$ → $L_1 \cup L_2 \cup L_3 \cup L_4$

(d) L_1, L_4 → $L_1 \cup L_4$; L_2, L_3 → $L_2 \cup L_3$ → $L_1 \cup L_2 \cup L_3 \cup L_4$

(e) How should the four lists be merged so that the total number of comparisons is a minimum? It is not sufficient to simply examine parts (a)–(d) since there are other ways to merge the lists.

12. Let L_1, L_2, L_3, L_4, L_5, L_6 be sorted lists having 5, 6, 9, 22, 29, 34 elements, respectively.
   (a) Show how the six lists should be merged so that the total number of comparisons can be kept to a minimum.
   (b) How many comparisons might be needed in your procedure?

13. Repeat Exercise 12 for seven lists having 2, 5, 8, 12, 16, 22, 24 elements, respectively.

14. Let T be an optimal binary tree whose weights satisfy $w_1 < w_2 < \cdots < w_t$. Show that the corresponding level numbers satisfy $l_1 \geq l_2 \geq l_3 \geq \cdots \geq l_t$.

15. Look at Exercise 1 again, and note that whenever a vertex of weight $w_1 + w_2$ in a tree T' is replaced by a subtree with weights w_1 and w_2, then the weight increases by $w_1 + w_2$. That is, the new tree T has weight $W(T') + w_1 + w_2$, just as in the proof of the theorem.

CHAPTER HIGHLIGHTS

As always, one of the best ways to use this material for review is to follow the suggestions at the end of Chapter 0. Ask yourself: What does it mean? Why is it here? How can I use it?

Concepts

tree, subtree, forest
   leaf, parent, child, descendant
spanning tree, minimal spanning tree, spanning forest
directed tree
root, rooted tree, ordered rooted tree
binary/m-ary tree
   regular/full m-ary tree
preorder, postorder, inorder traversals
Polish, reverse Polish, infix notations
weighted tree, weight of a tree
optimal binary tree
   merge of lists, prefix code

Facts

The following statements about a graph with $n \geq 1$ vertices and no loops are equivalent:

   (a) G is a tree.
   (b) There is just one simple path between any two distinct vertices.
   (c) G is connected, but won't be if an edge is removed.
   (d) G is acyclic, but won't be if an edge is added.
   (e) G is acyclic and has $n - 1$ edges [as many as possible].
   (f) G is connected and has $n - 1$ edges [as few as possible].

Any vertex of a tree can be used as a root for a unique corresponding rooted tree.
Rooted trees and height can be defined recursively.
Rooted trees are natural data structures for random access to data and for branching and recursive programs.
An ordered rooted tree cannot always be recovered from its preorder, inorder or postorder listing.
It *can* be recovered from its preorder or postorder listing given knowledge of how many children each vertex has.

Methods

KRUSKAL'S and PRIM'S algorithms to construct minimal spanning trees [or forests] for weighted graphs.
Depth-first search to traverse a tree.
PREORDER, POSTORDER and INORDER to list vertices of an ordered rooted tree.

TREESORT and TOPSORT to generate a topologically sorted labeling of an acyclic digraph.

TREE and FOREST to build a spanning forest in an undirected graph.

TREE+ and FOREST+ to keep track of the trees in a spanning forest and to check for cycles.

Method of charges for estimating running time of an algorithm.

Construction and deciphering of Polish and reverse Polish forms for algebraic expressions.

HUFFMAN'S recursive algorithm to find an optimal binary tree with given weights.

Use of binary weighted trees to determine efficient merging patterns and efficient prefix codes.

10

BOOLEAN ALGEBRA

The term "Boolean algebra" is used to describe a variety of related topics, ranging from symbolic logic and truth tables to the arithmetic performed by electrical relay networks or electronic computers. This chapter begins by developing abstract structures called Boolean algebras, starting from posets and lattices. The algebra involved is reminiscent of the truth tables of Chapter 2. The switching functions associated with electronic logic networks turn out to give important examples of Boolean algebras, and we are able to study the networks and their functions using the algebraic tools we develop. The chapter concludes with a brief discussion of Karnaugh maps, a tool somewhat like Venn diagrams, which can be useful in analyzing moderately complicated logical expressions.

§ 10.1 Lattices

Recall that in § 7.1 we called a poset $(P, \leqq)$ a **lattice** provided every finite subset has both a least upper bound and a greatest lower bound. We also introduced two binary operations $\vee$ and $\wedge$ on P where

$$x \vee y = \text{lub}\{x, y\} \quad \text{and} \quad x \wedge y = \text{glb}\{x, y\}.$$

We will see in Theorem 1 that these binary operations satisfy the following properties:

1La. $x \vee y = y \vee x$
   b. $x \wedge y = y \wedge x$ commutative laws

2La. $(x \vee y) \vee z = x \vee (y \vee z)$
   b. $(x \wedge y) \wedge z = x \wedge (y \wedge z)$ associative laws

3La. $x \vee (x \wedge y) = x$
   b. $x \wedge (x \vee y) = x$ absorption laws

Theorem 2 will say that these six properties characterize lattices, in the sense that a set closed under two binary operations $\vee$ and $\wedge$ satisfying the properties can be given a natural partial ordering which makes it a lattice. Thus every general fact about lattices must somehow be a consequence of these six properties, and whenever we encounter a set with two binary operations satisfying these conditions we can view it as a lattice and immediately conclude that it has whatever general properties lattices have. One of our goals, then, will be to determine what can be proved just using the six listed properties. Let us call any set L with two binary operations $\vee$ and $\wedge$ that satisfy these six properties an **algebraic lattice**. We will sometimes write $(L, \vee, \wedge)$ to stress our interest in the operations $\vee$ and $\wedge$. As in § 7.1 we read $x \vee y$ as "x join y" and $x \wedge y$ as "x meet y."

Theorem 1   Consider a lattice $(P, \leq)$ and define $x \vee y = \mathrm{lub}\{x, y\}$ and $x \wedge y = \mathrm{glb}\{x, y\}$. Then $(P, \vee, \wedge)$ is an algebraic lattice.

   Proof. The commutative laws are clear. To check property 2La, consider $x, y, z \in P$ and let $u = (x \vee y) \vee z$ and $v = x \vee (y \vee z)$. Since $y \leq x \vee y \leq u$ and $z \leq u$, the element u is an upper bound for $\{y, z\}$. Since $y \vee z$ is the *least* upper bound, we have $y \vee z \leq u$. Also $x \leq x \vee y \leq u$, so u is an upper bound for $\{x, y \vee z\}$ and hence $x \vee (y \vee z) \leq u$. In other words, $v \leq u$. Similar reasoning shows that $u \leq v$ and therefore $u = v$.

   To check property 3La, consider $x, y \in P$. Since $x \leq x \vee w$ for any w, we have $x \leq x \vee (x \wedge y)$. Since $x \leq x$ and $x \wedge y \leq x$, the element x is an upper bound for $\{x, x \wedge y\}$ and so $x \vee (x \wedge y) \leq x$. Thus $x \vee (x \wedge y) = x$, as desired.

   Properties 2Lb and 3Lb have similar proofs. ■

   Before we prove results for algebraic lattices, we mention a **duality principle**. If we consider one of the six properties defining an algebraic lattice and we replace each $\vee$ by $\wedge$ and each $\wedge$ by $\vee$, then we obtain another one of the properties, which we might call the dual property. Thus 1La and 1Lb are duals for each other, etc. It follows that any theorem or identity that holds for an algebraic lattice will remain true if we dualize it, that is, if we replace each $\vee$ by $\wedge$ and each $\wedge$ by $\vee$.

Proposition Let $(L, \vee, \wedge)$ be an algebraic lattice.
    (a) The operations $\vee$ and $\wedge$ satisfy the idempotent laws: $x \vee x = x$ and $x \wedge x = x$ for all $x \in L$.
    (b) For $x, y \in L$ we have $x \vee y = y$ if and only if $x \wedge y = x$.

Proof. (a) With $y = x \vee x$, absorption law 3La gives the identity $x = x \vee [x \wedge (x \vee x)]$. But $[x \wedge (x \vee x)] = x$ by absorption law 3Lb, so $x = x \vee x$. The other idempotent law follows by duality.
    (b) Suppose that $x \vee y = y$. Then

$$x = x \wedge (x \vee y) \quad \text{absorption law 3Lb}$$
$$= x \wedge y \quad \text{since } x \vee y = y \text{ by supposition.}$$

The other implication follows by duality, interchanging the roles of x and y. ∎

Theorem 1 shows that a lattice $(P, \leq)$ gives rise to an algebraic lattice. Theorem 2 provides a converse.

Theorem 2 Let $(L, \vee, \wedge)$ be an algebraic lattice. We define a relation $\leq$ on L as follows:

$$x \leq y \quad \text{if and only if} \quad x \vee y = y.$$

Then $\leq$ is a partial order, and $(L, \leq)$ is a lattice in which $\text{lub}\{x, y\} = x \vee y$ and $\text{glb}\{x, y\} = x \wedge y$ for all $x, y \in L$.

The proposition above shows that we could just as well define $x \leq y$ if and only if $x \wedge y = x$.

Proof. We check first that $\leq$ is a partial order.
    (R)   Since $x \vee x = x$ by the idempotent law, we have $x \leq x$.
    (AS) Suppose $x \leq y$ and $y \leq x$, so that $x \vee y = y$ and $y \vee x = x$. Then $x = y \vee x = x \vee y = y$, using the commutative law.
    (T)   Suppose that $x \leq y$ and $y \leq z$, so that $x \vee y = y$ and $y \vee z = z$. Then

$$x \vee z = x \vee (y \vee z) \quad \text{since } y \vee z = z$$
$$= (x \vee y) \vee z \quad \text{associative law}$$
$$= y \vee z \quad \text{since } x \vee y = y$$
$$= z \quad \text{since } y \vee z = z.$$

That is, $x \leq z$ as desired.
    Next we verify $\text{lub}\{x, y\} = x \vee y$. Since $x \vee (x \vee y) = (x \vee x) \vee y = x \vee y$, the definition of $\leq$ shows that $x \leq x \vee y$. Similarly, $y \leq x \vee y$ and so $x \vee y$ is an upper bound for $\{x, y\}$. To show $x \vee y$ is the *least* upper bound, consider another upper bound u for $\{x, y\}$. Then $x \leq u$ and $y \leq u$, so that $x \vee u = u$ and $y \vee u = u$ and hence

$$(x \vee y) \vee u = x \vee (y \vee u) = x \vee u = u.$$

Therefore $x \vee y \leq u$. This shows that $x \vee y$ is the least upper bound for $\{x, y\}$, i.e., $x \vee y = \text{lub}\{x, y\}$. A similar [dual!] argument shows that $x \wedge y = \text{glb}\{x, y\}$. ∎

Theorem 1 shows that if $(L, \leq)$ is a lattice with $\vee$ and $\wedge$ defined by $x \vee y = \text{lub}\{x, y\}$ and $x \wedge y = \text{glb}\{x, y\}$ relative to the partial order $\leq$, then $(L, \vee, \wedge)$ is an algebraic lattice. Theorem 2 shows that if we go on to define $\underline{\leq}$ on L by $x \underline{\leq} y$ if and only if $x \vee y = y$, then $(L, \underline{\leq})$ is a lattice. Is $(L, \underline{\leq})$ the lattice we started with? That is, is $\underline{\leq}$ the same as $\leq$? If $x \leq y$ then $y = \text{lub}\{x, y\} = x \vee y$ and so $x \underline{\leq} y$. Conversely, if $x \underline{\leq} y$ then $y = x \vee y = \text{lub}\{x, y\}$, so $x \leq y$. Thus $\underline{\leq}$ is the partial order $\leq$ we started with. A similar analysis shows that starting with an algebraic lattice, forming its poset and then forming its corresponding algebraic lattice gives us back our original algebraic lattice.

EXAMPLE 1   Consider any set S and let $\mathscr{P}(S)$ be the set of all subsets of S. With the operations $\cup$ and $\cap$, $\mathscr{P}(S)$ is an algebraic lattice. Motivated by Theorem 2, for A, B in $\mathscr{P}(S)$ we define

$$A \underline{\leq} B \quad \text{if and only if} \quad A \cup B = B.$$

As noted back in Exercise 12 of § 1.2, $A \cup B = B$ if and only if $A \subseteq B$. So the relations $\underline{\leq}$ and $\subseteq$ are the same. We have not defined a new relation, but we have given a slightly different and more algebraic definition of $\subseteq$ that is in terms of $\cup$ alone. ∎

EXAMPLE 2   Let $\mathscr{S}$ be a set of compound propositions such that $P \wedge Q$ and $P \vee Q$ belong to $\mathscr{S}$ whenever P, Q belong to $\mathscr{S}$. [Here $\vee$ and $\wedge$ are the usual logical connectives.] We want to regard propositions in $\mathscr{S}$ as the same if they are logically equivalent, i.e., we want to view $\Leftrightarrow$ as an equivalence relation on $\mathscr{S}$ as in Example 7 of § 7.3.

Let $[\mathscr{S}]$ be the set of equivalence classes of $\mathscr{S}$. If P_1 and P_2 are logically equivalent and Q_1 and Q_2 are logically equivalent, then $P_1 \vee Q_1$ and $P_2 \vee Q_2$ are also logically equivalent. Consequently, the following definition on equivalence classes is unambiguous:

$$[P] \vee [Q] = [P \vee Q].$$

[See the end of § 7.3 for a discussion concerning this issue of unambiguous definitions on equivalence classes.] Similarly, the definition

$$[P] \wedge [Q] = [P \wedge Q]$$

is unambiguous. It is believable and easy to check that, with these operations $\vee$ and $\wedge$, $[\mathscr{S}]$ is an algebraic lattice. For example, we know $P \vee Q \Leftrightarrow Q \vee P$ from Table 1 of § 2.2, and so

$$[P] \vee [Q] = [P \vee Q] = [Q \vee P] = [Q] \vee [P].$$

Theorem 2 provides an order on $[\mathscr{S}]$:

$$[P] \leq [Q] \quad \text{if and only if} \quad [P] \vee [Q] = [Q].$$

Now $[P] \vee [Q] = [Q]$ means $[P \vee Q] = [Q]$ so that $P \vee Q \Leftrightarrow Q$, which holds if and only if $P \Rightarrow Q$ since

$$((P \vee Q) \leftrightarrow Q) \leftrightarrow (P \rightarrow Q)$$

is a tautology. We conclude that $[P] \leq [Q]$ if and only if $P \Rightarrow Q$. ∎

In Examples 1 and 2, we began with an algebraic lattice and obtained a poset structure. Sometimes it is more natural to start with the poset.

EXAMPLE 3   Consider the set FUN($\mathbb{R}$, $\mathbb{R}$) of all functions from $\mathbb{R}$ to $\mathbb{R}$. As in Example 4 of § 7.2, we define $f \leq g$ provided $f(x) \leq g(x)$ for all $x \in \mathbb{R}$. This poset is a lattice, so it is an algebraic lattice by Theorem 1. For f and g in FUN($\mathbb{R}$, $\mathbb{R}$), $f \vee g$ denotes the least upper bound for $\{f, g\}$, that is, the smallest function that is greater than or equal to both f and g. It is easy to show that

$$(f \vee g)(x) = \max\{f(x), g(x)\} \quad \text{for all} \quad x \in \mathbb{R}.$$

Likewise, the greatest lower bound $f \wedge g$ for $\{f, g\}$ is the function given by

$$(f \wedge g)(x) = \min\{f(x), g(x)\} \quad \text{for all} \quad x \in \mathbb{R}. ∎$$

Consider again an algebraic lattice $(L, \vee, \wedge)$. An element in L that is greater than or equal to all other elements in L is called a **universal upper bound**. It is usually written 1; thus $x \leq 1$, $x \vee 1 = 1$ and $x \wedge 1 = x$ for all $x \in L$. A **universal lower bound** is an element 0 such that $0 \leq x$ for all $x \in L$; note that $0 \vee x = x$ and $0 \wedge x = 0$. A lattice might or might not have universal upper and lower bounds.

Recall from § 7.1 that an element y **covers** an element x if $x \prec y$ and there is no u with $x \prec u \prec y$. If L has a universal lower bound 0 then the elements that cover 0 are called **atoms**.

EXAMPLE 4   (a) Consider the lattice $\mathscr{P}(S)$ of all subsets of S. The empty set $\varnothing$ is a universal lower bound, since $\varnothing \subseteq A$ for all $A \in \mathscr{P}(S)$. Likewise, S itself is a universal upper bound. The atoms are the one-element subsets of S.

(b) Consider the set FUN(S, $\mathbb{B}$) of functions from a set S to $\mathbb{B} = \{0, 1\}$, made into a lattice by $f \leq g$ provided $f(x) \leq g(x)$ for all $x \in S$. As in Example 3, $(f \vee g)(x) = \max\{f(x), g(x)\}$, so

$$(f \vee g)(x) = \begin{cases} 1 & \text{if} \quad f(x) = 1 \quad \text{or} \quad g(x) = 1 \\ 0 & \text{otherwise} \end{cases}$$

and similarly,

$$(f \wedge g)(x) = \begin{cases} 1 & \text{if} \quad f(x) = 1 \quad \text{and} \quad g(x) = 1 \\ 0 & \text{otherwise.} \end{cases}$$

The atoms in FUN(S, $\mathbb{B}$) are the functions which have the value 1 at exactly one member of S and the value 0 at all other members, that is, the characteristic functions of the one-element subsets of S. ∎

An element x in an algebraic lattice is **join-irreducible** if $x = y \vee z$ implies $x = y$ or $x = z$, i.e., if x cannot be written as the join of two elements different from itself. We show that atoms are join-irreducible. Consider an atom a and assume $a = y \vee z$ where $a \neq y$ and $a \neq z$. Then $y \neq 0$ since otherwise a would equal z. Thus $0 \prec y$; and $y \prec a$ since $y \vee a = y \vee (y \vee z) = (y \vee y) \vee z = y \vee z = a$. This shows that y is between 0 and a so that a cannot be an atom. An induction-like argument establishes the following theorem; see Exercise 17.

Theorem 3 In a finite algebraic lattice, every element can be written as the join of join-irreducible elements.

EXAMPLE 5 Consider $\mathscr{P}(S)$ where S is finite. If A is a subset of S, say $A = \{a_1, \dots, a_m\}$, then

$$A = \{a_1\} \cup \{a_2\} \cup \cdots \cup \{a_m\}.$$

This shows that A is the join [union!] of atoms, which are join-irreducible. So Theorem 3 is nearly obvious in this case. ∎

Let $(L, \vee, \wedge)$ be an algebraic lattice. A subset M of L is a **sublattice** provided

$$x, y \in M \quad \text{imply} \quad x \vee y \in M \quad \text{and} \quad x \wedge y \in M,$$

i.e., provided M is closed under the operations $\vee$ and $\wedge$ of L.

EXAMPLE 6 Consider the lattice $(L, \leq)$ with Hasse diagram shown in Figure 1. The figure also shows Hasse diagrams of four subposets of $(L, \leq)$. The subset $M_1 = \{t, u, v, x\}$ is a sublattice of L, since the meet and join in L of any two members of M_1 belong to M_1. The subset $M_2 = \{t, u, v, w, x\}$ is not a

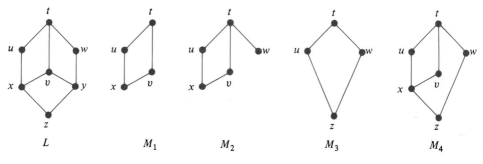

Figure 1

sublattice of L since, for example, $v \wedge w = y$ in L but $y \notin M_2$. In fact, $\{v, w\}$ has no lower bound in $(M_2, \leq)$ and so $(M_2, \leq)$ is not a lattice. The subset $M_3 = \{t, u, w, z\}$ is a sublattice of L since $u \vee w = t$ and $u \wedge w = z$ are in M_3. The subset $M_4 = \{t, u, v, w, x, z\}$ is not a sublattice of L. As was the case for M_2, M_4 does not contain $v \wedge w = y$. Notice, though, that the poset $(M_4, \leq)$ is a lattice in its own right, since z is the greatest lower bound of v and w in $(M_4, \leq)$. Thus *a subposet which is a lattice need not be a sublattice.* ∎

EXERCISES 10.1

1. Complete the proof of Theorem 1 by verifying properties 2Lb and 3Lb.

2. Verify the equalities in Example 3.

3. (a) Verify that $[\mathscr{S}]$ in Example 2 is a lattice.
   (b) If $\mathscr{S}$ contains a contradiction c, then $[c]$ is a universal lower bound. Why?
   (c) If $\mathscr{S}$ contains a tautology t, then $[t]$ is a universal upper bound. Why?

4. Write the duals of the following lattice equations [which are not always true].
   (a) $x \vee (y \wedge z) = (x \vee y) \wedge z$
   (b) $x \vee (y \wedge z) = (x \vee y) \wedge (x \vee z)$

5. Let L be an algebraic lattice with universal lower bound 0 and universal upper bound 1.
   (a) Is 0 join-irreducible? Explain.
   (b) Is 1 join-irreducible? Explain.

6. Consider a finite lattice $(P, \leq)$ and its Hasse diagram. Explain why an element is join-irreducible if and only if it covers at most one element.

7. Consider the lattice L in Figure 1.
   (a) List the atoms of the lattice.
   (b) List the join-irreducible elements of the lattice.
   (c) Write each element of the lattice as a join of join-irreducible elements.

8. Repeat Exercise 7 for Figure 2(a).

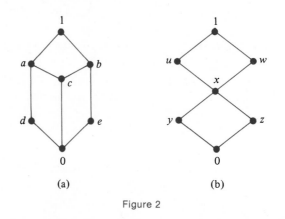

(a)　　　　　　　　(b)

Figure 2

9. Consider the lattice $(\mathbb{P}, |)$; recall $m|n$ if m divides n.
   (a) What is the universal lower bound?
   (b) Does $\mathbb{P}$ have a universal upper bound?
   (c) Describe the atoms of $\mathbb{P}$.
   (d) Describe the join-irreducible elements of $\mathbb{P}$.

10. Repeat Exercise 7 for Figure 2(b).

11. Let D_{90} be the set of all divisors of 90, including 1 and 90. D_{90} is a lattice with the order $|$, where $m|n$ means m divides n.
    (a) Draw the Hasse diagram for this lattice.
    (b) Calculate $6 \vee 10$, $6 \wedge 10$, $9 \vee 30$ and $9 \wedge 30$.
    (c) List the atoms of D_{90}.
    (d) List the join-irreducible elements of D_{90}.
    (e) Write 90, 18 and 5 as joins of join-irreducible elements.

12. Find all sublattices of D_{90} that have four elements including 1 and 90.

13. Consider the lattices T and U in Figure 4 of § 7.1, page 257. Each has a universal lower bound.
    (a) List the atoms for each lattice.
    (b) Do either of these lattices have join-irreducible elements that are not atoms?
    (c) How many [nonempty] sublattices are there of T? of U?

14. For $x, y \in \mathbb{R}$ we define $x \vee y = \max\{x, y\}$ and $x \wedge y = \min\{x, y\}$.
    (a) Show that $(\mathbb{R}, \vee, \wedge)$ is an algebraic lattice.
    (b) What is the order on $\mathbb{R}$ given by Theorem 2?
    (c) Explain why all elements of $\mathbb{R}$ are join-irreducible.

15. Two algebraic lattices $(L_1, \vee, \wedge)$ and $(L_2, \cup, \cap)$ are **isomorphic** if there is a one-to-one correspondence $\phi : L_1 \to L_2$ such that

$$\phi(x \vee y) = \phi(x) \cup \phi(y) \quad \text{and} \quad \phi(x \wedge y) = \phi(x) \cap \phi(y)$$

for all $x, y \in L_1$.
    (a) Show that in this case $\phi(x) \leq \phi(y)$ if and only if $x \leq y$.
    (b) Show that if $(L_1, \vee, \wedge)$ and $(L_2, \cup, \cap)$ are isomorphic, then x is an atom for L_1 if and only if $\phi(x)$ is an atom for L_2.
    (c) Show that the lattice L in Figure 1 is not isomorphic to the lattice in Figure 2(a).
    (d) Show that the lattice D_{30} of divisors of 30 is isomorphic to a lattice $\mathscr{P}(S)$ where $|S| = 3$. *Suggestion:* Use $S = \{2, 3, 5\}$.

16. Two algebraic lattices $(L_1, \vee, \wedge)$ and $(L_2, \cup, \cap)$ are **anti-isomorphic** if there is a one-to-one correspondence $\phi : L_1 \to L_2$ such that

$$\phi(x \vee y) = \phi(x) \cap \phi(y) \quad \text{and} \quad \phi(x \wedge y) = \phi(x) \cup \phi(y)$$

for all $x, y \in L_1$. Show that the lattice L in Figure 1 is anti-isomorphic to the lattice in Figure 2(a). *Hint:* Turn one of their Hasse diagrams upside down.

17. Prove Theorem 3. *Hint:* If Theorem 3 fails for some finite algebraic lattice L, let B be the subset consisting of elements which aren't joins of join-irreducible

elements. Consider a minimal member m of the finite poset $(B, \leq)$; see Exercise 10 of § 7.1.

18. Let $(P, \leq)$ be a lattice. Show that if $x \leq y$, then $x \vee (z \wedge y) \leq (x \vee z) \wedge y$ for all $z \in P$.

§ 10.2   Distributive and Boolean Lattices

We are leading up to a discussion of Boolean algebras, which are probably the most important lattices in computer science. Their applications range from the analysis and design of electrical networks to abstract arguments in the theory of computational complexity. To begin the account we examine lattices which enjoy some, but perhaps not all, properties of Boolean algebras.

An algebraic lattice L is called a **distributive lattice** if the operations $\vee$ and $\wedge$ distribute over each other, that is, if

$$x \vee (y \wedge z) = (x \vee y) \wedge (x \vee z) \quad \text{and} \quad x \wedge (y \vee z) = (x \wedge y) \vee (x \wedge z)$$

for all $x, y, z \in L$.

EXAMPLE 1   (a) The lattice $\mathcal{P}(S)$ of subsets of a set S is a distributive lattice in view of the distributive laws in Table 1 of § 1.2.

(b) The lattice $[\mathscr{S}]$ of Example 2 of § 10.1 is a distributive lattice. This is a consequence of the distributive laws in Table 1 of § 2.2.

(c) Any chain, like $\mathbb{R}$ or $\mathbb{N}$ with the usual ordering, is a distributive lattice with $x \vee y = \max\{x, y\}$ and $x \wedge y = \min\{x, y\}$. One of the distributive laws says

$$\max\{x, \min\{y, z\}\} = \min\{\max\{x, y\}, \max\{x, z\}\}.$$

To verify this, we first suppose that $y \leq z$. Then $\max\{x, y\} \leq \max\{x, z\}$ and the asserted equality reduces to the identity

$$\max \{x, y\} = \max \{x, y\}.$$

The case $y \geq z$ has a similar verification. The other distributive law has a similar proof. ∎

EXAMPLE 2   (a) Consider the lattice $(\mathbb{P}, |)$ where $m|n$ means m divides n. This lattice is distributive. Here is the idea of the proof. If we write an integer k as a product of primes, each prime will occur a certain number of times, possibly 0. In particular, 2 will occur a certain number of times. Let's write $k = 2^u etc$ to signify that there are u factors of 2 in k. For examples,

$$12 = 2^2 \cdot 3 = 2^2 etc, \quad 62 = 2 \cdot 31 = 2^1 etc, \quad 64 = 2^6 \cdot 1 = 2^6 etc,$$

$$73 = 2^0 etc.$$

If $k = 2^u etc$, $m = 2^v etc$ and $n = 2^w etc$, then

$$k \vee m = \text{1cm } (k, m) = 2^{\max\{u, v\}} etc$$

and

$$k \wedge m = \gcd (k, m) = 2^{\min\{u, v\}} etc.$$

With this notation the distributive law $k \vee (m \wedge n) = (k \vee m) \wedge (k \vee n)$ becomes

$$2^{\max\{u, \min\{v, w\}\}} etc = 2^{\min\{\max\{u, v\}, \max\{u, w\}\}} etc.$$

Example 1(c) shows that the exponents of 2 above are equal. Exactly the same argument works for counting the numbers of factors of the other primes. So when $k \vee (m \wedge n)$ and $(k \vee m) \wedge (k \vee n)$ are written as products of primes, they are identical. That is, $k \vee (m \wedge n) = (k \vee m) \wedge (k \vee n)$. The other distributive law has a similar proof.

(b) For an integer $m \geq 2$, consider the lattice $(D_m, |)$ where D_m is the set of all divisors of m. Since $(D_m, |)$ is a sublattice of $(\mathbb{P}, |)$ it is also a distributive lattice. The distributive lattice D_{12} is drawn in Figure 1(a). ∎

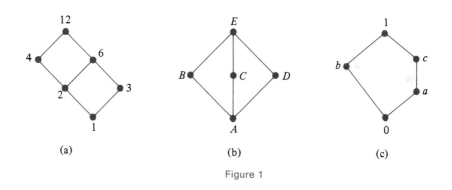

(a)   (b)   (c)

Figure 1

EXAMPLE 3   (a) The lattice in Figure 1(b) is not distributive. For example,

$$B \vee (C \wedge D) = B \vee A = B$$

while

$$(B \vee C) \wedge (B \vee D) = E \wedge E = E.$$

(b) The lattice in Figure 1(c) is not distributive either. For example,

$$a \vee (b \wedge c) = a \vee 0 = a$$

while

$$(a \vee b) \wedge (a \vee c) = 1 \wedge c = c. \quad ∎$$

It is interesting to note that it can be proved that a lattice is distributive if and only if it contains no sublattice that looks like either of the lattices in Figure 1(b) or 1(c). The next theorem shows that it is enough to check only one of the distributive laws, because then the other one holds automatically.

Theorem 1   For an algebraic lattice L the following properties imply each other:

(a) $x \vee (y \wedge z) = (x \vee y) \wedge (x \vee z)$ for all $x, y, z \in L;$

(b) $x \wedge (y \vee z) = (x \wedge y) \vee (x \wedge z)$ for all $x, y, z \in L.$

Proof. Suppose (a) holds. Then

$$(x \wedge y) \vee (x \wedge z) = [(x \wedge y) \vee x] \wedge [(x \wedge y) \vee z] \quad \text{property (a)}$$
$$= [x \vee (x \wedge y)] \wedge [z \vee (x \wedge y)] \quad \text{commutativity}$$
$$= x \wedge [z \vee (x \wedge y)] \quad \text{absorption}$$
$$= x \wedge [(z \vee x) \wedge (z \vee y)] \quad \text{property (a)}$$
$$= [x \wedge (z \vee x)] \wedge (z \vee y) \quad \text{associativity}$$
$$= [x \wedge (x \vee z)] \wedge (y \vee z) \quad \text{commutativity}$$
$$= x \wedge (y \vee z) \quad \text{absorption.}$$

So (a) implies (b); (b) implies (a) by the duality principle. ∎

For the remainder of this section we assume that L is an algebraic lattice with universal upper and lower bounds 1 and 0. Elements x and y in L are said to be **complements** if

$$x \vee y = 1 \quad \text{and} \quad x \wedge y = 0.$$

An element x in L might have no complements, might have a single unique complement or it might have more than one complement. It is easy to check that 0 and 1 are complements and that they are unique complements for each other. We call a lattice **complemented** if every element has at least one complement.

EXAMPLE 4   (a) The lattice in Figure 1(b) is complemented. The complements are not necessarily unique: both C and D are complements for B.

(b) The lattice in Figure 1(c) is also complemented. Again complements are not unique: both a and c are complements for b.

(c) The rather uninteresting lattice in Figure 2(a) illustrates a couple of important points. In the first place, it is not complemented. In fact, if we had

$$x \vee y = 1 \quad \text{and} \quad x \wedge y = 0,$$

then the first equality would force $y = 1$ while the second equality would force $y = 0$.

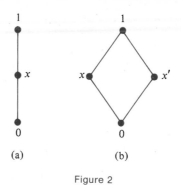

Figure 2

The second point is that this little lattice is distributive, as noted in Example 1(c). So distributive lattices with 1 and 0 need not be complemented. On the other hand, the next theorem shows that complements are unique in a complemented distributive lattice. ∎

Theorem 2   In a distributive lattice L with 1 and 0, if an element x has a complement then it is unique.

Proof. We assume that $x \vee y = 1$, $x \wedge y = 0$, $x \vee z = 1$, $x \wedge z = 0$, and will show $y = z$. Indeed

$$
\begin{aligned}
y &= y \vee 0 & &\text{since } 0 \leq y \\
&= y \vee (x \wedge z) & &\text{since } x \wedge z = 0 \\
&= (y \vee x) \wedge (y \vee z) & &\text{distributive law} \\
&= 1 \wedge (y \vee z) & &\text{since } y \vee x = x \vee y = 1 \\
&= (x \vee z) \wedge (y \vee z) & &\text{since } x \vee z = 1 \\
&= (x \wedge y) \vee z & &\text{distributive law} \\
&= 0 \vee z & &\text{since } x \wedge y = 0 \\
&= z & &\text{since } 0 \leq z. \quad \blacksquare
\end{aligned}
$$

For a general lattice in which unique complements exist, we will often write x' for the complement of x. Then

$$x \vee x' = 1 \quad \text{and} \quad x \wedge x' = 0.$$

Note that in such a lattice, each x belongs to a sublattice that looks like Figure 2(b) where $(x')' = x$. It follows that the correspondence $x \to x'$ assigning each element to its complement is a one-to-one function from L onto itself.

A distributive lattice that is complemented is called a **Boolean lattice**. Such a lattice is a set with two binary operations, $\vee$ and $\wedge$, and a "unary operation," $x \rightarrow x'$, that satisfy a number of properties which we have already discussed. Boolean lattices are sometimes called Boolean algebras. They are usually called lattices when the emphasis is on the underlying partial order, while they are called algebras when the stress is on the algebraic operations $\vee$, $\wedge$ and $'$. In the next section we will define Boolean algebras in a purely algebraic way and we will show that the new definition and our present one coincide.

EXAMPLE 5 (a) Recall that $\mathscr{P}(S)$ is a distributive lattice with universal upper and lower bounds S and $\varnothing$. For any A in $\mathscr{P}(S)$,

$$A \cup A^c = S \quad \text{and} \quad A \cap A^c = \varnothing$$

and so $\mathscr{P}(S)$ is a complemented lattice with $A' = A^c$, and hence is a Boolean lattice. The complement of the element A in the lattice $\mathscr{P}(S)$ is the complement of the set A. This is clearly where complemented lattices got their name.

(b) The lattice $[\mathscr{S}]$ of Example 2 of § 10.1 is a distributive lattice with universal upper and lower bounds $[t]$ and $[c]$, provided $\mathscr{S}$ contains a tautology t and a contradiction c. See Example 1(b) and also Exercise 3 of § 10.1. $[\mathscr{S}]$ is a Boolean lattice if

$$P \in \mathscr{S} \quad \text{implies} \quad \neg P \in \mathscr{S}.$$

This is because

$$[P] \vee [\neg P] = [t] \quad \text{and} \quad [P] \wedge [\neg P] = [c]$$

which follow from rule 7 in Table 1 of § 2.2.

(c) Of course $(\mathbb{P}, |)$ is not a Boolean lattice; it doesn't have a universal upper bound. Some of its sublattices $(D_m, |)$ are Boolean and some aren't. For example, in the lattice D_{12} of Figure 1, the elements 2 and 6 have no complements. See Exercise 3 for more about this. ∎

We now have a general setting in which to prove the **DeMorgan laws**.

Theorem 3 If L is a Boolean lattice, then

$$(x \vee y)' = x' \wedge y' \quad \text{and} \quad (x \wedge y)' = x' \vee y'$$

for all $x, y \in L$.

Proof. We have

$$(x \vee y) \vee (x' \wedge y') = [(x \vee y) \vee x'] \wedge [(x \vee y) \vee y'] \quad \text{distributivity}$$

$$= [y \vee (x \vee x')] \wedge [x \vee (y \vee y')]$$
$$\text{associativity and commutativity}$$

$$= [y \vee 1] \wedge [x \vee 1] = 1 \wedge 1 = 1.$$

Similarly $(x \vee y) \wedge (x' \wedge y') = 0$, and so $(x \vee y)' = x' \wedge y'$.

The other DeMorgan law has an analogous proof. Since $(z')' = z$ for all z, we can also derive the other law from the one already proved:

$$x \wedge y = (x')' \wedge (y')' = (x' \vee y')'$$

and so

$$(x \wedge y)' = (x' \vee y')'' = x' \vee y'. \quad \blacksquare$$

Suppose we interchange $\vee$ and $\wedge$ in some formula valid in a Boolean lattice L. Since 1 is defined by $x \vee 1 = 1$ for all x, and 0 by $x \wedge 0 = 0$ for all x, the elements 0 and 1 also switch roles. The conditions

$$x \vee x' = 1 \quad \text{and} \quad x \wedge x' = 0$$

become

$$x \wedge x' = 0 \quad \text{and} \quad x \vee x' = 1$$

and so the defining properties of x' are not changed by the switch. These remarks show that the following **duality principle** holds for Boolean lattices: if $\vee$ and $\wedge$ are interchanged and 0 and 1 are interchanged throughout a valid formula, another valid formula is obtained. In terms of the partial order $\leq$ on L, the duality principle holds because the inverse partial order $\geq$ also makes L a Boolean lattice.

EXERCISES 10.2

1. Consider the lattice L_1 with Hasse diagram in Figure 3(a).
   (a) List all the atoms of L_1.
   (b) List all join-irreducible elements of L_1.
   (c) Write 1 as the join of join-irreducible elements.
   (d) Give the complements, if they exist, for the following: $a, b, d, 0$.
   (e) Is L_1 a complemented lattice? Explain.
   (f) Is L_1 a distributive lattice? *Hint*: Use Theorem 2.

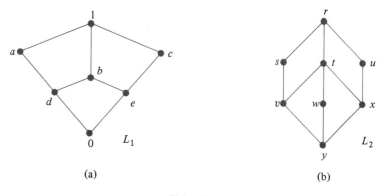

(a)   (b)

Figure 3

2. Consider the lattice L_2 with Hasse diagram in Figure 3(b).
   (a) Give the universal upper and lower bounds for L_2.
   (b) Find $v \vee x$, $s \vee v$ and $u \wedge v$.
   (c) Is L_2 a complemented lattice? Explain.
   (d) Find an element that has two complements.
   (e) Is L_2 a distributive lattice? Explain.

3. (a) Show that the elements 2 and 6 in the lattice D_{12} have no complements.
   (b) Consider any integer $m \geq 2$. Prove that D_m is complemented if and only if m is a product of *distinct* primes, i.e., no prime appears more than once when m is written as a product of primes.

4. (a) Draw the Hasse diagram for the lattice $(D_{24}, |)$.
   (b) Give the complements, if they exist, for the following: 2, 3, 4, 6.
   (c) Is D_{24} a complemented lattice? Explain.
   (d) Is D_{24} a distributive lattice? Explain.
   (e) Is D_{24} a Boolean lattice? Explain.

5. (a) Show that Figure 4(a) is the Hasse diagram for the lattice $(D_{36}, |)$.
   (b) Is the lattice distributive?
   (c) Is it complemented?

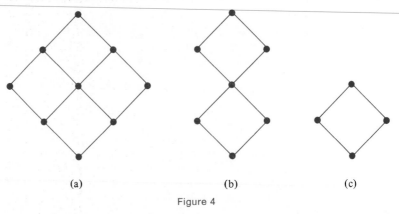

(a)   (b)   (c)

Figure 4

6. (a) Show that Figure 4(b) is the Hasse diagram for a distributive lattice.
   (b) Is it complemented? Explain.

7. Repeat Exercise 6 for Figure 4(c).

8. (a) Find the atoms for the lattice in Figure 1(a).
   (b) Find the join-irreducible elements.
   (c) Write the universal upper bound as the join of join-irreducible elements.

9. Repeat Exercise 8 for $(D_{36}, |)$ shown in Figure 4(a).

10. (a) Which chains have universal upper and lower bounds?
    (b) Which chains are distributive?
    (c) Which chains are complemented?

11. Let L be a distributive lattice. Prove that if x, y in L satisfy $x \vee a = y \vee a$ and $x \wedge a = y \wedge a$ for some $a \in L$, then $x = y$.

12. Use Theorem 2 to show that the lattices in Figures 1(b) and 1(c) are not distributive.

13. Let L be a lattice with 1 and 0. Show that 0 is the unique complement for 1 and vice versa.

14. Prove or disprove.
    (a) Every finite lattice is distributive.
    (b) Every finite lattice has a universal lower bound.

15. Let L be a Boolean lattice.
    (a) Prove that if $x \leq y$ then $y' \leq x'$.
    (b) Prove that if $y \wedge z = 0$ then $y \leq z'$.
    (c) Prove that if $x \leq y$ and $y \wedge z = 0$ then $z \leq x'$.

§ 10.3   Boolean Algebras

Consider a Boolean lattice $(L, \leq)$ with associated operations $\vee$, $\wedge$ and correspondence $x \to x'$. The lattice properties, the distributivity, and the existence of universal upper and lower bounds and complements translate into the following algebraic laws.

1Ba. $x \vee y = y \vee x$ }
  b. $x \wedge y = y \wedge x$ }   commutative laws

2Ba. $(x \vee y) \vee z = x \vee (y \vee z)$ }
  b. $(x \wedge y) \wedge z = x \wedge (y \wedge z)$ }   associative laws

3Ba. $x \vee (y \wedge z) = (x \vee y) \wedge (x \vee z)$ }
  b. $x \wedge (y \vee z) = (x \wedge y) \vee (x \wedge z)$ }   distributive laws

4Ba. $0 \vee x = x$ }
  b. $x \wedge 1 = x$ }   identity laws

5Ba. $x \vee x' = 1$
  b. $x \wedge x' = 0.$

We define a **Boolean algebra** to be a set with two binary operations $\vee$ and $\wedge$, a unary operation $'$ and distinct elements 0 and 1 satisfying these laws. Note that Boolean algebras satisfy the same principle of duality as stated for Boolean lattices at the end of § 10.2 [and for the same reasons].

Theorem 1   The following properties hold in a Boolean algebra:

6Ba. $x \vee x = x$ }
  b. $x \wedge x = x$ }   idempotent laws

7Ba. $x \vee 1 = 1$ }
  b. $x \wedge 0 = 0$ }   more identity laws

8Ba. $x \vee (x \wedge y) = x$ }
  b. $x \wedge (x \vee y) = x$ }   absorption laws

Proof. Here is 6Ba:

$$
\begin{aligned}
x \vee x &= (x \vee x) \wedge 1 && \text{identity law 4Bb} \\
&= (x \vee x) \wedge (x \vee x') && \text{5Ba} \\
&= x \vee (x \wedge x') && \text{distributive law 3Ba} \\
&= x \vee 0 && \text{5Bb} \\
&= 0 \vee x && \text{commutative law 1Ba} \\
&= x && \text{identity law 4Ba.}
\end{aligned}
$$

For 7Ba, observe

$$
\begin{aligned}
x \vee 1 &= x \vee (x \vee x') && \text{5Ba} \\
&= (x \vee x) \vee x' && \text{associative law 2Ba} \\
&= x \vee x' && \text{idempotent law 6Ba} \\
&= 1 && \text{5Ba.}
\end{aligned}
$$

And for 8Ba, we have

$$
\begin{aligned}
x \vee (x \wedge y) &= (x \wedge 1) \vee (x \wedge y) && \text{identity law 4Bb} \\
&= x \wedge (1 \vee y) && \text{distributive law 3Bb} \\
&= x \wedge (y \vee 1) && \text{commutative law 1Ba} \\
&= x \wedge 1 && \text{identity law 7Ba} \\
&= x && \text{identity law 4Bb.}
\end{aligned}
$$

Now 6Bb, 7Bb and 8Bb follow by duality. ▪

It turns out that the associative laws are redundant in the definition of a Boolean algebra. They are a consequence of the other defining laws for a Boolean algebra. In fact, Theorem 1 can be proved without using the associative laws. Proofs of these facts are tedious and not very informative, so we omit them.

Theorem 2   Boolean lattices are Boolean algebras and Boolean algebras are Boolean lattices.

Proof. We have already seen that Boolean lattices satisfy the laws 1Ba through 5Bb which define Boolean algebras. Now consider a Boolean algebra A. The absorption laws 8Ba and 8Bb proved in Theorem 1 together with 1Ba through 2Bb show that A is an algebraic lattice. Properties 3Ba through 5Bb show that A is distributive and complemented. Thus A is a Boolean lattice. ▪

The theorem just proved shows that Boolean algebras satisfy any properties established for Boolean lattices. In particular, the DeMorgan laws hold by Theorem 3 of § 10.2.

$$
\left.
\begin{aligned}
\text{9Ba.} \quad & (x \vee y)' = x' \wedge y' \\
\text{b.} \quad & (x \wedge y)' = x' \vee y'
\end{aligned}
\right\}
\quad \text{DeMorgan laws.}
$$

EXAMPLE 1   (a) The Boolean lattices in Example 5 of § 10.2 are Boolean algebras. In particular, $\mathscr{P}(S)$ is a Boolean algebra with respect to the operations $\cup, \cap$ and complementation.

(b) We will see that every finite Boolean algebra, regardless of the setting in which it arises or the notation used, looks like [that is, is isomorphic to] $\mathscr{P}(S)$ for some finite set S. A very simple but important Boolean algebra is $\mathbb{B} = \{0, 1\}$ where $\vee$ and $\wedge$ are the usual Boolean operations "or" and "and" [Table 1 of § 7.5], $0' = 1$ and $1' = 0$.

(c) In Example 4(b) of § 10.1 we observed that the set $\text{FUN}(S, \mathbb{B})$ of functions from S to $\mathbb{B} = \{0, 1\}$ is a lattice. Note that

$$(f \vee g)(x) = f(x) \vee g(x)$$

and

$$(f \wedge g)(x) = f(x) \wedge g(x)$$

for all $x \in S$; the operations $\vee$ and $\wedge$ on the right side are the operations in $\mathbb{B}$ mentioned in part (b). If we also define

$$(f')(x) = f(x)' \quad \text{for all} \quad x \in S,$$

then $\text{FUN}(S, \mathbb{B})$ is a Boolean algebra. The universal lower bound is the constant function on S equal to 0 everywhere, and the universal upper bound is the constant function equal to 1 everywhere.

If $S = \{1, 2, \ldots, n\}$ we may write $\text{FUN}(S, \mathbb{B})$ as $\mathbb{B}^n$ and view $\mathbb{B}^n$ as the set of all n-tuples of elements from $\mathbb{B} = \{0, 1\}$. The Boolean operations are then defined coordinatewise. For example [with $n = 3$], we have

$$\langle 1, 0, 1 \rangle \vee \langle 0, 1, 1 \rangle = \langle 1, 1, 1 \rangle,$$

$$\langle 1, 0, 1 \rangle \wedge \langle 0, 1, 1 \rangle = \langle 0, 0, 1 \rangle \quad \text{and} \quad \langle 1, 0, 1 \rangle' = \langle 0, 1, 0 \rangle.$$

The universal upper and lower bounds are $\langle 1, 1, 1 \rangle$ and $\langle 0, 0, 0 \rangle$. The atoms have exactly one entry equal to 1 and so the atoms are $\langle 1, 0, 0 \rangle$, $\langle 0, 1, 0 \rangle$ and $\langle 0, 0, 1 \rangle$. ∎

We next show that the atoms of a finite Boolean algebra are the building blocks for the algebra. Compare the result with Theorem 3 of § 10.1.

Theorem 3   Let A be a finite Boolean algebra with set of atoms $S = \{a_1, \ldots, a_n\}$. Each x in A can be written as a join of distinct atoms:

(1) $$x = a_{i_1} \vee \cdots \vee a_{i_k}.$$

Moreover, such an expression is unique, except for the order of the atoms. Indeed, the atoms $a_{i_1}, \ldots, a_{i_k}$ are precisely the atoms $\preceq x$.

Proof. The proof that every element has the form (1) is a sort of induction. First we explain the idea. If $x = 0$ or if x is an atom, we're done. Otherwise, there is a y in A such that $0 \prec y \prec x$. Then

$$x = x \vee y = (x \vee y) \wedge (y' \vee y) = (x \wedge y') \vee y.$$

Moreover, we have $x \wedge y' \prec x$ because otherwise $x \wedge y' = x$, $y \prec x = x \wedge y' \leq y'$ and $y \wedge y' = y$, which is impossible. So x is the join of the two smaller elements y and $x \wedge y'$. [Note that this shows that only atoms and 0 are join-irreducible.] If both y and $x \wedge y'$ are atoms, we're done. Otherwise we decompose them further into joins of smaller elements. Since A is finite, this process eventually stops and we've split x into a join of atoms. A recursive algorithm for finding an expression as a join of atoms can be based on this method.

A formal proof of (1) proceeds along the following lines. Assume some elements do not have the form (1) and let B be the nonempty subset of all such x. The poset $(B, \leq)$ has a minimal element. As in the last paragraph, the minimal element can be decomposed as the join of smaller elements. These minimal elements are not in B, so each is the join of atoms. Then the smallest element of B is also a join of atoms, a contradiction.

We next show that each x in A satisfies

(2) $$x = \bigvee \{a \in S : a \leq x\};$$

the notation on the right represents the join of all the elements in the set $\{a \in S : a \leq x\}$. We assume $x \neq 0$, since we view 0 as the join of the empty set of atoms. By (1), we know that 1 is the join of a set of atoms. It follows that

$$1 = \bigvee \{a \in S : a \leq 1\} = a_1 \vee \cdots \vee a_n,$$

since we could add the remaining atoms without destroying the identity. [Actually, in view of the theorem we are proving, there won't be any remaining atoms, but we don't know this yet.] Now

$$x = x \wedge 1 = x \wedge (a_1 \vee \cdots \vee a_n) = (x \wedge a_1) \vee \cdots \vee (x \wedge a_n).$$

Since $x \wedge a_i = a_i$ if $a_i \leq x$ and $x \wedge a_i = 0$ otherwise, this establishes the equality (2).

To check uniqueness, suppose $x = b_1 \vee \cdots \vee b_k$ is an expression for x as a join of atoms. Then $b_i \leq x$ for each i and so all the b_i's belong to $\{a \in S : a \leq x\}$. On the other hand, if $a \in S$ and $a \leq x$, then

$$0 \neq a = a \wedge x = a \wedge (b_1 \vee \cdots \vee b_k) = (a \wedge b_1) \vee \cdots \vee (a \wedge b_k).$$

Some $a \wedge b_i$ must be different from 0 and so $a \wedge b_i = a = b_i$ since a and b_i are atoms. Thus a is one of the b_i's. Consequently $\{b_1, \ldots, b_k\}$ is precisely the set of atoms $\leq x$. ∎

The next theorem tells us that a Boolean algebra is completely determined [up to isomorphism] by the number of atoms it has.

Theorem 4   If A is a finite Boolean algebra with set of atoms $S = \{a_1, \ldots, a_n\}$ and if B is a finite Boolean algebra with set of atoms $T = \{b_1, \ldots, b_n\}$, then there is a one-to-one correspondence ϕ of A onto B such that $\phi(a_i) = b_i$ for each i,

(1)
$$\phi(x \vee y) = \phi(x) \vee \phi(y),$$

(2)
$$\phi(x \wedge y) = \phi(x) \wedge \phi(y)$$

and

(3)
$$\phi(x') = \phi(x)'$$

for all $x, y \in A$.

A one-to-one correspondence ϕ between Boolean algebras which satisfies properties (1), (2) and (3) is called a **Boolean algebra isomorphism**.

Proof. By Theorem 3, every x in A can be written uniquely in the form

$$x = a_{i_1} \vee \cdots \vee a_{i_k}.$$

We define

$$\phi(x) = b_{i_1} \vee \cdots \vee b_{i_k}.$$

In particular,

$$\phi(a_i) = b_i \quad \text{for} \quad i = 1, 2, \ldots, n.$$

By our definition and Theorem 3,

$$\phi(x) = \bigvee \{\phi(a) : a \in S \text{ and } a \leq x\}$$

and also

$$\phi(x) = \bigvee \{b : b \in T \text{ and } b \leq \phi(x)\}.$$

Since the expression for $\phi(x)$ is unique, we conclude that

(4)
$$a \leq x \quad \text{if and only if} \quad \phi(a) \leq \phi(x)$$

for $a \in S$. To verify (1), consider x, y in A and note that for $a \in S$:

$$\phi(a) \leq \phi(x \vee y) \Leftrightarrow a \leq (x \vee y) \Leftrightarrow a \leq x \text{ or } a \leq y \quad \text{[see Exercise 9(a)]}$$
$$\Leftrightarrow \phi(a) \leq \phi(x) \text{ or } \phi(a) \leq \phi(y).$$

That is, for $b \in T$:

$$b \leq \phi(x \vee y) \Leftrightarrow b \leq \phi(x) \text{ or } b \leq \phi(y) \Leftrightarrow b \leq \phi(x) \vee \phi(y).$$

It follows from Theorem 3 applied to B that $\phi(x \vee y) = \phi(x) \vee \phi(y)$. Assertion (2) has a similar proof. Now

$$\phi(x) \vee \phi(x') = \phi(x \vee x') = \phi(1)$$

and

$$\phi(x) \wedge \phi(x') = \phi(x \wedge x') = \phi(0),$$

so $\phi(x') = \phi(x)'$.   ∎

If a set S has n elements, then the Boolean algebra $\mathscr{P}(S)$ [with operations $\cup$, $\cap$ and complementation] has exactly n atoms, namely the one-element subsets of S. So we have the following corollary.

Corollary   Every finite Boolean algebra with n atoms is isomorphic to the Boolean algebra $\mathscr{P}(S)$ of all subsets of an n-element set S, and hence has exactly 2^n elements.

As we have already indicated, finite Boolean algebras often arise in contexts different from $\mathscr{P}(S)$. Again let $\mathbb{B} = \{0, 1\}$ and for $n = 1, 2, \ldots$, let $\mathbb{B}^n$ be the Boolean algebra of Example 1. An n-**variable Boolean function** is a function

$$f : \mathbb{B}^n \to \mathbb{B}.$$

We write $\mathscr{F}_n$ for the set of all n-variable Boolean functions, and if $f \in \mathscr{F}_n$ and $\langle a_1, \ldots, a_n \rangle \in \mathbb{B}^n$ with each $a_i = 0$ or 1 we write $f(a_1, \ldots, a_n)$ for the value $f(\langle a_1, \ldots, a_n \rangle)$.

EXAMPLE 2   A 3-variable Boolean function is an f such that $f(a, b, c) = 0$ or 1 for each of the 2^3 choices of a, b and c. We could think of setting each of three switches to one of two positions. Then f behaves like a black box which produces an output of 0 or 1 depending on the settings of the switches and the internal structure of the box. Since there are 8 ways to set the switches and each setting can lead to either of 2 outputs, depending on the function, there are $2^8 = 256$ 3-variable Boolean functions. That is, $|\mathscr{F}_3| = 256$.

A 3-variable Boolean function f can be viewed as a column in a truth table, for example:

a	b	c	f
0	0	0	1
0	0	1	1
0	1	0	0
0	1	1	1
1	0	0	0
1	0	1	0
1	1	0	0
1	1	1	1

This is just one of the 256 possible tables, since column f can be any arrangement of 0's and 1's.   ∎

The counting argument in Example 2 works in general, so that $|\mathscr{F}_n| = 2^{(2^n)}$, a very big number unless n is very small. Just as in Example 1(c), $\mathscr{F}_n$ is a Boolean algebra with the Boolean operations defined coordinatewise.

EXAMPLE 3   We illustrate the Boolean operations in $\mathscr{F}_3$ in the following truth table for the indicated functions f and g:

a	b	c	f	g	$f \vee g$	$f \wedge g$	f'	$f' \wedge g$
0	0	0	1	0	1	0	0	0
0	0	1	1	1	1	1	0	0
0	1	0	0	0	0	0	1	0
0	1	1	1	0	1	0	0	0
1	0	0	0	1	1	0	1	1
1	0	1	0	0	0	0	1	0
1	1	0	0	1	1	0	1	1
1	1	1	1	0	1	0	0	0

Note that, since $f \wedge g$ takes the value 1 at exactly one point in $\mathbb{B}^3$, it is an atom of the Boolean algebra $\mathscr{F}_3$. There are seven other atoms in $\mathscr{F}_3$. As usual, it is of interest to write any member of a Boolean algebra, in this instance $\mathscr{F}_3$, as a join of atoms. In the next section we will indicate how this is done using Boolean expressions. ∎

EXERCISES 10.3

1. (a) Verify that $\mathbb{B} = \{0, 1\}$ in Example 1(b) is a Boolean algebra by checking some of the laws 1Ba through 5Bb.
   (b) Do the same for FUN(S, $\mathbb{B}$) in Example 1(c).

2. (a) Let $S = \{a, b, c, d, e\}$ and write $\{a, c, d\}$ as a join of atoms in $\mathscr{P}(S)$.
   (b) Write $\langle 1, 0, 1, 1, 0 \rangle$ as a join of atoms in $\mathbb{B}^5$.
   (c) Let f be the function in FUN(S, $\mathbb{B}$) that maps a, c and d to 1 and b and e to 0. Write f as a join of atoms in FUN(S, $\mathbb{B}$).

3. The lattice $(D_{30}, |)$ is a Boolean lattice [Exercise 3 of § 10.2] and hence a Boolean algebra.
   (a) Draw a Hasse diagram for this poset.
   (b) List the atoms of D_{30}.
   (c) Find all Boolean subalgebras of D_{30}. Note that the subalgebras must contain 1 and 30.
   (d) Find a sublattice with four elements that is not a Boolean subalgebra.

4. The lattice $(D_{210}, |)$ is a Boolean algebra. Find a set S so that $\mathscr{P}(S)$ and D_{210} are isomorphic Boolean algebras and exhibit an isomorphism between them.

5. Find a set S so that $\mathscr{P}(S)$ and $\mathbb{B}^5$ are isomorphic Boolean algebras. Exhibit a Boolean algebra isomorphism from $\mathbb{B}^5$ to $\mathscr{P}(S)$.

6. Describe the atoms of FUN(S, $\mathbb{B}$) in Example 1(c). Is your description valid even if S is infinite?

7. (a) Is there a Boolean algebra with 6 elements? Explain.
   (b) Is every finite Boolean algebra isomorphic to a Boolean algebra $\mathscr{F}_n$ of Boolean functions? Explain.

8. (a) Describe the atoms of the lattice $\mathscr{P}(\mathbb{N})$.
   (b) Is every member of the lattice the join of atoms? Discuss.

9. Let x, y be elements of a Boolean algebra, and let a be an atom.
   (a) Show that $a \leq x \vee y$ if and only if $a \leq x$ or $a \leq y$.
   (b) Show that $a \leq x \wedge y$ if and only if $a \leq x$ and $a \leq y$.
   (c) Show that either $a \leq x$ or $a \leq x'$ and not both.

10. Let x and y be elements of a finite Boolean algebra, each written as a join of atoms:

$$x = a_1 \vee \cdots \vee a_n \quad \text{and} \quad y = b_1 \vee \cdots \vee b_m.$$

   (a) Explain how to write $x \vee y$ and $x \wedge y$ as joins of distinct atoms. Illustrate with examples.
   (b) How would you write x' as the join of distinct atoms?

11. Show that if ϕ is a Boolean algebra isomorphism between Boolean algebras A and B, then $x \leq y$ if and only if $\phi(x) \leq \phi(y)$.

12. Let $S = [0, 1)$ and let $\mathscr{A}$ consist of the empty set $\varnothing$ and all subsets of S that can be written as a finite union of intervals of the form $[a, b)$.
   (a) Show that each member of $\mathscr{A}$ can be written as a finite *disjoint* union of intervals of the form $[a, b)$.
   (b) Show that $\mathscr{A}$ is a Boolean algebra with respect to the operations $\cup$, $\cap$ and complementation.
   (c) Show that $\mathscr{A}$ has no atoms whatever.

§ 10.4   Boolean Expressions

The main purpose of this section is to introduce the mathematical terminology and ideas which are used in applying Boolean algebra methods to circuit design and logical analysis. The next section will discuss the applications themselves more fully.

A Boolean expression is a string of symbols involving 0 and 1, some variables and the Boolean operations. To be more precise, we define **Boolean expressions in n variables** $x_1, x_2, \ldots, x_n$ recursively by:

(B) The symbols 0 and 1 and $x_1, x_2, \ldots, x_n$ are Boolean expressions in $x_1, \ldots, x_n$.

(R) If E_1 and E_2 are Boolean expressions in $x_1, x_2, \ldots, x_n$ so are $(E_1 \vee E_2)$, $(E_1 \wedge E_2)$ and E_1'.

As usual, in practice we will normally omit the outside parentheses and will freely use the associative laws.

EXAMPLE 1   (a) Here are four Boolean expressions in the three variables x, y, z:

$$(x \vee y) \wedge (x' \vee z) \wedge 1; \quad (x' \wedge z) \vee (x' \wedge y) \vee z'; \quad x \vee y; \quad z.$$

The first two obviously involve all three variables. The last two don't. Whether we regard $x \vee y$ as an expression in two or three or more variables often doesn't matter. When it does matter and the context doesn't make the variables clear, we will be careful to say how we are viewing the expression.

The Boolean expressions 0 and 1 can be viewed as expressions in any number of variables, just as constant functions can be viewed as functions of one or of several variables.

(b) An example of a Boolean expression in n variables is

$$(x_1 \wedge x_2 \wedge \cdots \wedge x_n) \vee (x_1' \wedge x_2 \wedge \cdots \wedge x_n) \vee (x_1 \wedge x_2' \wedge x_3 \wedge \cdots \wedge x_n). \quad \blacksquare$$

The usage of both symbols $\vee$ and $\wedge$ leads to bulky and awkward Boolean expressions, so we will usually replace the connective $\wedge$ by a dot or no symbol at all.

EXAMPLE 2   (a) With this new convention for $\wedge$, the first two Boolean expressions in Example 1(a) can be written as

$$(x \vee y) \cdot (x' \vee z) \cdot 1 \quad \text{and} \quad (x'z) \vee (x'y) \vee z'$$

or, more simply, as

$$(x \vee y)(x' \vee z)1 \quad \text{and} \quad x'z \vee x'y \vee z';$$

just as in ordinary algebra, the "product" $\wedge$ or $\cdot$ takes precedence over the "sum" $\vee$.

(b) The Boolean expression in Example 1(b) is

$$x_1 x_2 \cdots x_n \vee x_1' x_2 \cdots x_n \vee x_1 x_2' x_3 \cdots x_n.$$

(c) The expression $xyz \vee xy'z \vee x'z$ is shorthand for

$$(x \wedge y \wedge z) \vee (x \wedge y' \wedge z) \vee (x' \wedge z). \quad \blacksquare$$

If we substitute 0 or 1 for each occurrence of each variable in a Boolean expression, we get an expression involving 0, 1, $\vee$, $\wedge$ and $'$ which has a meaning as a member of the Boolean algebra $\mathbb{B} = \{0, 1\}$. For example, replacing x by 0, y by 1 and z by 1 in the Boolean expression $x'z \vee x'y \vee z'$ gives

$$0'1 \vee 0'1 \vee 1' = (1 \wedge 1) \vee (1 \wedge 1) \vee 0 = 1 \vee 1 \vee 0 = 1.$$

In general, if E is a Boolean expression in the n variables $x_1, x_2, \ldots, x_n$, then E defines a **Boolean function** mapping $\mathbb{B}^n$ into $\mathbb{B}$ whose function value at

$\langle a_1, a_2, \ldots, a_n \rangle$ is the element of $\mathbb{B}$ obtained by replacing x_1 by a_1, x_2 by $a_2, \ldots$, and x_n by a_n in E.

EXAMPLE 3   The Boolean function mapping $\mathbb{B}^3$ into $\mathbb{B}$ that corresponds to $x'z \vee x'y \vee z'$ is given in the following table where, just as with truth tables for propositions, we first calculate the Boolean functions for some of the subexpressions. The fourth entry in the last column is the value we calculated a moment ago. Note that the Boolean expression z' corresponds to the function on $\mathbb{B}^3$ that maps each triple $\langle a, b, c \rangle$ to c' where a, b, $c \in \{0, 1\}$. Similarly, z corresponds to the function that maps each $\langle a, b, c \rangle$ to c. ▮

x	y	z	$x'z$	$x'y$	z'	$x'z \vee x'y \vee z'$
0	0	0	0	0	1	1
0	0	1	1	0	0	1
0	1	0	0	1	1	1
0	1	1	1	1	0	1
1	0	0	0	0	1	1
1	0	1	0	0	0	0
1	1	0	0	0	1	1
1	1	1	0	0	0	0

We will regard two Boolean expressions as **equivalent** provided their corresponding Boolean functions are the same. For instance, $x(y \vee z)$ and $(xy) \vee (xz)$ are equivalent, since each corresponds to the function with value 1 at $\langle a, b, c \rangle = \langle 1, 1, 0 \rangle, \langle 1, 0, 1 \rangle$ or $\langle 1, 1, 1 \rangle$ and 0 otherwise. We will write $x(y \vee z) = (xy) \vee (xz)$, and in general we will write $E = F$ if the two Boolean expressions E and F are equivalent. The usage of "$=$" to denote this equivalence relation is customary and seems to cause no confusion.

The use of notation in this way is familiar from our experience with algebraic expressions and algebraic functions on $\mathbb{R}$. Technically, the algebraic expressions $(x + 1)(x - 1)$ and $x^2 - 1$ are different [because they *look* different] but the functions f and g defined by

$$f(x) = (x + 1)(x - 1) \quad \text{and} \quad g(x) = x^2 - 1$$

are equal. We regard the two expressions as equivalent and commonly use either $(x + 1)(x - 1)$ or $x^2 - 1$ as a name for the function they define. Similarly, we will often use Boolean expressions as names for the Boolean functions they define.

EXAMPLE 4   The function in $\mathscr{F}_3$ named xy is defined by $xy(\langle a, b, c \rangle) = ab$ for all $\langle a, b, c \rangle$ in $\mathbb{B}^3$, so

$$xy(\langle a, b, c \rangle) = \begin{cases} 1 & \text{if } a = b = 1, \\ 0 & \text{otherwise.} \end{cases}$$

Similarly the functions named $x \vee z'$ and $xy'z$ satisfy

$$(x \vee z')(\langle a, b, c \rangle) = a \vee c' = \begin{cases} 1 & \text{if} \quad a = 1 \quad \text{or} \quad c = 0, \\ 0 & \text{otherwise,} \end{cases}$$

and

$$xy'z(\langle a, b, c \rangle) = ab'c = \begin{cases} 1 & \text{if} \quad a = 1, b = 0, c = 1, \\ 0 & \text{otherwise.} \end{cases}$$

Since $xy'z$ takes the value 1 at exactly one point in $\mathbb{B}^3$, it is an atom of the Boolean algebra $\mathscr{F}_3$. The other seven atoms in $\mathscr{F}_3$ are

$$xyz, \; xyz', \; xy'z', \; x'yz, \; x'yz', \; x'y'z \text{ and } x'y'z'. \quad \blacksquare$$

Suppose that E_1, E_2, E_3 are Boolean expressions in n variables. Since $\mathscr{F}_n$ is a Boolean algebra, the Boolean expressions $E_1(E_2 \vee E_3)$ and $(E_1 E_2) \vee (E_1 E_3)$ define the same function. Thus the two expressions are equivalent, and we can write the distributive law

$$E_1(E_2 \vee E_3) = (E_1 E_2) \vee (E_1 E_3).$$

In the same way, Boolean expressions satisfy all the other laws of a Boolean algebra as well, so long as we are willing to write equivalences as if they were equations.

Boolean expressions such as x or y' consisting of a single variable or its complement are called **literals.** The functions which correspond to them have the value 1 at half of the elements of $\mathbb{B}^n$. For example, the literal y' for $n = 3$ corresponds to the function with value 1 at all points $\langle a, 0, c \rangle$ in $\mathbb{B}^3$ and value 0 at all points $\langle a, 1, c \rangle$.

We saw in Example 4(b) of § 10.1 and Example 1(c) of § 10.3 that the atoms of $\mathscr{F}_n$ are the functions which have the value 1 at exactly one member of $\mathbb{B}^n$. Each atom corresponds to a Boolean expression of a special form, called a minterm. A **minterm** in n variables is a meet [i.e., product] of n literals, each involving a different variable.

EXAMPLE 5   (a) The expressions $xy'z'$ and $x'yz'$ are minterms in the three variables x, y, z. The corresponding functions in $\mathscr{F}_3$ have the value 1 only at $\langle 1, 0, 0 \rangle$ and $\langle 0, 1, 0 \rangle$, respectively.

(b) The expression xz' is a minterm in the two variables x, z. It is not a minterm in the three variables x, y, z; the corresponding function in $\mathscr{F}_3$ has value 1 both at $\langle 1, 0, 0 \rangle$ and at $\langle 1, 1, 0 \rangle$.

(c) The expression $xyx'z$ is not a minterm since it involves the variable x in more than one literal. In fact, this expression is equivalent to 0. The expression $xy'zx$ is not a minterm either; it is equivalent to the minterm $xy'z$ in x, y, z, however.

(d) In the following table we list the eight elements of $\mathbb{B}^3$ and the corresponding minterms that take the value 1 at the indicated elements.

Note that the literals corresponding to 0 entries are complemented while the other literals are not.

$\langle a, b, c \rangle$	Minterm with value 1 at $\langle a, b, c \rangle$
$\langle 0, 0, 0 \rangle$	$x'y'z'$
$\langle 0, 0, 1 \rangle$	$x'y'z$
$\langle 0, 1, 0 \rangle$	$x'yz'$
$\langle 0, 1, 1 \rangle$	$x'yz$
$\langle 1, 0, 0 \rangle$	$xy'z'$
$\langle 1, 0, 1 \rangle$	$xy'z$
$\langle 1, 1, 0 \rangle$	xyz'
$\langle 1, 1, 1 \rangle$	xyz

According to Theorem 3 of § 10.3, every member of $\mathscr{F}_n$ can be written as a join of atoms. Since atoms in $\mathscr{F}_n$ correspond to minterms, every Boolean expression in n variables is equivalent to a join of distinct minterms. Moreover, such a representation as a join is unique, apart from the order in which the minterms are written. We call the join of minterms equivalent to a given Boolean expression E the **minterm canonical form** of E. [Another popular term, which we will not use, is **disjunctive normal form**, or DNF.] Parts (b) and (c) of the next example illustrate two different procedures for finding minterm canonical forms.

EXAMPLE 6   (a) The Boolean expression

$$x'yz' \vee xy'z' \vee xy'z \vee xyz'$$

is a join of minterms in x, y, z as it stands, so this expression is its own minterm canonical form. The corresponding Boolean function has the values shown in the righthand column of the table. The 1's in the column tell which atoms in $\mathscr{F}_3$ are involved, and hence determine the corresponding minterms. For instance, the 1 in the $\langle 1, 1, 0 \rangle$ row corresponds to xyz'.

x	y	z	$x'yz'$	$xy'z'$	$xy'z$	xyz'	$x'yz' \vee xy'z' \vee xy'z \vee xyz'$
0	0	0	0	0	0	0	0
0	0	1	0	0	0	0	0
0	1	0	1	0	0	0	1
0	1	1	0	0	0	0	0
1	0	0	0	1	0	0	1
1	0	1	0	0	1	0	1
1	1	0	0	0	0	1	1
1	1	1	0	0	0	0	0

(b) The Boolean expression $(x \vee yz')(yz)'$ is not written as a join of minterms. To get its minterm canonical form we can calculate the values of the corresponding Boolean function. For instance, $x = 0$, $y = 0$, $z = 0$ gives the value

$$(0 \lor 01)(00)' = (0 \lor 0)0' = 01 = 0$$

and $x = 1$, $y = 0$, $z = 1$ gives

$$(1 \lor 00)(01)' = (1 \lor 0)0' = 11 = 1.$$

When we calculate all eight values of the function we get the righthand column in the table in part (a). Thus $(x \lor yz')(yz)'$ is equivalent to the join of minterms in part (a), i.e., its minterm canonical form is $x'yz' \lor xy'z' \lor xy'z \lor xyz'$.

(c) We can attack $(x \lor yz')(yz)'$ directly and try to convert it into a join of minterms using Boolean algebra laws. Recall that we write $E = F$ in case the Boolean expressions E and F are equivalent. By the Boolean algebra laws,

$(x \lor yz')(yz)' = (x \lor yz')(y' \lor z')$	DeMorgan law
$\quad = (x(y' \lor z')) \lor ((yz')(y' \lor z'))$	distributive law
$\quad = (xy' \lor xz') \lor (yz'y' \lor yz'z')$	distributive law twice
$\quad = (xy' \lor xz') \lor (0 \lor yz')$	$yy' = 0$, $z'z' = z'$
$\quad = xy' \lor xz' \lor yz'$	associative law and property of 0.

We first applied the DeMorgan laws to get all complementation down to the level of the literals. Then we distributed $\lor$ across meets as far as possible.

Now we have an expression as a join of meets of literals, but not as a join of minterms in x, y, z. Consider the subexpression xy' which is missing the variable z. Since $z \lor z' = 1$ we have $xy' = xy'1 = xy'(z \lor z') = xy'z \lor xy'z'$, which is a join of minterms. We can do the same sort of thing to the other two terms and get

$$xy' \lor xz' \lor yz' = (xy'z \lor xy'z') \lor (xyz' \lor xy'z') \lor (xyz' \lor x'yz')$$

which is a join of minterms. Deleting repetitions gives the minterm canonical form

$$xy'z \lor xy'z' \lor xyz' \lor x'yz'$$

for the expression $(x \lor yz')(yz)'$ we started with. This is of course the same as the answer obtained in part (b). ∎

The methods illustrated in this example work in general. Given a Boolean expression, we can calculate the values of the Boolean function it defines—in effect, find its truth table. Then each value of 1 corresponds to a minterm in the canonical form of the expression. This is the method of Example 6(b). From this point of view the minterm canonical form is just another way of looking at the Boolean function.

Alternatively, we can obtain the minterm canonical form as in Example 6(c). First use the DeMorgan laws to move all complementation to the literals. Then distribute $\lor$ over products wherever possible. Then replace xx

by x and xx' by 0 as necessary and insert missing variables using $x \vee x' = 1$. Finally, eliminate duplicates.

It is not always clear which technique is preferable for a given Boolean expression. One would not want to do a lot of calculations by hand using either method. Fortunately, the minterm canonical form is primarily useful as a theoretical tool, and when calculations do arise in practice they can be performed by machine using simple algorithms.

From a theoretical point of view, the minterm canonical form of a Boolean expression is very valuable, since it gives the expression in terms of its basic parts, namely minterms or atoms. As we will illustrate in § 10.5, Boolean expressions can be realized as electronic circuits and equivalent Boolean expressions correspond to electronic circuits that perform identically, i.e., give the same outputs for given inputs. Hence it is of interest to be able to "simplify" Boolean expressions, thereby obtaining "simplified" electronic circuits.

There are various ways to measure the simplification. It would be impossible to describe here all methods which have practical importance, but we can at least discuss one simple criterion. Let's say that a join of products [i.e., meets] of literals is **optimal** if there is no equivalent Boolean expression which is a join of fewer products and if among all equivalent joins of the same number of products there are none with fewer literals. Our task is to find an optimal join of products equivalent to a given Boolean expression. We can suppose that we have already found *one* equivalent join of products, namely the minterm canonical form.

EXAMPLE 7   (a) Consider the expression $(xy)'z$. The table shows the values of the Boolean function it defines. The minterm canonical form is thus $x'y'z \vee x'yz \vee xy'z$. This expression is not optimal. By the Boolean algebra laws, $(xy)'z = (x' \vee y')z = x'z \vee y'z$, which is a join of only two terms with four literals. We will be able to show in § 10.6 that $x'z \vee y'z$ is optimal [or see Exercise 12].

x	y	z	xy	$(xy)'$	$(xy)'z$
0	0	0	0	1	0
0	0	1	0	1	1
0	1	0	0	1	0
0	1	1	0	1	1
1	0	0	0	1	0
1	0	1	0	1	1
1	1	0	1	0	0
1	1	1	1	0	0

This example illustrates a problem which can arise in practice. It seems plausible that a circuit to produce $x'z \vee y'z$ might be simpler than one to produce $x'y'z \vee x'yz \vee xy'z$, but perhaps a circuit to produce the original expression $(xy)'z$ would be simplest of all. We return to this point in § 10.6.

(b) Consider the join of products $E = x'z' \vee x'y \vee xy' \vee xz$. Is it optimal? We use Boolean algebra calculations, including the $x \vee x' = 1$ trick, to find its minterm canonical form:

$$E = x'yz' \vee x'y'z' \vee x'yz \vee x'yz' \vee xy'z \vee xy'z' \vee xyz \vee xy'z$$

$$= x'yz' \vee x'y'z' \vee x'yz \vee xy'z \vee xy'z' \vee xyz.$$

This has just made matters worse—more products and more literals. We want to repackage the expression in some clever way. Observe that we can group the six minterms together in pairs $x'yz'$ and $x'y'z'$, $x'yz$ and xyz, $xy'z$ and $xy'z'$, so that two minterms in the same pair differ in exactly one literal. Since

$$x'yz' \vee x'y'z' = x'(y \vee y')z' = x'z',$$

$$x'yz \vee xyz = yz \quad \text{and} \quad xy'z \vee xy'z' = xy',$$

$E = x'z' \vee yz \vee xy'$. A different grouping gives

$$x'yz' \vee x'yz = x'y, \qquad x'y'z' \vee xy'z' = y'z' \quad \text{and} \quad xy'z \vee xyz = xz,$$

so that $E = x'y \vee y'z' \vee xz$. Each of these product joins $x'z' \vee yz \vee xy'$ and $x'y \vee y'z' \vee xz$ will be shown to be optimal in Example 2(c) of §10.6. That is, no join of products which is equivalent to E has fewer than three products, and no join with three products has fewer than six literals. Whether or not we believe this, each of these expressions looks simpler than the join of four products we started with. ∎

There is a method, called the **Quine-McCluskey procedure**, which builds optimal expressions by systematically grouping products together which differ in only one literal. The algorithm is tedious to use by hand but is readily programmed for computer calculation. Among other references, the textbooks *Applications-Oriented Algebra* by J. L. Fisher and *Modern Applied Algebra* by G. Birkhoff and T. C. Bartee contain readable accounts of the method.

Another procedure for finding optimal expressions, called the method of **Karnaugh maps,** has a resemblance to Venn diagrams. The method works pretty well for Boolean expressions in three or four variables, where the problems are fairly simple anyway, but is less useful for more than four variables. The textbook *Computer Hardware and Organization* by M. E. Sloan devotes several sections to Karnaugh maps and discusses their advantages and disadvantages in applications. We will illustrate the method in § 10.6, after we have described the elements of logical circuitry.

EXERCISES 10.4

1. Let $f : \mathbb{B}^3 \to \mathbb{B}$ be the Boolean function such that $f(0, 0, 0) = f(0, 0, 1) = f(1, 1, 0) = 1$ and $f(a, b, c) = 0$ for all other $\langle a, b, c \rangle \in \mathbb{B}^3$. Write the corresponding Boolean expression in minterm canonical form.

2. Give the Boolean function corresponding to the Boolean expression in Example 7(b).

3. For each of the following Boolean expressions in x, y, z describe the corresponding Boolean function and write the minterm canonical form.
   (a) xy (b) z' (c) $xy \vee z'$ (d) 1

4. Consider the Boolean expression $x \vee yz$ in x, y, z.
   (a) Determine the corresponding Boolean function $f : \mathbb{B}^3 \to \mathbb{B}$.
   (b) Write the expression in minterm canonical form.

5. Find the minterm canonical form for the four-variable Boolean expressions
   (a) $(x_1 x_2 x_3') \vee (x_1' x_2 x_3 x_4')$ (b) $(x_1 \vee x_2) x_3' x_4$

6. Use the method of Example 6(c) to find the minterm canonical form of the 3-variable Boolean expression $((x \vee y)' \vee z)'$.

7. (a) Find a join of products involving a total of three literals which is equivalent to the expression
$$xz \vee [y' \vee y'z] \vee xy'z'.$$
   (b) Repeat part (a) for $[(xy \vee xyz) \vee xz] \vee z$.

8. The Boolean function $f : \mathbb{B}^3 \to \mathbb{B}$ is given by $f(a, b, c) = a +_2 b +_2 c$ for $\langle a, b, c \rangle \in \mathbb{B}^3$.
   (a) Determine a Boolean expression corresponding to f.
   (b) Write the expression in minterm canonical form with variables x, y, z.

9. Find an optimal expression equivalent to
$$(x \vee y)' \vee z \vee x(yz \vee y'z').$$

10. Group the three minterms in $xyz \vee xyz' \vee xy'z$ in pairs to obtain an equivalent expression as a join of two products with two literals each.

11. There is a notion of maxterm dual to the notion of minterm. A **maxterm** in $x_1, \ldots, x_n$ is a join of n literals, each involving a different one of $x_1, \ldots, x_n$.
   (a) Use the DeMorgan laws to show that every Boolean expression in variables $x_1, \ldots, x_n$ is equivalent to a product of maxterms.
   (b) Write $xy' \vee x'y$ as a product of maxterms in x, y.

12. (a) Show that $x'z \vee y'z$ is not equivalent to a product of literals.
   (b) Show that $x'z \vee y'z$ is not equivalent to a join of products of literals in which one "product" is a single literal. [Parts (a) and (b) together show that $x'z \vee y'z$ is optimal.]

13. Prove that if E_1 and E_2 are Boolean expressions in $x_1, \ldots, x_n$ then $E_1 \vee E_2$ and $E_2 \vee E_1$ are equivalent.

§ 10.5 Logic Networks

Computer science at the hardware level involves designing devices to produce appropriate outputs from given inputs. For inputs and outputs which are 0's and 1's, this becomes a problem of designing circuitry to transform input data according to the rules for Boolean functions. In this section we will look briefly at ways in which Boolean algebra methods can be applied to logical design.

The basic building blocks of our logic networks are small units called **gates** which correspond to simple Boolean functions. Hardware versions of these units are available from manufacturers, packaged in a wide variety of configurations. Figure 1 shows the standard ANSI/IEEE symbols for the six

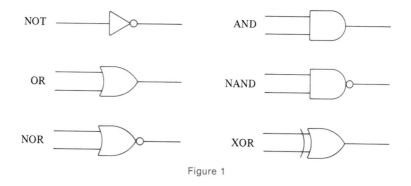

Figure 1

most elementary gates. We use the convention that the lines entering the symbol from the left are input lines, and the line on the right is the output line. Placing a small circle on an input or output line complements the signal on that line. The table shows the Boolean function values associated with these six gates and gives the corresponding Boolean function names for inputs x and y. AND, OR, NAND and NOR gates are also available with more than two input lines.

| | | x' | $x \vee y$ | $(x \vee y)'$ | xy | $(xy)'$ | $x \oplus y$ |
| | | NOT | OR | NOR | AND | NAND | XOR |
x	y						
0	0	1	0	1	0	1	0
0	1	1	1	0	0	1	1
1	0	0	1	0	0	1	1
1	1	0	1	0	1	0	0

EXAMPLE 1   (a) The gate shown in Figure 2(a) corresponds to the Boolean function $(x \vee y')'$, or equivalently $x'y$.

(b) The 3-input AND gate in Figure 2(b) goes with the function $x'yz$.

(c) The gate in Figure 2(c) gives $(x'y')'$ or $x \vee y$, so it acts like an OR gate. ∎

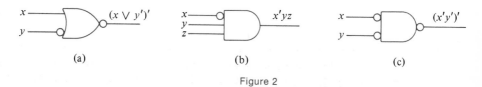

Figure 2

We consider the problem of designing a network of gates to produce a given complicated Boolean function of several variables. One major consideration is to keep the number of gates small. Another is to keep the length of the longest chain of gates small. Still other criteria arise in concrete practical applications.

EXAMPLE 2   Consider the foolishly designed network shown in Figure 3(a). [Dots indicate points where input lines divide.] There are four gates in the network. Reading from left to right there are two chains which are three gates long. We calculate the Boolean functions at A, B, C and D:

$$A = (x \vee y')'; \qquad B = x \vee z;$$

$$C = (A \vee B)' = ((x \vee y')' \vee (x \vee z))';$$

$$D = C \vee y = ((x \vee y')' \vee (x \vee z))' \vee y.$$

Boolean algebra laws give

$$D = ((x \vee y')(x \vee z)') \vee y = ((x \vee y')x'z') \vee y$$

$$= (xx'z' \vee y'x'z') \vee y = y'x'z' \vee y = (y' \vee y)(x'z' \vee y)$$

$$= x'z' \vee y.$$

The network shown in Figure 3(b) produces the same output, since

$$E = (x \vee z)' = x'z' \quad \text{and} \quad F = E \vee y = x'z' \vee y. \quad \blacksquare$$

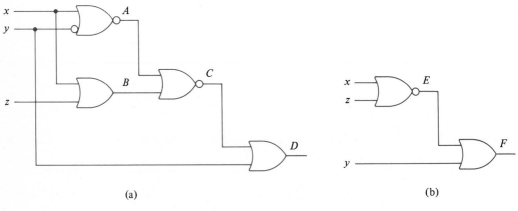

(a)   (b)

Figure 3

This simple example shows how it is sometimes possible to redesign a complicated network into one using fewer gates. One reason for trying to reduce the lengths of chains of gates is that in many situations, including programmed simulations of hard-wired circuits, the operation of each gate takes a fixed basic unit of time, and the gates in a chain must operate one after the other. Long chains mean slow operation.

It happens that the expression $x'z' \vee y$ which we obtained for the complicated expression D in the last example is an optimal expression for D in the sense of § 10.4. Optimal expressions do not always give the simplest networks. For example, one can show [Exercise 7(a) of § 10.6] that $xz \vee yz$ is an optimal expression in x, y, z. Now $xz \vee yz = (x \vee y)z$, which can be implemented with an OR gate and an AND gate, whereas to implement $xz \vee yz$ directly would require two AND gates to form xz and yz and an OR gate to finish the job. What this means is that in practical situations our definition of "optimal" should change to match the hardware available.

In some settings it is desirable to have all gates of the same type or of at most two types. It turns out that we can do everything just with NAND or just with NOR. Which of these two types of gates is more convenient to use may depend on the particular transistor technology being employed. The table in Figure 4(a) shows how to write NOT, OR and AND in terms of NAND. Figure 4(b) shows the corresponding networks. This table also answers Exercise 12 of § 2.3, since NAND is another name for the Sheffer stroke referred to in that exercise. Exercise 2 asks for a corresponding table and figure for NOR. The network for OR in Figure 4 could also have been written as a single NAND gate with both inputs complemented. Complementing may or may not take separate gates in a particular application, depending on the technology involved and the source of the inputs. In most of our discussion we proceed as if complementation can be done at no cost.

Combinations of AND and OR such as those which arise in joins of products can easily be done entirely with NAND's.

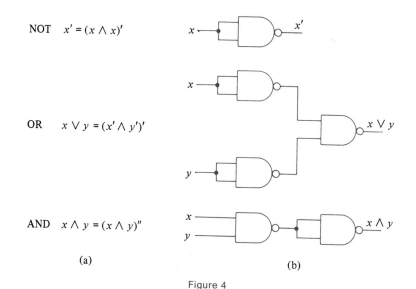

NOT $x' = (x \wedge x)'$

OR $x \vee y = (x' \wedge y')'$

AND $x \wedge y = (x \wedge y)''$

(a)   (b)

Figure 4

EXAMPLE 3   Figure 5 shows a simple illustration. Just replace all AND's and OR's by NAND's in an AND-OR 2-stage network to get an equivalent network. An OR-AND 2-stage network can be replaced by a NOR network in a similar way [Exercise 4]. ∎

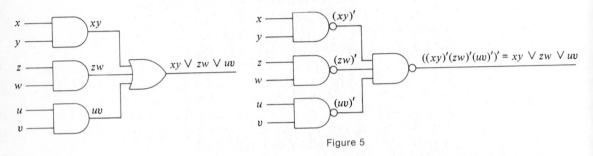

Figure 5

Logical networks can be viewed as acyclic digraphs with the sources labeled by variables $x_1, x_2, \ldots$, the other vertices labeled with $\vee$, $\wedge$ and $\oplus$, and some edges labeled $\neg$ for complementation. Each vertex then has an associated Boolean expression in the variables which label the sources.

EXAMPLE 4   The network of Figure 6(a) yields the digraph of Figure 6(b), with all edges directed from left to right. If we insert a 1-input $\wedge$-vertex in the middle of the edge from z to $(x \wedge y)' \vee z \vee (x \wedge z' \wedge w)$ we don't change the logic, and we get a digraph in which the vertices appear in columns—first a variable column, then an $\wedge$ column, then an $\vee$ column—and in which edges go only from one column to the next. ∎

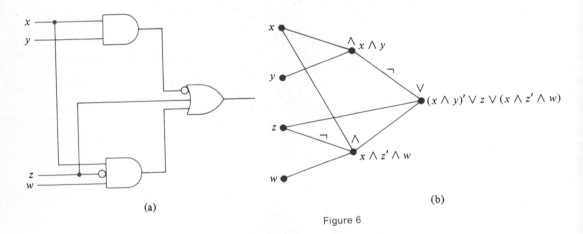

(a)   (b)

Figure 6

The labeled digraph in Example 4 describes a computation of the Boolean function $(x \wedge y)' \vee z \vee (x \wedge z' \wedge w)$. In a similar way, any such labeled digraph describes computations for the Boolean functions which are associated with its sinks. Since every Boolean expression can be written as a join of products of literals, every Boolean function has a computation which can be described by a digraph like the one in Example 4, with a variable column, an $\wedge$ column and an $\vee$ column [consisting of a single vertex]. Indeed, as we saw in Example 3, all corresponding gates can be made NAND gates, so the $\vee$ vertex can be made an $\wedge$ vertex.

In a digraph of this sort no path has length greater than 2. The interpretation is that the associated computation takes just 2 units of time. The price we pay may be an enormous number of gates.

EXAMPLE 5 Consider the Boolean expression $E = x_1 \oplus x_2 \oplus \cdots \oplus x_n$. The corresponding Boolean function on $\mathbb{B}^n$ takes the value 1 at $\langle a_1, a_2, \ldots, a_n \rangle$ if and only if an odd number of the entries $a_1, a_2, \ldots, a_n$ are 1. The corresponding minterms are the ones with an odd number of uncomplemented literals. Hence the minimal canonical form for E uses half of all the possible minterms and is a join of 2^{n-1} terms.

We next show that the minterm canonical form for this E is optimal, i.e., whenever E is written as a join of products of literals, there will be at least 2^{n-1} terms. This is because such a join will have to include each of the minterms mentioned in the last paragraph. Otherwise some term would be a product of fewer than n literals; say x_1 and x_1' were both missing. Some choice of values of $a_1, a_2, \ldots, a_n$ will make this term have value 1. Note that an odd number of the entries $a_1, a_2, \ldots, a_n$ will be 1. If we change a_1 [from 0 to 1 or from 1 to 0], the term will still have value 1 but an even number of the entries $a_1, a_2, \ldots, a_n$ will be 1. No term of E can have this property, so each term for E must involve all n variables.

The observations of the last two paragraphs show that a length two digraph associated with $E = x_1 \oplus x_2 \oplus \cdots \oplus x_n$ must have at least $2^{n-1} + 1$ $\wedge$ and $\vee$ vertices, a number which grows exponentially with n. ∎

If we are willing to let the paths grow in length, we can divide and conquer to keep the total number of vertices manageable. Figure 7 shows the

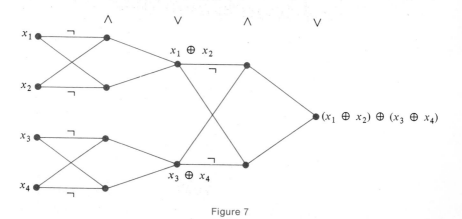

Figure 7

idea for $x_1 \oplus x_2 \oplus x_3 \oplus x_4$. This digraph has 9 $\wedge$ and $\vee$ vertices. So does the digraph associated with the join of products computation of $x_1 \oplus x_2 \oplus x_3 \oplus x_4$, since $2^3 + 1 = 9$; we have made no improvement. But how about $x_1 \oplus x_2 \oplus \cdots \oplus x_8$? The join-of-products digraph has $2^7 + 1 = 129$ $\wedge$ and $\vee$ vertices, while the analogue of the Figure 7 digraph only has $9 + 9 + 2 + 1 = 21$ $\wedge$ and $\vee$ vertices [Exercise 11].

For $x_1 \oplus x_2 \oplus \cdots \oplus x_n$ in general, the comparison is $2^{n-1} + 1$ gates for the 2-stage computation versus only $3(n-1)$ gates for the divide-and-conquer scheme, while the maximum path length increases from 2 to at most $2 \log n$ [Exercise 12]. Thus doubling the number of inputs increases path length by at most 2.

EXAMPLE 6   Circuits to perform operations in binary arithmetic are an important class of logical networks. Consider the problem of adding two binary integers. The example shown in Figure 8 illustrates the method, starting with the two integers 25 and 11 written in binary form and arranged one above the other with their digits in columns. Starting from the right we add the digits in a column. If the sum is 0 or 1 we write the sum in the answer line and carry a digit 0 one column to the left. If the sum is 2 or 3 [i.e., 10 or 11 in binary], we write 0 or 1, respectively, and go to the next column with a carry digit 1.

Carries $\longrightarrow$ 1 1 0 1 1

Numbers being $\Big\}\!\!\longrightarrow$ 1 1 0 0 1 $25 =$ $16 + 8$ $+ 1$
added $\longrightarrow$ $+$ 1 0 1 1 $+11 =$ 8 $+2 + 1$

Answer $\longrightarrow$ 1 0 0 1 0 0 $36 = 32$ $+ 4$

Figure 8

For the digit on the right which has no incoming carry digit, we would use the logical network called a **half-adder** shown in Figure 9. The value of S is the sum of x and y mod 2, and C gives the output carry digit. For the more general case with a carry input, C_I, as well as a carry output, C_O, we can combine two half-adders and an OR gate to get the network of Figure 10, called a **full-adder**. Here C_O is 1 if and only if both x and y are 1 or exactly

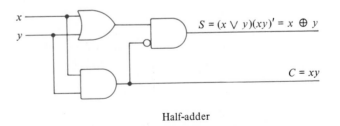

$S = (x \vee y)(xy)' = x \oplus y$

$C = xy$

Half-adder

Figure 9

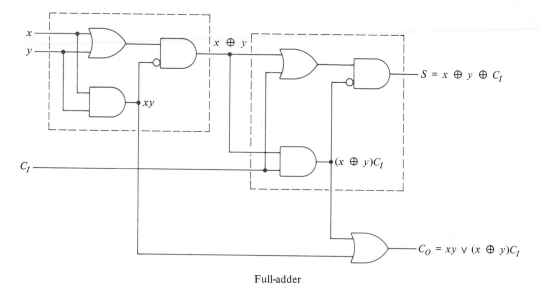

Full-adder

Figure 10

one of them is 1 and C_I is also 1, i.e., if and only if at least two of x, y and C_I are 1. Several full-adders can be combined into a network for adding n-digit binary numbers, or a single full-adder can be used repeatedly with suitable delay devices to feed the inputs in sequentially. In practice, each of these two schemes is slower than necessary. Fancy networks have been designed to add more rapidly and to perform other arithmetic operations. Our purpose in including this example was simply to illustrate the use of logical networks in hardware design. ∎

The full-adder also shows how networks to implement two or more Boolean functions can be blended together. The minterm canonical form of $S = x \oplus y \oplus C_I$ is

$$xyC_I \lor xy'C_I' \lor x'yC_I' \lor x'y'C_I$$

which turns out to be optimal [Example 5, or Exercise 7(c) of § 10.6]. It can be implemented with a logic network using four AND gates and one OR gate if we allow four input lines. The optimal sum of products expression for C_O is $xy \lor xC_I \lor yC_I$, which can be produced with three AND gates and one OR gate. To produce S and C_O separately would appear to require $4 + 1 + 3 + 1 = 9$ gates, yet Figure 10 shows we can get by with 7 gates if we want both S and C_O at once. Moreover, each gate in Figure 10 has only two input lines. The message which we hope will be clear from this discussion is that the design of economical logic networks is not an easy problem.

EXERCISES 10.5

Note. In these exercises, inputs may be complemented unless otherwise specified.

1. (a) Describe the Boolean function which corresponds to the logical network shown in Figure 11.

   (b) Sketch an equivalent network consisting of two 2-input gates.

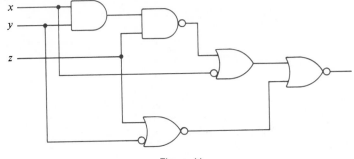

Figure 11

2. Write logical equations and sketch networks as in Figure 4 which show how to express NOT, OR and AND in terms of NOR without complementation of inputs.

3. Sketch logical networks equivalent to those in Figure 12, and composed entirely of NAND gates.

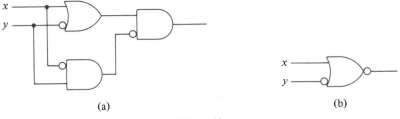

(a)   (b)

Figure 12

4. Sketch logical networks equivalent to those in Figure 13, and composed entirely of NOR gates.

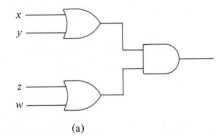

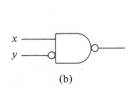

(a)   (b)

Figure 13

5. Sketch a logical network for the function XOR using
   (a) two AND gates and one OR gate.
   (b) two OR gates and one AND gate.

6. Sketch a logical network which has output 1 if and only if
   (a) exactly one of the inputs x, y, z has the value 1.
   (b) at least two of the inputs x, y, z, w have value 1.

7. Calculate the values of S and C_O for a full-adder with the given input values.
   (a) $x = 1$, $y = 0$, $C_I = 0$ (b) $x = 1$, $y = 1$, $C_I = 0$
   (c) $x = 0$, $y = 1$, $C_I = 1$ (d) $x = 1$, $y = 1$, $C_I = 1$

8. Find all values of x, y and C_I which produce the following outputs from a full-adder.
   (a) $S = 0$, $C_O = 0$ (b) $S = 0$, $C_O = 1$ (c) $S = 1$, $C_O = 1$

9. Consider the "triangle" and "circle" gates whose outputs are as shown in Figure 14. Show how to make a logic network from these two types of gates without complementation on input or output lines to produce the Boolean function
   (a) x' (b) xy (c) $x \lor y$

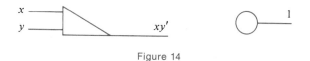

Figure 14

10. AND-OR-INVERT gates are available commercially which produce the same effect as the logical network shown in Figure 15. What inputs should be used to make such a gate into an XOR gate?

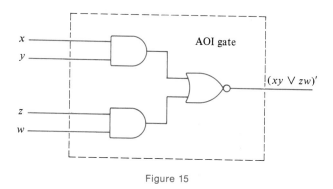

Figure 15

11. (a) Draw a digraph like the one in Figure 7 for a divide-and-conquer computation of $x_1 \oplus x_2 \oplus \cdots \oplus x_8$.
    (b) Draw the digraph for the 2-stage join-of-products computation of $x_1 \oplus x_2 \oplus x_3 \oplus x_4$.
    (c) How many $\land$ vertices are there in the digraph of the join-of-products computation of $x_1 \oplus x_2 \oplus \cdots \oplus x_8$?
    (d) Would you like to draw the digraph in part (c)?

12. Show by induction that for $n \geq 2$ there is a digraph for the computation of $x_1 \oplus x_2 \oplus \cdots \oplus x_n$ which has $3(n-1)$ $\wedge$ and $\vee$ vertices and is such that if $2^m \geq n$ then every path has length at most $2m$. *Suggestion*: Consider k with $2^{k-1} < n \leq 2^k$ and combine digraphs for 2^{k-1} and $n - 2^{k-1}$ variables.

13. Draw a digraph for the computation of $x_1 \oplus x_2 \oplus \cdots \oplus x_6$ with 15 $\wedge$ and $\vee$ vertices and all paths of length at most 6. *Suggestion*: See Exercise 12.

§ 10.6   Karnaugh Maps

Instead of trying to find the most economical or "best" logic network possible, we may decide or be forced to settle for a solution which just seems reasonably good. Optimal solutions in the sense of § 10.4 can be considered to be approximately best, so a technique for finding optimal solutions is worth having. The method of **Karnaugh maps**, which we now discuss briefly, is such a scheme. We can think of it as a sort of Boolean algebra mixture of the Venn diagrams and truth tables that we used earlier to visualize relationships between sets and between propositions.

We consider first the case of a three-variable Boolean function in x, y, z. The Karnaugh map of such a function is a 2×4 table, such as the ones in Figure 1. Each of the eight squares in the table corresponds to a minterm. The plus marks tell which minterms are involved in the function described by the table. The columns of a Karnaugh map are arranged so that neighboring columns differ in just one literal. If we wrap the table around and sew the left edge to the right edge, then we get a cylinder whose columns still have this property.

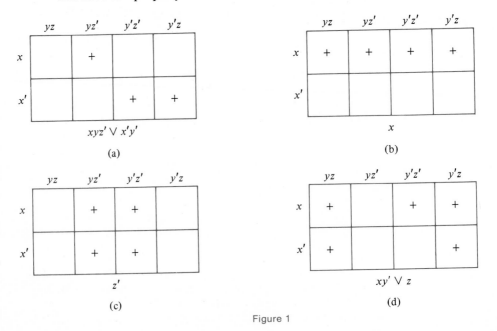

Figure 1

EXAMPLE 1   (a) In Figure 1(a) the minterm canonical form is $xyz' \vee x'y'z' \vee x'y'z$. Since $x'y'z' \vee x'y'z = x'y'(z' \vee z) = x'y'$, the function can also be written as $xyz' \vee x'y'$.

(b) The Karnaugh maps for literals are particularly simple. The map for x in Figure 1(b) has the whole first row marked; x' has the whole second row marked. The map for y has the left 2×2 block marked and the one for y' has the right 2×2 block marked. The map for z' has just the entries in the middle 2×2 block marked; see Figure 1(c). If we sew the left edge to the right edge, the columns involving z also form a 2×2 block.

(c) The map in Figure 1(d) has all entries involving z marked, and also the minterm $xy'z'$. Since both xy' boxes are marked, the function can be written as $xy' \vee z$. ∎

We now have a cylindrical map on which the literals x and x' correspond to 1×4 blocks, the literals y, z, y' and z' correspond to 2×2 blocks, products of two literals correspond to 1×2 or 2×1 blocks and products of three literals correspond to 1×1 blocks.

To find an optimal expression for a given Boolean function in x, y, and z, we outline blocks corresponding to products by performing the following steps.

Step 1. Mark the squares on the Karnaugh map corresponding to the function.

Step 2. (a) Outline all marked blocks with 8 squares. [If all 8 boxes are marked, the function is 1. Yawn.]
(b) Outline all marked blocks with 4 squares which are not contained in larger outlined blocks.
(c) Outline all marked blocks with 2 squares which are not contained in larger outlined blocks.
(d) Outline all marked squares which are not contained in larger outlined blocks.

Step 3. Select a set of outlined blocks which
(a) has every marked square in at least one selected block
(b) has as few blocks as possible and
(c) among all sets satisfying (b) gives an expression with as few literals as possible. ∎

We will need to say more about how to satisfy (b) and (c) in Step 3, but first we consider some examples.

EXAMPLE 2   (a) Consider the Boolean function with Karnaugh map in Figure 2(a). The "rounded rectangles" outline three blocks, one with four squares, corresponding to y, and two with two squares, corresponding to xz' and $x'z$. The $x'z$ block is made from squares on the two sides of the seam where we sewed the left and right edges together. Since it takes all three outlined

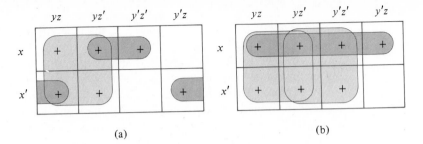

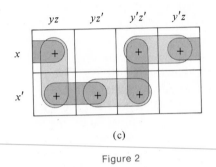

(c)

Figure 2

blocks to cover all marked squares, we must use all three blocks in Step 3. The resulting optimal expression is $y \vee xz' \vee x'z$.

(b) The Boolean function $(x'y'z)'$ is mapped in Figure 2(b). Here the outlined blocks go with x, y and z'. Again, it takes all three to cover the marked squares, so the optimal expression is $x \vee y \vee z'$.

(c) The Karnaugh map in Figure 2(c) has six outlined blocks, each with two squares. The marked squares can be covered with either of two sets of three blocks, corresponding to

$$x'y \vee xz \vee y'z' \quad \text{and} \quad x'z' \vee yz \vee xy'.$$

Since no fewer than three of the blocks can cover six squares, both of these expressions are optimal. We saw this Boolean function in Example 7(b) of § 10.4. ∎

Example 2(c) shows a situation in which more than one choice is possible. To illustrate the problems choices may cause in selecting the blocks in Step 3 we increase the number of variables to four, say w, x, y, z. Now the map is a 4×4 table, such as the ones in Figure 3, and we can think of sewing the top and bottom edges together to form a tube and then the left and right edges together to form a doughnut-shaped surface. The three-step procedure is the same as before, except that in Step 2 we start by looking for blocks with 16 squares.

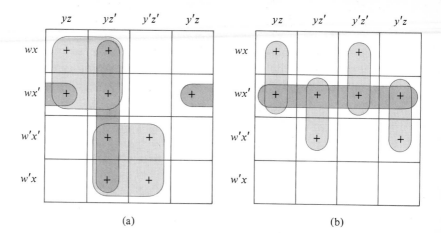

(a)   (b)

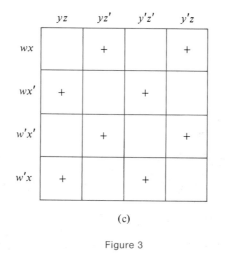

(c)

Figure 3

EXAMPLE 3   (a) The map in Figure 3(a) has four outlined blocks, three with four squares corresponding to wy, yz' and $w'z'$, and one with two squares corresponding to $wx'z$. The two-square block is the only one containing the marked $wx'y'z$ square, and the blocks for wy and $w'z'$ are the only ones containing the squares for $wxyz$ and $w'x'y'z'$, respectively, so these three blocks must be used. Since they cover all the marked squares, they meet the conditions of Step 3. The optimal expression is $wx'z \lor wy \lor w'z'$.

(b) Bigger is not always better. The map in Figure 3(b) has five blocks. Each two-square block is essential, since each is the only block containing one of the marked squares. The big four-square wx' block is superfluous, since its squares are already covered by the other blocks. The optimal expression is $wyz \lor x'yz' \lor wy'z' \lor x'y'z$.

(c) The checkerboard pattern in Figure 3(c) describes the Boolean function $w \oplus x \oplus y \oplus z$ in w, x, y, z. In this case all eight blocks are 1×1 and the optimal expression is just the minterm canonical form. A similar conclusion holds for $x_1 \oplus x_2 \oplus \cdots \oplus x_n$ in general, as we saw in Example 5 of § 10.5.   ▮

EXAMPLE 4   The maps in Example 3 offered no real choices; the essential blocks already covered all marked squares. The map of Figure 4(a) offers the opposite extreme. Every marked square is in at least two blocks. Clearly we must use at least one two-square block to cover $wx'yz'$. Suppose we choose the $wx'z'$ block. We can finish the job by choosing the four additional blocks shown in Figure 4(b). The resulting expression is

$$wx'z' \vee wy' \vee w'y \vee w'z \vee w'x.$$

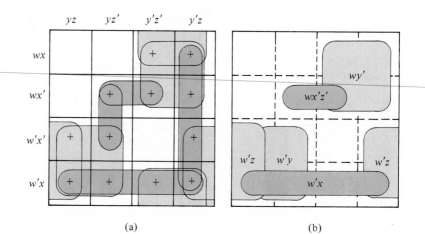

(a)   (b)

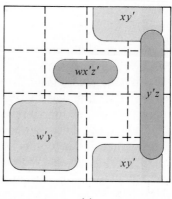

(c)

Figure 4

Figure 4(c) shows another choice of blocks, this time with only four blocks altogether. The corresponding expression is

$$wx'z' \lor w'y \lor xy' \lor y'z.$$

This expression is better, but is it optimal? The only possible improvement would be to reduce to one two-square and two four-square blocks. Since there are twelve squares to cover, no such improvement is possible; the expression is optimal. ∎

The rules for deciding which blocks to choose in situations like this are fairly complicated. It is not enough simply to choose the essential blocks because we are forced to, and then cover the remaining squares with the largest blocks possible. Such a procedure could lead to the nonoptimal solution of Figure 4(b).

For hand calculations, the tried-and-true method is to stare at the picture until the answer becomes clear. For machine calculations, which are necessary for more than five variables in any case, the Karnaugh map procedure is logically the same as the Quine-McCluskey method, for which software exists.

EXERCISES 10.6

For each of the Karnaugh maps in Exercises 1 through 4, give the corresponding minterm canonical form and an optimal expression.

1.

	yz	yz'	$y'z'$	$y'z$
x	+	+	+	+
x'	+			+

2.

	yz	yz'	$y'z'$	$y'z$
x	+	+		+
x'	+		+	+

3.

	yz	yz'	$y'z'$	$y'z$
x	+	+		+
x'			+	+

4.

	yz	yz'	$y'z'$	$y'z$
x	+			+
x'			+	

5. Draw the Karnaugh maps and outline the blocks for the given Boolean functions of x, y, z using the three-step procedure.
  (a) $x \lor x'yz$ (b) $(x \lor yz)'$
  (c) $y'z \lor xyz$ (d) $y \lor z$

6. Suppose the Boolean functions E and F each have Karnaugh maps consisting of a single block, and suppose the block for E contains the block for F.
   (a) How are the optimal expressions for E and F related?
   (b) Give examples of E and F related in this way.

7. Draw the Karnaugh map of each of the following Boolean expressions in x, y, z and show that the expression is optimal.
   (a) $xz \vee yz$ (b) $xy \vee xz \vee yz$
   (c) $xyz \vee xy'z' \vee x'yz' \vee x'y'z$

8. Repeat Exercise 7 for the following expressions in x, y, z, w.
   (a) $x' \vee yzw$ (b) $x'z' \vee xy'z \vee w'xy$
   (c) $wxz \vee wx'z' \vee w'x'z \vee w'xz'$

9. Find optimal expressions for the Boolean functions with these Karnaugh maps.

(a)

	yz	yz'	y'z'	y'z
wx	+	+	+	
wx'		+	+	+
w'x'	+	+	+	+
w'x	+	+	+	

(b)

	yz	yz'	y'z'	y'z
wx	+			
wx'				
w'x'	+	+	+	+
w'x		+	+	+

(c)

	yz	yz'	y'z'	y'z
wx	+	+	+	
wx'	+		+	+
w'x'		+	+	
w'x			+	+

(d)

	yz	yz'	y'z'	y'z
wx	+			+
wx'	+			+
w'x'		+	+	
w'x	+			+

CHAPTER HIGHLIGHTS

As usual: What does it mean? Why is it here? How can I use it? Think of examples.

Concepts

algebraic lattice
   sublattice
   join-irreducible, atom
   distributive lattice
   complement
   Boolean lattice, Boolean algebra
      duality principle
Boolean function, Boolean expression
   logical network, gate
   equivalent Boolean expressions, equivalent logical networks
      minterm canonical form
      optimal join of products
   labeled digraph
   Karnaugh map, block

Facts

Each lattice determines an algebraic lattice, and conversely. Boolean lattices correspond to Boolean algebras.
Every element of a finite algebraic lattice is a join of join-irreducible elements. This expression is essentially unique in a Boolean algebra.
Complements are unique in distributive lattices with 0 and 1.
The DeMorgan laws hold in Boolean algebras.
Any two Boolean algebras with n atoms are isomorphic.
Every logical network is equivalent to one using just NAND gates or just NOR gates.
Boolean expressions and logical networks correspond to labeled digraphs [§ 10.5].
Optimal Boolean expressions may not correspond to the simplest networks.
Choosing essential blocks first in a Karnaugh map and then greedily choosing the largest remaining blocks to cover may not give an optimal expression.

Methods

Use of the "truth table" of a Boolean expression to get its minterm canonical form.
Or use of the DeMorgan and distributive laws to get it.
Use of a Karnaugh map to find all optimal expressions equivalent to a given Boolean expression.

11

ALGEBRAIC SYSTEMS

A nonempty set which is closed under one or more binary operations is called an **algebra** or **algebraic system**. In Chapter 4 we considered semigroups, monoids and groups, which are algebras with one operation, and in Chapter 10 we studied lattices and Boolean algebras, which have more than one operation. In this chapter we will take a closer look at semigroups and groups. At the end, we will also briefly discuss rings and fields, algebras which can be thought of as generalizations of $\mathbb{Z}$ and $\mathbb{R}$ and of the set $\mathfrak{M}_{n,n}$ of $n \times n$ matrices.

Given an algebra, it turns out to be useful to look at its subalgebras and their relationship to each other and to the whole algebra. It also is helpful to consider the functions into other algebras from the given algebra which are compatible with the binary operations in a natural way. We first develop some of the basic facts about subgroups and about the appropriate functions for groups, and then show how to use the results to count the number of essentially different objects in collections with a high degree of symmetry.

§ 11.1 Generators

Chapter 4 contained an introduction to semigroups, with examples such as matrices, $\mathbb{Z}(p)$, Σ^* and $\mathcal{P}(\Sigma^*)$. Since then we have seen many other examples, including Boolean algebras with $\vee$ or $\wedge$ as operations. Since most common binary operations are associative, we encounter semigroups virtually every time we consider binary operations.

In this section we consider how to build a semigroup or group out of a handful of its elements. If a semigroup is at all large it may be impractical to keep a list of its elements, even in a computer, and of course there is no hope of listing all members of an infinite semigroup. Even if the elements *can* be stored, it may be impractical to examine them one at a time to test for a particular property. For a large semigroup S our goal is to find somehow a relatively small subset A of S from which all the elements of S can be built. In some sense the subset A will contain all the information about S.

The atoms of a Boolean algebra give an example of the sort of subset we want. By Theorem 3 of § 10.3, every element of a finite Boolean algebra is a join of atoms, i.e., every member of the algebra can be formed from atoms by joining elements repeatedly. If the algebra has 2^n members, we can build it out of its n atoms, and n is typically much smaller than 2^n.

Consider now a general semigroup $(S, \square)$ and a nonempty subset A of S. We define the set A^+ **generated by** A recursively as follows.

(B) $A \subseteq A^+$.
(R) If $x, y \in A^+$ then $x \square y \in A^+$.

These conditions imply, for example, that if $a_1, a_2, a_3, a_4 \in A$ then these four elements are in A^+, and so are $a_1 \square a_2$, $a_3 \square a_1 \square a_2$, $a_4 \square a_3 \square a_1 \square a_2$ and $a_1 \square a_2 \square a_3 \square a_1 \square a_2$. We call the members of A^+ the **products** formed from **factors** in A. Here "product" is a generic term for the result of combining elements of S using the operation $\square$. If the operation is $+$ or $\vee$ we will usually replace the word "product" by "sum" or "join," respectively. By (B) a product may have just one factor.

We say that A **generates** S if $A^+ = S$. Whether or not A generates S, A^+ is a subsemigroup of S by the following fact.

Theorem 1 Let A be a nonempty subset of the semigroup $(S, \square)$. Then A^+ is the unique smallest subsemigroup of S which contains A.

Proof. By (B), A^+ contains A. By (R), A^+ is closed under the operation $\square$, so by definition A^+ is a subsemigroup of S.

Now consider an arbitrary subsemigroup T of S which contains A. We'll show $A^+ \subseteq T$, from which it will follow that A^+ is the unique smallest subsemigroup containing A. We want to show that $x \in T$ for every x in A^+. Since A^+ is defined recursively, it's natural to use the Generalized Principle of Mathematical Induction introduced in § 6.4. Let $p(x)$ be the proposition "$x \in T$." To show that $p(x)$ is true for all x in A^+ we must establish that:

(B') $p(x)$ is true for the members of A^+ specified in the basis.
(I) If a member z of A^+ is specified by (R) in terms of previously defined members, say $z = x \square y$, then $p(x) \wedge p(y) \to p(z)$ is true.

Now (B') is true since $A \subseteq T$ by the choice of T. Condition (I) just says "if $z = x \square y$ and if $x \in T$ and $y \in T$ then $z \in T$," which is true because T is a subsemigroup. The conditions for the Generalized Principle are met, so $x \in T$ for all x in A^+. ∎

Theorem 1 helps us describe the members of A^+ without recursion, as follows.

Theorem 2   Let A be a nonempty subset of the semigroup $(S, \square)$. Then A^+ consists of all elements of S of the form $a_1 \square \cdots \square a_n$ for $n \in \mathbb{P}$ and $a_1, \ldots, a_n \in A$.

Proof. Let X be the set of all products $a_1 \square \cdots \square a_n$. If $a_1, \ldots, a_n$, $b_1, \ldots, b_m$ are in A, then $a_1 \square \cdots \square a_n \square b_1 \square \cdots \square b_m$ is a product of this same form. Thus X is closed under $\square$, i.e., it's a subsemigroup of S. The case $n = 1$ gives $A \subseteq X$, so $A^+ \subseteq X$ by Theorem 1.

To show $X \subseteq A^+$, we use ordinary induction on n. Since $A \subseteq A^+$, $a_1 \in A^+$ whenever $a_1 \in A$. Assume inductively that $a_1 \square \cdots \square a_n \in A^+$ for some n and some $a_1, \ldots, a_n$ in A and let $a_{n+1} \in A$. Since $a_{n+1} \in A^+$, we have $a_1 \square \cdots \square a_n \square a_{n+1} \in A^+$ by (R) in the recursive definition of A^+. By induction, every member of X is in A^+. ∎

EXAMPLE 1   (a) Consider the semigroup $(\mathbb{Z}, +)$. The subsemigroup $\{2\}^+$ consists of all "products" of members of $\{2\}$. Since the operation here is $+$, $\{2\}^+$ consists of all sums $2 + 2 + \cdots + 2$, i.e., of all positive even integers. Thus $\{2\}^+ = 2\mathbb{P} = \{2n : n \in \mathbb{P}\}$.

(b) More generally, the subsemigroup of $(S, \square)$ generated by a single element x is the set

$$\{x\}^+ = \{x^n : n \in \mathbb{P}\}$$

[or $\{x\}^+ = \{nx : n \in \mathbb{P}\}$ if $\square$ is $+$]. Such a single-generator semigroup is called a **cyclic semigroup**.

(c) The subsemigroup $\{2\}^+$ of $(\mathbb{Z}, \cdot)$ is $\{2^n : n \in \mathbb{P}\} = \{2, 4, 8, 16, \ldots\}$. Notice that our notation is deficient; we can't tell from the expression $\{2\}^+$ alone whether we mean this subsemigroup or the one in part (a).

(d) The subsemigroup $\{2, 7\}^+$ of $(\mathbb{Z}, \cdot)$ generated by $\{2, 7\}$ is the set $\{2^m 7^n : m, n \in \mathbb{N}$ and $m + n \geq 1\}$.

(e) We show that the cyclic subsemigroup of $(\mathfrak{M}_{2,2}, \cdot)$ generated by the matrix

$$\mathbf{M} = \begin{bmatrix} 1 & 1 \\ 0 & 1 \end{bmatrix}$$

is

$$\{\mathbf{M}\}^+ = \left\{ \begin{bmatrix} 1 & n \\ 0 & 1 \end{bmatrix} : n \in \mathbb{P} \right\}.$$

By part (b) it will be enough to show that $\mathbf{M}^n = \begin{bmatrix} 1 & n \\ 0 & 1 \end{bmatrix}$ for $n \in \mathbb{P}$. This is clear for $n = 1$. Assume inductively that $\mathbf{M}^n = \begin{bmatrix} 1 & n \\ 0 & 1 \end{bmatrix}$ for some $n \in \mathbb{P}$. Then matrix multiplication gives

$$\mathbf{M}^{n+1} = \mathbf{M}^n \cdot \mathbf{M} = \begin{bmatrix} 1 & n \\ 0 & 1 \end{bmatrix} \begin{bmatrix} 1 & 1 \\ 0 & 1 \end{bmatrix} = \begin{bmatrix} 1 & 1+n \\ 0 & 1 \end{bmatrix}.$$

The result now follows by mathematical induction.

(f) Let Σ be an alphabet and let A be a language in $\mathscr{P}(\Sigma^*)$, i.e., a subset of Σ^*. In § 4.4 we defined the positive closure A^+ to be $\bigcup\limits_{n=1}^{\infty} A^n$. In fact A^+ is the subsemigroup of Σ^* generated by A using the concatenation operation on Σ^*. ∎

Now consider a group $(G, \square)$, i.e., a semigroup with an identity, say e, and in which inverses exist. If A is a nonempty subset of G we can still consider the subsemigroup A^+ of G generated by A, but it is often more natural to take advantage of the inverses and consider the **subgroup generated by** A, denoted by $\langle A \rangle$ and defined recursively as follows.

(B) $A \subseteq \langle A \rangle$.
(R_1) If $x, y \in \langle A \rangle$ then $x \square y \in \langle A \rangle$.
(R_2) If $x \in \langle A \rangle$ then $x^{-1} \in \langle A \rangle$.

Conditions (R_1) and (R_2) show that $\langle A \rangle$ is closed under the operations $\square$ and inversion. Consider an $a \in A$. Then $a \in \langle A \rangle$ by (B), so $a^{-1} \in \langle A \rangle$ by (R_2) and thus the identity $e = a \square a^{-1}$ also belongs to $\langle A \rangle$ by (R_1). Hence $\langle A \rangle$ is a subgroup of G. In fact, $\langle A \rangle$ is the unique smallest subgroup of G which contains A [Exercise 13].

EXAMPLE 2   (a) Consider the group $(\mathbb{Z}, +)$, in which the identity is 0 and the inverse of n is $-n$. The subgroup $\langle \{4\} \rangle$ must contain $4, 4 + 4, 4 + 4 + 4$, etc. by (R_1), so it must contain all positive multiples of 4. It also contains the inverses [i.e., negatives] of these elements by (R_2), so it must contain all integer multiples of 4. That is, $\langle \{4\} \rangle \supseteq 4\mathbb{Z}$. Since $4\mathbb{Z}$ is a subgroup containing 4, and since $\langle \{4\} \rangle$ is the smallest such subgroup, $\langle \{4\} \rangle = 4\mathbb{Z}$. Note that $\{4\}^+ = 4\mathbb{P}$, which is not the same as $\langle \{4\} \rangle$.

(b) Consider the subgroup $\langle \{4, -6\} \rangle$ of $(\mathbb{Z}, +)$. It contains $4 + (-6) = -2$ by (R_1), so it contains $-(-2) = 2$ by (R_2). As in part (a), this means that $\langle \{4, -6\} \rangle \supseteq \langle \{2\} \rangle = 2\mathbb{Z}$. Now every number we can form from 4 and -6 by adding two numbers or taking negatives is still going to be even, so $\langle \{4, -6\} \rangle$ consists only of even numbers. Thus $\langle \{4, -6\} \rangle \subseteq 2\mathbb{Z}$ and so $\langle \{4, -6\} \rangle = 2\mathbb{Z}$. ∎

We say that a subset A of a group G **generates** G [as a group, now], or that A is a set of **generators** for G, if $\langle A \rangle = G$. The subgroup $\langle A \rangle$ must contain elements such as $a_1 \square a_3 \square a_2$, $a_1 \square a_2^{-1}$ and $a_3^{-1} \square a_2 \square a_1^{-1} \square a_2$ for $a_1, a_2, a_3 \in A$. The next theorem says that products like these are all there are in $\langle A \rangle$.

Theorem 3   Let A be a nonempty subset of the group $(G, \square)$. Then $\langle A \rangle$ consists of all products formed from members of A and their inverses.

Proof. For convenience, denote $\{a^{-1} : a \in A\}$ by A^{-1}. By Theorem 2, the set of products described in this theorem is the subsemigroup $(A \cup A^{-1})^+$. Thus we want to show that

$$\langle A \rangle = (A \cup A^{-1})^+.$$

By (B) and (R_2), $\langle A \rangle$ contains $A \cup A^{-1}$. Since $\langle A \rangle$ is closed under $\square$, and since $(A \cup A^{-1})^+$ is the smallest subsemigroup of G which contains $A \cup A^{-1}$, we have $\langle A \rangle \supseteq (A \cup A^{-1})^+$.

To show the reverse inclusion, we use the Generalized Principle of Induction. Let $p(x)$ be the proposition "$x \in (A \cup A^{-1})^+$." We want to show that $p(x)$ is true for every x in $\langle A \rangle$.

If $x \in A$ then $p(x)$ is true; thus $p(x)$ is true for the members x of $\langle A \rangle$ specified in (B).

Now each member of $\langle A \rangle$ specified by (R_1) is of the form $x \square y$. We must check that

(I_1) $p(x) \wedge p(y) \rightarrow p(x \square y)$ is true,

i.e., that if x and y are in $(A \cup A^{-1})^+$ then so is $x \square y$. Since $(A \cup A^{-1})^+$ is closed under $\square$, this condition is met.

Finally, we must check

(I_2) $p(x) \rightarrow p(x^{-1})$ is true.

Now if $p(x)$ is true, then $x \in (A \cup A^{-1})^+$, so by Theorem 2 we have $x = x_1 \square x_2 \square \cdots \square x_n$ for some $x_1, x_2, \ldots, x_n \in A \cup A^{-1}$. It is easy to check by multiplication by x that $x^{-1} = x_n^{-1} \square \cdots \square x_2^{-1} \square x_1^{-1}$. Since $(a^{-1})^{-1} = a$, the elements $x_1^{-1}, x_2^{-1}, \ldots, x_n^{-1}$ are also in $A \cup A^{-1}$, and so x^{-1} belongs to $(A \cup A^{-1})^+$ by Theorem 2. That is, $p(x^{-1})$ is true.   ∎

EXAMPLE 3   (a) If A has just one member, say $A = \{a\}$, we usually write $\langle a \rangle$ instead of $\langle \{a\} \rangle$. The subgroup $\langle a \rangle$ consists of all products of factors a and a^{-1}. We can cancel an a with an a^{-1}; for instance

$$a \square a \square a^{-1} \square a^{-1} \square a^{-1} \square a \square a^{-1} = a \square e \square a^{-1} \square e \square a^{-1}$$

$$= a \square a^{-1} \square a^{-1}$$

$$= e \square a^{-1} = a^{-1}.$$

So each such product is equal to a power of a or a power of a^{-1}. That is, $\langle a \rangle = \{a^k : k \in \mathbb{Z}\}$, where as always $a^0 = e$ and $a^{-k} = (a^{-1})^k$ for $k \in \mathbb{P}$. [If the operation is $+$, we have $\langle a \rangle = \{ka : k \in \mathbb{Z}\}$ with $0a = 0$ and $(-k)a = k(-a) = -(ka)$ for $k \in \mathbb{P}$.]

A group $\langle a \rangle$ generated by a single element is called a **cyclic group**. Such groups are always commutative.

(b) Consider the group $(\mathbb{Z}, +)$. For $n \in \mathbb{Z}$ we have $\langle n \rangle = n\mathbb{Z} = \{nk : k \in \mathbb{Z}\}$. In particular $\langle 0 \rangle = \{0\}$, $\langle 1 \rangle = \mathbb{Z}$ and $\langle -1 \rangle = \mathbb{Z}$.

(c) The subgroup $\langle \{3, 5\} \rangle$ of $(\mathbb{Z}, +)$ is $\mathbb{Z}$ itself, since we know that $\langle \{3, 5\} \rangle$ contains $3 + 3 - 5 = 1$ so $\langle \{3, 5\} \rangle \supseteq \langle 1 \rangle = \mathbb{Z}$. Notice that neither 3 nor 5 by itself generates $\mathbb{Z}$. ∎

In fact, all subgroups of $(\mathbb{Z}, +)$ turn out to be cyclic, which is the simplest possible situation.

Theorem 4   Every subgroup of $(\mathbb{Z}, +)$ is of the form $n\mathbb{Z}$ for some $n \in \mathbb{N}$.

Proof. Consider a subgroup H of $(\mathbb{Z}, +)$. If $H = \{0\}$ then $H = 0\mathbb{Z}$, which is of the required form. Suppose $H \neq \{0\}$. If $0 \neq m \in H$ then also $-m \in H$. Thus $H \cap \mathbb{P}$ is nonempty, so it has a smallest element, say n. We show that $H = n\mathbb{Z}$. Since $n \in H$ and H is a subgroup, we have $n\mathbb{Z} \subseteq H$. Consider an element m of H. By the Division Algorithm $m = qn + r$ with $0 \leq r < n$. Since $n\mathbb{Z} \subseteq H$, $qn \in H$ and thus $r = m - qn \in H$. Since $r < n$ and n is the smallest positive member of H, we must have $r = 0$. That is, $m = qn \in n\mathbb{Z}$. Since m was arbitrary in H, $H \subseteq n\mathbb{Z}$ as claimed. ∎

In a general group we expect that a set A with more than one member will generate a subgroup which is not cyclic.

EXAMPLE 4   Consider the set G of all one-to-one functions of $\{1, 2, 3\}$ onto itself, i.e., all permutations of $\{1, 2, 3\}$. There are $3! = 6$ functions in G, and they form a group under composition, with identity e defined by $e(x) = x$ for $x = 1, 2, 3$. This group can be generated by two of its elements:

$$f \text{ defined by } f(1) = 2, \quad f(2) = 1, \quad f(3) = 3,$$

and

$$g \text{ defined by } g(1) = 2, \quad g(2) = 3, \quad g(3) = 1.$$

The following table gives the function values of six different products formed from f, g and g^{-1}.

	1	2	3
$f \circ f = e$	1	2	3
f	2	1	3
g	2	3	1
g^{-1}	3	1	2
$f \circ g$	1	3	2
$f \circ g^{-1}$	3	2	1

These must be the six different members of G, so $G = \langle\{f, g\}\rangle$. All products formed from f, g and g^{-1} must be somewhere in this list. For instance, one can check that $g \circ f = f \circ g^{-1}$, $g \circ f \circ g = f$ and $f \circ g \circ f \circ g = e$. Since $f \circ g \neq g \circ f$, G is not commutative, so it can't possibly be cyclic. ∎

One common use for generators is to show that all members of a group have a certain property by showing that the members of a generating set have the property.

EXAMPLE 5   (a) Consider a group G which consists of some permutations of $\{1, 2, 3, \ldots, 100\}$ onto itself, with composition as operation. In practice the members of G may not be known very well, but we may know a set A of generators of G. Suppose that we know that each a in A satisfies $a(1) = 1$, $a(2) = 3$ and $a(3) = 2$. We can conclude that:

$$g(1) = 1 \quad \text{for every} \quad g \in G,$$

and

$$\{g(2), g(3)\} = \{2, 3\} \quad \text{for every} \quad g \in G.$$

Here's how.

First look at $g(1)$. We claim that $\{g \in G : g(1) = 1\}$ is a subgroup of G, for if $g, h \in G$ with $g(1) = 1$ and $h(1) = 1$ then $(g \circ h)(1) = g(h(1)) = g(1) = 1$, and $g^{-1}(1) = g^{-1}(g(1)) = 1$. This subgroup contains A by assumption, so it contains $\langle A \rangle$. Since $\langle A \rangle = G$, we have $G = \{g \in G : g(1) = 1\}$, as claimed.

Similarly, one can check that $\{g \in G : \{g(2), g(3)\} = \{2, 3\}\}$ is a subgroup of G containing A, so it too must be G itself.

Note that we don't claim that $g(2) = 3$ for every $g \in G$, even though every generator in A has this property. The reason is that $\{g \in G : g(2) = 3\}$ is not a subgroup. In fact, it doesn't contain the identity.

(b) Consider a group G of isomorphisms of a graph onto itself, with composition as operation. If there is some vertex, edge or component which is sent into itself by all members of a generating set for G, then that vertex, edge or component is sent into itself by all members of G. The argument is essentially the argument in part (a). In general we say that the function f **fixes** the point x, the set S, the vertex $v, \ldots$ if $f(x) = x$, $f(S) = S$, $f(v) = v, \ldots$. If G is a group of permutations of some set S and $x \in S$ the set $\{g \in G : g(x) = x\}$ is always a subgroup of G, called the **subgroup fixing** x [Exercise 14]. If this subgroup contains a set of generators for G, it must be G itself. ∎

We saw in earlier examples that the subsemigroup $\{2\}^+ = 2\mathbb{P}$ and the subgroup $\langle 2 \rangle = 2\mathbb{Z}$ are different subsets of the infinite group $(\mathbb{Z}, +)$. For finite groups, the next theorem shows that A^+ and $\langle A \rangle$ must be the same.

Theorem 5   If $(G, \square)$ is a group, then every finite subsemigroup of G is a subgroup.

> *Proof.* Consider a finite subsemigroup H of G. We just need to show that if $a \in H$ then $a^{-1} \in H$. Since H is finite, the function $n \to a^n$ mapping $\mathbb{P}$ to H cannot be one-to-one. Hence $a^n = a^m$ for some $n, m \in \mathbb{P}$ with $n < m$. Then
>
> $$e = a^n \square (a^{-1})^n = a^m \square (a^{-1})^n = a^{m-n}.$$
>
> If $m - n = 1$, then $a = e$, so $a^{-1} = a \in H$. Otherwise $m - n \geq 2$ and so
>
> $$a^{-1} = e \square a^{-1} = a^{m-n} \square a^{-1} = a^{m-n-1} \in H. \quad \blacksquare$$

This result is a time saver in practice. It means that to check to see if a finite subset of a known group is a subgroup it's only necessary to check closure under the operation of the group.

EXERCISES 11.1

1. Describe each of the following subsemigroups of $(\mathbb{Z}, +)$.
   (a) $\{1\}^+$ (b) $\{0\}^+$ (c) $\{-1, 2\}^+$
   (d) $\mathbb{P}^+$ (e) $\mathbb{Z}^+$ (f) $\{2, 3\}^+$
   (g) $\{6\}^+ \cap \{9\}^+$

2. Describe each of the following subsemigroups of $(\mathbb{Z}, \cdot)$.
   (a) $\{1\}^+$ (b) $\{0\}^+$ (c) $\{-1, 2\}^+$
   (d) $\mathbb{P}^+$ (e) $\mathbb{Z}^+$ (f) $\{2, 3\}^+$

3. Which of the semigroups in Exercise 1 are cyclic semigroups? Justify your answers.

4. Which of the semigroups in Exercise 2 are cyclic semigroups? Justify your answers.

5. Describe each of the following subgroups of $(\mathbb{Z}, +)$.
   (a) $\langle 1 \rangle$ (b) $\langle 0 \rangle$ (c) $\langle \{-1, 2\} \rangle$
   (d) $\langle \mathbb{Z} \rangle$ (e) $\langle \{2, 3\} \rangle$ (f) $\langle 6 \rangle \cap \langle 9 \rangle$

6. Which of the subgroups in Exercise 5 are cyclic groups? Justify your answers.

7. Recall that a monoid is a semigroup with an identity element. If A is a subset of a monoid $(M, \square)$ with identity e, the **submonoid generated by** A is defined to be $A^+ \cup \{e\}$. Find each of the following.
   (a) The submonoid of $(\mathbb{Z}, +)$ generated by $\{2\}$.
   (b) The submonoid of $(\mathbb{Z}, +)$ generated by $\{1, -1\}$.
   (c) The submonoid of $(\mathbb{Z}, +)$ generated by $\{0\}$.
   (d) The submonoid of $(\mathbb{Z} \cdot)$ generated by $\{1\}$.
   (e) The submonoid of Σ^* generated by Σ, using concatenation on Σ^*.

8. (a) Give an example of a cyclic semigroup and a subsemigroup of it which is not cyclic.
   (b) Give an example of a cyclic group which is not a cyclic semigroup.
   (c) Give an example of a cyclic group which *is* a cyclic semigroup.

9. List the members of each of the following finite subsemigroups of $(\mathfrak{M}_{3,3}, \cdot)$.

(a) $\left\{ \begin{bmatrix} 0 & 1 & 0 \\ 1 & 0 & 0 \\ 0 & 0 & 1 \end{bmatrix} \right\}^{+}$

(b) $\left\{ \begin{bmatrix} 0 & 1 & 0 \\ 1 & 0 & 0 \\ 0 & 0 & 1 \end{bmatrix}, \begin{bmatrix} 0 & 0 & 1 \\ 1 & 0 & 0 \\ 0 & 1 & 0 \end{bmatrix} \right\}^{+}$

(c) $\left\{ \begin{bmatrix} 0 & 2 & 3 \\ 0 & 0 & 4 \\ 0 & 0 & 0 \end{bmatrix} \right\}^{+}$

10. (a) Which of the subsemigroups in Exercise 9 are groups?
(b) Which are commutative?
(c) Which are cyclic?

11. The set S_4 of all one-to-one functions of $\{1, 2, 3, 4\}$ onto itself is a group under composition of functions. Which of the following are subgroups of $(S_4, \circ)$? Justify your answers.
(a) $\{f \in S_4 : f(4) = 4\}$
(b) $\{f \in S_4 : f(1) = 2\}$
(c) $\{f \in S_4 : f(1) \in \{1, 2\}\}$
(d) $\{f \in S_4 : f(1) \in \{1, 2\} \text{ and } f(2) \in \{1, 2\}\}$

12. (a) Find a 24-element subset of S_4 which generates the group $(S_4, \circ)$ in Exercise 11. [There are actually some 2-element generating sets for S_4, but they are harder to find.]
(b) Is $(S_4, \circ)$ a cyclic group? Justify your answer.

13. Use the Generalized Principle of Mathematical Induction to show that $\langle A \rangle$ is the smallest subgroup of the group G containing A.

14. Consider a group G of permutations [i.e., one-to-one functions] of a set S onto itself, with composition of functions as operation. Show that $\{g \in G : g(x) = x\}$ is a subgroup of G for each x in S. *Suggestion*: See Example 5(a).

§ 11.2   Subsemigroups, Subgroups and Cosets

In § 11.1 we formed subsemigroups by building them up from the inside, using sets of elements to generate them. Once we have several subsemigroups on hand, we can form others by intersection.

Theorem 1   Let $(S, \square)$ be a semigroup. The intersection of any collection of subsemigroups of S is either empty or is itself a subsemigroup of S.

Proof. If s and t belong to the intersection, then they both belong to each subsemigroup in the collection, so their product $s \square t$ does too. That is, the intersection is closed under $\square$. ∎

Corollary 1 If S is a monoid, the intersection of any nonempty collection of submonoids of S is a submonoid of S.

Proof. A submonoid is a subsemigroup of S which contains the identity element e of S. If each member of the collection contains e, so does the intersection. ∎

Corollary 2 The intersection of any nonempty collection of subgroups of a group is a subgroup.

Proof. Every member of the intersection belongs to each subgroup, so its inverse does too. Thus its inverse also belongs to the intersection. Hence the intersection is a submonoid closed under taking inverses, i.e., a subgroup. ∎

EXAMPLE 1 (a) Both $2\mathbb{Z}$ and $3\mathbb{Z}$ are subsemigroups of $(\mathbb{Z}, \cdot)$, so their intersection $6\mathbb{Z}$ is too.

(b) The intersection $\mathbb{P} \cap 2\mathbb{Z} = \{2k : k \in \mathbb{P}\}$ is a subsemigroup of $(\mathbb{Z}, \cdot)$.

(c) Both $\mathbb{P}$ and $\{-k : k \in \mathbb{P}\}$ are subsemigroups of $(\mathbb{Z}, +)$. Their intersection is empty.

(d) Consider a group G of permutations of a set X, with composition as operation. For elements x and y in X the sets $\{g \in G : g(x) = x\}$ and $\{g \in G : g(y) = y\}$ are subgroups of G [Example 5(b) of § 11.1]. Their intersection $\{g \in G : g(x) = x \text{ and } g(y) = y\}$ is the subgroup of G fixing both x and y. More generally, if $Y \subseteq X$ the set $\{g \in G : g(x) = x \text{ for all } x \in Y\}$ is the intersection of the subgroups fixing the members of Y, so it is itself a subgroup of G. ∎

EXAMPLE 2 (a) Consider a nonempty subset A of a semigroup S. By Theorem 1 the intersection of all of the subsemigroups which contain A [including S itself, naturally] is a subsemigroup which contains A. It must be the smallest subsemigroup containing A, namely A^+. This construction gives a way to define A^+ without describing its elements in terms of A.

(b) Consider a nonempty subset A of a group G. By Corollary 2, the intersection of all subgroups of G containing A is $\langle A \rangle$, the unique smallest subgroup of G containing A. ∎

Semigroups are only required to be closed under an associative operation, so it is not particularly surprising that we can't say much of a general nature about the relationship between subsemigroups and the larger semigroup in which they are contained.

EXAMPLE 3 To see a sample of how wild things can get, consider an arbitrary nonempty set S, and define $\square$ on S by $x \square y = y$ for all $x, y \in S$; i.e., the product of two elements is always just the second element. Then $(x \square y) \square z = z = x \square z = x \square (y \square z)$, so $\square$ is associative and $(S, \square)$ is a semigroup. *Every* nonempty subset of S is a subsemigroup, since it's closed under $\square$. ∎

As this example shows, the subsemigroups of a semigroup may be pretty arbitrarily embedded in the whole semigroup. So it is particularly striking that the subgroups of a group are quite strongly influenced by the structure of the group as a whole. Each subgroup defines a partition of the group in a natural way, which in the case of finite groups allows us to use arithmetic to draw significant group-theoretic conclusions. To state our main result we need to introduce some notation.

Consider a subgroup H of a group $(G, \square)$ with identity element e. A **coset** of H in G is a subset of the form

$$H \square x = \{h \square x : h \in H\}$$

for some x in G. The coset $H \square e$ is H itself, and indeed $H \square h = H$ for every h in H [Exercise 13(a)]. Since $e \in H$, the coset $H \square x$ contains $e \square x = x$. Thus every x in G belongs to at least one coset. In fact, each x belongs to just one coset.

Theorem 2   The cosets of a subgroup of a group form a partition of the group.

Proof. Consider a group G with subgroup H. We just showed that G is the union of the various cosets $H \square x$, so we only need to show that any two distinct cosets are disjoint. Consider first an element $z \in H \square x$. Then $z = h \square x$ for some $h \in H$. For each $k \in H$, $k \square z = k \square (h \square x) = (k \square h) \square x$ is in $H \square x$, because $k \square h$ is also in the subgroup H. Thus $H \square z \subseteq H \square x$. Moreover, $x = h^{-1} \square h \square x = h^{-1} \square z \in H \square z$, because h^{-1} is in the subgroup H, so the roles of x and z can be reversed and we conclude that $H \square x \subseteq H \square z$. [This is where the proof breaks down for semigroups.] We've shown that if $z \in H \square x$ then $H \square z = H \square x$.

Now suppose $z \in (H \square x) \cap (H \square y)$. Then by the argument above $H \square z = H \square x$ and $H \square z = H \square y$, so $H \square x = H \square y$. In other words, cosets which overlap are identical. ∎

EXAMPLE 4   Consider the subgroup $3\mathbb{Z}$ of the group $(\mathbb{Z}, +)$. The cosets are the sets of the form $3\mathbb{Z} + n = \{3k + n : k \in \mathbb{Z}\}$. There are just three of them, namely $3\mathbb{Z}$, $3\mathbb{Z} + 1$ and $3\mathbb{Z} + 2$. Every n in $\mathbb{Z}$ can be written as $n = 3q + r$ with $r \in \{0, 1, 2\}$, so $\mathbb{Z}$ is the union of these three disjoint sets. Similarly, for every $m \in \mathbb{P}$ the cosets of $m\mathbb{Z}$ in $(\mathbb{Z}, +)$ are the sets $m\mathbb{Z} + n$ where $n = 0, 1, 2, \ldots,$ $m - 1$. ∎

Instead of the cosets $H \square x$ which we have been considering, we could just as easily have looked at **left cosets**, of the form $x \square H = \{x \square h : h \in H\}$. If G is commutative, then $H \square x = x \square H$, but in general the left coset $x \square H$ and right coset $H \square x$ are different subsets of G. The proof of Theorem 2 is still valid for left cosets, with the obvious right-left switches.

EXAMPLE 5 Consider the permutation group $\mathrm{PERM}(X)$ of all one-to-one functions of the nonempty set X onto itself, with composition as operation. Choose an element x_0 in X, and define the equivalence relation $\sim$ on $\mathrm{PERM}(X)$ by letting $f \sim g$ if and only if $f(x_0) = g(x_0)$. The function e defined by $e(x) = x$ for all x in X is the identity of $\mathrm{PERM}(X)$, and $f \sim e$ if and only if $f(x_0) = x_0$. As in Example 5(b) of § 11.1, $\{f \in \mathrm{PERM}(X) : f(x_0) = x_0\}$ is a subgroup of $\mathrm{PERM}(X)$; call this subgroup FIX. Thus FIX is the equivalence class of e.

Now consider the equivalence class of some function g in $\mathrm{PERM}(X)$. We have $f \sim g$ if and only if $f(x_0) = g(x_0)$ if and only if $(g^{-1} \circ f)(x_0) = g^{-1}(f(x_0)) = g^{-1}(g(x_0)) = x_0$ if and only if $g^{-1} \circ f \in \mathrm{FIX}$. But $g^{-1} \circ f = h \in \mathrm{FIX}$ if and only if $f = g \circ g^{-1} \circ f = g \circ h \in g \circ \mathrm{FIX}$. That is, the left coset $g \circ \mathrm{FIX}$ is the equivalence class of g, consisting of all functions f in $\mathrm{PERM}(X)$ with $f(x_0) = g(x_0)$. In this case, the equivalence classes which partition $\mathrm{PERM}(X)$ are actually the left cosets of the subgroup FIX.

If $g \notin \mathrm{FIX}$ and if X has at least three elements we can show that the right coset $\mathrm{FIX} \circ g$ is not the same as $g \circ \mathrm{FIX}$. Since $g \notin \mathrm{FIX}$, $g(x_0) \neq x_0$. Choose an f in $\mathrm{PERM}(X)$ with $f(x_0) = x_0$ but $f(g(x_0)) \neq g(x_0)$. Then $f \in \mathrm{FIX}$, so $f \circ g \in \mathrm{FIX} \circ g$, but $(f \circ g)(x_0) \neq g(x_0)$, so $f \circ g \notin g \circ \mathrm{FIX}$.

Exercise 9 deals with this example in detail for a three-element set X. ∎

Not only do the cosets of H in G partition G, they all have the same size, namely the size of H.

Theorem 3   Let H be a subgroup of the group $(G, \square)$ and let $x \in G$. The function ϕ given by $\phi(h) = h \square x$ is a one-to-one correspondence of H onto the coset $H \square x$.

Proof. The function ϕ certainly maps H into $H \square x$. The function ψ given by $\psi(k) = k \square x^{-1}$ is its inverse, since if $k = h \square x \in H \square x$ with $h \in H$ then $k \square x^{-1} = h \square x \square x^{-1} = h$. Since ϕ has an inverse on $H \square x$, it is a one-to-one correspondence of H onto $H \square x$. ∎

Our next result is one of the basic workhorses of finite group theory. To state it we need some notation. For H a subgroup of G, let G/H be the set of [right] cosets of H in G, and let $|H|$, $|G|$ and $|G/H|$ be the numbers of elements in H, G and G/H, respectively.

Corollary   Let H be a subgroup of the finite group G. Then
(Lagrange's
Theorem)
$$|G| = |G/H| \cdot |H|.$$

In particular, $|H|$ and $|G/H|$ divide $|G|$.

Proof. There are $|G/H|$ cosets of H, each of which has $|H|$ members, by Theorem 3. They partition G by Theorem 2. ∎

Theorem 3 and Lagrange's theorem have valid analogues for left cosets, so G has the same number of left cosets of H as right cosets. Exercise 15 asks for a direct proof of this fact.

EXAMPLE 6   (a) A group with 10 members can only have subgroups with 1, 2, 5 or 10 members. Contrast this with Example 3; a semigroup with 10 members can have subsemigroups with n members for $n = 1, 2, \ldots, 10$.

(b) A group with 81 members can only have subgroups with 1, 3, 9, 27 or 81 members.

(c) Suppose the set X has n elements. Then the group PERM(X) of Example 5 has $n!$ elements, and the subgroup FIX fixing x_0 has $(n-1)!$ elements. There are n left cosets $g \circ$ FIX, one for each possible value of $g(x_0)$. Lagrange's theorem in this case takes the form

$$n! = n \cdot (n-1)!.$$

(d) Let G be the group of isomorphisms of the graph in Figure 1 back onto itself, with composition as operation. One can use Lagrange's theorem to help check that G has 10 members. We see at once the five "rotations," through angles of $0°$, $72°$, $144°$, $216°$ and $288°$, which form a subgroup, call it R. The correspondence g defined by $g(p) = p$, $g(q) = t$, $g(r) = s$, $g(s) = r$ and $g(t) = q$ is in G but not in R, so $|G| > |R| = 5$. By Lagrange's theorem $|G|$ is a multiple of 5, so it's at least 10. Now for an element f of G there are only five possible choices for $f(p)$, and then only two possible choices for $f(q)$ [one on each side of $f(p)$], after which the rest of the action of f is completely determined. So G has at most $5 \cdot 2 = 10$ members, and thus has exactly 10.

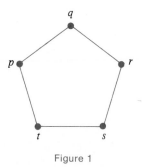

Figure 1

Half of the members of G are in R. The rest form the single coset $R \circ g$, where g is defined above. In this case $R \circ g = g \circ R$. [See also Exercise 14.]

The subgroup $\langle g \rangle$ consists just of e and g, since $g \circ g = e$, and so $|\langle g \rangle| = 2$. Since $10 = |G| = |G/\langle g \rangle| \cdot |\langle g \rangle|$, there are 5 cosets of $\langle g \rangle$ in G. In fact, $\langle g \rangle$ is the subgroup of G fixing the point p. As in part (c) above, the cosets of $\langle g \rangle$ are simply the five sets of isomorphisms in G taking p to each of its five possible images, namely $\{f \in G : f(p) = p\}$, $\{f \in G : f(p) = q\}$, $\{f \in G : f(p) = r\}$, etc. ■

The **order** of an element in a finite group is defined to be the number of elements in the cyclic subgroup it generates. By Lagrange's theorem, the order of every element of G divides $|G|$.

It is not hard to give examples of groups for which some divisors of $|G|$ are not orders of elements. Indeed, any noncyclic group is an example. An example of a group G which has no subgroup with n elements for some divisor n of $|G|$ is harder to describe, but in fact such groups are in a sense the most common.

EXERCISES 11.2

1. (a) Find the intersection of the subsemigroups $2\mathbb{P}$ and $3\mathbb{P}$ of the semigroup $(\mathbb{P}, +)$.
   (b) Is the intersection in part (a) cyclic? Explain.
   (c) Repeat part (b) with the semigroup $(\mathbb{P}, \cdot)$.

2. (a) Find the intersection of the subsemigroups of $(\mathbb{P}, \cdot)$ generated by 2 and 3, respectively.
   (b) Is the intersection in part (a) cyclic? Explain.

3. (a) Find the intersection of the three subsemigroups $4\mathbb{P}$, $6\mathbb{P}$, $10\mathbb{P}$ of the semigroup $(\mathbb{P}, \cdot)$.
   (b) Is the intersection in part (a) cyclic? Explain.
   (c) Repeat part (b) with the semigroup $(\mathbb{P}, +)$.

4. (a) Find the intersection of all the subgroups $n\mathbb{Z}$ of $(\mathbb{Z}, +)$, where $n \in \mathbb{P}$.
   (b) Is the intersection in part (a) cyclic? Explain.

5. (a) Give an example of a one-to-one correspondence between $4\mathbb{Z}$ and the coset $4\mathbb{Z} + 3$ in $(\mathbb{Z}, +)$.
   (b) Give another example.

6. Consider the group $(\mathbb{Z}, +)$. Write $\mathbb{Z}$ as a disjoint union of five cosets of a subgroup.

7. Give an example of a group G and subgroups H and K such that $H \cup K$ is not a subgroup of G.

8. (a) Show that the subgroup R in Example 6(d) is cyclic, and describe a generator for it.
   (b) Show that the group G in the example contains elements of orders 1, 2 and 5, but not 10.

9. Let $X = \{1, 2, 3\}$ and $x_0 = 1$ in Example 5.
   (a) Find $|\text{PERM}(X)|$.
   (b) Find $|\text{FIX}|$ and describe each of the functions in FIX.
   (c) The function g defined by $g(1) = 2$, $g(2) = 3$ and $g(3) = 1$ is not in FIX. Describe the members of the coset $\text{FIX} \circ g$.
   (d) Show that $\text{FIX} \circ g \neq g \circ \text{FIX}$.
   (e) Show that in fact $\text{FIX} \circ g$ is not a left coset at all.
   (f) How many cosets does FIX have in $\text{PERM}(X)$?

10. The table below describes a binary operation • on the set $G = \{a, b, c, d, e\}$ with e as identity element.

•	e	a	b	c	d
e	e	a	b	c	d
a	a	e	c	d	b
b	b	d	a	e	c
c	c	b	d	a	e
d	d	c	e	b	a

(a) Show that the set $\{e, a\}$ is a group under • as operation.

(b) Without doing any calculations, use the result of part (a) and Lagrange's theorem to conclude that $(G, •)$ is not a group.

11. The following table gives the binary operation for a group $(G, •)$ with elements a, b, c, d, e, f.

•	e	a	b	c	d	f
e	e	a	b	c	d	f
a	a	b	e	d	f	c
b	b	e	a	f	c	d
c	c	f	d	e	b	a
d	d	c	f	a	e	b
f	f	d	c	b	a	e

(a) List the members of the subgroup $\langle a \rangle$.

(b) Show that $\langle a \rangle • c = c • \langle a \rangle$.

(c) Find all the subgroups with two members.

(d) Find $|G/\langle d \rangle|$.

(e) Describe the right cosets of $\langle d \rangle$.

12. Repeat Exercise 11 for the group with the following table.

•	e	a	b	c	d	f
e	e	a	b	c	d	f
a	a	b	e	d	f	c
b	b	e	a	f	c	d
c	c	d	f	a	b	e
d	d	f	c	b	e	a
f	f	c	d	e	a	b

13. (a) Show that if H is a subgroup of a group $(G, \square)$, and if $g \in G$, then $H \square g = H$ if and only if $g \in H$. Try to be clever and use Theorem 2.

(b) Show by example that the conclusion in part (a) can fail if G is just a semigroup and H a subsemigroup. *Suggestion*: Let $G = H = \mathbb{P}$.

14. Consider a finite group $(G, \square)$ with a subgroup H such that $|G| = 2|H|$. Show that $g \square H = H \square g$ for every g in G. *Suggestion*: Consider the two cases $g \in H$ and $g \notin H$ separately.

15. Let H be a subgroup of the group $(G, \square)$.
   (a) Show that for each g in G the right coset $H \square g^{-1}$ consists of the inverses of the elements in the left coset $g \square H$.
   (b) Describe a one-to-one correspondence between the set of left cosets of H in G and the set of right cosets.

16. Prove Theorem 3 without mentioning the inverse correspondence.

17. Let H be a subgroup of a group $(G, \square)$ and for $x, y \in G$ define $x \sim y$ if $y \square x^{-1} \in H$.
   (a) Show that $\sim$ is an equivalence relation on G.
   (b) Show that the partition in Theorem 2 is precisely the partition of equivalence classes for $\sim$ described in Theorem 1 of § 7.3.

18. Show that if H is a subgroup of $(G, \bullet)$, then $x \bullet H \bullet x^{-1}$ is also a subgroup for each x in G.

19. The set $\mathscr{L}(S)$ of all subsemigroups of a semigroup S is partially ordered by $\subseteq$.
   (a) Show that the least upper bound $K \vee L$ of two members K, L of $\mathscr{L}(S)$ is $(K \cup L)^{+}$.
   (b) Under what conditions on K and L does a greatest lower bound $K \wedge L$ exist? What is it when it exists?
   (c) Show that if S is a finite group, then $(\mathscr{L}(S), \subseteq)$ is a lattice. *Suggestion*: Appeal to Theorem 5 of § 11.1.

§ 11.3   Homomorphisms

This section introduces the functions that are most important in the study of semigroups and groups. After Example 4 the discussion focuses on groups and on a special kind of subgroup that plays a key role. There is a lot of new material that needs time to sink in. Perhaps the best strategy is to read through the whole section fairly quickly and then come back and master a piece at a time.

There is really only one natural way in which a function from one semigroup to another can take their algebraic structures into account: it must take products of elements to products of their images. In symbols, an "algebra-compatible" function h from a semigroup S with operation $\bullet$ to a semigroup T with operation $\square$ must satisfy

$$h(s_1 \bullet s_2) = h(s_1) \square h(s_2) \quad \text{for all} \quad s_1, s_2 \in S.$$

Such a function is called a **homomorphism**. If S and T are not just semigroups but are groups, we may emphasize the fact by calling h a **group homomorphism**.

EXAMPLE 1   (a) Let $(S, \bullet)$ and $(T, \square)$ both be $(\mathbb{Z}, +)$. The homomorphism condition is that

$$h(m + n) = h(m) + h(n) \quad \forall m, n \in \mathbb{Z}.$$

The function h defined by $h(n) = 5n$ for all n is an example of a group homomorphism, since

$$h(m + n) = 5 \cdot (m + n) = 5m + 5n = h(m) + h(n).$$

There is nothing special about 5; any other integer would define a group homomorphism in the same way.

(b) Let $(S, \bullet)$ be $(\mathbb{P}, +)$, let $(T, \square)$ be $(\mathbb{P}, \cdot)$, and let $h(m) = 2^m$ for m in $\mathbb{Z}$. Since

$$h(m + n) = 2^{m+n} = 2^m \cdot 2^n = h(m) \cdot h(n),$$

h is a homomorphism of $(\mathbb{P}, +)$ into $(\mathbb{P}, \cdot)$.

(c) As in Exercise 8 of § 3.1, let $\text{trace}(\mathbf{A}) = \sum_{i=1}^{n} a_{ii}$ for $\mathbf{A} = [a_{ij}]$ in $\mathfrak{M}_{n,n}$, the set of $n \times n$ matrices. It is easy to see that

$$\text{trace}(\mathbf{A} + \mathbf{B}) = \text{trace}(\mathbf{A}) + \text{trace}(\mathbf{B})$$

for $\mathbf{A}, \mathbf{B} \in \mathfrak{M}_{n,n}$. Thus the function trace from $(\mathfrak{M}_{n,n}, +)$ to $(\mathbb{R}, +)$ is a group homomorphism.

(d) Recall that $a = \log_2 b$ if and only if $b = 2^a$. The function h from $(\{x \in \mathbb{R} : x > 0\}, \cdot)$ to $(\mathbb{R}, +)$ given by $h(x) = \log_2 x$ is a group homomorphism since $\log_2(xy) = \log_2(x) + \log_2(y)$. ∎

EXAMPLE 2   Consider a positive integer p. For $n \in \mathbb{Z}$, let $\text{Rem}_p(n)$ be the remainder when n is divided by p. Then Rem_p is a function from $\mathbb{Z}$ into $\mathbb{Z}(p)$, by the Division Algorithm in § 4.3. In fact, Rem_p is a group homomorphism of $(\mathbb{Z}, +)$ onto $(\mathbb{Z}(p), +_p)$, and also a homomorphism of $(\mathbb{Z}, *)$ onto $(\mathbb{Z}(p), *_p)$. To see that Rem_p is an additive homomorphism, we need to show

$$\text{Rem}_p(m + n) = \text{Rem}_p(m) +_p \text{Rem}_p(n) \qquad \text{for} \quad m, n \in \mathbb{Z}.$$

So it suffices to show

$$\text{Rem}_p(m + n) \equiv \text{Rem}_p(m) + \text{Rem}_p(n) \pmod{p}.$$

Since $m \equiv \text{Rem}_p(m) \pmod{p}$ and $n \equiv \text{Rem}_p(n) \pmod{p}$, Theorem 2(b) of § 4.3 shows that

$$m + n \equiv \text{Rem}_p(m) + \text{Rem}_p(n) \pmod{p}.$$

Since $m + n \equiv \text{Rem}_p(m + n) \pmod{p}$, part (T) of Theorem 2(a) of § 4.3 shows that

$$\text{Rem}_p(m + n) \equiv \text{Rem}_p(m) + \text{Rem}_p(n) \pmod{p},$$

as desired.

Simply change each $+$ to $*$ and each $+_p$ to $*_p$ in this argument to show that Rem_p is a multiplicative homomorphism. ∎

If h is a homomorphism from $(S, \bullet)$ to $(T, \square)$, then the image $h(S)$ is a subsemigroup of T, since if $h(s_1)$ and $h(s_2)$ are members of $h(S)$ then $h(s_1) \square h(s_2) = h(s_1 \bullet s_2) \in h(S)$, and so $h(S)$ is closed under $\square$. If $(S, \bullet)$ and $(T, \square)$ are groups, we can show that $h(S)$ is a subgroup of $(T, \square)$, as follows. Suppose e is the identity of S. Then $h(e) \square h(s) = h(e \bullet s) = h(s)$ for every s in S. So $h(e) = h(e) \square h(s) \square h(s)^{-1} = h(s) \square h(s)^{-1}$, which is the identity of T. That is, h sends the identity of S to the identity of T. Moreover, $h(e) = h(s^{-1} \bullet s) = h(s^{-1}) \square h(s)$, so

$$h(s)^{-1} = h(e) \square h(s)^{-1} = h(s^{-1}) \square h(s) \square h(s)^{-1} = h(s^{-1}).$$

Thus h takes inverses to inverses, and $h(S)$ contains the inverses of all of its elements. Since $h(S)$ is closed under products and taking inverses, it is a subgroup of T.

EXAMPLE 3 Let $(G, \square)$ be a group and let $g \in G$. Recall from Example 3(a) of § 11.1 that the cyclic group $\langle g \rangle$ generated by g is $\{g^n : n \in \mathbb{Z}\}$, where $g^{-k} = (g^{-1})^k$ for $k \in \mathbb{P}$. One can check by looking at cases that

$$g^{n+m} = g^n \square g^m \quad \forall n, m \in \mathbb{Z}.$$

That is, the function h from $(\mathbb{Z}, +)$ to $(G, \square)$ given by $h(n) = g^n$ is a group homomorphism. Its image $h(\mathbb{Z})$ is $\langle g \rangle$. ∎

We next say what it means for two semigroups to look just alike. A semigroup homomorphism which is a one-to-one correspondence between two semigroups is called a **semigroup isomorphism**. We have seen isomorphisms before; graph isomorphisms were one-to-one correspondences preserving the graph structure, and Boolean algebra isomorphisms preserved the operations $\vee$, $\wedge$ and $'$. Semigroup isomorphisms preserve products. If h is a semigroup isomorphism of $(S, \bullet)$ onto $(T, \square)$ then

$$h(s_1 \bullet s_2) = h(s_1) \square h(s_2),$$

and since h is one-to-one

$$h(s_3) = h(s_1) \square h(s_2) \Leftrightarrow s_3 = s_1 \bullet s_2.$$

If there is a semigroup isomorphism of S onto T, we say S and T are **isomorphic** and write $S \simeq T$ [or $(S, \bullet) \simeq (T, \square)$ if the operations need to be mentioned].

EXAMPLE 4 Let X be any set. We define the "complementation function" h from $\mathscr{P}(X)$ into $\mathscr{P}(X)$ by $h(A) = A^c$ for $A \in \mathscr{P}(X)$. By a DeMorgan law in Table 1 of § 1.2,

$$h(A \cup B) = (A \cup B)^c = A^c \cap B^c = h(A) \cap h(B),$$

so h is a homomorphism from the semigroup $(\mathscr{P}(X), \cup)$ into the semigroup $(\mathscr{P}(X), \cap)$. The function h is a one-to-one correspondence of $\mathscr{P}(X)$ onto

$\mathscr{P}(X)$, so it is an isomorphism of $(\mathscr{P}(X), \cup)$ onto $(\mathscr{P}(X), \cap)$. In fact, h is its own inverse [why?] and so h is also an isomorphism of $(\mathscr{P}(X), \cap)$ onto $(\mathscr{P}(X), \cup)$. ∎

Isomorphisms get used in two different ways, as we saw when we looked at graph isomorphisms. Sometimes we want to call attention to the fact that two apparently different semigroups are actually identical in structure. At other times the identical structure is obvious, but we want to examine the various isomorphisms which are possible, for instance from S back onto itself, to see how much symmetry is present. We will explore symmetry when we study groups in § 11.4.

Imagine now that we have a homomorphism h from a semigroup S to a semigroup T and that we know very little about S but all about T. If h is an isomorphism, then S is essentially identical with T and h tells us all about S. If h is one-to-one but not onto T, then at least we know that S looks just like the subsemigroup $h(S)$ of T. We can analyze S by looking at $h(S)$, which we know pretty well because it's inside the known semigroup T. Even if h is not one-to-one, we can still perhaps learn something about S by looking at $h(S)$. If S is complicated but $h(S)$ isn't, this may be a good way to begin to learn about the structure of S. Such an approach works especially well if S is a group, as we will see next.

In the remainder of this section we will look at what happens when h is a homomorphism from a group $(G, \bullet)$ to a group $(H, \square)$. Let e be the identity element of G and let e' be the identity of H. Then, as we saw after Example 2, $h(e) = e'$ and $h(x^{-1}) = h(x)^{-1}$ for each x in G.

Theorem 1   With notation as above, let $K = \{x \in G : h(x) = e'\}$. Then
    (a) K is a subgroup of G.
    (b) $K \bullet x = x \bullet K$ for each x in G.
    (c) $K \bullet x = \{z \in G : h(z) = h(x)\}$ for each x in G.

The subgroup K is called the **kernel** of the homomorphism h.

Proof. (a) If $x, y \in K$, then $h(x \bullet y) = h(x) \square h(y) = e' \square e' = e'$, so $x \bullet y \in K$. Moreover, $h(x^{-1}) = h(x)^{-1} = (e')^{-1} = e'$, so $x^{-1} \in K$. Since K is closed under products and inverses, it is a subgroup of G.

(b) To show that $K \bullet x \subseteq x \bullet K$, it is enough to consider $k \in K$ and show that $k \bullet x \in x \bullet K$. Since $k \bullet x = x \bullet (x^{-1} \bullet k \bullet x)$, it suffices to show that $x^{-1} \bullet k \bullet x \in K$. But $h(x^{-1} \bullet k) = h(x^{-1}) \square h(k) = h(x)^{-1} \square e' = h(x)^{-1}$ and so $h(x^{-1} \bullet k \bullet x) = h(x^{-1} \bullet k) \square h(x) = h(x)^{-1} \square h(x) = e'$. A similar argument shows that $x \bullet K \subseteq K \bullet x$.

(c) For $k \in K$, $h(k \bullet x) = h(k) \square h(x) = e' \square h(x) = h(x)$, and so $K \bullet x \subseteq \{z \in G : h(z) = h(x)\}$. In the other direction, if $h(z) = h(x)$ then $h(z \bullet x^{-1}) = h(z) \square h(x)^{-1} = e'$, which means $z \bullet x^{-1} \in K$ and so $z = z \bullet x^{-1} \bullet x \in K \bullet x$. ∎

We saw in Theorem 3 of § 11.2 that all cosets $K \bullet x$ have the same number of elements, namely $|K|$, and this fact gives us a useful test for one-to-oneness of homomorphisms.

Corollary   A group homomorphism is one-to-one if and only if its kernel is just the identity element.

Proof. Since K is a subgroup, it contains the identity e for sure. Moreover, by Theorem 1(c), h is one-to-one if and only if all cosets $K \bullet x$ have exactly one element, which is true if and only if K itself has just one member. ∎

EXAMPLE 5   (a) Define the homomorphism h from $(\mathbb{Z}, +)$ to $(\mathbb{Z}, +)$ by $h(n) = 5n$. The kernel of h is $\{n \in \mathbb{Z} : 5n = 0\} = \{0\}$, and h is one-to-one.

(b) Consider Rem_6 from $(\mathbb{Z}, +)$ to $(\mathbb{Z}(6), +_6)$, where $\mathrm{Rem}_6(n)$ is the remainder on dividing n by 6, as in Example 2. Then

$$K = \{n \in \mathbb{Z} : n \text{ is a multiple of } 6\} = 6\mathbb{Z}. ∎$$

A subgroup K of a group $(G, \bullet)$ with the property that $K \bullet x = x \bullet K$ for every x in G is called a **normal subgroup** of G. Theorem 1(b) shows that kernels of homomorphisms are normal. If G is commutative, every subgroup is normal, but in the general case there will be nonnormal subgroups. When K is normal, G/K has a natural group structure, which we describe in the next theorem. Note that in this case it doesn't matter whether we think of left or right cosets.

Theorem 2   Let K be a normal subgroup of the group $(G, \bullet)$. Then

(a) $(K \bullet x) \bullet (K \bullet y) = K \bullet (x \bullet y)$ for all $x, y \in G$.

(b) The set G/K of cosets of K in G is a group under the operation $*$ defined by

$$(K \bullet x) * (K \bullet y) = K \bullet (x \bullet y).$$

(c) The function $v : G \to G/K$ defined by $v(x) = K \bullet x$ is a homomorphism with kernel K.

Proof. (a) By $(K \bullet x) \bullet (K \bullet y)$ we mean the set

$$K \bullet x \bullet K \bullet y = \{k_1 \bullet x \bullet k_2 \bullet y : k_1, k_2 \in K\}.$$

Since K is a subgroup of G, $K = K \bullet e \subseteq K \bullet K \subseteq K$, so $K = K \bullet K$. Since K is normal, we have $x \bullet K = K \bullet x$, and so

$$K \bullet x \bullet K \bullet y = K \bullet K \bullet x \bullet y = K \bullet (x \bullet y) \in G/K.$$

(b) According to (a), $(K \bullet x) * (K \bullet y)$ is $(K \bullet x) \bullet (K \bullet y)$ and thus $*$ is a well-defined binary operation on G/K. It is easy to check that it is associative, that K is the identity and that $K \bullet x^{-1} = (K \bullet x)^{-1}$.

(c) We have $v(x \bullet y) = K \bullet (x \bullet y) = (K \bullet x) * (K \bullet y) = v(x) * v(y)$ by definition of $*$, so v is a homomorphism. If $v(x) = v(e)$ then $x \in K \bullet x = K \bullet e = K$, and if $x \in K$ then $v(x) = K \bullet x = K = v(e)$. Thus K is the kernel of v. ∎

Theorems 1 and 2 together tell us that kernels of group homomorphisms are normal subgroups and, conversely, every normal subgroup is the kernel of some homomorphism.

EXAMPLE 6 Let $(G, \bullet) = (\mathbb{Z}, +)$ and let $K = 6\mathbb{Z}$. Theorem 2(b) tells us that $\mathbb{Z}/6\mathbb{Z}$ is a group under the operation

$$(6\mathbb{Z} + k) * (6\mathbb{Z} + m) = 6\mathbb{Z} + k + m.$$

The identity of $\mathbb{Z}/6\mathbb{Z}$ is $6\mathbb{Z}$.

Theorem 2(c) tells us that if $v(k) = 6\mathbb{Z} + k$, then v maps $\mathbb{Z}$ onto $\mathbb{Z}/6\mathbb{Z}$ and the kernel of v is $6\mathbb{Z}$, i.e.,

$$\{k \in \mathbb{Z} : v(k) = 6\mathbb{Z}\} = \{k \in \mathbb{Z} : 6\mathbb{Z} + k = 6\mathbb{Z}\} = 6\mathbb{Z}.$$

In Example 5(b) we observed that $6\mathbb{Z}$ is also the kernel of $\text{Rem}_6 = h$, which maps $(\mathbb{Z}, +)$ onto $(\mathbb{Z}(6), +_6)$. If we define

$$h^*(6\mathbb{Z} + k) = k \quad \text{for} \quad k \in \{0, 1, 2, 3, 4, 5\} = \mathbb{Z}(6),$$

we obtain a one-to-one correspondence of $\mathbb{Z}/6\mathbb{Z}$ onto $\mathbb{Z}(6)$. Moreover, h^* is a group isomorphism because

$$\begin{aligned} h^*((6\mathbb{Z} + k) * (6\mathbb{Z} + m)) &= h^*(6\mathbb{Z} + k + m) = h^*(6\mathbb{Z} + k +_6 m) \\ &= k +_6 m = h^*(6\mathbb{Z} + k) +_6 h^*(6\mathbb{Z} + m). \end{aligned}$$

Thus the groups $\mathbb{Z}/6\mathbb{Z}$ and $\mathbb{Z}(6)$ are isomorphic. Using our G and K notation, this says that G/K and $h(G)$ are isomorphic, and illustrates the next theorem. ∎

Theorem 3 Let h be a homomorphism from the group $(G, \bullet)$ to the group $(H, \square)$, with kernel K. Then G/K is isomorphic to $h(G)$ under the isomorphism h^* defined by

$$h^*(K \bullet x) = h(x).$$

Proof. If $K \bullet x = K \bullet y$ then $h(x) = h(y)$, so the mapping h^* is well-defined. That is, $h^*(K \bullet x)$ doesn't depend on the representative we use for this coset.

To see that h^* is one-to-one, suppose that $h^*(K \bullet x) = h^*(K \bullet y)$. Then $h(x) = h(y)$, and so $K \bullet x = K \bullet y$ by Theorem 1(c). So h^* is one-to-one. Its image is clearly $h(G)$, so all we need to do is to observe that h^* is a homomorphism:

$$\begin{aligned} h^*((K \bullet x) * (K \bullet y)) &= h^*(K \bullet (x \bullet y)) && \text{definition of } * \\ &= h(x \bullet y) && \text{definition of } h^* \\ &= h(x) \square h(y) && h \text{ is a homomorphism} \\ &= h^*(K \bullet x) \square h^*(K \bullet y) && \text{definition of } h^* \text{ again.} \quad \blacksquare \end{aligned}$$

If G is finite, then Lagrange's theorem in § 11.2 says that $|G/K| = |G|/|K|$. Since $|h(G)| = |G/K|$ by Theorem 3, we have the following.

Corollary   Let h be a homomorphism defined on a finite group G, with kernel K. Then $|h(G)| = |G|/|K|$. In particular, $|h(G)|$ divides $|G|$.

EXAMPLE 7   (a) Consider the mapping $h: \mathbb{Z}(30) \to \mathbb{Z}(30)$ defined by $h(n) = 6 *_{30} n$. Then h is an additive homomorphism with image $h(\mathbb{Z}(30)) = 6 *_{30} \mathbb{Z}(30) = \{0, 6, 12, 18, 24\}$, which has 5 elements. The kernel $K = \{0, 5, 10, 15, 20, 25\}$ has 6 elements and

$$|h(\mathbb{Z}(30))| = 5 = 30/6 = |\mathbb{Z}(30)|/|K|.$$

(b) The mapping $h: \mathbb{Z}(6) \to \mathbb{Z}(15)$ defined by $h(n) = 5 *_{15} n$ is an additive homomorphism, with $h(0) = h(3) = 0$, $h(1) = h(4) = 5$, $h(2) = h(5) = 10$. The image is $\{0, 5, 10\} = 5 *_{15} \mathbb{Z}(15)$. The kernel is $\{0, 3\}$. Sure enough, we have $|5 *_{15} \mathbb{Z}(15)| = 3$ while also $|\mathbb{Z}(6)|/|\{0, 3\}| = 6/2 = 3$.

(c) The group $\text{PERM}(\{1, 2, 3\})$, defined in Example 5 of § 11.2, has $3! = 6$ members. By the last corollary, its homomorphic images can only have 1, 2, 3 or 6 members. Actually, 3 is impossible, since no subgroup K of $\text{PERM}(\{1, 2, 3\})$ with just two members satisfies $K \circ g = g \circ K$ for all g. This fact is essentially Exercise 9(d) of § 11.2. ∎

The message of Theorem 3 is that to study homomorphic images of G, it is enough to look at the various groups G/K which can be made from cosets of certain subgroups K of G. The message of Theorem 2 is that we can identify the interesting subgroups; they are the ones satisfying the equations $K \bullet x = x \bullet K$ for all x.

Instead of comparing left and right cosets, we can test a subgroup for normality by looking at its **conjugates**, sets of the form $x \bullet K \bullet x^{-1}$. If K is a subgroup of G, then so is $x \bullet K \bullet x^{-1}$ for each x in G [Exercise 18 of § 11.2]. Moreover, $x \bullet K = K \bullet x$ if and only if $x \bullet K \bullet x^{-1} = K \bullet x \bullet x^{-1} = K$, so K is normal if and only if $K = x \bullet K \bullet x^{-1}$ for every x, i.e., if and only if all conjugates of K are equal to K. In fact, to show K normal it is enough to show $x \bullet K \bullet x^{-1} \subseteq K$ for all x in G, for in that case we have

$$K = x^{-1} \bullet (x \bullet K \bullet x^{-1}) \bullet x \subseteq x^{-1} \bullet K \bullet x \subseteq K$$

since x^{-1} is also in G.

EXERCISES 11.3

1. Which of the following functions h from $(\mathbb{Z}, +)$ to $(\mathbb{Z}, +)$ are homomorphisms?
(a) $h(n) = 6n$
(b) $h(n) = n + 1$
(c) $h(n) = -n$
(d) $h(n) = n^2$
(e) $h(n) = (6n^2 + 3n)/(2n + 1)$

2. Which of the following functions h are homomorphisms from $(\mathbb{P}, +)$ to $(\mathbb{P}, \cdot)$?

(a) $h(n) = 6^n$ (b) $h(n) = n$

(c) $h(n) = (-6)^n$ (d) $h(n) = n^2$

(e) $h(n) = 2^{n+1}$

3. Which of the homomorphisms in Exercise 1 are isomorphisms? Explain briefly.

4. Which of the homomorphisms in Example 1 are isomorphisms? Explain briefly.

5. Let $F = \text{FUN}(\mathbb{R}, \mathbb{R})$, and define $+$ on F by

$$(f + g)(x) = f(x) + g(x) \quad \text{for all} \quad x \in \mathbb{R}.$$

Define h from F to $\mathbb{R}$ by $h(f) = f(73)$.

(a) Show that h is a homomorphism of $(F, +)$ onto $(\mathbb{R}, +)$.

(b) Define $*$ on F by $(f * g)(x) = f(x) \cdot g(x)$ for all x. Show that h is a homomorphism of $(F, *)$ onto $(\mathbb{R}, \cdot)$.

6. Let Σ be the English alphabet. Define h on Σ^* by $h(w) = $ length of w. Explain why h is a homomorphism from Σ^* with its usual operation to $(\mathbb{N}, +)$.

7. Find the kernel of each of the following group homomorphisms h.

(a) From $(\mathbb{Z}, +)$ to $(\mathbb{Z}, +)$, defined by $h(n) = 73n$.

(b) From $(\mathbb{Z}, +)$ to $(\mathbb{Z}, +)$, defined by $h(n) = 0$ for all n.

(c) From $(\mathbb{Z}, +)$ to $(\mathbb{Z}(5), +_5)$, defined by $h(n) = \text{Rem}_5(n)$, the remainder on division of n by 5.

(d) From $(\mathbb{Z}, +)$ to $(\mathbb{Z}, +)$, defined by $h(n) = n$.

8. For each homomorphism in Exercise 7 describe the coset of the kernel which contains 73.

9. Suppose that h is a homomorphism defined on a group G and that $|G| = 12$ and $|h(G)| = 3$.

(a) Find $|K|$ where K is the kernel of h.

(b) How many members of G does h map onto each member of $h(G)$?

(c) What is $|G/K|$?

10. Define $+$ on $\text{FUN}(\mathbb{P}, \mathbb{Z})$ by

$$(f + g)(n) = f(n) + g(n) \quad \text{for all} \quad n \in \mathbb{P}.$$

Define h from $\text{FUN}(\mathbb{P}, \mathbb{Z})$ to $\mathbb{Z}$ by $h(f) = f(73)$.

(a) Verify that h is a homomorphism from $(\text{FUN}(\mathbb{P}, \mathbb{Z}), +)$ to $(\mathbb{Z}, +)$.

(b) Find the kernel of h.

11. The **direct product** of two groups $(G, \bullet)$ and $(H, \square)$ is the group $(G \times H, \triangle)$ with operation $\triangle$ defined by

$$\langle g_1, h_1 \rangle \triangle \langle g_2, h_2 \rangle = \langle g_1 \bullet g_2, h_1 \square h_2 \rangle \quad \forall g_1, g_2 \in G, \forall h_1, h_2 \in H.$$

(a) Describe the identity element of $G \times H$ and the inverse of an element $\langle g, h \rangle$ in $G \times H$.

(b) Verify that the mapping $f : G \times H \to G$ defined by $f(\langle g, h \rangle) = g$ is a homomorphism.

(c) Find the kernel of the homomorphism f in part (b).

(d) Find a normal subgroup of $G \times H$ which is isomorphic to H.

12. Define h from $(\mathbb{Z}, +)$ to $\mathbb{Z}(2) \times \mathbb{Z}(3)$ [see Exercise 11] by $h(n) = \langle n_2, n_3 \rangle$, where $n_k = \text{Rem}_k(n)$, the remainder on division of n by k, and the operation on $\mathbb{Z}(k)$ is $+_k$ for $k = 2, 3$.
    (a) Verify that h is a homomorphism.
    (b) Verify that $h(\mathbb{Z}) = \mathbb{Z}(2) \times \mathbb{Z}(3)$.
    (c) Find the kernel of h.
    (d) Show that $\mathbb{Z}(2) \times \mathbb{Z}(3) \simeq \mathbb{Z}(6)$.

13. Let H be the set of 2×2 matrices of the form $\begin{bmatrix} 1 & x \\ 0 & 1 \end{bmatrix}$, with matrix multiplication as operation.
    (a) Verify that H is a group, with

$$\begin{bmatrix} 1 & x \\ 0 & 1 \end{bmatrix}^{-1} = \begin{bmatrix} 1 & -x \\ 0 & 1 \end{bmatrix}.$$

    (b) Verify that

$$\begin{bmatrix} 0 & 1 \\ 1 & 0 \end{bmatrix} \cdot H \neq H \cdot \begin{bmatrix} 0 & 1 \\ 1 & 0 \end{bmatrix}.$$

    This shows that H is not a normal subgroup of the multiplicative group G of all 2×2 matrices that have inverses.
    (c) Show that H is a normal subgroup of the group T of all 2×2 matrices of the form

$$\begin{bmatrix} y & z \\ 0 & 1/y \end{bmatrix}, \quad y \neq 0,$$

    with multiplication as operation.
    (d) The mapping h from T to $(\mathbb{R} \setminus \{0\}, \cdot)$ defined by

$$h\left(\begin{bmatrix} y & z \\ 0 & 1/y \end{bmatrix}\right) = y$$

    is a group homomorphism. Find its kernel.
    (e) Show that T/H is isomorphic to the group of non-0 real numbers under multiplication.

14. An **antihomomorphism** from $(S, \bullet)$ to $(T, \square)$ is a function k such that

$$k(s_1 \bullet s_2) = k(s_2) \square k(s_1) \quad \text{for all} \quad s_1, s_2 \in S.$$

    (a) Use the result of Exercise 19 of § 4.1 to show that the transpose mapping k given by $k(\mathbf{A}) = \mathbf{A}^T$ is an antihomomorphism from $\mathfrak{M}_{n,n}$ under multiplication into itself. [This k could be called an **anti-isomorphism** since it is one-to-one and onto.]
    (b) Show that the mapping $x \to x^{-1}$ is always an antihomomorphism of a group onto itself.
    (c) Show that if k_1 and k_2 are antihomomorphisms for which the composition $k_1 \circ k_2$ is defined, then $k_1 \circ k_2$ is a homomorphism.
    (d) Give an example of an antihomomorphism which is also a homomorphism.

15. Show that if h is a homomorphism defined on the group G and if $h(g)$ has just one pre-image under h for some g in G, then h is one-to-one.

16. Show that if H and K are normal subgroups of a group $(G, \bullet)$, then $H \cap K$ is a normal subgroup. *Suggestion*: Consider $x \bullet (H \cap K) \bullet x^{-1}$.

17. Let H be a subgroup of $(G, \bullet)$.
    (a) Show that $\{x \in G : x \bullet H \bullet x^{-1} = H\}$ is a subgroup of G.
    (b) Conclude that if G is generated by a subset A and if $x \bullet H \bullet x^{-1} = H$ for all $x \in A$, then H is a normal subgroup.

18. Consider a finite group G with a subgroup H such that $|G| = 2|H|$. Show that H must be a normal subgroup of G. *Suggestion*: See Exercise 14 of § 11.2.

19. (a) Show that if S is a semigroup, if A generates S and if h is a homomorphism defined on S, then $h(A)$ generates $h(S)$.
    (b) Is the corresponding statement true for groups and group homomorphisms? Justify your answer.

20. (a) Show that if f is a homomorphism from a semigroup $(S, \bullet)$ to a semigroup $(T, \square)$, and if g is a homomorphism from $(T, \square)$ to a semigroup $(U, \triangle)$, then $g \circ f$ is a homomorphism.
    (b) Show that if h is an isomorphism of $(S, \bullet)$ onto $(T, \square)$, then h^{-1} is an isomorphism of $(T, \square)$ onto $(S, \bullet)$.
    (c) Use the results of parts (a) and (b) to show that the relation $\simeq$ is reflexive, symmetric and transitive.

21. An element z of a semigroup $(S, \bullet)$ is called a **zero element** or **zero** of S in case

$$z \bullet s = s \bullet z = z \quad \text{for all } s \text{ in } S.$$

    (a) Show that a semigroup cannot have more than one zero element.
    (b) Give an example of an infinite semigroup which has a zero element.
    (c) Give an example of a finite semigroup which has at least two members and which has a zero element.

22. Let z be a zero element of a semigroup $(S, \bullet)$ and let h be a homomorphism from $(S, \bullet)$ to a semigroup $(T, \square)$.
    (a) Show that $h(z)$ is a zero element of $h(S)$.
    (b) Must $h(z)$ be a zero element of $(T, \square)$? Justify your answer.

23. Suppose that $(S, \bullet)$ is a monoid with identity e and that h is a semigroup homomorphism from $(S, \bullet)$ to $(T, \square)$.
    (a) Show that $(h(S), \square)$ is a monoid with identity $h(e)$.
    (b) Must $h(e)$ be an identity of $(T, \square)$? Justify your answer.

§ 11.4   Symmetry and Groups of Permutations

Commutative groups frequently arise in settings in which the binary operation is addition. Noncommutative groups, on the other hand, commonly occur where the operation is composition of functions. They are especially useful in describing the symmetry of other objects, such as graphs or digraphs.

EXAMPLE 1 Recall from § 8.2 that an automorphism of a digraph D is a graph isomorphism of D onto itself. One way to describe the symmetry of a digraph, such as the ones shown in Figure 1, is to list its automorphisms. We noted in § 8.2 that the composition of two digraph isomorphisms is also an isomorphism. And it's easy to check that the inverse of an isomorphism is also an isomorphism. This means that the set Aut(D) of automorphisms of a digraph

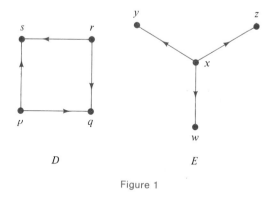

Figure 1

D is a group, with composition of functions as operation. In the case of the digraph D in Figure 1, which has two sources and two sinks, the automorphisms must send sources to sources and sinks to sinks. We can check that there are four different automorphisms, whose function values are given in Figure 2(a). The table in Figure 2(b) gives the products in the group (Aut(D), ∘). This group happens to be commutative.

	p	q	r	s
e	p	q	r	s
f	p	s	r	q
g	r	q	p	s
h	r	s	p	q

	e	f	g	h
e	e	f	g	h
f	f	e	h	g
g	g	h	e	f
h	h	g	f	e

Aut(D)

(a)   (b)

Figure 2

This section contains several tables, such as the ones in Figure 2, which contain information about groups. It is a good idea on first reading to look quickly at the tables to see what kinds of information they contain, and perhaps to verify a few entries. Later on, especially when you are working the exercises, you can use the tables instead of making the calculations yourself.

The digraph E in Figure 1 has another kind of symmetry and a different automorphism group. We see that every automorphism of E must send x to itself but otherwise can permute the remaining vertices in any way. Figure 3 gives the function values of the automorphisms of E and the table of $(\text{Aut}(E), \circ)$. The group $\text{Aut}(E)$ is essentially the same as—i.e., is isomorphic to—the noncommutative group $\text{PERM}(\{y, z, w\})$, whose function values we get by suppressing the x-column in the table of values of $\text{Aut}(E)$. We saw the table for $(\text{Aut}(E), \circ)$ before in Exercise 11 of § 11.2. Except for the names of the vertices, the group $\text{PERM}(\{y, z, w\})$ appears in Exercise 9 of § 11.2 and elsewhere. Every member of $\text{Aut}(E)$ fixes the vertex x. The subgroup of $\text{Aut}(E)$ fixing y is $\{e, c\}$. The subgroups $\{e, d\}$ and $\{e, f\}$ fix w and z, respectively. ∎

	x	y	z	w
e	x	y	z	w
a	x	z	w	y
b	x	w	y	z
c	x	y	w	z
d	x	z	y	w
f	x	w	z	y

$\circ$	e	a	b	c	d	f
e	e	a	b	c	d	f
a	a	b	e	d	f	c
b	b	e	a	f	c	d
c	c	f	d	e	b	a
d	d	c	f	a	e	b
f	f	d	c	b	a	e

$$\text{Aut}(E) \simeq \text{PERM}(\{y, z, w\})$$

Figure 3

The groups which arise in analyzing symmetry are generally groups of permutations of sets. For instance, the automorphism group $\text{Aut}(D)$ in Example 1 consisted of certain permutations of the set $V(D)$ of vertices of D. The group $\text{PERM}(V(D))$ has $4! = 24$ members, but only four of them are in its subgroup $\text{Aut}(D)$.

Groups of permutations may appear to be rather special kinds of groups. Our first theorem says they are not; in a sense, every group is a permutation group.

Theorem 1 Let $(G, \square)$ be a group. Then G is isomorphic to a group of permutations of
(Cayley's the set G itself.
Theorem)

Proof. We need to set up a one-to-one homomorphism λ of $(G, \square)$ into $(\text{PERM}(G), \circ)$. Then G will be isomorphic to the image $\lambda(G)$, which will be a subgroup of $\text{PERM}(G)$.

To each g in G we associate a function g^* from G to G defined by

$$g^*(x) = g \square x \quad \forall x \in G.$$

Since G is closed under $\square$, $g \square x$ is in G, so g^* does map G into G. Moreover

$$((g^{-1})^* \circ g^*)(x) = (g^{-1})^*(g^*(x)) = g^{-1} \square (g \square x) = x$$

for every x, and thus $(g^{-1})^* \circ g^*$ is the identity function on G. So is $g^* \circ (g^{-1})^*$ for a similar reason. Thus $(g^{-1})^*$ is the inverse of the function g^*. Since g^* has an inverse, it follows from the theorem of § 3.2 that g^* must be one-to-one and onto. Thus g^* belongs to PERM(G).

We now define $\lambda: G \to \text{PERM}(G)$ by $\lambda(g) = g^*$ for all g in G. If g and h are in G, then $\lambda(g \square h) = (g \square h)^*$. But

$$(g \square h)^*(x) = (g \square h) \square x = g \square (h \square x) = g^*(h^*(x)) = (g^* \circ h^*)(x)$$

for every x in G, so $(g \square h)^* = g^* \circ h^*$. Thus

$$\lambda(g \square h) = g^* \circ h^* = \lambda(g) \circ \lambda(h),$$

and hence λ is a homomorphism from G into PERM(G).

To check that λ is one-to-one, observe that if e is the identity element of G then

$$(\lambda(g))(e) = g^*(e) = g \square e = g,$$

so if $\lambda(g) = \lambda(h)$ then

$$g = (\lambda(g))(e) = (\lambda(h))(e) = h. \quad \blacksquare$$

Groups can be pretty complicated, and so Cayley's theorem can be viewed as saying that permutation groups must be pretty complicated. There is often some benefit to be gained, however, from taking a permutation group perspective.

Consider a group G of permutations, so that G is a subgroup of (PERM(S), $\circ$) for some set S. We sometimes say in this case that G **acts on** S. For g in G and s in S, it is common to write gs instead of $g(s)$. For each s in S, the subset $\{gs : g \in G\}$ of S is called the **orbit** of s under G, and is denoted by Gs. It consists of all members of S that s can be taken to by the various members of G. Since the identity e of PERM(S) belongs to G, we see that $s = es \in Gs$. In particular, S is the union of all of the orbits under G.

EXAMPLE 2 (a) The group Aut(E) of Example 1 is a subgroup of PERM($\{x, y, z, w\}$). The orbit of y is

$$\text{Aut}(E)y = \{e(y), a(y), b(y), c(y), d(y), f(y)\}$$
$$= \{y, z, w, y, z, w\} = \{y, z, w\}.$$

Similarly Aut(E)z = Aut(E)w = $\{y, z, w\}$, and

$$\text{Aut}(E)x = \{e(x), a(x), b(x), c(x), d(x), f(x)\}$$
$$= \{x, x, x, x, x, x\} = \{x\}.$$

(b) The group Aut(D) of Example 1 acts on the set $\{p, q, r, s\}$ with orbits

$$\text{Aut}(D)p = \text{Aut}(D)r = \{p, r\} = \{x : x \text{ is a source of } D\}$$
$$\text{Aut}(D)q = \text{Aut}(D)s = \{q, s\} = \{x : x \text{ is a sink of } D\}. \quad \blacksquare$$

Suppose now that G acts on the set S and that T is a subset of S. Then the set $G_T = \{g \in G : g(T) = T\}$ is a subgroup of G, since if $g, h \in G_T$ then $(g \circ h)(T) = g(h(T)) = g(T) = T$, and also $g^{-1}(T) = g^{-1}(g(T)) = T$ so that G_T is closed under composition and inverses. If T consists of a single element s of S, then we write G_s in place of $G_{\{s\}}$. Thus $G_s = \{g \in G : gs = s\}$ is a subgroup of G. There is a link between the subgroup G_s of G and the orbit $Gs \subseteq S$.

Theorem 2 Suppose the group G acts on the finite set S. For each s in S we have

$$|Gs| = |G/G_s|.$$

Hence $|Gs|$ divides $|G|$.

Proof. Recall that $|G/G_s|$ is the number of cosets of the subgroup G_s in G. Consider first a chosen element g in G. For h in G we have

$$gs = hs \Leftrightarrow s = g^{-1}(gs) = g^{-1}(hs) = (g^{-1} \circ h)s$$
$$\Leftrightarrow g^{-1} \circ h \in G_s \Leftrightarrow h = g \circ (g^{-1} \circ h) \in g \circ G_s.$$

That is, h in G takes s to gs precisely when $h \in g \circ G_s$. If we switch to a different coset $g' \circ G_s$, all of its members take s to $g's$, which must be an element of the orbit Gs different from gs. In this way, each coset $g \circ G_s$ corresponds to a unique element gs of Gs, so Gs and G/G_s have the same number of elements. By Lagrange's theorem of § 11.2, $|G/G_s|$ divides $|G|$. $\quad \blacksquare$

Corollary Suppose the group G acts on S as in Theorem 2. Choose $s_1, \ldots, s_m$ in S with exactly one s_i in each orbit Gs of G on S. Then we have

$$|S| = \sum_{i=1}^{m} |Gs_i| = \sum_{i=1}^{m} |G/G_{s_i}|.$$

Proof. The second equality follows from Theorem 2, so we just need to show that the orbits partition S. Suppose $Gs \cap Gt \neq \varnothing$ and let $u \in Gs \cap Gt$. Then $u = gs = ht$ for some $g, h \in G$, so $s = g^{-1}(ht) = (g^{-1} \circ h)t \in Gt$ and thus $Gs \subseteq G(Gt) \subseteq (G \circ G)t \subseteq Gt$. Similarly $Gt \subseteq Gs$. In other words, $Gs = Gt$ if $Gs \cap Gt \neq \varnothing$. $\quad \blacksquare$

This corollary can be useful in the following sort of situation. Suppose that we know, for some reason, that every subgroup of G [except G itself] has an even number of cosets, and that $|S|$ is odd. The numbers $|G/G_{s_i}|$ cannot all be even, since by the corollary their sum is odd, so some G_{s_i} must be G itself. That is, there is an s_i in S such that $Gs_i = \{s_i\}$, which means that every g in G fixes s_i.

Here is another setting in which the corollary can be useful. Consider a group G of permutations on a finite set S. Then S breaks up into the disjoint union $S = Gs_1 \cup \cdots \cup Gs_m$ of orbits under G. We can view G as acting on each orbit separately, as follows. For g in G and $i = 1, 2, \ldots, m$, define the function g_i on Gs_i by $g_i(t) = gt$ for each t in Gs_i. Then $g_i(Gs_i) = gGs_i \subseteq Gs_i$, so g_i maps the orbit Gs_i back into itself; we call g_i the **restriction** of g to Gs_i. Since g is a permutation of S, each g_i is one-to-one. Since Gs_i is finite, g_i must be onto Gs_i and hence $g_i \in \text{PERM}(Gs_i)$. Moreover, for $g, h \in G$ and $t \in Gs_i$ we have

$$(g \circ h)_i(t) = (g \circ h)t = g(ht) = g_i(h_i(t)) = (g_i \circ h_i)(t),$$

so $(g \circ h)_i = g_i \circ h_i$. In other words, the mapping $\lambda_i : g \to g_i$ is a homomorphism of G into $\text{PERM}(Gs_i)$. In a typical case, the sets Gs_i will be smaller than S, and we will be able to analyze the action of $\lambda_i(G)$ on each orbit Gs_i more easily than the action of G on S itself. Moreover, we will have lost nothing by breaking S up in this way, since the action of any member of G on S can be reconstructed from a knowledge of how its restrictions act on the orbits.

EXAMPLE 3 Consider the graph H in Figure 4. Let $G = \text{Aut}(H)$. Since graph automorphisms must preserve degrees of vertices, the orbit of w under G is contained in $\{u, w, x, y\}$. We can see by inspection that there actually are automorphisms of H taking w to any one of u, w, x, y, so the orbit is $Gw = \{u, w, x, y\}$. Similarly, we have $Gs = \{s, t\}$ and $Gr = \{r\}$.

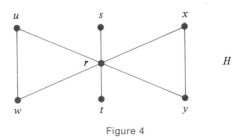

Figure 4

Consider an automorphism g of H. There are 4 choices for $g(w)$. Each such choice determines $g(u)$, but there are still two possibilities for $g(x)$. All told, there are 8 possible permutations g can produce on the orbit $\{u, w, x, y\}$. Moreover, for each of these 8 choices there are still two choices for $g(s)$. Thus $|G| = 16$. By Lagrange's theorem, the number of cosets of every subgroup of G divides 16, so all subgroups except G itself have an even number of cosets. The graph H has an odd number of vertices and sure enough, just as the corollary of Theorem 2 predicts, there is a vertex which is sent to itself by all members of $\text{Aut}(H)$.

The table in Figure 5(a) lists the sixteen members of G. The tables in Figures 5(b), 5(c) and 5(d) list their restrictions to the three orbits of G. Note the repetitions of blocks of rows in Figures 5(b), 5(c) and 5(d).

r	s	t	u	w	x	y
r	s	t	u	w	x	y
r	s	t	w	u	x	y
r	s	t	u	w	y	x
r	s	t	w	u	y	x
r	s	t	x	y	u	w
r	s	t	x	y	w	u
r	s	t	y	x	u	w
r	s	t	y	x	w	u
r	t	s	u	w	x	y
r	t	s	w	u	x	y
r	t	s	u	w	y	x
r	t	s	w	u	y	x
r	t	s	x	y	u	w
r	t	s	x	y	w	u
r	t	s	y	x	u	w
r	t	s	y	x	w	u

r	s	t	u	w	x	y
r	s	t	u	w	x	y
r	s	t	w	u	x	y
r	s	t	u	w	y	x
r	s	t	w	u	y	x
r	s	t	x	y	u	w
r	s	t	x	y	w	u
r	s	t	y	x	u	w
r	s	t	y	x	w	u
r	t	s	u	w	x	y
r	t	s	w	u	x	y
r	t	s	u	w	y	x
r	t	s	w	u	y	x
r	t	s	x	y	u	w
r	t	s	x	y	w	u
r	t	s	y	x	u	w
r	t	s	y	x	w	u

$G = \text{Aut}(H)$ Restrictions to the orbits

(a)   (b)   (c)   (d)

Figure 5

The first eight automorphisms in this table form the subgroup $G_s = G_t$. The first four listed form $G_s \cap G_{\{u, w\}}$. We see from Figures 5(c) and 5(d) that there are just two different possible restrictions of an element of Aut(H) to $\{s, t\}$, and just eight to the set $\{u, w, x, y\}$.

Let λ_1, λ_2, and λ_3 be the homomorphisms of G into PERM($\{r\}$), PERM($\{s, t\}$) and PERM($\{u, w, x, y\}$) induced by restriction. Then λ_1 is boring: everything maps to the identity. The mapping λ_2 is less dull; its kernel is the normal subgroup $G_s \cap G_t = G_s$. The kernel of λ_3 is

$$G_u \cap G_w \cap G_x \cap G_y = G_u \cap G_x,$$

which has just two elements, shown in the first and ninth rows of Figure 5. Thus

$$|\lambda_1(G)| = 1, \quad |\lambda_2(G)| = \tfrac{16}{8} = 2 \quad \text{and} \quad |\lambda_3(G)| = \tfrac{16}{2} = 8.$$

An element g in G is completely determined by the triple $\langle \lambda_1(g), \lambda_2(g), \lambda_3(g) \rangle$ in

$$\text{PERM}(\{r\}) \times \text{PERM}(\{s, t\}) \times \text{PERM}(\{u, w, x, y\}).$$

There are $1 \cdot 2 \cdot 24 = 48$ such triples, but only 16 of them correspond to automorphisms of H. ∎

Much more can be said about permutation groups, but to go farther we would need to introduce a good deal of specialized notation. This section has aimed primarily at introducing groups as ways of describing symmetry and at giving a bit of the flavor of permutation actions of groups on sets. In the next section we pursue the ideas of the corollary to Theorem 2. For more on the general subject one can consult an introductory textbook on abstract algebra which begins with an account of groups.

EXERCISES 11.4

1. (a) How many automorphisms does the binary tree T_1 in Figure 6 have?
   (b) Repeat part (a) for the tree T_2 in Figure 6.
   (c) Show that every subgroup of Aut(T_2), except Aut(T_2) itself, has an even number of cosets.
   (d) Find a vertex of T_2 which is sent into itself by every automorphism of T_2. Is such a vertex guaranteed by the corollary to Theorem 2?

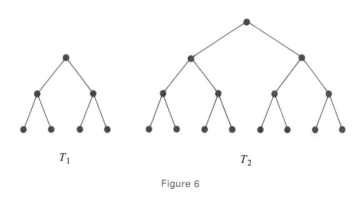

T_1 T_2

Figure 6

2. Consider the group $G = \text{Aut}(E)$ in Example 2, and note that Gy, Gz, Gw and Gx are given in that example.
   (a) Use Figure 3 to verify that $G_y = \{e, c\}$.
   (b) As in part (a), determine G_z, G_w and G_x.
   (c) Verify Theorem 2 for $s = y$, z, w and x.
   (d) Verify the corollary to Theorem 2 in this case.

3. Consider the tree T shown in Figure 7. Observe that a table for the group Aut(T) of automorphisms can be obtained from Figure 5(a) by adding suitable columns for the vertices p, q, v and z.
   (a) How would this table change if a new vertex m were added in the middle of the edge from p to q?
   (b) The graph obtained in part (a) has an even number of vertices. Are any of them fixed by all automorphisms of the graph? Explain.
   (c) What are the orbits of Aut(T) acting on the set $V(T)$ of vertices of T?

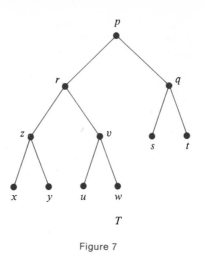

Figure 7

4. (a) Show how to construct, for each n in $\mathbb{P}$, a tree T such that $|\text{Aut}(T)| = 2^n$. *Hint*: Keep attaching suitable graphs with two automorphisms.
(b) Show how to construct, for each n in $\mathbb{P}$, a digraph D such that $|\text{Aut}(D)| = n$.

5. (a) Find all subgroups of the group $\text{Aut}(D)$ described in Figure 2.
(b) Which of these subgroups are normal subgroups of $\text{Aut}(D)$?
(c) Find the kernel of each restriction mapping λ_i of $\text{Aut}(D)$ on its orbits in $V(D)$.
(d) What is the intersection of the kernels in part (c)?

6. Repeat Exercise 5 for the digraph E of Figure 1; its automorphism group is given in Figure 3.

7. The group G whose members are listed in Figure 5(a) has a subgroup G_u with four elements.
(a) Make a table showing the members of G_u.
(b) Let g be the fourth automorphism in Figure 5(a). Make a table showing the members of the coset $g \circ G_u$.
(c) Find the image of u under each member of the coset in part (b).

8. A group G acting on a set S is said to act **transitively** on S if $S = Gs$ for some s in S. Suppose G acts transitively on S.
(a) Show that $S = Gs$ for every s in S.
(b) Show that if G is finite then so is S, and $|G| = |G_s| \cdot |S|$ for each s in S.

9. What can you say about the sizes of the orbits of a group with 27 members?

10. Suppose the group G acts on the set S and H is a subgroup of G. Then H also acts on S. Show that each orbit of G on S is a union of orbits of H.

11. Let G be a group acting on a set S. Define

$$R = \{\langle s, t \rangle \in S \times S : gs = t \text{ for some } g \in G\}.$$

(a) Show that R is an equivalence relation on S.
(b) Describe the partition π of S corresponding to R.

12. This exercise refers to Example 3.

(a) Determine the numbers $|G_u|, |G_x|, |G_u \cap G_x|$ and $|G_{\{u, x\}}|$.

(b) Give an example of a triple in

$$\text{PERM}(\{r\}) \times \text{PERM}(\{s, t\}) \times \text{PERM}(\{u, w, x, y\})$$

that doesn't correspond to an automorphism of H.

13. This exercise shows that every monoid is isomorphic to a monoid of functions from a set into itself. Consider a monoid $(M, \square)$ with identity e. For each m in M, define $m^*(x) = m \square x$ for each x in M.

(a) Show that $m^* \in \text{FUN}(M, M)$.

(b) Define $\lambda : M \to \text{FUN}(M, M)$ by $\lambda(m) = m^*$. Show that λ is one-to-one.

(c) Show that λ is a homomorphism of $(M, \square)$ into $(\text{FUN}(M, M), \circ)$.

14. Consider a semigroup $(S, *)$ and an element z not in S. Define $\square$ on $S \cup \{z\}$ by

$$x \square y = x * y \quad \text{for} \quad x, y \in S,$$

$$x \square z = z \square x = x \quad \text{for} \quad x \in S,$$

$$z \square z = z.$$

Show that $(S \cup \{z\}, \square)$ is a monoid with identity z.

This exercise and Exercise 13 together show that every semigroup is isomorphic to a semigroup of functions from a set to itself. First stick S into $S \cup \{z\}$, and then map $S \cup \{z\}$ into $\text{FUN}(S \cup \{z\}, S \cup \{z\})$.

15. For any set $S, x * y = y$ defines a semigroup $(S, *)$, as noted in Example 3 of § 11.2. Let $S = \{a, b, c\}$ and apply Exercise 14 to find a semigroup of functions isomorphic to $(S, *)$.

§ 11.5 Applications of Groups to Counting

Consider the problem of manufacturing logical circuits to compute all the different Boolean functions of four variables. On the one hand, since there are $2^4 = 16$ rows in a truth table for such a function, there are $2^{16} = 65,536$ such functions. On the other hand, just by switching the input leads around we can make one circuit compute various different functions, so we don't need as many circuits as there are functions. If we are willing to use external gates to complement inputs and outputs, we need to manufacture still fewer circuits. How many can we get by with?

Or consider coloring the faces of a cube, with only three colors allowed. If one face is red and the rest are blue, it doesn't really matter which face is red, since we can rotate the cube to put the red face wherever we want. Given two ways of coloring the faces with one red face, three blue faces and two green faces, it may or may not be possible to rotate the cube so that the two colorings are really the same. How many essentially different ways are there to color the cube?

Questions like these ask us to take symmetry into consideration somehow in counting distinct possibilities. In this section we see how groups, which describe symmetry, can help us find answers. We begin with a somewhat simpler situation and develop methods which we can then apply to problems such as the Boolean function and cube-coloring questions.

This section contains some fairly complicated formulas and calculations. We have done a number of the calculations for you, but there are still places where it is pretty slow going. Have courage. Careful reading of the examples will pay off in understanding of the key ideas.

Consider a group G acting [as permutations] on a set S. In § 11.4 we looked at the set G_s of elements of G which fixed a given member s of S. Now we switch perspective and for each g in G consider the set

$$F(g) = \{s \in S : gs = s\}$$

consisting of all members of S which g fixes. These sets are connected with the orbits of G on S in a surprising way.

Theorem 1   Let G be a finite group acting on a set S. The number of orbits of G on S is

$$\frac{1}{|G|}\left(\sum_{g \in G} |F(g)|\right).$$

The sum here has one term for each g in G. Before discussing the proof we look at what the theorem means.

EXAMPLE 1   (a) Consider the graph shown in Figure 1(a). Its symmetry is described by its group G of automorphisms, which is listed in the table in Figure 1(b). The automorphism group has eight members, and acts on the

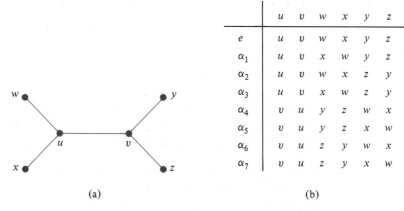

	u	v	w	x	y	z
e	u	v	w	x	y	z
α_1	u	v	x	w	y	z
α_2	u	v	w	x	z	y
α_3	u	v	x	w	z	y
α_4	v	u	y	z	w	x
α_5	v	u	y	z	x	w
α_6	v	u	z	y	w	x
α_7	v	u	z	y	x	w

(a)   (b)

Figure 1

6-element set $V = \{u, v, w, x, y, z\}$. To see which members of V each auto-morphism fixes, we look in the table.

$$|F(e)| = |\{u, v, w, x, y, z\}| = 6,$$

$$|F(\alpha_1)| = |\{u, v, y, z\}| = 4,$$

$$|F(\alpha_2)| = |\{u, v, w, x\}| = 4,$$

$$|F(\alpha_3)| = |\{u, v\}| = 2,$$

$$|F(\alpha_4)| = |F(\alpha_5)| = |F(\alpha_6)| = |F(\alpha_7)| = |\varnothing| = 0.$$

According to Theorem 1 the automorphism group has

$$\tfrac{1}{8}(6 + 4 + 4 + 2 + 0 + 0 + 0 + 0) = 2$$

orbits. In fact, we can see from the picture that there are exactly two orbits, namely $\{w, x, y, z\}$ and $\{u, v\}$, so Theorem 1 confirms our observation.

(b) The group G in part (a) has actions on other sets besides V. For example, G can be viewed as a group of permutations of the set E of edges of the graph; see Figure 2. From the table we find that $|F(e)| = 5$, $|F(\alpha_1)| = |F(\alpha_2)| = 3$ and $|F(\alpha_i)| = 1$ for $i \geq 3$. Theorem 1 says that E consists of

$$\tfrac{1}{8}(5 + 3 + 3 + 1 + 1 + 1 + 1 + 1) = 2$$

orbits under G. Indeed, the orbits are $\{e_3\}$ and $\{e_1, e_2, e_4, e_5\}$.

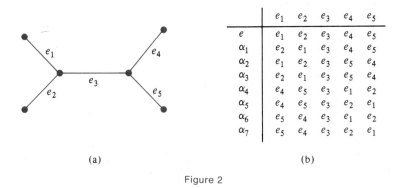

	e_1	e_2	e_3	e_4	e_5
e	e_1	e_2	e_3	e_4	e_5
α_1	e_2	e_1	e_3	e_4	e_5
α_2	e_1	e_2	e_3	e_5	e_4
α_3	e_2	e_1	e_3	e_5	e_4
α_4	e_4	e_5	e_3	e_1	e_2
α_5	e_4	e_5	e_3	e_2	e_1
α_6	e_5	e_4	e_3	e_1	e_2
α_7	e_5	e_4	e_3	e_2	e_1

(a)   (b)

Figure 2

(c) The group G also acts on the set T of all two-element subsets of V if we define $g(\{a, b\}) = \{g(a), g(b)\}$ for each graph automorphism g and for all vertices $a, b \in V$. This time $|F(e)| = |T| = \binom{6}{2} = 15$. Now α_1 sends each of the 2-element subsets of $\{u, v, y, z\}$ back to itself and also fixes $\{w, x\}$, since $\alpha_1(\{w, x\}) = \{\alpha_1(w), \alpha_1(x)\} = \{x, w\} = \{w, x\}$.   Thus $|F(\alpha_1)| = \binom{4}{2} + 1 = 7$. Similarly, $|F(\alpha_2)| = 7$. Since α_3 only fixes $\{u, v\}$, $\{w, x\}$ and $\{y, z\}$, we have $|F(\alpha_3)| = 3$. In the same way, we find that $F(\alpha_4) = \{\{u, v\}, \{w, y\}, \{x, z\}\}$, $F(\alpha_5) = \{\{u, v\}\} = F(\alpha_6)$, and $F(\alpha_7) = \{\{u, v\}, \{w, z\}, \{x, y\}\}$.

Theorem 1 then says that T consists of

$$\tfrac{1}{8}(15 + 7 + 7 + 3 + 3 + 1 + 1 + 3) = 5$$

orbits under G. This was not so obvious to begin with. Now we can observe that $\{w, u\}, \{w, v\}, \{w, x\}, \{w, y\}$ and $\{u, v\}$ belong to five different orbits. [Look at Figure 1(a) to see that none of these subsets can be mapped to another one by automorphisms of the graph.] If we had been asked originally to find a representative of each orbit, we would know we were done when we had exhibited these five subsets. ∎

Proof of Theorem 1. We have G acting on S, and we need to show that the number of orbits of G on S equals $\dfrac{1}{|G|} \sum_{g \in G} |F(g)|$. To do this, we consider the set

$$W = \{\langle g, s \rangle \in G \times S : gs = s\},$$

which we count in two different ways.

First, for each g in G we count pairs $\langle g, s \rangle$ with $gs = s$—there are $|F(g)|$ of them—and then add the answers. We get $|W| = \sum_{g \in G} |F(g)|$. We can also count the members of W by counting for each $s \in S$ the set $\{g \in G : gs = s\}$, and then adding the answers for all values of s. For a single value of s the set $\{g \in G : gs = s\}$ is the subgroup G_s of G fixing s. By Theorem 2 of § 11.4, the size of the orbit Gs of s is given by $|Gs| = |G/G_s| = |G|/|G_s|$. Thus $|G_s| = |G|/|Gs|$, and so

$$|W| = \sum_{s \in S} \frac{|G|}{|Gs|} = |G| \cdot \sum_{s \in S} \frac{1}{|Gs|}.$$

Now we group together the terms in the sum which come from a given orbit Gs. Since $Gs = Gs'$ for each s' in Gs,

$$\sum_{s' \in Gs} \frac{1}{|Gs'|} = \sum_{s' \in Gs} \frac{1}{|Gs|} = |Gs| \cdot \frac{1}{|Gs|} = 1.$$

That is, the orbit contributes a total value of 1 to $\sum_{s \in S} \dfrac{1}{|Gs|}$. Thus if there are m orbits then

$$\sum_{s \in S} \frac{1}{|Gs|} = 1 + \cdots + 1 = m,$$

and we get $|W| = |G| \cdot m$.

It follows that $|G| \cdot m = |W| = \sum_{g \in G} |F(g)|$, so that

$$m = \frac{1}{|G|} \sum_{g \in G} |F(g)|,$$

as claimed in the theorem. ∎

Since there are $|G|$ terms in the sum $\sum\limits_{g \in G} |F(g)|$, when we divide the sum by $|G|$ we obtain the average value of $|F(g)|$ over all members g of G. If some values are larger than average, then some must be smaller.

Corollary   If G acts transitively on S, i.e., if there is just one orbit of G on S, and if $|S| > 1$, then G contains an element g such that $gs \neq s$ for all $s \in S$.

Proof. By Theorem 1, the average value of $|F(g)|$ is 1, since there is just one orbit. Moreover, if e is the identity of G then $|F(e)| = |S| > 1$. Thus $|F(g)| < 1$ for at least one g, and for such a g $\{s \in S : gs = s\} = F(g) = \varnothing$. ∎

Now let us go back to the question of coloring a cube with colors red, blue and green. A natural group to consider is the group G of all rotations of space which send the cube back onto itself, since this group describes the rotational symmetry of the cube. The group acts on the set of faces of the cube, and it also acts on a more complicated set which we now describe. Imagine the cube held in some position. A **coloring** of the cube is then simply a way of assigning a color to each of the six faces of the cube, i.e., a *function f* from the set F of faces to the set C of colors. We want to regard two colorings as essentially the same in case rotating the cube takes one into the other. To describe mathematically what we mean by this statement, consider a coloring $f : F \to C$ and a rotation g in G. Then $f \circ g : F \to C$ defined by $(f \circ g)(X) = f(g(X))$ for each face X is also a coloring of the cube [simply rotate by g and then color faces according to f]. We regard f and f' as equivalent in case $f' = f \circ g$ for some rotation $g \in G$. We can view G as acting on the set $\mathrm{FUN}(F, C)$ of all colorings by having g take f to $f \circ g^{-1}$. [The inverse is necessary here for technical reasons, to give $(g_1 \circ g_2)(f) = g_1(g_2(f))$, but is irrelevant to the main ideas.] Then the equivalence classes of colorings are just the orbits of $\mathrm{FUN}(F, C)$ under G, and we can hope to count them using Theorem 1. To apply the theorem we need to count the colorings fixed by each member of G. The method we will use applies quite widely, so instead of answering the cube-coloring question just now, we first prove a general theorem and then apply it in Example 3 to the special case of the cube.

Theorem 2   Consider a finite group G acting on a set S. Then G also acts on $\mathrm{FUN}(S, C)$ for any set C if we define $g(f) = f \circ g^{-1}$ for $g \in G$ and $f : S \to C$. For each g in G, let $m(g)$ be the number of orbits of the cyclic group $\langle g \rangle$ on S. Then the number of orbits of G on $\mathrm{FUN}(S, C)$ is

$$\frac{1}{|G|} \sum_{g \in G} |C|^{m(g)}.$$

Proof. The action of G on $\mathrm{FUN}(S, C)$ is like the action of the rotation group on the set of face colorings of the cube which we just considered.

According to Theorem 1, to show the formula is correct we just need to show that for each $g \in G$ we have

$$|\{f \in \text{FUN}(S, C) : g(f) = f\}| = |C|^{m(g)}.$$

Now $g(f) = f \circ g^{-1}$, so $g(f) = f$ if and only if $(f \circ g^{-1})(s) = f(s)$ for every $s \in S$, which is true if and only if $f(g^{-1}(s)) = f(s)$ for every s. Replacing $g^{-1}(s)$ by t gives $g(f) = f$ if and only if $f(t) = f(g(t))$ for every $t \in S$, i.e., if and only if for each t the values $f(t), f(g(t)), f(g^2(t)), f(g^3(t)), \ldots$ and $f(g^{-1}(t)), f(g^{-2}(t)), \ldots$ are all the same. That is, $g(f) = f$ if and only if the function f is constant on each orbit $\{g^n(t) : n \in \mathbb{Z}\}$ under $\langle g \rangle$. To describe such an f we simply give its value on each $\langle g \rangle$-orbit. There are $|C|$ possible function values and $m(g)$ orbits under $\langle g \rangle$, so there are $|C|^{m(g)}$ functions f which are constant on $\langle g \rangle$-orbits. The theorem follows. ∎

A coloring of a set S with colors from C is just a function from S to C, and we regard two colorings as equivalent under the action of G if they belong to the same G-orbit $\{f \circ g : g \in G\}$ in $\text{FUN}(S, C)$. The G-orbits are the G-equivalence classes of colorings. With this terminology, Theorem 2 gives the following information.

Theorem 3 Consider a finite group G acting on a set S. For each positive integer k, let $C(k)$ be the number of [G-equivalence classes of] colorings of S using a set of k colors. Then

$$C(k) = \frac{1}{|G|} \sum_{g \in G} k^{m(g)}$$

where $m(g)$ is the number of orbits of $\langle g \rangle$ on S.

EXAMPLE 2 Before we get back to the cube, we color the vertices of a square with k colors, regarding two colorings as the same if we can turn one into the other by a suitable rotation of the square or by flipping it over. Figure 3(a) shows the

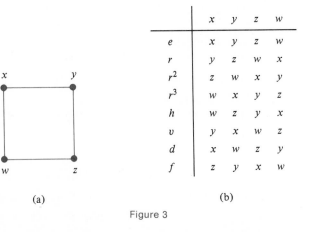

	x	y	z	w
e	x	y	z	w
r	y	z	w	x
r^2	z	w	x	y
r^3	w	x	y	z
h	w	z	y	x
v	y	x	w	z
d	x	w	z	y
f	z	y	x	w

(a)　　　　　　　　　　　　(b)

Figure 3

square, and Figure 3(b) lists the relevant group of permutations of its vertex set. One can check that $\langle e \rangle = \{e\}$, $\langle r \rangle = \langle r^3 \rangle = \{e, r, r^2, r^3\}$, $\langle r^2 \rangle = \{e, r^2\}$, $\langle h \rangle = \{e, h\}$, $\langle v \rangle = \{e, v\}$, $\langle d \rangle = \{e, d\}$ and $\langle f \rangle = \{e, f\}$. The table in Figure 4 lists the orbits of these cyclic subgroups. For example, since $\langle f \rangle = \{e, f\}$ the orbits of $\langle f \rangle$ are the sets $\{e(s), f(s)\}$ for s in $\{x, y, z, w\}$, so they are $\{e(x), f(x)\} = \{x, z\}$, $\{e(y), f(y)\} = \{y\}$, $\{e(z), f(z)\} = \{z, x\}$ and $\{e(w), f(w)\} = \{w\}$. There are just three different orbits; thus $m(f) = 3$.

$\langle e \rangle$	$\{x\}, \{y\}, \{z\}, \{w\}$	$m(e) = 4$
$\langle r \rangle$	$\{x, y, z, w\}$	$m(r) = 1$
$\langle r^2 \rangle$	$\{x, z\}, \{y, w\}$	$m(r^2) = 2$
$\langle r^3 \rangle$	$\{x, y, z, w\}$	$m(r^3) = 1$
$\langle h \rangle$	$\{x, w\}, \{y, z\}$	$m(h) = 2$
$\langle v \rangle$	$\{x, y\}, \{w, z\}$	$m(v) = 2$
$\langle d \rangle$	$\{x\}, \{z\}, \{y, w\}$	$m(d) = 3$
$\langle f \rangle$	$\{x, z\}, \{y\}, \{w\}$	$m(f) = 3$

Figure 4

According to Theorem 3 there are

$$C(k) = \tfrac{1}{8}(k^4 + k + k^2 + k + k^2 + k^2 + k^3 + k^3)$$
$$= \tfrac{1}{8}(k^4 + 2k^3 + 3k^2 + 2k)$$

different ways to color the vertices of the square with k colors. For $k = 1$, this number is of course 1. The table in Figure 5 gives the numbers of colorings

k	Number of ways to color
1	1
2	6
3	21
4	55
5	120
6	231
7	406

(a)

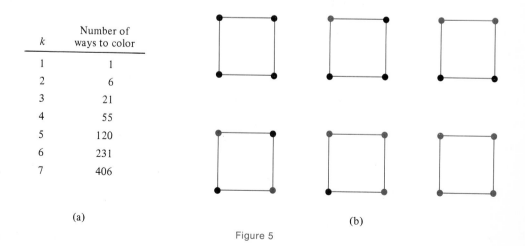

(b)

Figure 5

possible for the first few values of k. Figure 5(b) indicates the six different possibilities for two colors, including both of the one-color colorings. ▌

EXAMPLE 3   Now let us color the faces of the cube we started this section with. There are 24 rotations that send the cube back to itself. To list them all would take a fair amount of work, but in fact we only need to know their orbit sizes, and for that we can just count rotations of the five types illustrated in Figure 6(a). The table in Figure 6(b) tells how many there are of each type. It also gives the number $m(g)$ of orbits of their cyclic groups acting on the set of faces of the cube. For instance, consider the 90° rotation of type b. Each such rotation has an axis through the centers of two opposite faces. There are 6 faces, so there are 3 opposite pairs and hence 3 such axes. Figure 6(b) lists 6 rotations of type b: each axis gives two 90° rotations, one in each direction. A rotation g of type b has 3 $\langle g \rangle$-orbits: the faces the axes go through form orbits of size 1 and the other four faces form an orbit of size 4. Thus $m(g) = 3$. The remaining entries in Figure 6(b) have been determined by similar reasoning. Theorem 3 gives the formula

$$C(k) = \tfrac{1}{24}(6k^3 + 6k^3 + 3k^4 + 8k^2 + k^6)$$

for the number of colorings of the faces with k colors. Figure 6(c) lists the first few values of $C(k)$. ▌

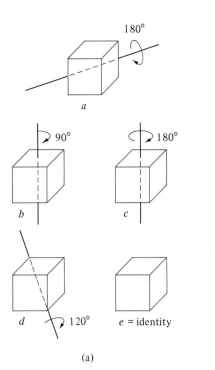

Type	Number of that type	$m(g)$
a	6	3
b	6	3
c	3	4
d	8	2
e	1	6

(b)

k	$C(k)$
1	1
2	10
3	57
4	240
5	800
6	2226

(c)

(a)

Figure 6

EXAMPLE 4 Consider now the problem of assigning different labels 1, 2, 3, 4, 5, 6 to the vertices of the graph in Figure 1 from Example 1(a), as illustrated in Figure 7. Two assignments are regarded as the same if there is a graph automorphism which takes one to the other. Thus the labelings shown in Figure 7 are all considered the same.

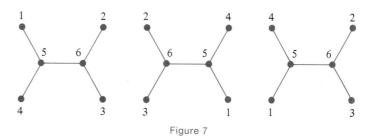

Figure 7

We can view this as a problem of coloring with exactly 6 colors. The formula from Theorem 3 gives us the number $C(k)$ of ways to color using at most k colors. Among the $C(6)$ colorings using at most 6 colors are some which use 5 or fewer colors, which we do not want to count. For $i = 1, 2, \ldots, 6$ let A_i be the set of colorings which do not use the ith color. Then $A_1 \cup A_2 \cup \cdots \cup A_6$ is the set of colorings using 5 or fewer of the colors, and the answer to our question is $C(6) - |A_1 \cup A_2 \cup \cdots \cup A_6|$. The Inclusion-Exclusion Principle of § 5.2 gives a formula for $|A_1 \cup A_2 \cup \cdots \cup A_6|$, namely

$$\sum_{i=1}^{6} |A_i| - \sum_{1 \le i < j \le 6} |A_i \cap A_j| + \sum_{1 \le i < j < k \le 6} |A_i \cap A_j \cap A_k| - \cdots,$$

where, for example, the third sum adds up the sizes of the intersections of three *distinct* sets from among $A_1, \ldots, A_6$. Now

$$|A_i| = C(5) \quad \text{for each} \quad i,$$

$$|A_i \cap A_j| = |\{\text{colorings not using } i\text{th and } j\text{th colors}\}|$$
$$= C(4) \text{ for } i < j,$$

$$|A_i \cap A_j \cap A_k| = C(3) \quad \text{for} \quad i < j < k,$$

etc. Thus

$$|A_1 \cup A_2 \cup \cdots \cup A_6| = \tbinom{6}{1}C(5) - \tbinom{6}{2}C(4) + \tbinom{6}{3}C(3) - \tbinom{6}{4}C(2) + \tbinom{6}{5}C(1)$$

and so the number of colorings which use exactly 6 colors is

$$C(6) - \tbinom{6}{1}C(5) + \tbinom{6}{2}C(4) - \tbinom{6}{3}C(3) + \tbinom{6}{4}C(2) - \tbinom{6}{5}C(1).$$

The table in Figure 8(a) is based on the table of group elements in Figure 1(b). Theorem 3 gives the formula

$$C(k) = \tfrac{1}{8}(k^6 + k^5 + k^5 + k^4 + k^3 + k^2 + k^2 + k^3)$$
$$= \tfrac{1}{8}(k^6 + 2k^5 + k^4 + 2k^3 + 2k^2),$$

	Orbits		k	$C(k)$
$\langle e \rangle$	$\{u\},\{v\},\{w\},\{x\},\{y\},\{z\}$	$m(e) = 6$	1	1
$\langle \alpha_1 \rangle$	$\{u\},\{v\},\{w,\ x\},\{y\},\{z\}$	$m(\alpha_1) = 5$	2	21
$\langle \alpha_2 \rangle$	$\{u\},\{v\},\{w\},\{x\},\{y,\ z\}$	$m(\alpha_2) = 5$	3	171
$\langle \alpha_3 \rangle$	$\{u\},\{v\},\{w,\ x\},\{y,\ z\}$	$m(\alpha_3) = 4$	4	820
$\langle \alpha_4 \rangle$	$\{u,\ v\},\{w,\ y\},\{x,\ z\}$	$m(\alpha_4) = 3$	5	2850
$\langle \alpha_5 \rangle$	$\{u,\ v\},\{w,\ y,\ x,\ z\}$	$m(\alpha_5) = 2$	6	8001
$\langle \alpha_6 \rangle$	$\{u,\ v\},\{w,\ z,\ x,\ y\}$	$m(\alpha_6) = 2$	7	19306
$\langle \alpha_7 \rangle$	$\{u,\ v\},\{w,\ z\},\{x,\ y\}$	$m(\alpha_7) = 3$		

(a) (b)

Figure 8

whose first seven values are shown in Figure 8(b). The number of colorings using exactly six colors is

$$8001 - 6 \cdot 2850 + 15 \cdot 820 - 20 \cdot 171 + 15 \cdot 21 - 6 \cdot 1 = 90.$$

Similarly, the number of colorings which use exactly two colors is just $C(2) - \binom{2}{1}C(1) = 21 - 2 = 19$ and the number using exactly seven colors is

$$19306 - 7 \cdot 8001 + 21 \cdot 2850 - 35 \cdot 820 + 35 \cdot 171 - 21 \cdot 21 + 7 \cdot 1 = 0. \quad \blacksquare$$

Now let us return to the problem of building logical circuits, at first for just two inputs. There are $2^4 = 16$ Boolean functions of two variables, namely the members of $\text{FUN}(\mathbb{B} \times \mathbb{B}, \mathbb{B})$, where $\mathbb{B} = \{0, 1\}$. We may think of a circuit as a black box with two input leads, one for x_1 and one for x_2, and one output lead. Each element $\langle a_1, a_2 \rangle$ in $\mathbb{B} \times \mathbb{B}$ corresponds to a choice of values a_1 for x_1 and a_2 for x_2.

Interchanging the connections for x_1 and x_2 amounts to replacing $\langle a_1, a_2 \rangle$ by $\langle a_2, a_1 \rangle$, and corresponds to the permutation g of $\mathbb{B} \times \mathbb{B}$ that interchanges $\langle 0, 1 \rangle$ and $\langle 1, 0 \rangle$. We want to regard two black boxes as equivalent if one will produce the same results as the other, or will produce the same results if we interchange its input leads. That is, two boxes are equivalent if their Boolean functions f and f' are either the same or satisfy $f' = f \circ g$. Since $|\mathbb{B}| = 2$, the problem looks just like the 2-color question for a four-element set with two elements which are interchangeable. We apply Theorem 2 with $S = \mathbb{B} \times \mathbb{B}$, $C = \mathbb{B}$ and $G = \langle g \rangle = \{e, g\}$. The number of G-orbits is

$$\tfrac{1}{2}(2^4 + 2^3) = 12.$$

We can confirm this result using the table in Figure 9 which lists all sixteen Boolean functions from $\mathbb{B} \times \mathbb{B}$ to $\mathbb{B}$. Functions 2 and 4 can be performed with the same black box, as can functions 3 and 5, 10 and 12, and 11 and 13, so the number of orbits is $16 - 4 = 12$.

Function numbers

	0	1	2	3	4	5	6	7	8	9	10	11	12	13	14	15
$\langle 0, 0 \rangle$	0	1	0	1	0	1	0	1	0	1	0	1	0	1	0	1
$\langle 0, 1 \rangle$	0	0	1	1	0	0	1	1	0	0	1	1	0	0	1	1
$\langle 1, 0 \rangle$	0	0	0	0	1	1	1	1	0	0	0	0	1	1	1	1
$\langle 1, 1 \rangle$	0	0	0	0	0	0	0	0	1	1	1	1	1	1	1	1

Figure 9

Now suppose we allow ourselves to complement inputs. Complementing the value on the first lead corresponds to interchanging $\langle 0, 0 \rangle$ with $\langle 1, 0 \rangle$, and $\langle 0, 1 \rangle$ with $\langle 1, 1 \rangle$. We denote this permutation of $\mathbb{B} \times \mathbb{B}$ by c_1 and the permutation corresponding to complementing the input on the second lead by c_2. Altogether the permutations g, c_1 and c_2 generate the group of permutations of $\mathbb{B} \times \mathbb{B}$ described in Figure 10(a). [This fact is not expected to be obvious; take our word for it.] This group also turns out to be isomorphic to the groups of Examples 1 and 2. It acts on the 4-element set $\mathbb{B} \times \mathbb{B}$ in the same way that the group in Example 2 acts on the vertices of the square, as we see by comparing Figure 10(a) with Figure 10(b), which is just Figure 3(b) rewritten with some rows and columns interchanged. The correspondence $\langle 0, 0 \rangle \to x$, $\langle 0, 1 \rangle \to y$, $\langle 1, 0 \rangle \to w$, $\langle 1, 1 \rangle \to z$ converts one table into the other. From Figure 5 we know that there are $C(2) = 6$ ways to 2-color the square, so there are six orbits of Boolean functions under the action of the group $\langle \{c_1, c_2, g\} \rangle$, i.e., six essentially different black boxes. Using the function numbers from Figure 9, the orbits in FUN($\mathbb{B} \times \mathbb{B}$, $\mathbb{B}$) are

$$\{0\}, \quad \{1, 2, 4, 8\}, \quad \{3, 5, 10, 12\}, \quad \{6, 9\}, \quad \{7, 11, 13, 14\}, \quad \{15\}.$$

	$\langle 0, 0 \rangle$	$\langle 0, 1 \rangle$	$\langle 1, 0 \rangle$	$\langle 1, 1 \rangle$
e	$\langle 0, 0 \rangle$	$\langle 0, 1 \rangle$	$\langle 1, 0 \rangle$	$\langle 1, 1 \rangle$
c_1	$\langle 1, 0 \rangle$	$\langle 1, 1 \rangle$	$\langle 0, 0 \rangle$	$\langle 0, 1 \rangle$
c_2	$\langle 0, 1 \rangle$	$\langle 0, 0 \rangle$	$\langle 1, 1 \rangle$	$\langle 1, 0 \rangle$
$c_1 \circ c_2$	$\langle 1, 1 \rangle$	$\langle 1, 0 \rangle$	$\langle 0, 1 \rangle$	$\langle 0, 0 \rangle$
g	$\langle 0, 0 \rangle$	$\langle 1, 0 \rangle$	$\langle 0, 1 \rangle$	$\langle 1, 1 \rangle$
$c_1 \circ g$	$\langle 1, 0 \rangle$	$\langle 0, 0 \rangle$	$\langle 1, 1 \rangle$	$\langle 0, 1 \rangle$
$c_2 \circ g$	$\langle 0, 1 \rangle$	$\langle 1, 1 \rangle$	$\langle 0, 0 \rangle$	$\langle 1, 0 \rangle$
$c_1 \circ c_2 \circ g$	$\langle 1, 1 \rangle$	$\langle 0, 1 \rangle$	$\langle 1, 0 \rangle$	$\langle 0, 0 \rangle$

(a)

	x	y	w	z
e	x	y	w	z
h	w	z	x	y
v	y	x	z	w
r^2	z	w	y	x
d	x	w	y	z
r^3	w	x	z	y
r	y	z	x	w
f	z	y	w	x

(b)

Figure 10

To build circuits, it would be enough to compute one function from each orbit, say the functions 0, 1, 3, 6, 7 and 15.

If we also allow ourselves to complement the output of a circuit, then a circuit which computes the function numbered n will also compute $15 - n$ and we need even fewer black boxes. A circuit for 0 will also compute 15, and one for 1 will compute 14 and hence also 7, 11 or 13. A circuit for 3 will also compute 12, which we already knew, and similarly a circuit for 6 will compute 9. The classes of functions are now

$$\{0, 15\}, \quad \{1, 2, 4, 8, 7, 11, 13, 14\}, \quad \{3, 5, 10, 12\} \quad \text{and} \quad \{6, 9\}.$$

It still requires four different circuits to compute all 2-variable Boolean functions, allowing complementation on both input and output leads.

Our methods generalize in theory to count the number of black boxes needed for n-input Boolean functions. In practice, the detailed determination of orbits for all elements of G gets exceedingly complicated. For 4-input functions the answer is that there are 222 different circuits required, even if we allow free complementation on input and output lines. This number is considerably smaller than $2^{16} = 65,536$.

Knowing how many different circuits there are does not help find representative circuits, though it does tell us when we have found enough. For 4-input functions, $\text{FUN}(\mathbb{B}^4, \mathbb{B})$ is small enough so that all 2^{16} of its members can be stored in a computer and one can find representatives in the following way using the algorithm MERGE PARTITIONS of § 7.7. As in the 2-input problem, the functions can be numbered $0, 1, \ldots, 2^{16} - 1$ so that n and $2^{16} - 1 - n$ are equivalent under complementation of outputs. It turns out that the group G which acts on $\mathbb{B}^4$ can be generated by three of its members. Each generator g partitions the set $\mathbb{B}^4$ into $\langle g \rangle$-orbits, and the join of these partitions with the $\{n, 2^{16} - 1 - n\}$ partition is the required 222-set partition. Exercises 15 and 16 illustrate the application of these ideas to 2-input functions.

Our methods have not taken systematic advantage of the symmetry of the group G itself. By using such symmetry one can obtain a formula for the number of G-orbits in $\text{FUN}(\mathbb{B}^n, \mathbb{B})$ whose members have exactly k of their values equal to 0 for $k = 1, 2, \ldots$. The textbook *Applied Modern Algebra* by Dornhoff and Hohn contains an account of such methods and the resulting formulas.

EXERCISES 11.5

1. The graph in Figure 11(a) has two automorphisms, which are described in Figure 11(b).
   (a) What is the average number of vertices fixed by the automorphisms of this graph?
   (b) Which of the automorphisms of this graph fix the average number of vertices?
   (c) Find the number of ways to color the vertices of this graph with k colors.

2. Verify Theorem 1 for
   (a) the group of automorphisms of the graph in Figure 11(a) acting on the set of vertices of the graph.
   (b) the group in part (a) acting on the set of edges of the graph in Figure 11(a).
   (c) the group of rotations in Example 3 acting on the set of faces of the cube.

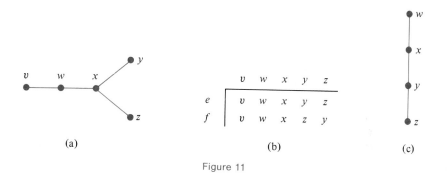

(a)   (b)   (c)

Figure 11

3. The graph in Figure 11(c) has two automorphisms.
   (a) How many ways are there to color the vertices of this graph with k colors?
   (b) How many ways are there to label the vertices of this graph with four different labels?

4. Show that the group $\langle\{g, c_1, c_2\}\rangle$ described in Figure 10(a) is also generated by $\{g, c_1\}$. *Suggestion:* Compute $g \circ c_1 \circ g$.

5. The graph in Figure 12(a) has six automorphisms, which are described in Figure 12(b).
   (a) Find the number of ways to color the vertices of this graph with k colors.
   (b) Find the number of ways to color the edges of this graph with k colors.

6. (a) Use the Inclusion-Exclusion Principle and the answer to Exercise 5(a) to find the number of ways to color the vertices of the graph in Figure 12(a) with exactly 3 colors.

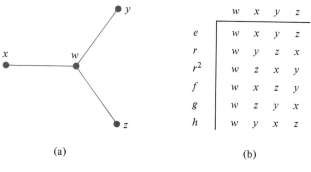

(a)   (b)

Figure 12

(b) Describe all the different colorings in part (a), using the colors red, blue and green.

(c) Find the number of ways to color the vertices of this graph with exactly 4 colors.

7. (a) How many ways are there to color the vertices of the square in Example 2 with exactly four colors?

(b) List the different colorings which use all 4 of the colors red, blue, green, yellow.

8. How many different circular necklaces can be made from 5 beads of k different colors? Consider two necklaces to be the same if one looks just like the other when it is rotated or flipped over. *Hint*: The group here consists of e, four nontrivial rotations and five flips. See Example 2 for the 4-bead case.

9. Consider the group G of Example 1 acting on the set $S = \{u, v, w, x, y, z\}$.
(a) How many pairs $\langle g, s \rangle$ in $G \times S$ are there with $gs = s$?
(b) How many such pairs are there with $g = \alpha_3$?
(c) How many such pairs are there with s in the G-orbit of x?
(d) How many such pairs are there with s in the G-orbit of u?

10. Consider the problem of coloring the faces of a cube using crayons which are red, green and blue, as in Example 3.
(a) How many colorings use exactly two of the three colors?
(b) Use any method to find how many colorings have four red faces and two blue faces.
(c) How many colorings have exactly four red faces?
(d) Would you like to use the method of inspection to find all colorings with exactly two faces of each color?

11. (a) How many ways are there to color the vertices of a cube with k colors?
(b) How many ways are there to color the edges of a cube with k colors?
Two colorings in parts (a) and (b) are considered the same if one can be turned into the other by a rotation of the cube. *Suggestion*: Use Figure 6(a) to create new tables like Figure 6(b) for the actions on vertices and edges.

12. Consider a group G acting on an n-element set S. Show that if $|F(g)| \geq 1$ for each $g \in G$, then G has at least $1 + (n-1)/|G|$ orbits in S. For $n > 1$ this implies the corollary to Theorem 1. *Hint*: Treat the element e of G separately from the others.

13. (a) How many different 2-input logical circuits are there if we only regard two circuits as the same if they have the same or complementary outputs?
(b) How many are there if we also consider two circuits to be the same if they produce the same function when the input leads on one are interchanged?

14. Exhibit an isomorphism between the group described in Figure 1(b) and the group described in Figure 3(b). *Hint*: Neighboring leaves of the tree correspond to diagonally opposite corners of the square.

15. Let G_1 and G_2 be groups of permutations of a finite set S, and let G_0 be the group generated by $G_1 \cup G_2$. Let π_1 be the partition of S consisting of the G_1-orbits, with similar definitions for π_2 and π_0. Show that $\pi_0 = \pi_1 \vee \pi_2$. *Hint*: Let R_1 be

the equivalence relation on S where $\langle s, t \rangle \in R_1$ if $gs = t$ for some $g \in G_1$, as in Exercise 11 of §11.4. If R_2 and R_0 are defined analogously, then the equality $\pi_0 = \pi_1 \vee \pi_2$ is equivalent to $R_0 = R_1 \vee R_2$.

16. Let S be the set FUN($\mathbb{B} \times \mathbb{B}$, $\mathbb{B}$) which is illustrated in Figure 9 and used to count logical circuits. Let G_1 be the group $\langle \{c_1, c_2, g\} \rangle$, which corresponds to complementing or interchanging inputs, and let G_2 be the two-element group corresponding to complementation of the output. Let G_0, π_1, π_2 and π_0 be as in Exercise 15.

   (a) Convince yourself that

$$\pi_1 = \{\{0\}, \{1, 2, 4, 8\}, \{3, 5, 10, 12\}, \{6, 9\}, \{7, 11, 13, 14\}, \{15\}\}$$

   and

$$\pi_2 = \{\{0, 15\}, \{1, 14\}, \{2, 13\}, \{3, 12\}, \{4, 11\}, \{5, 10\}, \{6, 9\}, \{7, 8\}\}.$$

   (b) Explain why G_0 is the group that corresponds to complementation of inputs and outputs.

   (c) Determine π_0.

§ 11.6   Other Algebraic Systems

So far in this chapter we have been looking at sets with just one binary operation on them. A number of important and familiar algebraic structures have two binary operations, usually written $+$ and $\cdot$. The "additive" operation $+$ is typically very well-behaved, while the "multiplicative" operation $\cdot$ is generally less so, and there are usually distributive laws relating the two operations to each other. In this section we briefly introduce rings and fields, which are the two main kinds of algebraic structures with two operations, and we give some examples and discuss the basic facts about homomorphisms of such systems.

EXAMPLE 1   (a) The sets $\mathbb{Z}$, $\mathbb{Q}$ and $\mathbb{R}$ are each closed under ordinary addition and multiplication. Both $+$ and $\cdot$ are commutative and associative for each of these sets. Moreover, these operations satisfy the distributive laws:

$$a \cdot (b + c) = (a \cdot b) + (a \cdot c) \quad \text{and} \quad (a + b) \cdot c = (a \cdot c) + (b \cdot c).$$

(b) The set $\mathfrak{M}_{n,n}$ of $n \times n$ matrices with real entries is closed under matrix addition and also under matrix multiplication. Both operations are associative, and addition is commutative, but matrix multiplication is not commutative. For instance

$$\begin{bmatrix} 2 & 2 \\ 1 & 1 \end{bmatrix}\begin{bmatrix} 3 & -1 \\ -3 & 1 \end{bmatrix} = \begin{bmatrix} 0 & 0 \\ 0 & 0 \end{bmatrix} \neq \begin{bmatrix} 5 & 5 \\ -5 & -5 \end{bmatrix} = \begin{bmatrix} 3 & -1 \\ -3 & 1 \end{bmatrix}\begin{bmatrix} 2 & 2 \\ 1 & 1 \end{bmatrix}.$$

The distributive laws

$$\mathbf{A}(\mathbf{B} + \mathbf{C}) = (\mathbf{A}\mathbf{B}) + (\mathbf{A}\mathbf{C}) \quad \text{and} \quad (\mathbf{A} + \mathbf{B})\mathbf{C} = (\mathbf{A}\mathbf{C}) + (\mathbf{B}\mathbf{C})$$

are valid.

(c) For $p \geq 2$ the operations $+_p$ and $*_p$ on $\mathbb{Z}(p)$ are commutative and associative by Theorem 3 of § 4.3, and the distributive laws hold [Exercise 4].

(d) Consider any commutative group $(G, +)$ with identity 0, and define the operation $\cdot$ trivially on G by $a \cdot b = 0$ for all $a, b \in G$. It follows at once that $\cdot$ is commutative and associative, and that $(G, +, \cdot)$ satisfies the distributive laws. ∎

Structures like the ones in Example 1 come up frequently enough to deserve a name. A **ring** is a set R closed under two binary operations, generally denoted $+$ and $\cdot$, such that

(a) $(R, +)$ is a commutative group,
(b) $(R, \cdot)$ is a semigroup,
(c) $a \cdot (b + c) = (a \cdot b) + (a \cdot c)$ and $(a + b) \cdot c = (a \cdot c) + (b \cdot c)$ for all $a, b, c \in R$.

If $(R, \cdot)$ is commutative, we say the ring $(R, +, \cdot)$ is **commutative**. The ring $(\mathfrak{M}_{n,n}, +, \cdot)$ in Example 1(b) is not commutative if $n > 1$. The other rings in Example 1 are commutative. A ring always has an additive identity element, denoted 0. If it has a multiplicative identity which is different from 0, we usually call the multiplicative identity 1 and we say the ring is a **ring with identity**. Each of the rings in Examples 1(a), (b) and (c) is a ring with identity. The **trivial ring** in Example 1(d) has no multiplicative identity unless G just consists of 0; since we require $1 \neq 0$, we do not consider $(\{0\}, +, \cdot)$ to be a ring with identity. The ring $(2\mathbb{Z}, +, \cdot)$ of even integers, with the usual sum and product, is another example of a commutative ring without identity.

The distributive laws make calculations in a ring behave very much like the arithmetic we are used to in $\mathbb{Z}$, allowing for the obvious fact that we need to watch out for noncommuting elements. For example, we have

$$(a + b)^2 = (a + b) \cdot (a + b) = [a \cdot (a + b)] + [b \cdot (a + b)]$$

$$= (a \cdot a) + (a \cdot b) + (b \cdot a) + (b \cdot b) = a^2 + a \cdot b + b \cdot a + b^2,$$

but this is not $a^2 + 2(a \cdot b) + b^2$ unless $a \cdot b = b \cdot a$. As another example, we get

$$(0 \cdot a) + (0 \cdot a) = (0 + 0) \cdot a = 0 \cdot a,$$

and so

$$0 \cdot a = (0 \cdot a) + (0 \cdot a) - (0 \cdot a) = (0 \cdot a) - (0 \cdot a) = 0.$$

Similarly $a \cdot 0 = 0$ for all a in a ring. One can also show that $(-a) \cdot b = -(a \cdot b) = a \cdot (-b)$ [Exercise 8].

We can never hope to divide by 0, but in the rings $(\mathbb{Q}, +, \cdot)$ and $(\mathbb{R}, +, \cdot)$ we can divide by every non-0 element. In $(\mathbb{Z}, +, \cdot)$ we can divide 6 by 3 successfully but cannot divide 6 by 5 to get an answer which is still in $\mathbb{Z}$. A

field is a commutative ring $(R, +, \cdot)$ in which the non-0 elements form a group under multiplication. In a field the inverse of a non-0 a is usually written a^{-1} or $1/a$, and it has the property that $a^{-1} \cdot a = a \cdot a^{-1} = 1$. We also often write b/a for $b \cdot a^{-1}$, a notation which we justify by the fact that $(b/a) \cdot a = b \cdot a^{-1} \cdot a = b$.

If $a \cdot b = a \cdot c$ in a field and if $a \neq 0$, then $b = a^{-1} \cdot a \cdot b = a^{-1} \cdot a \cdot c = c$; i.e., a can be canceled in the equation $a \cdot b = a \cdot c$. Such cancellation is sometimes possible even if inverses do not exist—for example in $(\mathbb{Z}, +, \cdot)$. A commutative ring with identity in which each non-0 element is cancelable in this way is called an **integral domain**. These rings form an important class intermediate between fields and more general commutative rings. One can show [Exercise 14(c)] that every finite integral domain is a field.

EXAMPLE 2   (a) Groups are always nonempty, so the multiplicative identity in a field is always non-0. The smallest possible field is $(\mathbb{Z}(2), +_2, *_2)$ which has just two elements 0 and 1 and operations as shown in the tables.

$+_2$	0	1		$*_2$	0	1
0	0	1		0	0	0
1	1	0		1	0	1

(b) The set $\mathrm{FUN}(\mathbb{R}, \mathbb{R})$ is a ring with $f + g$ and $f \cdot g$ defined by

$$(f + g)(x) = f(x) + g(x) \quad \text{and} \quad (f \cdot g)(x) = f(x) \cdot g(x)$$

for all $x \in \mathbb{R}$. The zero element is the constant function c_0 defined by $c_0(x) = 0$ for all x. This ring is commutative but is not an integral domain, even though it gets its multiplication from the field $\mathbb{R}$. For example, let $f(x) = 0$ for $x \leq 0$ and $f(x) = 1$ for $x > 0$, and let $g(x) = 1$ for $x \leq 0$ and $g(x) = 0$ for $x > 0$. Then $(f \cdot g)(x) = 0$ for every x, so $f \cdot g = 0 = f \cdot 0$ but $f \neq 0$ and $g \neq 0$. ∎

A **subring** of a ring R is simply a subset of R which is itself a ring under the two operations of R. A **subfield** of a field F is a subring of F which is itself a field; this means in particular that it contains the multiplicative identity 1 of F and is closed under taking inverses.

EXAMPLE 3   (a) The ring $(\mathbb{Z}, +, \cdot)$ has subrings $2\mathbb{Z}$, $73\mathbb{Z}$ and $\{0\}$ among others. In fact, the subrings of $\mathbb{Z}$ are precisely the rings $n\mathbb{Z}$ for n an integer in view of Theorem 4 of § 11.1. The ring $(\mathbb{Z}(p), +_p, *_p)$ is *not* a subring of $\mathbb{Z}$; in fact, the two rings have quite different structure. For example, for each a in $\mathbb{Z}(p)$ we have $a +_p a +_p \cdots +_p a = 0$ if there are p terms in the sum, whereas $a + a + \cdots + a = pa$ in $\mathbb{Z}$.

(b) Every subring R of a field F is an integral domain, because if $a, b, c \in R$ with $a \cdot b = a \cdot c$ and $a \neq 0$ then $b = c$ in F and hence also in R. In particular, the subring $(\mathbb{Z}, +, \cdot)$ of the field $(\mathbb{R}, +, \cdot)$ is an integral domain. It is not a field, since only 1 and -1 have multiplicative inverses in $\mathbb{Z}$. The field $(\mathbb{Q}, +, \cdot)$ is a subfield of $(\mathbb{R}, +, \cdot)$.   ∎

The appropriate mappings to use in studying rings are the ones which are compatible with both the additive and multiplicative structures. A **ring homomorphism** from a ring $(R, +, \cdot)$ to a ring $(S, +, \cdot)$ is a function $h : R \to S$ such that

$$h(a + b) = h(a) + h(b) \quad \text{and} \quad h(a \cdot b) = h(a) \cdot h(b)$$

for all $a, b \in R$. The operations on the left sides of these equations are the operations in R; those on the right are in S. Thus a ring homomorphism is just a function which is both a group homomorphism from $(R, +)$ to $(S, +)$ and a semigroup homomorphism from $(R, \cdot)$ to $(S, \cdot)$.

EXAMPLE 4   (a) The function Rem_p from $\mathbb{Z}$ to $\mathbb{Z}(p)$, defined by $\text{Rem}_p(m) = $ the remainder when m is divided by p, is a ring homomorphism. This is because

$$m + n \equiv \text{Rem}_p(m) +_p \text{Rem}_p(n) \ (\text{mod } p)$$

and

$$m \cdot n \equiv \text{Rem}_p(m) *_p \text{Rem}_p(n) \ (\text{mod } p)$$

for all $m, n \in \mathbb{Z}$ by definition of $+_p$ and $*_p$.

(b) The function h from $\mathbb{Z}$ to $\mathbb{Z}$ defined by $h(m) = 3m$ is an additive group homomorphism but is not a ring homomorphism because $h(m \cdot n) = 3mn$, while $h(m) \cdot h(n) = 3m \cdot 3n = 9mn$.

(c) The mapping h of $\text{FUN}(\mathbb{R}, \mathbb{R})$ into $\mathbb{R}$ given by $h(f) = f(73)$ is a ring homomorphism, since

$$h(f + g) = (f + g)(73) = f(73) + g(73) = h(f) + h(g)$$

and

$$h(f \cdot g) = (f \cdot g)(73) = f(73) \cdot g(73) = h(f) \cdot h(g).$$

More generally, for any set S and ring R we can make $\text{FUN}(S, R)$ into a ring just as we did for $\text{FUN}(\mathbb{R}, \mathbb{R})$, and each s in S gives rise to an evaluation homomorphism h from $\text{FUN}(S, R)$ to R defined by $h(f) = f(s)$ for all f in $\text{FUN}(S, R)$.

(d) The set $\text{POLY}(\mathbb{R})$ of all polynomials p, where $p(x) = a_0 + a_1 x + \cdots + a_n x^n$ with $n \in \mathbb{N}$ and with real coefficients $a_0, \ldots, a_n$, is a ring using $+$ and $\cdot$ defined in the usual way. We can think of $\text{POLY}(\mathbb{R})$ as a subring of $\text{FUN}(\mathbb{R}, \mathbb{R})$, since each polynomial defines a unique function and since one can show that two different polynomials must give different functions [i.e., have different graphs]. An evaluation homomorphism such as $f \to f(73)$ from $\text{FUN}(\mathbb{R}, \mathbb{R})$ to $\mathbb{R}$ yields a homomorphism from $\text{POLY}(\mathbb{R})$ to $\mathbb{R}$, defined in this instance by $p \to p(73)$. The evaluation homomorphism $p \to p(0)$ assigns to each polynomial p its constant coefficient a_0.

(e) As noted in Example 1(c) of § 11.3, the trace function from $\mathfrak{M}_{2,2}$ to $\mathbb{R}$ is an additive group homomorphism. Trace is not a ring homomorphism, however, because trace($\mathbf{AB}$) is not generally trace($\mathbf{A}$) $\cdot$ trace($\mathbf{B}$). ∎

Since ring homomorphisms are additive group homomorphisms, they have kernels. Suppose h is a homomorphism from $(R, +, \cdot)$ to $(S, +, \cdot)$. The kernel of h is the set $K = \{a \in R : h(a) = h(0)\}$, and for $a, b \in R$ we have

$$h(a) = h(b) \Leftrightarrow a - b \in K \Leftrightarrow a + K = b + K.$$

Everything is in additive dress here, so $a + K$ is the coset $\{a + k : k \in K\}$ of the subgroup K of $(R, +)$. As before, h is one-to-one if and only if $K = \{0\}$. The kernel of h is a special kind of subring of R. It is an additive subgroup of course, but it is also closed under multiplication not only by its own elements but even under multiplication by other elements in R:

$$a \in K, \quad r \in R \quad \text{imply} \quad a \cdot r, r \cdot a \in K.$$

The reason is that $h(a) = h(0)$ and if $r \in R$ then

$$h(a \cdot r) = h(a) \cdot h(r) = h(0) \cdot h(r) = h(0 \cdot r) = h(0)$$

and likewise $h(r \cdot a) = h(0)$.

An additive subgroup I of a ring $(R, +, \cdot)$ is called an **ideal** of R if $r \cdot a \in I$ and $a \cdot r \in I$ for all $a \in I$ and $r \in R$. Kernels of ring homomorphisms are ideals, as we noted in the last paragraph, and one can show [Exercise 10] that every ideal is the kernel of a homomorphism.

EXAMPLE 5   (a) The subgroup $\{0\}$ and the ring R itself are always ideals of R. These obvious ideals are the only ones there are in $\mathfrak{M}_{n,n}$; a proof of this fact takes some work.

(b) If the ring R is commutative and if $a \in R$, then the set $R \cdot a = \{r \cdot a : r \in R\}$ is an ideal of R. To check this we observe that $r \cdot a + s \cdot a = (r + s) \cdot a \in R \cdot a$ and $-(r \cdot a) = (-r) \cdot a \in R \cdot a$ for every $r, s \in R$, so $R \cdot a$ is an additive subgroup of R. Since $s \cdot (r \cdot a) = (s \cdot r) \cdot a \in R \cdot a$ for every $r, s \in R$, $R \cdot a$ is closed under multiplication by elements of R. An ideal of the form $R \cdot a$ is called a **principal ideal**.

All of the subgroups of $(\mathbb{Z}, +)$ are of the form $n\mathbb{Z}$ by Theorem 4 of § 11.1, so every ideal of $(\mathbb{Z}, +, \cdot)$ is principal. So are the ideals of POLY($\mathbb{R}$), as it turns out, but such a situation is very special. For example, in the commutative ring POLY($\mathbb{Z}$) consisting of polynomials with integer coefficients, the set of all polynomials $a_0 + a_1 x + \cdots + a_n x^n$ in which a_0 is even is an ideal which is not principal [Exercise 17].

(c) Ideals of fields are boring. For suppose I is a non-0 ideal of a field F. Let $0 \neq a \in I$. For every $b \in F$ we have $b = (b \cdot a^{-1}) \cdot a \in F \cdot a \subseteq I$, so $I = F$. That is, F has only the obvious ideals $\{0\}$ and F.   ∎

If R is a ring with ideal I, then the group R/I consisting of additive cosets $r + I = \{r + i : i \in I\}$ can be made into a ring in a natural way. We define

$$(r + I) + (s + I) = (r + s) + I$$

and

$$(r + I) \cdot (s + I) = r \cdot s + I.$$

We've already seen in § 11.3 that the addition on R/I is well-defined; we check multiplication. If $r + I = r' + I$ and $s + I = s' + I$, then $r - r' \in I$ and $s - s' \in I$ and hence

$$r \cdot s - r' \cdot s' = r \cdot s - r \cdot s' + r \cdot s' - r' \cdot s'$$

$$= r \cdot (s - s') + (r - r') \cdot s' \in r \cdot I + I \cdot s' \subseteq I.$$

Thus $r \cdot s + I = r' \cdot s' + I$ and our definition of product is independent of the choice of representatives we take in the cosets $r + I$ and $s + I$. The rest of the properties of a ring are easy to check.

The fundamental homomorphism theorem for groups leads to a corresponding result for rings. Consider a ring homomorphism h from R to S with kernel I. Then h is an additive group homomorphism of $(R, +)$ into $(S, +)$, so we already know from Theorem 3 of § 11.3 that the mapping h^* from R/I to $h(R)$ defined by $h^*(r + I) = h(r)$ for $r \in R$ is a group isomorphism. Since

$$h^*((r + I) \cdot (r' + I)) = h^*((r \cdot r') + I) = h(r \cdot r')$$

$$= h(r) \cdot h(r') = h^*(r + I) \cdot h^*(r' + I),$$

h^* is in fact a ring homomorphism. Therefore h^* is a **ring isomorphism** between R/I and $h(R)$, i.e., a ring homomorphism which is one-to-one and onto. We have shown the following.

Theorem 1   Let h be a homomorphism with kernel I from the ring R to the ring S. Then the mapping $r + I \rightarrow h(r)$ is an isomorphism of the ring R/I onto $h(R)$.

The ring R/I may have ideals of its own. They correspond to ideals of $h(R)$, by Theorem 1, but we can also associate them with ideals of R itself.

Theorem 2   Let h be a homomorphism from the ring R to the ring $h(R)$.

(a) If I is an ideal of R, then $h(I)$ is an ideal of $h(R)$.
(b) Every ideal of $h(R)$ is of the form $h(I)$ for some ideal I of R containing the kernel of h.

Proof. (a) Since I is a subgroup of $(R, +)$, $h(I)$ is a subgroup of $(h(R), +)$. If $h(r) \in h(R)$ and $h(a) \in h(I)$, then $h(a) \cdot h(r) = h(a \cdot r) \in h(I)$ since $a \cdot r$ is in the ideal I. Similarly, $h(r) \cdot h(a) \in h(I)$, so $h(I)$ is closed under multiplication by elements of $h(R)$.

(b) Suppose J is an ideal of $h(R)$. Let $I = h^{-1}(J) = \{a \in R : h(a) \in J\}$. If $a, b \in I$ then $h(a), h(b) \in J$, so $h(a + b) = h(a) + h(b)$ is in J and therefore $a + b \in I$. Similarly, $-a \in I$ for $a \in I$, and both $a \cdot r$ and $r \cdot a$ are in I for all $a \in I$ and $r \in R$. Thus I is an ideal of R containing the kernel $\{a \in R : h(a) = 0\}$ of h. Clearly $h(I) = J$. ∎

If K is an ideal of R, then the natural mapping $v : a \to a + K$ of R to R/K is a homomorphism with kernel K. Theorem 2, with v and R/K in place of h and $h(R)$, gives the following.

Corollary   Let K be an ideal of the ring R. The ideals of R/K are precisely the sets of the form I/K for I an ideal of R containing K. Moreover, for every ideal I of R the set $I + K = \{a + b : a \in I, b \in K\}$ is an ideal of R.

Proof. The second assertion is true because $v(I) = (I + K)/K$. ∎

EXAMPLE 6   (a) The kernel of the ring homomorphism Rem_p from $\mathbb{Z}$ onto $\mathbb{Z}(p)$ defined in Example 4(a) is the ideal $p\mathbb{Z}$. The ideals in $\mathbb{Z}$ are of the form $n\mathbb{Z}$, and their images are of the form $\mathrm{Rem}_p(n\mathbb{Z}) = \mathrm{Rem}_p(n) *_p \mathrm{Rem}_p(\mathbb{Z}) = \{\mathrm{Rem}_p(n) *_p a : a \in \mathbb{Z}(p)\}$. For example, there are just four such ideals in $\mathbb{Z}(6)$, namely

$$\mathrm{Rem}_6(\mathbb{Z}) = \mathbb{Z}(6) = \{0, 1, 2, 3, 4, 5\},$$

$$\mathrm{Rem}_6(2\mathbb{Z}) = 2 *_6 \mathbb{Z}(6) = \{0, 2, 4\},$$

$$\mathrm{Rem}_6(3\mathbb{Z}) = 3 *_6 \mathbb{Z}(6) = \{0, 3\},$$

$$\mathrm{Rem}_6(6\mathbb{Z}) = 0 *_6 \mathbb{Z}(6) = \{0\}.$$

To see that these are all, we observe that an ideal $n\mathbb{Z}$ of $\mathbb{Z}$ contains another ideal $m\mathbb{Z}$ if and only if m is a multiple of n. According to Theorem 2(b), the ideals of $\mathbb{Z}(6)$ correspond to the ideals of $\mathbb{Z}$ which contain $6\mathbb{Z}$, namely $\mathbb{Z}$, $2\mathbb{Z}$, $3\mathbb{Z}$ and $6\mathbb{Z}$.

(b) We can make $\mathbb{Z}(2) \times \mathbb{Z}(3)$ into a ring by defining

$$\langle m, n \rangle + \langle j, k \rangle = \langle m +_2 j, n +_3 k \rangle$$

and

$$\langle m, n \rangle \cdot \langle j, k \rangle = \langle m *_2 j, n *_3 k \rangle.$$

The mapping $h: m \to \langle \text{Rem}_2(m), \text{Rem}_3(m) \rangle$ is a ring homomorphism from $\mathbb{Z}$ into $\mathbb{Z}(2) \times \mathbb{Z}(3)$. Its kernel is

$$\{m \in \mathbb{Z} : m \equiv 0 \ (\text{mod } 2) \text{ and } m \equiv 0 \ (\text{mod } 3)\}$$

$$= \{m \in \mathbb{Z} : m \equiv 0 \ (\text{mod } 6)\} = 6\mathbb{Z}.$$

Thus by Theorem 1 $h(\mathbb{Z})$ is isomorphic to $\mathbb{Z}/6\mathbb{Z}$, i.e., is isomorphic to $\mathbb{Z}(6)$. Since $|h(\mathbb{Z})| = |\mathbb{Z}(6)| = 6$ and $|\mathbb{Z}(2) \times \mathbb{Z}(3)| = 2 \cdot 3 = 6$, h must map $\mathbb{Z}$ onto $\mathbb{Z}(2) \times \mathbb{Z}(3)$. That is,

$$\mathbb{Z}(2) \times \mathbb{Z}(3) \text{ is ring-isomorphic to } \mathbb{Z}(6).$$

One can check that the correspondence

$$\langle 0, 0 \rangle \to 0 \qquad \langle 1, 1 \rangle \to 1 \qquad \langle 0, 2 \rangle \to 2$$

$$\langle 1, 0 \rangle \to 3 \qquad \langle 0, 1 \rangle \to 4 \qquad \langle 1, 2 \rangle \to 5$$

is an isomorphism. The ideals of $\mathbb{Z}(2) \times \mathbb{Z}(3)$, which correspond to the ideals of $\mathbb{Z}(6)$ obtained in part (a), are the four subsets $\{\langle 0, 0 \rangle\}$, $\{\langle 0, 0 \rangle, \langle 0, 1 \rangle, \langle 0, 2 \rangle\} = \{0\} \times \mathbb{Z}(3)$, $\{\langle 0, 0 \rangle, \langle 1, 0 \rangle\} = \mathbb{Z}(2) \times \{0\}$, and $\mathbb{Z}(2) \times \mathbb{Z}(3)$ itself. ∎

The ideas in Example 6(b) generalize. If $R_1, \ldots, R_n$ is a list of rings, not necessarily distinct, we make the product $R_1 \times \cdots \times R_n$ into a ring by defining

$$\langle r_1, \ldots, r_n \rangle + \langle s_1, \ldots, s_n \rangle = \langle r_1 + s_1, \ldots, r_n + s_n \rangle$$

and

$$\langle r_1, \ldots, r_n \rangle \cdot \langle s_1, \ldots, s_n \rangle = \langle r_1 \cdot s_1, \ldots, r_n \cdot s_n \rangle,$$

where the operations in each coordinate are the operations defined on the corresponding ring. If $h_1, \ldots, h_n$ are homomorphisms from some ring R to $R_1, \ldots, R_n$, respectively, then one can check [Exercise 12] that the mapping h from R to $R_1 \times \cdots \times R_n$ defined by $h(r) = \langle h_1(r), \ldots, h_n(r) \rangle$ is a homomorphism. Its kernel is $\{r \in R : h_i(r) = 0 \text{ for } i = 1, \ldots, n\}$, i.e., it is the intersection of the kernels of $h_1, \ldots, h_n$.

Suppose now that $I_1, \ldots, I_n$ are ideals of R, and that for $j = 1, \ldots, n$ each h_j is the natural homomorphism from R onto R/I_j given by $h_j(r) = r + I_j$. Then the homomorphism h described in the last paragraph is defined by $h(r) = \langle r + I_1, \ldots, r + I_n \rangle$ for $r \in R$. Since I_j is the kernel of h_j, we obtain the following.

Theorem 3   Let R be a ring with ideals $I_1, \ldots, I_n$. Then $I_1 \cap \cdots \cap I_n$ is an ideal of R and $R/(I_1 \cap \cdots \cap I_n)$ is isomorphic to a subring of $(R/I_1) \times \cdots \times (R/I_n)$.

In Example 6(b), with $I_1 = 2\mathbb{Z}$ and $I_2 = 3\mathbb{Z}$, the ring $\mathbb{Z}/(2\mathbb{Z} \cap 3\mathbb{Z})$ was isomorphic to the whole ring $(\mathbb{Z}/2\mathbb{Z}) \times (\mathbb{Z}/3\mathbb{Z})$, but in general $h(R)$ is only a subring of $R_1 \times \cdots \times R_n$. For example, in $\mathbb{Z}$ we have $6\mathbb{Z} \cap 10\mathbb{Z} \cap 15\mathbb{Z} = 60\mathbb{Z}$ [check this] and so $\mathbb{Z}/(6\mathbb{Z} \cap 10\mathbb{Z} \cap 15\mathbb{Z}) = \mathbb{Z}/60\mathbb{Z}$ has 60 members, while $(\mathbb{Z}/6\mathbb{Z}) \times (\mathbb{Z}/10\mathbb{Z}) \times (\mathbb{Z}/15\mathbb{Z})$ has $6 \cdot 10 \cdot 15 = 900$ elements. Exercise 9 gives another example.

There is a great deal more which can be said about rings and fields. We have only introduced the most basic ideas and a few examples, but we hope to have given some feeling for the kinds of questions it might be reasonable to ask about these systems and the kinds of answers one might get. The study of groups, rings and fields makes up a large part of the area of mathematics called abstract algebra. At this point you are in a good position to read an introductory book in this area.

EXERCISES 11.6

In these exercises, the words "homomorphism" and "isomorphism" mean "ring homomorphism" and "ring isomorphism."

1. Which of the following sets are subrings of $(\mathbb{R}, +, \cdot)$?
   (a) $2\mathbb{Z}$
   (b) $2\mathbb{R}$
   (c) $\mathbb{N}$
   (d) $\{m + n\sqrt{2} : m, n \in \mathbb{Z}\}$
   (e) $\{m/2 : m \in \mathbb{Z}\}$
   (f) $\{m/2^a : m \in \mathbb{Z}, a \in \mathbb{P}\}$

2. (a) For each subset in Exercise 1 which is a subring of $\mathbb{R}$ verify closure under addition and multiplication.
   (b) For each subset in Exercise 1 which is not a subring of $\mathbb{R}$ give a property of subrings which the subset does not satisfy.

3. Which of the following functions are ring homomorphisms? Justify your answer in each case.
   (a) $h : \text{FUN}(\mathbb{R}, \mathbb{R}) \to \mathbb{R}$ defined by $h(f) = f(0)$.
   (b) $h : \mathbb{R} \to \mathbb{R}$ defined by $h(r) = r^2$.
   (c) $h : \mathbb{R} \to \text{FUN}(\mathbb{R}, \mathbb{R})$ defined by $(h(r))(x) = r$. I.e., $h(r)$ is the constant function on $\mathbb{R}$ having value r at every x.
   (d) $h : \mathbb{Z} \to \mathbb{R}$ defined by $h(n) = n$.
   (e) $h : \mathbb{Z}/3\mathbb{Z} \to \mathbb{Z}/6\mathbb{Z}$ defined by $h(n + 3\mathbb{Z}) = 2n + 6\mathbb{Z}$.

4. Verify that the distributive laws hold in $(\mathbb{Z}(p), +_p, *_p)$.

5. Describe the ideals in the trivial ring G of Example 1(d).

6. Find the kernel of the homomorphism h from $\text{FUN}(\mathbb{R}, \mathbb{R})$ to $\mathbb{R}$ in Example 4(c).

7. Consider the ring $\mathbb{Z}$. Write each of the following in the form $n\mathbb{Z}$ with $n \in \mathbb{N}$.
   (a) $6\mathbb{Z} \cap 8\mathbb{Z}$
   (b) $6\mathbb{Z} + 8\mathbb{Z}$
   (c) $3\mathbb{Z} + 2\mathbb{Z}$
   (d) $6\mathbb{Z} + 10\mathbb{Z} + 15\mathbb{Z}$

8. Show that in a ring $(-a) \cdot b = -(a \cdot b) = a \cdot (-b)$ for every a and b.

9. (a) Verify that the mapping h from $\mathbb{Z}(12)$ to $\mathbb{Z}(4) \times \mathbb{Z}(6)$ given by $h(m) = \langle \mathrm{Rem}_4(m), \mathrm{Rem}_6(m) \rangle$ is a well-defined homomorphism.
   (b) Find the kernel of h.
   (c) Find an element of $\mathbb{Z}(4) \times \mathbb{Z}(6)$ which is not in the image of h.
   (d) Which elements in $\mathbb{Z}(12)$ are mapped to $\langle 1, 3 \rangle$ by h?

10. Let I be an ideal of a ring R.
    (a) Show that the mapping $r \to r + I$ is a ring homomorphism of R onto R/I.
    (b) Find the kernel of this homomorphism.

11. (a) Show that $(\mathbb{Z}(6), +_6, *_6)$ is not a field.
    (b) Show that $(\mathbb{Z}(5), +_5, *_5)$ is a field.
    (c) Show that if F and K are fields, then $F \times K$ is not a field.

12. If $R_1, \ldots, R_n$ are rings and if $h_1, \ldots, h_n$ are homomorphisms from a ring R into $R_1, \ldots, R_n$, respectively, then the mapping h defined by $h(r) = \langle h_1(r), \ldots, h_n(r) \rangle$ is a homomorphism of R into $R_1 \times \cdots \times R_n$. Verify this fact for $n = 2$.

13. (a) Show that if h is a ring homomorphism from a field F to a ring R, then either $h(a) = 0$ for all $a \in F$ or h is one-to-one.
    (b) Show that if I is an ideal of the ring R such that R/I is a field, then I and R are the only ideals of R which contain I.

14. Let R be a commutative ring with identity.
    (a) Show that R is an integral domain if and only if the mapping $a \to r \cdot a$ from R to R is one-to-one for each non-0 r in R.
    (b) Show that R is a field if and only if this mapping is a one-to-one correspondence of R onto itself for each non-0 r in R.
    (c) Show that every finite integral domain is a field.

15. (a) Find an ideal I of $\mathbb{Z}$ for which $\mathbb{Z}/I$ is isomorphic to $\mathbb{Z}(3) \times \mathbb{Z}(5)$.
    (b) Describe an isomorphism between $\mathbb{Z}(12)$ and $\mathbb{Z}(3) \times \mathbb{Z}(4)$.
    (c) Show that $\mathbb{Z}$ has no ideal I with $\mathbb{Z}/I$ isomorphic to $\mathbb{Z}(2) \times \mathbb{Z}(2)$.

16. (a) Draw a Hasse diagram for the set of ideals of $\mathbb{Z}(12)$, partially ordered by inclusion $\subseteq$.
    (b) Repeat part (a) for the ring $\mathbb{Z}(3) \times \mathbb{Z}(4)$.

17. Consider the ring $R = \mathrm{POLY}(\mathbb{Z})$ of polynomials in x with integer coefficients.
    (a) Describe the members of the ideals $R \cdot 2$, $R \cdot x$ and $R \cdot 2 + R \cdot x$.
    (b) Show that there is no polynomial p in R for which $R \cdot p = R \cdot 2 + R \cdot x$.

18. The set $\mathbb{B} \times \mathbb{B}$ can be made into a ring in another way besides the one described in the text. Define $+$ and $\cdot$ by the tables:

$+$	$\langle 0, 0 \rangle$	$\langle 1, 0 \rangle$	$\langle 0, 1 \rangle$	$\langle 1, 1 \rangle$
$\langle 0, 0 \rangle$	$\langle 0, 0 \rangle$	$\langle 1, 0 \rangle$	$\langle 0, 1 \rangle$	$\langle 1, 1 \rangle$
$\langle 1, 0 \rangle$	$\langle 1, 0 \rangle$	$\langle 0, 0 \rangle$	$\langle 1, 1 \rangle$	$\langle 0, 1 \rangle$
$\langle 0, 1 \rangle$	$\langle 0, 1 \rangle$	$\langle 1, 1 \rangle$	$\langle 0, 0 \rangle$	$\langle 1, 0 \rangle$
$\langle 1, 1 \rangle$	$\langle 1, 1 \rangle$	$\langle 0, 1 \rangle$	$\langle 1, 0 \rangle$	$\langle 0, 0 \rangle$

$\cdot$	$\langle 0, 0 \rangle$	$\langle 1, 0 \rangle$	$\langle 0, 1 \rangle$	$\langle 1, 1 \rangle$
$\langle 0, 0 \rangle$	$\langle 0, 0 \rangle$	$\langle 0, 0 \rangle$	$\langle 0, 0 \rangle$	$\langle 0, 0 \rangle$
$\langle 1, 0 \rangle$	$\langle 0, 0 \rangle$	$\langle 1, 0 \rangle$	$\langle 0, 1 \rangle$	$\langle 1, 1 \rangle$
$\langle 0, 1 \rangle$	$\langle 0, 0 \rangle$	$\langle 0, 1 \rangle$	$\langle 1, 1 \rangle$	$\langle 1, 0 \rangle$
$\langle 1, 1 \rangle$	$\langle 0, 0 \rangle$	$\langle 1, 1 \rangle$	$\langle 1, 0 \rangle$	$\langle 0, 1 \rangle$

Verify that $\mathbb{B} \times \mathbb{B}$ is an additive group and that the non-0 elements form a group under multiplication. *Suggestion*: Save work by exhibiting known groups isomorphic to your alleged groups. [Finite fields such as this are important in algebraic coding to minimize the effects of noise on transmission channels.]

CHAPTER HIGHLIGHTS

Concepts

algebraic system
  semigroup, monoid, group
  ring, integral domain, field
    identity, zero
generate, set of generators
order of an element
coset, right coset
homomorphism
  isomorphism
  kernel, normal subgroup, ideal
  natural operation on G/K
  natural isomorphism of G/K onto $h(G)$
group of permutations acting on a set
  subgroup fixing something
  orbit
  coloring, G-equivalent colorings

Facts

Intersections of subsemigroups [subgroups, subrings, normal subgroups, ideals] are subsemigroups [subgroups, etc.].

The subsemigroup [subgroup] generated by A consists of all products of members of A [and their inverses].

The subgroups of $(\mathbb{Z}, +)$ are the sets $n\mathbb{Z}$, all of which are cyclic groups. They are the ideals of $(\mathbb{Z}, +, \cdot)$.

The cosets of a subgroup partition a group into subsets of equal size.

Lagrange's theorem: $|G| = |G/H| \cdot |H|$.

The order of an element of G divides $|G|$.

Cosets of normal subgroups [of ideals] can be combined in a natural way to form a group [ring].

Fundamental Theorem: If a group homomorphism h on a group G has kernel K, then $G/K \simeq h(G)$.

A group or ring homomorphism is one-to-one if and only if its kernel consists just of the identity element.

A homomorphism takes identities, inverses, subgroups, ideals to corresponding objects in the image.

Every subgroup [ideal] of the image of h is the image of a subgroup [ideal] which contains the kernel of h.

Cayley's theorem: Every group is isomorphic to a group of permutations.

The size of the G-orbit of s is the number of cosets of the subgroup fixing s, i.e., $|Gs| = |G/G_s|$.

$$|S| = \sum_{i=1}^{m} |G/G_{s_i}| \text{ with one } s_i \text{ from each orbit.}$$

The number of orbits of G acting on S is the average number of elements of S fixed by members of G.

The number of orbits of G on FUN(S, C) is $\dfrac{1}{|G|} \sum_{g \in G} |C|^{m(g)}$.

$$C(k) = \frac{1}{|G|} \sum_{g \in G} k^{m(g)}.$$

If $I_1, \ldots, I_n$ are ideals of R, then $R/(I_1 \cap \cdots \cap I_n)$ is isomorphic to a subring of $(R/I_1) \times \cdots \times (R/I_n)$.

DICTIONARY

The words listed below are of three general sorts: English words with which the reader may not be completely familiar, common English words whose usage in mathematical writing is specialized, and technical mathematical terms which are assumed background for this book. For technical terms introduced in this book, see the index.

absurd. Clearly impossible, being contrary to some evident truth.

all. See *every.*

ambiguous. Capable of more than one interpretation or meaning.

anomaly. Something which is, or appears to be, inconsistent.

any. See *every.*

assume. "assume" and "suppose" mean the same thing and ask that we imagine a situation for the moment.

axiom. An assertion that is accepted and used without a proof. Obvious or self-evident axioms are preferred.

bona fide. Genuine or legitimate.

cf. Compare.

chicanery. Trickery.

class. See *set.*

collapse. To fall or come together.

collection. See *set.*

common factor. The integer m is a "common factor" or "common divisor" of two integers if it divides them both. See *divisible by.*

comparable. Capable of being compared.

conjecture. A guess or opinion, preferably based on some experience or other source of wisdom.

corollary. See *theorem.*

define. Often this looks like an instruction [as in "Define $f(x) = x^2$"] when it is merely a [bad] mathematical way of saying "We define" or "Let."

disprove. The instruction "prove or disprove" means that the assertion should either be proved true or proved false. Which you do, of course, depends on whether the assertion is actually true or not.

distinct. Different.

distinguishable. A collection of objects is regarded as "distinguishable" if there is some property or characteristic that makes it possible to tell different objects apart. In contrast, we would regard ten red marbles of the same size as "indistinguishable."

divisible by. Consider integers m and n. We say that "n is divisible by m," that "m divides n," that "n is a multiple of m," or that "m is a factor of n" if $n = mk$ for some integer k. We write $m|n$ to signify any of these statements. For example, $3|6$, $4|20$ and $8|8$. Also $m|0$ for all m since $0 = mk$ for $k = 0$.

e.g. For example.

entries. The individual numbers or objects in ordered pairs, in matrices or in sequences.

even number. An integer that is exactly divisible by 2, i.e., any integer that can be written as $2k$ where $k \in \mathbb{Z}$. Note that 0 is an even number.

every. The expressions "for every," "for any" and "for all" mean the same thing. They all mean "for all choices of the variable in question," and so they correspond to the quantifier $\forall$ in § 6.1. The expression "for some" means "for at least one" and corresponds to the quantifier $\exists$ in § 6.1. It is generally good practice to avoid the use of "any," which is sometimes misunderstood. For example, "If $p(n)$ is true for any n" usually means "If $p(n)$ is true for some n," not "If $p(n)$ is true for every n."

factor. See *divisible by.*

family. See *set.*

fictitious. Imaginary, not actual.

gcd. If m and n are positive integers, then $\gcd(m, n)$ is the "greatest common divisor" of m and n, that is, the largest integer that m and n are both divisible by. $\operatorname{lcm}(m, n)$ is their "least common multiple," i.e., the smallest positive integer that is a multiple of both m and n. For example $\gcd(10, 25) = 5$ and $\operatorname{lcm}(10, 25) = 50$. gcd and lcm are often most easily calculated by writing m and n as products of primes. For example, for $m = 168$ and $n = 450$ we have $m = 2^3 \cdot 3 \cdot 7$ and $n = 2 \cdot 3^2 \cdot 5^2$, and so

$$\gcd(m, n) = 2 \cdot 3 = 6 \quad \text{and} \quad \operatorname{lcm}(m, n) = 2^3 \cdot 3^2 \cdot 5^2 \cdot 7 = 12{,}600.$$

greatest common divisor. See *gcd.*

inclusion. We sometimes refer to the relation $A \subseteq B$ as an "inclusion" just as we may refer to $A = B$ as an "equality."

indices. Plural of index.

inspection. Something can be seen "by inspection" if it can be seen directly without calculation or modification.

invertible. Having an inverse.

irrational. An "irrational number" is a real number that is not rational, i.e., cannot be written as m/n for $m, n \in \mathbb{Z}$, $n \neq 0$. Examples include $\sqrt{2}$, $\sqrt{3}$, $\sqrt[3]{2}$, π and e.

lcm. See *gcd*.

least common multiple. See *gcd*.

lemma. See *theorem*.

loop. A part of a computer program that is used repeatedly.

matrices. Plural of matrix.

max. For two real numbers a and b, we write $\max\{a, b\}$ for the larger of the two. If $a = b$, then $\max\{a, b\} = a = b$.

min. For two real numbers a and b, we write $\min\{a, b\}$ for the smaller of the two. If $a = b$, then $\min\{a, b\} = a = b$.

multiple of. See *divisible by*.

necessary. We say a property p is "necessary" for a property q if p must hold whenever q holds, i.e., $q \Rightarrow p$. The property p is "sufficient" for q if p is enough to guarantee q, i.e., $p \Rightarrow q$. So p is necessary and sufficient for q provided $p \Leftrightarrow q$.

odd number. An integer that is not an even integer. An odd number can be written as $2k + 1$ for some $k \in \mathbb{Z}$.

permute. To change the order of a sequence of elements.

prime number. An integer ≥ 2 that cannot be written as the product of two integers that are both ≥ 2. The first few primes are $2, 3, 5, 7, 11, 13, 17, 19$.

proposition. See *theorem*.

redundant. Unnecessary or excessive.

relatively prime. Two positive integers m and n are "relatively prime" if they have no common factors, i.e., if $\gcd(m, n) = 1$. See *gcd*.

sequence. A list of things following one another. A formal definition is given in § 3.3.

set. The terms "set," "collection," "class" and "family" are used interchangeably. We tend to refer to families of sets, for example, to avoid the expression "sets of sets."

some. See *every*.

sufficient. See *necessary*.

suppose. See *assume*.

synonym. A word having the same meaning as another word. Two words that have the same meaning are *synonymous*.

theorem. A "theorem," "proposition," "lemma" or "corollary" is some assertion that has been or can be proved. The term "proposition" also has a special use in logic; see § 2.1. Theorems are usually the most important facts. Lemmas are usually not of primary interest and are used to prove later theorems or propositions. Corollaries are usually easy consequences of theorems or propositions just presented.

truncate. To shorten by cutting.

unambiguous. Not ambiguous. See *ambiguous.*

underlying set. The basic set on which the objects [like functions or operations] are defined.

vertices. Plural of vertex.

ANSWERS AND HINTS

Wise students will only look at these answers and hints after seriously working on the problems. When only hints are given, you should write out the details to check understanding as well as to get practice in communicating mathematical ideas.

Section 0.1, page 7

1. All but (b) and (d).
3. (c) is the only cycle listed.
5. (a) $s\, u\, t\, v$ is one. (c) $u\, s\, t\, u\, v\, w\, y\, x\, z\, y$. There are no longest paths connecting vertices. For example, one can go from s to v as follows: $s\, u\, v\, u\, v\, u\, v \cdots u\, v$.
7. (a) is true and (b) is false. To see the failure of (b), consider the graph in Figure 9(a).
9. (a) and (b) are now both true.
11. Use a graph like Figure 3(b) of § 0.2.
13. If a cycle contains a loop at v, it must have vertex sequence $v\, v$, since otherwise v would be repeated in the middle of the cycle.

Section 0.2, page 14

1. Only Figure 9(b) has an Euler circuit. Only Figures 9(a) and 9(c) have Hamilton circuits.
3. It won't break down until the second visit to the other vertex of degree 3, namely t.
5. Figures 2(a) and 2(b) have Hamilton circuits.
7. Only the edges e_6 and e_7 are interchanged. Of course the new spanning tree has the same total weight, namely 1330, since they are both *minimal* spanning trees.
9. (b) A tree with n vertices has exactly $n - 1$ edges.
11. (b) For each pair of vertices in a tree, there is exactly one path connecting them that repeats no edges.
13. (a) $t\, u,\, t\, v,\, u\, v,\, u\, w,\, v\, w,\, x\, y,\, x\, z$ and $y\, z$. (b) $s\, t$ and $w\, x$.
15. (a) If the tree has n vertices, it must have $n - 7$ of degree 1. So we must have

$$5 + 5 + 3 + 3 + 3 + 2 + 2 + 1 \cdot (n - 7) = 2n - 2$$

by Exercise 14. Solve for n.

Section 0.3, page 22

Here are some answers for the trees labeled as indicated.

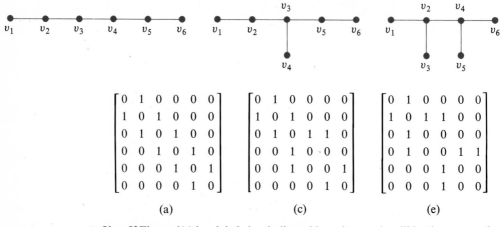

$$\begin{bmatrix} 0 & 1 & 0 & 0 & 0 & 0 \\ 1 & 0 & 1 & 0 & 0 & 0 \\ 0 & 1 & 0 & 1 & 0 & 0 \\ 0 & 0 & 1 & 0 & 1 & 0 \\ 0 & 0 & 0 & 1 & 0 & 1 \\ 0 & 0 & 0 & 0 & 1 & 0 \end{bmatrix} \qquad \begin{bmatrix} 0 & 1 & 0 & 0 & 0 & 0 \\ 1 & 0 & 1 & 0 & 0 & 0 \\ 0 & 1 & 0 & 1 & 1 & 0 \\ 0 & 0 & 1 & 0 & 0 & 0 \\ 0 & 0 & 1 & 0 & 0 & 1 \\ 0 & 0 & 0 & 0 & 1 & 0 \end{bmatrix} \qquad \begin{bmatrix} 0 & 1 & 0 & 0 & 0 & 0 \\ 1 & 0 & 1 & 1 & 0 & 0 \\ 0 & 1 & 0 & 0 & 0 & 0 \\ 0 & 1 & 0 & 0 & 1 & 1 \\ 0 & 0 & 0 & 1 & 0 & 0 \\ 0 & 0 & 0 & 1 & 0 & 0 \end{bmatrix}$$

$$\text{(a)} \qquad\qquad\qquad \text{(c)} \qquad\qquad\qquad \text{(e)}$$

3. Yes. If Figure 6(c) is relabeled as indicated here, its matrix will be the same as for Figure 6(b).

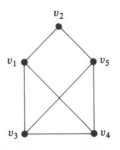

5. There are nine. "All" here means that every such graph is isomorphic to one of the nine, and no two of the nine are isomorphic to each other.

7. Consider loops to guess which two they are, and draw both. Then relabel the vertices on one of them.

9. (a) Either work it out pair by pair or multiply

$$\begin{bmatrix} 0 & 1 & 0 & 0 \\ 1 & 1 & 0 & 0 \\ 0 & 0 & 1 & 0 \\ 0 & 0 & 0 & 0 \end{bmatrix}$$

by itself and then draw the graph.

(b) No. (c) $\begin{bmatrix} 1 & 1 & 0 & 0 \\ 1 & 2 & 0 & 0 \\ 0 & 0 & 1 & 0 \\ 0 & 0 & 0 & 0 \end{bmatrix}.$ (d) $\begin{bmatrix} 1 & 1 & 0 & 0 \\ 1 & 1 & 0 & 0 \\ 0 & 0 & 1 & 0 \\ 0 & 0 & 0 & 0 \end{bmatrix}.$

11. (b) No.   (c) $\begin{bmatrix} 1 & 0 & 1 & 0 & 0 & 0 \\ 0 & 2 & 0 & 1 & 1 & 0 \\ 1 & 0 & 3 & 0 & 0 & 1 \\ 0 & 1 & 0 & 1 & 1 & 0 \\ 0 & 1 & 0 & 1 & 2 & 0 \\ 0 & 0 & 1 & 0 & 0 & 1 \end{bmatrix}$.   (d) $\begin{bmatrix} 1 & 1 & 1 & 1 & 1 & 1 \\ 1 & 1 & 1 & 1 & 1 & 1 \\ 1 & 1 & 1 & 1 & 1 & 1 \\ 1 & 1 & 1 & 1 & 1 & 1 \\ 1 & 1 & 1 & 1 & 1 & 1 \\ 1 & 1 & 1 & 1 & 1 & 1 \end{bmatrix}$.

13. The degree of vertex v_i is $2 \cdot \mathbf{M}[i, i]$ plus the sum of the other numbers in the ith row. If this doesn't look like a formula, how's this:

$$\text{degree}(v_i) = \mathbf{M}[i, i] + \mathbf{M}[i, 1] + \mathbf{M}[i, 2] + \cdots + \mathbf{M}[i, n]?$$

Section 1.1, page 32

1. (a) 0, 5, 10, 15, 20, say.   (c) $\varnothing$, $\{1\}$, $\{2, 3\}$, $\{3, 4\}$, $\{5\}$, say.   (e) 1, 1/2, 1/3, 1/4, 1/73, say.   (g) 1, 2, 4, 16, 18, say.
3. (a) ϵ, a, ab, cab, ba, say.   (c) $aaaa$, $aaab$, $aabb$, etc.
5. (a) 0.   (c) 138.   (e) 73.   (g) 0.   (i) ∞.   (k) ∞.   (m) ∞.   (o) ∞.
7. $A \subseteq A$, $B \subseteq B$, C is a subset of A and C, D is a subset of A, B and D.
9. (a) aba is in all three and has length 3 in each.   (c) cba is in Σ_1^* and length(cba) = 3.   (e) $caab$ is in Σ_1^* with length 4 and is in Σ_2^* with length 3.
11. (a) Yes.   (c) Delete first letters from the string until no longer possible. If ϵ is reached, the original string is in Σ^*. Otherwise, it isn't.

Section 1.2, page 39

1. (a) $\{1, 2, 3, 5, 7, 9, 11\}$.   (c) $\{1, 5, 7, 9, 11\}$.   (e) $\{3, 6, 12\}$.   (g) 16.
3. (a) $[2, 3]$.   (c) $[0, 2)$.   (e) $(-\infty, 0) \cup (3, \infty)$.
5. (a) $\varnothing$.   (c) $\varnothing$.   (e) $\{\epsilon, ab, ba\}$.
7. $A \oplus A = \varnothing$ and $A \oplus \varnothing = A$.
9. $(A \cap B \cap C)^c = (A \cap B)^c \cup C^c$ and $(A \cap B)^c = A^c \cup B^c$ by DeMorgan law 9b. Now substitute.
11. (a) False. Try $A = \varnothing$.   (c) True. Show $x \in B$ implies $x \in C$ by considering two cases: $x \in A$ and $x \notin A$. Similarly, $x \in C$ implies $x \in B$.   (e) The hint to part (c) also applies here.

13.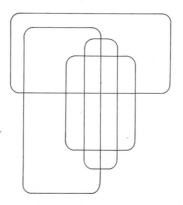

See *Mathematics Magazine 58* (Sept. 1985), p. 251 for five rectangles forming a Venn diagram.

Section 1.3, page 48

1. (a) 42.   (c) 1.   (e) 154.
3. (a) 3, 12, 39 and 120.   (c) 3, 9 and 45.
5. (a) $-2, 2, 0, 0$ and 0.
7. (a) D_{10}.   (c) D_4.   (e) $\{k \in \mathbb{P} : k$ is a multiple of 4 or of 6 but not of 12$\} = \{4, 6, 8, 16, 18, 20, 28, \ldots\}$.
9. (a) $A_2 = \{2, 3, 4, \ldots\}$, $B_2 = \{0, 1, 2, 3, 4\}$,
   $A_4 = \{4, 5, 6, \ldots\}$, $B_4 = \{0, 1, 2, 3, 4, 5, 6, 7, 8\}$.
   (c) $A_1 \cap B_1 = \{1, 2\}$, $A_2 \cap B_2 = \{2, 3, 4\}$, $A_3 \cap B_3 = \{3, 4, 5, 6\}$, $A_7 \cap B_7 = \{7, 8, 9, \ldots, 14\}$, $A_1^c = \{0\}$, $A_2^c = \{0, 1\}$, $A_3^c = \{0, 1, 2\}$ and $A_7^c = \{0, 1, 2, 3, 4, 5, 6\}$.
   (e) A_3, $\mathbb{N}$, B_3 and $\varnothing$.
11. (a) $\{w \in \Sigma^* : \text{length}(w) \leq n\}$.
13. Imitate the proof in Example 3.
15. Divide $11^1 - 4^1 = 7$, $11^2 - 4^2 = 105$ and $11^3 - 4^3 = 1267$ by 7.
17. $p(n)$ is true for all $n \geq 4$.

Section 1.4, page 57

1. (a) $\langle a, a \rangle, \langle a, b \rangle, \langle a, c \rangle, \langle b, a \rangle$, etc. There are nine altogether.
   (c) $\langle a, a \rangle, \langle b, b \rangle$.
3. (a) $\langle 0, 0 \rangle, \langle 1, 1 \rangle, \langle 2, 2 \rangle, \ldots, \langle 6, 6 \rangle$, say.
   (c) $\langle 6, 1 \rangle, \langle 6, 2 \rangle, \langle 6, 3 \rangle, \ldots, \langle 6, 7 \rangle$, say.
   (e) The set has exactly five elements. List them.
5. (a)

7. (a) To find $A \cap B$, you need all $\langle x, y \rangle$ satisfying $2x - y = 4$ *and* $x + 3y = 9$. Solve this system of two equations in two unknowns. Answer: $A \cap B = \{\langle 3, 2 \rangle\}$.   (c) $\{\langle 9/7, 18/7 \rangle\}$.
9. (a) $\langle \epsilon, \epsilon \rangle, \langle a, a \rangle, \langle b, b \rangle, \langle a, b \rangle, \langle aa, bc \rangle$, etc.
   (b) There are eight, including $\langle a, b, a \rangle$ and $\langle b, a, a \rangle$.

11. (a) $\begin{bmatrix} -1 & 1 & 4 \\ 0 & 3 & 2 \\ 2 & -2 & 3 \end{bmatrix}$. (c) $\begin{bmatrix} 5 & 8 & 7 \\ 5 & 1 & 5 \\ 7 & 3 & 5 \end{bmatrix}$. (e) $\begin{bmatrix} 5 & 5 & 7 \\ 8 & 1 & 3 \\ 7 & 5 & 5 \end{bmatrix}$.

(g) $\begin{bmatrix} 12 & 12 & 8 \\ 12 & -4 & 8 \\ 8 & 8 & 4 \end{bmatrix}$. (i) $\begin{bmatrix} 4 & 8 & 9 \\ 6 & 4 & 3 \\ 11 & 5 & 8 \end{bmatrix}$.

13. (a) $\begin{bmatrix} 1 & -1 & 1 & -1 \\ -1 & 1 & -1 & 1 \\ 1 & -1 & 1 & -1 \end{bmatrix}$. (c) Not defined. (e) $\begin{bmatrix} 3 & 2 & 5 & 4 \\ 2 & 5 & 4 & 7 \\ 5 & 4 & 7 & 6 \end{bmatrix}$.

15. (a) $\begin{bmatrix} 1 & 0 & 0 \\ 0 & 1 & 0 \\ 0 & 0 & 1 \end{bmatrix}$, $\begin{bmatrix} 1 & 0 & 0 \\ 0 & 0 & 1 \\ 0 & 1 & 0 \end{bmatrix}$, $\begin{bmatrix} 0 & 1 & 0 \\ 1 & 0 & 0 \\ 0 & 0 & 1 \end{bmatrix}$, $\begin{bmatrix} 0 & 1 & 0 \\ 0 & 0 & 1 \\ 1 & 0 & 0 \end{bmatrix}$, $\begin{bmatrix} 0 & 0 & 1 \\ 1 & 0 & 0 \\ 0 & 1 & 0 \end{bmatrix}$,

$\begin{bmatrix} 0 & 0 & 1 \\ 0 & 1 & 0 \\ 1 & 0 & 0 \end{bmatrix}$.

(b) Four of them equal their transposes.

17. (a) $\begin{bmatrix} 1 & 0 \\ n & 1 \end{bmatrix}$. (c) $\{n \in \mathbb{N} : n \text{ is odd}\}$.

19. Clearly $s_1 = s_2$ and $t_1 = t_2$ imply $\langle s_1, t_1 \rangle = \langle s_2, t_2 \rangle$. If $\langle s_1, t_1 \rangle = \langle s_2, t_2 \rangle$, then

$$\{s_1\} = \{s_2\} \quad \text{and} \quad \{s_1, t_1\} = \{s_2, t_2\}.$$

Show that these equalities force $s_1 = s_2$ and $t_1 = t_2$.

Section 2.1, page 66

1. (a) $p \wedge q$. (c) $\neg p \to (\neg q \wedge r)$. (e) $\neg r \to q$.
3. (b) In Example 2, parts (b) and (d) are true and parts (a) and (c) are false.
5. The proposition is true for all $x, y \in [0, \infty)$, but is false when applied to all $x, y \in \mathbb{R}$.
7. (a) $\neg r \to \neg q$. (c) If it is false that $x = 0$ or $x = 1$, then $x^2 \neq x$.
9. (a) $3^3 < 3^3$ is false.
11. (a) $(-1 + 1)^2 = 0 < 1 = (-1)^2$.
   (c) No. If $x \geq 0$, then $(x + 1)^2 = x^2 + 2x + 1 > x^2$. In fact, if $x \geq -\frac{1}{2}$, then $2x + 1 \geq 0$ and $(x + 1)^2 \geq x^2$.
13. (a) $\langle 0, -1 \rangle$
15. (a) $p \to q$. (b) $p \to r$. (c) $\neg r \to p$. (d) $q \to p$. (e) $r \to q$.
   (f) $r \to (q \vee p)$ or $(r \to q) \vee p$.

Section 2.2, page 76

1. (a) Converse: $(q \wedge r) \to p$
   Contrapositive: $\neg(q \wedge r) \to \neg p$.
   (c) Converse: If $3 + 3 = 8$, then $2 + 2 = 4$.
   Contrapositive: If $3 + 3 \neq 8$, then $2 + 2 \neq 4$.
3. (a) $q \to p$. (c) $p \to q$, $\neg q \to \neg p$, $\neg p \vee q$.
5. (a) 0. (c) 1.

Note. For truth tables, only the final columns are given.

7. (a)

p	q	¬(p ∧ q)
0	0	1
0	1	1
1	0	1
1	1	0

(c)

p	q	¬p ∧ ¬q
0	0	1
0	1	0
1	0	0
1	1	0

9.

p	q	r	final column
0	0	0	1
0	0	1	1
0	1	0	1
0	1	1	0
1	0	0	1
1	0	1	1
1	1	0	1
1	1	1	0

11.

p	q	r	part (a)	part (b)
0	0	0	0	0
0	0	1	1	0
0	1	0	1	1
0	1	1	1	0
1	0	0	1	1
1	0	1	1	0
1	1	0	1	1
1	1	1	1	0

15. (a) One need only consider rows in which $[(p \land r) \to (q \land r)]$ is false, i.e., $(p \land r)$ is true and $(q \land r)$ is false. This leaves one row to consider:

p	q	r
1	0	1

(c) One need only consider rows in which $[(p \land r) \to (q \land s)]$ is false, i.e., $(p \land r)$ is true and $(q \land s)$ is false. This leaves three rows to consider:

p	q	r	s
1	0	1	0
1	0	1	1
1	1	1	0

17. Let p = "He finished dinner" and q = "He was sent to bed." Then p is true and q is true, so the logician's statement $\neg p \to \neg q$ has truth value True. She was logically correct, but not very nice.

Section 2.3, page 82

1. We obtain successive equivalences using the indicated rules and suitable substitutions:

$[(p \lor r) \land (q \to r)]$	
$[(p \lor r) \land (\neg q \lor r)]$	rule 10a
$[(r \lor p) \land (r \lor \neg q)]$	rule 2a
$r \lor (p \land \neg q)$	rule 4a
$(p \land \neg q) \lor r$	rule 2a
$\neg(\neg p \lor \neg\neg q) \lor r$	rule 8d
$\neg(\neg p \lor q) \lor r$	rule 1
$\neg(p \to q) \lor r$	rule 10a
$(p \to q) \to r$	rule 10a

3. For example, rule 23 corresponds to the rule of inference

$$P \leftrightarrow Q$$
$$Q \leftrightarrow R$$
$$\therefore \ P \leftrightarrow R$$

5. (b)

p	$p \oplus p$
0	0
1	0

(d)

p	$(p \oplus p) \oplus p$
0	0
1	1

7. Use truth tables.

9. (a) The propositions are not equivalent, as can be verified by comparing the first or third rows of their truth tables.   (b) and (c) are true. It is probably easiest to verify them using truth tables.

11. (a) See rules 11a and 11b.   (c) No. Any proposition involving only p, q, $\wedge$ and $\vee$ will have truth value 0 whenever p and q both have truth values 0. See Exercise 11 of §6.4.

13. Consider the row of the truth table where p is false and q is true.

15. Both possible truth values, true or false, lead to a contradiction. So this is not a proposition. It is an example of what is called a **paradox**.

Section 2.4, page 87

1. (a)

1. $p \to (q \vee r)$	hypothesis
2. $q \to s$	hypothesis
3. $r \to t$	hypothesis
4. $(q \vee r) \to (s \vee t)$	2, 3; rule of inference based on 26a
5. $p \to (s \vee t)$	1, 4; hypothetical syllogism (rule 33)

(c)

1. $p \to (q \to r)$	hypothesis
2. $p \vee \neg s$	hypothesis
3. q	hypothesis
4. $\neg s \vee p$	2; commutative law 2a
5. $s \to p$	4; implication (rule 10a)
6. $s \to (q \to r)$	1, 5; hypothetical syllogism (rule 33)
7. $(s \wedge q) \to r$	6; exportation (rule 14)
8. $q \to [s \to (q \wedge s)]$	rule 22
9. $s \to (q \wedge s)$	3, 8; modus ponens (rule 30)
10. $s \to (s \wedge q)$	9; commutative law 2b
11. $s \to r$	7, 10; hypothetical syllogism (rule 33)

3. Lines 1, 2, 3 are hypotheses. Line 4 is the negation of the conclusion. Line 5 follows from lines 2 and 4 and the rule of inference 31 (modus tollens). Line 6 follows from line 5 and a DeMorgan law. Line 7 follows from line 6 and the rule of inference 29 (simplification). Etc.

5. (a) Let c = "my computations are correct," b = "I pay the electric bill," r = "I run out of money," and p = "the power stays on." Then the theorem is: if

$(c \wedge b) \to r$ and $\neg b \to \neg p$, then $(\neg r \wedge p) \to \neg c$. Here is a formal proof; you should supply the missing explanations. Or give your own proof.

1. $(c \wedge b) \to r$
2. $\neg b \to \neg p$
3. $\neg r \to \neg(c \wedge b)$
4. $\neg r \to (\neg c \vee \neg b)$
5. $p \to b$
6. $(\neg r \wedge p) \to ((\neg c \vee \neg b) \wedge b)$ 4, 5; rule of inference corresponding to rule 26b
7. $(\neg r \wedge p) \to (b \wedge (\neg c \vee \neg b))$
8. $(\neg r \wedge p) \to ((b \wedge \neg c) \vee (b \wedge \neg b))$
9. $(\neg r \wedge p) \to ((b \wedge \neg c) \vee \text{contradiction})$
10. $(\neg r \wedge p) \to (b \wedge \neg c)$
11. $(\neg r \wedge p) \to (\neg c \wedge b)$
12. $(\neg c \wedge b) \to \neg c$ simplification (rule 17)
13. $(\neg r \wedge p) \to \neg c$

(c) Let $j =$ "I get the job," $w =$ "I work hard," $p =$ "I get promoted," and $h =$ "I will be happy." Then the theorem is: if $(j \wedge w) \to p$, $p \to h$ and $\neg h$, then $\neg j \vee \neg w$. You should supply the formal proof.

7. (a) Here is a formal proof; you should supply all the explanations. Or give your own proof.

1. $p \to q$
2. $r \to (p \wedge s)$
3. $(q \wedge s) \to (p \wedge t)$
4. $\neg t$
5. $\neg(p \to \neg r)$
6. $p \wedge r$
7. $(p \wedge s) \to s$
8. $r \to s$

9. $(p \wedge r) \to (q \wedge s)$
10. $q \wedge s$
11. $p \wedge t$
12. $t \wedge p$
13. t
14. $t \wedge \neg t$
15. contradiction

9. (a) With suggestive notation, the hypotheses are $\neg b \to \neg s$, $s \to p$ and $\neg p$. We can infer $\neg s$ using the contrapositive. We cannot infer either b or $\neg b$. Of course, we can infer more complex propositions, like $\neg p \vee s$ or $(s \wedge b) \to p$.
(c) The hypotheses are $(m \vee f) \to c$, $n \to c$ and $\neg n$. No interesting conclusions, such as m or $\neg c$, can be inferred.

Section 2.5, page 95

1. Give a direct proof using the following fact. If m and n are even integers, then there exist j and k in $\mathbb{Z}$ so that $m = 2j$ and $n = 2k$.
3. This can be done via four cases: see Example 5.
5. This can be done via three cases: (i) $n = 3k$ for some $k \in \mathbb{N}$; (ii) $n = 3k + 1$ for some $k \in \mathbb{N}$; (iii) $n = 3k + 2$ for some $k \in \mathbb{N}$.
7. (a) Give a direct proof, as in Exercise 1.   (c) False. Give an example.
(e) False. Finding an example will be easy; the sum of four consecutive integers is *never* divisible by 4.
9. (a) Trivially true.   (c) Vacuously true.

11. Since $\{n \in \mathbb{N} : n$ is prime and $n \leq 10^{21}\}$ is a finite set, Example 3 shows that $\{n \in N : n$ is prime and $n > 10^{21}\}$ is nonempty. As in Example 10, this set has a least element. The proof relies on Example 3, which is nonconstructive.

13. (a) None of the numbers in the set

$$\{k \in \mathbb{N} : (n + 1)! + 2 \leq k \leq (n + 1)! + (n + 1)\}$$

is prime, since if $2 \leq m \leq n + 1$, then m divides $(n + 1)!$ and so m also divides $(n + 1)! + m$.

(b) Yes. Since $7! = 5040$, the proof shows that all the numbers from 5042 to 5047 are nonprime.

(c) Simply add 5048 to the list in part (b). Incidentally, a sequence of seven nonprimes starts with 90.

Section 2.6, page 104

Induction proofs should be written carefully and completely. These answers will serve only as guides, not as models.

1. Check the basis. For the inductive step, assume the equality holds for n. Then

$$\sum_{k=1}^{n+1} k^2 = \sum_{k=1}^{n} k^2 + (n + 1)^2 = \frac{n(n + 1)(2n + 1)}{6} + (n + 1)^2.$$

Some algebra shows that the right-hand side equals $\dfrac{(n + 1)(n + 2)(2n + 3)}{6}$, and so the equality holds for $n + 1$ whenever it holds for n.

3. The algebra for the inductive step is

$$\sum_{k=0}^{n+1} a^k = \sum_{k=0}^{n} a^k + a^{n+1} = \frac{a^{n+1} - 1}{a - 1} + a^{n+1} = \frac{a^{n+2} - 1}{a - 1}.$$

5. (a) 2, 5, 10, 17.   (b) $S = n^2 + 1$.

7. Show that $11^{n+1} - 4^{n+1} = 11 \cdot (11^n - 4^n) + 7 \cdot 4^n$. Imitate Example 5.

11. (a) The inequality holds for all $n \geq 4$.

(b) Prove $3n < n^2 - 1$ for $n \geq 4$ by induction. For the inductive step, assume $3n < n^2 - 1$; you want $3(n + 1) < (n + 1)^2 - 1$, i.e., $3n + 3 < n^2 + 2n$. Since $3n + 3 < (n^2 - 1) + 3 = n^2 + 2$, it suffices to observe that $2 \leq 2n$.

13. (a) Assume $p(n)$ is true. Then $(n + 1)^2 + 5(n + 1) + 1 = (n^2 + 5n + 1) + (2n + 6)$. $n^2 + 5n + 1$ is even by assumption and $2n + 6$ is clearly even, so $p(n + 1)$ is true.

(b) All propositions $p(n)$ are false. *Moral*: The basis of induction is crucial for mathematical induction.

15. *Hint*: $5^{n+1} - 4(n + 1) - 1 = 5(5^n - 4n - 1) + 16n$.

17. Here $p(n)$ is the proposition "$|\sin nx| \leq n|\sin x|$ for all $x \in \mathbb{R}$." Clearly $p(1)$ holds. By algebra and trigonometry,

$$|\sin (n + 1)x| = |\sin (nx + x)| = |\sin nx \cos x + \cos nx \sin x|$$

$$\leq |\sin nx| \cdot |\cos x| + |\cos nx| \cdot |\sin x| \leq |\sin nx| + |\sin x|.$$

Now assume $p(n)$ is true and show $p(n + 1)$ is true.

Section 3.1, page 116

1. (a) $f(3) = 27$, $f(1/3) = 1/3$, $f(-1/3) = 1/27$, $f(-3) = 27$.
   (c) $\text{Im}(f) = [0, \infty)$.

3. (a) No; S is bigger than T.   (c) Yes. For example, let $f(1) = a$, $f(2) = b$,
   $f(3) = c$, $f(4) = f(5) = d$.   (e) No.

5. (a) $f(\langle 2, 1 \rangle) = 2^2 3^1 = 12$, $f(\langle 1, 2 \rangle) = 2^1 3^2 = 18$, etc.
   (b) If not, $2^m 3^n = 2^{m'} 3^{n'}$ for some $\langle m, n \rangle \neq \langle m', n' \rangle$. Then $m \neq m'$ [why?]. Say
   $m < m'$. Divide both sides by 2^m to get a number that is both odd and even, a
   contradiction.   (c) Consider 5, for instance.

7. (a) Pick b_0 in B. For every $a \in A$, $\text{PROJ}(\langle a, b_0 \rangle) = a$ and so every a in A is in the
   image of PROJ. That is, PROJ maps $A \times B$ onto A.

9. $\{n \in \mathbb{Z} : n \text{ is even}\}$.

11. (a) $f \circ g \circ h(x) = (x^8 + 1)^{-3} - 4(x^8 + 1)^{-1}$.
    (c) $h \circ g \circ f(x) = [(x^3 - 4x)^2 + 1]^{-4}$.
    (e) $g \circ g(x) = (x^2 + 1)^2/[1 + (x^2 + 1)^2]$.
    (g) $g \circ h(x) = (x^8 + 1)^{-1}$.

13. Since $g \circ f : S \to U$ and $h : U \to V$, the composition $h \circ (g \circ f)$ is defined and maps S
    to V. A similar remark applies to $(h \circ g) \circ f$. Show that the functions' values agree
    at each $x \in S$.

15. (a) 1, 0, -1 and 0.
    (c) $g \circ f$ is the characteristic function of $\mathbb{Z} \backslash E$. $f \circ f(n) = n - 2$ for all $n \in \mathbb{Z}$.

Section 3.3, page 125

1. (a) $f^{-1}(y) = (y - 3)/2$.   (c) $h^{-1}(y) = 2 + \sqrt[3]{y}$.

3. (a) All of them; verify this.
   (c) $\text{SUM}^{-1}(4)$ has 5 elements, $\text{PROD}^{-1}(4)$ has 3 elements, $\text{MAX}^{-1}(4)$ has 9
   elements, and $\text{MIN}^{-1}(4)$ is infinite.

5. (b) $g(0) = 0$, $g(1) = 0$, $g(2) = 1$, $g(3) = 2$, etc.
   (c) One-to-oneness is easy. Note that $0 \notin \text{Im}(f)$.
   (e) Evaluate $f \circ g$ at 0.

9. Since f and g are invertible, the functions $f^{-1} : T \to S$, $g^{-1} : U \to T$ and
   $f^{-1} \circ g^{-1} : U \to S$ exist. So it suffices to show $(g \circ f) \circ (f^{-1} \circ g^{-1}) = 1_U$ and
   $(f^{-1} \circ g^{-1}) \circ (g \circ f) = 1_S$.

11. (a) Prove the contrapositive: if $s_1, s_2 \in S$ and $f(s_1) = f(s_2)$, then $s_1 = s_2$. The
    proof will be very short.

13. (a) Suppose $t \in f(f^{-1}(B))$. Then $t = f(s)$ for some $s \in f^{-1}(B)$. $s \in f^{-1}(B)$ means
    that $f(s) \in B$. So $t \in B$. This works for any t, so $f(f^{-1}(B)) \subseteq B$.

15. (a) The first sentence shows that the f in the example cannot be one-to-one. At
    the other extreme, if f is constant, $f \circ g = f \circ h$ for all g and h. Provide a specific
    example.
    (c) f maps $\text{Dom}(f)$ onto $\text{Dom}(g) = \text{Dom}(h)$.

Section 3.3, page 130

1. (a) 0, 1/3, 1/2, 3/5, 2/3, 5/7.
   (c) Note that $a_{n+1} = \dfrac{(n + 1) - 1}{(n + 1) + 1} = \dfrac{n}{n + 2}$ for $n \in \mathbb{P}$.

3. (a) $\begin{bmatrix} 0 & -1 \\ 1 & 0 \end{bmatrix}$, $\begin{bmatrix} 1 & 0 \\ 2 & 1 \end{bmatrix}$, $\begin{bmatrix} 2 & 1 \\ 3 & 2 \end{bmatrix}$, $\begin{bmatrix} 3 & 2 \\ 4 & 3 \end{bmatrix}$.

5. (a) 0, 0, 2, 6, 12, 20, 30.

   (b) Just substitute the values into both sides. Induction is not needed.

   (c) Same comment.

7. (a) 0, 0, 1, 1, 2, 2, 3, 3, ...

   (c) $\langle 0, 0 \rangle$, $\langle 0, 1 \rangle$, $\langle 1, 1 \rangle$, $\langle 1, 2 \rangle$, $\langle 2, 2 \rangle$, $\langle 2, 3 \rangle$, $\langle 3, 3 \rangle$, ...

9. (a) $k = 2$

   (c) $k = -1$ Hint: $n - 1 \geq n - (n/2) = n/2$ for $n \geq 2$.

   (e) $k = 2$ Hint: $(n^3 + 2n - 1)/(n + 1) < (n^3 + 2n - 1)/n$.

   (g) $k = 1$ Hint: $n^2 + 1 \leq 2n^2$.

11. (a) If $f(n) \leq C_1 g(n)$ and $g(n) \leq C_2 h(n)$ for constants C_1, C_2, then $f(n) \leq C_1 C_2 h(n)$.

13. (a) Let $\text{DIGIT}(n) = m$. Then $10^{\text{DIGIT}(n)} = 10^m$ is written as a 1 followed by m 0's, so it's larger than any m-digit number, such as n. And 10^{m-1} is a 1 followed by $m - 1$ 0's so it's the smallest m-digit number.

   (b) Since $n < 10^{\text{DIGIT}(n)}$ by part (a),

   $$\log_{10} n < \log_{10} 10^{\text{DIGIT}(n)} = \text{DIGIT}(n)$$

   for every $n \in \mathbb{P}$.

   (c) Use part (a) and consider $n \geq 10$.

Section 3.4, page 138

1. (a) 1, 2, 1, 2, 1, 2, 1, 2, ... (b) $\{1, 2\}$.

3. (a) $\text{SEQ}(n) = 3^n$.

   (b) (B) $\text{SEQ}(0) = 1$,

   (R) $\text{SEQ}(n + 1) = 3 * \text{SEQ}(n)$ for $n \geq 1$.

5. No. It's okay up to $\text{SEQ}(100)$, but $\text{SEQ}(101)$ is not defined, since we cannot divide by zero. If, in (R), we restricted n to be ≤ 100, we would obtain a recursively defined *finite* sequence.

7. (a) 1, 3, 8. (b) $s_n = 2s_{n-1} + 2s_{n-2}$ for $n \geq 2$.

9. (a) 1, 1, 2, 4. (b) $t_n = 2^{n-1}$ for $n \geq 1$.

11. (a) $a_6 = a_5 + 2a_4 = a_4 + 2a_3 + 2a_4 = 3(a_3 + 2a_2) + 2a_3 = 5(a_2 + 2a_1) + 6a_2 = 11(a_1 + 2a_0) + 10a_1 = 11 \cdot 3 + 10 = 43$. This calculation uses only two intermediate value addresses at any given time. Other recursive calculations are possible that use more.

   (b) Since $2a_{n-2}$ is always even, the oddness of a_{n-1} implies the oddness of $a_{n-1} + 2a_{n-2} = a_n$. Give an induction proof.

13. $\text{SEQ}(n) = 2^{n-1}$ for $n \geq 1$.

15. (a) $A(1) = 1$. $A(n) = n \cdot A(n - 1)$.

    (b) $A(6) = 6 \cdot A(5) = 6 \cdot 5 \cdot A(4) = \cdots = 6! = 720$.

    (c) Yes.

17. $f^{(1)} = f^{(0)} \circ f$ by (R) with $n = 0$

    $= 1_S \circ f$ by (B)

    $= f$ by a property of 1_S.

    Also [supply reasons]

    $$f^{(2)} = f^{(1)} \circ f = f \circ f \quad \text{and} \quad f^{(3)} = f^{(2)} \circ f = (f \circ f) \circ f.$$

(a) (B) UNION$(1) = A_1$,

    (R) UNION$(n) = A_n \cup$ UNION$(n - 1)$ for $n \geq 2$.

(d) Empty intersection should be the universe, in this case S.
For the inductive step, assume FACT$(n) = n!$ Then

$$
\begin{aligned}
\text{FACT}(n + 1) &= (n + 1) * \text{FACT}(n) && \text{definition of FACT} \\
&= (n + 1) * n! && \text{inductive assumption} \\
&= (n + 1) * 1 * 2 * \cdots * n && \text{definition of } n! \\
&= (n + 1)! && \text{commutative law and} \\
& && \text{definition of } (n + 1)!.
\end{aligned}
$$

Section 3.5, page 144

1. $s_n = 3 \cdot (-2)^n$ for $n \in \mathbb{N}$.

3. We prove this by induction. $s_n = a^n \cdot s_0$ holds for $n = 0$ because $a^0 = 1$. If it holds for some n, then $s_{n+1} = as_n = a(a^n \cdot s_0) = a^{n+1} \cdot s_0$ and so the result holds for $n + 1$.

5. $s_0 = 3^0 - 2 \cdot 0 \cdot 3^0 = 1$. $s_1 = 3^1 - 2 \cdot 1 \cdot 3^1 = -3$. For $n \geq 2$,

$$
\begin{aligned}
6s_{n-1} - 9s_{n-2} &= 6[3^{n-1} - 2(n - 1) \cdot 3^{n-1}] - 9[3^{n-2} - 2(n - 2) \cdot 3^{n-2}] \\
&= 2[3^n - 2(n - 1) \cdot 3^n] - [3^n - 2(n - 2) \cdot 3^n] \\
&= 3^n[2 - 4(n - 1) - 1 + 2(n - 2)] \\
&= 3^n[1 - 2n] = s_n.
\end{aligned}
$$

7. This time $c_1 = 3$ and $c_2 = 0$ and so $s_n = 3 \cdot 2^n$ for $n \in \mathbb{N}$.

9. Solve $1 = c_1 + c_2$ and $2 = c_1 r_1 + c_2 r_2$ for c_1 and c_2 to obtain $c_1 = (1 + r_1)/\sqrt{5}$ and $c_2 = -(1 + r_2)/\sqrt{5}$. Hence

$$
s_n = \frac{1}{\sqrt{5}} (r_1^n + r_1^{n+1} - r_2^n - r_2^{n+1}) \text{ for all } n,
$$

where r_1, r_2 are as in Example 3.

11. (a) $r_1 = -3, r_2 = 2, c_1 = c_2 = 1$ and so $s_n = (-3)^n + 2^n$ for $n \in \mathbb{N}$.

(c) Here the characteristic equation has one solution $r = 2$. $c_1 = 1$ and $c_2 = 3$ and so $s_n = 2^n + 3n \cdot 2^n$ for $n \in \mathbb{N}$.

(e) $s_{2n} = 1, s_{2n+1} = 4$ for all $n \in \mathbb{N}$.

(g) $s_n = (-3)^n$ for $n \in \mathbb{N}$.

Section 3.6, page 150

1. (a) 4, -3 and 5 are polynomial functions by (B). $I(x) = x$ is a polynomial function by (B). So $-3x$ is a polynomial function by the (fg) part of (R). Also $I^2 = I \cdot I$ where $I^2(x) = x^2$ is a polynomial function by the (fg) part of (R). And so $I^2 \cdot I = I^3$ where $I^3(x) = x^3$ is a polynomial function by the (fg) part of (R).

    Now $5x^3$ is a polynomial function by the (fg) part of (R). So $-3x + 5x^3$ is a polynomial function by the $(f + g)$ part of (R). Finally, $4 + (-3x + 5x^3)$ is a polynomial function by the $(f + g)$ part of (R). Tedious, isn't it?

3. (a) $((x + y) + z)$ or $(x + (y + z))$.

(c) $((xy)z)$ or $(x(yz))$.

5. (a) By (B), x, y and 2 are wff's. By the (f^g) part of (R), we conclude that (x^2) and (y^2) are wff's. So by the ($f + g$) part of (R), $((x^2) + (y^2))$ is a wff.

(c) By (B), X and Y are wff's. By the ($f + g$) part of (R), $(X + Y)$ is a wff. By the ($f - g$) part of (R), $(X - Y)$ is a wff. Finally, by the ($f * g$) part of (R), $((X + Y) * (X - Y))$ is a wff.

7. Simply add "$(P \oplus Q)$," to the recursive clause (R).

9. (a) First note that

$$g_1(x) = \begin{cases} -x & \text{for} \quad x < 0 \\ x & \text{for} \quad x \geq 0 \end{cases}.$$

Now x and $-x$ are piecewise polynomial functions (ppf's) by (B). So by (R) with $a = 0$, we see that g_1 is also a ppf.

11. (a) (B) $\langle 0, 0 \rangle \in S$,

(R) if $\langle m, n \rangle \in S$, then $\langle m + 1, n + 3 \rangle \in S$.

(b) By (B), $\langle 0, 0 \rangle \in S$. So by (R), $\langle 1, 3 \rangle \in S$. Again by (R), $\langle 2, 6 \rangle \in S$. And again by (R), $\langle 3, 9 \rangle \in S$.

13. (a) (B) ϵ is in S,

(R) if $w \in S$, then $aw \in S$ and $wb \in S$.

(b) $\epsilon \in S$ by (B). Now repeated use of (R) yields $a\epsilon \in S$, i.e., $a \in S$, so $ab \in S$, so $abb \in S$, so $abbb \in S$.

15. (a) 2. (c) 2. (e) 4.

Section 4.1, page 159

1. (a) $\begin{bmatrix} -8 & 13 \\ 2 & 9 \end{bmatrix}$. (c) $\begin{bmatrix} 31 & -16 & -6 \\ 29 & 4 & 26 \end{bmatrix}$. (e) $\begin{bmatrix} -1 & 7 \\ 8 & 0 \\ 16 & 0 \end{bmatrix}$.

3. The products written in parts (a), (c) and (e) do not exist.

5. (a) $\begin{bmatrix} 1 & 10 \\ 11 & 19 \end{bmatrix}$. 7. (a) $\begin{bmatrix} 7 & 14 \\ 8 & 11 \\ 2 & -6 \end{bmatrix}$.

9. (a) The (i, j) entry of $a\mathbf{A}$ is $a\mathbf{A}[i, j]$. Similarly for $b\mathbf{B}$ and so the (i, j) entry of $a\mathbf{A} + b\mathbf{B}$ is $a\mathbf{A}[i, j] + b\mathbf{B}[i, j]$. So the (i, j) entry of $c(a\mathbf{A} + b\mathbf{B})$ is $ca\mathbf{A}[i, j] + cb\mathbf{B}[i, j]$. A similar discussion shows that this is the (i, j) entry of $(ca)\mathbf{A} + (cb)\mathbf{B}$. Since their entries are equal, the matrices $c(a\mathbf{A} + b\mathbf{B})$ and $(ca)\mathbf{A} + (cb)\mathbf{B}$ are equal.

(c) The (j, i) entries of both $(a\mathbf{A})^T$ and $a\mathbf{A}^T$ equal $a\mathbf{A}[i, j]$. Here $1 \leq i \leq m$ and $1 \leq j \leq n$. So the matrices are equal.

11. (a) $\mathbf{B}$. (c) $\mathbf{0}$. (e) $\mathbf{A}$.

15. (a) In fact, $\mathbf{AB} = \mathbf{BA} = a\mathbf{B}$ for all $\mathbf{B}$ in $\mathfrak{M}_{2,2}$.

(b) $\mathbf{AB} = \mathbf{BA}$ with $\mathbf{B} = \begin{bmatrix} 1 & 0 \\ 0 & 0 \end{bmatrix}$ forces $\begin{bmatrix} a & 0 \\ c & 0 \end{bmatrix} = \begin{bmatrix} a & b \\ 0 & 0 \end{bmatrix}$ and so $b = c = 0$. So $\mathbf{A} = \begin{bmatrix} a & 0 \\ 0 & d \end{bmatrix}$. Now try $\mathbf{B} = \begin{bmatrix} 0 & 1 \\ 0 & 0 \end{bmatrix}$.

17. Since $(\mathbf{A} + \mathbf{B})(\mathbf{A} - \mathbf{B}) = \mathbf{A}^2 + \mathbf{BA} - \mathbf{AB} - \mathbf{B}^2$, it is enough to find $\mathbf{A}$ and $\mathbf{B}$ where $\mathbf{BA} \neq \mathbf{AB}$.

For $1 \le k \le p$ and $1 \le i \le m$,

$$(\mathbf{B}^T\mathbf{A}^T)[k, i] = \sum_{j=1}^{n} \mathbf{B}^T[k, j]\mathbf{A}^T[j, i] = \sum_{j=1}^{n} B[j, k]A[i, j].$$

Compare with the (k, i) entry of $(\mathbf{AB})^T$.

21. (a) Consider $1 \le i \le m$ and $1 \le k \le p$ and compare the (i, k) entries of $(\mathbf{A} + \mathbf{B})\mathbf{C}$ and $\mathbf{AC} + \mathbf{BC}$.

23. (a) This is an easy induction proof on k. At the inductive step, argue $\mathbf{BA}^{k+1} = \mathbf{BA}^k\mathbf{A} = \mathbf{A}^k\mathbf{BA} = \mathbf{A}^k\mathbf{AB} = \mathbf{A}^{k+1}\mathbf{B}$, with reasons, of course.
(b) Apply induction and part (a).

Section 4.2, page 168

1. (a) Yes. (b) Yes, 1. (c) No. Only 1 itself has an inverse. (d) It is a commutative monoid, but not a group.

3. Same answers as for Exercise 1.

5. (b) Yes, the empty set $\varnothing$, since $A \oplus \varnothing = A$ for all $A \in \mathscr{P}(S)$. (c) Yes, $A \oplus A = \varnothing$. Elements of $\mathscr{P}(S)$ are their own inverses with respect to this operation. (d) $(\mathscr{P}(S), \oplus)$ is a group.

7. See Example 8(b).

9. (a) No. (c) No, the zero matrix has no inverse, for example. Exercise 12 tells which matrices in $\mathfrak{M}_{2,2}$ have inverses.

11. (a) *break, fast, fastfood, lunchbreak, foodfood.*

13. (a) $\mathbf{I}^{-1} = \mathbf{I}$. (c) Not invertible. (e) $\mathbf{D}^{-1} = \mathbf{D}$.

15.

k	$g(k)$	$h(k)$
1	2	2
2	3	3
3	1	4
4	4	5
5	5	1

17. (b) $(\mathbb{N}, \max)$ is a monoid because 0 is an identity. $(\mathbb{N}, \min)$ has no identity and so it is not a monoid. Check these claims.

19. (a) $f(\langle f(\langle s_1, s_2\rangle), s_3\rangle) = f(\langle s_1, f(\langle s_2, s_3\rangle)\rangle)$ for all $s_1, s_2, s_3 \in S$.

21. Suppose that t and u are both inverses for s and consider tsu.

23. (a) For an element x, show $\chi_{A \cap B}(x) = \chi_A(x) \cdot \chi_B(x)$ by considering cases: $x \in A \cap B$; $x \notin A \cap B$.

Section 4.3, page 175

1. (a) $q = 6, r = 2$. (c) $q = -7, r = 1$. (e) $q = 5711, r = 31$.

3. (a) $-4, 0, 4$. (c) $-2, 2, 6$. (e) $-4, 0, 4$.

5. (a) 1. (c) 1. (e) 0.

7. (a) 3 and 2. (c) $m *_{10} k$ is the last [decimal] digit of $m * k$.

9.

$+_4$	0	1	2	3
0	0	1	2	3
1	1	2	3	0
2	2	3	0	1
3	3	0	1	2

$*_4$	0	1	2	3
0	0	0	0	0
1	0	1	2	3
2	0	2	0	2
3	0	3	2	1

11. Solutions are 1, 3, 2 and 4, respectively.

13. It's a semigroup by Theorem 3. Since $m *_p 1 = m$ for all $m \in \mathbb{Z}(p)$, 1 is a multiplicative identity for $\mathbb{Z}(p)$. Thus $(\mathbb{Z}(p), *_p)$ is a monoid. Commutativity is obvious, since $m *_p n$ is the remainder on dividing mn by p, $n *_p m$ is the remainder on dividing nm by p, and $mn = nm$.

15. *Hint*: If $a - b$ is a multiple of p, then so is $(a - b)(a + b)$.

Section 4.4, page 180

1. (a) $\{abab, abba, baab, baba\}$. (c) $\{ab, ba, bbab, bbba\}$.
   (e) $\{aba, abb, baa, bab\}$.

3. (a) $B^* = \{\epsilon, bb, bbbb, \ldots\} = \{b^{2n} : n \in \mathbb{N}\}$. (c) aa is one.

5. (b) Clearly $A^* \subseteq (A^+)^* \subseteq (A^*)^*$, and $(A^*)^* = A^*$ by Theorem 1(b). So all these sets are equal. A similar argument shows that $A^* = (A^*)^+$.

7. (a) False. (c) True. (e) False. (g) False. (i) True. (k) False.
   (m) True. Don't forget to provide examples for the false statements.

9. Since $A^* = A^+ \cup \{\epsilon\}$, this amounts to the statement "$\epsilon \in A^+ \Leftrightarrow \epsilon \in A$." Suppose $\epsilon \in A^+$. Then $\epsilon \in A^*A$ by Theorem 1(c), so $\epsilon = uv$ for $u \in A^*$, $v \in A$. But then $u = v = \epsilon$, so $\epsilon \in A$. The reverse implication is clear.

11. Parts (a) and (c) involve straightforward induction arguments. For (b), use the ideas in Exercise 8(a) to show

$$(BA)^+ = \bigcup_{n \in \mathbb{N}} (BA)^{n+1} = \bigcup_{n \in \mathbb{N}} (B(AB)^n A)$$

$$= B\left(\bigcup_{n \in \mathbb{N}} (AB)^n \right) A = B(AB)^* A.$$

13. (a) and (b). For $n = 0$ the claim is $\{\epsilon\} \subseteq \{\epsilon\}$. For $n = 1$,

$$AB \subseteq AA \cup AB \cup BA \cup BB = (A \cup B)^2.$$

For the inductive step, $(AB)^n AB \subseteq (A \cup B)^{2n}(A \cup B)^2$.
   (c) By part (b), $(AB)^* = \bigcup_{n \in \mathbb{N}} (AB)^n \subseteq \bigcup_{n \in \mathbb{N}} (A \cup B)^{2n} \subseteq (A \cup B)^*$.

15. (a) The smallest monoid containing $\{a, ab\}$ consists of ϵ and all words that start with a and have no consecutive b's.

Section 5.1, page 192

1. (a) 56. (c) 56. (e) 1.

3. (a) Draw a Venn diagram and work from the inside out. The 10 is given as $|S \cap B \cap J|$. The 17 is calculated from $27 = |J \cap S|$. Etc. Answer = 15. (b) 71.

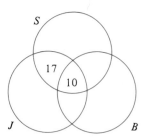

5. (a) 126. (b) 105.

7. (a) 0. (b) 840. (c) 2401.

9. (a) This is the same as the number of ways of drawing ten cards so that the *first* one is not a repetition. Hence $52(51)^9$. (b) $52^{10} - 52(51)^9$.

11. (a) $13 \cdot \binom{4}{4} \cdot \binom{48}{1} = 624$. (b) 5108.

   (c) $13 \cdot \binom{4}{3} \cdot \binom{12}{2} \cdot 4 \cdot 4 = 54{,}912$. (d) 1,098,240.

15. (a) It is the $n \times n$ matrix with 0's on the diagonal and 1's elsewhere.

17. (b) Counting all the r-element subsets of an n-element set is the same as counting all their $(n - r)$-element complements.

19. (c) If true for n, then

$$\sum_{r=0}^{n+1} \binom{n+1}{r} = 1 + \sum_{r=1}^{n} \binom{n+1}{r} + 1 = 1 + \sum_{r=1}^{n} \binom{n}{r-1} + \sum_{r=1}^{n} \binom{n}{r} + 1$$

$$= \sum_{r=1}^{n+1} \binom{n}{r-1} + \sum_{r=0}^{n} \binom{n}{r} = 2 \sum_{r=0}^{n} \binom{n}{r} = 2 \cdot 2^n = 2^{n+1}.$$

21. (a) $\sum_{k=3}^{5} \binom{k}{3} = \binom{3}{3} + \binom{4}{3} + \binom{5}{3} = 15 = \binom{6}{4}$.

   (c) Let $p(n)$ be " $\sum_{k=m}^{n} \binom{k}{m} = \binom{n+1}{m+1}$ whenever $1 \le m \le n$." Check that $p(1)$ is true. Assume that $p(n)$ is true for some $n \ge 1$. Then whenever $1 \le m \le n$

$$\sum_{k=m}^{n+1} \binom{k}{m} = \left(\sum_{k=m}^{n} \binom{k}{m} \right) + \binom{n+1}{m}$$

$$= \binom{n+1}{m+1} + \binom{n+1}{m} \qquad \text{[by the inductive assumption]}$$

$$= \binom{n+2}{m+1} \qquad \text{[give reasons]}.$$

   Check the case $m = n + 1$.

Section 5.2, page 204

1. 125.

3. 466; remember $D_4 \cap D_6 = D_{12}$ not D_{24}.

5. (a) $\dfrac{15!}{3! \, 4! \, 5! \, 3!}$. (b) $\binom{15}{3}\binom{15}{4}\binom{15}{5}$.

7. (a) $\binom{15}{3} = 455$. (b) 35.

9. 1260.

11. (a) 625. (b) 505. (c) 250. (d) 303.

13. There are $\frac{1}{2}\binom{2n}{n}$ unordered such partitions and $\binom{2n}{n}$ ordered partitions.

15. (a) 165. (b) 56.

17. (a) 59,049. (b) 252. (c) 120. (d) 15,360. (e) 4200. (f) 55,980.

19. (a)

(b) 01010100, etc.

Section 5.3, page 211

1. (a) Apply the Pigeon-Hole Principle to the partition $\{A_0, A_1, A_2\}$ of the set S of four integers where

$$A_i = \{n \in S : n \equiv i \pmod{3}\}.$$

Alternatively, apply the second version of the Pigeon-Hole Principle to the function $f : S \rightarrow \mathbb{Z}(3)$ defined so that $f(n) \equiv n \pmod{3}$ for each $n \in S$.
(b) Apply the Pigeon-Hole Principle to the function $f : \{1, 2, \ldots, p + 1\} \rightarrow \mathbb{Z}(p)$ defined so that $f(j) \equiv a_j \pmod{p}$.

3. Here $|S| = 73$ and $73/8 > 9$, so some box has more than 9 marbles.

5. For each 4-element subset B of A, let $f(B)$ be the sum of the numbers in B. Explain why f maps the set of 4-element subsets of A into $\{10, 11, 12, \ldots, 194\}$.

Note that A has $\binom{10}{4} = 210$ 4-element subsets. Apply the second version of the Pigeon-Hole Principle to f.

7. For each 2-element subset T of A, let $f(T)$ be the sum of the two elements. Then f maps the 300 2-element subsets of A into $\{3, 4, 5, \ldots, 299\}$.

9. The repeating blocks are various permutations of 142857.

11. (a) Look at the six blocks

$$\langle n_1, n_2, n_3, n_4 \rangle, \quad \langle n_5, n_6, n_7, n_8 \rangle, \quad \ldots, \quad \langle n_{21}, n_{22}, n_{23}, n_{24} \rangle.$$

(b) Use Example 3(b) of § 2.6.
(d) Look at the five blocks

$$\langle n_1, \ldots, n_5 \rangle, \quad \langle n_6, \ldots, n_{10} \rangle, \quad \langle n_{11}, \ldots, n_{15} \rangle, \quad \langle n_{16}, \ldots, n_{20} \rangle, \quad \langle n_{21}, \ldots, n_{24} \rangle.$$

13. If $0 \in \text{Im}(f)$, then n_1, $n_1 + n_2$ or $n_1 + n_2 + n_3$ is divisible by 3. Otherwise f is not one-to-one and there are three cases:

$$n_1 \equiv n_1 + n_2 \pmod{3}; \quad n_1 \equiv n_1 + n_2 + n_3 \pmod{3};$$

$$n_1 + n_2 \equiv n_1 + n_2 + n_3 \pmod{3}.$$

15. (a) 262,144. (b) 73,502. (c) 20,160. (d) 60.

17. (a) Show that S must contain both members of some pair $\langle 2k - 1, 2k \rangle$.
(b) For each $m \in S$, write $m = 2^k \cdot n$ where n is odd and let $f(m) = n$. Then $f : S \rightarrow \{1, 3, 5, \ldots, 2n - 1\}$.

Section 5.4, page 220

1. (a) True. (c) False. (e) True. Compare Exercise 3 of § 3.1.
3. (a) A function of the form $f(x) = ax + b$ will work if you choose a and b so that $f(0) = -1$ and $f(1) = 1$. Sketch your answer to see that it works.
   (b) Use g where $g(x) = 1 - x$.
   (c) Modify hint to part (a).
   (d) $1/x$.
   (e) Map $(1, \infty)$ onto $(0, \infty)$ using $h(x) = x - 1$ and compose with your answer in part (d).
   (f) $f(x) = 2^x$, say. Sketch f to see that it works.
5. (a) Use the data

x	.1	.2	.3	.4	.5	.6	.7	.8	.9
$f(x)$	-8.89	-3.75	$--1.90$	$-.83$	0	.83	1.90	3.75	8.89

7. Only the sets in (b) and (c) are countably infinite.
9. (a) Let $f : S \to T$ be a one-to-one correspondence where T is a countable set. There is a one-to-one correspondence $g : T \to \mathbb{P}$ [why?]. Then $g \circ f$ is a one-to-one correspondence of S onto $\mathbb{P}$.
11. (a) Apply part (b) of the theorem to $S \times T = \bigcup_{t \in T} (S \times \{t\})$. Explain why each $S \times \{t\}$ is countable.
    (b) For each $t \in T$, let $g(t)$ be an element in S such that $f(g(t)) = t$. Show that g is one-to-one and apply part (a) of the theorem.
13. (a) For each f in $\mathrm{FUN}(\mathbb{P}, \{0, 1\})$, define $\phi(f)$ to be the set $\{n \in \mathbb{P} : f(n) = 1\}$. Show ϕ is a one-to-one function from $\mathrm{FUN}(\mathbb{P}, \{0, 1\})$ onto $\mathscr{P}(\mathbb{P})$.
    (b) Use Example 2(a) and Exercise 9.
15. For the inductive step, use the identity $S^n = S^{n-1} \times S$.

Section 6.1, page 229

1. (a) 0. (b) 0. (c) 1. (d) 0. (e) 0.
3. (a) $\forall x\, \forall y\, \forall z[((x < y) \wedge (y < z)) \to (x < z)]$; universes $\mathbb{R}$.
   (c) $\forall m\, \forall n\, \exists p[(m < p) \wedge (p < n)]$; universes $\mathbb{N}$.
   (e) $\forall n\, \exists m[m < n]$; universes $\mathbb{N}$.
5. (a) $\forall w_1\, \forall w_2\, \forall w_3[(w_1 w_2 = w_1 w_3) \to (w_2 = w_3)]$.
   (c) $\forall w_1\, \forall w_2[w_1 w_2 = w_2 w_1]$.
7. (a) x, z are bound; y is free. (c) Same answers as part (a).
9. (a) x, y are free; there are no bound variables.
   (b) $\forall x\, \forall y[(x - y)^2 = x^2 - y^2]$ is false.
   (c) $\exists x\, \exists y[(x - y)^2 = x^2 - y^2]$ is true.
11. (a) No. $\exists m[m + 1 = n]$ is false for $n = 0$. (b) Yes.
13. (a) $\exists! x\, \forall y[x + y = y]$. (c) $\exists! A\, \forall B[A \subseteq B]$. Here A, B vary over the universe of discourse $\mathscr{P}(\mathbb{N})$. Note that $\forall B[A \subseteq B]$ is true if and only if $A = \varnothing$.
    (e) "$f : A \to B$ is a one-to-one function" $\to \forall b\, \exists! a[f(a) = b]$. Here a ranges over A and b ranges over B. One way to make this clear is to write $\forall b \in B\, \exists! a \in A[f(a) = b]$.

15. (a) True.   (c) False; e.g., 3 is in the right-hand set.   (e) False; the right-hand set is empty.   (g) True.   (i) True.   (k) True.   (m) True.

17. (a) 0.   (c) 1.   (e) 0.

Section 6.2, page 238

1. (a) Every club member has been a passenger on every airline if and only if every airline has had every club member as a passenger.   (c) If there is a club member who has been a passenger on every airline, then every airline has had a club member as a passenger.

3. (a) See rule 8c.

5. $\exists n[\neg\{p(n) \to p(n+1)\}]$ or $\exists n[p(n) \wedge \neg p(n+1)]$.

7. (a) $\exists x\, \exists y[(x < y) \wedge \forall z\{(z \leq x) \vee (y \leq z)\}]$.
   (c) 0; for example, $[x < y \to \exists z\{x < z < y\}]$ is false for $x = 3$ and $y = 4$.

9. One can let $q(x, y)$ be the predicate "$x = y$." Another way to handle $\exists x\, p(x, x)$ is to let $r(x)$ be the 1-place predicate $p(x, x)$. Then $\exists x\, r(x)$ is a compound predicate.

11. $\exists N\, \forall n[p(n) \to (n < N)]$

Section 6.3, page 244

3. Show that $n^5 - n$ is always even. Then use the identity $(n + 1)^5 = n^5 + 5n^4 + 10n^3 + 10n^2 + 5n + 1$ [from the binomial theorem].

5. Yes. The oddness of a_n depends only on the oddness of a_{n-1}, since $2a_{n-2}$ is even whether a_{n-2} is odd or not.

7. (b) $a_n = n + 1$ for all $n \in \mathbb{N}$.
   (c) The basis needs to be checked for $n = 0$ and $n = 1$. For the inductive step, consider $n \geq 2$ and assume $a_k = k + 1$ for $0 \leq k < n$. Then $a_n = 2a_{n-1} - a_{n-2} = 2n - (n - 1) = n + 1$. This completes the inductive step, and so $a_n = n + 1$ for all $n \in \mathbb{N}$ by the Second Principle of Induction.

9. (b) $a_n = 2^n$ for all $n \in \mathbb{N}$.
   (c) The basis needs to be checked for $n = 0$ and $n = 1$. For the inductive step, consider $n \geq 2$ and assume that $a_k = 2^k$ for $0 \leq k < n$.

11. (b) The basis needs to be checked for $n = 0$, 1 and 2. For the inductive step, consider $n \geq 3$ and assume that a_k is odd for $0 \leq k < n$. Then $a_{n-1}, a_{n-2}, a_{n-3}$ are all odd. Since the sum of three odd integers is odd [if not obvious, prove it], a_n is also odd.
   (c) Since the inequality is claimed for $n \geq 1$ and since you will want to use the identity $a_n = a_{n-1} + a_{n-2} + a_{n-3}$ in the inductive step, you will need $n - 3 \geq 1$ in the inductive step. So check the basis for $n = 1$, 2 and 3. For the inductive step, consider $n \geq 4$ and assume that $a_k \leq 2^{k-1}$ for $1 \leq k < n$.

13. (a) 2, 3, 4, 6.
   (b) The inequality must be checked for $n = 3$, 4 and 5 before applying the Second Principle of Mathematical Induction to $b_n = b_{n-1} + b_{n-3}$.
   (c) The inequality must be checked for $n = 2$, 3 and 4. Then use the Second Principle of Mathematical Induction and part (b).

15. Check for $n = 0$ and 1 before applying induction. It may be simpler to prove "SEQ$(n) \leq 1$ for all n" separately from "SEQ$(n) \geq 0$ for all n."

17. (a) 1, 3, 4, 7, 11, 18, etc.

(b) First check for $n = 2$ and 3. Then apply induction:

$$\text{LUC}(n) = \text{LUC}(n - 1) + \text{LUC}(n - 2)$$

$$= [\text{FIB}(n - 1) + \text{FIB}(n - 3)] + [\text{FIB}(n - 2) + \text{FIB}(n - 4)]$$

$$= [\text{FIB}(n - 1) + \text{FIB}(n - 2)] + [\text{FIB}(n - 3) + \text{FIB}(n - 4)]$$

$$= \text{FIB}(n) + \text{FIB}(n - 2).$$

Be sure to supply explanations.

19. For $y > 0$, let $L(y)$ be the largest integer 2^k with $2^k \le y$. Show that $L(n) = T(n)$ for all n by showing first that $L(\lfloor n/2 \rfloor) = L(n/2)$ for $n \ge 2$ and then using the Second Principle of Induction.

21. Show that $S(n) \le n$ for every n by the Second Principle of Induction.

Section 6.4, page 250

1. *Hints:*

$$\frac{1}{n + 2} + \cdots + \frac{1}{2n + 2} = \left(\frac{1}{n + 1} + \cdots + \frac{1}{2n} \right) + \left(\frac{1}{2n + 1} + \frac{1}{2n + 2} - \frac{1}{n + 1} \right)$$

and

$$\frac{1}{2n + 1} + \frac{1}{2n + 2} - \frac{1}{n + 1} = \frac{1}{2n + 1} - \frac{1}{2n + 2}.$$

Alternatively, to avoid induction, let $f(n) = \sum_{k=1}^{n} \frac{1}{k}$ and write both sides in terms of f.

3. *Hints:* $5^{n+2} + 2 \cdot 3^{n+1} + 1 = 5(5^{n+1} + 2 \cdot 3^n + 1) - 4(3^n + 1)$. Show that $3^n + 1$ is always even.

5. (a) $\langle 2, 3 \rangle$, $\langle 4, 6 \rangle$, etc.

(b) (B) is clear since 5 divides $0 + 0$. For (R) you need to check

"if 5 divides $m + n$, then 5 divides $(m + 2) + (n + 3)$."

Alternatively, prove that every member of S is of the form $\langle 2k, 3k \rangle$ for $k \in \mathbb{N}$.

7. (a) Obviously $A \subseteq \mathbb{N} \times \mathbb{N}$. To show $\mathbb{N} \times \mathbb{N} \subseteq A$, apply the ordinary First Principle of Mathematical Induction to the propositions

$$p(k) = \text{"if } \langle m, n \rangle \in \mathbb{N} \times \mathbb{N} \text{ and } m + n = k, \text{ then } \langle m, n \rangle \in A."$$

9. (a) For w in Σ^*, let

$$p(w) = \text{"length}(\bar{w}) = \text{length}(w)."$$

Since $\bar{\epsilon} = \epsilon$, $p(\epsilon)$ is clearly true. You need to show $p(w) \rightarrow p(wx)$:

$$\text{length}(\bar{w}) = \text{length}(w) \text{ implies length}(\overline{wx}) = \text{length}(wx).$$

(b) Fix w_1, say, and work with $p(w) = \text{"}\overline{w_1 w} = \bar{w} \bar{w}_1."$

11. (b) There are too many p's and P's around, so let's use $r(P)$ for the proposition-valued function on $\mathscr{F}$, to which we wish to apply the general principle of induction. Then you need to prove all $r(P)$ are true where

$$r(P) = \text{"if } p, q \text{ are false, then } P \text{ is false."}$$

Section 7.1, page 262

1. (a)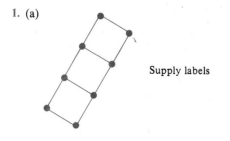

Supply labels

3. (a) h, o, p, q, r, z.   (c) B and C.   (e) f, z, p, does not exist.
5. (a) Only $\leqq$. $<$ is not reflexive and $\leqq$ is not antisymmetric.
   (c)

7. See Figures 2 and 6 or Exercise 6 for two different sorts of failure.
9. (a) $a \vee b = \max\{a, b\}$, $a \wedge b = \min\{a, b\}$.   (c) 73.   (e) $\sqrt{73}$.
13. Not if Σ has more than one element. Show that antisymmetry fails.
15. (a) Yes.   (b) The relation is not transitive.
17. (a) No. Every finite subset of $\mathbb{N}$ is a subset of a larger finite subset of $\mathbb{N}$.
   (c) $\text{lub}\{A, B\} = A \cup B$. Note that $A \cup B \in \mathscr{F}(\mathbb{N})$ for all $A, B \in \mathscr{F}(\mathbb{N})$.
   (e) Yes; see parts (c) and (d).
19. (a) Show that b satisfies the definition of $\text{lub}\{x, y, z\}$, i.e., $x \leqq b, y \leqq b, z \leqq b$ and if $x \leqq c, y \leqq c, z \leqq c$ then $b \leqq c$.
   (c) Use commutativity of $\vee$ and part (a).

Section 7.2, page 272

1. (a) $\{\varnothing, \{1\}, \{1, 4\}, \{1, 4, 3\}, \{1, 4, 3, 5\}, \{1, 4, 3, 5, 2\}\}$ is one.
3. (a) 501, 502, ..., 1000.   (c) Yes. Think of primes or see Exercise 16.
5. Yes. If $a \leqq b$ then $a \vee b = b$ and $a \wedge b = a$.
7. (a) Transitivity, for example. If $f \leqslant g$ and $g \leqslant h$, then $f(t) \leqq g(t)$ and $g(t) \leqq h(t)$ in S, for all t in T. Since $\leqq$ is transitive on S, $f(t) \leqq h(t)$ for all t, so $f \leqslant h$ in $\text{FUN}(T, S)$.
   (c) $f(t) \leqq f(t) \vee g(t) = h(t)$ for all t, so $f \leqslant h$. Similarly $g \leqslant h$, so h is an upper bound for $\{f, g\}$. Show that if $f \leqslant k$ and $g \leqslant k$ then $h \leqslant k$, so that h is the least upper bound for $\{f, g\}$.

9. Supply labels.   (a)   (c)

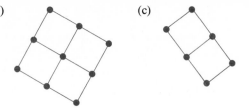

11. $|T| = 1$ and S is a chain, or $|S| = 1$. If S is not a chain, $\text{FUN}(T, S)$ contains constant functions which are not comparable. If $|S| > 1$ and $|T| > 1$, select $s_1 \neq s_2$ and $t_1 \neq t_2$ and define f and g so that $f(t_1) = s_1$, $f(t_2) = s_2$, $g(t_1) = s_2$, $g(t_2) = s_1$.

13. (a) $\langle 0, 0 \rangle, \langle 0, 1 \rangle, \langle 0, 2 \rangle, \langle 1, 0 \rangle, \langle 1, 1 \rangle, \langle 1, 2 \rangle, \langle 2, 0 \rangle, \langle 2, 1 \rangle, \langle 2, 2 \rangle.$
    (c) $\langle 3, 0 \rangle, \langle 3, 1 \rangle, \langle 3, 2 \rangle, \langle 4, 0 \rangle, \langle 4, 1 \rangle, \langle 4, 2 \rangle.$

15. (a) Yes. Both $(\mathbb{N}, \leqq)$ and $(\mathbb{N}, \geqq)$ are chains. Apply Theorem 1.
    (c) No. Consider $\{\langle 0, n \rangle : n \in \mathbb{N}\}$, for example. Note that $\langle 0, n + 1 \rangle \prec \langle 0, n \rangle$.

17. (a) 000, 0010, 010, 10, 1000, 101, 11.

19. (a) in of the list this order words sentence standard increasing.

21. Antisymmetry is immediate. For transitivity consider cases. Suppose

$$\langle s_1, \ldots, s_n \rangle \prec \langle t_1, \ldots, t_n \rangle \text{ and } \langle t_1, \ldots, t_n \rangle \prec \langle u_1, \ldots, u_n \rangle.$$

If $s_1 \prec_1 t_1$ then $s_1 \prec_1 t_1 \leqq_1 u_1$, so $\langle s_1, \ldots, s_n \rangle \prec \langle u_1, \ldots, u_n \rangle$.
If $s_1 = t_1, \ldots, s_{r-1} = t_{r-1}, s_r \prec_r t_r$ and $t_1 = u_1, \ldots, t_{p-1} = u_{p-1}, t_p \prec_p u_p$ and $r < p$, then $s_1 = u_1, \ldots, s_{r-1} = u_{r-1}$ and $s_r \prec_r t_r = u_r$ and again $\langle s_1, \ldots, s_n \rangle \prec \langle u_1, \ldots, u_n \rangle$. The remaining cases are similar.

Section 7.3, page 283

1. (a) is an equivalence relation.   (c) is also an equivalence relation, unless one is concerned about those rare individuals who maintain residences in more than one state.   (e) is not an equivalence relation. Why?

3. Very much so.

5. One needs to show:
    (R) For each (s_n), the set $\{n \in \mathbb{N} : s_n \neq s_n\}$ is finite, i.e., $(s_n) \sim (s_n)$.
    (S) If $\{n \in \mathbb{N} : s_n \neq t_n\}$ is finite, so is $\{n \in \mathbb{N} : t_n \neq s_n\}$, i.e., $(s_n) \sim (t_n)$ implies $(t_n) \sim (s_n)$.
    (T) If $\{n \in \mathbb{N} : s_n \neq t_n\}$ and $\{n \in \mathbb{N} : t_n \neq u_n\}$ are finite, so is $\{n \in \mathbb{N} : s_n \neq u_n\}$, i.e., $(s_n) \sim (t_n)$ and $(t_n) \sim (u_n)$ imply $(s_n) \sim (u_n)$.

7. One needs to show:
    (R) For each $x \in S$, $f(x) = x$ for some $f \in G$.
    (S) If $f(x) = y$ for some $f \in G$, then $y = g(x)$ for some $g \in G$.
    (T) If $f(x) = y$ for some $f \in G$ and $g(y) = z$ for some $g \in G$, then $h(x) = z$ for some $h \in G$.

9. (a) Verify directly, or apply Theorem 2(a) with $f(m) = m^2$ for $m \in \mathbb{Z}$.

11. (a) It consists of all finite subsets of S.

13. (a) $m \equiv n \pmod 1$ for all $m, n \in \mathbb{Z}$. There is only one equivalence class.

15. Apply Theorem 2, using the length function. The equivalence classes are the sets $\Sigma^k, k \in \mathbb{N}$.

17. Fifteen.

19. (a) Not well-defined: depends on the representative. For example $[3] = [-3]$ and $-3 \leq 2$. If the definition made sense, we would have $[3] = [-3] \leq [2]$ hence $3 \leq 2$.
   (b) Trouble. For example, $[2] = [-2]$ but $(2)^2 + (2) + 1 \neq (-2)^2 + (-2) + 1$.
   (c) Nothing wrong. If $[m] = [n]$ then $m^4 + m^2 + 1 = n^4 + n^2 + 1$.

Section 7.4, page 290

1. (a) R_1 satisfies (AR) and (S).
   (c) R_3 satisfies (R), (AS) and (T).
   (e) R_5 satisfies only (S).
3. R_3 is the only partial order, and R_2 is the only equivalence relation. Explain why each of the other relations isn't of one of these types.
5. Two of the relations are reflexive. Five of them are symmetric.
7. (a) $\{\langle 0, 5 \rangle, \langle 1, 4 \rangle, \langle 2, 3 \rangle, \langle 3, 2 \rangle, \langle 4, 1 \rangle, \langle 5, 0 \rangle\}$.
9. (a) This is an equivalence relation with 4 equivalence classes.
   (c) This relation is not reflexive, since 0 is not related to 0. It is symmetric and transitive.
11. (a) The relation satisfies (R) and (T). Note that antisymmetry fails because $A^* = B^*$ need not imply $A = B$.
13. (a) The empty relation satisfies (AR), (S), (AS) and (T). The last three properties hold vacuously.
15. (a) To show that $R \cup E$ is a partial order, show
   (R) $\langle x, x \rangle \in R \cup E$ for all $x \in S$,
   (AS) $\langle x, y \rangle \in R \cup E$ and $\langle y, x \rangle \in R \cup E$ imply $x = y$,
   (T) $\langle x, y \rangle \in R \cup E$ and $\langle y, z \rangle \in R \cup E$ imply $\langle x, z \rangle \in R \cup E$.
   To verify (T), consider cases.
17. (a) Yes. This is clear: if $E \subseteq R_1$ and $E \subseteq R_2$ then $E \subseteq R_1 \cup R_2$.
   (c) No. For a small example, let $S = \{a, b, c\}$, $R_1 = \{\langle a, b \rangle\}$ and $R_2 = \{\langle b, c \rangle\}$.
19. (a) Suppose R is symmetric. If $\langle x, y \rangle \in R$, then $\langle y, x \rangle \in R$ by symmetry and so $\langle x, y \rangle \in R^{-1}$. Similarly $\langle x, y \rangle \in R^{-1}$ implies $\langle x, y \rangle \in R$ [check] so that $R = R^{-1}$. For the converse, suppose that $R = R^{-1}$ and show R is symmetric.

Section 7.5, page 302

1. (a) $\mathbf{A} * \mathbf{A} = \begin{bmatrix} 1 & 1 & 1 \\ 1 & 1 & 1 \\ 1 & 1 & 1 \end{bmatrix}$. Since $\mathbf{A} * \mathbf{A} \leq \mathbf{A}$ is not true, R is not transitive.
   (c) Not transitive.
3. (b) Boolean matrices for $(R_1 \cap R_2)R_1^{-1}$ and $R_1 R_1^{-1} \cap R_2 R_1^{-1}$ are

$$[\mathbf{A}_1 \wedge \mathbf{A}_2] * \mathbf{A}_1^T = \begin{bmatrix} 0 & 0 & 0 \\ 0 & 0 & 0 \\ 0 & 0 & 0 \end{bmatrix} \text{ and } \mathbf{A}_1 * \mathbf{A}_1^T \wedge \mathbf{A}_2 * \mathbf{A}_1^T = \begin{bmatrix} 0 & 0 & 0 \\ 0 & 0 & 0 \\ 1 & 0 & 0 \end{bmatrix}.$$

   (c) The Boolean matrices are $\mathbf{A}_2 * [\mathbf{A}_1^T \vee \mathbf{A}_2^T]$ and $(\mathbf{A}_2 * \mathbf{A}_1^T) \vee (\mathbf{A}_2 * \mathbf{A}_2^T)$. They are equal.

5. (a) The matrix for R is $\mathbf{A} = \begin{bmatrix} 0 & 0 & 0 \\ 1 & 0 & 1 \\ 0 & 1 & 0 \end{bmatrix}$. The matrices for R^{-1} and R^2 are $\mathbf{A}^T$ and $\mathbf{A} * \mathbf{A}$.

(c) No; compare A and $A * A$ and note that $A * A \leqq A$ fails.

(e) Yes.

7. (a) Matrix for R^0 is the identity matrix. Matrix for R^1 is A, of course. Matrix for R^n is $A * A$ for $n \geq 2$ as should be checked by induction. For $n < 0$, use transposes.

(b) R is reflexive, but not symmetric or transitive.

9. (a) $A_f = \begin{bmatrix} 0 & 0 & 1 & 0 \\ 0 & 1 & 0 & 0 \\ 0 & 1 & 0 & 0 \\ 0 & 1 & 0 & 0 \end{bmatrix}$ and $A_g = \begin{bmatrix} 0 & 0 & 0 & 1 \\ 0 & 0 & 1 & 0 \\ 0 & 1 & 0 & 0 \\ 1 & 0 & 0 & 0 \end{bmatrix}$.

(b) They will be different, since the Boolean matrix for $R_f R_g$ is $A_f * A_g$; this is the Boolean matrix for $R_{g \circ f}$ but not for $R_{f \circ g}$.

(c) One does and one doesn't.

11. Don't use Boolean matrices; the sets S, T, U might be infinite.

(a) $R_1 R_3 \cup R_1 R_4 \subseteq R_1 (R_3 \cup R_4)$ by Example 2(a). For the reverse inclusion, consider $\langle x, z \rangle$ in $R_1 (R_3 \cup R_4)$ and show $\langle x, z \rangle$ is in $R_1 R_3$ or $R_1 R_4$.

(c) $R_1 (R_3 \cap R_4) \subseteq R_1 R_3 \cap R_1 R_4$. As in part (b), equality need not hold. For example, consider R_1, R_3, R_4 with Boolean matrices

$$A_1 = \begin{bmatrix} 1 & 1 \\ 0 & 0 \end{bmatrix}, \quad A_3 = \begin{bmatrix} 0 & 0 \\ 0 & 1 \end{bmatrix}, \quad A_4 = \begin{bmatrix} 0 & 1 \\ 0 & 0 \end{bmatrix}.$$

13. (c) One is and one isn't.

15. Say S, T and U have m, n and p elements. The problem is equivalent to showing $(A * B)^T = B^T * A^T$ where A and B are $m \times n$ and $n \times p$ Boolean matrices. Why? So compare (k, i)-entries of the two matrices where $1 \leqq i \leqq m$, $1 \leqq k \leqq p$.

17. Given $m \times n$, $n \times p$ and $p \times q$ Boolean matrices A_1, A_2, A_3, they correspond to relations R_1, R_2, R_3 where R_1 is a relation from $\{1, 2, \ldots, m\}$ to $\{1, 2, \ldots, n\}$, etc. The matrices for $(R_1 R_2) R_3$ and $R_1 (R_2 R_3)$ are $(A_1 * A_2) * A_3$ and $A_1 * (A_2 * A_3)$ by four applications of Theorem 1.

Section 7.6, page 310

1. (a) $\begin{bmatrix} 1 & 1 & 0 \\ 0 & 1 & 0 \\ 0 & 0 & 1 \end{bmatrix}$. (c) $\begin{bmatrix} 1 & 1 & 0 \\ 1 & 1 & 0 \\ 0 & 0 & 1 \end{bmatrix}$. (e) $\begin{bmatrix} 1 & 1 & 0 \\ 1 & 1 & 0 \\ 0 & 0 & 1 \end{bmatrix}$.

3. $\{1, 2\}$, $\{3\}$.

5. (a) $\begin{bmatrix} 1 & 1 & 0 & 1 \\ 1 & 1 & 1 & 0 \\ 0 & 1 & 1 & 0 \\ 1 & 0 & 1 & 1 \end{bmatrix}$. (c) $\begin{bmatrix} 1 & 1 & 0 & 1 \\ 1 & 1 & 1 & 0 \\ 0 & 1 & 1 & 1 \\ 1 & 0 & 1 & 1 \end{bmatrix}$. (e) $\begin{bmatrix} 1 & 1 & 1 & 1 \\ 1 & 1 & 1 & 1 \\ 1 & 1 & 1 & 1 \\ 1 & 1 & 1 & 1 \end{bmatrix}$.

7. (a) $r(R)$ is the usual order $\leqq$.

(c) $rs(R)$ is the universal relation on $\mathbb{P}$.

(e) R is already transitive.

9. $\langle h_1, h_2 \rangle \in st(R)$ if $h_1 = h_2$ or if one of h_1, h_2 is the High Hermit. On the other hand, $ts(R)$ is the universal relation on F.O.H.H.

11. (a) Compare $(R \cup E) \cup (R \cup E)^{-1}$ and $(R \cup R^{-1}) \cup E$; see Exercise 20 of § 7.4.

13. Any relation that contains R will include the pair $\langle 1, 1 \rangle$. This violates antireflexivity.

15. (a) For any relation R, let $\{R_i : i \in I\}$ be the set of all relations containing R that have property p. Show that $\bigcap_{i \in I} R_i$ is the smallest such relation.

(c) $S \times S$ is not antireflexive.

(d) Intersect the onto relations on $\{1, 2\}$ having Boolean matrices $\begin{bmatrix} 1 & 1 \\ 1 & 0 \end{bmatrix}$ and $\begin{bmatrix} 1 & 0 \\ 1 & 1 \end{bmatrix}$. Is the intersection an onto relation?

Section 7.7, page 318

1. $\pi_1 \wedge \pi_2$ has 3 sets; $\pi_1 \vee \pi_2 = \{\{\text{all marbles}\}\}$.
3. $\pi_1 \vee \pi_2$ has 2 sets.
5. (b) $\pi_3 \wedge \pi_5$ has 15 sets.
   (c) Since $\langle 1, 7 \rangle$ and $\langle 7, 2 \rangle$ are in $R_3 \cup R_5$, $\langle 1, 2 \rangle$ is in $R_3 \vee R_5$. To deal with $\langle 47, 73 \rangle$, note that $\langle 47, 2 \rangle$ is in $R_3 \cup R_5$ and you already know that $\langle 2, 1 \rangle$ and $\langle 1, 73 \rangle$ are in $R_3 \vee R_5$.
7. (a) For example, $\alpha(1) = \alpha(3) = \alpha(5) = 1$, $\alpha(2) = \alpha(6) = 2$, $\alpha(4) = 4$.
   (c) $\alpha(n) = n$ for $n = 1, 2, 3, 4, 5, 6$.
   (e) The "equality relation."
9. The final γ's should be

j	1	2	3	4	5	6	7	8
$\gamma(j)$ for $\pi_1 \wedge \pi_2$	1	1	3	4	5	6	7	8
$\gamma(j)$ for $\pi_1 \vee \pi_2$	5	5	6	5	5	6	6	5

11. The final γ's should be

j	1	2	3	4	5	6	7	8
$\gamma(j)$ for $\pi_1 \wedge \pi_2$	1	1	3	4	5	3	7	4
$\gamma(j)$ for $\pi_1 \vee \pi_2$	8	8	6	8	5	6	6	8

13. (a) Antisymmetry: If $\pi_1 \leq \pi_2$ and $\pi_2 \leq \pi_1$ and if A is a set in π_1 there are sets B in π_2 and C in π_1 with $A \subseteq B$ and $B \subseteq C$. The sets in π_1 are disjoint, so $A \subseteq C$ implies $A = C$. Thus $A = B \in \pi_2$. Similarly every set in π_2 is in π_1.
15. (a) $\gamma(j) = 0$ to start. If $\gamma(j)$ is still 0 when $k = j$, then $\gamma(j)$ changes to j at that point.
   (c) The value changes at least once by part (a). If the first change occurs at $k = k_0$, then by part (b) the algorithm skips to Step 4 for later values $k = k'$.
   (e) For any j, the final value of $\gamma(j)$ is the smallest k for which $\alpha(j) = \alpha(k)$ and $\beta(j) = \beta(k)$, by parts (b) and (c). If $\alpha(i) = \alpha(j)$ and $\beta(i) = \beta(j)$, then this value is the same for i and j.

Section 8.1, page 330

1. (a)

e	a	b	c	d	e	f
$\gamma(e)$	$\langle x, v \rangle$	$\langle v, x \rangle$	$\langle v, w \rangle$	$\langle w, y \rangle$	$\langle w, y \rangle$	$\langle y, x \rangle$

(c)

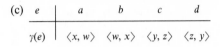

e	a	b	c	d
$\gamma(e)$	$\langle x, w \rangle$	$\langle w, x \rangle$	$\langle y, z \rangle$	$\langle z, y \rangle$

3. (a) Yes. (c) No. (e) Yes.

5. (a) $x\,w\,y$ is one. (c) $v\,x\,w$ is one. (e) $z\,y\,x\,w\,v$.

7. (a) $\text{SUCC}(t) = \varnothing$, $\text{SUCC}(u) = \{t, w, x\}$, $\text{SUCC}(v) = \{t, y\}$, $\text{SUCC}(w) = \{y\}$, etc. (c) t and z.

9. (a) $R(s) = \{s, t, u, w, x, y, z\} = R(t)$, $R(u) = \{w, x, y, z\}$, etc.

11. (a)

Supply labels.

(c) s, u, x, z.

13. Transitivity is a general property of reachability: string together a path from u to v and a path from v to w to get a path from u to w. Antisymmetry follows from acyclicity.

15. (a) In the proof of Theorem 1, choose a shortest path consisting of edges of the given path.

17. (a) Show that $\hat{G}$ is also acyclic. Apply the lemma to $\hat{G}$. A sink for $\hat{G}$ is a source for G.

19. (a) They are the sources of H.

21. If SINK chooses the source for v in Step 1, then it must choose each of the remaining vertices before it is done.

Section 8.2, page 342

1. (a) Here is one way.

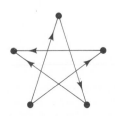

3. (a) One example is α where $\alpha(p) = v$, $\alpha(q) = y$, $\alpha(r) = w$, $\alpha(s) = z$, $\alpha(t) = x$.

5. The isomorphism is $\alpha(u) = u$ for all u if the digraphs are labeled as follows.

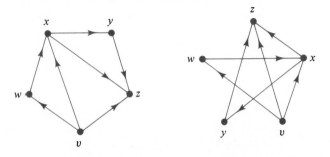

7. (a) One of them is the identity function and the other interchanges p and r. Define them explicitly.

9. (a) $D_{0,3} = \{v\}$, $D_{1,2} = \{s, y, z\}$, $D_{2,1} = \{t, w, x\}$, $D_{3,0} = \{u\}$.

11. It is terrible. Isomorphism is a binary relation; it makes no sense to say that "G is related." If you were asked "Are $1 + 1$ and 2 equal?" you wouldn't respond "$1 + 1$ is, but 2 isn't."

13. v is a sink if and only if outdeg(v) = 0, and outdeg(v) = outdeg($\alpha(v)$).

15. Use the second equation in the theorem.

17. (a) See the answer to Exercise 13; v is a source if and only if $\alpha(v)$ is a source.

Section 8.3, page 351

1.

W^*	A	B	C	D
A	∞	1.0	1.4	1.2
B	.4	∞	.4	.2
C	.7	.3	∞	.5
D	.8	.5	.2	∞

3.

W	m	q	r	s	w	x	y	z
m	∞	6	∞	2	∞	4	∞	∞
q	∞	∞	4	∞	4	∞	∞	∞
r	∞	∞	∞	∞	∞	∞	∞	3
s	∞	3	∞	∞	5	1	∞	∞
w	∞	∞	2	∞	∞	∞	2	5
x	∞	∞	∞	∞	3	∞	6	∞
y	∞	∞	∞	∞	∞	∞	∞	1
z	∞	∞	∞	∞	∞	∞	∞	∞

W^*	m	q	r	s	w	x	y	z
m	∞	5	8	2	6	3	8	9
q	∞	∞	4	∞	4	∞	6	7
r	∞	∞	∞	∞	∞	∞	∞	3
s	∞	3	6	∞	4	1	6	7
w	∞	∞	2	∞	∞	∞	2	3
x	∞	∞	5	∞	3	∞	5	6
y	∞	∞	∞	∞	∞	∞	∞	1
z	∞	∞	∞	∞	∞	∞	∞	∞

5. (a) Here is one example.

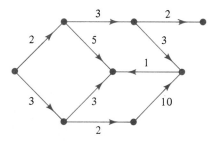

7. (a) $s\,v\,x f$ is one. There is another.

9. (a) Labeling as shown gives critical edges $\langle s, u \rangle$, $\langle u, w \rangle$, $\langle s, t \rangle$, $\langle t, w \rangle$, $\langle w, x \rangle$, $\langle x, y \rangle$ and $\langle y, f \rangle$.

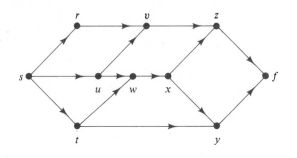

(c) 2.

11. (a) Shrink the 0-edges to make their two endpoints the same.

13. (a) $A(u) = M(s, u) =$ weight of a max-path from s to u. If there is an edge $\langle w, u \rangle$, a max-path from s to w followed by that edge has total weight at most $M(s, u)$. That is, $A(w) + W(w, u) \leq A(u)$. If $\langle w, u \rangle$ is an edge in a max-path from s to u, then $A(w) + W(w, u) = A(u)$.

15. (a) $FF(u, v) = A(v) - A(u) - W(u, v)$.   (c) The slack time at v.

Section 8.4, page 361

1. (a)

e	a	b	c	d	e	f	g	h	k
$\gamma(e)$	$\{w, x\}$	$\{x, u\}$	$\{t, u\}$	$\{t, v\}$	$\{u, v\}$	$\{u, y\}$	$\{v, z\}$	$\{x, y\}$	$\{y, z\}$

3. (a) In the notation of the proof $x_0 = y_0 = v$, $x_n = x_3 = x$, $y_m = y_4 = x$. Also $i = 1$ since $x_1 \neq y_1$. The pair $\langle j, k \rangle$ could be either $\langle 1, 3 \rangle$ [since $x_1 = y_3 = u$] or $\langle 2, 2 \rangle$ [since $x_2 = y_2 = y$]. The cycle is $v\,u\,y\,z\,v$ in either case.

(c) The cycle is $x\,u\,t\,v\,z\,y\,x$.

5. (a)

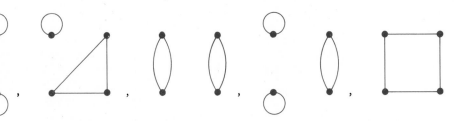

(c) See Exercise 12 or Theorem 3.

7. (a), (c) and (d) are regular, but (b) is not. Or count edges. (a) and (c) have cycles of length 3, but (d) does not. (a) and (c) are isomorphic; exhibit an isomorphism between them.

9. (a) $\binom{8}{5} = 56$.   (c) $8 \cdot 7 + 8 \cdot 7 \cdot 6 + 8 \cdot 7 \cdot 6 \cdot 6 = 2{,}408$.

11. (a) 19. [Use Theorem 3.]

13. (a)   (e)

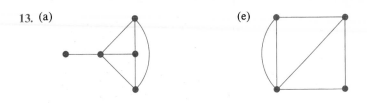

(c) See Exercise 12 or Theorem 3. (g) K_4.

15. Assume no loops or parallel edges. Consider a longest path $v_1 \cdots v_m$ with distinct vertices. There is another edge at v_m. Adjoint it to the path to get a closed path and use Proposition 1.

17. Use $|V(G)| = D_0(G) + D_1(G) + D_2(G) + \cdots$ and Theorem 3.

Section 8.5, page 370

1. (a) $v_3 \, v_1 \, v_2 \, v_3 \, v_6 \, v_2 \, v_4 \, v_6 \, v_5 \, v_1 \, v_4 \, v_5 \, v_3 \, v_4$ is one.

3. No. The edges and corners form a graph with 8 vertices, each of degree 3. It has no Euler path. See Figure 6(c) of § 8.6.

5. The ones with n odd.

7. $\{0, 1\}^3$ consists of 3-tuples of 0's and 1's, which we may view as binary strings of length 3. The graph is then

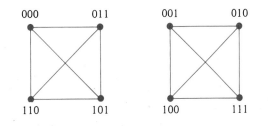

(a) 2. (c) No.

9. (a) Join the odd-degree vertices in pairs, with k new edges. The new graph has an Euler circuit, by Theorem 2. The new edges do not appear next to each other in the circuit and they partition the circuit into k simple paths of G.

(c) Imitate the proof. That is, add edges $\{v_2, v_3\}$ and $\{v_5, v_6\}$, say, create an Euler circuit, and then remove the two new edges.

Section 8.6, page 378

1. (a) Yes. Try $v_1 \, v_2 \, v_6 \, v_5 \, v_4 \, v_3 \, v_1$, for example.

(c) No. If v_1 is in V_1 then $\{v_2, v_3, v_4, v_5\} \subseteq V_2$, but v_2 and v_3 are joined by an edge.

3. (a) $2(n!)^2$. (c) m and n even, or m odd and $n = 2$, or $m = n = 1$.

5. One possible Hamilton circuit has vertex sequence 000, 001, 011, 111, 110, 101, 010, 100, 000 corresponding to the circular arrangement 0 0 0 1 1 1 0 1. Find another example.

7. No, for both questions. Why?

9. (a) K_n^+ has n vertices and just one more edge than K_{n-1} has, so it has exactly $\frac{1}{2}(n - 1)(n - 2) + 1$ edges.

11. (a)

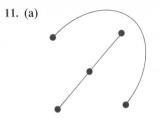

(c) Choose two vertices u and v in G. If they are *not* joined by an edge in G they are joined by an edge in the complement. If they *are* joined by an edge in G, they are in the same component of G. Choose w in some other component. Then $u\,w\,v$ is a path in the complement. In either case, u and v are joined by a path in the complement.

(e) No. Give an example.

13. Given G_{n+1}, consider the subgraph H_0 where $V(H_0)$ consists of all binary $(n+1)$-tuples with 0 in the $(n+1)$-st digit and $E(H_0)$ is the set of all edges of G_{n+1} connecting vertices in $V(H_0)$. Define H_1 similarly. Show H_0, H_1 are isomorphic to G_n, and so have Hamilton circuits. Use these to construct a Hamilton circuit for G_{n+1}. For $n = 2$, see how this works in Figure 6.

Section 8.7, page 385

1. (a) $u\,A\,v$ if there is an edge e in $E(G)$ with $\gamma(e) = \{u, v\}$.

3. $\mathbf{M}_A = \begin{bmatrix} 0 & 1 & 0 & 0 & 0 & 0 \\ 1 & 0 & 1 & 0 & 0 & 0 \\ 0 & 1 & 0 & 1 & 0 & 1 \\ 0 & 0 & 1 & 0 & 1 & 0 \\ 0 & 0 & 0 & 1 & 0 & 0 \\ 0 & 0 & 1 & 0 & 0 & 0 \end{bmatrix}$, $\mathbf{M}_{A^4} = \begin{bmatrix} 1 & 0 & 1 & 0 & 1 & 0 \\ 0 & 1 & 0 & 1 & 0 & 1 \\ 1 & 0 & 1 & 0 & 1 & 0 \\ 0 & 1 & 0 & 1 & 0 & 1 \\ 1 & 0 & 1 & 0 & 1 & 0 \\ 0 & 1 & 0 & 1 & 0 & 1 \end{bmatrix}$, $\mathbf{M}_{\bar{R}}$ has all 1's.

5. (a) $\mathbf{M}_A = \begin{bmatrix} 0 & 1 & 1 & 0 & 0 & 0 \\ 0 & 0 & 1 & 0 & 0 & 0 \\ 0 & 0 & 0 & 1 & 0 & 0 \\ 0 & 0 & 0 & 0 & 1 & 1 \\ 0 & 0 & 0 & 0 & 0 & 0 \\ 0 & 0 & 0 & 0 & 1 & 1 \end{bmatrix}$ and $\mathbf{M}_R = \begin{bmatrix} 0 & 1 & 1 & 1 & 1 & 1 \\ 0 & 0 & 1 & 1 & 1 & 1 \\ 0 & 0 & 0 & 1 & 1 & 1 \\ 0 & 0 & 0 & 0 & 1 & 1 \\ 0 & 0 & 0 & 0 & 0 & 0 \\ 0 & 0 & 0 & 0 & 1 & 1 \end{bmatrix}$.

7. (a) 3. (c) 4.

9. (a) $S = \{\langle v_i, v_j \rangle : i - j \text{ is even}\}$.

(c) Yes. Check the reflexive, symmetric, transitive properties.

11. (a) It has 1 vertex and no edges.

13. (a) Yes, if we allow parallel edges. No, if we don't, since in that case $\mathbf{M}_A$ completely determines $E(G)$.

(c) The indegree of v is the sum of the entries in column v of $\mathbf{M}_A$. Outdegrees are row sums.

15. (a) See Theorem 2 of § 8.1.

Section 8.8, page 396

1. (a) $\mathbf{W}^* = \begin{bmatrix} \infty & 1 & 2 & 3 & 4 & 5 & 7 \\ \infty & \infty & \infty & 4 & 3 & 4 & 6 \\ \infty & \infty & \infty & 1 & 4 & 5 & 7 \\ \infty & \infty & \infty & \infty & 3 & 4 & 6 \\ \infty & \infty & \infty & \infty & \infty & 1 & 3 \\ \infty & \infty & \infty & \infty & \infty & \infty & 3 \\ \infty & \infty & \infty & \infty & \infty & \infty & \infty \end{bmatrix}$

3. (a)

L	$D(1)$	$D(2)$	$D(3)$	$D(4)$	$D(5)$	$D(6)$	$D(7)$
{1}	∞	1	2	∞	∞	∞	∞
{1, 2}	∞	1	2	5	4	∞	∞
{1, 2, 3}	∞	1	2	3	4	9	∞
{1, 2, 3, 4}	∞	1	2	3	4	9	∞
{1, 2, 3, 4, 5}	∞	1	2	3	4	5	7
{1, 2, 3, 4, 5, 6}	∞	1	2	3	4	5	7

No change now

5. (a) $\mathbf{W}_2 = \begin{bmatrix} \infty & \infty & \infty & \infty & 1 & \infty & \infty \\ \infty & \infty & \infty & \infty & \infty & \infty & 1 \\ \infty & \infty & \infty & 1 & \infty & 1 & \infty \\ \infty & \infty & 1 & \infty & 1 & \infty & \infty \\ 1 & \infty & \infty & 1 & 2 & \infty & \infty \\ \infty & \infty & 1 & \infty & \infty & \infty & 1 \\ \infty & 1 & \infty & \infty & \infty & 1 & 2 \end{bmatrix}$

$\mathbf{W}_4 = \begin{bmatrix} \infty & \infty & \infty & \infty & 1 & \infty & \infty \\ \infty & \infty & \infty & \infty & \infty & \infty & 1 \\ \infty & \infty & 2 & 1 & 2 & 1 & \infty \\ \infty & \infty & 1 & 2 & 1 & 2 & \infty \\ 1 & \infty & 2 & 1 & 2 & 3 & \infty \\ \infty & \infty & 1 & 2 & 3 & 2 & 1 \\ \infty & 1 & \infty & \infty & \infty & 1 & 2 \end{bmatrix}$

$\mathbf{W}_7 = \begin{bmatrix} 2 & 6 & 3 & 2 & 1 & 4 & 5 \\ 6 & 2 & 3 & 4 & 5 & 2 & 1 \\ 3 & 3 & 2 & 1 & 2 & 1 & 2 \\ 2 & 4 & 1 & 2 & 1 & 2 & 3 \\ 1 & 5 & 2 & 1 & 2 & 3 & 4 \\ 4 & 2 & 1 & 2 & 3 & 2 & 1 \\ 5 & 1 & 2 & 3 & 4 & 1 & 2 \end{bmatrix}$

7. $\mathbf{W^*} = \begin{bmatrix} \infty & 8 & 7 & 5 & 2 \\ \infty & \infty & \infty & \infty & \infty \\ \infty & 1 & \infty & \infty & \infty \\ \infty & 3 & 2 & \infty & \infty \\ \infty & 6 & 5 & 3 & \infty \end{bmatrix}.$

9. (a) $\mathbf{D}_0 = \mathbf{D}_1 = [-\infty \quad 1 \quad 2 \quad -\infty \quad -\infty \quad -\infty \quad -\infty]$

$\mathbf{D}_2 = [-\infty \quad 1 \quad 2 \quad 5 \quad 4 \quad -\infty \quad -\infty]$

$\mathbf{D}_3 = [-\infty \quad 1 \quad 2 \quad 5 \quad 7 \quad 9 \quad -\infty]$

$\mathbf{D}_4 = [-\infty \quad 1 \quad 2 \quad 5 \quad 8 \quad 13 \quad -\infty]$

$\mathbf{D}_5 = [-\infty \quad 1 \quad 2 \quad 5 \quad 8 \quad 13 \quad 11]$

$\mathbf{D}_6 = [-\infty \quad 1 \quad 2 \quad 5 \quad 8 \quad 13 \quad 16]$

11. (a) The algorithm would give

L	$D(1)$	$D(2)$	$D(3)$	$D(4)$
$\{1\}$	$-\infty$	5	6	$-\infty$
$\{1, 3\}$	$-\infty$	5	6	9

No change

whereas $M(1, 3) = 9$ and $M(1, 4) = 12$.

(c) Both algorithms would still fail to give correct values of $M(1, 4)$.

Section 8.9, page 402

1. (a) $\mathbf{P}_0 = \begin{bmatrix} 0 & 2 & 3 & 0 & 0 & 0 & 0 \\ 0 & 0 & 0 & 4 & 5 & 0 & 0 \\ 0 & 0 & 0 & 4 & 5 & 6 & 0 \\ 0 & 0 & 0 & 0 & 5 & 6 & 0 \\ 0 & 0 & 0 & 0 & 0 & 6 & 7 \\ 0 & 0 & 0 & 0 & 0 & 0 & 7 \\ 0 & 0 & 0 & 0 & 0 & 0 & 0 \end{bmatrix}$

$\mathbf{P}_{\text{final}} = \begin{bmatrix} 0 & 2 & 3 & 3 & 2 & 2 & 2 \\ 0 & 0 & 0 & 4 & 5 & 5 & 5 \\ 0 & 0 & 0 & 4 & 4 & 4 & 4 \\ 0 & 0 & 0 & 0 & 5 & 5 & 5 \\ 0 & 0 & 0 & 0 & 0 & 6 & 7 \\ 0 & 0 & 0 & 0 & 0 & 0 & 7 \\ 0 & 0 & 0 & 0 & 0 & 0 & 0 \end{bmatrix}$

3. (a)

k	$D(1)$	$D(2)$	$D(3)$	$D(4)$	$D(5)$	$P(1)$	$P(2)$	$P(3)$	$P(4)$	$P(5)$
1	∞	2	1*	7	∞	0	1	1	1	0
3	∞	2*	1	7	4	0	1	1	1	3
2	∞	2	1	6	4*	0	1	1	2	3
5	∞	2	1	5*	4	0	1	1	5	3

* in column $D(k)$ marks the time that k is chosen for L and $D(k)$ is frozen.

5. (a) $\mathbf{W}_0 = \mathbf{W}_1 = \begin{bmatrix} \infty & 1 & \infty & 7 & \infty \\ \infty & \infty & 4 & 2 & \infty \\ \infty & \infty & \infty & \infty & 3 \\ \infty & \infty & 1 & \infty & 5 \\ \infty & \infty & \infty & \infty & \infty \end{bmatrix}$ $\mathbf{P}_0 = \mathbf{P}_1 = \begin{bmatrix} 0 & 2 & 0 & 4 & 0 \\ 0 & 0 & 3 & 4 & 0 \\ 0 & 0 & 0 & 0 & 5 \\ 0 & 0 & 3 & 0 & 5 \\ 0 & 0 & 0 & 0 & 0 \end{bmatrix}$

$\mathbf{W}_2 = \begin{bmatrix} \infty & 1 & 5 & 3 & \infty \\ \infty & \infty & 4 & 2 & \infty \\ \infty & \infty & \infty & \infty & 3 \\ \infty & \infty & 1 & \infty & 5 \\ \infty & \infty & \infty & \infty & \infty \end{bmatrix}$ $\mathbf{P}_2 = \begin{bmatrix} 0 & 2 & 2 & 2 & 0 \\ 0 & 0 & 3 & 4 & 0 \\ 0 & 0 & 0 & 0 & 5 \\ 0 & 0 & 3 & 0 & 5 \\ 0 & 0 & 0 & 0 & 0 \end{bmatrix}$

$\mathbf{W}_3 = \begin{bmatrix} \infty & 1 & 5 & 3 & 8 \\ \infty & \infty & 4 & 2 & 7 \\ \infty & \infty & \infty & \infty & 3 \\ \infty & \infty & 1 & \infty & 4 \\ \infty & \infty & \infty & \infty & \infty \end{bmatrix}$ $\mathbf{P}_3 = \begin{bmatrix} 0 & 2 & 2 & 2 & 2 \\ 0 & 0 & 3 & 4 & 3 \\ 0 & 0 & 0 & 0 & 5 \\ 0 & 0 & 3 & 0 & 3 \\ 0 & 0 & 0 & 0 & 0 \end{bmatrix}$

$\begin{aligned}\mathbf{W}_4 = \\ \mathbf{W}_5 = \mathbf{W}^* = \end{aligned} \begin{bmatrix} \infty & 1 & 4 & 3 & 7 \\ \infty & \infty & 3 & 2 & 6 \\ \infty & \infty & \infty & \infty & 3 \\ \infty & \infty & 1 & \infty & 4 \\ \infty & \infty & \infty & \infty & \infty \end{bmatrix}$ $\begin{aligned}\mathbf{P}_4 = \\ \mathbf{P}_5 = \mathbf{P}^* = \end{aligned} \begin{bmatrix} 0 & 2 & 2 & 2 & 2 \\ 0 & 0 & 4 & 4 & 4 \\ 0 & 0 & 0 & 0 & 5 \\ 0 & 0 & 3 & 0 & 3 \\ 0 & 0 & 0 & 0 & 0 \end{bmatrix}$

7. (a) Create a row matrix $\mathbf{P}$ with $\mathbf{P}[j] = j$ initially if there is an edge from v_1 to v_j and $\mathbf{P}[j] = 0$ otherwise. Add the line

$$\text{Replace } \mathbf{P}[j] \text{ by } \mathbf{P}[k].$$

9. (a) $i\,t(R)\,j$ if $i < j$.
11. (a) Initialization of $\mathbf{P}$ takes time $O(n)$, and the replacement step still only takes constant time during each pass through the For loop.

Section 9.1, page 416

1. There are fourteen, including the trivial tree with one vertex. Six of them are in Figure 3 of § 0.3.
3. (a) One example is

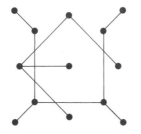

5. Use $\mathbb{Z}$ for the set of vertices.

7. (a)

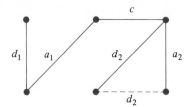

Edges a_1 and a_2 can be chosen in either order. So can d_1 and d_2. Edge d_2' can be chosen instead of d_2. The weight is 10.

9. Use the corollary to Theorem 1.

11. One possible graph is

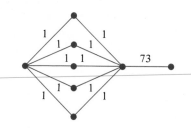

13. (a) Apply Theorem 1 of § 8.4.

15. (a) Suppose the components have n_1, n_2, ..., n_m vertices, so that altogether $n_1 + n_2 + \cdots + n_m = n$. The ith component is a tree, so it has $n_i - 1$ edges, by Theorem 3. The total number of edges is $(n_1 - 1) + (n_2 - 1) + \cdots + (n_m - 1) = n - m$.

17. 1687 miles.

19. There exist minimal spanning trees S and T with $e_k \in T \backslash S$. Imitate the fourth paragraph of the proof of Theorem 4 to obtain $e_i \in S \backslash T$ where $i < k$. This will contradict your choice of k.

Section 9.2, page 428

1. (c) 3.

3. (a) Rooted trees in Figures 5(b) and 5(c) have height 2; the one in 5(d) has height 3.

 (b) Only the rooted tree in Figure 5(c) is a regular 3-ary tree.

5. (a) There are seven of them.   (c) 1.

9. Use the formulas for p and t in Example 5.

11. There are 2^k words of length k.

13. (a) One such mapping is α where $\alpha(r) = a$, $\alpha(s) = b$, $\alpha(p) = d$, $\alpha(q) = c$, $\alpha(v) = e$, $\alpha(u) = f$, $\alpha(w) = g$, $\alpha(x) = j$, $\alpha(y) = k$, $\alpha(z) = h$.

15. (a) Suppose α were an isomorphism of the rooted tree in Figure 5(b) onto the rooted tree in Figure 5(c). Then $\alpha(v) = x$. Since α is one-to-one, v and x must have the same number of children. But v has four and x has three.

17. Say

Section 9.3, page 439

1. Postorder: $v, y, w, z, x, t, p, u, q, s, r$.
3. Preorder: $r, t, x, v, y, z, w, u, p, q, s$.
5. $T = T_r = r, T_w, T_u$
   $\qquad = r, w, T_v, T_x, u, T_t, T_s$
   $\qquad = r, w, v, x, T_y, T_z, u, t, s, T_p, T_q$
   $\qquad = r, w, v, x, y, z, u, t, s, p, q$.
7. (a) By (B), the leaves represent trivial trees. By (R), T_w, T_r and T_s are rooted trees. Use (R) with T_w and T_r to see that T_x is a rooted tree. Then use (R) with T_x and T_s to see that $T = T_u$ is a rooted tree. Draw the trees, as in Figure 1.
   (b) The trivial trees have height 0. So by (R), height$(T_w) = 1 + \max\{0, 0\} = 1$. Similarly, height$(T_r)$ = height$(T_s) = 1$. Then

   $$\text{height}(T_x) = 1 + \max\{1, 1\} = 2 \quad \text{and} \quad \text{height}(T) = 1 + \max\{2, 1\} = 3.$$

9. After the trivial trees are listed one obtains:

   $$T_w: \quad v, y, w; \quad T_r: \quad z, t, r; \quad T_s: \quad p, q, s;$$

   $$T_x: \quad T_w, T_r, x, \quad \text{i.e.,} \quad v, y, w, z, t, r, x; \quad \text{etc.}$$

15. (a)   (c)

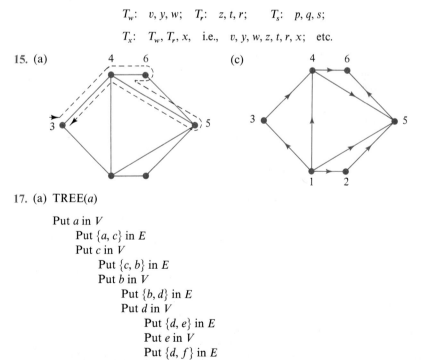

17. (a) TREE(a)

   Put a in V
       Put $\{a, c\}$ in E
       Put c in V
           Put $\{c, b\}$ in E
           Put b in V
               Put $\{b, d\}$ in E
               Put d in V
                   Put $\{d, e\}$ in E
                   Put e in V
                   Put $\{d, f\}$ in E
                   Put f in V

   See the drawing on the next page.

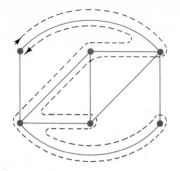

19. Since the digraph is acyclic there can be at most one edge joining each of the $n(n-1)/2$ pairs of distinct vertices.

Section 9.4, page 447

1. Reverse Polish: $x\,4\,2 \uparrow\, - y * 2\,3\,/ +$.
3. (a) Polish: $- * + a\,b - a\,b - \uparrow a\,2 \uparrow b\,2$;
    Infix: $(a+b)*(a-b) - ((a \uparrow 2) - (b \uparrow 2))$.
5. (a) 20.   (b) 10.
7. (a) $3\,x * 4 - 2 \uparrow$.
    (c) The answer depends on how the terms are associated. For the choice $(x - x^2) + (x^3 - x^4)$, the answer is $x\,x\,2 \uparrow - x\,3 \uparrow x\,4 \uparrow - +$.
9. (a) $a\,b\,c * *$ and $a\,b * c *$
    (c) The associative law is $a\,b\,c * * = a\,b * c *$.
11. (b) Reverse Polish: $p\,q\,p \neg \lor \rightarrow$; Polish: $\rightarrow p \lor q \neg p$.
13. (a) Infix: $(p \land (p \rightarrow q)) \rightarrow q$.
15. (a) Both give $a\,/\,b + c$.
17. (a) **(B)** Numerical constants and variables are wff's.
    (R) If f and g are wff's, so are $+ f\,g$, $- f\,g$, $* f\,g$, $/ f\,g$ and $\uparrow f\,g$.
19. (a) **(B)** Variables, such as p, q, r, are wff's.
    (R) If P and Q are wff's, so are $P\,Q \lor$, $P\,Q \land$, $P\,Q \rightarrow$, $P\,Q \leftrightarrow$ and $P \neg$.
    (b) Argue, in turn, that $q \neg$, $p\,q \neg \land$ and $p\,q \neg \land \neg$ are wff's. Likewise, $p\,q \neg \rightarrow$ is a wff. Thus $p\,q \neg \land \neg p\,q \neg \rightarrow \lor$ is a wff.

Section 9.5, page 459

1. (a) 35, 56, 70, 82.
3. (a) Weight = 84.   (c) Weight = 244.
5. All but (b) are prefix codes. In (b), 01 consists of the first two digits of 0111.
7. (b) 269.
9. (a) The vertex labeled 0 has only one child, 00.   (b) Consider any string beginning with 01.
11. (a) 484.   (c) 373.   (e) It will involve at most 354 comparisons.
13. (b) 221.

Section 10.1, page 470

3. (a) The commutative and associative laws follow, as in Example 2, from the corresponding laws in Table 1 of § 2.2. For the absorption laws, see Exercise 4 of § 2.3.

(b) Because $c \Rightarrow P$ for all P. This can be verified by looking at the truth table of $c \to p$ or by applying rules 6c, 2b and 17 of § 2.2.

5. (a) Yes.   (b) Maybe, maybe not. In Figure 4 of § 7.1, T and U are lattices. The universal upper bound for T is join-irreducible but the one for U is not.

7. (a) x, y.   (b) u, w, x, y, z.   (c) $t = u \vee w$, $v = x \vee y$ and the others are already join-irreducible.

9. (a) 1.   (c) They are the primes.   (d) They are the powers of primes, i.e., all integers of the form p^k where p is prime and $k \in \mathbb{P}$.

11. (a) Note $D_{90} = \{1, 2, 3, 5, 6, 9, 10, 15, 18, 30, 45, 90\}$.   (c) 2, 3, 5.
    (e) $90 = 2 \vee 9 \vee 5$; $18 = 2 \vee 9$; 5 is already join-irreducible.

13. (c) T has 15 and U has 19.

15. (a) $x \leq y \Leftrightarrow x \vee y = y \Leftrightarrow \phi(x \vee y) = \phi(y) \Leftrightarrow \phi(x) \cup \phi(y) = \phi(y) \Leftrightarrow$
    $\phi(x) \leq \phi(y)$. Supply reasons.
    (c) Use part (b).

17. Since m isn't itself join-irreducible, $m = x \vee y$ for some x and y different from m. Then $x \prec m$ and $y \prec m$. Hence x and y are not in B and so they are joins of join-irreducible elements. Use these facts to show that m is also a join of join-irreducible elements, a contradiction.

Section 10.2, page 477

1. (a) d, e.   (c) $1 = a \vee c$.   (e) No. b has no complement.

3. (b) It is straightforward to show D_m is complemented if m is a product of distinct primes; do it. Next assume m is not the product of distinct primes and show D_m is not complemented. To simplify notation, assume 2 is a prime that appears more than once so that $m = 2^u etc$ where $u \geq 2$. Then 2 has no complement. In fact, if k were a complement for 2, we would have $\gcd(2, k) = 1$ and $\text{lcm}(2, k) = m$. Then k would be odd and so $\text{lcm}(2, k)$ would have only one factor of 2, a contradiction.

5. (b) See Example 2(b).   (c) See Exercise 3(b).

7. The figure represents $(D_6, |)$.

9. (a) 2, 3.   (b) 1, 2, 3, 4, 9.

11. Supply reasons:

$$x = x \wedge (x \vee a) = x \wedge (y \vee a)$$

$$= (x \wedge y) \vee (x \wedge a) = (x \wedge y) \vee (y \wedge a)$$

$$= (y \wedge x) \vee (y \wedge a) = y \wedge (x \vee a) = y \wedge (y \vee a) = y.$$

13. Assume $1 \vee x = 1$ and $1 \wedge x = 0$, and show $x = 0$. Then assume $0 \vee y = 1$ and $0 \wedge y = 0$, and show $y = 1$.

15. (a) Use $y' \vee x' = (y \wedge x)' = x'$.
    (b) Show $y = y \wedge z'$, starting with $y = y \wedge 1 = y \wedge (z \vee z')$.
    (c) Use parts (a) and (b).

Section 10.3, page 485

1. (a) Since the operations $\vee$ and $\wedge$ treat 0 and 1 just as if they represent truth values, checking the laws 1Ba through 5Bb for all cases amounts to checking corresponding truth tables. Do enough until the situation is clear to you.

3. (a) Note that $D_{30} = \{1, 2, 3, 5, 6, 10, 15, 30\}$.
    (c) $\{1, 30\}$, $\{1, 2, 15, 30\}$, $\{1, 3, 10, 30\}$, $\{1, 5, 6, 30\}$, D_{30}.

5. One solution is to set $S = \{1, 2, 3, 4, 5\}$ and define

$$\phi(\langle a_1, a_2, a_3, a_4, a_5 \rangle) = \{i \in S : a_i = 1\}.$$

7. (a) No. A finite Boolean algebra has 2^n elements for some n.

9. (a) If $a \leq x$ or $a \leq y$, then surely $a \leq x \vee y$. Suppose $a \leq x \vee y$. Then $a = a \wedge (x \vee y) = (a \wedge x) \vee (a \wedge y)$. One of $a \wedge x$ and $a \wedge y$, say $a \wedge x$, must be different from 0. But $0 \prec a \wedge x \leq a$, so $a \wedge x = a$ and $a \leq x$.

(c) $a \leq 1 = x \vee x'$, so $a \leq x$ or $a \leq x'$ by part (a). Both $a \leq x$ and $a \leq x'$ would imply $a \leq x \wedge x' = 0$ by part (b), a contradiction.

11. $x \leq y \Leftrightarrow x \vee y = y \Leftrightarrow \phi(x \vee y) = \phi(y) \Leftrightarrow \phi(x) \vee \phi(y) = \phi(y) \Leftrightarrow \phi(x) \leq \phi(y)$.

Section 10.4, page 493

1. $x'y'z' \vee x'y'z \vee xyz'$.

3. (c)

x	y	z	$xy \vee z'$
0	0	0	1
0	0	1	0
0	1	0	1
0	1	1	0
1	0	0	1
1	0	1	0
1	1	0	1
1	1	1	1

$x'y'z' \vee x'yz' \vee xy'z' \vee xyz' \vee xyz$.

5. (a) $x_1 x_2 x_3' x_4 \vee x_1 x_2 x_3' x_4' \vee x_1' x_2 x_3 x_4'$.

7. (a) $xz \vee y'$.

9. $y' \vee z$.

11. (a) Find the minterm canonical form for E'. Then find $E = (E')'$ using DeMorgan laws, first on joins and then on products.

(b) $(x' \vee y')(x \vee y)$.

13. Show that the values of the corresponding Boolean functions are the same.

Section 10.5, page 502

1. (a) xyz.

3. (a)

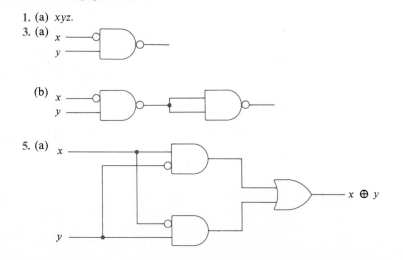

(b)

5. (a)

$x \oplus y$

7. (a) $S = 1, C_O = 0.$ (c) $S = 0, C_O = 1.$

9. (a)

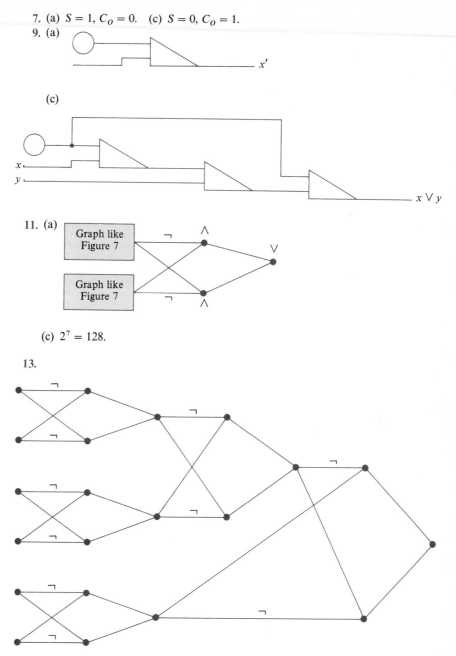

(c)

x'

$x \lor y$

11. (a)

Graph like Figure 7

Graph like Figure 7

(c) $2^7 = 128.$

13.

Vertices need labels.

Section 10.6, page 509

1. $xyz \lor xyz' \lor xy'z' \lor xy'z \lor x'yz \lor x'y'z = x \lor z.$

3. $xyz \lor xyz' \lor xy'z \lor x'y'z' \lor x'y'z = xz \lor xy \lor x'y' = xy \lor y'z \lor x'y'.$

5. (a)

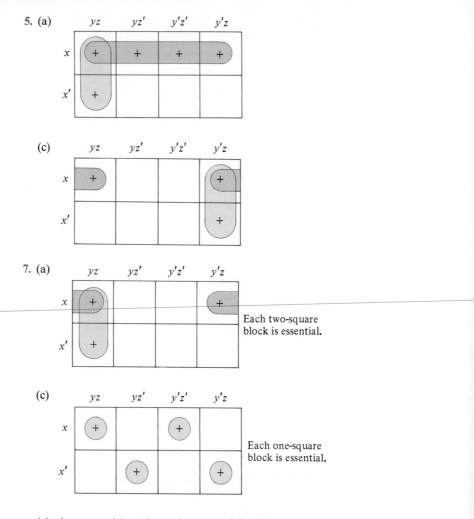

(a) $z' \lor xy \lor x'y' \lor w'y$ or $z' \lor xy \lor x'y' \lor w'x'$.
(c) $w'x'z' \lor w'xy' \lor wxy \lor wx'z \lor y'z'$, not $w'x'z' \lor w'xy' \lor wx'y' \lor wyz \lor wxz'$, which also has five product terms but one more literal.

Section 11.1, page 519

1. (a) $\mathbb{P}$. (c) $\mathbb{Z}$. (e) $\mathbb{Z}$. (g) $18\mathbb{P} = \{18k : k \in \mathbb{P}\}$.
3. $\mathbb{P} = \{1\}^+$, $\{0\} = \{0\}^+$, $18\mathbb{P} = \{18\}^+$.
5. (a) $\mathbb{Z}$. (c) $\mathbb{Z}$. (e) $\mathbb{Z}$.
7. (a) $2\mathbb{N} = \{2k : k \in \mathbb{N}\}$. (c) $\{0\}$. (e) Σ^*.

9. (a) $\begin{bmatrix} 0 & 1 & 0 \\ 1 & 0 & 0 \\ 0 & 0 & 1 \end{bmatrix}$ and $\begin{bmatrix} 1 & 0 & 0 \\ 0 & 1 & 0 \\ 0 & 0 & 1 \end{bmatrix}$.

(c) $\begin{bmatrix} 0 & 2 & 3 \\ 0 & 0 & 4 \\ 0 & 0 & 0 \end{bmatrix}$, $\begin{bmatrix} 0 & 0 & 8 \\ 0 & 0 & 0 \\ 0 & 0 & 0 \end{bmatrix}$ and $\begin{bmatrix} 0 & 0 & 0 \\ 0 & 0 & 0 \\ 0 & 0 & 0 \end{bmatrix}$.

11. (a) and (d). Check closure under $\circ$. For (b) and (c) give counterexamples.
13. Let H be a subgroup containing A. To show that $\langle A \rangle \subseteq H$, let $p(x)$ be "$x \in H$" and show that

    (B) $p(x)\,\forall x \in H$,
    (I_1) $p(x) \wedge p(y) \to p(x \,\square\, y)$,
    (I_2) $p(x) \to p(x^{-1})$

are true.

Section 11.2, page 525

1. (a) $6\mathbb{P} = \{6k : k \in \mathbb{P}\}$.
    (c) No. For example, 6 and 12 are not both powers of the same member of $6\mathbb{P}$, so cannot both lie in the same cyclic subgroup.
3. (a) $60\mathbb{P}$. (c) Yes. Give a generator.
5. (a) $f(n) = n + 3$ defines one.
7. Any example in which neither H nor K contains the other will work.
9. (a) $6 = 3!$. (c) g and h, where $h(1) = 3$, $h(2) = 2$ and $h(3) = 1$. (e) It contains g so could only be $g \circ \text{FIX}$. Apply part (d).
11. (a) $\langle a \rangle = \{e, a, b\}$. (c) $\langle c \rangle, \langle d \rangle, \langle f \rangle$. (e) $\{e, d\}, \{a, c\}, \{b, f\}$.
13. (a) $H \,\square\, g$ contains $e \,\square\, g = g$. By Theorem 2, either $H \,\square\, g = H$ or else $H \,\square\, g$ and H are disjoint.
15. (a) For $h \in H$, $(g \,\square\, h)^{-1} = h^{-1} \,\square\, g^{-1} \in H \,\square\, g^{-1}$, so

$$\{x^{-1} : x \in g \,\square\, H\} \subseteq H \,\square\, g^{-1}.$$

Moreover, $h \,\square\, g^{-1} = (g \,\square\, h^{-1})^{-1}$ is in $\{x^{-1} : x \in g \,\square\, H\}$ for $h \in H$, so

$$H \,\square\, g^{-1} \subseteq \{x^{-1} : x \in g \,\square\, H\}.$$

    (b) Use the result of part (a).
17. (a) (R) Note that $x \,\square\, x^{-1} = e \in H$.
    (S) Note that $x \,\square\, y^{-1} = (y \,\square\, x^{-1})^{-1}$.
    (T) Note that $z \,\square\, x^{-1} = (z \,\square\, y^{-1}) \,\square\, (y \,\square\, x^{-1})$.
19. (a) See Example 2(a). (c) By Theorem 5 of § 11.1, $\mathscr{L}(S)$ consists of subgroups of S. Any two subgroups intersect nontrivially by Corollary 2. Apply part (b).

Section 11.3, page 533

1. (a), (c), (e).
3. (a) Not an isomorphism, since it doesn't map $\mathbb{Z}$ onto $\mathbb{Z}$.
    (c) Is an isomorphism, being a one-to-one and onto homomorphism.
    (e) Not an isomorphism, since $h(n) = 3n$ does not map $\mathbb{Z}$ onto $\mathbb{Z}$.
5. (a) $h(f + g) = (f + g)(73) = f(73) + g(73) = h(f) + h(g)$.
7. (a) $\{0\}$. (c) $5\mathbb{Z} = \{5n : n \in \mathbb{Z}\}$.
9. (a) 4. (c) 3.
11. (a) The identity is $\langle e_G, e_H \rangle$, where e_G and e_H are the respective identities of G and H; $\langle g, h \rangle^{-1} = \langle g^{-1}, h^{-1} \rangle$.
    (c) $\{\langle e_G, h \rangle : h \in H\}$.

13. (c) $\begin{bmatrix} y & z \\ 0 & 1/y \end{bmatrix}\begin{bmatrix} 1 & x \\ 0 & 1 \end{bmatrix}\begin{bmatrix} 1/y & -z \\ 0 & y \end{bmatrix} = \begin{bmatrix} 1 & xy^2 \\ 0 & 1 \end{bmatrix}$ is in H.

   (e) Use the result of part (d) and Theorem 3.

15. The preimage of $h(g)$ is $K \bullet g$ where K is the kernel of h. By assumption $|K \bullet g| = 1$. So $|K| = 1$ and h is one-to-one as noted in the corollary to Theorem 1.

17. (a) Use the identity
$$(x \bullet y) \bullet H \bullet (x \bullet y)^{-1} = x \bullet (y \bullet H \bullet y^{-1}) \bullet x^{-1}.$$

   Also, if $H = x \bullet H \bullet x^{-1}$, then
$$x^{-1} \bullet H \bullet x = x^{-1} \bullet (x \bullet H \bullet x^{-1}) \bullet x = H.$$

19. (a) More generally, if $A \subseteq S$ then
$$h(A^+) = h(\{x : x \text{ is a product } a_1 \cdots a_n \text{ of members of } A\})$$
$$= \{h(x) : x = a_1 \cdots a_n, \quad a_1, \ldots, a_n \in A\}$$
$$= \{y : y = h(a_1) \cdots h(a_n), \quad a_1, \ldots, a_n \in A\} = h(A)^+.$$

   (b) Yes. Indeed, $h(\langle A \rangle) = \langle h(A) \rangle$.

21. (a) If z' is also a zero, then $z' = z \bullet z' = z$.

   (b) $(\mathbb{Z}, \cdot)$.

   (c) $(\{0, 1\}, \cdot)$. Another example is $(\mathscr{P}(X), \cap)$, with zero element $\varnothing$.

23. (a) $h(S)$ is closed under products because $h(s) \square h(s') = h(s \bullet s') \in h(S)$. And $h(e)$ is an identity because $h(s) \square h(e) = h(s \bullet e) = h(s) = h(e \bullet s) = h(e) \square h(s)$ for all $h(s) \in h(S)$.

Section 11.4, page 543

1. (a) 8.   (c) Apply Lagrange's theorem.

3. (a) There would be a new column with all entries m. (c) There are 6 orbits.

5. (a) $\{e\}, \{e, f\}, \{e, g\}, \{e, h\}, \{e, f, g, h\}$.

   (c) Kernel on $\{p, r\}$ is $\{e, f\}$ and kernel on $\{q, s\}$ is $\{e, g\}$.

7. (a) Rows of Figure 5(a) with u in the u-column.   (c) w in each case.

9. They can only have 1, 3, 9 or 27 elements.

11. (b) The sets in π are the G-orbits in S.

13. Imitate the proof of Cayley's theorem.

15. The functions a^*, b^*, c^* will be defined on $\{a, b, c, z\}$.

Section 11.5, page 556

1. (a) 4.   (c) $(k^5 + k^4)/2$.

3. (a) $C(k) = (k^4 + k^2)/2$.

5. (a) $C(k) = (k^4 + 2k^2 + 3k^3)/6$.

7. (a) 3.

9. (a) 16.   [Count entries in Figure 1(b) which match their column headings.]

   (c) 8.

11. (a) $C(k) = (k^8 + 17k^4 + 6k^2)/24$.

13. (a) 8.

15. As noted in § 7.7, $R_1 \vee R_2$ is the transitive closure of $R_1 \cup R_2$. We easily have $R_1 \cup R_2 \subseteq R_0$, so $R_1 \vee R_2 \subseteq R_0$. If $\langle s, t \rangle \in R_0$, then we have $h(s) = t$ for some $h \in \langle G_1 \cup G_2 \rangle$. Apply Theorem 3 of § 11.1 to conclude that

$$h = g_m \circ g_{m-1} \circ \cdots \circ g_2 \circ g_1$$

where each g_i is in $G_1 \cup G_2$. If $s_0 = s$ and $s_i = g_i(s_{i-1})$ for $i = 1, \ldots, m$, then $s_m = t$ and each pair $\langle s_{i-1}, s_i \rangle$ belongs to $R_1 \cup R_2$. Hence $\langle s, t \rangle = \langle s_0, s_m \rangle$ is in $R_1 \vee R_2$ by Theorem 2 of § 7.6.

Section 11.6, page 567

1. (a), (b), (d), (f).
3. All but (b); (b) is not an additive homomorphism.
5. Every subgroup is an ideal.
7. (a) $24\mathbb{Z}$.   (c) $\mathbb{Z} = 1\mathbb{Z}$, since $1 = 3 \cdot 1 + 2 \cdot (-1) \in 3\mathbb{Z} + 2\mathbb{Z}$.
9. (a) Verify well-definedness directly, or apply Theorem 1 to the homomorphism $m \to \langle \text{Rem}_4(m), \text{Rem}_6(m) \rangle$ from $\mathbb{Z}$ to $\mathbb{Z}(4) \times \mathbb{Z}(6)$, as in Example 6(b).
   (c) $\langle 1, 4 \rangle$ is one of the twelve; find another one. '
11. (a) Since $2 *_6 3 = 0$, 2 has no inverse.
    (b) Exhibit an inverse for each non-0 element. [In fact, $(\mathbb{Z}(p), +_p, *_p)$ is a field if and only if p is prime. In case p is not prime, the idea in part (a) works. If p is prime, the mapping $m \to k *_p m$ is a permutation of $\mathbb{Z}(p)$ for each non-0 $k \in \mathbb{Z}(p)$. The inverse permutation corresponds to the inverse of k.]
    (c) Consider the inverse of $\langle 1, 0 \rangle$ for example.
13. (a) The kernel of h is either F or $\{0\}$ by Example 5(c).
    (b) Part (a) and the corollary to Theorem 2 apply.
15. (a) $I = 15\mathbb{Z}$.   (b) See Example 6(b).
    (c) It would have to be $4\mathbb{Z}$ [why?], but $\mathbb{Z}/4\mathbb{Z}$ has an element a with $a + a \neq 0$.
17. (a) $R \cdot 2 = \{a_0 + a_1 x + \cdots + a_n x^n \in R : \text{every } a_i \text{ is even}\}$,
       $R \cdot x = \{a_0 + a_1 x + \cdots + a_n x^n \in R : a_0 = 0\}$,
       $R \cdot 2 + R \cdot x = \{a_0 + a_1 x + \cdots + a_n x^n \in R : a_0 \text{ is even}\}$.
    (b) Suppose $R \cdot p = R \cdot 2 + R \cdot x$ for some $p \in R$. Since $2 \in R \cdot p$, p must be constant, and since $x \in R \cdot p$, p is 1 or -1. But then $R \cdot p = R$, a contradiction.

INDEX

Note: Where page references for an entry are not in regular numerical sequence, the first reference listed is the *primary* reference.

A

associativity *(cont.)*
  of ⊕, 39
assume, 571
atoms, 468, 481
automorphism of graphs, 341
axiom, 571

B

basis for induction, 98
  for recursion, 133, 145
biconditional ↔, 71
big-oh notation, 128
binary code, 455
binary operation, 163
binary relation, 285, 288
binary search tree, 419–420
binary tree, 424
  optimal, 452
binary up-down counter, 384
binds a variable, 227
binomial coefficients, 187
binomial theorem, 190
bipartite graph, 377
bona fide, 571
Boolean algebra, 479
Boolean algebra
    isomorphism, 483
Boolean expressions, 486
Boolean function, 484, 487–488
Boolean lattice, 475
Boolean matrices, 296
Boolean operations, 295
Boolean product, 296
bound variable, 225, 227
branch node, 423
bridge deal, 199
bridge hand, 199

C

cancellation laws for
    matrices, 162
Cantor's diagonal procedure, 216
cards, 188–189
cases, proof by, 92
Cayley's theorem, 538
cf., 571
chain, 265–266
characteristic equation, 141

characteristic function, 113
charge, 434
chicanery, 571
child, children, 423
circuit, 324
class, 571
closed interval notation, 29
closed path in a digraph, 324
  in a graph, 4, 355
closed under, 163
closed under intersections, 311
closure operator, 179, 304
codes, 455
codomain of a function, 107
coefficient matrix, 54
collapse, 571
collection, 571
coloring cubes, etc., 549–552
column vectors, 56
combinations, 187
common factor, 571
commutative law for
    matrices, 154
commutative laws for logic, 74
  for sets, 36
commutative ring, 560
commutative semigroup, 164
comparable, 572
complement, 35
  of a graph, 378
complemented lattice, 474
complements in lattices, 474
complete bipartite graph, 377
complete graph, 188, 360
complete set of invariants, 337
component of a graph, 364
composite relation, 292–293
composition of functions, 113
  associativity of, 115
compound predicates, 232
compound propositions, 63, 232
concatenation, 166
conclusion, 80
conditional implication →, 69
congruence modulo p, 170, 278
congruent triangles, 275
conjecture, 572
conjugates of subgroups, 533

conjunction, 81
connected digraph, 384
connected graph, 6, 364
connectives (*see* logical
    connectives)
constant function, 113
constructive dilemmas, 75
constructive proof, 93
contained in, 28
contradiction, 72
contradiction proof, 85, 90
contrapositive, 64, 74
contrapositive proof, 90
converse, 64
corollary, 572
coset of a subgroup, 522
countable set, 213
countably infinite, 213
counterexample, 65, 236
counting rules:
  inclusion-exclusion, 195
  lemma, 197
  objects in boxes, 201
  ordered partitions, 198
  permutations, 197
  pigeon-hole, 206, 207
  power, 186
  product, 184
  union, 183
covers (in a poset), 255, 468
critical edge or path, 349
cube coloring, 552
curry recipe, 347
cycle in a digraph, 324
  in a graph, 4, 355
cyclic group, 517
cyclic semigroup, 514

D

DAG, 325
decimal expansions of
    rationals, 209
decreasing order, 208
define, 572
degree of a polynomial, 149
  of a vertex, 5, 338, 358
degree sequence, 358
DeMorgan laws for lattices, 476, 480
  for logic, 74, 233, 234
  for sets, 36, 45
depth of a wff, 151
depth-first search
    algorithms, 434
descendant, 423

Traveling Salesperson
  Problem, 11
traversal algorithms, 431–432
tree, 12, 146, 407
  for counting, 185
TREE algorithm, 437, 438
tree traversal algorithm, 430
TREESORT algorithm, 435
trivial proof, 93
trivial ring, 560
trivial tree, 429
trivially true, 93
true, false, 61
true theorem, 80
truncate, 574
truth tables, 70

U

unambiguous, 574
uncountable set, 214
underlying set, 574
undirected graph, 354
undirected path, 424
union counting rules, 183
union of sets, 33, 44
universal lower bound, 468
universal quantifier, 224
universal relation, 291
universal set, 35
universal upper bound, 468

universe, 35
  of discourse, 224
up-down counter, 384
upper bound, 259
  universal, 468

V

vacuous proof, 93
vacuously true, 93
valid proof or argument, 83
valid proposition, 80
values of a function, 107
variables of a truth table, 70
variables, free and bound,
  225–227
variables, logical, 70, 231
vector, 56, 57
vector addition, 56
Venn diagram, 34
vertex sequence, 4, 324, 355
vertices, 574
vertices of a digraph, 322
  of a graph, 3, 112, 354

W

WARSHALL'S algorithm,
  392, 400
weight of an edge or path,
  344

of a graph, 11, 411
of a leaf, 449
of a subgraph, 411
of a tree, 449
weighted digraph, 344
weighted graph, 11, 411
weighted tree, 449
well-defined function, 149,
  283
well-formed-formula (*see* wff)
well-ordered set, 267
well-ordering principle, 94,
  243
wff for algebra, 148
  for Polish notation, 446
  for propositional calculus,
  148
with or without replacement,
  184–185
word, 30, 146
words, multiplying, 166

X

XOR gate, 495

Z

zero element, 536
zero matrix, 153